FORESTRY
PRINCIPLES AND APPLICATIONS

Editors & Authors

ANTONY JOSEPH RAJ Ph.D.

Associate Professor (Forestry)
Department of Land Resources Management and Environmental Protection
Mekelle University, Ethiopia
& Associate Dean (On Leave)
College of Forestry and Environment
Sam Higginbottom Institute of Agriculture, Technology and Sciences
(Formerly Allahabad Agricultural Institute) Deemed University
Allahabad, Uttar Pradesh, India

S.B. LAL Ph.D.

Pro-Vice Chancellor and Dean
College of Forestry & Environment of SHIATS-Deemed University
(Formerly Allahabad Agricultural Institute)
Allahabad

Published by:

Scientific Publishers (India)
5 A, New Pali Road, P.O. Box 91
Jodhpur 342 001 (India)

Branch Office
Scientific Publishers (India)
4806/24, Ansari Road, Daryaganj
New Delhi - 110 002 (India)

E-mail: info@scientificpub.com
Website: www.scientificpub.com

ISBN: 978-81-7233-810-7

Typeset by Rajesh Ojha
Printed in India

FOREWORD

Forests are a critical link in the transition to a green economy – one that promotes sustainable development and poverty eradication as we move towards a low carbon and more equitable future. Forests supply wood, fibre, fuelwood and non-wood forest products for industrial and non-industrial uses. Biologically-rich forest ecosystems provide a wide range fo provisioning, regulating, cultural and supporting services for human well being collectively known as ecosystem services.

The sustainability of forest ecosystems depends on sustained management, efficient utilization and effective protection measures against deforestation and forest degradation. The role of ecosystem services are equally important for sustaining livelihoods and maintaining environmental conditions. The benefits of reducing deforestation for climate change alone is estimated to be in the trillions. Yet despite these huge ecological, economical, social and health benefits, forest are still being destroyed at an alarming rate of 13 million hectares annually for limited and short-term gains.

This comprehensive textbook "Forestry: Principles and Applications" by Dr. Antony Joseph Raj and Prof. (Dr.) S.B. Lal will provide a significant contribution in university and college lecture halls. This book forms a comprehensive and thoroughly up-to-date text on forestry and a detailed exploration of specific and modern technologies in the field of forestry. By bringing current knowledge on forestry and natural resources together in one place, this book will advance our ability to manage the remaining forest resources and safeguard their continuing contribution to human beings. It is my wish that students, teachers, scientists and professional foresters will find the information on this textbook valuable. In future, the acquired knowledge will be helpful not only for the development of sustainable forest management in India but also as an input to policy formulation with respect to Indian forests.

I welcome this latest book which is first of its kind in India as an editorial textbook on forestry. I believe this book, with its high standard, will serve the students for the preparation of competitive exams like UPSC Civil Service Exam, UPSC Indian Forest Service Exam, ICAR Scientists/NET Exam, etc. I would like to congratulate the authors, Dr. Antony Joseph Raj and Prof. (Dr.) S.B. Lal, for their meticulous and hard work in bringing this important textbook into being.

[Rev.(Prof. Dr.) RAJENDRA B. LAL]

Vice-Chancellor

Sam Higginbottom Institute of Agriculture, Technology and Sciences

(Formerly Allahabad Agricultural Institute) Deemed University

Allahabad, Uttar Pradesh, India

September, 2012

PREFACE

The material resources and life supporting functions provided by forest lands are particularly important in developing countries where some people still rely on them directly as sources of food, fodder, fuelwood, medicine, shelter, building materials and as centers of certain cultural practices. In recent years, the capability of forests to mitigate climate change, provide renewable products and energy, maintain biological diversity, protect land and water resources, provide recreation facilities, improve air quality, help alleviate poverty and contribute to developing a greener economy have received increasing attention. But at the same time, these benefits are threatened by land and forest degradation, global warming and extreme weather events. An ever increasing population places enormous demands on the forest and land resources. Carbon sequestration in forest ecosystems has become an important issue both in the political discussion about abrupt climate change and forest ecosystem research.

National Forest Policy 1988 mandates expanding forest cover from 23.81 to 33 per cent of India's territory and the overarching objective of the Green India Mission is to increase forest and tree cover in 5 million hectares and improve quality of forest cover in another 5 million hectares. There is a challenge to devise comprehensive work plan for sustainable development of forests in India in next twenty years. To cope with challenging task, forest managers have to integrate modern knowledge resulting from all the disciplines of forestry into the management plans. We are expected to shift towards modern trends and latest concepts to upgrade the knowledge and skills in forestry. This book is the first in the forestry text book series to synthesize information on relevant processes, factors and causes of carbon turnover in forest ecosystem and the technical and economic potential of carbon sequestration. Accordingly the book is able to fill an important gap between the needs of global forest and environmental policies and local forest management. In fact, the book makes a valuable contribution to the knowledge of students, academicians, research scientists, modern foresters and policy makers, which will, in turn, guide efforts to manage the remaining forests and new forests in the millennia to come.

There are many forestry textbooks available in India which provide a more balanced account of subjects. The classic examples are textbooks authored by pioneers like P.K.R. Nair, D.N. Tewari, L.S. Khanna, A.P. Dwivedi, A.N. Chaturvedi, Tribhawan Mehta, Ram Prakash, S.S. Negi, R.K. Luna and others. In many aspects, the topics and structure of this textbook is highly meritorious and unique. All the individual chapters are contributed by Subject Matter Specialists/ Experts of high repute carefully selected based on their experience and practical knowledge from throughout India and abroad (few chapters).

In fact, the book collates valuable knowledge on forestry and natural resources which will be useful to students for their regular semester exams, University entrance examinations for admission to M.Sc. & Ph.D. programmes, preparation of competitive exams like Civil Service Exams (UPSC), Indian Forestry Service (UPSC), ICAR-ARS Scientist Exam, ICFRE Forestry Scientist Exams, NET Exam, State Public Service Commission Exams, etc. This compilation will be most useful for the people in Universities and Colleges, Research Institutions dealing with biological sciences, agricultural sciences and forestry sciences and Forest Departments of all the States of India. The general subject matters available in the book will be beneficial to persons from forestry, agriculture and natural resources field in other countries.

This textbook will contribute significantly to academic teaching and scientific research. Additional information or suggestions that will improve this book are invited from experts. Finally we would like to express our deep gratitude to all our friends and well wishers for providing numerous comments and suggestions for enhancing the quality of the book.

ANTONY JOSEPH RAJ
S.B. LAL

September 2012

LIST OF CONTRIBUTORS

Prof. S.B. Lal
Pro-Vice Chancellor and Dean, College of Forestry and Environment, SHIATS Deemed University, (Formerly Allahabad Agricultural Institute), Allahabad, Uttar Pradesh

Shri. K.C.A. Arun Prasad, IFS
Director & Dy. CWLW (Keoladeo National Park) in Rajasthan Forest Department (Earlier) and Presently on Deputation as Associate Professor, Indira Gandhi National Forest Academy, Dehradun, Uttarakhand

Shri. L.Chandrasekar, IFS
Divisional Forest Officer, Kerala Forest Department (Earlier) and Presently on Deputation in Central Academy for State Forest Service, Coimbatore, Tamil Nadu

Shri. Rajesh G. Nair, IFS
Additional Chief Executive, Chilika Development Authority, Bhubaneswar, Odisha

Shri. M. Senthil Kumar, IFS
Divisional Forest Officer,
Dharmapuri Division,
Dharmapuri, Tamil Nadu

Shri. N. Satheesh, IFS
Divisional Forest Officer, Satyamangalam Division, Sathyamangalam,
Tamil Nadu

Prof. M. Paramathma
Executive Director, National Oilseeds and Vegetable Oil Development (NOVOD) Board, Ministry of Agriculture, Department of Agriculture and Cooperation, Government of India, Gurgaon, Haryana

Prof. S. Chellamuthu
Director, Water Technology Centre,
Tamil Nadu Agricultural University,
Coimbatore, Tamil Nadu

Prof. N.B. Singh
Director of Extension Education,
Dr.Y.S.Parmar University of Horticulture and Forestry,
Nauni, Solan, Himachal Pradesh

Prof. S.T. Naik
Associate Director of Extension, University of Agricultural Sciences, Dharwad, Karnataka

Prof. C. Sekhar
Professor and Head, Department of Forest Resource Management, Forest College and Research Institute (FC&RI), Mettupalayam, Tamil Nadu

Prof. C.T. Ashok Kumar
Department of Entomology, college of agriculture, U.A.S., G.K.V.K., Bangalore, Karnataka

Dr. K. Sasikumar, IFS
Divisional Forest Officer,
North Tripura District,
Kumarghat, Tripura

Shri. Sanjayan Kumar, IFS
Deputy Director,
Periyar Tiger Reserve,
Kerala

Shri. R. Manikanda Ramanujam, IFS
Deputy Conservator of Forest
Wadsa Forest Division
Maharashtra

Shri. Bheemsingh Mohandaas, IFS
Assistant Conservator of Forest
(Working Plan),
J&K Forest Department,
Jammu and Kashmir

Shri. T. Mohan Raj, IFS
Assistant Conservator of Forest,
Tonk, Rajasthan

Dr. A.V. Santhoshkumar
Associate Professor & Head, Department of Tree Physiology and Breeding, College of Forestry
Kerala Agricultural University, Thrissur,
Kerala

Dr. Munish Kumar
Associate Professor, Department of Soil Conservation and Water Management/Forestry, C.S. Azad University of Agriculture and Technology, Kanpur, Uttar Pradesh

Dr. Sanjeev Thakur
Senior Scientist, Department of Tree Improvement and Genetic Resources, College of Forestry, Dr.Y.S.Parmar University of Horticulture and Forestry,
Nauni, Solan, Himachal Pradesh

Dr. Ramakrishna Hegde
Associate Professor & Head, Department of Silviculture and Agroforestry, College of Forestry, Ponnampet (UAS, Bangalore),
Karnataka

Dr. K.T. Parthiban
Associate Professor (Forestry), Forest College and Research Institute (FC&RI),
Mettupalayam, Tamil Nadu

Dr. P. Masilamani
Associate Professor (Seed Science &Technology)
Agricultural Engineering College and Research Institute, Kumulur, Tiruchirappalli,
Tamil Nadu

Dr. C. Buvaneswaran
Scientist, Division of Forestry, Land Use and Climate Change, Institute of Forest Genetics and Tree Breeding, Coimbatore,
Tamil Nadu

Dr. A. Venkatesh
Principal Scientist (Forestry), ICAR Research Complex for NEH Region,
Umiam (Barapani), Meghalaya

Dr. Sharad Nema
Associate professor (Forestry) & Head, School of Studies (Forestry & Wildlife), Bastar University,
Jagdalpur, Chhattisgarh

Dr. J. Jayaprakash
Scientist, Central Soil and Water Conservation Research and Training Institute (ICAR),
Dehradun

Mr. Etefa Guyassa
Head, Department of Land Resources Management and Environmental Protection,
Mekelle University,
Mekelle, Ethiopia

Dr. P. Ratha Krishnan
Senior Scientist (Forestry),
Central Arid Zone Research Institute,
Jodhpur, Rajasthan

Shri. S. Gopakumar
Assistant Professor, Dept. of Forest Management &Utilization, College of Forestry,
Kerala Agricultural University,
Thrissur, Kerala

Dr. S. Umesh Kanna
Assistant Professor (Forestry), Forest College and Research Institute (FC&RI),
Mettupalayam, Tamil Nadu

Shri. Ashutosh Pandey
Oilseeds Division, National Oilseeds and Vegetable Oil Development (NOVOD) Board,
Ministry of Agriculture, Department of Agriculture and Cooperation,
Government of India, Gurgaon, Haryana

Shri. Jadegowda
Assistant Professor, College of Forestry,
Ponnampet (UAS, Bangalore),
Karnataka

Mrs. Smitha Rajesh
Department of Forestry,
Orissa University of Agriculture and Technology,
Bhubaneswar, Odisha

Ms. S. Vennila
Forest College and Research Institute (FC&RI),
Mettupalayam,
Tamil Nadu

Dr. S. K. Uttam
Assistant Professor, Department of Soil Conservation and Water Management/Forestry,
C.S. Azad University of Agricultural & Technology, Kanpur, Uttar Pradesh

Dr. Manmohan J.R. Dobriyal
Head In-charge, Department of Forest Products and Utilization, College of Horticulture and Forestry, MPUAT,
Jhalawar, Rajasthan

Dr. Munesh Kumar
Assistant Professor, Department of Forestry,
HNB Garhwal University, Srinagar Garhwal,
Uttarakhand

Dr. Rainer W. Bussmann
William L. Brown Center,
Missouri Botanical Garden,
St. Louis, MO 63110, USA

Dr. A. Vidhyavathi
Assistant Professor, Department of Forest Resource Management, Forest College and Research Institute (FC&RI), Mettupalayam,
Tamil Nadu

Dr. Afaq Majid Wani
Assistant Professor, College of Forestry and Environment, SHIATS Deemed University,
(Formerly Allahabad Agricultural Institute),
Allahabad, Uttar Pradesh

Shri. R. Ezhumalai
Scientist -Wood Science, Institute of Wood Science and Technology (IWST),
Bangalore, Karnataka

CONTENTS

PART 1:
NATURAL RESOURCES

PART 2:
TREE AND FOREST

LIST OF TABLES

LIST OF FIGURES

ABBREVIATIONS AND ACRONYMS

AAUs : Assigned Amount Units
ABA : Abscisic Acid
ACF : Assistant Conservator of Forests
AI : Aridity Index
AOSA : Association of Official Seed Analysts
APCCF : Additional Principal Chief Conservator of Forests
APFC : Asia-Pacific Forestry Commission
ASEAN : Association of South East Asian Nations
ATO : African Timber Organization
AYUSH : Department of Ayurveda, Yoga & Naturopathy, Unani, Siddha & Homeopathy
BAF : Basal Area Factor
BC : Buffering Capacity
BDA : Biological Diversity Act
B-I Process : Bhopal-India Process
BMC : Biodiversity Management Committees
BR : Biosphere Reserve
BSI : Botanical Survey of India
C&I : Criteria and Indicators
C:N Ratio : Carbon Nitrogen Ratio
CAI : Current Annual Increment
CAZRI : Central Arid Zone Research Institute
CBD : Convention on Biological Diversity
CBF : Central Board of Forestry
CBP : Conservation Breeding Programme
CBT : Conservation Bench Terraces
CCF : Chief Conservator of Forests
CCS : Carbon Capture and Sequestration

CDM	: Clean Development Mechanism
CEC	: Cation Exchange Capacity
CERs	: Certified Emission Reductions
CF	: Conservator of Forests
CGIAR	: Consultative Group on International Agricultural Research
CIFOR	: Centre for International Forestry Research
CITES	: Convention on International Trade in Endangered Species of Wild Fauna and Flora
CMS	: Convention on the Conservation of Migratory Species of Wild Animals
COC	: Chain of Custody
COFO	: Committee on Forestry
COP	: Conference of the Parties
CPF	: Collaborative Partnership on Forests
CR	: Contingent Ranking
CSD	: United Nations Commission on Sustainable Development
CSIRO	: Commonwealth Scientific and Industrial Research Organization
CSO	: Clonal Seed Orchard
CVM	: Contingent Valuation Method
CWLW	: Chief Wild Life Warden
CZA	: Central Zoo Authority
DANIDA	: Danish International Developmental Agency
dbh	: diameter at breast height
DBHOB	: Diameter at Breast Height Over Bark
DCF	: Deputy Conservator of Forests
DF	: Diesel Fuel
DFO	: Divisional Forest Officer
DoLR	: Department of Land Resources
ECOSOC	: United Nations Economic and Social Council
EIA	: Environmental Impact Assessment
EMC	: Equilibrium Moisture Content
ENSO	: El Nino - Southern Oscillation
ERUs	: Emission Reduction Units
ESP	: Exchangeable Sodium Percent
FAO	: Food and Agricultural Organization of the United Nations
FDA	: Forest Development Agency
FFA	: Free Fatty Acids
FMU	: Forest Management Unit
FORWORKNET	: Forestry Workforce Network

FPC : Forest Protection Committee
FRA : Global Forest Resources Assessment
FRI : Forest Research Institute
FRM : Forest Reproductive Material
FSC : Forest Stewardship Council
FSI : Forest Survey of India
FSO : Forest Settlement Officer
FSP : Fibre Saturation Point
FYM : Farm Yard Manure
FYP : Five Year Plan
GA : Gibberellic Acid
GA : Genetic Advance
GATT : General Agreement on Tariffs and Trade
GCA : General Combining Ability
GEF : Global Environment Facility
GFMC : Global Fire Monitoring Center
GHGs : Green House Gases
GIM : Green India Mission
GIS : Geographic Information System
GLASOD : Global Assessment of Soil Degradation
GM : Genetically Modified
GPS : Global Positioning System
GtC : Gigatons of Carbon
GWP : Global Warming Potential
HYV : High Yielding Variety
IAA : Indole Acetic Acid
IBA : Indole Butyric Acid
IBWL : Indian Board for Wildlife
ICAR : Indian Council of Agricultural Research
ICBN : International Code of Botanical Nomenclature
ICFRE : Indian Council of Forestry Research and Education
ICIMOD : International Centre for Integrated Mountain Development
ICRAF : World Agroforestry Centre
IFAD : International Fund for Agricultural Development
IFF : Intergovernmental Forum on Forests
IIFM : Indian Institute of Forest Management
IIRS : Indian Institute of Remote Sensing

IPCC : Intergovernmental Panel on Climate Change
IPF : Intergovernmental Panel on Forests
IPIRTI : Indian Plywood Industries Research and Training Institute
IPM : Integrated Pest Management
ISRO : Indian Space Research Organisation
ISTA : International Seed Testing Association
ITTA : International Tropical Timber Agreement
ITTO : International Tropical Timber Organization
IUCN : International Union for Conservation of Nature
IUDZG : International Union of Directors of Zoological Gardens
IUFRO : International Union of Forest Research Organizations
IVI : Importance Value Index
IWRM : Integrated Water Resources Management
JFM : Joint Forest Management
JFMC : Joint Forest Management Committees
JI : Joint Implementation
KFRI : Kerala Forest Research Institute
LaCONES : Laboratory for Conservation of Endangered Species
LADA : Land Degradation Assessment in Drylands
LULUCF : Land Use, Land Use Change and Forestry
M ha : Million Hectare
MA : Millennium Ecosystem Assessment
MAB : The Man and the Biosphere Programme
MAI : Mean Annual Increment
MAP : Medicinal and Aromatic Plants
MC : Moisture Content
MDGs : Millennium Development Goals
MOA : Ministry of Agriculture
MoEF : Ministry of Environment and Forests
MOP : Meetings of Parties of the Kyoto Protocol
MoRD : Ministry of Rural Development
MPT : Multipurpose Tree
MUSLE : Modified Universal Soil Loss Equation
MW : Megawatt
MXD : Maximum Latewood Density
NAEB : National Afforestation and Eco-Development Board
NAPCD : National Action Programme to Combat Desertification

NBA : National Biodiversity Authority
NBSS&LUP : National Bureau of Soil Survey and Land Use Planning
NCA : National Commission on Agriculture
NCDMA : National CDM Authority
NFAP : National Forestry Action Programme
NFT : Nitrogen Fixing Tree
NLBI : Non-Legally Binding Instrument on All Types of Forests
NMPB : National Medicinal Plants Board
NPV : Nuclear Polyhedrosis Virus/ Net Present Value
NRCAF : National Research Centre for Agroforestry
NRSA : National Remote Sensing Agency
NRSC : National Remote Sensing Center
NTFP : Non-Timber Forest Products
NWDB : National Wasteland Development Board
NWFP : Non-Wood Forest Products
OECD : Organisation for Economic Co-operation and Development
OM : Organic Matter
PA : Protected Area
PCCF : Principal Chief Conservator of Forests
PCR : Polymerase Chain Reaction
PIP : Pugmark Impression Pads
POPs : Convention on Persistent Organic Pollutants
PRA : Participatory Rural Appraisal
PROFOR : Programme on Forests
PTGs : Primitive Tribal Groups
PVTGs : Particularly Vulnerable Tribal Groups
PWPR : Preliminary Working Plan Report
RAPD : Random Amplified Polymorphic DNA
REDD : Reducing Emissions from Deforestation and forest Degradation
RFLP : Restriction Fragment Length Polymorphism
RFO : Range Forest Officer
RIL : Reduced Impact Logging
RMUs : Removal Units
RSG : Re-introduction Specialist Group
RUPFOR : Resource Unit for Participatory Forestry
RWE : Round Wood Equivalent
SBA : Stand Basal Area

SBB	: State Biodiversity Boards
SCA	: Specific Combining Ability
SFM	: Sustainable Forest Management/ Soil Fertility Management
SIA	: Social Impact Assessment
SLEM	: Sustainable Land and Ecosystem Management
SMPB	: State Medicinal Plants Board
SOC	: Soil Organic Carbon
SOFO	: State of the World's Forests
SOM	: Soil Organic Matter
SPA	: Seed Production Area
SSO	: Seedling Seed Orchard
SSR	: Simple Sequence Repeat
TBA	: Tree Basal Area
TBOs	: Tree Borne Oilseeds
TEV	: Total Economic Value
TIES	: The International Ecotourism Society
TOF	: Trees Outside Forests
UNCCD	: United Nations Convention to Combat Desertification
UNCED	: United Nations Conference on Environment and Development
UNCHE	: United Nations Conference on the Human Environment
UNCOD	: United Nations Conference on Desertification
UNCTAD	: United Nations Conference on Trade and Development
UNDESA	: United Nations Department of Economic and Social Affairs
UNDP	: United Nations Development Programme
UNEP	: United Nations Environment Programme
UNESCO	: United Nations Educational, Scientific and Cultural Organization
UNFCCC	: United Nations Framework Convention on Climate Change
UNFF	: United Nations Forum on Forests
UNISDR	: United Nations International Strategy for Disaster Reduction
UNPFII	: United Nations Permanent Forum on Indigenous Issues
UNWTO	: United Nations World Tourism Organization
USEPA	: United States Environmental Protection Agency
USLE	: Universal Soil Loss Equation
VA	: Village Associations
VERs	: Verified Emission Reductions
VFC	: Village Forest Committee
WAZA	: World Association of Zoos and Aquariums

WBG : World Bank Group
WCFSD : World Commission on Forests and Sustainable Development
WF : Weightage Factor
WHC : World Heritage Convention
WII : Wildlife Institute of India
WMO : World Meteorological Organization
WPO : Working Plan Officer
WSSD : World Summit on Sustainable Development
WTA : Willingness To Accept
WTO : World Trade Organization
WTP : Willingness To Pay
WWF : World Wide Fund for Nature

PART 1:

NATURAL RESOURCES

1 SOIL RESOURCES

WHAT IS SOIL?

Soil is a natural body consisting of layers (soil horizons) of primarily mineral constituents of variable thicknesses, which differ from the parent materials in their texture, structure, consistence, colour, chemical, biological and other physical characteristics.

DOKUCHAEV DEFINITION

Pedology or soil science was established as a new academic discipline more than a hundred years ago by the prominent Russian scientist Vasily Dokuchaev (1846-1903), who has been recognized worldwide as its founder. Dokuchaev defined soil as an independent natural body developed over time under the influence of five soil-forming factors: parent rock, living organisms, climate, relief, and time. This has become a classic definition of soil.

Soils are applied solely to those superficial or nearly superficial horizons of rocks, that have been more or less modified naturally by the interaction of water, air and various kinds of organisms, either living or dead; this being reflected in a certain manner in composition, structure and colour of such formations. Where these conditions are absent, there are no natural soils.

VELAYUTHAM AND BHATTACHARYA (2000) DEFINITION

Soil is a dynamic natural body developed as a result of pedogenic processes through weathering of rocks, consisting of mineral and organic constituents, possessing definite physical, chemical and biological properties having a variable depth over the surface of the earth and providing a medium for plant growth.

HILGARD DEFINITION

Soil is the more or less loose and crumby part of the outer earth crust in which, by means of their roots, plants may or do find foot hold and nourishment as well as all other conditions essential to their growth.

BUCKMAN AND BRADY DEFINITION

Soil may be defined as a dynamic natural body on the surface of the earth in which plants grow and composed of mineral, organic materials and living forms.

FOREST SOILS

Forest soils are soils which have developed under the influence of forest growth and are capable of supporting forest ecosystem. It is the natural medium for growth of tree roots and associated forest vegetation. They have a major influence on trees and forest stands. Forest soils have a fundamental influence on forest management and forest operations. Similarly forest operations have the potential to maintain, improve, or degrade the condition of forest soils.

SOIL COMPONENTS

Soils consist of four major components i.e. mineral matter, organic matter (OM), water and air. The relative proportions of these four soil components vary with soil type and climatic conditions. A review on the approximate proportions (by volume) of the four soil components in a mineral soil under optimum conditions for plant growth reveals that soil contains about 50 per cent solid space and 50 per cent pore space. Total solid space of soil is occupied by mineral matter and OM by about 45 and 5 per cent respectively. Total pore space of the soil is occupied by air and water by 25 and 25 per cent respectively. OM content is very small and varies from about 3-5 per cent by weight in a top soil.

MINERAL COMPONENTS

Mineral is the inorganic component of soil which includes soil particles such as sand, silt and clay. Mineral particles are inorganic materials derived from rocks and minerals. They are extremely variable in size and composition. The larger soil particles, which include stones, gravel and coarse sands, are generally rock fragments of various kinds. These larger particles are often aggregates of several different minerals. Soil composition depends on types of primary and secondary minerals, weathering, most common elements in soils, particle and bulk density.

ORGANIC MATTER

Organic matter is the organic component of soil which includes the residue of dead plants, animals and organisms. It consists of nutrients necessary for plant growth such as nitrogen, phosphorous and potassium. Soils which contain 30 per cent or more organic matter are considered organic soils; all other soils are identified as mineral soils. A generous amount of organic matter is essential for successful crop production as it is critical to many soil processes. Organic components include non-humic substances and humic substances.

SOIL WATER

Water is the major transport agent for fluxes within and between terrestrial ecosystems. It is a prerequisite for all active life, and participates in geochemical cycles by weathering geological substrates, by leaching materials to groundwater and by moving ions and particles through the soil profile. The following soil water parameters determine the quality of soils viz. soil water content, soil solution, soil water flow, qualitative description of soil wetness.

SOIL AIR

Soil pores, the voids between minerals, organic matter, and living organisms, are filled with air or water. There is a dynamic equilibrium between water and air content within a soil. When water enters the soil, it displaces air from some of the pores. Soil air is studied under the following parameters viz. composition of soil air, movement of gasses within soil, soil porosity.

SOIL TEXTURE

Soil texture is determined by identifying the proportions of the sand, silt, clay and organic matter. Sand is the large sized (0.05 - 2 mm diameter) particles and they can be seen with the naked eye. Silt is the medium sized (0.002 - 0.05 mm diameter) particles and cannot be individually seen with the naked eye. Clay is the small sized (< 0.002 mm diameter) particles. Relative proportions of these minerals can be estimated by visually observing and feeling the soil. Determination of soil texture takes much experience and skill, but can be improved with practice.

Sand particles do not retain water well, therefore soils containing much sand tend to allow water to flow through it or evaporate quickly. At elevated flow rates, water could also flush the nutrients in the soil beyond the plant's root zone making it unavailable for plant uptake. Silt and clay particles retain water well but do not allow water to flow through it very quickly. Therefore, it is very hard for a plant to extract water and nutrients from soil if it contains much silt and clay.

Organic matter contains nutrients available for plant uptake, improves water infiltration, decreases evaporation and increases the water holding capacity. Soils which contain a well-balanced mixture of sand, silt, clay and organic matter retain water well and allow it to flow through the soil at an appropriate rate such that it does not inhibit plant growth. Largely because of the organic matter, it also contains essential nutrients for plant growth. This type of soil is called loam which is the best for plant growth.

SOIL STRUCTURE

Soil structure is the physical arrangement of all soil components. It is this arrangement which determines how much air and water can be present in, absorbed onto, or flow through the soil. The small fragments of mineral soil are often held together in clumps of specific form. The size, shape and strength of these clumps determine the soil structure. Soils which contain more air space allow greater water infiltration and root penetration. Air spaces can be formed through the movement of earthworks in the soil. Soils with less air space can naturally form compacted layers known as hardpans. This type of soil tends to create drought sensitive conditions as many crops cannot successfully root in the zone below the hardpan resulting in poor plant growth.

SOILS OF INDIA

India, situated between the latitudes of 08°04' and 37°06' N and longitudes of 68°07' to 97°25' E, has a geographical area of 329 M ha. The soil temperature and moisture regimes play an important role in classifying soils and crop planning, as each plant species requires, specific soil temperature and moisture conditions for its optimal growth. India is dominantly represented by thermic and hyper-thermic temperatures and Ustic, Aridic and Udic moisture regimes. India with a wide variety of landforms, geological formations and climatic conditions, exhibits a large variety of soils. India represents all the major soils of the world. The grouping of these soils can be achieved by using either of the two systems, viz. genetic and soil taxonomy. The genetic system is based on genetic factors and processes. The soil taxonomy is based on the properties of soils which are the reflection of soil genesis. The soils of India, according to the genetic factors have been classified into eleven major soil groups.

- Alluvial soil
- Black cotton soil
- Red soil
- Laterite soil
- Desert soil
- Tarai soil
- Saline and sodic soil
- Organic soil
- Forest soil
- Podzole soil
- Skeletal soil

ALLUVIAL SOIL

The name alluvial is given to the soils that have developed in alluvium, irrespective of their place of occurrence and degree of development. They are one of the important groups of soils for agricultural production. They are extensively distributed in the states of Rajasthan, Haryana, Punjab, Uttar Pradesh and in valley of Krishna, Kaveri, Brahmaputra, Godawari, Narmada, Tapti and Surma rivers.

Order : New alluvial soil - Entisols order

: Old alluvial soil - Inceptisols order

Area : 75 M ha (1st largest group)

Rainfall : 350 mm - 1700 mm annually

Main crops : All crops

Properties :

- These soils are sandy loam to clay loam in textured
- Soil colour is light brown to dark brown

- Soil pH is 7-8
- Deficient in N, P and humus; moderate - high in K and excess in lime
- Generally friable in nature
- Dominancy of illite mineral
- Sometimes, Kankar layers are found in 'B' horizon

BLACK COTTON SOIL

The name black is given to soils that are very dark in colour and turn very hard on drying and sticky and plastic on wetting, and hence are very difficult to cultivate and manage. These soils are comparable with the Grumosols of the USA. In view of their black colour, they are comparable with the Chernozems of Russia and Prairie soils of the USA, but differ in their physiochemical properties. These soils are dominantly distributed in the central, western and southern states of India like, part of Madhya Pradesh, Gujarat, Maharashtra, West part of Andhra Pradesh, plateau of Deccan and Tamil Nadu.

Order : Vertisols order
Area : 72 M ha (2nd largest group)
Climate : Semi-arid to sub-humid climate
Main crops : Cotton, Bajra, Jowar, Ragi, Wheat and Millets
Properties :

- Highly clay textured with clay content between 30-80 per cent
- Soil colour is dark brown to black
- Soil pH is 7.8 - 9.5
- Deficient in N,P, OM and rich in K, Fe, Al, $CaCO_3$, $MgCO_3$, Montmorillonite
- Silica: sesquioxide ratio is 3 - 3.25
- Have high water holding capacity
- High sticky in nature
- CEC is 30-60 Cmol (P^+)/kg
- Swelling and shrinking property occurs

RED SOIL

The name Red is given to soils rich in sesquioxides that have developed on rocks of Archean origin (granite, gneiss) and on well drained, stable and higher land forms under hot, semi-arid to humid sub-tropical climatic conditions. Under such conditions, the weathering is moderately intense and leads to enhanced decalcification. These soils are predominantly observed in whole part of Chennai, Mysore, part of Uttar Pradesh (Mirzapur, Hamirpur, Jhansi), Bihar (Santhal Pargana, Chhotanagpur), West Bengal and eastern half part of Rajasthan, Madhya Pradesh, Andhra Pradesh and Odisha.

Order : Alfisols order

Area : 70 M ha (3rd largest group)

Climate : Sub-humid climate

Main crops : Cotton, Bajra, Jowar, Ragi, Wheat and Pulses

Properties :

- Loamy sand - clayey textured, poor in clay
- Soil colour is red due to presence of Hematite
- Neutral - acidic in nature
- Deficient in NPK and low in OM and lime
- CEC is < 20 Cmol (P^+) / kg
- Silica: sesquioxide ratio is 2

LATERITE SOIL

Found in typical monsoon conditions, under conditions of high temperature and heavy rainfall with alternate wet and dry periods. The alterations of wet and dry season leads to the leaching away of siliceous matter and lime of the rocks and a soil rich in oxides of iron and aluminum compounds is left behind. These are found in hill tops of south, hills of Rajasthan, plain part of Odisha, Maharashtra, Malabar and Central India.

Order : Oxisols order

Area : 25 M ha

Climate : Tropical and sub-tropical regions with high rainfall areas

Main crops : Low land-Paddy and Up land - Tea, Coffee, Rubber, and Cinchona

Properties :

- High clay content in soil
- Soil pH is 5-6
- Essential plant nutrients are less
- CEC is < 16 Cmol (P^+)/ kg
- Silica: sesquioxide ratio is 1-1.25
- Characterized by compact, honey combed or vasicular structure in sub-soil
- Deficient in lime

DESERT SOIL

Location : North-West part of India i.e. Rajasthan, Kathiabar, south Punjab and in west district of Uttar Pradesh

Order : Aridisols order

Area : 29 M ha

Climate : Arid regions

Main crops : Sorghum, Bajra, Pulses

Properties :

- Sandy - loamy sand textured with clay content between 3.5 to less than 10 per cent
- Soil colour is light brown-yellowish brown
- Soil pH is 7.8 - 9.2
- Poor soil fertility status
- Organic matter content of the soils is very low
- Poor water holding capacity
- High soluble salts

TARAI SOIL

Location : Tarai area of Punjab, Uttar Pradesh, Bihar, West Bengal and Himalaya

Order : Mollisols order

Area : 5.7 M ha

Climate : Wet regime and high water table condition for most part of the year

Main crops : Sorghum, Bajra, Pulses

Properties :

- Silty loam - loam textured soils
- Soil colour is black to grey black
- Soil pH is 4.7-5.8
- Waterlogged soil has poor fertility
- Well drained soil has high fertility
- In lower horizons, stones and gravels are dominant

SALINE AND SODIC SOIL

Location : Saline soils are found in Punjab, Uttar Pradesh, Rajasthan, Madhya Pradesh and Bihar, however, sodic soils are found in Gujarat, Maharashtra and Karnataka.

Order : Aridisols order

Area : 7 M ha

Climate : Arid to semi - arid climate

Main crops : Paddy dominant

Properties :

- Soil pH is > 7.3
- Poor soil fertility status
- Poor physical condition of soil
- In 'B' horizon, kankar layer is found
- Soil colour is white - brown (saline) and black (sodic soil)
- Excess salts are present
- Growth of most of the crop plant is reduced

ORGANIC SOIL

Location : Tarai of Himalaya, middle part of Bihar, North East part of Uttar Pradesh, Kerala, Cochin and Sundarban (West Bengal)

Order : Histosols order

Area : 2720 km^2

Climate : Humid climate with low temperature

Main crops : Paddy dominant

Properties :

- Soil colour is blue to black
- Soil pH is about 4
- OM is found 15-50 per cent and sometimes up to 80 per cent
- CEC is 50-100 Cmol (P^+) / kg

- Soil horizons are very clear and mature

FOREST SOIL

The forest soils may be defined as a part of the earth surface which serves as a medium for the substances of forest vegetation. It is a living, dynamic and biological entity which undergoes changes all the time in the course of forest vegetation and silvicultural systems. The formation of these soils is mainly governed by the characteristic deposition of organic matter derived from forest growth and found in Jammu & Kashmir, Uttarakhand, Himchal Pradesh, Uttar Pradesh, Bihar, Madhya Pradesh, Maharashtra, Kerala and North East regions.

Order : Deciduous forest - Histosols order
Evergreen forest - Spodosols order

Area : 75 M ha

Dominant crops : Apple, Pear

CHARACTERISTICS

- These soils are characterized with the presence of litter with an associated micro flora and fauna
- Root activities from the surface to the subsurface horizons are common phenomena in forest soils
- Forest soil has natural well distinct horizons
- Forest soils are characterized by higher porosity, higher permeability, more soil aggregates and excellent water holding capacity
- In forest soil, the C/N ratio is higher than agricultural soil
- Typical well developed forest soil profile consists of 4 horizons viz. organic horizon (O), alluvial horizon (A), zone of accumulation (B) and unaltered parent materials (C)

PODZOL SOIL

Location : Himalaya mountain and in some part of south India

Order : Spodosols order

Area : Very less area in India (<1 per cent)

Climate : Humid temperate region with high rainfall areas

Properties :

- pH of upper most of soil is low and increases with increase in soil depth
- CEC of 'A' horizon is less than the 'B' horizon

- 'Si' is dominant in 'A' horizon and 'Al' is less and vice versa in 'B' horizon
- Good crops cannot be grown due to absence of sufficient available plant nutrients
- Sesquioxide and humus work as cement and produce hard layer and kankar in 'B' horizon

SKELETAL SOIL

Location : Ladakh area of Himalaya Mountain and Vindhyan region of central India

Order : Entisols order

Area : 24 M ha

Climate : Cold climate

Properties :

- Silty loam - loam textured soil
- Soil colour is light yellow-brown.
- Unfertile soil
- Very shallow soils (6-15 cm deep)
- Natural soil profile and horizons are absent

NUTRIENT CYCLING

A nutrient cycle is the movement and exchange of organic and inorganic matter back into the production of living matter. The process is regulated by food web pathways that decompose matter into mineral nutrients. Nutrient cycles occur within ecosystems. Ecosystems are interconnected systems where matter and energy flows and is exchanged as organisms feed, digest and migrate. Minerals and nutrients accumulate in varied densities and uneven configurations across the planet. Ecosystems recycle locally, converting mineral nutrients into the production of biomass, and on a larger scale they participate in a global system of inputs and outputs where matter is exchanged and transported through a larger system of biogeochemical cycles.

Biogeochemical cycle is the movement (or cycling) of matter through a system (atmosphere, lithosphere, hydrosphere, biosphere). Biogeochemical cycles include carbon cycle, nitrogen cycle, sulfur cycle, phosphorus cycle, iron cycle, water cycle, oxygen cycle and others that continually recycle along with other mineral nutrients into productive ecological nutrition. There are two types of nutrient cycling viz. geochemical nutrient cycling, biological nutrient cycling.

GEOCHEMICAL NUTRIENT CYCLING

It is an open system which concerns the import export relationship of nutrients into and out of the ecosystem. It includes nutrient imports from such sources as nitrogen and fertilization. The outputs include leaching and erosion losses in drainage water, volatile losses from fire and desertification and removal in harvests.

BIOLOGICAL NUTRIENT CYCLING

It is the closed system which involves plant-soil exchanges of nutrients. It involves the transfer of nutrients between forest floor soil, associated plants and animal communities. It may also include the internal transfer of nutrients among organs within trees. Biological nutrient cycling deals with the translocation of nutrients within standing tree mass and nutrient transfer between soils and trees.

MAJOR PATHWAYS IN NUTRIENT CYCLING

Out of 126 naturally occurring chemical elements only 18 have been identified as essential elements (or nutrients) without which plants cannot grow and complete their life cycles. Elements essential for plant growth include: C, H, O, N, P, K, S, Ca, Mg (elements used by plants in relatively large amounts called as macronutrients), Fe, Mn, B, Zn, Cu, Cl, Co, Mo, Ni (elements used by plants in small amounts called as micronutrients). These essential nutrients play various important roles in the structure and metabolism of plants. They are continuously cycled through the atmosphere-soil-plant system according to specific transformation and transport processes. Some of these processes occur in soils and make elements available to plant roots and soil organisms.

CARBON CYCLE

The carbon cycle is a bio-geochemical cycle which exchanges carbon between various reservoirs or storage places within the earth system viz. the land, the oceans and the atmosphere.

NITROGEN CYCLE

It is the biogeochemical cycle that describes the transformations of nitrogen and nitrogen containing compounds in nature. It is the transformation of nitrogen from atmosphere (air) to soil, plants, animal life and returns back to air or soil through decay or denitrification.

PHOSPHORUS CYCLE

The phosphorus cycle describes the movement of phosphorus through the lithosphere, hydrosphere and biosphere. The atmosphere does not play a significant role, because phosphorus and phosphorus-based compounds are usually solids at the typical ranges of temperature and pressure found on Earth.

SULFUR CYCLE

It is the cyclic movement of sulfur in different chemical forms from the environment to organisms and then back to the environment from organisms. It includes mineralization of organic sulfur into inorganic forms such as hydrogen sulfide (H_2S), elemental sulfur, as well as sulfide minerals and then oxidation of hydrogen sulfide, sulfide, and elemental sulfur (S) to sulfate (SO_4^{2-}). Finally reduction of sulfate to sulfide and incorporation sulfide into organic compounds occur.

IRON CYCLE

The iron cycle is the biogeochemical cycle of iron through landforms, atmosphere and oceans. The cycling of Iron consist of largely of oxidation-reducion (redox) reactions that reduce ferric ion (Fe^{3+}) to ferrous ion (Fe^{2+}) and similarly oxidizes Fe^{2+} to Fe^{3+}.

NUTRIENT CYCLING IN FOREST ECOSYSTEM

Nutrient cycle in forest ecosystem is a dynamic and complex system of geological, chemical and biological cycling through which the soil organic matter and nutrient supplies are replenished and maintained. It ensures continued productivity of the site.

The processes encompassing nutrient cycling are of critical importance to the functioning and evolution of forest ecosystems. Cycling of nutrients in forest ecosystems is an integrating set of processes involving the transfer of energy and nutrients within an ecosystem, including inter-system and intra-system interactions within and between the atmosphere, biosphere, geosphere, and hydrosphere. Nutrient cycling involves a series of interrelated processes, such as nitrogen fixation, atmospheric deposition of nutrients, organic matter decomposition and mineralization, rock weathering and nutrient uptake. Since one process functions as the precursor to another, the flow of nutrients is linked in a set of specific interconnected steps that ultimately lead to a series of cyclic pathways.

FOREST NUTRIENT CYCLING

Forest nutrient cycling is defined as the processes of nutrient uptake, incorporation of mineral nutrients into biological tissues of plants and trees, litterfall and the decomposition of organic matter with the concomitant release of nutrients to soil by microorganisms.

To maintain all the biochemical reactions necessary for plant growth, at least 18 essential mineral nutrients are required from the geosphere. These nutrients are extracted from the soil by plant uptake and returned to the soil directly or indirectly as organic matter. The organic matter in turn is a source of energy for heterotrophs which further oxidize the organic compounds by decomposition processes (i.e., by respiration), simultaneously releasing the mineral nutrients back to the soil. Thus nutrient cycling is essential to the cyclic flow of nutrients in forest ecosystems.

NUTRIENT POOL IN VEGETATION: The quantity of nutrients contained in the living biomass of an ecosystem is termed as nutrient pool in vegetation. The total biomass and nutrient pools vary widely between contrasting forest types. Vegetation pool is typically segregated by individual species within an ecosystem and within a given species into various components: foliage, woody tissue, bark and roots. Several rotations of nutrient removal by forest harvest from low nutrient sites may result in a loss of productivity in subsequent rotations.

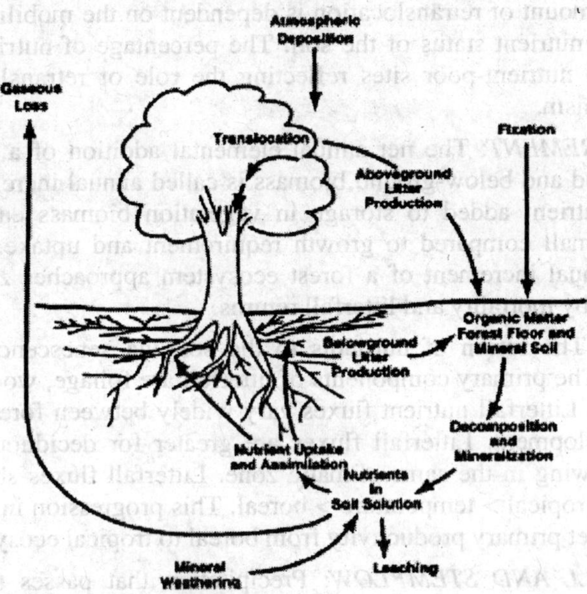

Fig. 1.1: Overview of Forest Nutrient Cycling

PLANT UPTAKE: The nutrients are taken into a plant by root and foliar uptake. But foliar uptake generally represents a very small proportion of the total nutrient uptake. Mycorrhizae fungi associations with plant roots can enhance the availability and uptake of several essential nutrients (e.g., phosphorus, copper, calcium and iron). Foliar uptake consists of ion uptake from wet deposition across the cell membrane or direct absorption of gases (e.g., SO_2 and NO_2).

The annual uptake (or nutrient demand) changes greatly over the course of development of a forest from establishment to maturity. During the period of early establishment, nutrient uptake will increase with increasing gross ecosystem production. Uptake rates will peak at approximately the time of canopy closure. Following canopy closure, gross production slows due to competition and mortality and more carbon is allocated to woody materials having a lower nutrient content (e.g., stem). Thus, uptake rates will decline from their maximum level at canopy closure and maintain a relatively constant value.

GROWTH REQUIREMENT OF NUTRIENTS: The growth requirement is defined as the total amount of a nutrient that is required each year by vegetation to meet its demands for nutrients associated with the annual increment, litterfall and canopy leaching. The requirement is obtained by uptake and retranslocation processes:

Growth Requirement = Uptake + Retranslocation

Growth Requirement = Annual Increment + litterfall + Canopy Leaching

Canopy Leaching = Throughfall + Stemflow - Atmospheric Deposition

RETRANSLOCATION: Nutrient removal from plant tissue into the perennial part of the plant prior to senescence is termed as retranslocation. Retranslocation occurs primarily from foliage and fine roots. Retranslocated nutrients are available for reuse by the plant in the next growth cycle. The amount of retranslocation is dependent on the mobility of the element, the tree species and the nutrient status of the soil. The percentage of nutrients retranslocated is generally greater on nutrient-poor sites reflecting the role of retranslocation as a nutrient conservation mechanism.

ANNUAL INCREMENT: The net annual elemental addition of a nutrient incorporated into the above-ground and below-ground biomass is called annual increment. In other words, the amount of a nutrient added to storage in vegetation biomass each year. The annual increment is very small compared to growth requirement and uptake. As a forest reaches steady state, the annual increment of a forest ecosystem approaches zero as the growth of biomass is balanced by mortality and litterfall returns.

LITTERFALL: The return of nutrients to the soil via senescence of plant tissues is termed as litterfall. The primary components of litterfall are foliage, woody tissues, roots and reproductive tissues. Litterfall nutrient fluxes vary widely between forest types and with the stage of forest development. Litterfall fluxes are greater for deciduous trees compared to coniferous trees growing in the same climatic zone. Litterfall fluxes show an increase with decreasing latitude: tropical > temperature > boreal. This progression in litterfall corresponds with an increase in net primary productivity from boreal to tropical ecosystems.

THROUGHFALL AND STEMFLOW: Precipitation that passes through a vegetation canopy reaches the soil surface as either canopy throughfall or stemflow. As the water penetrates the canopy and flows down the stems, nutrients may be leached from or taken up by the plant tissue. Thus, nutrient concentrations may either increase or decrease relative to the amount input by atmospheric deposition (e.g., wetfall and dryfall).

Net Canopy Exchange = Total Deposition (Wet + Dry) - (Throughfall + Stemflow)

HERBIVORY: The consumption of living tissue by herbivores is termed as herbivory. Herbivory short-circuits the nutrient cycle by consuming living tissues prior to senescence and litterfall. Many of the nutrients may be returned directly to the ecosystem by herbivore defecation and death or the nutrients may be transferred to other ecosystems by herbivore migration. In most forest ecosystems, herbivores consume less than 10 per cent of the annual net primary productivity.

DECOMPOSITION: Decomposition is defined as the interrelated processes by which organic matter is broken down to CO_2 and humus with a simultaneous release of nutrients. These processes are a critical link responsible for recycling of nutrients in the intra-system nutrient cycling. The decomposition process often begins with soil macrofauna which physically break down the particle-size of organic matter and predigesting organic materials. Action by microorganisms results in the progressive breakdown of organic matter.

Litter → partially decomposed organic matter → Humus

FACTORS AFFECTING DECOMPOSITION: The rate of decomposition is dependent on litter quality, microbial activity and environmental conditions. Litter quality includes nutrient content (e.g., C/N ratio), composition of organic matter especially lignin concentrations (lignin/nitrogen ratio) and concentrations of polyphenols (including tannins). Soil temperature and moisture content are very important factors affecting decomposition rates.

IMMOBILIZATION: Immobilization is the conversion of an element from an inorganic to organic form by microorganisms.

MINERALIZATION: Mineralization refers specifically to those decomposition processes that release inorganic compounds from organic matter. The amount of a nutrient released by mineralization that is available for plant utilization (uptake) is referred to as net mineralization:

Net Mineralization = Gross Mineralization – Immobilization

GASEOUS INPUTS AND OUTPUTS

NITROGEN FIXATION: Microorganisms may contribute significantly to nutrient inputs and losses from an ecosystem through oxidation and reduction reactions. The primary processes of importance to forest nutrient cycling are N_2 fixation and denitrification. Several genera of non-leguminous shrubs and trees have root nodules with symbiotic microorganisms which may fix atmospheric nitrogen. Nitrogen fixation is performed by a wide variety of microorganisms including symbiotic microorganisms (*Rhizobium*) and free-living organisms (organotrophs such as *Azotobacter* and phototrophs such as *Cyanobacteria*), Actinomycete (such as *Frankia* common to *Alnus, Ceanothus, Casuarina*) and lichens (phototrophic association of algae with fungi). Nitrogen fixation rates vary widely among species and ecosystems with a reported range of 2 to 300 kg/ha/year.

DENITRIFICATION: It is an important process in nitrate-rich ecosystems experiencing periodic anaerobic conditions. Under anaerobic conditions, NO_3^- is used by certain microorganisms as an electron receptor. The process involves microbial reduction of NO_3^- to NO_2^- and then to gaseous N_2O and N_2, which are subsequently lost to the atmosphere. Within the context of the global nitrogen cycle, the amount of biological nitrogen fixation (140 x 10^{12} g N per year) in terrestrial ecosystems nearly balances the estimated loss of nitrogen from terrestrial ecosystems by denitrification (130x 10^{12} g N per year).

NUTRIENT DEPOSITION

DRY DEPOSITION: Gravitational sedimentation of particles during periods without precipitation. These particles include: aerosols, sea salts, particulate material and adsorbed/ reacted gases captured by vegetation.

WET DEPOSITION: The dissolved constituents in precipitation reaching forest ecosystem is called wet deposition. Wet deposition may contribute relatively large annual inputs of nutrients into forest ecosystems, especially in regions affected by air pollution.

WEATHERING

Weathering is defined as the breaking down of rocks and minerals by physical, chemical and biological agencies. It is the breakdown and decomposition of earth material, namely rocks. Weathering of rocks-forming minerals can create new products from pre-existing rocks. The physical disintegration of rocks affects soil development and texture. Weathering releases chemical compounds that become available for biological processes. The weathering of carbonate minerals releases carbon to the atmosphere which impacts on atmospheric chemistry and temperature. Weathering of rocks and minerals means destructive process. Weathering process are mostly slow (hundreds to thousands of years).

Earth weathering occurs in three ways. Physical weathering, also called mechanical weathering, involves the disintegration of rock materials. Physical weathering incurs no change in the chemistry of the material being altered. Instead, it simply breaks large pieces into smaller ones. Chemical weathering involves the decomposition of rocks and sediment. In this case, a chemical change occurs and a new product is created from the material that has undergone weathering. In biological weathering, biological or living organisms help in weathering to some extent. Weathering processes are determined by the climate and vegetation of a place. Dry locations tend to be dominated by physical weathering and moist places by chemical weathering.

Table 1.1: Three Types of Weathering

Physical Weathering/ Mechanical Weathering	Chemical Weathering	Biological Weathering
It means disintegration of rocks and minerals	It means decomposition	It means disintegration and decomposition of rocks and minerals by biological or living agents
Primary minerals and rocks are splitted in fragment	Chemical weathering is not possible without the presence of water	Biological agent are man and animals, higher plants and their roots, microorganisms
Mechanical action of various weathering agents i.e. water, wind, ice, hail storm, frost, temperature and glacier.	Weathering takes place mainly at the surface of rock minerals and formation of secondary products	
Consolidated massive rocks converted into smaller bites without any chemical change	Chemical agents help in weathering to some extent through the processes such as solution, hydration, hydrolysis, oxidation, reduction, carbonation etc.	
In physical weathering, size and shape of rocks and minerals may be changed		

SOIL FORMATION

Soil formation is a long slow process. It is estimated that an inch of soil takes 500 to 1000 years to form. Soil is constantly being formed and eroded. The soil formation is the process of weathering of rocks and minerals into regolith and subsequently regolith is transformed to true soil. Soil formation is constructive process which is slow and continuous. Pedogenesis is the study of soil formation. Soil formation is a process of two clear stages i.e. weathering of rocks and minerals (destructive process) and development of true soil (constructive process). Soil formation is the result of biological process.

FACTORS OF SOIL FORMATION

There are five major factors that influence soil formation viz. parent material, climate, living organisms (especially native vegetation), topography and time. The combined influence of these soil-forming factors determines the properties of a soil and their degree of expression. Hans Jenny (1941) expressed the relationship of these five factors by equation:

$$S = f\,(cl, r, o, p, t, \ldots)$$

Where,

S = soil properties (organic matter content, pH, soil texture etc.)
cl = regional climate
o = potential biota
r = topography
p = parent material
t = time

ACTIVE FACTORS

CLIMATE: Climate affects soil formation largely through precipitation and temperature. Water percolates and moves from one place of parent material to another, it carries with it substances in solution and in suspension. Rainfall affects profile development through erosion producing these soils on slopes and deposition of soil material downhill. With increasing moisture, N and C content, clay content, aggregation, saturation capacity and exchangeable H tend to increase. The depth of $CaCO_3$ horizon (calcic) in pedocals soils increase with increasing moisture.

With increasing temperature, the depth of weathering and clay content show an increase. Under tropical and sub tropical climate with high rainfall, hilly leached soils have developed i.e. laterite soil. Low temperature increases leaching by reducing evaporation and favours the accumulation of organic matter by slowing down the process of decomposition.

BIOSPHERE: Biosphere consists of vegetation and organisms. Vegetation shows main effect on soil formation through the amount and nature of organic matter, it adds to the soils. Vegetation also helps to control erosion. Burrowing animals, rodents, earthworms, ants etc. are highly important in soil formation. Burrowing animals cause constant mixing within the soil profile. The role of microorganisms as soil formers is intimately related to humification and mineralization.

PASSIVE FACTORS

RELIEF OR TOPOGRAPHY: With the same kind of climate and parent material , soils that have developed on steep hill sides have thinner A and B horizons due to the surface erodes quite rapidly and less water moves downward within the profile. Soil materials on gently sloping topography have more water passing through them and the profile is generally deeper.

PARENT MATERIAL OR DEVELOPMENT OF SOIL PROFILE: The presence of limestone in parent material shows the development of acidity in humid regions. Trees growing in lime stone material produce leaf litter that is relatively rich in Ca.

TIME: The length of time required for a soil to develop horizons depends upon many factors i.e. climate, nature of PM, burrowing animals and relief.

SOIL FORMING PROCESSES

Podzolization: When humus, sesquioxide and clay goes down in 'B' horizon by percolation from 'A' horizon. As a result, increases the ratio of silica: sesquioxide in 'A' horizon and accumulation of humus, sesquioxide and clay in 'B' horizon, the soil so formed is called podzol soil and process is known as podzolization.

Laterization: When mobile silica (SiO_2) is deposited in 'B' horizon and residue sesquioxide on 'A' horizon. As a result, it decreases the ratio of silica: sesquioxide (< 2) and 'A' horizon is brick like due to presence of Fe_2O_3 (Hematite). The soil formed in this process is called laterite soil or latosols and process is called laterization.

Pedoturbation: Inter mixing of soil horizon is called pedoturbation or it is the process of mixing in soil.

Calcification: Calcification is the process of precipitation and accumulation of carbonates in the solum or in some part of the profile.

Decalcification: Reactions that remove $CaCO_3$ from one or more soil horizons or movement of $CaCO_3$ out of a section of the profile is called decalcification.

Eluviation: Leaching or removing from upper horizon to lower horizon is known as eluviation. When salts, elements, organic matter , small particles etc. are dissolved or coincide in water or suspension with water and goes down in lower horizons by percolation from the upper surface, the process is called eluviation and the upper horizons are called eluvial horizon. Generally, 'A' horizon is also called eluvial horizon.

Illuviation: When precipitation and deposition of salts, elements, OM, small particles etc. occur in lower surface, the process is called illuviation and lower horizons are called illuvial horizon. Generally, 'B' horizon is also called illuvial horizon.

Soil Erosion: Soil erosion is defined as the detachment and transportation of soil mass from one place to another through action of water or wind or beating action of raindrops.

Leaching: Washing out soluble materials from the solum is called leaching. Leaching is possible when water table is down. Leaching process is very useful in salt affected areas.

Enrichment: Addition of material to a soil body is called enrichment.

Cumulization: Cumulization is the process of accumulation of mineral material on the soil surface either by air or water. Process occurs in where material eroded from upper slope has been accumulated in lower slope resulting heap the lower portion.

Clay Migration/ Lessivage: Removal of fine clay in suspension form from the upper part of soil profile and its accumulation in the lower part is called clay migration. Physical downward movement of clay mineral is termed as clay migration.

Silication: Accumulation of silica is called silication.

Resilication: Formation of Kaolinite from Gibbsite or formation of Smectite from Kaolinite in presence of large amount of $Si(OH)_4$ (Silicic acid) at higher pH values.

Desilication: The chemical migration of silica out of a section of soil profile.

Decomposition: The breakdown of mineral and organic materials.

Synthesis: The formation of new particles of mineral and organic species.

Melanization: The process refers to changes in colour value in soil whether caused by addition or losses of organic matter. Melanization develops Mollic horizon due to incorporation of organic matter.

Littering: The accumulation of organic litter and associated humus on the mineral soil surface to a depth of <30 cm.

Humification: The transformation of raw organic material into humus. Humification is a complex process involving various organisms i.e. fungi, bacteria, actinomycetes, earthworms, termites. Humification helps in the formation of surface humus layer.

Mineralization: Decomposition of OM and subsequent release of oxide solids.

Gleization: The reduction of Fe under anaerobic waterlogged soil condition with the production of bluish-greenish grey colour with or without yellowish brown, brown and black mottles and ferric or maganiferrous concretions is called gleization. Gleization occurs in flood plain areas.

Braunification/ Rubification: Release of Iron from primary minerals.

Salinization: The process of accumulation of soluble salts i.e. SO_4^-, Cl^- and $NO3^-$ of Ca, Mg, Na, and K in soil is called salinization and the soil so formed is called saline soil.

Alkalinization: The accumulation of Na^+ ions on soil colloidal complex is known as alkalinization and the soil so formed is called alkali or sodic soil. It develops natric horizon.

Desalinization: Removal of soluble salts from the surface soil is known as desalinization. It develops salic horizon.

Degradation: Leaching of Na^+ and salts from surface soil and pH is about 6, the process is known as degradation and the soil so formed is called degraded sodic soil.

Brunisolic: Fulvic acid reacts with Fe and makes complex compound and may cause its eluviations which is known as brunisolic. The material is brown in colour.

Ferrugination: Iron sesquides adhere firmly to sand grains and gravels giving them red colour resulting cement like. It leads to development of brown, reddish brown or reddish soil colour.

Audification: Accumulation of H^+ ions in soil body.

Chelation: Forming complexes with metals by organic agents.

Compaction: The physical reduction of the air content resulting in an increased BD.

Dehydration: Loss of water reverting the compound to the original status.

Deposition: Deposition is the sedimentation of transported material. Accumulation of soil particles, nutrients and enriched of medium sized particles.

Disaggregation: Breaking down of aggregates.

Induration: Hardening of a section of the profile caused by iron pans and plinthite and other cementing agents i.e. Si or Ca.

Decolouration/ Leucinization: Lighting colour in a section of profile and formation of albic endopedon.

Oxidation: Formation of an oxide or the release of electrons.

Solifluction: Slow flow of saturated soil on a permanent frozen soil.

Upward Movement: Movement of dissolved or suspended matter by capillarity.

Carbonization: The conversion of OM into charcoal is called carbonization.

Horizonation: The development of all horizons in soil is called horizonation.

Paludization: Paludiation is the process of accumulation of a thicker mass of organic materials under poorly drained soil or anaerobic soil condition.

Sulfofication: The biological oxidation of sulphur and its compound.

STAGES OF SOIL FORMATION

There are five stages of soil formation (Edward Mohr and Van Baren, 1954).

Initial stage	: Un-weathered parent material
Juvenile stage	: Weathering start, but much of the original material still un-weathered.
Verile stage	: Easily weatherable minerals fairly decomposed, clay content increased, and slowly weatherable minerals still.
Senile stage	: Decomposition reaches at final stage, only most resistant minerals survive.
Final stage	: Soil development completed.

ORGANIC MATTER

The residue of animals, microorganisms, plants, vegetation etc. which are found on the upper layer of soil surface or added from outside, is called organic matter (OM). It consists of a wide range of organic (carbonaceous) substances including living organisms, carbonaceous remains of organisms, organic compounds produced by current and past metabolism in the soil. Organic matter is lost from the soil as CO_2 produced by microbial respiration. Because of such loss, repeated additions of new plant and animal residues are necessary to maintain soil organic matter. More carbon is stored in the world's soil than in plant biomass and atmosphere combined. There are three sources of organic matter viz.

Plant sources	: Straw, green leaves, stem, root stubbles, twigs
Animal sources	: Dead animals, birds, insects, orgainisms, bones
Other sources	: FYM, compost, green manuring, fish meal, dried blood etc.

The influence of organic matter on soil properties and consequently on plant growth is far greater than the low amount it presents in soil. Organic matter binds mineral particles into a granular soil structure that is largely responsible for the loose, easily managed condition of productive soil. The soil's mineral particles are clustered into aggregates by the binding effect of the polysaccharide gums produced by soil microorganisms, networks of fungal hyphae, some humus compounds, by the activity of earthworms and by plant root systems. The aggregates encapsulate a portion of the humus and protect it from microbial degradation. Almost 60 per cent of the soil organic matter on a dry basis is made of soil organic carbon (SOC).

$$SOC = SOM \times 0.575$$

ROLE OF ORGANIC MATTER

It makes good physical condition of soil and makes P and K more available in acidic soil. Organic matter maintains favourable condition of aeration, infiltration and permeability and maintains optimum temperature. OM increases water holding capacity and buffering capacity of soil and decreases water erosion, wind erosion and alkalinity of soil. It also prevents leaching loss and helps in growth of microorganisms. It supplies food for microorganisms and serves as a reservoir of mineral nutrients.

DECOMPOSITION OF ORGANIC MATTER

DECOMPOSITION OF SOLUBLE SUBSTANCES: Sugar and water soluble nitrogenous compound are the first to be decomposed and they produce very readily available sources of C, N and energy for the microorganisms. Water soluble organic substances are break down to NH_4^+ and NO_3^- by the process of ammonification and nitrification.

AMMONIFICATION: The transformation of organic nitrogenous compounds into NH_4^+ forms by enzymatic hydrolysis through some intermediate steps. Amonification occurs as a result of action of enzymes produced by microorganisms. Ammonifying bacteria is Nitrosococcus. Nitrogen enters the soil through the decomposition of protein in dead organic matter. This process liberates a lot of energy which can be used by the saprotrophic microbes.

Organic nitrogenous compounds $\rightarrow$ Polypeptides $\rightarrow$ Amino acids $\rightarrow$ NH_3 or salt of NH_4^+

Organic nitrogen $\longrightarrow NH_3$

NITRIFICATION: The process of conversion of NH_4 to NO_2^- by *Nitrosomonas* bacteria and then to NO_3^- by *Nitrobactor* bacteria is known as nitrification. This involves two oxidation processes. The ammonia produced by ammonification is an energy rich substrate for *Nitrosomonas* bacteria. They oxidise it to nitrite:

$$2NH_4^+ + 3O_2 \rightarrow 2NO_2^- + 2H_2O + 4H^+ + 66 \text{ K Cal energy}$$

This in turn provides a substrate for *Nitrobacter* bacteria which oxidise the nitrite to nitrate:

$$2NO_2^- + O_2 \rightarrow 2NO_3^- + 18 \text{ K Cal energy}$$

This energy is the only source of energy for these prokaryotes and they are chemoautotrophs.

DENITRIFICATION: The process of conversion of soil NO_3^- (nitrates) into gaseous nitrogen or nitrous oxide by *Pseudomonas, Clostridium* or Bacillus bacteria is called denitrification. Denitrification is also called nitrate respiration.

$2NO_3^- \rightarrow 3O_2 + N_2\uparrow$ providing up to 2385kJ

$2NO_2^- \rightarrow 2O_2 + N_2\uparrow$

The liberated oxygen is used as an electron acceptor in the processes that oxidise organic molecules, such as glucose. These microbes are heterotrophs. Denitrification occurs in poor drainage or waterlogged (anaerobic) soils. In denitrification, loss of N is as high as 87 per cent of added nitrogen. The rate of denitification is very low in acid soils with pH 3.6-4.8. Denitrification occurs by reductase enzyme.

BREAKDOWN OF PROTEIN: The conversion of proteins into amino acids, amines and Urea by the action of heterotrophic bacteria (*Rhizobium, Azotobactor, Clostridium, Amonifers, Denitrifers*) or fungi is known as Aminization. Bacteria in alkaline soil and neutral soils while fungi dominate in acidic soils. Amine ($R\text{-}NH_2$) is formed by Aminization.

BREAKDOWN OF CELLULOSE: Cellulose is converted to cellubiose by cellulase enzyme and then to glucose by cellubiase enzymes in the process called enzymatic hydrolysis. Glucose is oxidised to CO_2 in aerobic condition and converted to organic acids and achocols under anaerobic condition. Decomposition of cellulose is quite rapid in well aerated soils. It is slower in acid soils than in neutral and alkaline soils.

BREAKDOWN OF HEMICELLULOSE: Hemicellulose (polysaccharides) is converted to oligosaccharides by hemicellulase enzyme and then to simple sugars in the process called hydrolysis. Simple sugars are oxidised to CO_2 in aerobic condition and converted to organic acids and achocols under anaerobic condition.

BREAKDOWN OF STARCH: Starch (glucose of polymer) is converted to maltose by amylase enzyme and then to glucose by maltase and α-glucosidase, in the process called hydrolysis.

DECOMPOSITION OF LIGNIN: Lignin decomposers slowly and much slower than cellulose. Complete oxidation of lignin gives rise to CO_2 and water.

SIMPLE DECOMPOSITION OF PRODUCTS (By Microorganisms)

Carbon ⟶ CO_2, CO_3, HCO_3, elemental C

Nitrogen ⟶ NH_4^+, NO_2, NO_3, gaseous N

Sulphur ⟶ S, H_2S, SO_4, SO_3, CS_2 (carbon disulphide)

Phosphorus ⟶ H_2PO_4, HPO_4, PO_4

MINERALIZATION OF ORGANIC SULFUR (S)

Organic sulfur compounds ⟶ Decay products ⟶ SO_{4^-} and $2H^+$

MINERALIZATION OF ORGANIC PHOSPHORUS (P)

$$\text{Organic P} \longrightarrow H_2PO_4^- \xrightarrow{6.5-7.5\text{ pH}} HPO_4^-$$

FACTORS AFFECTING AVAILABILITY OF ORGANIC MATTER (OM)

CLIMATE: In cold climates, soil has higher OM due to slow decomposition of OM. With increase rainfall, soil has higher OM due to more growth of vegetation. In warmer climate, soil has Poor OM due to fast decomposition of OM.

NATURAL VEGETATION: Grass land areas and trees areas has higher OM. Similarly organic matter content of grass land soil is generally higher than forest soils mainly in surface soil.

TEXTURE: Light textured soil has less OM due to fast decomposition of OM and heavy textured soil has more OM due to slow decomposition.

STRUCTURE: Spheroidal structure has rich OM due to good soil physical condition.

DRAINAGE: Poorly drained soil has higher OM due to poor aeration and well drained soil has less OM due to good aeration.

pH: Best pH is 6-8 for higher OM.

MOISTURE: Waterlogged soil condition has slow decomposition of OM whereas dry soil has fast rate of decomposition.

TILLAGE: Continuous tillage results in poor OM whereas zero tillage conserves OM.

CROPPING SYSTEM: Cultivated soils have lower OM whereas, forest soils and grass land soils have higher OM. Most of cultivated soils contain 1-5 per cent OM which is available in mostly top 25 cm (10") of the soil. Under Indian conditions, most cultivated soils contain about 1-2 per cent OM in surface soil.

CARBON NITROGEN RATIO (C:N RATIO)

The ratio of weight of organic carbon (C) to the weight of total nitrogen (N) in a soil or OM is known as C:N ratio. The importance of C:N ratio is in mineralization and conserve N and OM in soil.

Table 1.2: C:N Ratio of Organic Matter and Soils

Organic Matter/ Soil	C:N Ratio
Soil microbes	4:1-9:1
Bacteria	5:1
Actinomycates	6:1
Fungi	10:1
Humus	10-12:1
Green legumes	12:1
Alfalfa	13:1
Clovers	20:1
Manures	20:1
FYM	20-30:1
Berseem	21:1
Hay (Legume grasses)	25:1
Blue grass	30:1
Green rye	36:1
Green grasses	40:1

Maize stalk, Sugarcane Bagashi	40-50:1
Grain straw, dry grasses	80:1
Rice stover	89:1
Wheat straw	124:1
Jowar stalk	93:1
Wood burada	350:1
Saw dust	400-500:1
General soil, cultivated soil /arable land	8-15:1
Mineral soil, waterlogged soil	10-12:1
Organic matter (OM)	40:1-12:1
Tropical soils	10:1

KEY POINTS

If C:N ratio of organic matter is < 20, it will result more N mineralization and more gain of available N in soil. If C:N ratio of OM is 20-30, resulting no gain of available N. If C:N ratio of OM is > 30, resulting less N-mineralization and more N- immobilization and finally loss of available N in soil. If decomposition of manure is slow, then wider/higher C:N ratio is present. If decomposition of manure is fast, then lower/narrow C:N ratio is obseved. Similarly, C:N ratio of sub-surface soil is narrow than surface soil. It is reported that C:N ratio of dry areas soil is narrower than humid areas soil. C:N ratio of mature residues i.e. legumes or non-legumes is higher than younger succulent materials. C:N ratio of warmer regions (arid climate) is lower than colder regions (humid climate). As decomposition occurs, C:N ratio decreases due to loss of carbon and conservation of nitrogen. C:N ratio of OM with less carbohydrate is narrow/lower than the OM with higher carbohydrate. Generally C:N ratio of Indian soils is observed as 14:1. Lower C:N ratio means greater amount of mineralization of N to NH^{+}_{4} and NO_3^- form.

HUMUS

Humus is defined as a complex and generally resistant mixture of brown or dark brown, amorphous and colloidal material modified from the original tissues or synthesized by the various soil organisms. Humus is the principal component of the soil carbon store which is a stable form of organic carbon. Humus is the final product of the decomposition of fresh organic matter. All of the carbon in the soil originally comes from the atmosphere when atmospheric CO_2 is captured by plant photosynthesis and converted into plant materials. The plant materials fall into the ground and forms organic matter. The fresh organic matter is decomposed by the action of soil microorganisms and rapidly releases most of its carbon within a couple of years. But part resists biological action and remains in the soil as humus which persists for, on average, hundreds to thousands of years, forming the soil carbon store.

HUMUS PROPERTIES

Humus is the colloidal complex of stable organic compounds. Humus contains higher proportion of organic compounds that are resistant to decomposition. The soil humus level represents the underlying level of the soil carbon store. As the fresh organic matter is the source of humus, more the OM, the more the soil carbon level. The colour of humus is brown, dark brown to black and the C:N ratio of humus is 10-12:1. CEC of humus is 100-300 Cmol(P^+) / kg soil and the density is between 1.3-1.5 Mg/m^3. Humus absorbs 80-90 per cent water under saturated condition.

CLASSIFICATION OF HUMUS

According to Molder, humus is classified into three groups on the basis of colour and solubility in water and alkali solution viz.

(i) Soluble in water : Crenic, Apocrenic acid

(ii) Soluble in alkali : Almic (brown), Humic acid (black)

(iii) Insoluble in alkali : Almin, Humin

According to Oden, humus is classified into four categories viz.

(i) Humus coal

(ii) Hematomelanic acid

(iii) Humic acid

(iv) Fulvic acid

Table 1.3: Composition of Humus

Material Basis		Weight basis	
Lignin	55%	Carbon	50-60%
Amino acid	35%	Hydrogen	3-4%
Carbohydrate	15%	Oxygen	30-40%
Hemicellullose	7%	Nitrogen	4-5%
Resin, fat, wax	3%	Phosphorous	0.4-0.5%
Others	5%		

2 WATER RESOURCES

INTRODUCTION

Water is essential for sustaining all forms of life, food production, economic development and for general well being. It is impossible to substitute for most of its uses, difficult to de-pollute, expensive to transport and it is truly a unique gift to mankind from nature. Water is also one of the most manageable of the natural resources as it is capable of diversion, transport, storage and recycling. India receives annual precipitation of about 4,000 km^3, including snowfall. Out of this, monsoon rainfall is of the order of 3,000 km^3. Rainfall in India is dependent on the south-west and north-east monsoons, on shallow cyclonic depressions and disturbances and on local storms. Most of it takes place under the influence of south-west monsoon between June and September except in Tamil Nadu, where it is under the influence of north-east monsoon during October and November.

Variations in climatic characteristics both in space and time are responsible for uneven distribution of precipitation in India. This uneven distribution of the precipitation results in highly uneven distribution of available water resources both in space and time, which leads to floods and drought affecting the vast areas of the country. Better and scientific structural and non-structural measures are required for mitigating the floods and droughts. The availability of water resources may be further enhanced by rejuvenation of dying lakes, ponds and tanks and increasing the artificial means of groundwater recharge. In addition to these measures, interbasin transfer of water provides one of the options for mitigating the problems of the surplus and deficit basins.

The surface water and groundwater resources of the country play a major role in agriculture, hydropower generation, livestock production, industrial activities, forestry, fisheries, navigation, recreational activities, etc. According to National Water Policy in the planning and operation of systems, water allocation priorities should be broadly as drinking water, irrigation, hydropower, ecology, agro-industries and non-agricultural industries and navigation. India is gifted with a river system comprising more than 20 major rivers with several tributaries. Many of these rivers are perennial and some of these are seasonal. The rivers like Ganges, Brahmaputra and Indus originate from the Himalayas and carry water throughout the year. The snow and ice melt of the Himalayas and the base flow contribute the flows during the lean season. Apart from the water available in the various rivers of the country, the groundwater is also an important source of water for drinking, irrigation, industrial uses, etc. It accounts for about 80 per cent of domestic water requirement and more than 45 per cent of the total irrigation in the country.

As per the international norms, if per capita water availability is less than 1,700 m^3 per year then the country is categorized as water stressed and if it is less than 1,000 m^3 per capita per year then the country is classified as water scarce. In India, per capita surface water availability in the years 1991 and 2001 were 2,309 and 1,902 m^3 and these are projected to reduce to 1,401 and 1,191 m^3 by the years 2025 and 2050 respectively. Hence, there is a need for proper planning, development and management of the greatest assets of the country, viz. water and land resources for raising the standards of living of the millions of people, particularly in the rural areas.

WATER RESOURCES OF INDIA

Although India occupies only 3.29 million km^2 geographical area, which forms 2.4 per cent of the world's land area, it supports over 15 per cent of the world's population. India supports about 1/6th of world population, 1/50th of world's land and 1/25th of world's water resources. India also has a livestock population of 500 million, which is about 20 per cent of the world's total livestock population. More than half of these are cattle, forming the backbone of Indian agriculture. The total utilizable water resources of the country are assessed as 1,086 km^3.

SURFACE WATER RESOURCES

In the past, several organizations and individuals have estimated water availability for the nation. Recently, the National Commission for Integrated Water Resources Development estimated the basin-wise average annual flow in Indian river systems as 1,953 km^3. The details are given in Table 2.1. Utilizable water resource is the quantum of withdrawal water from its place of natural occurrence. The utilizable annual surface water of the country is 690 km^3. There is considerable scope for increasing the utilization of water in the Ganga–Brahmaputra basins by construction of storages at suitable locations in neighbouring countries.

Table 2.1: Basin-wise Average Flow and Utilizable Water (in km^3/year)

S. No.	River Basin	Average Annual Flow	Utilizable Flow
1.	Indus	73.31	46.00
2.	Ganga – Brahmaputra-Meghna Basin		
	2a. Ganga	525.02	250.00
	2b. Brahmaputra sub-basin	629.05	24.00
	2c. Meghna (Barak) sub-basin	48.36	NA
3.	Subarnarekha	12.37	6.81
4.	Brahmni-Baitarani	27.48	18.30
5.	Mahanadi	66.88	49.99
6.	Godavari	110.54	76.30
7.	Krishna	69.81	58.00

8.	Pennar	6.32	6.86
9.	Cauvery	21.36	19.00
10.	Tapi	14.88	14.50
11.	Narmada	45.64	34.50
12.	Mahi	11.02	3.10
13.	Sabarmati	3.81	1.93
14.	West-flowing rivers of Kachchh and Saurashtra including Luni	15.10	14.98
15.	West flowing rivers south of Tapi	200.94	36.21
16.	East-flowing rivers between Mahanadi and Godavari	17.08	NA
17.	East-flowing river between Godavari and Krishna	1.81	13.11
18.	East-flowing river between Krishna and Pennar	3.63	NA
19.	East-flowing river between Pennar and Cauvery	9.98	16.73
20.	East-flowing rivers south of Cauvery	6.48	NA
21.	Area for North Ladakh not draining into Indus	0	NA
22.	Rivers draining into Bangladesh	8.57	NA
23.	Rivers draining into Myanmar	22.43	NA
24.	Drainage areas of Andaman, Nicobar and Lakshadweep Islands	0	NA
	Total (rounded)	1,953	690

GROUNDWATER RESOURCES

The annual potential natural groundwater recharge from rainfall in India is about 342.43 km^3, which is 8.56 per cent of total annual rainfall of the country. The annual potential groundwater recharge augmentation from canal irrigation system is about 89.46 km^3. Thus, total replenishable groundwater resource of the country is assessed as 431.89 km^3. After allotting 15 per cent of this quantity for drinking and 6 km^3 for industrial purposes, the remaining can be utilized for irrigation purposes. Thus, the available groundwater resource for irrigation is 361 km^3, of which utilizable quantity (90 per cent) is 325 km^3. The estimates by the Central Groundwater Board (CGWB) of total replenishable groundwater resource, provision for domestic, industrial and irrigation uses and utilizable groundwater resources for future use are given in Table 2.2 The basinwise per capita water availability varies between 13,393 m^3 per annum for the Brahmaputra–Barak basin to about 300 m^3 per annum for the Sabarmati basin. The state-wise estimates of dynamic groundwater (fresh) resource made by the CGWB are given in Table 2.3. The basin-wise groundwater potential of the country is given in Table 2.4.

Table 2.2: Groundwater Resources of India (in km^3/year)

S.No.	Categories	Quantity
1.	Total replenishable groundwater resource	432
2.	Provision for domestic, industrial and other uses	71
3.	Available groundwater resource for irrigation	361
4.	Utilizable groundwater resource for irrigation (90% of the sl. no. 3)	325
5.	Total utilizable groundwater resource (Sum of s.nos. 2 and 4)	396

Table 2.3: State-wise Dynamic Fresh Groundwater Resource of India (in km^3/year)

S. No.	State	Replenishable Groundwater Resource from Normal Natural Recharge	Replenishment due to Recharge Augmentation from Canal Irrigation	Total Replenishable Groundwater Resource	Percentage Contribution of Recharge Augmentation to Total GW Resource
1.	Andhra Pradesh	20.03	15.26	35.29	43
2.	Arunachal Pradesh	1.44	0.00	1.44	0
3.	Assam	24.23	0.49	24.72	2
4.	Bihar	28.31	5.21	33.52	16
5.	Goa	0.18	0.03	0.21	14
6.	Gujarat	16.38	4.00	20.38	20
7.	Haryana	4.73	3.80	8.53	45
8.	Himachal Pradesh	0.29	0.08	0.37	22
9.	Jammu & Kashmir	2.43	2.00	4.43	45
10.	Karnataka	14.18	2.01	16.19	12
11.	Kerala	6.63	1.27	7.90	16
12.	Madhya Pradesh	45.29	5.60	50.89	11
13.	Maharashtra	33.40	4.47	37.87	12
14.	Manipur	3.15	0.00	3.15	0
15.	Meghalaya	0.54	0.00	0.54	0
16.	Mizoram	NOT ASSESSED			
17.	Nagaland	0.72	0.00	0.72	0
18.	Orissa	16.49	3.52	20.01	18
19.	Punjab	9.47	9.19	18.66	49
20.	Rajasthan	10.98	1.72	12.70	14
21.	Sikkim	NOT ASSESSED			
22.	Tamil Nadu	18.91	7.48	26.39	28
23.	Tripura	0.57	0.10	0.67	15
24.	Uttar Pradesh	63.43	20.39	83.32	24
25.	West Bengal	20.30	2.79	23.09	12
26.	Union territories	0.35	0.05	0.40	12
	Total	342.43	89.46	431.89	21

Table 2.4: Groundwater Potential in River Basins of India (pro-rata basis) (in km^3/year)

S. No.	Basin	Total Repleni-shable Ground water Resources	Provision for Domestic, Industrial and Other Cases	Available Ground water for Irrigation	Net Draft	Balance Ground water Potential	Level of Ground water Development (%)
1.	Brahmani with Baitarni	4.05	0.61	3.44	0.29	3.16	8.45
2.	Brahmaputra	26.55	3.98	22.56	0.76	21.80	3.37
3.	Chambal composite	7.19	1.08	6.11	2.45	3.66	40.09
4.	Cauvery	12.30	1.84	10.45	5.78	4.67	55.33
5.	Ganga	170.99	26.03	144.96	48.59	96.37	33.52
6.	Godavari	40.65	9.66	30.99	6.05	24.92	19.53
7.	Indus	26.49	3.05	23.43	18.21	5.22	77.71
8.	Krishna	26.41	5.58	20.83	6.33	14.50	30.39
9.	Kutch and Saurashtra Composite	11.23	1.74	9.49	4.85	4.64	51.14
10.	Madras and South Tamil Nadu	18.22	2.73	15.48	8.93	6.55	57.68
11.	Mahanadi	16.46	2.47	13.99	0.97	13.02	6.95
12.	Meghna	8.52	1.28	7.24	0.29	6.95	3.94
13.	Narmada	10.83	1.65	9.17	1.99	7.18	21.74
14.	Northeast composite	18.84	2.83	16.02	2.76	13.26	17.20
15.	Pennar	4.93	0.74	4.19	1.53	2.66	36.60
16.	Subarnarekha	1.82	0.27	1.55	0.15	1.40	9.57
17.	Tapi	8.27	2.34	5.93	1.96	3.97	33.05
18.	Western Ghat	17.69	3.19	14.50	3.32	11.18	22.88
	Total	431.43	71.08	360.35	115.21	245.13	31.97

WATER REQUIREMENTS OF INDIA

Traditionally, India has been an agriculture-based economy. Hence, development of irrigation to increase agricultural production for making the country self-sustained and for poverty alleviation has been of crucial importance for the planners. Accordingly, the irrigation sector was assigned a very high priority in the 5-year plans. Long-term planning has to account for the growth of population. Table 2.5 provides details of the population of India and per capita water availability as well as utilizable surface water for some of the years from 1951 to 2050 (projected). The availability of water in India shows wide spatial and temporal variations.

Table 2.5: Per-capita per year Availability and Utilizable Surface Water in India (in m^3)

S. No.	Year	Population (millions)	Per-capita Surface Water Availability	Per-capita Utilizable Surface Water
1.	1951	361	5,410	1,911
2.	1955	395	4,944	1,746
3.	1991	846	2,309	816
4.	2001	1027	1,902	672
5.	2025 (projected)	1,286 (low growth)	1,519	495
		1,333 (high growth)	1,495	–
6.	2050 (projected)	1,346 (low growth)	1,451	421
		1,581 (high growth)	1,235	–

DOMESTIC USE

Community water supply is the most important requirement and it is about 5 per cent of the total water use. About 7 km^3 of surface water and 18 km^3 of groundwater are being used for community water supply in urban and rural areas. Along with the increase in population, another important change from the point of view of water supply is higher rate of urbanization. According to the projections, the higher is the economic growth, the higher would be urbanization. It is expected that nearly 61 per cent of the population will be living in urban areas by the year 2050 in high-growth scenario as against 48 per cent in low-growth scenario. Based on these norms and projection of population, it is estimated that by 2050, water requirements per year for domestic use will be 90 km^3 for low-demand scenario and 111 km^3 for high-demand scenario. It is expected that about 70 per cent of urban water requirement and 30 per cent of rural water requirement will be met by surface water sources and the remaining from groundwater.

IRRIGATION

The irrigated area in the country was only 22.6 million hectare (M ha) in 1950–51. Since the food production was much below the requirement of the country, due attention was paid for expansion of irrigation. The ultimate irrigation potential of India has been estimated as 140 M ha. Out of this, 76 M ha would come from surface water and 64 M ha from groundwater sources. The quantum of water used for irrigation by the last century was of the order of 300 km^3 of surface water and 128 km^3 of groundwater, total 428 km^3. The estimates indicate that by the year 2025, the water requirement for irrigation would be 561 km^3 for low-demand scenario and 611 km^3 for high-demand scenario. These requirements are likely to further increase to 628 km^3 for low-demand scenario and 807 km^3 for high-demand scenario by 2050.

HYDROELECTRIC POWER

The hydropower potential of India has been estimated at 84,044 MW at 60 per cent load factor. At the time of independence, the installed capacity of hydropower projects was 508

MW. By the end of 1998, the installed hydropower capacity was about 22,000 MW. India has plans to develop 60,000 MW additional hydropower by the twelfth five-year plan. It includes 14,393 MW during the tenth five-year plan (2002–2007); 20,000 MW during eleventh (2007–2012) and 26,000 MW during the twelfth (2012–2017) five-year plans.

INDUSTRIAL WATER REQUIREMENT

The present water use in the industrial sector is of the order of 15 km^3. The water use by thermal and nuclear power plants with installed capacities of 40,000 MW and 1,500 MW (1990 figures) respectively, is estimated to be about 19 km^3. In view of shortage of water, the industries are expected to switch over to water efficient technologies. If the present rate of water use continues, the water requirement for industries in 2050 would be 103 km^3; this is likely to be nearly 81 km^3 if water saving technologies is adopted on a large scale.

TOTAL WATER REQUIREMENTS

The total annual requirement of water for various sectors has been estimated and its break up is given Table 2.6. With the increasing population as well as all round development in the country, the utilization of water has also been increasing at a fast pace. In 1951, the actual utilization of surface water was about 20 per cent and 10 per cent in the case of groundwater. The utilizable water in river basins is highly uneven. For example in the Brahmaputra basin, which contributes 629 billion m^3 of surface water of the country's total flow, only 24 billion m^3 is utilizable.

WATER RESOURCES MANAGEMENT IN INDIA

The water resources management practices may be based on increasing the water supply and managing the water demand under the stressed water availability conditions. Data monitoring, processing, storage, retrieval and dissemination constitute the very important aspects of the water resources management. These data may be utilized not only for management but also for the planning and design of the water resources structures. In addition to these, now-a-days decision support systems are being developed for providing the necessary inputs to the decision makers for water resources management. Also, knowledge sharing, people's participation, mass communication and capacity building are essential for effective water resources management.

FLOOD MANAGEMENT

The main causes of floods in India are inadequate capacity within river banks to contain high flows, river bank erosion and silting of river beds. The additional factors are as land slides leading to obstruction of flow and change of the river course, retardation of flow due to tidal and backwater effects, poor natural drainage in the flood-prone area, cyclone and associated heavy rain storms or cloud bursts, snowmelt and glacial outbursts and dam break flows. The committees and commissions constituted by the government have given valuable recommendations on different issues of flood management. The structural measures, such as the construction of reservoirs, embankments, levees, spurs, etc. have been implemented in some of the states to reduce the damages in the flood plains. The flood management measures have to be more focused and targeted towards the decided objectives within a stipulated time frame. The total live storage capacity of completed projects in India is about 174 km^3. A large flood storage space in reservoirs is required for a successful flood management programme. Flood management also calls for community participation. Farmers, professional bodies, industries and voluntary organizations have to be aware about flood management.

Table 2.6: Annual Water Requirement for Different Uses (in km^3)

Use	Year 1997-98	Year 2010			Year 2025			Year 2050		
		Low	High	%	Low	High	%	Low	High	%
					SURFACE WATER					
Irrigation	318	330	339	48	325	366	43	375	463	39
Domestic	17	23	24	3	30	36	5	48	65	6
Industries	21	26	26	4	47	47	6	57	57	5
Power	7	14	15	2	25	26	3	50	56	5
Inland navigation	–	7	7	1	10	10	1	15	15	1
Environment-Ecology	–	5	5	1	10	10	1	20	20	2
Evaporation losses	36	42	42	6	50	50	6	76	76	6
Total	399	447	458	65	497	545	65	641	752	64
					GROUNDWATER					
Irrigation	206	213	218	31	236	245	29	253	344	29
Domestic	13	19	19	2	25	26	3	42	46	4
Industries	9	11	11	1	20	20	2	24	24	2
Power	2	4	4	1	6	7	1	13	14	1
Total	230	247	252	35	287	298	35	332	428	36
Grand total	629	694	710	100	784	843	100	973	1180	100
					TOTAL WATER USE					
Irrigation	524	543	557	78	561	611	72	628	807	68
Domestic	30	42	43	6	55	62	7	90	111	9
Industries	30	37	37	5	67	67	8	81	81	7
Power	9	18	19	3	31	33	4	63	70	6
Inland navigation	0	7	7	1	10	10	1	15	15	1
Environment-Ecology	0	5	5	1	10	10	1	20	20	2
Evaporation losses	36	42	42	6	50	50	6	76	76	7
Total	629	694	710	100	784	843	100	973	1180	100

DROUGHT MANAGEMENT

The drought-prone area assessed in the country is of the order of 51.12 M ha. The traditional system of drought monitoring and estimating losses by crop cutting needs replacement with real time remote sensing, GIS, GPS and modelling techniques for ensuring transparency and quick response. Scope of losses may be extended to groundwater depletion, damage to perennial trees, plantations, orchards and depletion in fertility of livestock. Food, fodder, agricultural inputs and water banks may be established in vulnerable areas instead of their storage in surplus regions to avoid transport bottlenecks during the drought. An integrated basin development approach is required to be developed and implemented for preparing the drought management plan before, during and after the occurrence of the drought. In this regard, there is a need for the development of the decision support systems (DSS) for the monitoring and management of the drought on basin scale utilizing the advanced capabilities of remote sensing, geographical information system and knowledge based systems.

GROUNDWATER MANAGEMENT

To protect the aquifers from overexploitation, an effective groundwater management policy oriented towards promotion of efficiency, equity and sustainability is required. The exploitation of groundwater resources should be regulated so as not to exceed the recharging possibilities, as well as to ensure social equity. Overexploitation of groundwater should be avoided, especially near the coasts to prevent ingress of seawater into freshwater aquifers. Artificial recharge measures need to be urgently implemented in the overexploited areas. Promotion of participatory action in rehabilitating tanks for recharging would go a long way in augmenting groundwater supply. It is desirable that the irrigation needs for fulfilling crop water requirements should be satisfied by judicious utilization of available canal water in conjunction with groundwater so as to keep the water table within the acceptable range. Thus, the optimal conjunctive use of the region's surface and groundwater resources would help in minimizing the problems of water logging and groundwater mining.

WATER CONSERVATION

Water conservation implies improving the availability of water through augmentation by means of storage of water in surface reservoirs, tanks, soil and groundwater zone. There is a great potential for better conservation and management of water resources in its various uses. Since agriculture accounts for about 69 per cent of all water withdrawn, the greatest potential for conservation lies in increasing irrigation efficiencies. An important supplement to conservation is to minimize the wastage of water. Prices of water for all uses should be fixed, keeping in mind its economic value, control of wastage and the ability of users to pay. As water is becoming scarcer, pricing will be an important factor in avoiding wastage and ensuring optimal use.

WATERSHED MANAGEMENT

Watershed management refers to the conservation, regeneration and the judicious use of all the resources – natural (land, water, plants, animals) and human – within the watershed area. Watershed Management tries to bring about the best possible balance in the environment between natural resources on the one side and man and animals on the other. Since it is the man who is primarily responsible for degradation of environment, regeneration and conservation can only be possible by promoting awakening and participation among the people who inhabit the watersheds. Man and his environment are interdependent. The changes in the environment directly affect the lives of the people depending on it. A degraded environment means a degraded quality of life of the people. Environmental degradation can be tackled effectively through the holistic development of the watershed. A watershed provides a natural geo-hydrological unit for planning any developmental initiative.

Watershed is the unit of management in Integrated Water Resources Management (IWRM), where surface water and groundwater are inextricably linked and related to land use and management. Watershed management aims to establish a workable and efficient framework for the integrated use, regulation and development of land and water resources in a watershed for socio-economic growth. Local communities play a central role in the planning, implementation and funding of activities within participatory watershed development programmes. In these initiatives, people use their traditional knowledge, available resources, imagination and creativity to develop watershed and implement community-centered programme. Currently, many programmes, campaigns and projects are underway in different parts of India to spread mass awareness and mobilize the general population in managing water resources.

RAINWATER HARVESTING

Rainwater harvesting is the process to capture and store rainfall for its efficient utilization and conservation to control its runoff, evaporation and seepage. Different methods of rainwater harvesting were developed to suit the geographical and meteorological conditions of the region in various parts of the country. Traditional rainwater harvesting, which is still prevalent in rural areas, is done by using surface storage bodies like lakes, ponds, irrigation tanks, temple tanks, etc. There is a need to recharge aquifers and conserve rainwater through water harvesting structures. In urban areas, rainwater will have to be harvested using rooftops and open spaces. Harvesting rainwater not only reduces the possibility of flooding, but also decreases the community's dependence on groundwater for domestic uses. Apart from bridging the demand–supply gap, recharging improves the quality of groundwater, raises the water table in wells/bore-wells and prevents flooding and choking of drains. One can also save energy to pump groundwater as water table rises.

WATER QUALITY CONSERVATION AND ENVIRONMENT RESTORATION

Implementation of water pollution prevention strategies and restoration of ecological systems are integral components of all developmental plans. To preserve our water and environment, we need to make systematic changes in the way we grow our food, manufacture the goods and dispose off the waste. Manufacturers may reduce water pollution by reusing materials and chemicals and switching over to less toxic alternatives. Environmental improvement and restoration should be planned and implemented such that the freshwater resources are protected and their quality is maintained and/or enhanced.

RECYCLE AND REUSE OF WATER

Another way through which we can improve freshwater availability is by recycle and reuse of water. Use of water of lesser quality, such as reclaimed wastewater, for cooling and fire fighting is an attractive option for large and complex industries to reduce their water costs, increase production and decrease the consumption of energy. This conserves better quality waters for potable uses. Currently, recycling of water is not practiced on a large scale in India and there is considerable scope and incentive to use this alternative.

LEGAL RESTRICTIONS ON WATER USE

One of the active strategies could include provisions of legal restrictions on use of water, mainly during the period of scarcity. In India a national water policy has been adopted, which includes policy directions for development and management of water resources. Also, provision of legal restrictions on proper utilization of groundwater resources has been advocated at various fora in the country. In fact, Gujarat has already enacted such legislation and other states may also follow the suit.

LAND USE PLANNING AND CROPPING PATTERN

Another strategy that could be adopted refers to planning of land use especially in new land developments. Areas where water supply priorities are low can be planted with drought-resistant varieties of trees. The selection of cropping pattern as per availability of water will reduce adverse impacts of drought on potential water consuming crops. The plants suitable for water scare areas can be (i) with shorter growth period (ii) high-yielding plants requiring no increase in water supply (iii) plants that can tolerate saline irrigation water (iv) plants with low transpiration rates and (v) plants with deep and well-branched roots.

WATER PRICING

Another strategy, which needs consideration, is changes in water pricing structures. Mostly water rates are based only on a portion of what it costs to obtain, develop, transport, treat and deliver water to the consumer. The costs are usually paid for in part by subsidies from the state using tax payers' money and therefore are not accurately reflected in an individual's water bill. The rates can be thought to vary with the availability either by an increase in price during period of abundant supply. The policy of peak demand pricing can also be effective in conserving water in which higher rates can be proposed for water use beyond a prescribed limit. Moreover, some economic incentives for using small amount of water can be given to consumers for encouraging water conservation.

APPLICATIONS OF DECISION SUPPORT SYSTEM IN WATER RESOURCES

Knowledge based Decision Support System (DSS) focuses on the interaction between the use and the data, models and computers. Automating the process with a DSS could effectively improve the water resources planning and management. The aim has been to create a system in which the mechanics of linking one component with another are largely transparent to the user. Although time and effort is required in customizing the system for a particular river basin, its use has been designed to be as simple as possible. Communication is by means of a user-friendly interface, which makes extensive use of hypertext for guidance and colour graphics in presenting the results.

PEOPLE PARTICIPATION AND CAPACITY BUILDING

For making the people of various sections of the society aware about the different issues of water resources management, a participatory approach may be adopted. Mass communication programmes may be launched using the modern communication means for educating the people about water conservation and efficient utilization of water. Capacity building should be perceived as the process whereby a community equips itself to become an active and well-informed partner in decision making. The process of capacity building must be aimed at both increasing access to water resources and changing the power relationships between the stakeholders. Capacity building is not only limited to officials and technicians but must also include the general awareness of the local population regarding their responsibilities in sustainable management of the water resources. Policy decisions in any water resources project should be directed to improve knowledge, attitude and practices about the linkages between health and hygiene, provide higher water supply service levels and to improve environment through safe disposal of human waste. Sustainable management of water requires decentralized decisions by giving authority, responsibility and financial support to communities to manage their natural resources and thereby protect the environment.

CONCLUSION

Integrated and coordinated development of surface water and groundwater resources and their conjunctive use should be envisaged right from the project planning stage and should form an integral part of the project implementation. There is a need for proper management of groundwater resources, which presently require adequate inputs including manpower, financial inputs, technologies, etc. Some of the important measures which may be taken up for sustainable development of groundwater resources include improving public water supply, use of energy pricing and supply to manage agricultural groundwater draft, increasing rain-water harvesting and groundwater recharge, transfer of surface water in lieu of groundwater pumping, increasing the economic growth and reduction in dependence on agriculture and formalizing the water sector. An accurate assessment of available surface and groundwater resources, considering the manmade changes, is needed for planning, design and operation of the water resources projects as well as for watershed management.

3 FOREST RESOURCES

INTRODUCTION

Forest ecosystem is a dynamic complex of plant, animal and microorganism communities and their abiotic environment interacting as a functional unit, where trees are a key component of the system. Humans, with their cultural, economic and environmental needs are an integral part of many forest ecosystems. Forest is an area with a high density of trees which provide multitude of benefits. Forests are a very crucial natural resource and have always been central in human life. Forests are unique because of the diversity of plant resources and land services they provide, and their potential to be renewable under sustainable management practices. The material resources and life supporting functions provided by forest lands are particularly important in developing countries where some people still rely on them directly as sources of food, fodder, fuelwood, medicine, shelter, building materials and as centers of certain cultural practices.

In recent years, the capability of forests to mitigate climate change, provide renewable products and energy, maintain biological diversity, protect land and water resources, provide recreation facilities, improve air quality, help alleviate poverty and contribute to developing a greener economy have received increasing attention. But at the same time, these benefits are threatened by global warming and extreme weather events.

Sustainably managed forests have multiple environmental and socio-economic functions which are important at the global, national and local scales. They play a vital part in sustainable economic development and improve the quality of life of people across the world. Forests are considered an essential part of the solution to many global challenges. Reliable and up-to-date information on the state of forest resources such as forest area and area change, growing stock, wood and non-wood products, carbon stock, protected areas, biological diversity, use of forest for recreation and other services, contribution to national economies are crucial to support decision-making for policies and programmes in forestry and sustainable development at all levels.

Forests provide a wide range fo provisioning, regulating, cultural and supporting services for human well being collectively known as ecosystem services. The sustainability of forest ecosystems depends on sustained management, efficient utilization and effective protection measures against deforestation and forest degradation. The role of ecosystem services are equally important for sustaining livelihoods and maintaining environmental conditions.

FOREST DEFINITION

Forest includes natural forests and forest plantations. The FAO definition of a forest is considered as the basic one, but many other useful definitions of forest exist in published form. The fact that "forest" has been defined in many ways is a reflection of the diversity of forests and forest ecosystems in the world and of the diversity of human approaches to forests.

Forests are generally conceived as areas having tree growth. The Forest Survey of India (FSI) has accordingly recognised dense, open and scrub forests based on the percentage of canopy cover. The Indian Forest Act, 1927 does not define the forests and the legal extent of forests depends upon the process of notifications. Hon'ble Supreme Court of India has taken a view regarding extending purview of the Forest Conservation Act, 1980 to all the areas falling in the "dictionary meaning" of forests irrespective of ownership and control. In addition to the protective and production functions of tree cover, there is a growing awareness of livelihood and life-sustaining values of non-wood forest products.

LEGAL DEFINITION

Some legal definitions of 'forest' are based on the actual vegetation on the ground, whereas other definitions are based on a defined land area which may have no vegetation on it at all but is legally under the jurisdiction of the national agency which manages forests and natural resources. Any land recorded as forest in any land record is legally forest land whether or not there is any vegetation on the land.

GOVERNMENT OF INDIA

Forest: Forest is defined as a land having tree crown cover value of minimum 15 per cent, land area value of minimum 0.05 hectare and tree height value of minimum 2 meters.

FOOD AND AGRICULTURAL ORGANIZATION

Forest: Forest includes natural forests and forest plantations. It is used to refer to land with a tree canopy cover of more than 10 per cent and area of more than 0.5 ha. Forests are determined both by the presence of trees and the absence of other predominant land uses. The trees should be able to reach a minimum height of 5 m. Young stands that have not yet but are expected to reach a crown density of 10 percent and tree height of 5 m are included under forest, as are temporarily unstocked areas. The term includes forests used for purposes of production, protection, multiple-use or conservation (i.e. forest in national parks, nature reserves and other protected areas), as well as forest stands on agricultural lands (e.g. windbreaks and shelterbelts of trees with a width of more than 20 m), and rubberwood plantations and cork oak stands. The term specifically excludes stands of trees established primarily for agricultural production, for example fruit tree plantations. It also excludes trees planted in agroforestry systems

Primary Forest: Forest of native species, in which there are no clearly visible indications of human activity and ecological processes are not significantly disturbed. A primary forest is a forest that has never been logged and has developed following natural disturbances and under natural processes, regardless of its age.

Forest Plantations: A forest established by planting or/and seeding in the process of afforestation or reforestation. It consists of introduced species or, in some cases, indigenous species

Natural Forest: A forest composed of indigenous trees and not classified as forest plantation

GLOBAL FOREST RESOURCE ASSESSMENT

Forest: Land spanning more than 0.5 hectares with trees higher than 5 metres and a canopy cover of more than10 per cent or trees able to reach these thresholds *in situ*. It does not include land that is predominantly under agricultural or urban land use.

Other Wooded Land: Land not classified as "Forest", spanning more than 0.5 hectares; with trees higher than 5 meters and a canopy cover of 5-10 per cent, or trees able to reach these thresholds *in situ*; or with a combined cover of shrubs, bushes and trees above 10 per cent. It does not include land that is predominantly under agricultural or urban land use.

Other Land: All land that is not classified as "Forest" or "Other wooded land"

Other Land with Tree Cover: Land classified as "Other Land", spanning more than 0.5 hectares with a canopy cover of more than 10 per cent of trees able to reach a height of 5 meters at maturity.

EUROPEAN COMMISSION

Forest: It has defined forestland as having at least 20 per cent canopy closure (10 per cent in Mediterranean forests) and a minimum area of 0.5 hectares

KYOTO PROTOCOL

Forest: Forest is a minimum area of land of 0.05-1.0 hectares with tree crown cover (or equivalent stocking level) of more than 10-30 per cent with trees with the potential to reach a minimum height of 2-5 meters at maturity *in situ*.

CONVENTION ON BIOLOGICAL DIVERSITY

Forest: A forest is a land area of more than 0.5 ha, with a tree canopy cover of more than 10 per cent, which is not primarily under agricultural or other specific non-forest land use. In the case of young forests or regions where tree growth is climatically suppressed, the trees should be capable of reaching a height of 5 m *in situ*, and of meeting the canopy cover requirement.

Forest Biome: This reflects the ecological and physiognomic characteristics of the vegetation and broadly corresponds to climatic regions of the Earth. In this document, it is used in reference to boreal, temperate and tropical forest biomes.

Forest Type: Within biomes, a forest type is a group of forest ecosystems of generally similar composition that can be readily differentiated from other such groups by their tree and undercanopy species composition, productivity and/or crown closure.

GLOBAL SCENARIO OF FOREST RESOURCES

The world's forests including largest rainforests, tropical forests, temperate forests, savannas and semi-arid biomes are critical for biodiversity conservation. The ecologically rich set of forests has extraordinary potential to provide services of economic significance, including water, food, wood, fiber, carbon sequestration, non-timber forest products (NTFP) and tourism destinations, as well as erosion control, flood mitigation, water purification, pollination, waste assimilation and disease regulation.

Despite the many links between forests and human welfare, current patterns of forest use are largely unsustainable. Extraction rates exceed the capacity of these forests to regenerate. Conversion of forest lands to other uses frequently involves lands that cannot sustain those uses and are soon abandoned, reverting to degraded forest. These approaches do not realize the long-term potential of forest-based ecosystem services to support income generation, development and social equity, through the potential of forests to sustain themselves. Forest types differ widely, determined by factors including latitude, temperature, rainfall patterns, soil composition and human activity.

Forest area provides the first indication of the relative importance of forests in a country or region. Estimates of the change in forest area over time provide an indication of the demand for land for forestry and other uses. Forest area is relatively easy to measure and has been selected as one of the 60 indicators for monitoring progress towards the Millennium Development Goals, the 2010 Biodiversity Target and Global Objectives on Forests contained in the Non-legally Binding Agreement on all types of forests. Year 2010 was celebrated as the International year of Biodiversity to remind the people that forests represent some of the most diverse ecosystems on Earth. Year 2011 was declared as the International year of Forests to promote awareness and understanding of forests and forestry issues and to draw attention to four key areas viz. regional trends on forest resources, the development of sustainable forest industries, climate change adaptation and mitigation and the local value of forests.

GLOBAL OBJECTIVES ON FORESTS (UNGA 2008)

In 2006, at its sixth session, the United Nations Forum on Forests (UNFF) agreed on four shared Global Objectives on Forests, providing clear guidance on the future work of the international arrangement on forests.

(i) Reverse the loss of forest cover worldwide through sustainable forest management, including protection, restoration, afforestation and reforestation, and increase efforts to prevent forest degradation.

(ii) Enhance forest-based economic, social and environmental benefits, including by improving the livelihoods of forest-dependent people.

(iii) Increase significantly the area of protected forests worldwide and other areas of sustainably managed forests, as well as the proportion of forest products from sustainably managed forests.

(iv) Reverse the decline in official development assistance for sustainable forest management and mobilize significantly increased, new and additional financial resources from all sources for the implementation of sustainable forest management.

WORLD FORESTRY DAY

World forestry day has been celebrated around the world for 30 years to increase public awareness of the huge array of benefits wooded areas offer and commemorates the importance of forests and forestry to communities around the globe. The concept of having a World Forestry Day originated at the 23rd General Assembly of the European Confederation of Agriculture in 1971. Later that year, the United Nations Food and Agriculture Organisation (FAO) gave support to the idea believing the event would contribute a great deal to public awareness of the importance of forests and agreed that it should be observed every year around the world. March 21, the autumnal equinox in the Southern Hemisphere and the vernal equinox in the Northern Hemisphere, was chosen as the day to be celebrated offering information about the three key facets of forestry: protection, production and recreation.

GLOBAL FOREST COVER ASSESSMENT

Growing awareness of forest's role in the global carbon cycle has created interest in world's forests to unprecedented heights. FAO, in cooperation with its member countries, has monitored the world's forest at five to ten year intervals since 1945. These global assessments provide valuable information to policy-makers, international negotiations, arrangements and organizations related to forests and to the general public. The Global Forest Resources assessment (FRA) is the most comprehensive assessment to date. It examines the current status and recent trends for all types of forests in 233 countries. It covers the all seven thematic elements of sustainable forest management viz. extent of forest resources, forest biological diversity, forest health and vitality, productive functions of forest resources, protective functions of forest resources, socio-economic functions of forests and legal, policy and institutional framework.

Table 3.1: Global Forest Cover by Region (Source: FRA 2010a)

Region	Forest Area	
	1000 ha	% of Land Area
Africa	674,419	23
Asia	592,512	19
Europe	1,005,001	45
North and Central America	705,393	33
South America	864,351	49
Oceania	191,384	23
World	4,033,060	31

World's total forest area is 4.03 billion hectares, corresponding to 31 per cent of the total land area or an average of 0.6 ha of per capita. The ten most forest-rich countries are Russian

Federation (809 million ha), Brazil (520 million ha), Canada (310 million ha), United States of America (304 million ha), China (207 million ha), Democratic Republic of Congo (154 million ha), Australia (149 million ha), Indonesia (94 million ha), Sudan (70 million ha) and India (68 million ha). The top five countries accounts for more than half of the total forest area.

The total area of other wooded land is estimated to be 1.1 billion hectares, equivalent to nine per cent of the total land area. The total area of other land with tree cover was reported to be 79 million hectares. High forest countries, in terms of percentage of forest cover with respect to total geographical area, are French Guiana (98%), Suriname (95%) and Federal States of Micronesia (92%).

Deforestation, mainly due to conversion of forests to agricultural land, shows signs of decreasing in several countries but continues at an alarmingly high rate in others. Globally around 13 million hectares of forest were converted to other uses or lost through natural causes each year from 2000-2010. The total net change in forest area in the period 2000-2010 is estimated at -5.2 million hectares per year or equivalent to a loss of more than 140 km^2 of forest per day (FRA 2010a). The five countries with largest annual net loss of forest area are Brazil (2.64 million ha), Australia (0.56 million ha), Indonesia (0.5 million ha), Nigeria (0.41 million ha) and Tanzania (0.4 million ha).

FOREST PROFILE OF INDIA

FOREST ASSESSMENT IN INDIA

The term 'Forest Area' (or recorded forest area) refers to all the geographical areas recorded as 'Forests' in government records. Recorded forest areas largely consists of reserved forests (RF) and protected forests (PF) which have been constituted under the provisions of the Indian Forest Act, 1927. Besides RF and PF, the recorded forest area may also include unclassed forests and all such areas which have been recorded as forests in the revenue records or have been constituted so under any State act or local law.

Forest cover refers to all lands, more than one hectare in area, with a tree canopy density of more than 10 per cent irrespective of ownership and legal status. Such lands may not necessarily be a recorded forest area. It also includes orchards, bamboo and palm. Forest area denotes the legal status of the land, whereas forest cover indicates presence of trees on any land irrespective of their ownership.

FOREST AREA: At the time of independence, the recorded forest area of the country was 39.94 million ha comprising the government owned forest (26.16 million ha) and community and privately owned forest (13.78 million ha). The area increased to 68.02 million ha in 1950-51 with the addition of ex-princely and ex-proprietary forests. Out of this, 53.82 million ha was government forests and 14.20 million ha with community and private forests. The recorded forest area further increased to 75.18 million ha due to consolidation by the early eighties. FSI (2011) has reported 76.95 million ha of recorded forest area in the country at present comprising reserved forest (42.25 million ha), protected forest (21.40 million ha) and unclassed forest (13.30 million ha).

FOREST COVER: India is one of the few countries of the world to have a robust and scientific system of periodic forest cover assessment. Forest Survey of India (FSI), an organization under the Ministry of Environment and Forests, has been carrying out forest and tree cover mapping of the entire country. The forest cover mapping was started over two decades back in the year 1987 and so far, twelve cycles of forest cover mapping have been carried out on a biennial basis.

The forest and tree cover mapping is carried out using remote sensing technology which captures the unique spectral reflectance of the electro-magnetic radiation. This is then used for the characterization of vegetation and other land covers. The forest cover mapping has been carried out at a scale of 1:50,000. The first ever forest cover mapping in India was carried out using Landsat imageries pertaining to period 1981-83 at 1:1 million scale. The first forest cover assessment and mapping reported about 64.2 million ha of forest cover in the country corresponding to 19.52 per cent in the year 1987.

CURRENT FOREST COVER STATISTICS

India is the seventh largest country in the world having an area of 328.7 million hectares. Forest and tree cover of the country is 78.29 million hectares which is 23.81 per cent of the geographical area (FSI 2011). India has the total forest cover of 21.05 per cent (69.20 million ha) and tree cover of 2.76 per cent (9.08 million ha) respectively. Area-wise, Madhya Pradesh has the largest forest cover (77,700 km^2) in the country followed by Arunachal Pradesh (67,410 km^2), Chhattisgarh (55,674 km^2), Maharashtra (50,646 km^2) and Orissa (48,903 km^2). In terms of percentage of forest cover with respect to total geographical area, Mizoram with 90.68% has the highest, followed by Lakshadweep (84.56%), Andaman & Nicobar Islands (81.51%) and Arunachal Pradesh (80.50%).

Table 3.2: Forest and Tree Cover of India (Source: FSI 2011)

Class	Area (Km^2)	% of Geographical Area
Forest Cover		
(a) Very Dense Forest	83,471	2.54
(b) Moderately Dense Forest	320,736	9.76
(c) Open Forest	287,820	8.75
Total Forest Cover*	692,027	21.05
Tree Cover	90,844	2.76
Total Forest and Tree Cover	782,871	23.81
Scrub	42,177	1.28
Non-Forest	2,553,059	77.67
Total Geographical Area	3,287,263	100.00

(* Includes 4662 km^2 area under mangroves)

The total growing stock of India's forest and TOF is estimated as 6,047.15 million cubic meters which comprises 4,498.73 million cubic meters inside the forests and 1,548.42 million cubic meters outside the forests. Among the States/UTs, the maximum growing stock in forest is reported from Arunachal Pradesh (493 million cubic meters) followed by Uttarakhand (460 million cubic meters) and Chhattisgarh (334 million cubic meters).

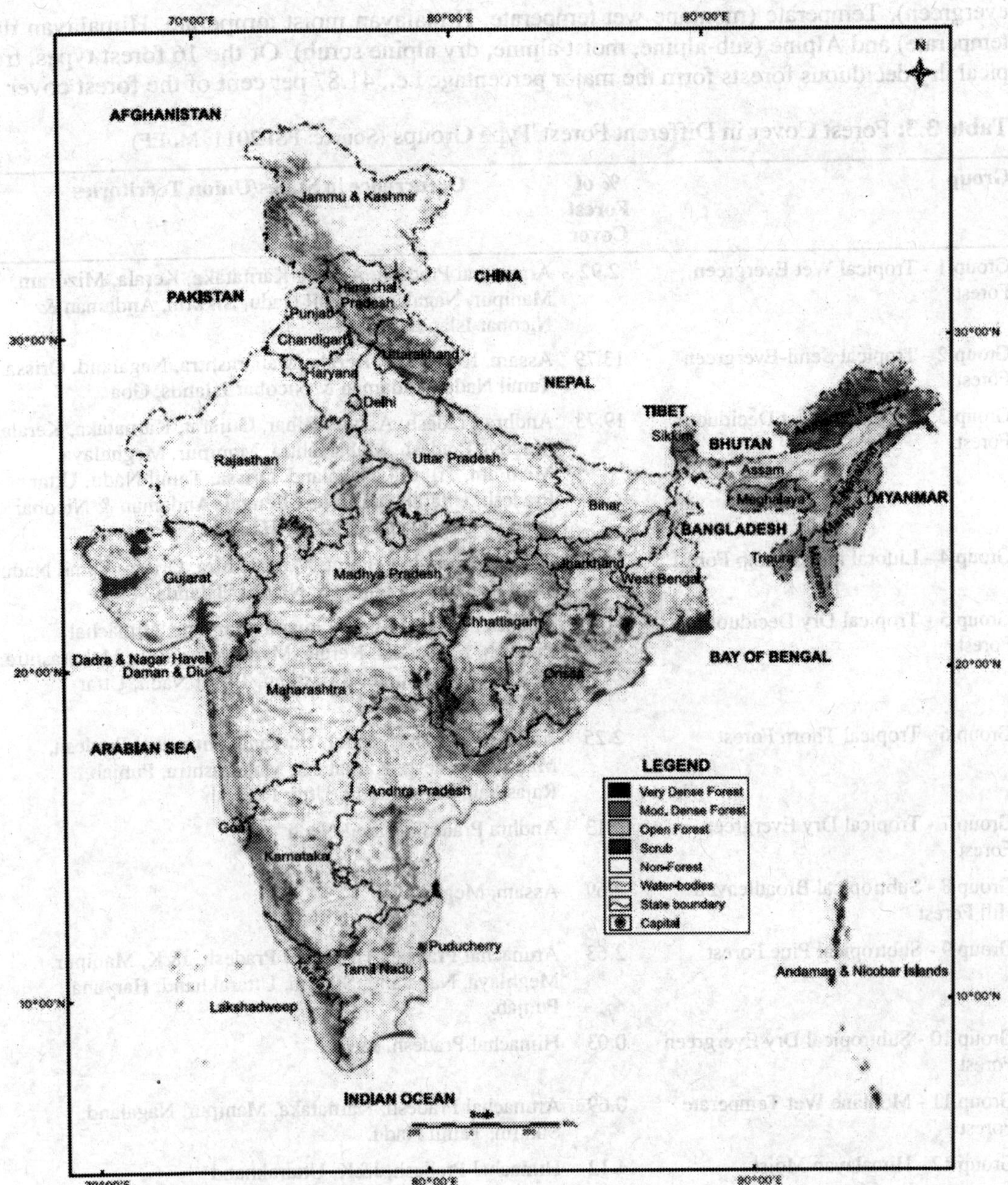

Fig. 3.1: Forest Cover Map of India (Source: FSI 2011)

CLASSIFICATION OF INDIAN FORESTS

Champion and Seth (1968) classified India's forests into four major ecosystem groups viz. tropical, sub-tropical, temperate and alpine. These major groups are further divided into 16 type groups: Tropical (wet evergreen, semi-evergreen, moist deciduous, littoral and swamp, dry deciduous, thorn, dry evergreen), Sub-tropical (broadleaved hill forests, pine, dry evergreen), Temperate (montane wet temperate, Himalayan moist temperate, Himalayan dry temperate) and Alpine (sub-alpine, moist alpine, dry alpine scrub). Of the 16 forest types, tropical dry deciduous forests form the major percentage i.e., 41.87 per cent of the forest cover.

Table 3.3: Forest Cover in Different Forest Type Groups (Source: FSI 2011; MoEF)

Group	% of Forest Cover	Occurrence in States/Union Territories
Group 1 - Tropical Wet Evergreen Forest	2.92	Arunachal Pradesh, Assam, Karnataka, Kerala, Mizoram, Manipur, Nagaland, Tamil Nadu, Sikkim, Andaman & Nicobar Islands, Goa
Group 2 - Tropical Semi-Evergreen Forest	13.79	Assam, Karnataka, Kerala, Maharashtra, Nagaland, Orissa, Tamil Nadu, Andaman & Nicobar Islands, Goa
Group 3 - Tropical Moist Deciduous Forest	19.73	Andhra Pradesh, Assam, Bihar, Gujarat, Karnataka, Kerala, Madhya Pradesh, Maharashtra, Manipur, Meghalaya, Mizoram, Tripura, Nagaland, Orissa, Tamil Nadu, Uttar Pradesh, Uttarakhand, West Bengal, Andaman & Nicobar Islands, Goa, Dadra & Nagar Haveli
Group 4 - Littoral and Swamp Forest	0.69	Andhra Pradesh, Gujarat, Maharashtra, Orissa, Tamil Nadu, West Bengal, Andaman & Nicobar Islands
Group 5 - Tropical Dry Deciduous Forest	41.87	Andhra Pradesh, Bihar, Gujarat, Haryana, Himachal Pradesh, Karnataka, Kerala, Madhya Pradesh, Maharashtra, J&K, Orissa, Punjab, Rajasthan, Tamil Nadu, Uttar Pradesh, Uttarakhand
Group 6 - Tropical Thorn Forest	2.25	Andhra Pradesh, Gujarat, Haryana, Himachal Pradesh, Madhya Pradesh, Karnataka, Maharashtra, Punjab, Rajasthan, Tamil Nadu, Uttar Pradesh
Group 7 - Tropical Dry Evergreen Forest	0.13	Andhra Pradesh, Tamil Nadu
Group 8 - Subtropical Broadleaved Hill Forest	2.69	Assam, Meghalaya
Group 9 - Subtropical Pine Forest	2.63	Arunachal Pradesh, Himachal Pradesh, J&K, Manipur, Meghlaya, Nagaland, Sikkim, Uttarakhand, Haryana, Punjab.
Group 10 - Subtropical Dry Evergreen Forest	0.03	Himachal Pradesh, J&K
Group 11 - Montane Wet Temperate Forest	0.69	Arunachal Pradesh, Karnataka, Manipur, Nagaland, Sikkim, Tamil Nadu.
Group 12 - Himalayan Moist Temperate Forest	4.12	Himachal Pradesh, J&K, Uttarakhand
Group 13 - Himalayan Dry Temperate Forest	0.84	J&K, Himachal Pradesh
Group 14, 15, 16 - Sub Alpine, Moist Alpine and Dry Alpine Forest	2.55	J&K, Himachal Pradesh, Uttarakhand
Plantation/TOF	5.07	All States
Total	100.00	

Over 45,000 species of plants exists in India and of this several thousands are endemic to the country. Bamboo in the country constitutes the largest bamboo resources of the world. The total bamboo bearing area in the country is estimated to be 13.96 million ha. The total carbon stock in the country's forests is estimated to be 6,663 million metric tons. Mangroves in India account for about 3 per cent of the world's mangrove vegetation and 0.14 per cent of the country's total geographical area and are spread over an area of 4,662.56 km^2 (FSI 2011). Sunderbans in West Bengal accounts for almost half of the total area under mangroves in India. Of the total geographical area, 2.54 per cent are classified as "very dense" (tree canopy density greater than 70 per cent), 9.76 per cent are "moderately dense" (tree canopy density from 40 to 70 per cent) and 8.75 per cent are "open forest" (tree canopy density from 10 to 40 per cent).

The forest vegetation ranges from that of the arid, semi-desert tracts of North-west India and the alpine scrub on the higher slopes of the Himalayas to the luxuriant evergreen forests of parts of Assam, Burma and the west coast of India, while the total number of woody plants comprises approximately 5,000 distinct species. The number of valuable timbers and other commercial products available in the forests is very large and many of them are as yet imperfectly known. Among the best known timber trees are teak (*Tectona grandis*), rosewood (*Dalbergia latifolia*), shisham (*Dalbergia sissoo*), sal (*Shorea robusta*), red cedar (*Toona ciliata*), deodar (*Cedrus deodara*), satinwood (*Chloroxylon swietenia*), padauk (*Pterocarpus dalbergioides*), bijasal (*Pterocarpus marsupium*), sandal (*Santalum album*), mahogany (*Swietenia* spp.), pine (*Pinus* spp.), axlewood (*Anogeissus latifolia*), benteak (*Lagerstoemia parviflora*), ironwood (*Mesua ferrea*), siris (*Albizia* spp.), arjun (*Terminalia arjuna*), oak (*Quercus* spp.), kathal (*Artocarpus heterophyllus*), aini (*Artocarpus hirsutus*), mango (*Mangifera indica*), babul (*Acacia nilotica*) The principal NTFPs are classified under bamboos, grass, fibres, oil-seeds, tanning materials, essential oils, gums, resins, rubber, drugs, lac etc.

CHRONOLOGICAL VIEW OF INDIAN FORESTRY

1800: The first step in Indian Forestry began in West Coast. A commission was appointed to enquire the availability of teak in Malabar forests

1805: A Forest Committee was constituted to enquire into the capacity of forests and the status of proprietary rights in them. As a result of the Forest Committee's report, a proclamation was made declaring 'royalty' rights over teak trees in the South and prohibiting unauthorized fellings of teak

1806: The Government of Madras appointed Captain Watson of the Police as the first Conservator of Forests. It was basically to organize the production of teak and other timber supply from the West Coast suitable for the Navy for ship building

1823: The post of Conservator of Forests was abolished

1842: Conolly, Collector of Malabar, initiated action that resulted in the world-famous first teak plantation in the country at Nilambur (Kerala). This was the first step towards the conservation forestry. This work was consolidated by the tireless effort of his assistant Sub-Conservator, Chathu Menon

1847: Gibson was appointed as Conservator of Forests by Bombay Government

August 3, 1855: The Charter of Indian Forests was promulgated for the first time and Lord Dalhousie, the Governor General of India, established the Indian Forest Department

1856: Cleghorn was appointed as the first regular Conservator of Forests in Madras Presidency

1858: Exotic *Acacia* and *Eucalyptus* species were introduced and resulted in the large scale plantations in Nilgiris

1861: Publication of book *Forests and Gardens of South India* authored by Cleghorn

1861: Government constituted the Central Provinces as a chief commissionership. Temple was appointed as first Chief Commissioner and Captain Pearson was appointed as Superintendent of Forests

1864: Lord Canning appointed Sir Dietrich Brandis (a German forester) as first Inspector General of Forests to the Government of India, a post he held from 1864 to 1883. He is the founder of Scientific Forestry in India. There was a shift from the exploitative principle towards conservation and better management of forest resources (systematic forest management)

1865: First Indian Forest Act enacted

1867: *Imperial Forest Service* was constituted

1869: Reorganization of Forest Department and Forest Service and forest officers classified into Conservators, Deputy Conservators and Assistant Conservators

1865-1870: Planting of Casuarina in the coastal areas of Madras for providing fuel needs of Railways

1871: The Forest Department was placed under the newly established Department of Revenue and Agriculture, itself under the umbrella of the Home Department

1874: Publication of book *Forest Flora of North West and Central India* authored by Sir Dietrich Brandis

1875: The *Indian Forester* was founded

1878: Formation of Indian Forest School at Dehradun for training forest rangers of all the Provinces

1878: A Revised Indian Forest Act came into existence. For the first time forests were classified into Reserve and Protected forests

1871 to 1900: Compilation of working plans and Forest Settlements

1881: Publication of book *Manual of Indian Timbers* authored by James Sykes Gamble

1882: Madras Forest Act was passed

1885: Forest training commenced at Cooper's Hill in England

1889-1896: Publication of book *Manual of Forestry* in five volumes authored by Sir Wilhelm Schlich

1891: Having realized the importance of a multi-tier forest administration in the federal and provincial Governments for effective management of forest resources the British India Government constituted *Provincial Forest Service* and Executive &

Subordinate Services, which were quite similar to the present day forest administrative hierarchy

1894: First National Forest Policy

June 5, **1906**: Establishment of Forest Research Institute (FRI) as Imperial Forest Research Institute and College with six research disciplines: silviculture, forest working plans, forest zoology, forest botany, forest economics and forest chemistry

1906: Publication of book *Indian Trees* authored by Sir Dietrich Brandis

1906: Appointment of R.S.Troup as first Imperial Forest Economist

1907: Appointment of Chief Conservator of Forests for each province

1909: R.S.Troup was appointed as first regular Silviculturist and served from 1909 to 1919. He laid the foundation of systematic research in the field of silviculture

1909: Royal Commission on Decentralization recommendations stated it was not considered necessary that Inspector General should be called as adviser over every working plan and it was considered sufficient that Chief Conservator of Forests will do the final scrutiny and check of the working plans

1910: *Board of Forestry*, comprising of representative chief conservators under the presidentship of the Inspector General of forests was set up, which met once in three years to discuss the forestry programmes

1912: Madras Forest College was started in Coimbatore for training forest rangers of South India

1917: The appointments of Chief Conservators of Forests began

1921: Publication of book *Silviculture of Indian Trees* in three volumes authored by R.S.Troup

1922: 40 per cent of the Imperial Forest Service was reserved for Indians

1927: New Indian Forest Act enacted

1935: The subject of "Forestry" which was managed by the Federal Government until then, was transferred to the "Provincial List" by the Government of India Act, 1935. National Character of forest administration was considerably diluted and forests became a provincial subject

1938: Indian Forest College (IFC) was established at Dehra Dun and the officers recruited to the Superior Forest Service by the provinces/states were trained there

1950: Central Board of Forestry was constituted

1952: New National Forest Policy

1952: Indian Board of Wildlife was formed

1966: The Indian Forest Service, one of the three All India Services, was constituted under the All India Services Act, 1951 by the Government of India

1972: The Indian Wildlife (Protection) Act

1973: Project Tiger, now renamed as the National Tiger Conservation Authority, was launched

1980: Forest (Conservation) Act

1981: Establishment of Forest Survey of India (FSI)

1982: Establishment of Wildlife Institute of India (WII)

1983: First National Wildlife Action Plan (NWAP) was adopted

1985: Creation of Ministry of Environment and Forests (MoEF) as a separate ministry

1986: Establishment of Indian Council of Forestry Research and Education (ICFRE)

1988: Current National Forest Policy

1990: Launch of Joint Forest Management Programme

1992: Project Elephant was launched

1999: Formulation of National Forestry Action Programme (NFAP)

2002-2016: New Wildlife Action Plan has been adopted

2003: National Forest Commission was set up

2006: The Forest Rights Act

2011: First Indian Forest Congress was organized

FORESTRY IN FIVE YEAR PLANS

Although every five year plans had specified objectives and programmes, the main activity under most of them was tree planting. The emphasis of the different FYPs regarding forestry was as follows.

- First and second FYPs: rehabilitation of degraded forests, introduction of economic species, survey and demarcation;
- Third and fourth FYPs: increasing productivity of forests through fast growing species plantation, scientific assessments and modern logging;
- Fifth FYP: large scale plantation, social forestry and forest conservation;
- Sixth FYP: social forestry and fuel-wood reserves to save natural forests;
- Seventh FYP: forest conservation, massive afforestation and wasteland development;
- Eight and Ninth FYP: preservation of biological and genetic diversity (both flora and fauna), protection of forest against biotic interference, utilization of wastelands, and promotion of people's participation through joint forest management (JFM).
- Tenth FYP: forest conservation, biodiversity conservation, sustainable development

The percentage share of Forestry Sector Outlay has always been low from 0.37 per cent of the total outlay in First FYP (1951-1956) to 0.27 per cent in Eleventh FYP (2007-2012). The highest allocation was in the Seventh FYP (1.03%). For the sustainable development of the sector, allocation to the forestry sector should be raised to about 4 to 5 per cent of the total outlay of the country.

CENTRAL BOARD OF FORESTRY

A Central Board of Forestry was constituted in 1950 to provide guidance to the government in the formulation of policy and programs. Central Board of Forestry (CBF) was reconstituted in the year 2006 with Hon'ble Prime Minister as Chairman and Union Minister of Environment & Forests as Vice-Chairman of the Board to guide the forestry sector.

THE DUTIES AND RESPONSIBILITIES OF THE CBF: The Board will advise the central and state/UT Governments on all matters concerning Forest conservation and Management as well as related matters. The Board will have special concern and focus on: Forest Policy, Biodiversity, Legislation, Inter-State matters, Inter-Ministerial matters, Identification of broad areas of R&D, Training and Education, Indian Forest Service, Appointment of Committees and Sub-Committees/ working Groups. CBF will meet at least once a year. Standing Committees will monitor the implementation of recommendation of CBF. Each state/UT is given an opportunity to present the policy issues pertaining to their area in CBF.

NATIONAL FORESTRY ACTION PROGRAMME (NFAP)

Ministry of Environment and Forests (MoEF) has formulated the National Forestry Action Programme (NFAP) and released it on 9 August 1999. It is a comprehensive work plan for sustainable development of forests in India in next twenty years as well as to achieve the national goal of 33 per cent geographic area of the country under the forest and tree cover as enshrined in the National Forest Policy, 1988. This exercise has been undertaken as a part of the programme recommended by the United Nations Conference for Environment and Development (UNCED) and its subsequent forum, the Commission on Sustainable Development (CSD) and Intergovernmental Panel on Forestry (IPF) for the launch of National Forest Programmes globally. The exercise has been financially supported by the UNDP.

The objective of NFAP is to enhance the contribution of forestry and tree resources to ecological stability and people cantered development through qualitative and quantitative improvements in the forest resources. It is to integrate the forestry development programmes in the country within the framework of national Five Year Plans (FYP). In order to have an integrated programme with the coordination of all states, the State Forest Departments were assigned to the task of preparation of State Forestry Action Programmes (SFAPs). NFAP has been formulated by integrating and amalgamating 26 State Forestry Action Programmes. Five inter-related basic issues have also been identified and these are the basis of the NFAP structure.

(i) Protect existing forest resources

(ii) Improve forest productivity

(iii) Reduce total demand

(iv) Strengthen policy and institutional frame work

(v) Expand forest area

TOP PRIORITIES IN NFAP

- For sustainability of forests, productivity of forest plantations to be increased at least 3 to 5 cubic meter per ha per year by promoting regeneration and enrichment of plantations.

- Hygiene of forests to be improved through suitable silvicultural practices.
- Efforts to be made to bring one-third geographic area of the country under forest and tree cover by plantations on all categories of Wastelands and Agroforestry.
- Protected area network to be expanded and managed for Bio-Diversity conservation.
- Plantation on non-forest wastelands to be done mostly with fuel-wood species, as 70% of the wood produce from forests are used as fuel-wood. Species of pulpwood and other industrial wood may be encouraged in farm forestry.
- Institution for the people's participation in protection and development of degraded and fringe forests to be strengthened.
- NWEP species to be developed and value addition may be promoted at the village level.
- Grazing in forests to be regulated as per carrying capacity and silvicultural needs.
- Infrastructure for forest inventory, research and development to be strengthened. HRD should also be improved.
- Investment for the sustainable development of forests should be rational and in proportion to the total production.

4 WILDLIFE RESOURCES

WILDLIFE DEFINITION

Wildlife is defined as free-living animals of major significance to man. Wildlife is a term that does not enjoy a precise meaning and various definitions exist in literatures such as,

- Animals living in the wilderness without human intervention.
- All living things outside direct human control
- Plants and animals that are not cultivated or domesticated
- A practical definition : any free-living vertebrate animals, often excluding fish

In the Wildlife (Protection) Act, 1972, the following terms related with the wildlife resources are legally defined under the Chapter- I Preliminary Section 2 as follows:

- "Animal" includes amphibians, birds, mammals and reptiles and their young, and also includes, in the cases of birds and reptiles, their eggs
- "Wild animal" means any animal found wild in nature specified in Schedule I, Schedule II, Schedule III and Schedule IV
- "Wildlife" includes any animal, aquatic or land vegetation which forms part of any habitat

India is one of the world's richest countries in terms of its biodiversity and identified on the basis of immense species diversity. Country has nearly about 91,797 species of animals and 46,340 species of plants. The important mammals include elephant (Assam, Kerala, Karnataka), camel and wild ass (arid areas of Rajasthan and Rann of Kutch), rhinoceros (Assam, N.W. Bengal), Indian bison, buffaloes, nilgiri, chousingha, gazel, deer, etc. Among the predator animals included are lion (Gujarat), tiger (Sunderbans in Bengal) and leopard of various types found in the Himalayan region. The Himalayan ranges are the home of several interesting animals such as wild sheep, mountain goats, ibex (wild goat), tapir (a hoofed animal like a rhinoceros), snow leopard etc. India also has several species of monkeys such as the langur, the lion-tailed macaque, etc. There is also variety of birds found in different parts of country. The chief among them are peacock, goose, pheasants, duck, mynah, pigeon, parrot, crane, hornbill and sunbird etc. India also possesses 2634 species of fish which inhabit in lakes, rivers, seas and oceans. Our national animal is tiger and our national bird is peacock.

VALUES OF WILDLIFE

Wildlife in general comprises of all living organism i.e. plants, animals, microorganisms found in their natural habitats which are neither cultivated or domesticated nor tamed. But in its strictest sense, it includes uncultivated mammals, reptiles, birds and fishes etc. which are generally hunted. Importances of wildlife are:

1. Ecological Balance: Wildlife maintains ecological balance by preserving the environment as a self sustaining system through,

   (a) Regulation of population of different species.

   (b) Food-chains.

   (c) Natural cycles between biotic and abiotic environments.
2. Wildlife serves as a gene banks for breeding improved varieties in plant and animal.
3. Wildlife helps in plant propagation e.g. pollination in certain plants is performed by wild animals like birds, insects, bats etc.
4. Wildlife serves as scavengers and decomposers for cleaning and sustainable of Environment.
5. Scientific and research importance for studies particularly related to anatomy, physiology, ecology, evolutionary aspects. The wildlife is useful as experimental animals e.g. monkey, rabbit, guinea-pig, rat, etc. These trials on wildlife species are used to sustain and save human life.
6. Plant products or NWFPs of plant origin and animal products of economic importance such as several drugs, silk, honey, hair, feathers, guano, leather, musk, ivory etc. are obtained from wildlife. Besides these, certain benefits are also obtained from tourism and exports of animal products such as hide, ivory, etc. after death of wild animals, though nowadays it is legally banned and not in practice.
7. Potential Uses: Food, drugs and other useful products may be obtained and derived from wildlife.
8. Soil erosion: It is prevented by plant cover, litter, mixing of litter by movement of wild animals and conversion to spongy humus by microorganisms.
9. Maintenance of climate and control of pollution by wild plants and trees.

Values of wildlife are categorized into direct and indirect values as described below:

DIRECT VALUES: Direct values are considered which have consumptive use value such as non-market value of firewood, game, etc. and productive use value like commercial value of timber, fish, etc.

INDIRECT VALUES: The indirect values are classified as follows:

- Non-consumptive use value: scientific research, bird-watching, etc.
- Option value: value of maintaining options available for the future.

- Existence value: value of ethical feelings of existence of wildlife.

Values of wildlife are also categorized into positive and negative values as described below:

POSITIVE VALUES: It pertains to the positive or beneficial aspects of wildlife. They have direct or indirect bearing on the life of human being. It may be ethical, cultural, ecological, environmental, scientific, asthetic, commercial or economic and game values.

- *Ethical value*: There are many examples of wildlife having ethical values in Hindu mythology e.g. Ramayana, Mahabharata, Panchatantra, Arthshastra, etc.
- *Cultural value*: Wildlife has a strong and intimate relationship with Indian culture, tradition and religion. Some of the important examples are elephant nose of Lord Ganesha, tiger associated with Goddess Durga, cobra snake with Lord Shiva, peacock with Goddess Saraswati, lion is the sign of bravery and ferocity, monkey helped to Lord Rama etc. which show religious and cultural values of wildlife in India.
- *Ecological value*: All wildlife plays an important role in maintaining ecological balance of our earth and they have symbiotic association with the nature.
- *Environmental value*: They help in protection of land degradation, soil erosion control and climate control.
- *Scientific value*: Wild plants and animals both have immense scientific values. The newly developed medicines and drugs are first tested on animals like apes, monkey, rats, mice, pig, etc. for various diseases in human beings. Every year thousands of animals are used for these experiments.
- *Aesthetic value*: Wild animals play an important role in aesthetic and recreational point of view. The beauty of nature gives immense pleasure to mankind. The wildlife sanctuaries and national park, zoos etc. are the important sources of recreation.
- *Commercial or economic value*: Wild animals also have great commercial or economic values e.g. fishing, trade of animal products etc.
- *Game value*: Hunting, fishing etc. are very popular sports in the past but nowadays it is restricted in almost all sanctuaries and national parks and other reserve forests.
- *Nutritional value*: Some wild animals have been a source of food for human beings. The word 'wild meat' is used to designate meat from wild animals, keeping in mind that terms vary widely according to regions and cultures.

NEGATIVE OR ADVERSE VALUES: The wildlife, however, may be seen as sometimes presenting negative values as described below:

- Casualties to people: Human beings attacked by wild animals, vehicle collisions, etc are examples of human-animal conflict in wild life areas.
- Predation: The predators such as leopard, bear, wolf or lynx, etc. are regularly responsible for attacks on sheep, cattle and other livestock.

- Crop damage: Several wildlife species are responsible for significant damage, both to crops (elephant, deer, wild boar, wild rabbit, hare, wild pigeon etc.) and to regenerating forests (cervids)
- Disease dissemination to domestic animals.
- Invasive pests have anti-values

WILDLIFE DISTRIBUTION

ZOOGEOGRAPHICAL REGION

Zoogeography is the study of distribution of animals on the earth, which occurs in different regions of the world in a distinct pattern. The study of the distribution of land animals, also called 'faunal regions'. The distribution of some animal species is so typical that it is difficult to explain their occurrence in a particular region. The presence or absence of an animal in a certain region is governed by ecological and historical factors. Zoogeography attempts to understand the distribution of animals with regard to evolutionary and environmental influences. Zoogeographical regions can usually be distinguished by the distribution of animals in certain physiographical regions. Sclater (1857) was the first one to give concepts of zoogeography and divided the continental masses into six *realms* based on his studies on the bird fauna under two *creatio* or centres of creation, namely, Palaeogeana (Old world) and Neogeana (New world).

Table 4.1: Zoogeographic Divisions of World (Sclater 1857)

Creatio	Realms	Area
Palaeogeana	Palaearctic	Temperate Eurasia
(Old world)	Ethiopian	West Paleotropic, Africa
	Indian	Middle Paleotropic, Tropical Asia
	Australian	East Paleotropic, Australia, New Guinea, Tasmania
Neogeana	Nearctic	North America, Greenland, North America up to Mexico
(New world)	Neotropical	South America south of Mexico

Table 4.2: Zoogeographic Regions of World (Wallace 1876; Darlington 1957)

S No	Region	Area
1	Nearctic	North America and Greenland
2	Palaearctic	Eurasia, without India
3	Ethiopian	Africa, south of the Sahara
4	Oriental	India and Indochina
5	Australian	Australia and New Zealand
6	Neotropical	South and Central America, and the Antilles

A.R. Wallace (1876), who is considered father of modern zoogeography agreed with Sclater's classification but proposed the name ORIENTAL instead of INDIAN and

AFRICAN instead of ETHIOPIAN regions because the earlier names represented countries and not the zoogeographical regions. The widely accepted modern classification of land masses into six regions is given below which is based on Wallace (1876) and Darlington (1957). On this basis, the world can be divided into six zoogeographical regions:

In the perspective of Earth's history, the present position of India is of recent origin and it forms a part of the Oriental region in the above Zoo-geographic classification.

INDIAN FAUNA AND THEIR DISTRIBUTION

India is one of the 12 mega biodiversity countries of the world. Out of the 25 unique biodiversity 'hotspots', India has two hotspots namely Eastern Himalaya (Indo-Burma) and the Western Ghats which contain about 20 per cent of the world's flora. India is divided into various ecological sub-divisions. The flora and fauna, specific to the environment are found in these sub-divisions because they can thrive well in the conditions prevailing there. Ecologically, India can be divided into three main regions:

- Himalayan Mountain system
- Peninsular India Sub-region (woodlands and desert)
- Tropical Evergreen Forest of Indo-Malayan Sub-region

HIMALAYAN MOUNTAIN SYSTEM

This region is characterized by altitudinal as well as East-West variations.

THE HIMALAYAN FOOTHILLS (Eastern frontiers of Kashmir to Assam)

These regions are characterized by bhaber and tarai tract and consist of thick boulder deposits. The natural monsoon forest extends well into this region and the dominant species is sal. Tarai is characterized by grassy meadow and savanna vegetation. This is one of the richest areas for the typical big mammals of Northern India.

Fauna distribution in the Himalayan foothills is unique e.g. elephant, panther, sambar, swamp deer, cheetal, hog deer, barking deer, wild boar, wild dog, hyena, jackal, black and the sloth bears, tiger, gangatic gharial, hispid hare and pygmy hog. Among rodents porcupine is the most prominent. The great Indian one horned rhinoceros is considered as a rare species. The wild buffalo and gaur are now considered as an endangered species. The brow antlered deer is another rare deer found in this region confined to Keibul Lamjo Sanctuary in Manipur.

WESTERN HIMALAYAS (from Kashmir including Ladakh to Kumaon)

This is a high altitudinal zone with a belt of coniferous alpine forests rich with rhododendron, dwarf hill with bamboo and birch forest mixed with the alpine pastures and extends upto the snow line. The typical fauna of the high altitude cold region includes wild ass which is mostly found in the cold desert regions, largest bovid-the yak, wild goats-tahr, markhor and ibex, three species of sheep-nayan or largest sheep, bharal or blue sheep and shapu or urial which is the smallest of the wild sheep. There are two kinds of antelopes viz. the chiru and the Tibetan gazelle. The deer species, hangul or Kashmir stag are mostly confined to the Dachingam Sanctuary in Jammu and Kashmir near Srinagar. Musk deer is another deer which inhabits the high altitude in Western Himalayas and are much valued for their musk pods. Due to high poaching scale, musk deer have now become threatened. The smaller mammals include marmot, pika or mouse hare. One of the most beautiful animals, snow leopard is found in Western Himalayas and now regarded as a threatened species.

EASTERN HIMALAYAS

The character of vegetation changes at the altitudes of 1525 m to 1830 m as go up from dense evergreen and semi evergreen vegetation of the foothills of Eastern Himalayas to higher altitudes on these mountain regions. Oaks magnolias, laurels and birches covered with mosses and ferns replace the sal, silk cotton trees and giant bamboos of the foothills. Fauna of this region includes red panda, hog badgers, ferret badgers, crestless porcupines and three wild goats-serow, goral and takins.

PENINSULAR INDIAN SUB-REGION

It is characterized by the raised plateau land of the Deccan extending northwards into the flood plains of the Indo-Gangetic basin and westwards into the great Thar desert of Rajasthan. Ecologically it may include the salt flats of the little Rann of Kutch. Overall, this region is divided into two broad zones. First one is the tropical deciduous woodlands and its extension into the drainage basin of the Ganges river system. The second is the desert region of Rajasthan or the Thar.

PENINSULAR INDIA AND THE DRAINAGE BASIN

The area has relatively higher rainfall and has sal as the predominant species while the Southern plateau is characterized by teak as the principal species. The Western Ghats and the central belt lying to the west of it, is a region of very high rainfall and is characterized by evergreen vegetation. The typical fauna of the area includes elephant, muntjak, sambar deer, wild boar, cheetal, hog deer, swamp deer, barasinga, nilgai, black bucks, chinkara gazelle, wild dog or dhole, tiger, leopard, cheetah, lion, stripped hyena and jackal. The cheetah has now become an extinct animal in India.

THE INDIAN DESERT

The Thar or the Indian desert is linked through Sind and Baluchistan in Pakistan to the deserts of Arabia and Middle East. Here the temperature rises very high reaching upto 48^0 C or more and the night temperature falls upto zero. Due to the scarcity of water, the desert animals have some modified physical characters and behaviours and have grown some remarkable physiological adaptations to conserve water. The fauna includes rodents which probably represent the largest group in the mammalian fauna of the desert. The Asiatic wild ass is one of the endangered species found in Rann of Kutch. The ungulate blackbucks inhabit the scrub forests and open savannah. These animals are fully protected by the Vishnoi tribes due to their religious beliefs in them. Desert cat and the caracal represent the carnivorous animals of these regions. Reptiles include desert lizards and snakes like saw scaled vipers. The great Indian bustard, one of the largest birds, is found in these deserts.

TROPICAL EVERGREEN FOREST OF INDO-MALAYAN SUB-REGION

It is covered with rain forest with a highly complex environment and contains a wealth of species of flora and fauna. This region consists of a heavy rainfall zones and comprises North Eastern India (Khasi Jaintia hills and the lower Himalayan slopes embracing areas of Arunachal Pradesh, Assam, Meghalaya, Nagaland and Tripura) and the Western Ghats region in the south including the Malabar coast. In the Nilgiris, an evergreen vegetation known as "Sholas" or Shola forests are also found.

There are some exclusively confined species such as hollock gibbon, red panda, tree shrew, giant squirrel and flying lemurs which are specific to these areas. Himalayan animals as the tahr, pine marten and the European otter occur in the Nilgiris. The Hollock Gibbon (the only ape found in India at present), golden langur, caped langur or leaf monkey, Assam macaque, pigtailed macaque, lion tailed macaque (which is now considered rare and endangered species), Nilgiri langur and slender loris, bat, giant squirrel, civet, binturong and red panda are characteristic of the evergreen forest in these regions. The flying squirrel, nilgiri mongoose and stripped necked mongoose and malabar civet are characteristic animals of the Western Ghats.

THE WILDLIFE IN THE ANDAMAN AND THE NICOBAR ISLANDS

The Andaman and the Nicobar islands lie in the Bay of Bengal. The climate of these islands is tropical and is the home of tropical rain forest. These islands posses about 35 species of mammals, number of interesting reptiles and a variety of marine fauna. The typical fauna of this area includes Andaman pig (one of the large mammals found in the Andaman), crab eating macaque, palm civet, sambar and spotted deer which are more prominent. Dugong, false killer whale and Dolphin are important marine mammals found around the island. The Dugong is the rare and threatened animal. The avifauna includes Narcondum hornbill, the Nicobar pigeon, megapode which has Australian affinities, white bellied sea eagle, white breasted swiftlet and a number of fruit pigeons. About 240 species and sub-species of birds are reported to found in these islands, of which 92 are migrants, 32 residents and 112 as resident endemic species. It has also salt water crocodile which is a threatened animal. A number of marine turtles are found, of which the green turtle is common. The coconut crab, green lizard, cobra, viper, coral and sea snakes, pythons are found in the Nicobar group.

WATER BIRDS IN THE BHARATPUR SANCTUARY

Bharatpur is the fresh water swamp of Keoladeo Ghana with the Kadam tree, Acacia tree and profuse aquatic vegetation extending over an area of 2,900 hectares. Over a million birds come here during winter season from far distance temperate regions of Eurasia and unique to its richness and variety. Nearly 328 species of birds have been identified in this region, of which 110 are migratory birds. Amongst the breeding birds in the Bharatpur Sanctuary are painted stork, egret, open billed stork, spoonbill, white ibis, heron, snake bird or Indian darter, cormorant. Among the winter visitors, there are a number of geese and duck, rosy or white pelican, dalmatian pelican, bar-headed geese, gray lag geese, pochard, pintails, sandpipers, large number of warblers, wagtail and pipit. The VIP visitor bird to this area is the Siberian crane which is one of the rarest birds. Water sedge, *Cyperus rotundus* found in the swamps of Bharatpur is the main attraction for the Siberian crane, which is a vegetarian. Its tuberous roots form the crane's main food.

THE MANGROVE FOREST OF THE SUNDARBANS

Mangrove forests are found in the Sunderbans delta of the river Ganges where both the Brahmaputra and the Ganges join and drain into the Bay of Bengal. Water logging, humidity, salinity and regular inundations are the salient features of this habitat. These areas are characterized by the pneumatophores roots of Sonerentians. The *Rhizophora* tree is found in these mangrove regions which possess the supporting roots or the stilt roots. Small crabs and fishes are found in the muds and mudskippers. Among the crabs, land crab and fiddler crab are the most famous ones because of their brilliantly coloured claws. Spotted deer, wild boar, monitor lizard and monkeys characterize the area. The most interesting animal found in the Sundarbans is the royal Bengal Tiger. The most peculiar tendency of these tigers is their tendency to man-eating.

WILD ANIMAL STATUS

Due to increasing anthropogenic pressures, many species are under threat. Human activities, both directly and indirectly, are responsible for current high rates of habitat loss resulting in biodiversity loss. The land fragmentation due to agricultural activities, extraction, developmental activities and environmental degradation are some of the main causes of serious threat to wildlife. Some of the major causes of threat to wildlife animals are:

- *Habitats Destruction* is the most serious threat. Loss of biodiversity and habitat occurs by following reasons,
    - Agricultural activities such as cropping land, shifting cultivation, grazing land, land conversion for agriculture, land fragmentation and degradation leads to the formation of isolated, small and scattered populations
    - Extraction activities including mining, fishing, logging and harvesting
    - Development activities like human settlements, industrialization, modernization and associated infrastructure (construction of dams, reservoirs, roads, rails, buildings, etc.)
    - Deforestation for timber, wood, fuelwood, Jhum cultivation, etc.
    - Natural calamities like flood, drought, earthquake, forest fire etc.
- Hunting for food, safety and pleasure
- Poaching, smuggling and illegal trade of wildlife and its derivatives
- Forest invasive species like lantana, parthenium etc.
- Disturbance in ecosystem and ecological balance
- Over-exploitation of wild resources
- Pollution of water, soil and environment, global warming and climate change

IUCN CATEGORIES OF WILD ANIMALS

The International Union for Conservation of Nature and Natural Resources (IUCN) maintains a "Red Data Book" about the identified threatened species for the purpose of conservation. The following categories have been identified for the purpose

- **Normal Species:** Species whose population levels are considered to be normal for their survival.
- **Endangered Species (E):** These are species (or taxa) which are in danger of extinction and whose survival is unlikely if the causal factors continue operating. Examples of such species are black buck, crocodile, Indian wild ass, Indian rhinoceros, Asiatic lion, tiger, lion tailed macaque, red panda, Gangetic dolphin, wild cat, etc.
- **Vulnerable Species (V):** These are species (or taxa) likely to move into the endangered category in the near future if the causal factors continue operating. Examples of such species are black buck, brown bear, Indian giant squirrel, Asiatic wild ass, sloth bear, tiger, Asiatic lion, etc.
- **Rare Species (R):** Species (or taxa) with small world populations that are not at present endangered or vulnerable, but are at risk. Examples of such species are the Himalayan brown bear, wild Asiatic buffalo, desert fox and hornbill, etc.
- **Threatened Species (T):** Threatened is used in the conservation context for species (or taxa) which are in one of the categories endangered, vulnerable and rare.
- **Out of Danger (O):** The species (or taxa) formerly included in one of the above categories, but which are now considered relatively secure because of effective conservation measures undertaken or the previous threat to their survival has been removed.
- **Intermediate (I):** The species (or taxa) that are suspected of belonging to one of the above categories, but for which insufficient information is currently available and these may also be referred as the category of Data Deficient.

In 1994, IUCN introduced new Red Data categories which are as follows:

- **Extinct (Ex):** A taxon is extinct when there is no reasonable doubt that the last individual has died.
- **Extinct in the wild (EW):** A taxon is extinct in the wild when it is known only to survive in cultivation, in captive well outside the past range.
- **Critically Endangered (CE):** A taxon is critically endangered when it is facing an extremely high risk of extinction in the wild in the immediate future.
- **Endangered (EN):** A taxon is endangered when it is not critically endangered but is facing a very high risk of extinction in the wild in the near future.
- **Vulnerable (VU):** A taxon is vulnerable when it is not critically endangered or endangered but is facing a high risk of extinction in the wild in the medium-term future.

- **Lower risk (LR):** A taxon is lower risk when it has been evaluated, does not satisfy the criteria for any of the categories critically endangered, endangered or vulnerable. Taxa are subcategorized:
  - (i) **Conservation dependent (CD):** Taxa which are the focus of a continuing taxon-specific or habitat-specific conservation programme targeted towards the taxon in question.
  - (ii) **Near Threatened (NT):** Taxa which do not qualify for conservation dependent, but which are close to qualifying for vulnerable.
  - (iii) **Least Concern (LC):** Taxa which do not qualify for conservation dependent, or near threatened.
- **Data Deficient (DD):** A taxon is data deficient when there is inadequate information to make a direct, or indirect, assessment of its risk of extinction based on its distribution and /or population status.
- **Not Evaluated (NE):** A taxon is not evaluated when it has not yet been assessed against the criteria for critically endangered, endangered and vulnerable.

Endemic Species: These are species which are only found in some particular areas usually isolated by natural or geographical barriers. Examples of such species are the Andaman teal, Nicobar pigeon, Andaman wild pig, etc.

Extinct Species: These are species which are not found after searches of known or likely areas where they may occur. A species may be extinct from a local area, region, country, continent or the entire earth. Examples of such species are the Asiatic cheetah, pink head duck.

ENDANGERED WILD ANIMALS OF INDIA

The following list includes wild animals rated as Critically Endangered (CR), Endangered (EN) or Vulnerable (VU) according to the Red Data Book of International Union for Conservation of Nature (IUCN)

Critically Endangered Birds

1. *Ardea insignis* (White-bellied Heron)
2. *Eurynorhynchus pygmeus* (Spoon-billed Sandpiper)
3. *Grus leucogeranus* (Siberian Crane)
4. *Gyps bengalensis* (White-rumped Vulture)
5. *Gyps indicus*
6. *Gyps tenuirostris* (Slender-billed Vulture)
7. *Heteroglaux blewitti* (Forest Owlet)
8. *Houbaropsis bengalensis* (Bengal Florican)
9. *Ophrysia superciliosa* (Himalayan Quail)
10. *Rhinoptilus bitorquatus (Jerdon's Courser)*, Endemic to India
11. *Sarcogyps calvus* (Red-headed Vulture)

12. *Vanellus gregarius* (Sociable Lapwing)
13. *Rhodonessa caryophyllacea* (Pink-headed Duck)
14. *Great Indian Bustard*

Critically Endangered Mammals, Reptiles, Amphibians and Fishes

1. *Anoxypristis cuspidata* (Knifetooth sawfish)
2. *Batagur baska* (Four-toed Terrapin)
3. *Batagur kachuga (*Red-crowned Roofed Turtle)
4. *Biswamoyopterus biswasi* (Namdapha Flying Squirrel)
5. *Carcharhinus hemiodon* (Pondicherry shark)
6. *Cremnomys elvira* (Large Rock-rat)
7. *Crocidura andamanensis* (Andaman White-toothed Shrew)
8. *Crocidura jenkinsi* (Jenkin's Shrew), Endemic to India
9. *Crocidura nicobarica* (Nicobar Shrew)
10. *Dermochelys coriacea* (Leatherback)
11. *Dicerorhinus sumatrensis* (Sumatran Rhinoceros)
12. *Eretmochelys imbricata* (Hawksbill Turtle)
13. *Fejervarya murthii*
14. *Gavialis gangeticus* (Fish-eating crocodile)
15. *Glyphis gangeticus* (Ganges Shark), Endemic to India
16. *Indirana gundia*
17. Himalayan Wolf
18. Kashmir Stag
19. *Indirana phrynoderma*
20. *Ingerana charlesdarwini*
21. *Labeo potail* (Deccan Labeo)
22. *Micrixalus kottigeharensis*
23. *Millardia kondana* (Kondana Rat)
24. *Philautus chalazodes*
25. *Philautus griet*
26. *Philautus ponmudi*
27. *Philautus sanctisilvaticus*
28. *Philautus shillongensis*
29. *Philautus* sp. nov. 'Amboli Forest'
30. *Philautus* sp. nov. 'Munnar'

31. *Porcula salvania* (Pygmy Hog or Sus salvanius)
32. *Pristis microdon* (Leichhardt's sawfish)
33. *Pristis zijsron* (Narrowsnout sawfish)
34. *Rhacophorus pseudomalabaricus*
35. *Rhinoceros sondaicus* (Javan Rhinoceros)
36. *Viverra civettina* (Malabar Large-spotted Civet)

Endangered Species

1. Andaman Shrew (*Crocidura andamanensis*), Endemic to India
2. Andaman Spiny Shrew (*Crocidura hispida*), Endemic to India
3. Asian Arowana (*Scleropages formosus*
4. Asiatic Black Bear (*Selenarctos thibetanus*)
5. Asiatic Lion (*Panthera leo persica*)
6. Asiatic Wild Dog/ Dhole (*Cuon alpinus*)
7. Banteng (*Bos javanicus*)
8. Blue Whale (*Balaenoptera musculus*)
9. Capped Leaf Monkey (*Trachypithecus pileatus*)
10. Chiru (Tibetan Antelope) (*Pantholops hodgsonii*)
11. Wild Cat (*Felis silvestris ornata*)
12. Fin Whale (*Balaenoptera physalus*)
13. Ganges River Dolphin (*Platanista gangetica gangetica*)
14. Golden Leaf Monkey (*Trachypithecus geei*)
15. Hispid Hare (*Caprolagus hispidus*)
16. Hoolock Gibbon (*Bunipithecus hoolock*) (Previously *Hylobates hoolock*)
17. Indian Elephant or Asian Elephant (*Elephas maximus*)
18. Indus River Dolphin (*Platanista minor*)
19. Kondana Soft-furred Rat (*Millardia kondana*), Endemic to India
20. Lion-tailed Macaque (*Macaca silenus*), Endemic to India
21. Loggerhead Sea Turtle (*Caretta caretta*)
22. Malabar Civet (*Viverra civettina*)
23. Markhor (*Capra falconeri*)
24. Narcondam Hornbill (*Rhyticeros narcondami*)
25. Nicobar Tree Shrew (*Tupaia nicobarica*), Endemic to India
26. Nilgiri Leaf Monkey (*Presbytis johni*)
27. Nilgiri Tahr (*Hemitragus hylocrius*), Endemic to India

28. Olive Ridley Turtle, Endemic to Odisha and Andhra Pradesh, India
29. Particoloured Flying Squirrel (*Hylopetes alboniger*)
30. Peter's Tube-nosed Bat (*Murina grisea*), Endemic to India
31. Red Panda (Lesser Panda) (*Ailurus fulgens*)
32. Royal Bengal Tiger (*Panthera tigris tigris*)
33. Sei Whale (*Balaenoptera borealis*)
34. Servant Mouse (*Mus famulus*), Endemic to India
35. Snow Leopard (*Uncia uncia*)
36. Wild Water Buffalo (*Bubalus bubalis*) (Previously *Bubalus arnee*)
37. Woolly Flying Squirrel (*Eupetaurus cinereus*)
38. Himalayan Tahr

Vulnerable Species

1. Andaman Horseshoe Bat (*Rhinolophus cognatus*), Endemic to India
2. Andaman Rat (*Rattus stoicus*), Endemic to India
3. Argali (*Ovis ammon*)
4. Himalayan W-toothed Shrew (*Crocidura attenuate*)
5. Sri Lankan Highland Shrew (*Suncus montanus*)
6. Asiatic Black Bear (*Ursus thibetanus*)
7. Asiatic Golden Cat (*Catopuma temminckii*)
8. Assamese Macaque (*Macaca assamensis*)
9. Back-striped Weasel (*Mustela strigidorsa*)
10. Barasingha (*Cervus duvauceli*)
11. Bare-bellied Hedgehog (*Hemiechinus nudiventris*), Endemic to India
12. Black buck (*Antilope cervicapra*)
13. Brow-antlered Deer (*Cervus eldi eldi*)
14. Brown Bear (*Ursus arctos*)
15. Brown fish owl (*Ketupa zeylonensis*), Endemic to India
16. Brown Palm Civet (*Paradoxurus jerdoni*)
17. Central Kashmir Vole (*Alticola montosa*), Endemic to India
18. Clouded Leopard (*Neofelis nebulosa*)
19. Day's Shrew (*Suncus dayi*), Endemic to India
20. Dhole (*Cuon alpines*)
21. Dugong (*Dugong dugon*)
22. Eld's Deer (*Cervus eldi*).

23. Elvira Rat (*Cremnomys Elvira*), Endemic to India
24. European Otter or Eurasian Otter (*Lutra lutra*)
25. Fishing Cat (*Prionailurus viverrinus*)
26. Four-horned Antelope (*Tetracerus quadricornis*)
27. Ganges River Dolphin (*Platanista gangetica*)
28. Gaur (*Bos gaurus*)
29. Golden Jackal (*Canis aureus*)
30. Goral (*Nemorhaedus goral*)
31. Himalayan Musk Deer (*Moschus chrysogaster*)
32. Himalayan Shrew (*Soriculus nigrescens*)
33. Himalayan Tahr (*Hemitragus jemlahicus*)
34. Humpback Whale (*Megaptera novaeangliae*)
35. Indian Fox (*Vulpes bengalensis*)
36. Indian Giant Squirrel (*Ratufa indica*), Endemic to India
37. Indian Wolf (*Canis lupus indica*)
38. Irrawaddy Squirrel (*Callosciurus pygerythrus*)
39. Jerdon's Palm Civet (*Paradoxurus jerdoni*), Endemic to India
40. Kashmir Cave Bat (*Myotis longipes*)
41. Kerala Rat (*Rattus ranjiniae*), Endemic to India
42. Khajuria's Leaf-nosed Bat (*Hipposideros durgadasi*), Endemic to India
43. Kolar Leaf-nosed Bat (*Hipposideros hypophyllus*), Endemic to India
44. Lesser Horseshoe Bat (*Rhinolophus hipposideros*)
45. Lesser Panda (*Ailurus fulgens*)
46. Mainland Serow (*Capricornis sumatraensis*)
47. Malayan Porcupine (*Hystrix brachyuran*)
48. Mandelli's Mouse-eared Bat (*Myotis sicarius*)
49. Marbled Cat (*Pardofelis marmorata*)
50. Mouflon (or Urial) (*Ovis orientalis*)
51. Nicobar Flying Fox (*Pteropus faunulus*), Endemic to India
52. Nilgiri Leaf Monkey (*Trachypithecus johnii*), Endemic to India
53. Nilgiri Marten (*Martes gwatkinsii*), Endemic to India
54. Nonsense Rat (*Rattus burrus*), Endemic to India
55. Asiatic Wild Ass (*Equus hemionus*)
56. Pale Grey Shrew (*Crocidura pergrisea*), Endemic to India
57. Palm Rat (*Rattus palmarum*) Endemic to India

58. Red Goral (*Naemorhedus baileyi*)
59. Rock Eagle-owl (*Bubo bengalensis*), Endemic to India
60. Rusty-spotted Cat (*Prionailurus rubiginosus*)
61. Sikkim Rat (*Rattus sikkimensis*)
62. Sloth Bear (*Melursus ursinus*)
63. Slow Loris (*Loris tardigradus*)
64. Smooth-coated Otter (*Lutrogale perspicillata*) (Previously *Lutra perspicillata*)
65. Sperm Whale (*Physeter macrocephalus*)
66. Sri Lankan Giant Squirrel (*Ratufa macroura*)
67. Stumptail Macaque (*Macaca arctoides*)
68. Takin (*Budorcas taxicolor*)
69. Wild Goat (*Capra aegagrus*)
70. Wild Yak (*Bos grunniens*)
71. Tiger (*Panthera tigris tigris*)
72. Asiatic Lions (*Panthera leo persica*)

Threatened

1. Indian Wild Ass (*Equus hemionus khur*)
2. Leopard (*Panthera pardus*)
3. Red Fox (*Vulpes vulpes montana*)
4. Kashmir Stag (*Praygnaa*)
5. Black Buck (in Uttar Pradesh)

PROTECTED AREAS

Protected Areas are ecological/biogeographically area where wildlife is conserved by maintaining habitats, natural resources and preventing poaching. They are delimited to protect biological diversity and includes national parks, sanctuaries and biosphere reserve. Protected Areas covering all the important habitats of the Country have been created, as per the provisions of the Wild Life (Protection) Act, 1972 to provide better protection to wildlife, including threatened species and their habitat. A network of 669 Protected Areas (PAs) has been established to provide better protection to wildlife, including threatened species and their habitat. PAs in India covered about 1,61,132 sq.km. (4.9 per cent of total geographic area), comprising 102 National Parks, 515 Wildlife Sanctuaries, 48 Conservation Reserves and 4 Community Reserves. The details of protected areas state-wise in the country are given in the Table 4.3. Till date, 40 Tiger Reserves and 28 Elephant Reserves have been designated for species specific management of tiger and elephant habitats in India.

Table 4.3: State-wise Details of the Protected Area Network of India

S No	State/UT	No. of National Park	No. of Wildlife Sanctuaries	No. of Conservation Reserves	No. of Community Reserves
1	Andhra Pradesh	6	21	0	0
2	Arunachal Pradesh	2	11	0	0
3	Assam	5	18	0	0
4	Bihar	1	12	0	0
5	Chhattisgarh	3	11	0	0
6	Goa	1	6	0	0
7	Gujarat	4	23	1	0
8	Haryana	2	8	2	0
9	Himachal Pradesh	5	32	0	0
10	Jammu &Kashmir	4	15	34	0
11	Jharkhand	1	11	0	0
12	Karnataka	5	22	2	1
13	Kerala	6	16	0	1
14	Madhya Pradesh	9	25	0	0
15	Maharashtra	6	35	1	0
16	Manipur	1	1	0	0
17	Meghalaya	2	3	0	0
18	Mizoram	2	8	0	0
19	Nagaland	1	3	0	0
20	Odisha	2	18	0	0
21	Punjab	0	12	1	2
22	Rajasthan	5	25	3	0
23	Sikkim	1	7	0	0
24	Tamil Nadu	5	21	1	0
25	Tripura	2	4	0	0
26	Uttar Pradesh	1	23	0	0
27	Uttarakhand	6	6	2	0
28	West Bengal	5	15	0	0
29	Andaman & Nicobar	9	96	0	0
30	Chandigarh	0	2	0	0
31	Dadar & Nagar Haweli	0	1	0	0
32	Lakshadweep	0	1	0	0
33	Daman & Diu	0	1	0	0
34	Delhi	0	1	0	0
35	Pondicherry	0	1	0	0
	Total	102	515	47*	4

*** Note: Recently one more is added, As on 2012, the total conservation reserves are 48**

National Parks: IBWL defines national parks as an area dedicated by statute for all time to conserve scenery, natural and historical objects of national significance and wildlife, and where provision is made for enjoyment of the same by the public. India's first national park (an IUCN category II protected area) was established in 1936 as Hailey National Park, now known as Jim Corbett National Park. The most important are Corbett (for Tiger Protection) in Uttarakhand; Dudhwa (Monkey) in UP; Gir (Lion) in Gujarat; Shivpuri (Wolf) and Kanha (Deer, Antelope) in Madhya Pradesh; Rajdewra (Monkey) in Jharkhand; Similipal (Tiger) in Odisha; Kanheri (Deer) and Todaba (Bison) in Maharashtra.

Wildlife Sanctuaries: IBWL defines wildlife sanctuaries as an area where killing or capturing of any species of bird or animal is prohibited except under orders of competent authority and whole boundaries and characters should be sacrosanct (free from outrage) as for as possible. The following are the most important sancturies: Dachigam in Jammu and Kashmir; Rajaji and Banbasa in Uttarakhand and Assam; Ranthambhore and Sariska in Rajasthan; Kawal in Maharashtra, Chandaka in Odisha, Pocharam and Etumagaram in Andhra Pradesh, Bandipur in Karnataka, Guindy and Mudumalai in Tamil Nadu, and Periyar in Kerala. The most important bird sanctuaries are: Bharatpur in Rajasthan, Rangnathaittoo in Karnataka and Vedanthangal in Tamil Nadu.

Table 4.4: National Parks and Wildlife Sanctuaries in India (Upto December 2006) (Source: National Wildlife Database, May 2007)

State/UTs	Area of State (km²)	Wildlife Sanctuaries			National Parks		
		Number	Area (km²)	% of State Area	Number	Area (km²)	% of State Area
Andhra Pradesh	2,75,068	22	12,600.09	4.58	4	373.23	0.14
Arunachal Pradesh	83,743	11	7,606.37	9.08	2	2,290.82	2.74
Assam	78,438	18	1,932.00	2.46	5	1,977.79	2.52
Bihar	94,163	12	2,856.06	3.03	1	335.65	0.36
Chhattisgarh	1,35,194	11	3,572.93	2.64	3	2,929.5	2.17
Goa	3,702	6	647.96	17.5	1	107.0	2.89
Gujarat	1,96,024	22	16,440.94	8.39	4	480.11	0.24
Haryana	44,212	10	225.37	0.58	2	48.25	0.11
Himachal Pradesh	55,673	32	5,770.85	10.37	2	1,429.4	2.57
Jammu & Kashmir	2,22,235	15	10,312.25	4.64	4	3,930.25	1.77
Jharkhand	79,714	11	1,962.72	2.48	1	231.67	0.29
Karnataka	1,91,791	21	3,888.22	2.03	5	2,472.18	1.29
Kerala	38,863	14	1,891.07	4.87	6	558.16	1.44
Madhya Pradesh	3,08,252	25	7,158.40	2.32	9	3,656.36	1.19
Maharashtra	3,07,690	35	14,152.69	4.6	6	1,273.6	0.41
Manipur	22,327	1	184.40	0.83	1	40.0	0.18

Meghalaya	22,429	3	34.20	0.15	2	267.48	1.19
Mizoram	21,081	7	840.75	3.99	2	150.0	0.71
Nagaland	16,579	3	20.34	0.12	1	202.02	1.22
Odisha	1,55,707	18	6,969.15	4.48	2	990.7	0.64
Punjab	50,362	11	317.89	0.63	0	0	0
Rajasthan	3,42,239	23	5,447.03	1.59	5	4,122.33	1.20
Sikkim	7,096	7	399.10	5.62	1	1,784.0	25.14
Tamil Nadu	1,30,058	20	2,997.60	2.3	5	307.84	0.24
Tripura	10,486	4	603.62	5.76	1	5.08	0.05
Uttar Pradesh	2,40,926	23	5,222.47	2.17	1	490.0	0.2
Uttarakhand	53,485	6	2,413.76	4.51	6	4,725.0	7.62
West Bengal	88,752	15	1,203.28	1.36	5	1,693.25	1.9
Union Territories							
Andaman & Nicobar	8,249	96	389.39	4.72	9	1,156.91	14.02
Chandigarh	114	2	26.10	22.89	0	0	0
Dadra & Nagar Haveli	491	1	92.16	18.77	0	0	0
Daman & Diu	112	1	2.18	1.95	0	0	0
Delhi	1,483	1	27.20	1.83	0	0	0
Lakshadweep	32	1	0.01	0.03	0	0	0
Pondicherry	493	0	0	0	0	0	0
India	32,87,263	508*	1,18,594.54	3.61	93*	38,029.18	1.16

* Note: As on 2012, there are 515 Wildlife Sanctuaries and 102 National Parks in the Country

Table 4.5: List of some important Wildlife Sanctuaries and National Parks in India

S No	Name of Sanctuary/ National Park	Place/ State	Area (Sq.Km.)	Wildlife conserved
1	Kaziranga Wildlife Sanctuary	Sibasaa, Assam	430	Rhinoceros, elephant, wild buffalo, bison, tiger, leopard, sloth bear, sambhar, pelican, stork, eagle, etc.
2	Manas Wildlife Sanctuary and Tiger Reserve	Kamrup, Assam	540	Tiger, panther, wild dog, bear, rhinoceros, gaur, golden langur etc.
3	Jaldapara Wildlife Sanctuary	Jalpaiguri, West Bengal	65	Rhinoceros, gaur, elephant, tiger, leopard, deer, birds and reptiles.
4	Kolameru Bird Sanctuary	Tadepallegudum, Andhra Pradesh	NA	A breeding place for pelican and other visiting marine birds.
5	Chilika Lake	Chilika, Odisha	100	Waterfowl, duck, cranes, ospreys, golden plover, sand piper, stone curlew, flamingo, etc.

6	Vendanthangal Bird Sanctuary	Chennai, Tamilnadu	0.30	Flamingo, pelican, black buck, chital, wild boar
7	Point Calimer Wildlife Sanctuary	Thanjavur, Tamirnadu	0.30	Panther, tiger, sambhar, chital
8	Mundanthurai Sanctuary	Tirunelveli, Tamilnadu	520	Elephant, gaur, sambhar, leopard, jackal, nilgai, langur, grey horn bill, giant squirrel, flying squirrel
9	Periyar Wildlife Sanctuary	Periyar, Kerala	777	Elephant, gaur, sambhar, leopard, Nilgiri langur, grey hornbill, egret
10	Bharatpur Bird Sanctuary	Bharatpur, Rajasthan	29	Cormorants, Spoon bill, white ibis, Indian darter, egret, open billed stork, geese, duck, Siberian crane, deer, black buck, blue bull, wild boar
11	Palamau National Park	Dattongunj, West Bengal	345	Tiger, panther, sloth bear, elephant, chital, gaur, nilgai, chinkara, chowsingha, mouse deer, leopard, deer, birds and reptiles.
12	Hazaribagh National Park	Hazaribagh, Bihar	184	Wild boar, sambhar, nilgai, tiger, leopard, hyena, gaur etc.
13	Similipal National park	Similipal, Odisha	2750	Tiger, elephant, deer, chital, peafowl, sambhar, panther, gaur, hyena and sloth bear
14	Guindy National Park	Chennai, Tamil nadu	NA	Albinos of black buck, chital
15	Kanha National Park	Banjar Valley, Madhya Pradesh	940	Tiger, chital, panther, sambhar, black buck etc.
16	Tanoba National Park	Chandrapur	166	Tiger, sambhar, sloth bear, barking deer, blue bull, chinkara, bison, pea fowl etc.
17	Corbett National Park	Nainital, UP	525	Tiger, panther, sloth bear, hyena, elephant, blue bull, barking deer, Indian antelope, porcupine, pecker barbet, crocodile, python etc.
18	Gir National Park	Ahmedabad, Gujarat	1295	Asiatic lion, spotted deer, blue bull (nilgai), fur-homed antelope, chinkara, striped hyena, wild boar, porcupine, langur, python, crocodiles and birds like green pigeon, partridge, rock grouse etc.
19	Sariska National Park	Alwar, Rajasthan	800	Tiger, leopard, spotted deer, jungle cat, four-horned antelope, langur, porcupine, hedgehog, peafowl etc.
20	Dachigam Wild Life Sanctuary	Srinagar, Kashmir	89	Hangul or Kashmir stag, musk deer, leopard, black buck, brown bear, porcupine, pangolin, flying-lizard, monitor lizard, rat snake, python and various birds.
21	Ranganthitoo Bird Sanctuary	Bangalore, Karnataka	166	Open bill stork, white ibis, egret, spoon bill, wild duck, pea fowl, night heron etc.

Tiger Reserves

Project Tiger was launched by the Government of India in the year 1973 to save the endangered species of tiger in the country. Initially, there were 9 tiger reserves in 1973-74 and now the number has grown upto 40 as on January, 2012 spread over in 17 States of the country. The total area covered by the project tiger areas is 46,600 km2, which is 1.42 per cent of the total geographical area of the country. The most important are Corbett National Park (Uttrakhand), Sunderbans National Park (West Bengal), Bandhavgarh National Park (Madhya Pradesh), Sariska Wildlife Santuary (Rajasthan), Manas Tiger Reserve (Assam) and Periyar Tiger Reserve (Kerala).

Project Elephant

Project Elephant was launched in February, 1992 to assist States having free ranging populations of wild elephants to ensure long term survival of identified viable populations of elephants in their natural habitats. The project is being implemented in 12 States, *viz.* Andhra Pradesh, Arunachal Pradesh, Assam, Jharkhand, Karnataka, Kerala, Meghalaya, Nagaland, Odisha, Tamil Nadu, Uttaranchal and West Bengal. States are being given financial as well as technical assistance in achieving the objectives of the project. Help is also provided to other states with small populations of elephants for the purpose of census, training of field staff and mitigation of human-elephant conflict.

Recently eight elephant reserves have been set up which include Dihing - Patkai, Dhansiri - Lungding and Kaziranga - Karbi Anglong in Assam; Nilgiri, Coimbatore, Anamalai and Srivilliputhur in Tamil Nadu and Royala in Andhra Pradesh. A total of 25 Elephant Reserves (ERs) of the country extending over about 58,000 square km have been formally notified by various State Governments till 2005 and the number has gone to 28 as on 2012.

Crocodile Breeding Project

Three species found in the country (the gharial, *Gavialis gangeticus*; the saltwater crocodile, *Crocodylus porosus* and the mugger or marsh crocodile, *C. palustris*) were on the verge of extinction due to large scale hunting for their skins. Crocodile Breeding and Management project was launched by Government of India in April, 1975 initially in Odisha. Ironically, it was the gharial, which is completely harmless to man and of relatively low value in terms of its hide, which was most endangered. Stabilization of river banks and dam construction had greatly reduced the gharial's natural environment of free-flowing rivers. In addition, the increased use of nylon fishing nets resulted in the accidental ensnaring and drowning of many gharials. In 1973, it was estimated that fewer than 100 individuals of Gharial continued to survive in the wild in India. While larger numbers of saltwater crocodile and mugger were known to exist, they were not enough to avoid the total extinction of the species in the short-term future. Presently, due to the overwhelming success in breeding programme, forest departments have concluded captive breeding of the mugger.

PART 2:

TREE AND FOREST

5 TREE GROWTH AND FOREST

INTRODUCTION

Forest is a complex ecosystem consisting mainly of trees that have formed a buffer for the earth to protect life forms (plants, animals, insects, microorganisms, etc). The FAO has defined forest as land with tree crown cover (or equivalent stocking level) of more than 10 per cent and area of more than 0.5 hectare with more than 5 m height trees. In general, forest is a land set aside for the production of timber and other forest products (Gulhati 1981).

The history of Indian forests is closely linked with political history. The Epics Ramayana and Mahabharata give an attractive description of forests. The Dandakaranya, the Khandavban and the Nandanvan come from them. Ancient Hindu culture is said to have evolved in Aranyas. Literatures are not, however, scientific treatises. Nevertheless, they give a glimpse into the dim and distant past. They are an indication of the level of significance the forests had in the cultural life of people. The earliest indication of forestry administration in India is found in 300 BC. It was during the reign of Chandragupta Maurya. A Superintendent of Forests looked after forests and wildlife. Later, Ashoka the Great continued the process and much importance was given for planting trees along roadsides and camping sites. Ashoka is said to have set up the first sanctuary to protect the forests and all life in it.

The Mughal policy on forests was one of indifference. Neither did they seem to pay much attention to the forests nor did they have any religious scruples to destroy them. The Mughal rulers were avid hunters and spent a great deal of time in the forests. The Mughals looked upon the forests as game reserves. They were interested in trees for gardening. They also showed interest in plantations on either side of avenues. They therefore, displayed an aesthetic and utilitarian approach to plants. They lacked any comprehensive understanding of forests and lacked definitive approach for their preservation, propagation, protection and improvement. The forests were reclaimed for agriculture and the State supported this through incentives. Parts of the farmer community were pushed back into the forests due to the Mughal invasion. They took up shifting cultivation and damaged the forests.

Heavy destruction of forests also occurred in the later part of the 18th and early part of the 19th century. Europeans carried away much of the produce. In the early years of British Raj, large indents were made on the timber wealth of the country. The teak forests along the coast of Malabar were over-exploited. The timber was supplied to meet the requirement of the

British Navy. But it was during the British rule that the first practical move towards conservation in modern times took place. In the post-independence period, the task of consolidation of forests, unification of forest laws and extension of scientific management on a reasonably uniform basis was the most important task of forest administration. Rapid development and progress saw large forest tracts fragmented by roads, canals and townships. There was an increase in the exploitation of forest wealth. It was only in the 1970s that the importance of conservation of forests was realized and the preservation of India's remaining forests and wildlife was given a front seat.

WHAT IS FORESTRY?

Forestry is the art and science of managing forests so as to yield, on a continuous basis, a maximum in quality and quantity of forest products and services. In broad sense, it is for the handling of forest land to satisfy the needs of man. It includes the logging, manufacturing, marketing and use of wood products. Forests can be managed for single or multiple purposes, to include protection of watersheds, production of timber, wildlife conservation, regulation of stream flow, recreation, erosion control and general aesthetics. Forest ecosystems have come to be seen as the most important component of the biosphere and forestry has emerged as a vital field of science, natural resource management, applied art and technology. The main goal of forestry is to create and implement socio-environmental systems that allow forests to continue a sustainable continuation of environmental supplies and services. The challenge of forestry is to create systems that are socially accepted while sustaining the resource and any other resources that might be affected.

BRANCHES OF FORESTRY

- *Agroforestry*: It is the integration of forests into agricultural systems in order to optimize the production and positive effects within the system and minimize negative side effects of farming
- *Dendrology*: It involves the study and identification of economically useful tree species
- *Forest Ecology*: It studies the patterns and processes of a forest ecosystem
- *Forest Economics*: It studies the impact of economics on forest management decisions
- *Forest Hydrology*: It embodies the effects of changes in forest land use on the movement, distribution, and quality of water in the ecosystem
- *Forest Management*: It comprises the overall administrative, economic, legal and social aspects of forest regulation
- *Forest Mensuration*: It incorporates quantitative measurements of the forest stand to determine stand timber volume and productivity/health and provides a basis of which management decisions can be made
- *Forest Protection*: It comprises both the biotic and abiotic maladies affecting the health of the forest or tree, primarily by fungal pathogens, insects, nematodes,

parasitic plants and climatic factors (stress, salinity, fire, etc.) and their management

- *Silviculture*: It is the art and science of controlling the establishment, growth, composition, health and quality of forests to meet specific objectives
- *Tree Breeding*: It is the method of genetically modifying/selecting forest stock for improved growth or vigour characteristics
- *Urban Forestry*: It entails the care and management of urban tree populations for the purpose of improving the urban environment
- *Social Forestry*: It addresses human-forest interactions and the importance of community based natural resource management
- *Farm Forestry*: The practice of cultivating and managing trees in compact blocks on agricultural lands
- *Community Forestry*: Combination of forest conservation with rural development and poverty reduction objectives, accomplished through instating a legal framework that favours profitable and sustainable forest management
- *Eco-forestry* or *Environmental Forestry*: It emphasizes tree planting practices which strive to protect and restore ecosystems
- *Energy Forestry*: It includes specifically managing forests for the production of energy from biomass or bio-fuel derived from a fast-growing species of tree or woody shrub
- *Short Rotation Forestry*: Managing a forest that utilizes fast-growing species as a bio-based energy crop for domestic use and commercial use in power stations, alone or in combination with other fuels such as coal
- *Analog Forestry*: A management focus that seeks to establish a tree-dominated ecosystem that is similar in architectural structure and ecological function to the naturally occurring climax and sub-climax vegetation community
- *Boreal Forestry*: It analyzes the particular challenges of forestry in the world's boreal regions

TREE STRUCTURE AND GROWTH

In general, trees are big plants having erect stem, branches and other parts as like plants. It is composed of an underground part: the roots and two aboveground parts: the trunk and the crown. Their size and form are extremely variable and this extends to species within genus, varieties within a species and even the same species in different environments. Understanding how the different structures making up the tree development gives insight into how this many variations arise (Owens and Lund 2006).

During each growing season, a tree grows by elongation of the top shoot, side branches, roots and also adding new layer of wood and bark, all over. This increase or growth is also termed as increment. It can be expressed by measurements of height, girth, basal area, volume, quality and price over period. Generally the increment can be classified as:

- Current Annual Increment (CAI): Increment put on by a tree in a single or current year
- Periodic Increment (PI): Increment put on over a fixed period of time
- Mean Annual Increment (MAI): Total increment put on upto a given age divided by age

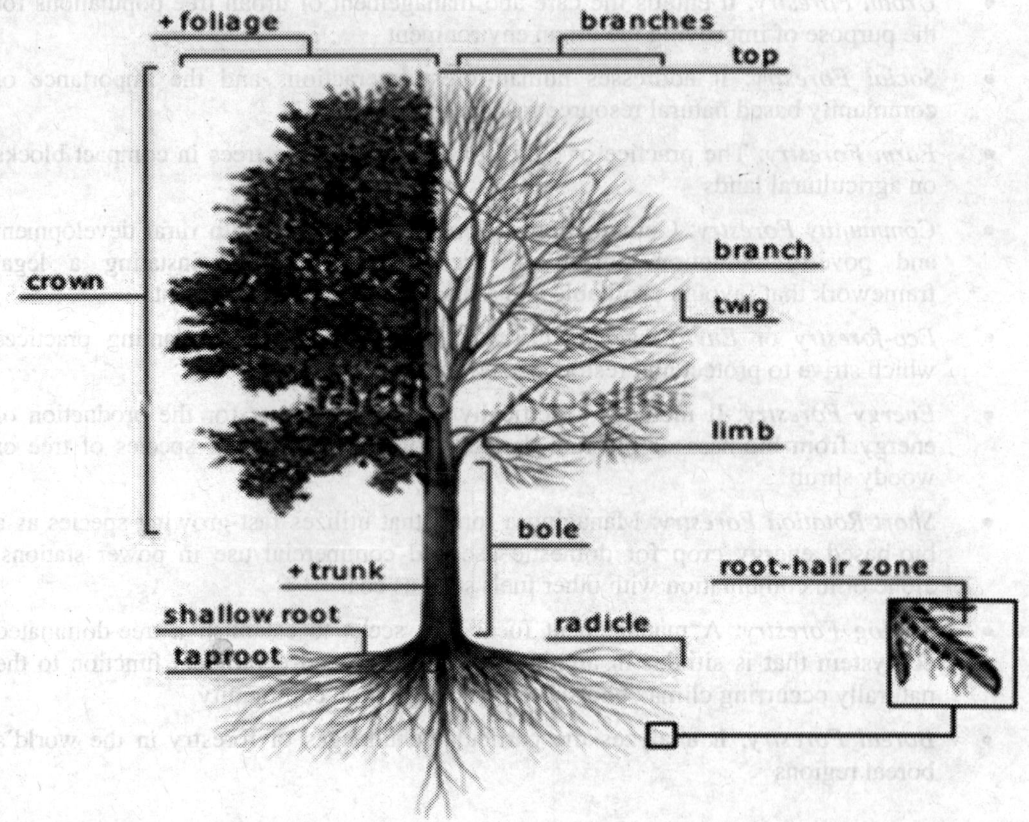

Fig. 5.1: Thematic Diagram of Tree

For describing the growth status of a tree, it is necessary to employ terms which are descriptive of some characteristic of the individual trees. The timber species are conveniently designated by certain size classes through their life development. Those commonly employed are:

(a) Seedling: from germination to the height of 3 feet.

(b) Small Sapling: from 3 to 10 feet tall.

(c) Large Sapling: from 10 feet tall to 4 inches d.b.h.

(d) Small Pole: from 4 to 8 inches d.b.h.

(e) Large Pole: from 8 to 12 inches d.b.h.

(f) Standard: from 12 to 24 inches d.b.h.

(g) Veteran: over 24 inches d.b.h.

CLASSIFICATION OF FORESTS

Forests can be broadly classified into many types, some of which are the taiga type (consisting of pines, spruce, etc.), the mixed temperate forests with both coniferous and deciduous trees, the temperate forests, the sub tropical forests, the tropical forests and the equatorial rainforests. It can be also be classified in different ways and to different degrees of specificity. One such way is in terms of the "biome" in which they exist, combined with leaf longevity of the dominant species (whether they are evergreen or deciduous). Another distinction is whether the forests composed predominantly of broadleaf trees, coniferous (needle-leaved) trees, or mixed. Accordingly world's forest can be grouped into 26 major types, which reflect climatic zones as well as the principal types of trees. These 26 major types can be reclassified into 6 broader categories:

(1) Temperate needle leaf forest

(2) Temperate broadleaf and mixed forest

(3) Tropical moist forest

(4) Tropical dry forests

(5) Sparse trees and parkland forests

(6) Forest plantations/ planted forest

In the tropical and subtropical region, forests are further subdivided into plantations and natural forests. Natural forests are forests composed of indigenous trees, regenerates naturally and not deliberately planted. Plantations are forest stands established by planting or/and seeding in the process of afforestation or reforestation or for some profit oriented purpose. In India, there are about 16 major types of forests (Table 5.1) from the tropical type to the dry type (Champion and Seth 1968).

Table 5.1: Vegetation Types of India (Champion and Seth 1968)

S No	Vegetation Type	General composition
1	Tropical wet evergreen forest	Dense tall forests, entirely evergreen or nearly so
2	Tropical semi evergreen forest	Dominants include deciduous species but evergreens are predominants
3	Tropical moist deciduous forest	Dominants mainly deciduous but sub-dominants and lower story largely evergreen, top canopy even and dense but 25 m high
4	Tropical dry deciduous forest	Entirely deciduous or nearly so, top canopy uneven rarely over 25 m high
5	Tropical thorn/ scrub forest	Deciduous with low thorny trees and xerophytes predominants, top canopy more or less broken, less than 10 m high

6	Tropical dry evergreen forest	Broad leaved evergreen trees predominates with some deciduous emergent, often dense but usually under 20 m high
7	Littoral and swamp forest	Mainly evergreens of varying density and height but always associated predominantly with wetness
8	Subtropical broad-leaved hill forest	Broad-leaved largely evergreen high forests
9	Subtropical pine forest	Pine associated predominant
10	Subtropical dry evergreen forest	Low xerophytic forest and scrubs
11	Montane wet temperate forest	Evergreen without coniferous species
12	Himalayan wet/ moist temperate forest	Evergreen forests mainly scleriphyllous oak and coniferous species
13	Himalayan dry temperate forest	Coniferous forests with sparse xerophytic undergrowth
14	Sub-alpine forest	Stunted deciduous or evergreen forests, usually close formation with or without confers
15	Moist alpine	Low but often dense scrub of evergreen species
16	Dry alpine	Xerophytic scrub in open formation mostly of deciduous in nature

DIRECT BENEFITS OF FORESTS

- They provide a sustainable source of timber (for cabinets, wardrobes, kitchens, desks/tables, chairs, doors, floors, paper, musical instruments, sports goods, boats, agricultural tools), food, fruit, fibre, fuelwood, charcoal, etc.
- They also provide NWFPs like essential oil, gum, resin, latex, tannin, dye, beads for jewellery, bio-fuel, bamboo, palm, rattan, etc
- Trees provide organic fertilizer, through the falling of leaves, branches and root segregates.
- Leaves provide a high-protein, palatable animal forage and cattle fodder
- They have provided medicines for centuries for malaria, heartburn, measles, lung diseases, rheumatism, flu, leprosy.
- They serve as national/ international borders.

INDIRECT BENEFITS OF FORESTS

- They improve living standards for local people
- They create microclimate conditions for life to flourish
- Through shade and by evaporating moisture from their leaves, trees can reduce daytime temperatures by about 5°F and buildings by 20°F
- The roots reduce soil erosion in rainy seasons and replenish the ground with nutrients from the air e.g. nitrogen fixation
- They retain groundwater supporting surrounding growth

- Tree roots stabilize river banks, reducing the risk of flooding
- They rebuild worn soils and improve agricultural productivity
- Cooling ground temperatures and sheltering from direct sunlight allows food crops and other growth to return where they otherwise could not grow
- They serve as homes for indigenous wildlife, encourages the return of endangered species. Hence, they sustain eco-system balance and biodiversity
- Home for bees, birds, bats, arboreal animals, etc.
- Tree and forest can act as natural wind breakers and separate different types of crop farms
- Forest is nature's gene bank and reservoir of variation and diversity

It would be impossible for our country to maintain the standard of living it enjoys without the products and services which forests provide. History teaches that the more prosperous nations are those which have productive forests. In every land and in all ages, trees have had an influence on the progress and welfare of man. His progress form a primitive cave dweller to his present civilized state cannot be told without frequent reference to trees and their products. Trees provided early man with food, medicines, fuel, shelter, protection, shade, tools and other needs. Today, over ten thousand products are reportedly made of and from wood. It is the raw material from which forest industries manufacture countless products for the home, factory and office.

IMPACT OF CLIMATIC FACTORS ON TREES AND FOREST

The different climate factors combined together affect the vegetation of a place. These factors influence or modify each other too. This is termed as the bio-climate. Meanwhile, physiography has a profound impact on vegetation, microclimatic types etc. Physiographic factors are as follows: configuration of the ground, altitude or elevation, slope, aspect and exposure.

It is likely that changing temperature and precipitation pattern will produce a strong direct impact on both natural and modified forests. A number of bio-geographical models demonstrate a polarward shift of potential vegetation for the 2 × CO_2 climate by 500 km or more for boreal zones (Cramer et al 2001; Foley et al 1998). The equilibrium models and some dynamic vegetation models project that this vegetation shift toward newly available areas with favourable climate conditions will eventually result in forest expansion and replacement of upto 50 per cent of current tundra area. There is, however, a concern that the lagged forest migration (compare the tree species migration rates after the last glacial period of few kilometers per decade or less to projected future climate zones shift rate of 50 km per decade) may lead to massive loss of natural forests with increased deforestation at the southern boundary of the boreal forests and a correspondent large carbon pulse (Malcolm et al 2002). At the same time, tree species migration rates can be much more rapid.

Meanwhile, increasing concentrations of the atmospheric CO_2, aside from modifying the temperature and precipitation pattern, may also increase the production through the "carbon fertilization effect". Experiments in closed or open-top chambers have demonstrated the very high potential for CO_2-induced growth enhancement, increase in stem growth and higher wood production. Regardless of the contradictory effects of variations in CO_2 concentration, insolation, nutrients availability, temperature and precipitation, the forest growth rate have been increasing since the middle of the 20th century.

The climate change-induced modifications of frequency and intensity of forest fires, outbreaks of insects and pathogens and extreme events such as high winds, may be more important than the direct impact of higher temperatures and elevated CO_2. At the same time, very few forest production models include these effects, which severely limit the reliability of the model results. Forest fire may be exclusion here. The last two decades demonstrated increasing burned areas in Canada, the western United States and Russia, because of both climatic conditions and other factors such as fuel conditions, ignition sources, land-use change and variations in fire protection (Westerling et al 2006; Gillet et al 2004). The potential losses of the timber, pulp and paper production, as well as the damage to health and non-timber forest products caused by elevated fire activity, are quite uncertain as much of the fire damage is expected to occur in less-accessible regions.

For many forest types, forest health questions are of great concern with pest and disease outbreaks as major sources of natural disturbance. The effects vary from defoliation and growth loss, to timber damage, to massive forest diebacks. Even without fires or insect damage, the change in frequency of extreme events, such as strong winds, winter storms, droughts, etc. can bring massive loss to commercial forestry. These effects of climate extremes on commercial forestry are region-specific. This assessment based on climate projections of Regional Climate Model of the Hadley Centre (HadRM3) using 740 ppm CO_2 and 575 ppm CO_2 scenarios of Special Report on Emissions Scenarios and the BIOME4 vegetation response model (Ravindranath et al 2006) gives the conclusion that, under the climate projection for the year 2085, 77 per cent and 68 per cent of the forested grids in India are likely to experience shift in forest types.

SITE QUALITY

The productivity of forest land is defined in terms of the maximum amount of volume that the land can produce over a given amount of time. Site quality is measured as an index related to this timber productivity. Site quality will vary with species, for instance a best quality site of teak may not be a poor site for Bamboo species. Of all the indirect measures investigated, the rate of height growth has been the most practical, consistent and useful indicator of site quality with respect to timber production. The standard practice has been to define site index in terms of the total height of the dominants - the largest, full-crowned trees in a stand.

Site index based on height growth is the most widely accepted method for estimating site quality. Height and age measurements from free-growing, uninjured, dominant and codominant trees are used with a family of height-age (site index) curves to estimate total tree height at a specified index age. For most northern species 50 years is the index age. The site index method is accurate and simple to use when (1) suitable trees are available for measuring site index, and (2) when accurate site index curves are available.

It is very important to understand that the productivity of timberlands varies greatly by site. On one site, very good growth may be observed, while on another site, the same species at the same age may grow very poorly. Site quality can be changed by fertilization, vegetation control, irrigation, or drainage. Only highly intensive treatment can make a poor site into productive site.

BIOCLIMATE AND MICROCLIMATE OF FORESTS

Climate is the set of characteristic temperature, humidity, sunshine, wind and other weather conditions that prevail over large areas of space for long periods of time. The smallest scale sub-climate is the microclimate defined as the climate that holds over a distance (in any direction) of less than 328 ft (100 m). Microclimate refers to a climate that holds over a very small area. Microclimates usually are slight modifications of the main background climate altered by features in the landscape. A forest creates a microclimate or bioclimate within the canopy of trees which is cooler, wetter and has altered soil chemistry compared to the area outside the forest. The altered climate found within forests can support organisms that cannot survive in the surrounding grassland. Microclimate frequently supports unique ecosystems. Mountain meadows, river valleys, tidal marshes and crop lands have one or several microclimates which help to determine the amount and type of organisms that thrive in these locations. Cooling of air temperature due to the effect of trees has been well documented in the past through various studies. A tree can be regarded as a natural "evaporative cooler" using up to 100 gallons of water a day (Kramer and Kozlowski 1960). The release of moisture into the atmosphere through the stomata of leaves is caused due to the evaporation of water at the stomata into the air, thus causing cooling of the air.

Microclimatic effects within a forest cover are explained in terms of morphological characteristics such as branching (bifurcation), growth periodicity (evergreen or deciduous), size, density, texture, orientation of leaves, height of trees, plant coverage, etc. Different forest types have different spatial organizations and so their effects on microclimate. Generalized microclimatic effects within a forest cover may be described as below:

- Modification of energy transfers: The canopy of a forest cover significantly changes the pattern of incoming and outgoing radiation that is reflected mainly as change in albedo, energy trapped and absorption of short-wave radiation and penetration of light.
- Modification of air flow
- Modification of air temperature
- Modification of precipitation
- Modification of humidity
- Modification of carbon-dioxide in air

STAND TYPES

Stand is an aggregation of trees or other growth occupying a specific area and sufficiently uniform in species composition, size, age, arrangement and condition as to be distinguished from the forest or other growth on adjoining areas (Burns and Honkala 1990). The composition of stands is conceived of as being either pure or mixed. These are defined as:

- *Pure Stand*: A stand in which at least 80 per cent of the trees in the main canopy are of single species.

- *Mixed Stand*: A stand in which less than 80 per cent of the trees in the canopy are of a single species.

The density of stocking expressed in number of trees, basal area, volume or other criteria, on a per-acre basis is called as stand density. In addition, stocking is further modified and defined as:

- *Fully Stocked Stands*: Stands in which all the growing space is effectively occupied but which still have ample room for development of the crop trees.
- *Overstocked Stands*: Stands in which the growing space is so completely utilized that growth has slowed down and many trees, including dominants, are being suppressed.
- *Understocked Stands*: Stands in which the growing space is not effectively occupied by crop trees.

Stands are usefully described and considered from the stand point of the age classes of which they are composed. Different stand forms generally available are:

- *Even-aged Stands*: Stands in which there exists relatively small age differences between individual trees.
- *Uneven-aged Stands*: Stands in which there exists relatively large age differences between individual trees. At least 3 age classes are present. A similar meaning is all-aged stand.
- *Two-aged Stands*: Stands in which there are two distinct age classes.

Mixed plantation systems seem to be the most appropriate for providing a broader range of options, such as production, protection, biodiversity conservation and restoration (Montagnini et al 1995; Guariguata et al 1995). Mixed plantations yield more diverse forest products than mono-specific stands, helping to diminish farmer's risks in unstable markets. If planned with consideration for each species' response to mixed conditions, mixed designs can be more productive than mono-specific systems. In addition, a mixture of species, each with different nutrient requirements and different nutrient cycling properties, may be overall less demanding on site nutrient than monoculture stands.

Mixed plantations can produce more biomass per unit area because competition among individuals is reduced and the site is used integrally (Montagnini et al 1995). The roots of different species may occupy different soil strata allowing more complete utilization of soil and water resources (Lamb and Lawrence 1993). More solar energy can be captured because different species have different light requirements and crowns are broadly distributed in the vertical plane (Guariguata et al 1995). However, the success of the establishment of mixed forest plantations depends on plantation design and an appropriate definition of the species to be used, taking into consideration ecological and silvicultural aspects (Wormald 1992).

NATURAL AND ARTIFICIAL REGENERATION

Natural regeneration is the process by which forests are restocked by vegetation that develops from seeds/ other mode of propagation originates from the available vegetation. Whereas artificial regeneration is the process of establishing seedlings in the nurseries and planting them in the field for gap filling or planting them in required spacing as plantation.

For the last four decades, forests were restocked and woodlands were created by using transplants grown in nurseries. Natural regeneration was rarely used until the late 1980's when a combination of factors, including a desire to "do things more naturally", and a change in the grant structure, lead to its use becoming more popular. Restocking by natural regeneration is often unsatisfactory, frequently for unknown reasons; which demonstrate the need for research to understand the processes occurring. The main oobjectives of artificial regeneration are:

SUPPLEMENTING NATURAL REGENERATION

The natural regeneration is a slow and difficult process and cannot be relied upon. Natural regeneration does not give adequate and uniform stocking over the area. Therefore the failed patches need to be filled upon by artificial methods. The coppice system of working also fails to give adequate natural regeneration and calls for artificial regeneration. The natural regeneration has always been a problem in Sal bearing moist deciduous forests in Uttar Pradesh, fir and spruce forests in Himachal Pradesh, tropical moist deciduous and evergreen forests in many States and dry deciduous forests containing teak in Madhya Pradesh, Maharashtra, Andhra Pradesh, etc.

REPLACING NATURAL REGENERATION

Due to increase in the biotic pressure, the natural regeneration in several areas is poor, slow and uncertain. In such areas, various operations e.g. shrub cutting, weeding, soil working, trenching, etc. increase the cost of natural regeneration but fail to give the desired result. Therefore, it is necessary to regenerate the forest with the help of artificial regeneration which must be done on a large scale where the labour supply is easily available. Artificial regeneration of fir-spruce forests in Himachal Pradesh, sal forests in Uttar Pradesh, West Bengal and Assam, teak forests in Madhya Pradesh, Maharashtra and Kerala, etc. are some of the examples, where the forests have been regenerated successfully. The artificial regeneration needs to be resorted where the forest fires often obliterate natural regeneration.

REFORESTATION OF DEGRADED FOREST AREA

Degradation of forest areas have occurred due to large scale firewood and poles requirement of poor villagers. They are not in a position to purchase firewood or any other fuel for running their hearth. The only easy way for them is to collect fuelwood from the nearby forests. A lot of damage is being done to our forests on account of this factor. The density of the forest is continuously going down especially in dry deciduous and thorn forests where the pressure of grazing, browsing, lopping, illicit removal and fire are heavy. There is no possibility of reclamation through natural regeneration. Such areas should be reforested by means of artificial regeneration.

INCREASING PROPORTION OF VALUABLE SPECIES

Many forests such as wet evergreen, moist deciduous and also the dry deciduous forests have more percentage of less valuable species. The more valuable species, on account of many factors, are unable to regenerate naturally. Such areas need to be enriched by intensively planting valuable species including exotics too. The value of the forests can be increased by introducing sal, teak, Dipterocarpus, rosewood, etc.

AFFORESTATION OF WASTELANDS

Wastelands have hardly any vegetation due to poor and unstable soil, lack of moisture, swampiness, etc. These areas can be afforested by selection of suitable species and proper technology of plantation development.

INTRODUCTION OF EXOTICS TO INCREASE PRODUCTION

The artificial regeneration can be taken up for introduction of valuable exotics. Some of the exotics which have been selected for plantations are: eucalypts, poplars, acacias, *Prosopis*, tropical pines, etc.

TO INCREASE PRODUCTION

Trees occupy land for many years. The land should be used in such a manner so as to give maximum economic returns during a given period of time. There will be a great loss due to low productivity of the area. The total growing stock in India is only 32 cubic m per ha. The present production level is 0.5 cubic m per ha per annum which is not sufficient to meet our growing demands. This can be solved by artificial regeneration of valuable and fast growing species in large scale.

TO PROMOTE FARM FORESTRY AND AGROFORESTRY

Growing trees simultaneously with the agricultural crop is always beneficial to trees as well as to the agricultural crop. If the land is not suited to good agriculture, only tree crop can be grown separately or alternately. The various farm forestry and agroforestry models viz. intercropping, alley cropping, etc. will certainly minimize the biotic pressure on the forests.

ENVIRONMENTAL CONSERVATION

Planting more and more trees will immensely help in ameliorating climate, conserving soil and water, retaining and developing water courses, controlling drought and floods and protection of fauna.

Artificial regeneration also has certain demerits:

(1) *Reduction in species richness and diversity*: The natural regeneration has the tendency to produce mixed crops which are rich in diversity whereas, the plantations are completely devoid of diversity. In a plantation, the canopy density is high and other species find it hard to come up except in failed patches. Due to lack of diversity, the population of wild animals, birds etc. are less as compared to their number in mixed stocking or natural forests.

(2) *Reduction in soil productivity*: The plantations are raised after removing the existing vegetation. As soon as the existing vegetation is removed, the soil gets exposed, site conditions are drastically changed and the net result is reduction in soil productivity. Loss in productivity has been observed for many species but no such reduction was found in acacia species. Hence, natural eco-system is the best for soil protection and climate amelioration.

(3) *Large damage by pests and diseases*: The outbreak of pests and diseases in the epidemic form may destroy the whole plantation. The various silvicultural practices

e.g. thinning, pruning, etc. may also invite various pests and diseases in the plantation area. The fungus attacks are transmitted mainly through roots rapidly in plantations whereas, their occurrence in natural forest is only sporadic.

ROTATION

The period which a forest crop takes between its formation and final felling is known as rotation or production period. Generally, rotation is expressed as an average age practice for harvesting to start a new crop. This term is also defined in various other ways by different authors and some of these definitions are given below:

- The planned number of years between the formation or regeneration of a crop and its final felling. In the case of a selection forest, the average age at which a tree is considered mature for felling (Forest Glossary)
- The number of years fixed by the working plan between the formation or regeneration and the final felling of a crop (Brasnett)
- Rotation age is the age of trees or crops at which they are felled and object of management for the time being are best served (Knuchel)
- Rotation or production period is the interval of time between the formation of a young crop by seeding, planting or other means and its final harvesting (Osmaston)
- Rotation is the period which elapses between the formation of a wood and the time when it is finally cut over (Jerram)

TYPES OF ROTATION

Rotation is an important factor in the regulation of yield and proper management of the forest as a whole. As stated earlier, it will depend mainly on the objects of management. Various types of rotation recognized in forestry are as follows:

PHYSICAL ROTATION: It is the rotation which coincides with the natural lease of life of a species on a given site. The natural life-span of trees varies greatly with species and the site factors. This rotation is applicable only in case of protection and amenity forests, park lands and in some cases roadside avenues. It is very variable, fairly long and also indefinite. Another interpretation of physical rotation is the age up to which the trees remain sound, or produce viable seed in high forests and, in coppice crops, can put forth reliable coppice shoots. This rotation is not of any relevance to economic forestry.

SILVICULTURAL ROTATION: It is the rotation through which a species retains satisfactory vigour of growth and reproduction on a given site. It can neither be lower than the age at which trees start producing fertile seed in sufficient quantity, nor beyond the age when they stop doing so. It is also necessary that soil conditions remain satisfactory for germination and establishment of seed. It is not only long but has also very wide range of limits, hence somewhat vague and may be used in combination with other rotations, such as Technical rotation.

TECHNICAL ROTATION: It is the rotation under which a species yields the maximum material of a specified size or suitability for economic conversion or for special use. It aims at producing the maximum material of specific dimension for specific purpose, such as railway sleepers, saw-logs, mine-props, transmission poles, match-wood, paper-wood.etc. Since one and the same tree may yield different assortments of material and the trees in a crop may attain given size at different times, the technical exploitable age offers no reliable fixed point for fixing the rotation. It does, however, allow for fixation of limits within which a tree or stand is adopted for the production of assortments in greater number or better quality. Technical rotation is adopted particularly by industrial firms which own forests for the purpose of supplying raw material for their plants (e.g., NEPA and West-coast paper mills, WIMCO match factory, ITC, etc.).

ROTATION OF MAXIMUM VOLUME PRODUCTION: It is the rotation that yields the maximum annual quantity of material i.e. the age at which the mean annual increment (M.A.I.) culminates. The M.A.I. refers to that of the stand and not that of individual trees. More often than not, the objective in forestry is quality or value production and rotation for higher value production is usually longer than for highest volume production. In turn, maximum volume rotation is usually longer than financial rotation. Common practice in forestry is to adopt a combination of rotation of maximum volume production and financial rotation.

ROTATION OF HIGHEST INCOME: It is the rotation which yields the highest average annual gross or net revenue irrespective of the capital value of the forest. It is calculated without interest and irrespective of the times when the items of income or expenditure occur. This rotation is important from the overall national point of view. With forestry in the public sector, attainment of highest gross revenue is more important than that of net income because larger expenditure and investment generates several social benefits and indirect advantages to the trade and industry. The private owner of a forest estate is interested in maximum net revenue by keeping the rotation period as short as possible.

The average net annual revenue or rental obtained from a stand of trees is expressed by the formula:

$$\text{Mean annual net revenue per unit area} = \left(Yr + \sum Tr - C - \sum E\right) / R$$

where,

Yr = Value of final felling per unit area

Tr = Value of all thinning during rotation period 'R' per unit area.

C = Cost of formation of stand per unit area.

E = Annual cost of administration per unit area

R = Rotation (years)

The two rotations will be about the same unless there is an appreciable increase in price for larger-sized timber which is, in fact, usual. If the size gradient is marked, then the rotation of highest net income will be comparatively longer. Again, if there is a special size of timber which fetches a particularly high price, the rotation which provides that price may be the rotation of highest income and possibly coincides with technical rotation.

FINANCIAL ROTATION: It is the rotation which yields the highest net return on the invested capital. It differs from the rotation of highest net income in that all items of revenue and expenditure are calculated with compound interest at an assumed rate, usually the rate at which the Govt. is able to borrow money. It is also defined as:

- The rotation which gives the highest discounted profit, usually at its commencement
- The rotation which is most profitable (Hiley)
- The rotation which gives the highest net return on capital value i.e. under which the soil expectation value calculated with a given rate of interest is the maximum (Brasnett)

Rotation ages of some important plantation species are as follows.

Teak: 60 - 120 yrs

Eucalypts, Poplars and Casuarinas: 4 - 8 yrs

Acacias: 8 - 12 yrs

Prosopis: 4 - 8 yrs

LENGTH OF ROTATION

The choice of the type of rotation will depend on the object of management, but the length of rotation of whatever type, will depend on the interaction of several physical and economic factors given below:

- *Rate of growth*: This will vary with species, site fertility, intensity of thinning, etc.
- *Silvicultural characteristics of the species*: For example, natural span of life, age of fertile seed production, age at which rate of growth culminates, age at which the quality of its timber is most desirable or begins to fall, etc.
- *Response of the soil*: Deterioration / exhaustion of soil by exposure, biotic influence etc.
- *Economic considerations*: Depending on a combination of factors of cost, prices of different sizes, time required to reach those sizes, etc. A factor tending to lengthen the rotation is the increase in value of large-sized timber, though not economically sound to grow large-sized timber over a long rotation.
- *Social conditions*: Socio-economic and employment policy of the State.

THINNING

Depending on the forest type and biophysical setting, hundreds to tens of thousands of seedlings per acre can naturally regenerate after a disturbance in a natural stand. Even with such high stand densities, at 60 to 150 years old only 100 to 200 stems per acre remain. This reduction is caused by intertree competition, drought, wind, frost, snow, diseases, insects, fire or a combination of these important mortality factors. These stocking reductions allowed the site's growth potential to be concentrated on fewer stems producing fewer but larger trees. The efficiency at which mortality factors reduce the number of stems on a site depends on the disturbance, forest type and biophysical setting. Individual trees in many forest areas do not readily succumb to intertree competition, often causing stagnated stands with thousands of stems per acre. Fire is one of the major mortality factors that causes natural thinning of dense stands.

In addition to natural events that reduce density of forest stands, forest management through application of thinnings also can alter species composition and stand structure. Depending on the objectives, thinnings can be applied to forest stands for various reasons. Classically, thinning is defined as cuttings made in immature stands in order to stimulate the growth of trees that remain and to increase the total yield of useful material from a stand. But, often any kind of partial cutting such as cleaning, weeding, liberation, improvement, sanitation and selection cuttings is termed thinning, especially outside the field of silviculture and all reduce the number of stems in a forest stand.

Thinning is defined as a felling made in an immature stand for the purpose of improving the growth and form of the trees that remain, without permanently breaking the canopy. It is a treatment of forest crop where by the number of trees growing in a stand is reduced. Thinning consists of a series of successive felling operation for a number of times before the crop matures. The interval between two successive felling may be fixed but it is dependent on the time required for canopy closure. Thinning is carried out in a crop after it reaches the sapling stage and continued upto the beginning of the regeneration period. The thinning principles are applicable to pure and even aged or nearly even aged crop or even aged groups of the trees in a crop. Thinning always increases spacing and decreases number of trees per unit area.

OBJECTIVES OF THINNING

- To improve the hygienic condition
- To create best conditions of growth
- Salvage the anticipated losses of the merchantable volume of timber
- To obtain desirable composition of crop
- Retaining seed bearers
- Improvement on wood quality
- To obtain intermediate yield and increasable net yield and financial out turn
- Decomposition of raw humus
- To reduce risk of disease and pests

TYPES OF THINNING

The category of trees to be removed in a thinning operation may depend on size or crown class distribution before and after the thinning. In low thinning or thinning from below, smaller trees are preferred for removing as compared to larger trees. But in high or crown thinning or thinning from above, only larger trees are removed. The thinning type is based on the relative dominance of trees within the stand (Vinod Kumar 1995). The classic objective of thinning is to redistribute growth potential to fewer trees past the sapling stage, leaving a stand with a desired structure and composition. In general, five methods of thinning are recognized:

(1) Low thinning or thinning from below

(2) Crown thinning or thinning from above

(3) Selection thinning or diameter-limit thinning

(4) Free thinning

(5) Mechanical thinning

Most often, forest stands do not develop with one canopy. Because of individual tree species, microsite differences and local disturbances, multiple crown classes usually develop. Four crown classes are commonly recognized and used to describe different stand structure (Smith 1962) *viz.* dominant, codominant, intermediate and suppressed. These crown classes are used to describe the trees removed in different types of thinnings.

LOW THINNING (thinning from below) is when trees are removed from the lower canopy, leaving large trees to occupy the site. This method mimics mortality caused by intertree competition or surface fires and concentrates site growth potential on dominant trees. Low thinnings primarily remove intermediate and suppressed trees, but heavy thinning also can remove many in the codominant crown class. Low thinnings not only remove understory canopies but also can alter species compositons. Usually, different tree species have characteristic development rates that result in individual species dominating specific canopy layers (Russel et al 1999).

Table 5.2: Trees Removed during different intensities of Low Thinning (Smith 1962)

Intensity	Trees Removed
Very Light	Poorest overtopped
Light	Overtopped and poorest intermediate
Moderate	Overtopped and intermediate
Heavy	Overtopped, intermediate and many codominants

CROWN THINNING (thinning from above) reduces crowding within the main canopy. Dominant and codominant trees are removed to favour residual trees in these same classes. This method is often used to remove selected species in the dominant and codominant crown classes that are competing with more desirable species (Nyland 1996). This method keeps vertical structure in place, which is often desirable for wildlife species. As with low thinning, crown thinning can create various stand structures and compositions while retaining vertical structure.

SELECTION THINNING removes dominant trees to favour smaller trees. This method is often applied by removing trees over a certain diameter. Diameter-limit cuts that continually remove largest trees may well be dysgenetic and can be a disguise for high grading (removing trees of high economic value). By removing the current value from a stand, future options often can be limited and the only recourse for the future may be to regenerate. Stand structures and species compositions created by using selection thinning are limited and in general, favour shade-tolerant species or trees occupying the intermediate and suppressed crown classes. Often the stands created by selection thinnings are prone to edpidemics of insects and diseases. Compared to other thinning methods, selection thinning is less useful because of the limited stand structures and compositions it can create (Russel et al 1999).

FREE THINNING sometimes called crop-tree thinning, primarily relesases selected trees. This method favours specific trees, whereas the remainder of the stand goes untreated. Depending on what is presented in various portions of a stand (tree spacing, species, vertical structure, etc.), the thinning criteria can be highly flexible, producing stands with large amounts of diversity. It can be used in any of the crown classes for releasing specific trees. This method has the most flexibility for creating various stand structures and compositions.

MECHANICAL THINNING removes trees based on specified spatial arrangements (Nyland 1996). This method is often applied in plantations where every other row (alternate row) or every other tree (alternate tree) in a row is removed. Such rigid thinning is easy to apply, but the stands created often lack diversity in either structure or composition. This method also resembles strip thinning, where a strip of trees are removed. Mechanical thinning is well suited for timber production on uniform sites but has limited value for producing conditions that meet other resources values.

THINNING CYCLE

It is defined as the planned interval in years which elapses between two successive or consecutive thinning in the same crop. Thinning cycles are not regular. The thinning cycles are shorter during early period of growth and longer when the crop is middle aged to mature.

THINNING INTENSITY

It indicates in numerical terms the extent to which the crop is thinned or the frequency at which thinning are carried out. It is a measure of the yield removed over a specified period of time. Thinning intensity may be "light" or "heavy" according to the number of trees to be removed. It can be expressed in volume per hectare per annum. The thinning cycle and thinning intensity are interdependent.

THINNING REGIME

It is the whole set of thinning carried out in a crop from earliest stage to maturity.

THINNING GRADE

It refers to the relative extent to which a crop is opened up in thinning.

THINNING SCHEDULE

It is the laying down of number of stems that should remain standing after thinning. For this, criteria of dimensions, site quality and age are taken into account. Fixing of thinning schedule requires data of yield and actual experience in the field. The rotation period depends on the objects of management.

6 FOREST ECOLOGY

INTRODUCTION

Forests comprise the most diverse ecosystems in the world, hosting a substantial part of all living species. Forests also provide a wide variety of essential services to society, ranging from natural resources such as timber, wildlife and other products to services such as water retention, climate regulation and carbon storage. A forest is more than trees and it includes the shrubs, wildflowers and grasses; the animals that depend on and live among these plants; the soil in which plant grow; the stream that flows through it; the insects, fungi, bacteria and organisms that cannot be seen readily; and the climate. In short, a forest is an ecosystem. An ecosystem is simply a specific area of the earth that includes all the living organisms and nonliving components of the environment that interact within its boundaries. An ecosystem can be any size, such as a log, pond, field, forest or the whole Earth's biosphere.

It is useful to understand and manage forest as an ecosystem because all the components influence one another. Certain aspects of the ecosystem such as climate cannot be controlled and one must learn to live within its constraints of temperature and precipitation. Forest management and activities are based on an understanding of ecology. Ecological concepts provide a basis for predicting the effects of our actions on ecosystems. Timber harvesting, grazing, wildlife habitat improvement and fire control intentionally or unintentionally change vegetation to favour or discourage certain ecological functions. Viewing management activities in an ecological context provides a basis for understanding the effects of those activities on ecosystem processes. Altering vegetation can change many things beyond the obvious and immediate changes in plant cover. Ecosystem approach would mean considering the effects of vegetation management on all comprehensible ecological functions and processes. Most of forest activities are aimed at improving the production of goods and services for human benefit. Evaluating those activities in an ecological context can provide insight into a larger picture, predict and avoid potential negative effects.

CONCEPTS OF ECOLOGY

Ecology is the study of the interactions that take place among living organisms and their abiotic environment. Interactions determine distribution and abundance of organisms. Ecology is historically an observational science, often descriptive about natural history.

Ecology and evolutionary biology are closely related sciences. Events that occur in the framework of ecological time (minutes, days, years) translate into effects over evolutionary time (decades, millennia). Ecological research scale ranges from individuals to the biosphere.

- *Organismal Ecology*: It is concerned about the way in which an individual interacts with its environment.
- *Population Ecology*: It is the study of a group of individuals of the same species. Population denotes all the organisms in an ecosystem that belong to the same species.
- *Community Ecology*: It deals with all interacting species within a particular area. Community denotes all the populations in an ecosystem. Plants and animals occurring together in a coherent group called as ecological community because of their adaptations to each other and the surrounding environment.

Habitat is the place in which an organism lives. It provides the kinds of food and shelter, the temperature, and the amount of moisture the organism needs to survive. Competition for food and space caused by population growth affects many organisms, including humans and it limits population size. Anything that restricts the number of individuals in a population is called limiting factor. It includes living and nonliving features of the ecosystem. Landscape ecology is the study of interactions among ecosystems. The biosphere is the global ecosystem. The part of Earth that supports life is called biosphere and it includes ecosphere, lithosphere, hydrosphere and atmosphere. Global climate research is an example of ecology at the biosphere scale. Biogeography is the study of past and present distribution of individual species. Species dispersal contributes to the distribution of organisms. Dispersal refers to the process of distribution of individuals within geographic population boundaries.

ECOSYSTEM

All living communities (biotic) and their associated nonliving environmental components (abiotic) in a defined area is called ecosystem. The self-sustaining structural and functional interactions between living and non-living components occur in the ecosystem. Ecosystem is classified as below,

NATURAL ECOSYSTEM

Terrestrial Ecosystems: grassland ecosystems, forest ecosystems, desert ecosystems, wetland ecosystems

Aquatic ecosystem:

- Lentic (Stagnant water) like lake, ponds, etc.
- Lotic (Flowing water) like stream, river, sea, etc.

ARTIFICIAL ECOSYSTEM: agriculture ecosystem, garden ecosystem, aquarium ecosystem, park ecosystem, kitchen garden ecosystem, etc.

FUNCTIONS AND COMPONENTS OF ECOSYSTEM

The producers, the green plants, fix radiant energy (solar energy) with the help of minerals taken from their edaphic (soil where they grow) or aerial environment and build up complex organic matter. These are their food. The carbon dioxide from air, water and minerals from water bodies and soil are taken up by green plants along with solar energy to make their food. This process is repeated again and again. This leads to continuous functioning of the ecosystem. So, with the help of solar energy they convert the chemical energy of the food to kinetic energy and finally heat energy. The animals eat up plants and other animals as food. So, the energy is transferred through food to animals. When plants and animals die, then decomposers (like certain bacteria and fungi) act on their dead bodies and decompose them into simple materials like carbon dioxide, water and minerals which go back to air, water bodies and soil from where they were taken.

An ecosystem consists of two main components:

Abiotic or Non-Living Components: inorganic substances, organic compounds, climatic factors

Biotic or Living Components: autotrophs or producers, heterotrophs or consumers, saprotrophs or decomposers

ENERGY FLOW

The transfer of energy and matter takes place in the process of predator and prey relationship in a food chain. The original source of energy is the energy from the sun. Out of the enormous amount of energy continuously radiated by the sun, most of it is reflected or refracted back (by atmosphere, earth surface and object like plants). Only a very small fraction, about one per cent, of the solar energy received by the plant is used through the process of photosynthesis. The transfer of energy from sun to producer, producer to primary consumer, primary consumer to secondary consumer and ultimately to tertiary consumer is termed as food chain. Energy pyramid shows that the amount of available energy decreases down the food chain. It takes a large number of producers to support a small number of primary consumers and it takes a large number of primary consumers to support a small number of secondary consumers. A system of interlocking and interdependent food chains is known as food web.

Autotrophs include all green plants, trees, shrubs along with some bacteria. Consumers are organisms that eat plants and animals, also known as herbivores and carnivores respectively. Bacteria, fungi, insects, or other organisms that break down organic material are called decomposers. The balance between the living beings and also with the non- living environment is called the ecological balance or simply a balance in nature. Unfortunately, however human activities have upset the ecological balance in nature, so it is more essential that balance in nature should be established soon because it is essential for existence of human beings.

FOREST ECOSYSTEM

A forest is a community of trees, shrubs, herbs, and associated plants and organisms that cover a considerable area that use oxygen, water and soil nutrients as the community attains maturity and reproduces itself. Forest ecology is the study of the complex interactions between the organic and inorganic elements of a forest ecosystem. A forest ecosystem is the organisms, soil, air and water associated with the forest. A forest ecosystem is interdependent which means every organism depends on every other living and nonliving element of the system. Forest Ecology is the study of the complex interactions between the living and nonliving elements of a forest ecosystem. Large geographic areas with similar climates and ecosystems are called biomes.

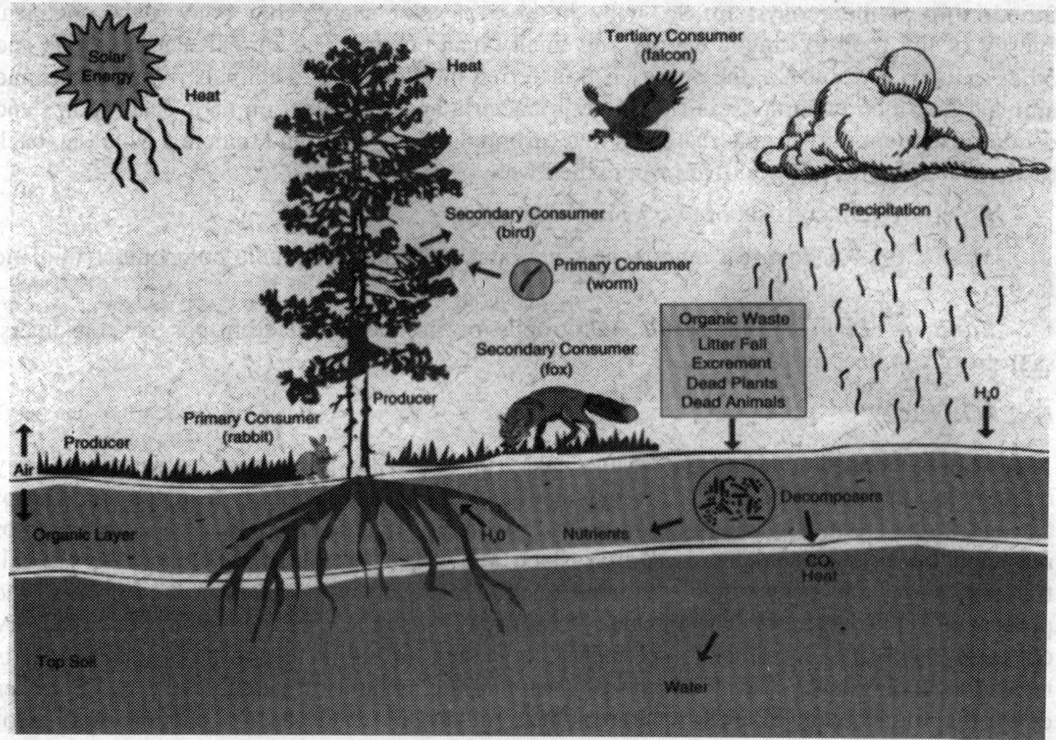

Fig. 6.1: Energy Flow in an Ecosystem (Source: Helgerson 2012)

All trees compete with each other to survive in a forest ecosystem. The limiting factor in a forest ecosystem is sunlight. Symbiosis literally means the living together of two dissimilar species of organisms, where one or the other, or both are affected. Symbiosis usually involves supply of food, protection, cleaning, transportation, etc. Mutualism is any relationship between two species of organisms that benefits both species. Commensalism is a relationship between two species where one species derives a benefit from the relationship and the second species is unaffected by it. Parasitism is a relationship between two species where one species benefits and the other is injured. Trees are parasitized by viruses, bacteria, fungi, nematodes and a few other plants. In the parasitism, parasites may or may not cause death of the host. In predation, one species (stronger predator) kill the other species (the prey) in order to obtain food. Allelopathy involves a plant's secretion of biochemical materials into the environment to inhibit germination or growth of surrounding vegetation. Allelopathy enhances tree survival and reproduction.

SUCCESSION

All trees compete for the same basic requirement of life: light, water, essential elements, oxygen and other necessities. The species that are better able to gather those requirements will outcompete others. Broadleaved or hardwood trees are better able to compete and grow on deeper, heavier, more productive soils. The needle-leaved trees or conifers do better on poorer, lighter textured soils. Forest ecosystems are dynamic in nature and keep on changing due to natural changes and man-made changes. Natural changes are caused by fire, storms, drought, flood, death and disease, while man-made changes are caused by harvesting, farming, trails, recreation and development. These changes are known as disturbances and spark the process called succession. Succession is the gradual change in plants and animal communities over time. It is a natural and gradual change in the types of species that live in an area. Primary succession occurs in an area that has no true soil and pioneer species are the first plants to grow at these barren sites. Pioneer species are the group of organisms, such as lichens, found in the primary stage of succession and that begin an area's soil-building process. Primary succession follows after the big disturbance in the forest ecosystem.

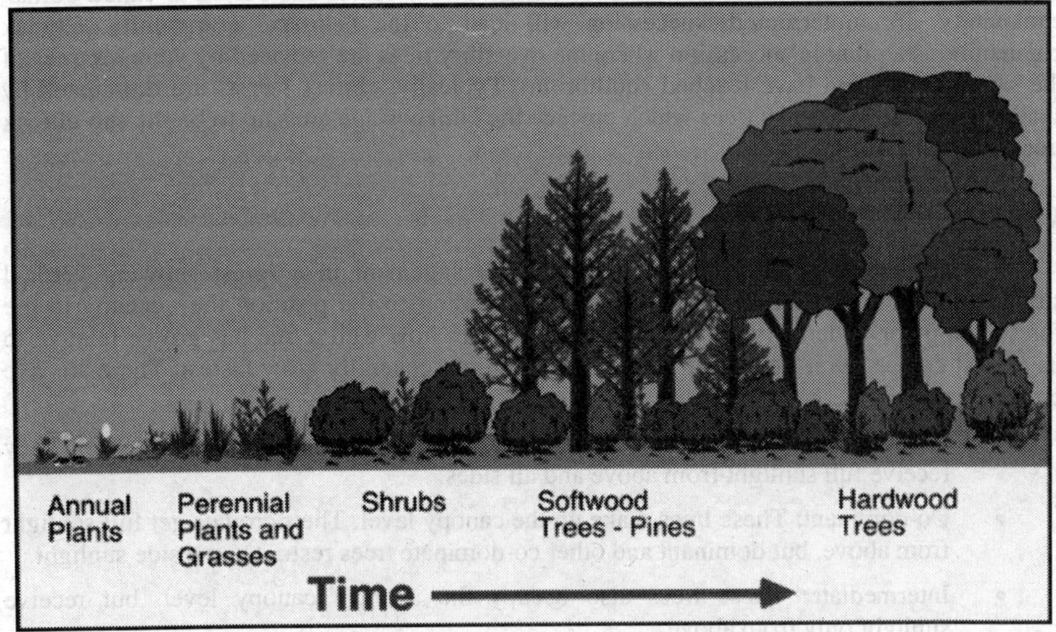

Fig. 6.2: Succession in Forest Ecosystem (Source: Delaware ENVIROTHON 2012)

Secondary succession occurs on landscapes where the natural vegetation has been removed or destroyed but the soil remains intact. Tree species that first colonize an area are usually shade-intolerant species and must have full sunlight to do well. Shade intolerant species are generally the "first in" after an event such as a clear-cut or a major fire that substantially opens the canopy. These trees are often called pioneer species. These trees tend to be fast growing, short-lived and have light seeds. Succession first begins in a disturbed area by means of seeds from bird droppings, animals carrying seeds in fur, wind blown seeds, existing seed bank in soil gets the needed sunlight and floods or high rains will leave behind seeds. Once they have developed some sun coverage, the shade-tolerant species will begin to grow and eventually take over the canopy. These trees normally are not the first to colonize open areas. Instead, they grow up into an existing canopy. Usually, these trees are found in the climax community. These trees tend to live a long time, grow slowly and have heavier seeds. Succession ensures the continuation of the forest by allowing other species the chance to grow and helps to increase the forest's biodiversity. Biodiversity is the species richness or variability among species in a given ecosystem. When an area is rich in native species, it encourages other native animals and insects to inhabit that area, this in turn makes the ecosystem stronger.

When forests are strong, they are better able to withstand and recover from stress the outside environment imposes. Trees occupy all layers of the forests. Forest canopy is formed by the branches and leaves from the tallest trees. Understory is the middle layer where shorter trees and shrubs grow. Forest floor is the third layer where trees begin as seedlings. Tree roots also make up this layer that hold onto to soil and organic matter so that grasses, ferns, fungi, decaying plants and logs, microorganisms and animals may flourish.

A community that has reached a stable stage of ecological succession is called climax community. If uninterrupted, succession will lead to the "climax" community. Climax community is a point in succession where the overstory trees are replaced by younger trees of the same species and have reached equilibrium. Typically, climax forests are dominated by shade-tolerant species. The trees which precede the climax stage are said to be the sub-climax trees.

VERTICAL STRUCTURE

Vertical structure refers to the ladder-like arrangement in a forest showing vertical stratification of trees. Trees can be classified according to the position they occupy in the forest canopy or understory. Such as system measures how well a tree has grown relative to its closest competitors. Trees that get the most sunlight generally grow fastest. There are five crown classes:

- Dominant: Dominant trees have tops that rise above the general canopy level. They receive full sunlight from above and all sides.
- Co-dominant: These trees make up the canopy level. Their crowns get full sunlight from above, but dominant and other co-dominate trees restrict some side sunlight.
- Intermediate: These trees also occupy the average canopy level, but receive sunlight only from above.
- Suppressed: These are trees that receive no direct overhead or side sunlight. They are usually slow growing and weak.
- Dead trees (snags): These can be found in the canopy, understory or forest floor.

FOREST COMMUNITY DYNAMICS

In nature different organisms grow with the association of other species. A group of many species i.e., plants and or animals living together with mutual tolerance and beneficial interactions are known as a community and more appropriately biotic community. The biotic community is defined as a group of one or more populations living together in time and space through the combined action of various environmental factors. In a biotic community, organisms always interact among themselves and with their physical environment. Thus, in a forest, the organisms like trees, shrubs, herbs, climbers, twiners, epiphytes, parasites, lizards, snakes, frogs, birds, insects, earthworms, lichens, microorganisms, etc. live together and form a community (Tiwari 1992).

Fig. 6.3: Vertical Structure in Forest Ecosystem (Source: MSU 2012)

A community has its own characteristics and these characteristics which have meaning only with reference to community level of organizations are as follows (Sharma 2009),

SPECIES DIVERSITY: Each community is made up of much different organisms–plant, animals, microbes, which differ taxonomically from each other.

GROWTH FORM AND STRUCTURE: Plant communities have several characteristics by which they can be described: structure, life form, spatial pattern, species composition, successional stages, biomass and functional processes. Community is described in terms of major growth forms as trees, shrubs, herbs etc. These different growth forms determine the structural pattern of a community. According to the mode of arrangement of the various growth forms, the communities exhibit zonation or horizontal layering as well as stratification or vertical layering.

The communities exhibit a structure or recognizable pattern in the spatial arrangement of their members. Thus structurally, a community may be divided horizontally into sub communities, which are units of homogeneous life-form and ecological relation. This horizontal division constitutes the zonation in the community. Latitudinal as well as altitudinal zonations of vegetation have also been shown in relation with climatic factors. In deep ponds and lakes, there may be recognized three zones i.e., littoral zone, limnetic zone and profundal zone. In each zone, organisms differ from each other. Stratification is another most common aspect of structure, which involves vertical rather than horizontal changes within the community. In each horizontal zone, there may be recognized distinct vertical stories.

DOMINANCE: In the community, all the species are not equally important. There are relatively only a few of these which determine the nature of the community. These few species exert a major controlling influence on the community. Such species are known as dominants.

SUCCESSION: Each community has its own developmental history. It develops as a result of a directional change in it with time.

TROPIC STRUCTURE: Nutritionally, in each community, a group of autotrophic plants as well as heterotrophic animals exist as self-sufficient, perfectly balanced assemblage of organism. Each community has its own composition structure and development history.

COMPOSITION: Community may be large or small. Larger ones such as forests extend over areas of several thousands of square kilometers. Others such as deserts, etc. are comparatively smaller with dimensions in hundreds of kilometers and still others such as meadows, rivers, ponds, rocky, plateaus, etc. occupying a more restricted area. Very small-sized communities are the groups of microorganisms in such microhabitats as leaf surface, fallen logs, litter soil, etc. The number of species and population abundance in communities vary greatly.

COMMUNITY STRUCTURE CHARACTERISTICS

As mentioned above, each community is characterized by its species diversity, growth forms and structure, dominance, succession trend, etc. To study the details of these aspects of any community, number of characters are considered and used to express the characteristics of the community. If a particular species is dominant in the community, then all the possible characteristics of that dominance is studied. Various characters used for studying community are classified into analytical characters and synthetic characters (Tiwari 1992; Sharma 2009).

ANALYTICAL CHARACTERS

Analytical characteristic features of a stand are directly observed or measured in the field. Analytic characters are either qualitative or quantitative.

QUALITATIVE CHARACTERISTICS: These characters are usually described and not measured.

- *Physiognomy*: The general appearance of vegetation or gross structure of a community is called physiognomy
- *Stratification*: The term stratification or layering is the distributions of organisms which live together and are adjusted in the vertical space, above and below ground.
- *Phenology*: The temporal behaviour of the species is known as phenology. Here each species in the community exhibits important events of its life-cycle. These events include: plant germinates, completes their vegetative growth, produces flowers, fruits and seeds, and further fruits and seed disperse etc.
- *Sociability*: Sociability indicates the closeness between members of a species in the community.
- *Life form spectrum* or *Biological spectrum* or *Phytoclimate spectrum*: This can be expressed in the form of percentage of different life-form classes in the total flora. In other words, it can also be defined as the statistical distribution of species among the life-forms of a flora. The biological spectrum of a region is related to the climate of the region.

QUANTITATIVE CHARACTERISTICS: These characters are measured quantitatively in the field. These include following characters:

(1) *Frequency*: Frequency expresses the distribution or dispersion of species in a community. The degree of dispersion is usually expressed in terms of percentage occurrence. The percentage frequency of each species is calculated as follows,

$$\text{Frequency (\%)} = \frac{\text{Number of quadrats in which the species occurred}}{\text{Total number of quadrats studied}} \times 100$$

(2) *Density*: The density is defined as the number of individuals per unit area. The density is calculated as,

$$\text{Density} = \frac{\text{Nunmber of individuals of each species in sampling units}}{\text{Total number of sampling units studied}}$$

(3) *Abundance*: It refers to density of individuals of a species in those sampling units only, in which a given species occurs. Thus abundance is the number of individuals per sampling unit of occurrence.

$$\text{Abundance} = \frac{\text{Total number of individuals}}{\text{Total number of sampling units, in which species occurred}}$$

(4) *IMPORTANCE VALUE INDEX (IVI)*: IVI is the sum of the relative frequency, relative density and relative dominance. The relative values are calculated as follows,

$$\text{Relative Frequency} = \frac{\text{Frequency of a species}}{\text{Frequency of all species}} \times 100$$

$$\text{Relative Density} = \frac{\text{Density of a species}}{\text{Density of all species}} \times 100$$

$$\text{Relative Dominance} = \frac{\text{Basal area of a species}}{\text{Basal area of all species}} \times 100$$

(5) *A/F RATIO*: Abundance to frequency ratio (A/F ratio) for different species is determined for eliciting the distribution pattern in terms of regular (<0.025), random (0.025-0.05) and contagious (>0.05), as follows (Curtis and Cottam 1956):

$$\text{A/F ratio} = \frac{\text{Abundnace}}{\text{Frequency}}$$

(6) *BASAL AREA AND BASAL COVER*: Girth of individual tree species observed in the quadrat are measured at breast height (1.37m) and basal area is calculated using following formula,

$$BA = \frac{\pi D^2}{4}$$

Where, D is the diameter of the tree at breast height

$$\text{Mean basal area (MBA)} = \frac{\text{Basal area of all individuals of species}}{\text{Total number of individuals of species}}$$

Total basal cover (TBC) refers to the mean basal area of species multiplied with respective density and expressed as m^2 per ha.

Total basal cover (TBC) = Mean basal area of species × Density of species

SYNTHETIC CHARACTERS

These characters are based upon the data recorded for the analytic characters. Thus data recorded for different stands are synthesized to understand various aspects of vegetation and for the comparison of the stands. Some important synthetic characters are: (1) Presence and Constance, (2) Fidelity, (3) Dominance, (4) Association Index, (5) Indices of Similarity and Dissimilarity and (vi) Indices of Diversity.

METHODS OF STUDYING FOREST COMMUNITIES

Various methods for studying of plant communities are broadly grouped into three major categories: floristic methods, physiognomic methods and phytosociological methods.

FLORISTIC METHODS

Here the flora is studied by listing various genera and species present in the communities. Thus vegetation is described in terms of the flora. These are descriptive methods and such method gives either little or no idea about the composition, structure, growth forms, etc. of the community.

PHYSIOGNOMIC METHODS

In this method, various species of the community are studied mainly of their life-forms (growth form), general stature (importance and value) and its distribution ranges, etc. Among various physiognomic methods, Raunkiaer's (1934) life-form method is most popular. According to Raunkiaer, moisture and heat conditions of the environment are of prime importance in determining the pattern of plant distribution in different zone of the earth. Further, the unfavourble period of the year again plays an important role and acts as limiting factor. Raunkiaer classified five major different life-forms. Here position and location of perennating organs or bud in relation to ground surface has been taken as a criterion for classification of species into different life forms.

PHYTOSOCIOLOGICAL METHODS

Above two methods do not provide detailed information on the species composition, structure, diversity, growth, trends of succession and other characteristics of the community. In phytosociological methods, there are generally three forms of sampling units: area, line and points. In area and line as sampling units, there is a definite area for sampling. Although in such communities which have forests with long grassy and shrubby belts, it becomes difficult to have a definite area. In such analysis, point is taken as sampling units. On the basis of the nature of sampling units, there are three popular methods for studying communities.

Table 6.1: Life-Form Classes (Raunkiaer 1934)

S No	Life-Form Class	Symbol	Description
1	Phanerophytes	P	Tree or shrubs, large perennial herbs, woody lianas, epiphytes, stem succulent in which perennating buds are well above ground (above 25 cm from ground surface) of twigs or branches. Predominating in tropical floras
2	Chamaephytes	Ch	Small shrubs, herbs and prostrate plants, in which perennating buds are above ground, but at less than 25 cm height. Predominating at high altitudes and latitudes
3	Hemicryptophytes	H	Biennial or perennial herbs and grasses in which perennating buds are just beneath or in soil surface covered with snow or litter during unfavourable period. Predominating in cold temperate zones
4	Cryptophytes	Cr	Perennating buds under water or below the soil surface as in hydrophytes, marsh plants, or terrestrial plants with underground rhizome, tuber, etc. Predominating in zones characterized by a long dry season.
5	Therophytes	Th	Usually annual herbs, which survive unfavourable period through seeds or spores. Predominating in cold and hot deserts.

QUADRAT METHOD: Here the sampling unit is definite area which is taken in the form of a square, rectangle or circle. Based on the vegetation type and purpose of study, quadrats may be three types:

- *List Quadrat*: Here species present in the area are just listed for recording.
- *List-Count Quadrat*: Here listing of species and numerical counts of individuals of each species are also made.
- *Chart Quadrat*: Here detailed growth and distribution of each species in space is recorded.

TRANSECT METHOD: Here, the sampling unit is line of suitable length or a segment of suitable length

POINT METHOD: The sampling unit is a point.

COMPETITIVE EXCLUSION PRINCIPLE

The competitive exclusion principle works rather well in simple laboratory experiment with two competing species but in natural ecosystem, it can find numerous examples. In 1959, Darwin in his book, *On the origin of species through natural selection*, has given the idea of competition or struggle between species. Darwin considered the competition as an important factor in evolution through survival of the fittest. The idea of replacement of individual species by stronger competitors was further developed by many scientists. Considering such studies, Hardin discussed the competitive exclusion principle in 1960 which states that two or more species having the same resource requirement cannot continue to exist in a stable environment and only one species utilizes the available resources by eliminating the other.

ECOLOGICAL NICHE CONCEPT

Varieties of organisms belonging to different species live together in the community. However, these organisms differ in their ecological requirements and functions. When two species with similar ecological requirements and functions occupy a particular habitat, it is always followed by a competition for the habitat resources and one of the two species is eliminated after a certain length of time. Thus it can be said that no two species with similar ecological requirement can exist together. Such organisms in the community occupy a specific habitat where it could survive and perform all life activities as best as it can. Considering such activities or behaviour of organisms in the community, Joseph Grinnel in 1917 used the term niche for the first time which refers to the sub-division of habitat (*Grinnellian Niche*). In 1927, Charles Elton in England used the term niche for the first time in the sense of the functional status of an organism in its community and defined the niche as an organism's place in the biotic environment, its relation to food, and enemies (*Eltonian niche*). In 1957, Hutchinson defined niche as the totality of the environmental variable, both of biotic and abiotic factors, to which a given species is uniquely adapted (*Hutchinsonian niche*). According to Hutchinson, a multidimensional space within which the environment allows an individual or species to survive indefinitely can be said as niche.

The term niche tells us how a species population is specialized within a community. Niche thus refers to the functional adaptation and distribution characteristics of a species. Species avoid competition by evolving different adaptations, tolerances, requirements, behaviour and so on. This observation led to the development of the concept of Ecological Niche. Based on the observations of Grinnell, Elton and Hutchinson, three aspects of the ecological niche are clearly recognized:

- *Spatial or Habitat Niche*: Grinnel considered the niche as sub-division or specific position of the habitat called as microhabitat.
- *Trophic Niche*: Elton considered the niche in the sense of functional status i.e., trophic position (energy relation) of an organism in its community.
- *Multidimensional or Hypervolume Niche*: Hutchinson considered the niche as the ranges of environmental factors representing environmental gradients in which the species can survive indefinitely.

In this context, the hypervolume niche represents the totality of the environmental interactions to which a given species is adapted. Hutchinson termed hypervolume niche as the fundamental niche. The *fundamental niche* represents the maximum niche that the species could occupy in the absence of competition from other species. The *realized niche* is the portion of the fundamental niche occupied by the species in the face of competition.

7 BIODIVERSITY AND CONSERVATION BIOLOGY

INTRODUCTION

Human population growth is clearly outstripping the Earth's non-renewable resources, biological fundaments and Earth's ability to revive. Ecological degradation and its corollary - biodiversity loss - pose a serious threat to development. Ecologically destructive economic activities are inefficient not merely because of the resulting resource misallocation but also because of the excessive scale of activity levels in relation to the limited availability of natural resources. In order to bring about sustainable resource conservation and management, it is essential to adopt several different approaches for managing our forests and biodiversity.

Future efforts for conservation and management of our natural resources must derive from a set of clear objectives, mechanisms for action and commitment from all stakeholders. Apart from this, halting the process of degradation and species loss requires specialized solutions and an understanding of ecological processes. Protecting biodiversity does not merely involve setting aside chunks of area as reserves. Instead, all the ecological processes that have maintained the area's biodiversity such as predation, pollination, parasitism, seed dispersal and herbivory, involving complex interactions between several species of plants and animal needs to be ensured. This is possible only if reserves are large enough to maintain these processes and some of the other crucial links in the web of life.

There is also the need for greater involvement of communities and for models which decentralize of management and conservation roles and responsibilities. Strengthening of information resources is important pertaining to forests, biodiversity - flora and fauna, causative factors for their degradation and major threats. Data on current status and ongoing losses/gains of biodiversity is very important for successfully implementing conservation strategies. More importantly, laws and policies governing natural resources need to be strengthened to tackle the scale of the problem and the insufficiencies should be addressed with a sense of urgency.

THE CONVENTION ON BIOLOGICAL DIVERSITY

The Convention on Biological Diversity (CBD), one of the key agreements adopted during the Earth Summit 1992, is the first comprehensive global agreement which was related to all aspects of biodiversity. The CBD has membership of 192 countries, sets out commitments for maintaining the world's ecological underpinnings, while pursuing economic development. The Convention, while reaffirming sovereign rights of nations over their biological resources, establishes three main goals: conservation of biological diversity, sustainable use of its components and the fair and equitable sharing of the benefits arising out of the use of genetic resources.

The year 1992 was a historical moment in the struggle for conserving the world's natural heritage which had set off the creation of many large protected areas. The 4th World Parks Congress was held in Caracas, Venezuela in February 1992 organized by the World Conservation Union (IUCN), the global alliance for conservation and wise use of natural resources. The world parks congress adopted the "Caracas Action Plan" with recommendations for strategic actions for protected areas creation and management over the decade from 1992 to 2002, calling for special attention to enhancing the capacity to manage marine protected areas and to include them as major components of national system plans. The Caracas congress devoted much of its attention to the need to build constituencies, involve major interest groups in planning and management, enhance revenue generation and financing and assess and quantify the benefits protected areas provide.

The United Nations Conference on Environment and Development or "Earth Summit" in Rio de Janeiro, Brazil was the largest-ever meeting of world leaders and a historic set of agreements was signed, including the CBD, which was the first global agreement on the integral conservation and sustainable use of biological diversity. The CBD was signed at the Earth Summit by over 150 Governments, and since then, more than 175 countries have ratified the agreement. The significance of the CBD was not only the commitment of almost all the countries of the world to promote biodiversity conservation, but for the first time in history, a funding mechanism was established that could at least launch a serious initiative for establishing biodiversity conservation programmes in countries that for compliance, would have to rely heavily on external financing and technical assistance. Launched in 1991 as an experimental financing facility, the Global Environment Facility (GEF) was restructured after the Earth Summit for the financing of a number of environmental issues among which biodiversity conservation is of primary importance. Two years after the Earth Summit, the IUCN published a "*Guide to the Convention on Biological Diversity*" which provides a solid basis on how to work with the CBD.

WHAT IS BIODIVERSITY?

DEFINITION

Biodiversity is a modern term which simply means "the variety of life on earth". This variety can be measured on several different levels. The most widely accepted definition for biodiversity is contained within the Convention on Biological Diversity. *'Biological diversity' means the variability among living organisms from all sources including, among other things, terrestrial, marine and other aquatic ecosystems and the ecological complexes of which they are part; this includes diversity within species, between species and of ecosystems.*

WORLD BIODIVERSITY

Estimates of global species diversity vary enormously because it is so difficult to guess how many species there may be in less well explored habitats such as untouched rain forest. Rain forest areas which have been sampled have shown such amazing biodiversity that the mind boggles over how many species there might remain to be discovered in unexplored rain forest areas and microhabitats. According to some biologists, the true numbers of living species on earth are much higher and any range from 3.6 million to more than 100 million. Ten million is probably nearer the mark. Only 1.4 million species have been named. New species are continually being discovered every year. The number of species present in little-known ecosystems such as the soil beneath our feet and the deep sea can only be guessed at. It has been estimated that the deep sea floor may contain as many as a million undescribed new species. It is absolutely difficult to mention the exact number of species available.

Earth is endowed with enormous biodiversity. About 3,00,000 green plants, 8,00,000 insects, 23,000 fishes, 9,000 birds, 6,500 reptiles, 4,100 mammals and few thousand microbes have been estimated to exist in the world. Biologists estimate that more than half of the species occur in the tropical rain forest on earth. Many world records of biodiversity have been reported from these natural greenhouse e.g., 425 kinds of trees in one hectare of Brazil's Atlantic forest and 1,300 butterfly species form a corner of Peru's Manu National Park. These figures are more than 10 times the number of species found in the comparable sites of World. On other side, McMurdo Dry valley of Antarctica has the poorest and the coldest soils in the world but still provides shelters to spares communities of bacteria, fungi and microscopic invertebrates (Chakraborty 2004).

INDIAN BIODIVERSITY

India is one of the 12 mega-biodiversity centers of the world. India is divided into 10 bio-geographic zones and 25 biotic provinces representing all the major ecosystems. India is recognized as a country rich in biodiversity because of its tropical location, varied physical features, altitude and climate. Various taxa showing the number of species reveals that India has nearly 30 per cent of the world's endemic flora and 62 per cent of the known amphibian species in the world. According to Zoological Survey of India, nearly 53,430 insects, 5,050 molluscs, 2,546 fishes, 204 amphibians, 456 reptiles, 1,228 birds and 372 mammals are reported from India.

The Indian flora is mainly concentrated in four hotspots of floristic diversity viz. Himalayas, Western Ghats (and Sri Lanka), NE India and Andaman Islands (Indo-Burma) and Nicobar Islands (Sundaland), which are identified amongst the thirty four global diversity hotspots. The recent group wise list of species in India and world recorded by Botanical Survey of India are presented in Table 7.1.

BIODIVERSITY CONSERVATION EFFORTS

At International level, several organizations took initiative to conserve global biodiversity since early 1970s. United Nations Environmental Programme (UNEP) was formulated by 113 member countries under the hood of United Nations on 5 June 1972 in Stockholm (Sweden). Through this work plan, different countries enacted laws on environment protection from time to time and these laws paved the way for biodiversity conservation as well. A global strategy for dealing with all aspects of biodiversity was prepared by the World Resources Institute (WRI), UNEP, FAO and IUCN. The detailed strategy was put forth at the United Nations Conference on Environment and Development 1992. The World Summit on Sustainable Development held in Johannesburg (South Africa) in 2002 formulated the guidelines for biodiversity conservation under a long-term development programme all over the globe.

Table 7.1: Group-wise Record of Species (Source: BSI 2010)

Group	No. of Species in India	No. of Species in World	Percentage
Virus/Bacteria	903	8,050	11.2
Algae	7,182	40,000	17.9
Fungi	14,588	72,000	20.2
Lichens	2,268	13,500	16.8
Bryophytes	2,451	16,600	14.7
Pteridophytes	1,236	10,000	12.3
Gymnosperms	69	650	10.6
Angiosperms	17,643	2,50,000	7.0
Total	46,340	4,10,800	11.2

TYPES OF BIODIVERSITY

Biological diversity is usually considered at three different levels: genetic diversity, species diversity and ecosystem diversity.

GENETIC DIVERSITY

Genetic diversity refers to the variation of genes within species or variation between individuals of the same species. This includes genetic variation between individuals in a single population, as well as variations between different populations of the same species. Genetic diversity can be measured using a variety of DNA-based and other techniques. New genetic variation is produced in populations of organisms that can reproduce sexually by recombination and in individuals by gene and chromosome mutations. The large differences in the amount and distribution of genetic variation can be attributed in part to the enormous variety and complexity of habitats, and the different ways organisms obtain their living. One estimate is that there are 10,000,000,000 different genes distributed across the world's biota, though they do not all make an identical contribution to overall genetic diversity. In particular, those genes which control fundamental biochemical processes are strongly conserved across different species groups (or taxa) and generally show little variation. Other more specialized genes display a greater degree of variation.

SPECIES DIVERSITY

Species diversity refers to the variety of species in a given region or area. This can either be determined by counting the number of different species present, or by determining taxonomic diversity. Taxonomic diversity is more precise and considers the relationship of species to each other. It can be measured by counting the number of different taxa (the main categories of classification) present. Aspects of species diversity can be measured in a number of ways. Most of these ways can be classified into three groups of measurement: species richness, species abundance and taxonomic or phylogenetic diversity.

SPECIES RICHNESS: Measures of species richness count the number of species in a defined area.

SPECIES ABUNDANCE: Measures of species abundance sample the relative numbers among species. A typical sample may contain several very common species, a few less common species and numerous rare species. Measures of species diversity that simplify information on species richness and relative abundance into a single index are in extensive use.

TAXONOMIC OR PHYLOGENETIC DIVERSITY: Another approach is to measure taxonomic or phylogenetic diversity, which considers the genetic relationships between different groups of species. These measures are based on analysis which results in a hierarchical classification usually represented by a tree that depicts the branching pattern which is thought to best represent the phylogenetic evolution of the taxa concerned.

On a broad scale, species diversity is not evenly distributed across the globe. The single most obvious pattern in the global distribution of species is that overall species richness is concentrated in equatorial regions and tends to decrease as one move from equatorial to polar regions. In general, there are more species per unit area in the tropics than in temperate regions and far more species in temperate regions than there are in polar regions. In addition, diversity in land ecosystems generally decreases with increasing altitude. Other factors which are generally believed to influence diversity on land are rainfall patterns and nutrient levels.

ECOSYSTEM DIVERSITY

Communities of plants and animals, together with the physical characteristics of their environment (e.g. geology, soil and climate) interlink together as an ecological system, or ecosystem. It is harder to define ecosystem diversity than species or genetic diversity because the boundaries of communities (associations of species) and ecosystems are more fluid. Ecosystem diversity is more difficult to measure because there are rarely clear boundaries between different ecosystems and they grade into one another. However, if consistent criteria are chosen to define the limits of an ecosystem, then their number and distribution can also be measured.

Ecosystem diversity encompasses the broad differences between ecosystem types, and the diversity of habitats and ecological processes occurring within each ecosystem type. Since the ecosystem concept is dynamic and thus variable, it can be applied at different scales, though for management purposes it is generally used to group broadly similar assemblages of communities, such as temperate rainforests, lakes or coral reefs etc.

THE VALUE OF BIOLOGICAL DIVERSITY

Generally, benefits arising from the components of biological diversity can be considered in three major groups: ecosystem services, biological resources and social benefits. The details of each components of biological diversity with examples are given below.

ECOSYSTEM SERVICES

PROTECTION OF WATER RESOURCES: The natural vegetation near catchments areas helps to maintain hydrological cycles, regulating and stabilizing water run-off, and act as a buffer against extreme events of flood and drought condition. However, removal of vegetation on these area results in siltation of catchment waterways, loss of water yield and quality and degradation of aquatic habitat.

SOILS FORMATION AND PROTECTION: Biological diversity helps in the formation and maintenance of soil structure and the retention of moisture and nutrient levels including productive capacity of the soil, prevention of landslides, safeguard coastlines and riverbanks and prevent the degradation of coral reefs and riverine and coastal fisheries by siltation. But the loss of biological diversity also contributes to losses such as salinization of soils, leaching of nutrients, accelerated erosion of topsoil and reducing the land's productivity.

NUTRIENT STORAGE AND CYCLING: Biological diversity provides the vital function of recycling nutrients. Plants are able to take up nutrients from the soil and these nutrients become basis of food chains, used by a wide range of other life forms.

POLLUTION BREAKDOWN AND ABSORPTION: Ecosystems and ecological processes play an important role in the breakdown and absorption of many pollutants created by humans and their activities. These include wastes such as sewage, garbage and oil spills. Components of ecosystems from bacteria to higher life forms are involved in these breakdown and assimilative processes. Excessive quantities of any pollutant, however, can be detrimental to the integrity of ecosystems and their biota. Some ecosystems, especially wetlands, have qualities that are particularly well suited to breaking down and absorbing pollutants. Natural and artificial wetlands are being used to filter effluents to remove nutrients, heavy metals and suspended solids, reduce the biochemical oxygen demand and destroy potentially harmful microorganisms.

CONTRIBUTION TO CLIMATE STABILITY: Vegetation influences climate at the macro and micro levels. Growing evidence suggests that undisturbed forest helps to maintain the rainfall in its immediate vicinity by recycling water vapour at a steady rate back into the atmosphere and through the canopy's effect in promoting atmospheric turbulence. At smaller scales, vegetation has a moderating influence on local climates and may create quite specific micro-climates. Some organisms are dependent on such micro-climates for their existence.

MAINTENANCE OF ECOSYSTEMS: Ecosystem relationships resemble a web of connections from one living thing to many other living and non-living things. They not only allow survival, but also maintain a balance between living things and the resources (such as food and shelter) they need to survive. Vegetation is integral to the maintenance of water and humidity levels and is essential for the maintenance of the oxygen/carbon dioxide balance of the atmosphere. Due to the complex nature of ecosystem relationships, the removal/ disturbance of the ecosystem could affect the functioning of other components of the ecosystem. Maintaining natural habitats helps ecosystem functions over a wider area.

BIOLOGICAL RESOURCES

FOOD: Biodiversity ensures food security. Animals and humans depend on plans for food. People have 7,000 plant species for food. Today only 20 plant species provide 90 per cent of world's food and just three plant species namely rice, wheat and maize supply more than 50 per cent food. Therefore only the conservation of biodiversity can ensure global food security for rapid increase human population as well as animal population. One of the important benefits of conservation of biodiversity is the wild plant gene pool which is available to augment the narrow genetic base of these established food crops, providing disease resistance, improved productivity and different environmental tolerances.

MEDICINAL: People have long used biological resources for medicinal purposes. Natural flora and fauna are the major sources of pharmaceuticals for curing various ailments. The Indian system of Ayurvedic medication is entirely dependent on plant biodiversity. Diverse plants have diverse medicinal value and provide useful drugs. For example, bark of the Cinchona plant contains an alkaloid called quinine which is prescribed by physicians for curing malaria. Some other plant species Rauwolfia, Aconite, Poppy are commonly used in pharmaceutical industries for extracting useful drugs.

WOOD PRODUCTS: Wood is a basic commodity used worldwide and is still largely harvested from the wild. It is a primary source of fuel, construction and forms the basis for paper production and other industrial uses.

FUTURE RESOURCES: There is also potential for further development of biotic resources for natural pesticides and other useful products such as fats and oils. The conservation of diversity is also essential for finding effective biological control organisms and for breeding disease resistant species. Genetic engineering of microorganisms promises further advances in the production of new compounds and processes.

SOCIAL BENEFITS

RESEARCH, EDUCATION AND MONITORING: There is still much to learn on how to get better use from biological resources, how to maintain the genetic base of harvested biological resources and how to rehabilitate degraded ecosystems. Natural areas provide excellent living laboratories for such studies, for comparison with other areas under different systems of use and for valuable research into ecology and evolution. Unaltered habitats are often essential for certain research approaches, providing controls against which the changes brought about by different management regimes may be measured and assessed.

RECREATION: Biological diversity is an intrinsic part of many areas for tourism and recreational purposes. People value such areas for a variety of recreational pursuits: film, photographs or literature based on or using wildlife, natural habitats and natural features; bird watching; and ecological field study and other scientific pursuits.

CULTURAL VALUES: The cultural value of biological diversity conservation for present and future generations is an important reason for conserving it today. Human cultures co-evolve with their environment and the conservation of biological diversity can be important for cultural identity. The natural environment provides for many of the inspirational, aesthetic, spiritual and educational needs of people now and in the future. The aesthetic values of our natural ecosystems and landscapes contribute to the emotional and spiritual well-being of a highly urbanized population. The conservation of biological diversity also has ethical benefits. The presence of a wide range of living organisms reminds people that they are but one interdependent part of Earth.

Aboriginal relationships to the land and sea, and its animals and plants are complex. To the people, the land and sea has a deep spiritual, economic, social, protective and recreational significance. By hunting and gathering, tribal aboriginal people are not only supplementing their diet with food very high in nutritional value; they are also confirming their self-sufficiency and, more importantly, educating their children in relationships to the land and to other aspects of their culture. Biological diversity conservation can contribute to the conservation of aboriginal cultural identity.

THREATS TO BIODIVERSITY

There are several factors for threat to biodiversity and biodiversity loss. Today we are losing two to five species per hour from tropical forests alone. This amounts to a loss of 16 million populations per year or 1800 populations per hour. If the trend of population loss remains continues, about 20 per cent of the total 2,50,000 higher plant species will be lost in the next few decades and another 25 per cent by the end of the 21st century (Uniyal et al 2010).

HABITAT LOSS AND DEGRADATION

Habitat loss and degradation is considered by conservation biologists to be the primary cause of biodiversity loss. Clearance of vegetation for food production, human settlement and timber leads to degradation of habitat. The destruction of resources leads fragmentation of the area into small segments. Habitat loss, alteration and fragmentation directly affect the species that rely on the habitat that is being changed.

INTRODUCTIONS OF INVASIVE ALIEN SPECIES

Invasive species considered second largest threat to biodiversity worldwide. Whether introduced on purpose or accidentally, non-native species can cause severe problems in the ecosystems they invade, from affecting individuals to causing huge changes in ecosystem functioning and the extinction of many species. Virtually all ecosystems worldwide have suffered invasion. In the USA, it is estimated that an economic loss of $79 billion was incurred because of 79 harmful alien species during 1906-91 (Stein and Flack 1996). In India, ICFRE has identified nearly 75 forest invasive species, out of which 61 are of plants, 12 are fungi and 14 are insects, which are a threat to the natural forest cover of the country. In India, some of the invasive species are *Parthenium hysterophours*, *Eupatorium adenophorum*, *Eupatorium odoratum*, *Mikania microantha*, *Ageratum conyzoides* and *Galinsoga poarviflora* which have already caused a huge loss to the local biodiversity (Raghubanshi et al 2005). In Indian flora, about 40 per cent of the species are alien of which 25 per cent are invasive.

OVER-EXPLOITATION OF NATURAL RESOURCES

Exploitation of biodiversity occurs for food, construction, industrial products, pet trade, fashion, traditional medicines etc. Selective removals of an individual species can imbalance ecosystems and all other organisms within them.

POLLUTION AND DISEASES

Pollution is currently poisoning all forms of life, both on land and water, and contributing to climate change. Any chemical in the wrong place or at the wrong concentration can be considered a pollutant. Transport, industry, construction, extraction, power generation etc. contribute pollutants to the air, land and water. These chemicals can directly affect biodiversity or lead to chemical imbalances in the environment that ultimately kill individuals, species and habitats. There are thousands of pollutants circulating through the earth's ecosystems, and many of these materials have significant, large-scale impacts on forests and aquatic ecosystems. Acid precipitation and pollution can also disrupt ecological processes.

HUMAN-INDUCED CLIMATE CHANGE

Human induced climate changes are due to emissions of greenhouse gases when fossil fuels are burnt. This can lead to a change in the abundance and distribution of individual species around the globe and will affect the crops we grow, cause a rise in sea levels and problems to many coastal ecosystems. In addition, the climate is becoming more unpredictable and extreme devastating events occur.

Habitat destruction is responsible for the extinction of thousands of flora and fauna each year. The forest area harboring about 50 per cent of the global species will be reduced to a tenth or less by 2020 AD (Raven 1987). The forest area has already been reduced to 55 per cent of its original cover and is being further reduced at a rate of more than 1,00,000 sq km per year (Wilson 1989). According to IUCN, if the current trend of species loss remains continue and no measure were taken, 25 per cent of the World's species may be lost by 2050 AD (Uniyal et al 2010). IUCN has reported that 70 per cent of plant species are under global threat along with a number of animal species (Table 7.2).

Table 7.2: Number of Global Species under Threat (Source: IUCN)

Taxon	Number of Species
Plants	12,151
Freshwater fish	3,120
Inbvertebrates	7,615
Ambhibians	6,285
Reptiles	1,678
Mammals	5,490
Birds	n/a

LOSS OF BIODIVERSITY

Extinction is a fact of life. Species have been evolving and dying out ever since the origin of life. One has to look at the fossil record to prove the fact. It has been estimated that surviving species constitute about 1 per cent of the species that have ever lived. However, species are now becoming extinct at an alarming rate, almost entirely as a direct result of human activities. Previous mass extinctions evident in the geological record are thought to have been brought about mainly by massive climatic or environmental shifts. Mass extinctions as a direct consequence of the activities of a single species are unprecedented in geological history.

The loss of species in tropical ecosystems such as the rain forests, is extremely well-publicized and of great concern. However, equally worrying is the loss of habitat and species closer to home. Predictions and estimates of future species losses calculate that a quarter of all species on earth are likely to be extinct, or on the way to extinction within 30 years. Another predicts that within 100 years, three quarters of all species will either be extinct, or in populations so small that they can be described as "the living dead". It must be emphasized that these are only predictions and based on modeling. The only possible conclusion is that unless human populations are substantially reduced, it is inevitable that biodiversity will suffer further major losses.

Loss of an individual species can have various different effects on the remaining species in an ecosystem. These effects depend upon the how important the species is in the ecosystem. Some species can be removed without apparent effect, while removal of others may have enormous effects on the remaining species. Species such as these are termed "keystone" species. Some species are more vulnerable to extinction than others and these include:

SPECIES AT THE TOP OF FOOD CHAINS

Large carnivores usually require fairly extensive territories in order to provide them with sufficient prey. As human populations increasingly encroach on wild areas and as habitats shrink in extent, the number of carnivores which can be accommodated in the area also decreases. These animals may also pose a threat to people, as populations expand into wilder areas inhabited by large carnivores. Protective measures, including elimination of offending animals in the area, further reduces numbers.

ENDEMIC LOCAL SPECIES

Endemic species found only in one geographical area with a very limited distribution. These are very vulnerable to local habitat disturbance or human development.

SPECIES WITH CHRONICALLY SMALL POPULATIONS

If populations become too small, then simply finding a mate, or interbreeding, can become serious problems.

MIGRATORY SPECIES

Species which need suitable habitats to feed and rest in widely spaced locations (which are often traditional and 'wired' into behaviour patterns) are very vulnerable to loss of these 'way stations'.

SPECIES WITH EXCEPTIONALLY COMPLEX LIFE CYCLES

If completion of a particular lifecycle requires several different elements to be in place at very specific times, then the species is vulnerable if there is disruption of any single element in the cycle.

SPECIALIST SPECIES

Species with very narrow requirements such as a single specific food source, e.g. a particular plant species.

BIODIVERSITY HOTSPOTS

The concept of biodiversity hotspots was drafted by British ecologist Norman Myers in 1988 as a means to address the dilemma of identifying the areas most important for preserving species. Myers recognized that despite their relatively small sizes, certain ecosystems account for a high percentage of global biodiversity. Many of these same areas face tremendous pressure from logging, agriculture, hunting and climate change. Myers reasoned that a prudent conservation strategy would be to target money and research at those regions where these threats are greatest to the greatest number of species. In the mid-1990s, together with partners at Conservation International, Myers ironed out a formula for hotspot designation: The region must support at least 1,500 plant species found nowhere else in the world and it must have lost at least 70 per cent of its original habitat.

Hotspots are generally defined as fragile area rich in biodiversity with high degree of endemism and higher incidence of rare and threatened species. Preliminary efforts were limited only to vascular plants present in 18 hotspots of the world (Myers 1988 and 1990). The number of hotspots has been increased to 25 (Myers et al 2000), out of these India has two hotspots i.e., Eastern Himalaya (Indo-Burma) and the Western Ghats. They are estimated to have about 7,000 and 2,180 endemic species of plants respectively. The selection of any area as hotspot should have the following:

- An area must contain at least 0.5 per cent or 1,500 of the world's 300,000 plant species as endemics
- A second determinant of hotspots status applied only after an area has met the plants criterion the degree of threat through habitat loss.
- A hotspot should have lost 70 per cent or more of its primary vegetation and form of habitat contains the most endemic species.

Around the world, 25 areas qualify as biodiversity hotspots under this definition, with nine others possible candidates. The biodiversity hotspots hold especially high numbers of endemic species, yet their combined area of remaining habitat covers only 2.3 per cent of the Earth's land surface. Each hotspot faces extreme threats and has already lost at least 70 per cent of its original natural vegetation. Over 50 per cent of the world's plant species and 42 per cent of all terrestrial vertebrate species are endemic to the 34 biodiversity hotspots.

Eleven hotspots have already lost at least 90 per cent vegetation and three have lost 95 per cent vegetation. Among the hotspots which are under greatest threat of extinction of species and habitats are termed as "hottest hotspots". Eight hotspots have been identified as "hottest hotspots". The failure to protect the hotspots would result in the loss of nearly 50 per cent of Earth's plants and terrestrial vertebrates.

Table 7.3: The Hotspots of the World (Myers et al 2000)

Hotspots of the Worlds	Original extent of primary vegetation (km^2)	Remaining primary vegetation (km^2) (% of original extent)	Area protected (km^2) (% of hotspot)	Plant species	Endemic plants (% of global plants, 300,000)	Verte-brate species	Endemic vertebrates (% of global vertebrates, 27,298)
Tropical Andes	1,258,000	314,500 (25.0%)	79,687 (25.3%)	45,000	20,000 (6.7%)	3,389	1,567 (5.7%)
Mesoamerica	1,155,000	231,000 (20.0%)	138,437 (59.9%)	24,000	5,000 (1.7%)	2,859	1,159 (4.2%)
**Caribbean	263,500	29,840 (11.3%)	29,840 (100.0%)	12,000	7,000 (2.3%)	1,518	779 (2.9%)
**Brazil's Atlantic Forest	1,227,600	91,930 (7.5%)	33,084 (35.9%)	20,000	8,000 (2.7%)	1,361	567 (2.1%)
Choc/Darien/ Western Ecuador	260,600	63,000 (24.2%)	16,471 (26.1%)	9,000	2,250 (0.8%)	1,625	418 (1.5%)
Brazil's Cerrado	1,783,200	356,630 (20.0%)	22,000 (6.2%)	10,000	4,400 (1.5%)	1,268	117 (0.4%)
Central Chile	300,000	90,000 (30.0%)	9,167 (10.2%)	3,429	1,605 (0.5%)	335	61 (0.2%)
California Floristic Province	324,000	80,000 (24.7%)	31,443 (39.3%)	4,426	2,125 (0.7%)	584	71 (0.3%)
**Madagascar*	594,150	59,038 (9.9%)	11,548 (19.6%)	12,000	9,704 (3.2%)	987	771 (2.8%)
**Eastern Arc and Coastal Forests of Tanzania/Kenya	30,000	2,000 (6.7%)	2,000 (100.0%)	4,000	1,500 (0.5%)	1,019	121 (0.4%)
Western African Forests	1,265,000	126,500 (10.0%)	20,324 (16.1%)	9,000	2,250 (0.8%)	1,320	270 (1.0%)
Cape Floristic Province	74,000	18,000 (24.3%)	14,060 (78.1%)	8,200	5,682 (1.9%)	562	53 (0.2%)
Succulent Karoo	112,000	30,000 (26.8%)	2,352 (7.8%)	4,849	1,940 (0.6%)	472	45 (0.2%)
Mediterranean Basin	2,362,000	110,000 (4.7%)	42,123 (38.3%)	25,000	13,000 (4.3%)	770	235 (0.9%)
Caucasus	500,000	50,000 (10.0%)	14,050 (28.1%)	6,300	1,600 (0.5%)	632	59 (0.2%)
**Sundaland	1,600,000	125,000 (7.8%)	90,000 (72.0%)	25,000	15,000 (5.0%)	1,800	701 (2.6%)

Wallacea	347,000	52,020 (15.0%)	20,415 (39.2%)	10,000	1,500 (0.5%)	1,142	529 (1.9%)
**Philippines	300,800	9,023 (3.0%)	3,910 (43.3%)	7,620	5,832 (1.9%)	1,093	518 (1.9%)
**Indo-Burma	2,060,000	100,000 (4.9%)	100,000 (100.0%)	13,500	7,000 (2.3%)	2,185	528 (1.9%)
South-Central China	800,000	64,000 (8.0%)	16,562 (25.9%)	12,000	3,500 (1.2%)	1,141	178 (0.7%)
**Western Ghats/Sri Lanka	182,500	12,450 (6.8%)	12,450 (100.0%)	4,780	2,180 (0.7%)	1,073	355 (1.3%)
SW Australia	309,850	33,336 (10.8%)	33,336 (100.0%)	5,469	4,331 (1.4%)	456	100 (0.4%)
New Caledonia	18,600	5,200 (28.0%)	526.7 (10.1%)	3,332	2,551 (0.9%)	190	84 (0.3%)
New Zealand	270,500	59,400 (22.0%)	52,068 (87.7%)	2,300	1,865 (0.6%)	217	136 (0.5%)
Polynesia/ Micronesia	46,000	10,024 (21.8%)	4,913 (49.0%)	6,557	3,334 (1.1%)	342	223 (0.8%)
Totals	17,444,300	2,122,891 (12.2%)	800,767 (37.7%)	†	133,149 (44%)	†	9,645 (35%)

* Madagascar includes the nearby islands of Mauritius, Reunion, Seychelles and Comores.

† These totals cannot be summed owing to overlapping between hotspots. **Hottest hotspots of the world

POPULATION DIVERSITY

There are four key components of population diversity: richness, the size of each population, spatial distribution and differentiation (Gary et al 2003).

POPULATION RICHNESS

Population richness is the number of populations of a species in a given area, which depends on the criteria used to delineate population boundaries. When the focus is on ecosystem services, it is proposed using as a criterion the spatial disjunctions in the services provided by conspecific individuals occupying a heterogeneous environment. If the required data are unavailable, standard genetic or demographic approaches would be the next best option.

POPULATION SIZE

Data about the number of individuals per population provide an indication of the frequency distribution of population sizes. Absolute numbers would be the most useful, but, owing to the difficulty of obtaining such information, orders of magnitude or some other representative measure might be more appropriate. It is important to document the distribution of population sizes to determine whether a species is characterized by, for example, a single large population and many small populations, or several similarly sized populations. This distribution has implications, not only for species conservation, but also for the contribution that each population makes to the functioning of ecosystems and raises the important issue of how variability in the number of individuals in a given population affects functionality.

POPULATION DISTRIBUTION

The third component of population diversity is the spatial distribution of the populations under study. The important measures here are the extent of the populations relative to their maximum possible extent in a defined area and population dispersion. Focusing on maximum possible extent facilitates comparisons between species with large geographical ranges and those with relatively restricted ranges. Geographical range is a measure of the maximum area in which a species can provide a given service. Population dispersion is a measure of the spatial aggregation of populations and how this can affect the delivery of services over a given area. For example, the services provided by fish populations might be extremely localized owing to natural (e.g. shorelines) or anthropogenic boundaries (e.g. regional fisheries) and the adaptation of populations to a narrow range of environmental variability.

GENETIC DIFFERENTIATION OF POPULATIONS

The final component of population diversity is genetic differentiation within and among populations. From both conservation and ecosystem service perspectives, more genetic variation within populations might confer greater resilience in the face of environmental change (such as the possible stabilizing influence of genetic diversity in key tree species enabling forests to track climate change. Genetic differentiation among populations can be associated with different types or levels of a given ecosystem service. Perhaps the most important ecosystem service in which study of genetic differentiation of populations is crucial in maintaining or augmenting crop yields, which depends on genetic material that often is only available from populations of wild relatives of crops.

MEASURMENT OF BIODIVERSITY

In spite of many tools and data sources, biodiversity remains difficult to quantify precisely. But precise answers are seldom needed to devise an effective understanding of where biodiversity is, how it is changing over space and time, the drivers responsible for such change, the consequences of such change for ecosystem services and human well-being and the response options available. Ideally, to assess the conditions and trends of biodiversity either globally or sub-globally, it is necessary to measure the abundance of all organisms over space and time, using taxonomy (such as the number of species), functional traits and the interactions among species that affect their dynamics and function (predation, parasitism, competition and facilitation such as pollination, for instance, and how strongly such interactions affect ecosystems).

Even more important would be to estimate turnover of biodiversity, not just point estimates in space or time. Currently, it is not possible to do this with much accuracy because the data are lacking. Even for the taxonomic component of biodiversity, where information is the best, considerable uncertainty remains about the true extent and changes in taxonomic diversity. Ecological indicators are scientific constructs that use quantitative data to measure aspects of biodiversity, ecosystem condition, services, or drivers of change, but no single ecological indicator captures all the dimensions of biodiversity.

SPECIES DIVERSITY

Diversity, which is synonymous with heterogeneity, comprises species richness and evenness. Indices that combine both richness and evenness (heterogeneity) into a single value are diversity indices. The total number of species in a community is referred to as species richness while species evenness or equitability explains as to how species abundance is distributed among species. Diversity has emerged as the most widely used criterion to assess the conservation potential and ecological value of a site.

DIVERSITY INDEX: The Heterogeneity or Diversity will be measured by Simpson's Diversity Index that is commonly referred to as the dominance measure and Shannon-Wiener Diversity Index, which is also referred to as the information statistic index.

$$\text{Simpsons Diversity Index (D)} = \sum_{i=1}^{s} P_i^2$$

$$\text{Shannon - Wieners Diversity Index (H)} = -\sum_{i=1}^{s} (P_i)(\log Pi)$$

where,

s = number of species

P_i = proportion of species i in the community

SPECIES RICHNESS: As a measure of species richness, the number of species in a community or sample will be calculated using Jackknife Index (J) and first Chao Index (C) which is the estimator of species richness.

$$\text{Jackknife Index (J)} = S + k\left[\frac{P-1}{P}\right]$$

$$\text{First Chao Index (C)} = S + \frac{a^2}{2b}$$

where,

S = Number of species found when all sample plots are pooled

P = Number of plots sampled

k = Number of species that occur in only one sample plot (unique)

a = Number of species that are represented by a single individual (singletons)

b = number of species that are represented by two individuals (doubletons)

SPECIES EVENNESS OR EQUITABILITY: Evenness measures attempts to quantify unequal representation against a hypothetical community in which all species are equally common. The Evenness or Equitability (E) will be quantified by expressing Simpson's index or Shannon index.

$$\text{Simpsons Measure of Evenness (E)} = \left(\frac{1}{D}\right) \times \left(\frac{1}{S}\right)$$

where,

D = Simpson's Diversity Index

S = Number of species found when all sample plots are pooled

$$\text{Shannon - Wieners Equitability or Evenness (J)} = \frac{H}{\log(S)}$$

where,

H = Shannon-Wiener's Diversity Index

S = Number of species found when all sample plots are pooled

BIODIVERSITY CONSERVATION STRATEGY

Many traditional cultural practices have served to protect important components of biodiversity. Similarly, a number of community-based resource management programmes have slowed the loss of biodiversity while contributing benefits to the people by placing community-level benefits as central objectives for sustainable management. Substantial investments have also been made by NGOs, Governments, and the private sector to reduce negative impacts on biodiversity, protect threatened biodiversity and use biodiversity sustainably.

Protected areas, including those managed primarily for biodiversity conservation and those managed for a wide range of sustainable uses, are extremely important, especially in environments where biodiversity loss is sensitive to changes in key drivers. PA systems are most successful if they are designed and managed in the context of an ecosystem approach, with due regard to the importance of corridors and interconnectivity of PAs and to external threats such as pollution, climate change and invasive species. At the global and regional scales, however, the current system of protected areas is not sufficient for conservation of all components of biodiversity.

IN SITU CONSERVATION

In situ conservation maintains not only the genetic diversity of species, but also the evolutionary adaptations that enable them to adapt continually to shifting environmental conditions, such as changes in pest populations or climate. *In situ* conservation also ensures that along with target species, a host of other interlinked species are also preserved as a by-product. It may often be the only conservation option, for example for species with recalcitrant seeds.

In situ conservation measures involve designating specific areas as protected sites. Protection may be offered at various levels, from complete protection and restriction of access, through various levels of permitted human use. In practice, complete protection is rarely necessary or advisable in a terrestrial context. Human beings have been a major part of the landscape for many thousands of years. Over the course of that time, human cultures have emerged and adapted to the local environment, discovering, using and altering biotic resources. Many areas that now appear 'natural' bear the hallmarks of millennia of human influence. Other species have evolved along with that influence and in many cases require the disturbance provided by humans to provide the necessary conditions for their survival. In other words, it is rarely advisable to relegate the countryside to the status of a museum piece.

Through this method biological species are conserved in their own habitats. Utmost care is taken to maintain the ecosystem. Human encroachment into the forest is restricted. Biosphere reserves, National parks and Sanctuaries are the examples of *in situ* conservation of biodiversity. In 1982, with help of United Nations Environment Programme (UNEP) and the World Wide Fund for Nature (WWF) has suggested 10 different protected areas for United Nations. In India Wild Life (Protection) Act came into existence in 1972. It includes the constitution and management of national parks, sanctuaries, game reserves and closed areas. This Act not only specifies the procedures of setting up of national parks but also specified the management parameters. The work and power of the officials have also been fixed. In India, there are about 100 National Parks and 515 Wildlife Sanctuaries, 43 Conservation Reserves and four Community Reserves (MoEF 2011).

BIOSPHERE RESERVES

The idea of 'Biosphere Reserves' was initiated by UNESCO in 1973-74 under its Man and Biosphere (MAB) Programme. The MAB, launched in 1970 by UNESCO, is a broad based ecological programme aimed to develop within the natural and social sciences a basis for the rational use and conservation of the resources of the biosphere and for the improvement of the relationship between man and the environment; to predict the consequences of today's actions on tomorrows world and thereby to increase man's ability to manage efficiently the natural resources of the biosphere.

Biosphere Reserves are thus special environments for both people and the nature and are living examples of how human beings and nature can co-exist while respecting each other's needs. Biosphere reserves are areas of terrestrial and coastal ecosystems promoting solutions to reconcile the conservation of biodiversity with its sustainable use. They are internationally recognized, nominated by national governments and remain under sovereign jurisdiction of the states where they are located. Biosphere reserves serve in some ways as 'living laboratories' for testing out and demonstrating integrated management of land, water and biodiversity. The list of Biosphere Reserves in India is given in Table 7.4.

STRUCTURE AND DESIGN OF BIOSPHERE RESERVES

Biosphere Reserves are demarcated into three inter-related zones: natural or core zone, manipulation or buffer zone and transition zone outside the buffer zone.

THE CORE ZONE: The core zone is kept absolutely undisturbed. It must contain suitable habitat for numerous plant and animal species, including higher order predators and may contain centres of endemism. Core areas often conserve the wild relatives of economic species and also represent important genetic reservoirs. The core zones also contain places of exceptional scientific interest. A core zone secures legal protection and management and research activities that do not affect natural processes and wildlife are allowed. Strict nature reserves and wilderness portions of the area are designated as core areas of BR. The core zone is to be kept free from all human pressures external to the system.

THE BUFFER ZONE: In the Buffer Zone, which adjoins or surrounds core zone, uses and activities are managed in ways that protect the core zone. These uses and activities include restoration, demonstration sites for enhancing value addition to the resources, limited recreation, tourism, fishing and grazing, which are permitted to reduce its effect on core zone. Research and educational activities are to be encouraged. Human activities, if natural within BR, are likely to be permitted to continue if these do not adversely affect the ecological diversity.

THE TRANSITION ZONE: The Transition Zone is the outermost part of a Biosphere Reserve. This is usually not delimited one and is a zone of cooperation where conservation, knowledge and management skills are applied and uses are managed in harmony with the purpose of the Biosphere Reserve. This includes settlements, crop lands, managed forests and area for intensive recreation and other economic uses characteristic of the region.

Table 7.4: List of Biosphere Reserves of India (MoEF 2011)

S No	Name of the BR & Total Geographical Area (km^2)	Date of Notification	Location in the State/ UT
1	Nilgiri (5,520)	01.08.1986	Part of Wynad, Nagarhole, Bandipur and Madumalai, Nilambur, Silent Valley and Siruvani hills (Tamil Nadu, Kerala and Karnataka)
2	Nanda Devi (6,497.03)	18.01.1988	Part of Chamoli, Pithoragarh & Almora Districts and Valley of Flowers (Uttarakhand)
3	Nokrek (820)	01.09.1988	Part of Garo Hills (Meghalaya)
4	Manas (2,837)	14.03.1989	Part of Kokrajhar, Bongaigaon, Barpeta, Nalbari, Kamprup and Darang Districts (Assam)
5	Sunderbans (9,630)	29.03.1989	Part of delta of Ganges & Brahamaputra river system (West Bengal)
6	Gulf of Mannar (10,500)	18.02.1989	Indian part of Gulf of Mannar between India and Sri Lanka (Tamil Nadu)
7	Great Nicobar (885)	06.01.1989	Southern most islands of Andaman and Nicobar (A&N Islands)
8	Similipal (4,374)	21.06.1994	Part of Mayurbhanj district (Orissa)
9	Dibru-Saikhowa (765)	28.07.1997	Part of Dibrugarh and Tinsukia districts (Assam)
10	Dehang debang (5,111.5)	02.09.1998	Part of Siang and Debang valley (Arunachal Pradesh)
11	Kanchanjunga (2,619.92)	07.02.2000	Parts of North and West Sikkim (Sikkim)
12	Pachmarhi (4,926.28)	03.03.1999	Parts of Betul, Hoshangabad and Chhindwara, district (Madhya Pradesh)
13	Agasthyamalai (3,500.36)	12.11.2001	Parts of Thirunelveli and Kanya Kumari (area expanded in Tamil Nadu on 30.3.2005) and Thiruvanthapuram, Kollam and Pathanmthitta in Kerala (TamilNadu & Kerala)

14	Achanakmar-Amarkantak (3835.51)	30.03.2005	Parts of Anuppur and Dindori districts of Madhya Pradesh and Parts of Bilaspur district of Chhattisgarh State (Madhya Pradesh & Chhattisgarh)
15	Kachchh (12,454)	29.01.2008	Parts of Kachchh, Rajkot, Surendranagar and Patan Civil Districts of Gujarat State
16	Cold Desert (7,770)	28.08.2009	Pin Valley National Park and surroundings; Chandratal and Sarchu & Kibber Wildlife Sancturary in Himachal Prachal.
17	Seshachalam (4,755.997)	20.09.2010	Parts of Chittoor and Kadapa districts of Andhra Pradesh

NATIONAL PARKS

The higher level of protection is given to the National parks. No grazing is permitted within the park. No private land holding or right is allowed within a National Park. No person is allowed to destroy, exploit or remove any wild life from a National Park or deprive any animals of its habitat within such National Park except when permission granted by the Chief Wildlife Warden and the State Government is available. The permission is only granted on the ground that such destruction, exploitation or removal of wildlife from the national park is essential for the betterment of wildlife there in (Section, 35 (6) of the Wildlife Protection Act, 1972).

WILDLIFE SANCTUARY

Wildlife sanctuaries are given a lesser level of protection. Certain type of activities may be permitted in Sanctuaries not only for the better protection of wildlife but also for any other good and sufficient reason. No person is allowed to hunt and wild animals in the sanctuaries or remove there from any wild animals, whether live or dead, or any trophy, or meat derived from such animals provided that the Chief Wild Life Warden is satisfied that such activity is for any other good or sufficient reason. The permission is granted on the previous approval by state Government (Section 29 (1) of the Wildlife Protection Act, 1972).

EX SITU CONSERVATION

Under *ex situ* conservation, various biological species collected from their habitats are maintained in a different place like botanical garden, zoo and aquarium. Similarly gene banks collect and maintain different species of plants and their varieties.

ZOOS

In the past, zoos were mainly display facilities for the purpose of public enjoyment and education. As large numbers of the species traditionally on display have become rarer in the wild, many zoos have taken on the additional role of building up numbers through captive breeding programmes. Although comparatively far more invertebrates than vertebrates face extinction, most captive breeding programmes in zoos focus on vertebrates. Threats to vertebrate extinction tend to be well publicized and vertebrates therefore, serve as a focus for public interest. Many of the most threatened vertebrates are large top carnivores, which the world stands to lose in disproportionate numbers. Such species require extensive ranges to provide sufficient prey to sustain them. In many cases, whole habitats for these predators have all but disappeared. Several species are now solely represented by animals in captivity. Captive breeding programmes are in place for numerous species. Many species have been reintroduced into the wild following such programs. In many cases, the species was actually extinct in the wild at the time of reintroduction. In some cases, all remaining individuals of a species, whose numbers are too low for survival in the wild, have been captured and the species has then been reintroduced after captive breeding.

The role of zoos in conservation is limited both by space and by expense. However, zoos are now shifting their emphasis from long-term holding of species, to returning animals to the wild after only a few generations. This frees up space for the conservation of other species. Genetic management of captive populations via stud records is essential to ensure genetic diversity is preserved as far as possible. There are now a variety of international computerized stud record systems which catalogue genealogical data on individual animals in zoos around the world. Mating can therefore be arranged by computer, to ensure that genetic diversity is preserved and in-breeding minimized.

Research has led to great advances in technologies for captive breeding. This includes techniques such as artificial insemination, embryo transfer and long-term cryogenic (frozen) storage of embryos. These techniques are all valuable because they allow new genetic lines to be introduced without having to transport the adults to new locations. The success of zoos in maintaining populations of endangered species is limited. According to World Resources Institute, only 26 of 274 species of rare mammals in captivity are maintaining self-sustaining populations.

AQUARIA

The role of aquaria has largely been as display and educational facilities. However, they are assuming new importance in captive breeding programmes. Growing threats to marine species and freshwater species in particular, are leading to the development of *ex situ* breeding programmes. The World Conservation Union (IUCN) is currently developing captive breeding programmes for endangered fishes. Natural habitats will be restored as part of the programme. Marine, as well as freshwater species are also the subject of captive breeding programmes.

BOTANICAL GARDEN

There are roughly 1,500 botanic gardens world-wide, holding 35,000 plant species (more than 15% of the world's flora). There is however, a general geographic imbalance. Only 230 of the world's 1,500 gardens are in the tropics. Considering the greater species richness of the tropics, this is an imbalance that needs to be addressed. A more serious problem with *ex situ* collections involves gaps in coverage of important species, particularly those of significant value in tropical countries. One of the most serious gaps is in the area of crops of regional importance, which are not widely traded on world markets. These often have recalcitrant seeds (unsuited to long-term storage) and are poorly represented in botanic collections. Wild crop relatives are also under-represented. These are a potential source of genes conferring resistance to diseases, pests and parasites and as such are a vital gene bank for commercial crops.

SEED BANKS

Plant genetic diversity can also be preserved *ex situ* through the use of seed banks. Seeds are small but tough and have evolved to survive all manner of adverse conditions and a host of attackers. Seeds can be divided into two main types, orthodox and recalcitrant. Orthodox seeds can be dried and stored at temperatures of -20^0 C. Almost all species in a temperate flora can be stored in this way. Surprisingly, many tropical seeds are also orthodox. Recalcitrant seeds, in contrast, die when dried and frozen in this manner. The result of storing seeds under frozen conditions is to slow down the rate at which they lose their ability to germinate. Seeds of crop plants such as maize and barley could probably survive thousands of years in such conditions, but for most plants, centuries is probably the norm. This makes seed banking an attractive conservation option, particularly when all others have failed. It offers an insurance technique for other methods of conservation.

INDIAN SCENARIO OF BIODIVERSITY CONSERVATION

In India, the concept of biodiversity conservation is not new. Several ancient literatures revealed the conservation efforts which were made in India. In past, Indian Emperor Ashok planted different trees for shade and shelter. The Bishnoi communities of Rajasthan (India) have projected antelopes as their blood brother. Saint Gautama Buddha attained enlightment meditating under a peepal tree. The women of the Mandal village of Uttarakhand protested against the deforestation through Chipko Movement under leadership of Sunderlal Bahuguna, Chandi Prasad Bhatt and Gour Devi in 1973. The Silent Valley Movement in Western Ghat mountain range and Narmada Bachao Movement for rehabilitation of affected people of dam project are important examples.

BIOLOGICAL DIVERSITY ACT

Taking cognizance of the International Convention on Biodiversity (CBD) and to address the excessive pressure on biodiversity, the Government of India has enacted Biological Diversity Act in 2002 (BDA 2002). Accordingly, the national, state and local level mechanisms have been provided for implementation of the Act. At the national level, National Biodiversity Authority (NBA) has been established by Government of India in October 2003 at Chennai (Tamil Nadu) under Section (8) of the Biological Diversity Act. The State Biodiversity Boards (SBB) have been established by the State Governments and Biodiversity Management Committees (BMC) constituted by the local bodies.

The NBA is a corporate body established in accordance with the provisions of Sec.8 of the Biological Diversity Act-2002. It is an autonomous, statutory and regulatory organization which is intended to implement the provisions of Biological Diversity Act-2002.Twenty four States have so far set up State Biodiversity Boards. Eleven states viz. Andhra Pradesh, Gujarat, Rajasthan, Karnataka, Madhya Pradesh, Manipur, Sikkim, West Bengal, Tripura, Uttar Pradesh and Maharashtra have also notified the state specific rules in accordance with the provisions of the Act. So far, over 30,000 BMCs have been constituted by the local bodies in 14 states - Andhra Pradesh, Himachal Pradesh, Karnataka, Madhya Pradesh, West Bengal, Kerala, Goa, Nagaland, Manipur, Uttar Pradesh, Gujarat, Punjab, Mizoram and Tripura. So far, 419 People's Biodiversity Registers have been documented in six states (MoEF 2011).

MEDICINAL PLANTS AND ITS CONSERVATION

Medicinal and aromatic plants (MAP) are of global significance, both within the context of the community and within the realm of international trade. Between 40,000 and 50,000 plant species are known to be used in traditional and modern medicinal systems on the global scale. The majority of these MAP species are provided by collection from the wild (Srivastava et al 1996). Many MAP species and populations are threatened by over-harvesting, land conversion and habitat destruction. This reality necessitates the development of proactive approaches for MAP collection and the incorporation of integrated principles to ensure the sustainability of the plant utilization.

India has 15 Agroclimatic zones and 6,000-7,000 plant species are estimated to have medicinal usage in folk and documented systems of medicine, like Ayurveda, Siddha, Unani and Homoeopathy. About 960 species of medicinal plants are estimated to be in trade of which 178 species have annual consumption levels in excess of 100 metric tons. The National Medicinal Plants Board (NMPB) was set-up in November 2000 by the Government of India with the primary mandate of coordinating all matters relating to medicinal plants and support policies and programmes for growth of trade, export, conservation and cultivation. The NMPB is placed under the Department of Ayurveda, Yoga & Naturopathy, Unani, Siddha & Homeopathy (AYUSH) of the Ministry of Health and Family Welfare.

Medicinal plants are not only a major resource base for the traditional medicine & herbal industry but also provide livelihood and health security to a large segment of Indian population. The domestic trade of medicinal plants in India is estimated to be Rs.80 to 90 billion. The Indian medicinal plants and their products also account of exports in the range of Rs.10 billion. There is global resurgence in traditional and alternative health care systems resulting in world herbal trade which stands at US$ 120 billion and is expected to reach US$ 7 trillion by 2050 (NMPB 2012). Indian share in the world trade, at present, however, is quite low.

There are 335-recorded threatened medicinal plants in India. They range from Critically Endangered (CR), Endangered (EN) and Vulnerable (VU) to Near Threatened (NT) as per IUCN Red List Categories and Criteria. In addition to National Medicinal Plants Board, State/Union Territory Medicinal Plants Boards (SMPBs/ UTMPBs) have been set up in all the States and UTs for promotion and conservation of medicinal plants. The National Medicinal Plants Board has been implementing various central sector schemes for conservation of medicinal plants viz. setting up medicinal plants board (10th Plan), Conservation, Development and Sustainable Management of Medicinal Plants (11th Plan), national mission on medicinal plants (11th Plan). Ministry of Environment Forests has prepared a State-wise list of threatened plants and animals along with guidelines for their rehabilitation in consultation with Botanical Survey of India and Zoological Survey of India.

Ministry of Environment & Forests undertook a UNDP Country Programme in the year 2006 for 9 States viz. Karnataka, Kerala, Tamil Nadu, Andhra Pradesh, Orissa, West Bengal, Madhya Pradesh and Rajasthan to promote conservation of Medicinal Plants and Traditional Knowledge for enhancing Health & Livelihood Security and mainstreaming these into the existing policies and programs of the forestry and health sector. The ministry has recently initiated another project with the support of UNDP and Global Environment Facility (GEF) namely GEF-UNDP Project on Mainstreaming Conservation and Sustainable Use of

Medicinal Plant Biodiversity implemented in three states namely Chhattisgarh in Central India, Arunachal Pradesh in North-East India and Uttaranchal in North-west India.

Table 7.5: List of Prioritized Medicinal Plants in India (NMPB 2012)

S No	Local Name/ Trade Name	Botanical Name	Family
1	Agar/Sanchi	*Aquilaria agallocha*	Thymelaeaceae
2	Amla	*Emblica officinalis*	Euphorbiaceae
3	Anantamool	*Hemidesmus indicus*	Periplocaceae
4	Arjun	*Terminalia arjuna*	Combretaceae
5	Ashok	*Saraca asoca*	Caesalpiniaceae
6	Ashwagandha	*Withania somnifera*	Solanaceae
7	Atis	*Aconitum heterophyllum*	Ranunculaceae
8	Bael	*Aegle marmelos*	Rutaceae
9	Bankakri	*Podophyllum hexandrum*	Podophyllaceae
10	Batsanabh	*Aconitum ferox*	Ranunculaceae
11	Bhomora	*Terminalia belerica*	Combretaceae
12	Bhumi Amlakhi	*Phyllanthus amarus*	Euphorbiaceae
13	Birmi	*Taxus baccata*	Taxaceae
14	Brahmi	*Bacopa monnieri*	Scrophulariaceae
15	Chandan	*Santalum album*	Santalaceae
16	Chirata	*Swertia chirata*	Gentianaceae
17	Cinnamomum	*Cinnamomum* spp.	Lauraceae
18	Daruhaldi	*Berberis aristata*	Berberidaceae
19	Dioscorea	*Dioscorea floribunda*	Dioscoreaceae
20	Foxglove	*Digitalis purpurea*	Scrophulariaceae
21	Giloe	*Tinospora cordifolia*	Menispermaceae
22	Ginko	*Ginkgo biloba*	Ginkgoaceae
23	Ginseng	*Panax pseudoginseng*	Araliaceae
24	Gritakumari	*Aloe vera*	Liliaceae
25	Gudmar	*Gymnema sylvestre*	Asclepiadaceae
26	Guggal	*Commiphora wightii*	Burseraceae
27	Haritaki	*Terminalia chebula*	Combretaceae
28	Isabgol	*Plantago ovata*	Plantaginaceae
29	Jatamansi	*Nardostachys jatamansi*	Valerianaceae
30	Jatropha/Ratanjot	*Jatropha curcas*	Euphorbiaceae
31	Kalihari	*Gloriosa superba*	Liliaceae
32	Kalmegh	*Andrographis paniculata*	Acanthaceae
33	Kesar/ Saffron	*Crocus sativus*	Iridaceae
34	Kokum	*Garcinia indica*	Clusinaceae (Guttiferae)
35	Kutaj	*Holarrhena antidysenterica*	Apocynaceae

36	Kuth	*Saussurea costus*	Asteraceae
37	Kutki	*Picrorhiza kurroa*	Scrophulariaceae
38	Makoy	*Solanum nigrum*	Solanaceae
39	Mulethi/Jastimadhu	*Glycyrrhiza glabra*	Fabaceae
40	Neem	*Azadirachta indica*	Meliaceae
41	Patharchur	*Coleus forskohlii*	Lamiaceae
42	Pippali	*Piper longum*	Piperaceae
43	Safed Musli	*Chlorophytum borivillianum*	Liliaceae
44	Sarpgandha	*Rauwolfia serpentina*	Apocynaceae
45	Senna	*Cassia angustifolia*	Caesalpiniaceae
46	Shatavari	*Asparagus racemosus*	Liliaceae
47	Stevia	*Stevia rebaudiana*	Asteraceae
48	Tulsi	*Ocimum sanctum*	Lamiaceae
49	Vach/Sweet flag	*Acorus calamus*	Araceae
50	Viavidang	*Embelia ribes*	Myrsinaceae

Medicinal plants are potential renewable natural resources. The conservation and sustainable utilization of medicinal plants must necessarily involve a long term, integrated, scientifically oriented action programme. This should involve the pertinent aspects of protection, preservation, maintenance, exploitation, conservation and sustainable utilization. A holistic and systematic approach envisaging interaction between social, economic and ecological systems will be a more desirable one. The most widely accepted scientific technologies of medicinal plants conservation are the *in situ* and *ex situ* methods.

INTERNATIONAL ORGANIZATIONS FOR MEDICINAL PLANTS CONSERVATION

- IUCN-The World Conservation Union
- Species Survival Commission
- Convention on Biological Diversity (CBD)
- Convention on International Trade in Endangered Species of Wild Flora and Fauna (CITES)
- Trade Records Analysis of Fauna And Flora in Commerce (TRAFFIC)
- World Wide Fund for Nature (WWF)
- United Nations Development Programme (UNDP)
- United Nations Environment Programme (UNEP)
- Global Environment Facility (GEF)

AUTHENTIC REFERENCE DOCUMENTS ON MEDICINAL PLANTS AND ITS CONSERVATION

- Red Lists of Threatened Species
- Red Data Books.

8 DENDROLOGY

INTRODUCTION

The teaching of every science begins by trying to identify and understand the basic units. This applies to all scientific disciplines, be it chemistry or biology. Taxonomy is the discipline which identifies names and classifies natural objects including plants, birds, reptiles, fungi and many others. The term taxonomy derives from the Latin word *taxis*, meaning arrangement or order and *nomos*, which refers to knowledge. In a more practical sense, plant taxonomy (also called plant systematics) attempts to teach the skill required to identify an unknown plant specimen.

WHAT IS DENDROLOGY?

In a forest landscape, the tree is the most visible and dominant biological organism. So, one of the important and early challenges of a forestry professional is to correctly identify these organisms, at least a few of them, by its exact name and family. The further application of the various forest management skills of a forest manager primarily rests on his fundamental knowledge about plant systematics. Plant taxonomy is an easy discipline if one could master some of the basic morphological characters of the plants and use it to unravel the identity of a plant. Dendrology is the study of trees and for some, it is also the taxonomy or systematics of woody plants.

THE APPLICATION OF DENDROLOGY

The Indian subcontinent is blessed with unique geographic position, distinct physiographic, edaphic and climatic zones and gradients. This has favoured the development of a very rich and diverse flora with high percentage of endemism. India is one of the top ten plant rich nations of the world and fourth among the Asian countries. Though with only 2.5 per cent of the world's land area, India accounts for 7.8 per cent of the recorded species of the world including 45,500 recorded species of plants. India is also considered as one of the world's 17 "megadiverse" countries in terms of biodiversity.

India is also recognized as one of the eight Vavilovian centres of origin and diversity of crop plants, having more than 300 wild ancestors and close relatives of cultivated plants, which are still evolving under natural conditions. India has four global biodiversity hot spots

(Eastern Himalaya, Indo-Burma, Western Ghats and Sri Lanka, and Sundaland). About 27 per cent of all the species of higher plants recorded in the Indian region can be found (about 4,000 of 15,000 species) in the Western Ghats. Further, almost 1,800 species are endemic to the region. Based on the uniqueness of the phyto-geographical zones and pattern of endemism, 25 micro hot spots centres of endemic flora have also been identified in India. The unique features of the plant diversity, among others, include 60 monotypic families and over 6,000 endemic species. It is widely believed that India also harbours over 256 globally threatened plant species. With over 16 major forest types and 251 subtypes, the total forest and tree cover is approximately 23.39 per cent of the geographical area.

This biological richness has prompted the establishment of the Botanical Survey of India, as early as in 1890. More recently in 1999, the Ministry of Environment and Forests, Government of India had launched the All India Coordinated Project on Capacity Building in Taxonomy (AICOPTAX) with the sole mission of enhancing India's capability for inventorying, monitoring, conserving and utilizing biodiversity as well as for establishing leadership in the field of taxonomy at regional and global levels. In 2003, India also set up a National Biodiversity Authority (NBA) with an explicit mandate of conservation of biological resources and associated knowledge as well as facilitating access to them in a sustainable manner.

India is also a party to CBD and therefore, committed to developing an inventory of its biodiversity resources. This documentation of indigenous knowledge with respect to biodiversity is a major task as our biodiversity is also under threat from Intellectual Property Rights issues. However, for an effective conservation, management and sustainable use of our tree resources, it is imperative that we first develop a sound taxonomic database of the same. In years to come, dendrologists must play an important role in getting the sovereign right of our country on our remaining tree wealth. In addition, when human induced climate change is altering species distribution and extinctions in the forest landscape, the role of a forester-dendrologist has more practical significance in biodiversity conservation.

HISTORY OF PLANT CLASSIFICATION

The illustrious Greek scientist, Theophrastus (ca.370–ca.286 B.C.), is considered the first to propose a system of plant classification, which he described in his two works *Historia Plantarum* and *De Causis Plantarum*. It is believed that Theophrastus proposed the four plant categories namely trees, shrubs, undershrubs and herbs. Dioscorides in his *De Materia Medica* (Latin) also listed and described the substances used for medical purposes and their therapeutic properties and uses.

The European naturalists in the 16th, 17th and 18th centuries increasingly began to discuss un-encountered plants and animals from their overseas voyages. So the need to develop a classification system so as to organize these unknown plants and animals became necessary. Herbalists like Otto Brunfels (1464–1534), Leonhard Fuchs (1501–1566), Matthias de L'Obel (1538–1616), John Gerard (1545–1612) and Charles de L'Ecluse (1526–1609) attempted various classification schemes. Andrea Caesalpino (1519–1603) organized plants by fruits and seeds. In the 17th century, Joseph Pitton de Tournefort (1656–1708) used a form classification that divided plants into groups based on petal characters. He also developed the modern "genus" concept. His contemporary John Ray (1628–1705) classified some 18,000 species in his *Methodus Plantarum* (London 1703), using a system based on form and gross morphology of plant structures. In the 18th century, Michel Adanson (1727–1806) grouped plants by affinities observed through multiple relationships of characters which helped to establish the concept of natural classification.

Carolus Linnaeus (1707-1778) decided to try his artificial system, in which he classified plants by counting their stamens and pistils. He created the sexual system consisting of 24 classes based on the number and arrangement of stamens in each flower. Within each class, the plant was then assigned to an order based on the number of pistils in each flower. Linnaeus, through his *Species Plantarum* (1753) also proposed the concept of binomial nomenclature.

Natural systems, based on morphological relationships emerged during the fag end of 18th century. Antoine Laurent de Jussieu (1748-1836), George Bentham (1800-1884) and Sir Joseph Dalton Hooker (1817-1911) developed natural systems of plant classification by placing together plants having similar combinations of characters. Both De Jussieu and Bentham and Hooker separated the seed plants into three classes namely, dicots, monocots and gymnosperms.

By the middle of the 19th century, new plant classifications emerged based on evolutionary development and phylogenetic concepts put forth by Alfred Wallace and Charles Darwin. Adolf Engler and Karl Prantl's system, though not truly phylogenetic, is one such system. The American taxonomist, Charles E Bessey (1845-1915) developed a more convincing phylogenetic plant classification system, by realigning Bentham and Hooker's natural classification. The plant classification schemes put forth by John Hutchinson, Fuller and Tippo (1954) and Cronquist (1981) are also examples of phylogenetic system.

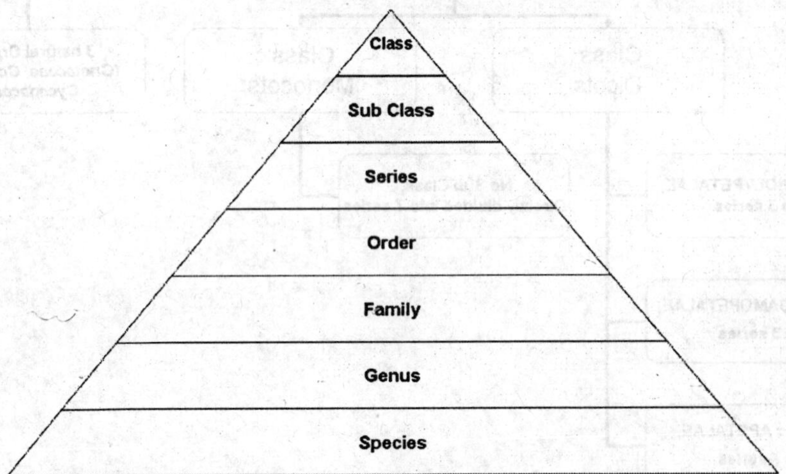

Fig. 8.1: Scheme of Classification under Bentham and Hooker's System

BENTHAM AND HOOKER'S NATURAL SYSTEM

George Bentham (1800-1884) and Sir Joseph Dalton Hooker (1817-1911), the two famous British botanists closely examined and placed 97,205 species of seed plants in 202 orders (now referred to as families) based on their morphological relationships. They produced a three volume book called *Genera Plantarum*. Their natural system is a refinement of those by de Candolle and by Lindley, which in turn were based directly that of de Jussieu. Most of the plants were classified into either class dicots or monocots, which were subsequently subdivided into sub class, series, orders, family, genus and finally the species (Fig. 8.1).

The refinements made by Bentham and Hooker can be seen in Polypetalae, where the new series Disciflorae are interpolated between the Thalamiflorae and the Calyciflorae. They also revised Monochlamydeae. The gymnosperms were treated as a third taxon collateral with and placed between the dicots and monocots. This system of Bentham and Hooker is being followed in India, United Kingdom and several other Commonwealth countries. It is also used in a number of herbaria and botanical gardens all over the world, including Kew herbarium.

The most important merit of Bentham and Hooker system is that an unknown plant can be easily identified by studying its morphology and by applying a logical process of elimination. For example: Dicots from Monocots and Polypetalae from Apetalae or Gamopetalae (Fig 8.2. to 8.6).

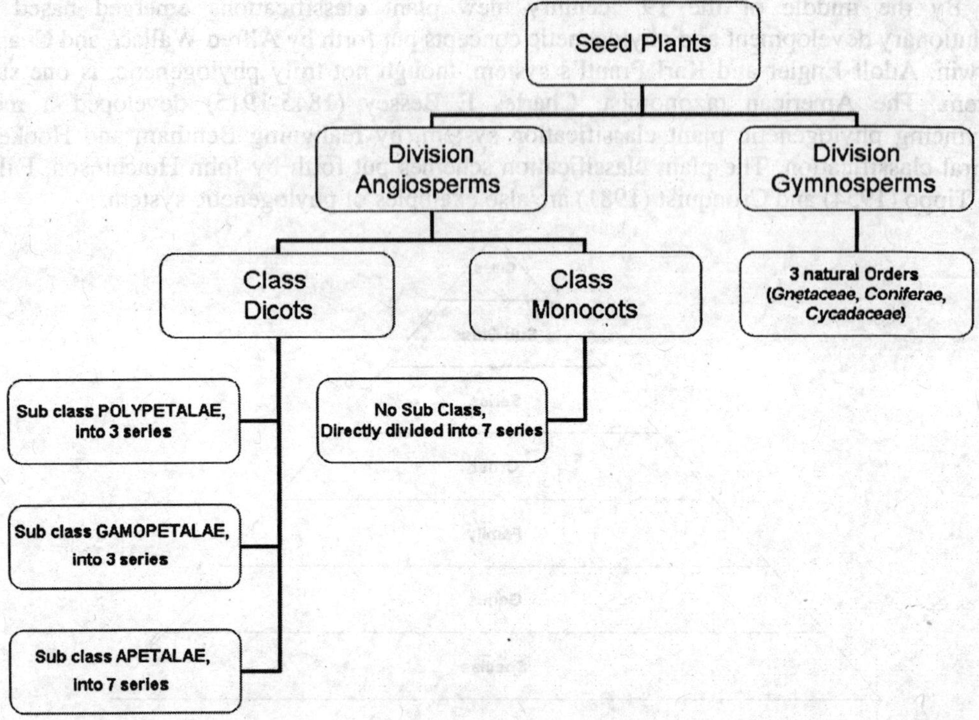

Fig. 8.2: Outline of Bentham and Hooker's Natural Classification

Further, this is a natural system which is based on actual examination of specimens. It is to be reasonably believed that the authors had some notion about evolutionary trends. For probably that reason, they placed Monocots, which are more advanced than Dicots, in the end of the classification scheme.

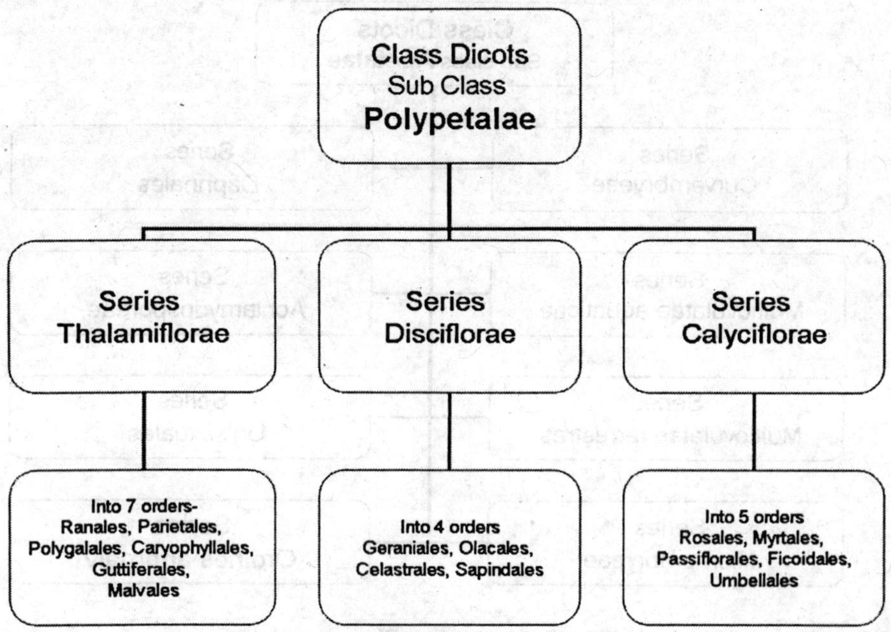

Fig. 8.3: Bentham and Hooker's Classification of Sub Class Polypetalae

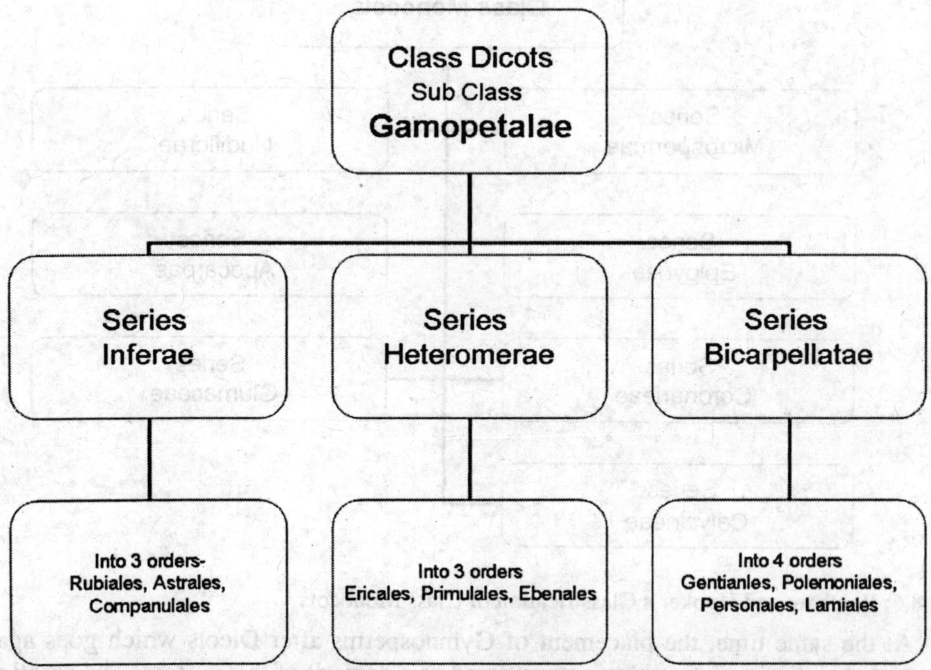

Fig. 8.4: Bentham and Hooker's Classification of Sub Class Gamopetalae

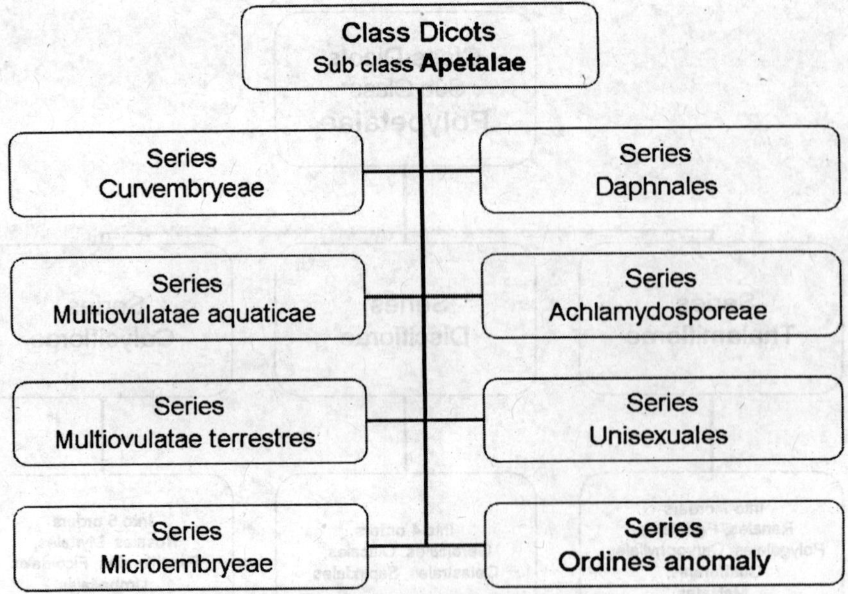

Fig. 8.5: Bentham and Hooker's Classification of Sub Class Apetalae

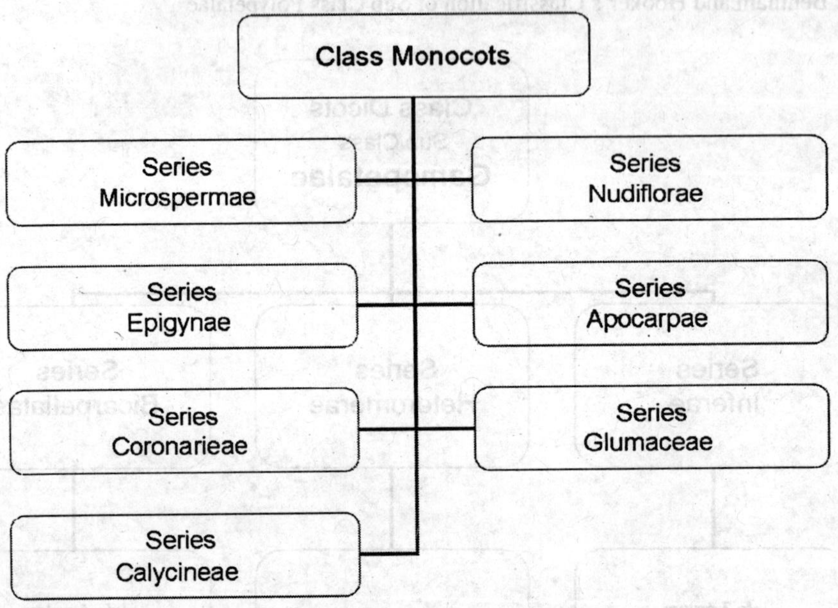

Fig. 8.6: Bentham and Hooker's Classification of Class Monocots

At the same time, the placement of Gymnosperms after Dicots which goes against the accepted evolutionary principles is considered as a demerit of this system. The creation of the sub-class Monochlamydae, which is actually a *mixed grouping of un-related plants,* is also considered as a demerit. The positioning of Orchidaceae, a group of highly advanced plants, as the first family in Monocots also defies evolutionary concepts.

INTERNATIONAL CODE OF BOTANICAL NOMENCLATURE

Use of botanical names and coinage of new botanical names is not to be done arbitrarily but must always follow standard procedures. The International Code of Botanical Nomenclature or ICBN codes have worldwide acceptance which basically outlines provisions for having a stable method of naming taxonomic groups, avoiding and rejecting the use of names which may cause error or ambiguity. The ICBN codes is intended to sort out errors and ambiguities arising from past misunderstandings and misidentifications and to ensure correct naming of new taxa. The ICBN sets the formal starting date of plant nomenclature at 01 May 1753, the publication of *Species Plantarum* by Linnaeus (or at later dates for specified groups and ranks).

The first attempt at developing an international agreement was made in Paris in 1867. Here, in the first International Botanical Congress called by the Swiss Botanist Alphonse de Candolle, the first set of International Rules of Botanical nomenclature was adopted along with the "Paris Code". Later, in 1892, a group of US botanists held a meeting in Rochester at which they presented some additions and modifications that they considered more objective. They proposed modifications of the Paris Code, but these modifications later presented at the second International Botanical Congress (Vienna 1905) were not accepted.

In 1930, taxonomists finally agreed on a single International Code of Botanical Nomenclature which will be revised every 6 years. The goals of all the revisions will always to achieve stability in scientific nomenclature. The revisions will be published in *Taxon*, the journal of the International Society of Plant Taxonomists. There are six principles that guide decisions concerning the code which include uniqueness principle (principle IV), type principle (principle II), priority principle (principle III), retroactivity principle (principle VI) and principles 1 and V. Proposals will become valid until they are voted on at the subsequent International Botanical Congress.

IDENTIFICATION OF TREES IN A TROPICAL FOREST LANDSCAPE

In a tropical forest, it is seldom possible to even remotely examine the various vegetative and reproductive characters of a tree species. The tree in question may be quite large and occupying a top canopy. Even with the help of a good binocular it will not be easy to grasp the morphology. In such a situation, a fallen leaf, flower, fruit or even a seed may be of much value to fix the identity of a tree species. By noticing the nature of the branching pattern, the characteristics of the bole or, if possible, by cutting a blaze, one can also achieve equal success in tree identification. Presented below are some typical diagnostic features which can be used to pin point a tree species in a tropical forest ecosystem.

LEAVES

Leaves can be simple (e.g. Many members of *Magnoliaceae, Dipterocarpaceae, Lauraceae* etc.) or compound (as in *Meliaceae, Fabaceae* etc.). Compound leaves may be digitate (*Ceiba, Bombax*), pinnate (*Meliaceae*), bipinnate (*Albizzia*), tripinnate (*Moringa*) or quadripinnate (*Oroxylum indicum*). The typical leaf of *Bauhinia* spp. is bifid. Lobed leaves are typical of *Sterculiaceae*. Leaves with a conspicuous intra-marginal vein may belong to *Myrtaceae* (*Syzygium* spp.). Leaves with glandular dots are very common in *Rutaceae* and *Myrtaceae*. In Zizyphus (*Rhamnaceae*), *Cinnamomum* (Lauraceae), *Grewia* (*Tiliaceae*), *Trema* (*Ulmaceae*) and *Strychnos* (*Loganiaceae*), 3-5 basal nerves can be found arising from the base of the leaves. The leaves of many *Elaeocarpus* spp., *Terminalia catappa* and *Ailatnthus triphysa* turn bright-red before falling.

The leaf-surface of many trees may be covered with various types of hairs or glands. The leaves of teak, when rubbed between the hands colour them red. The under-surface of the leaves of *Mallotus philippensis* has crimson glands. The leaves of many *Holigarna* spp. are allergic to a few.

BRANCHING

The nature of the branches of certain tree species is also quite distinctive. Trees of the genus *Garcinia*, *Bombax ceiba, Polyalthia fragrans* and *Terminalia catappa* have a characteristic whorled appearance and can be recognized at a glance. *Polyalthia longifolia*, *Cananga odorata* and most *Pinus* spp. have a drooping nature.

BOLE

The shape of the bole is also often very helpful in tree identification. The bottle shaped stem of the palm *Roystonea regia* is a classical example. Both *Bombax ceiba* and *Ceiba pentandra* has a round bole and also plank buttresses. Other tree species with prominent buttressing nature include *Tetrameles nudiflora, Palaquium ellipticum* and many *Ficus* and *Terminalia* spp. Trees like *Baccaurea courtallensis, Couroupita guianensis, Saraca indica* and many species of *Ficus* exhibit cauliflory. The bole of *Bombax ceiba, Xanthoxylum rhetsa, Zizyphus* spp., *Flacourtia* spp. and *Erythrina* spp. are covered with spines of various texture. The sheaths of the leaves of the climbing palms, *Calamus* spp. develop needle-like spines.

BARK

Generally the bark of trees in the evergreen forests will be smooth natured and light coloured with a green tinge. On the other hand, the deciduous tree species display an array of colour and textures. Also, the bark of a young tree is often smooth and of a different colour from that of the old tree. Trees with a smooth bark are many species of *Ficus*, *Tetrameles nudiflora*, *Garcinia* spp., *Hevea, Morus* and many members of *Lauraceae*, *Myristicaceae* and *Leguminosae*. Examples of trees with fissured bark are *Shorea, Pinus, Calophyllum, Erythrina, Moringa* etc. In *Aporosa* spp. and in *Bischofia javanica*, the bark is powdery in nature. In some species, the bark exfoliates. Examples include, *Gmelina arborea, Terminalia arjuna* and *Manihot glaziovii.* The tree species *Lagerstroemia parviflora* exfoliates and display its creamy white bark and so the nick name, "naked maiden of the forest". Using the "crocodile-skin bark", a *Terminalia tomentosa* could be quickly identified. The silvery smooth bark of *Sterculia urens* is a typical sight in a tropical dry deciduous forest.

ROOTS

Tree roots also have an important diagnostic value. For example, pneumatophores are very common among the various species of the mangrove genera, *Avicennia* and *Sonneratia.*

Carallia brachiata (*Rhizophoraceae*) also have aerial rootlets on stem and branches. Stilt roots are common in many *Rhizophora* spp., *Bruguiera sexangula* and *Elaeocarpus* spp. The aerial roots of certain *Ficus* species, especially the banyan tree is well known.

PLANT SAP/ LATEX

In some trees, when a blaze is taken, the nature of the juice (sap) oozing out is a key diagnostic character. White latex is produced by all the species of *Apocynaceae, Asclepiadaceae* and some members of *Euphorbiaceae*. The juice of *Anacardiaceae* is acrid in nature. Yellow latex is produced by most species of *Garcinia*. The red latex produced by *Bischofia javanica* and most species of *Myristicaceae* is a key spot character.

BLAZE COLOUR

The colour of the blaze also plays a key role in tree identification. The sulphur-yellow coloured blaze of *Zanthoxylum rhetsa* is quite typical. Red coloured blaze is characteristic of *Buchanania lanzan*, *Boswellia serrata*, *Bischofia javancia* and many members of *Myristicaceae*. The blaze is creamy white in *Alstonia scholaris*, *Dysoxylum malabaricum* and *Firmiana colorata*. The blaze of most *Lauraceae* members is aromatic, especially *Cinnamomum* spp.

FRUIT

Familiarity with the fruits of many trees may help us to identify the family, genera or species easily. The "pod" belongs to *Leguminosae* and comes in various sizes and shapes. Fruits are round in the case of *Hydnocarpus* and *Pterygota alata* (*Sterculiaceae*), while it is spherical in *Annona* (*A. squamosa* and *A. reticulata*), *Ficus* spp. and *Artocarpus hirsutus*. Capsules containing seeds enveloped in floss are found in *Malvaceae* and *Bombacaceae*. Follicle is the common form of fruit, especially in *Apocynaceae, Asclepiadaceae, Bignoniaceae, Moringaceae* and *Sterculiaceae*. Winged fruits can be seen in *Dipterocarpaceae*. In *Coniferae* and *Bignoniaceae*, winged seeds can be found.

BOTANICAL SPOT CHARACTERS AND UTILIZATION ASPECTS OF SELECTED TREE FAMILIES

MAGNOLIACEAE

Leaves alternate, mostly entire, simple; pinnately veined; stipules enclose the bud and leave conspicuous encircling scars on the twigs; terminal flowers often large and showy; numerous petals and sepals; androecium of numerous spirally arranged stamens at the base of a cone-like receptacle bearing numerous spirally arranged carpels; gynoecium of numerous spirally arranged pistils; woody, cone like aggregate of seed-bearing follicles; each seed has a fleshy red outer layer (aril) and hangs from its follicle by a threadlike stalk.

Magnolia champaca (L.) Baill. ex Pierre (Golden champa, Champak) - avenue tree for its fragrant flowers.

Magnolia nilagirica (Zenk.) Figlar (White champak) -avenue tree with fragrant white flowers.

ANNONACEAE

Trees, shrubs or vines; aromatic wood and foliage; simple leaves with alternate phyllotaxy; petiole short; stipules absent; flowers bisexual and drooping in appearance; actinomorphic; perianth usually tri-seriate; ovary superior; spirally arranged stamens; multi-pistillate gynoecium; Berry or an aggregate fruit; ruminate endosperm of the seeds.

Annona squamosa L. (Sweet sop)-fruit is edible

Cananga odorata (Lam.) Hook. f. & Thoms. (Ylang-ylang)-Ylang-Ylang oil of commerce.

Polyalthia longifolia (Sonner.) Thw. Enum. (Ashok; Mast tree)-beautiful avenue tree.

GUTTIFERAE (CLUSIACEAE)

Trees or shrubs; Simple large and leathery leaves; opposite or whorled leaves; Resinous sap (blaze exudes white or yellow sap); most species are highly endangered.

Calophyllum inophyllum L. (Alexandrian laurel)-non edible oil for bio-diesel production.

Garcinia gummi-gutta (L.) Robs (Malabar gamboge)-fruit has medicinal value.

Mesua ferrea L. (Ceylon ironwood)-durable timber.

DIPTEROCARPACEAE

Small to very lofty evergreen trees; Resinous wood- *dammar*; simple leaves with stellate hairs; winged fruit (Samara)-*persistent calyx becomes a wing*; tropical in distribution, most species produce highly valuable timber.

Hopea parviflora Bedd. (Iron wood of Malabar)-durable timber.

Vateria indica L. (White dammar)-resin is a valuable NTFP.

Shorea robusta (Sal)-durable timber.

MALVACEAE

Simple, entire (sometimes variously lobed) stipulate leaves with long petiole, alternate phyllotaxy; stipules present; Presence of mucilage in the vegetative parts; flowers bisexual; epicalyx present; ovary superior; monadelphous stamens; fruit a loculicidal capsule, seeds covered with hairs.

Thespesia populnea (L.) Soland. ex Correa (Portia tree)-avenue tree with yellow flowers.

Kydia calycina Roxb., (Roxburgh's Kydia)-deciduous tree.

STERCULIACEAE

Trees, shrubs or herbs; leaves alternate, simple entire or infrequently palmately lobed or digitate compound; monadelphous stamens with 2 celled anthers; fruit leathery or fleshy, dehiscent or indehiscent.

Firmiana colorata (Roxb.) (Coloured sterculia)-deciduous, avenue tree

Kleinhovia hospita L. (Guest tree)-avenue tree.

Pterospermum reticulatum Wight & Arn., -timber value

Sterculia guttata Roxb. ex DC., -avenue tree

Sterculia foetida L.,

Sterculia villosa Roxb. ex Smith

Theobroma cacao L., (Cacao)-commercially valuble fruits.

TILIACEAE

Alternate simple leaves, base oblique; margin dentate, serrate or lobed; cymose inflorescence; calyx 4-5 sepals, corolla 4-5 petals, stamens 10 or more; fruit, fleshy, berry like or dry capsule.

Grewia tiliifolia Vahl (Dhaman)-timber value.

Tilia americana L. (American basswood)-timber value.

BOMBACACEAE

Large deciduous trees with tall and stout boles; leaves alternate, simple or palmately compound; large, showy, bisexual flowers, often appearing before the leaves; fruit a loculicidal capsule; seed often embedded in floss.

Bombax ceiba L. (Red silk cotton)-floss used for stuffing mattress and pillows.

Ceiba pentandra (L.) Gaertn. (Silk cotton tree)-floss used for stuffing mattress and pillows.

Durio zibethinus Murray (Durian)-edible fruit

Adansonia digitata L. (Baobab)-timber

RUTACEAE

Herbs, shrubs and trees; mostly compound leaves, which are alternate; gland dotted/glandular; punctate/pellucid dotted leaves; winged petiole (lateral expansion of the petiole); sepals take the shape of a cup (cupular calyx); definite number of stamens (5+5) in 2 whorls (outer whorl opposite the petals).

Aegle marmelos (L.) Correa (Bhael, Indian wood apple)-sacred tree.

Chloroxylon swietenia DC. (Indian satinwood tree)-timber value.

Citrus limon (L.) Burm. f. (Lemon)-edible fruit.

MELIACEAE

Leaves usually alternate, pinnately compound; bitter taste of leaves; leaf base not isobilateral, but oblique; Fruit-berry/drupe/samara (winged); wood often scented.

Azadirachta indica A.Juss. (Neem)-medicinally important species.

Swietenia mahagoni (L.) Jacq., (Big leaf mahogany)-timber is durable.

SAPINDACEAE

Trees; Usually pinnate leaves; usually arillate seeds; pericarp of the nut contains a detergent like material (Saponin).

Filicium decipiens (Wight & Arn.) Thw. (Japanese fern tree)-avenue tree

Sapindus trifoliatus L. (Soapnut tree)-fruit is a NTFP.

Schleichera oleosa (Lour.) Oken (Macassar oil tree)-avenue tree.

ANACARDIACEAE

Simple, leathery leaves, often alternate; resinous bark which is often aromatic (*presence of acrid juice*); inflorescence is a highly branched panicle (male, female & bisexual flowers can be seen); stamens often double the number of petals (*Staminodes present*); drupaceous fruit.

Anacardium occidentale L. (Cashew nut tree)-commercially valuable nut.

Buchanania lanzan Spreng. (Almondette tree)-NTFP.

Gluta travancorica Bedd. (Red wood tree)-endemic to Western Ghats, India.

Mangifera indica L. (Mango)-fruits.

FABACEAE

Leaves alternate, pinnately compound and stipulate; In the exotic *Acacia* genus, the petioles are flattened into phyllodes; presence of stipular spines in some species of *Acacia* and *Robinia*; garden pea has large, leaf like stipules; leaves can alter their position at night by folding; leaves of *Mimosa pudica* can open by shock stimulus (pulvinus); in some species, tendrils or hooks are modified branches, leaves or leaflets.

Sub family **CAESALPINIACEAE**

Trees and shrubs; flowers zygomorphic; corolla caesalpiniaceous; posterior petal innermost; petals typically 5 and distinct; fruit a legume.

Amherstia nobilis Wall. (Pride of Burma)-avenue tree

Caesalpinia sappan L. (Brazil wood)-heartwood is valuable NTFP

Cassia fistula L. (Indian Laburnum)-avenue tree with golden yellow flowers

Delonix regia (Boj. ex Hook.) Rafin (Mayflower)-avenue tree with red flowers

Tamarindus indica L. (Tamarind)-fruit is edible.

Sub family **MIMOSAE**

Trees, shrubs, lianas; leaves mainly bipinnate and often with specialized glands; small, regular flowers aggregated into heads or spikes; actinomorphic; petals valvate in bud, sepals and petals generally united at base; stamens 10-many (sometimes over 100); fruit a loment.

Acacia catechu (L.f.) Willd. (Cutch tree)-NTFP

Albizia lebbeck (*L.*) *Benth (Siris)-forage and timber*

Albizia odoratissima (L. f.) (Ceylon rosewood; Kala siris)-timber value.

Samanea saman (Jacq.) Merr. (Rain tree)-avenue tree.

Sub family **PAPILIONACEAE**

Flowers zygomorphic; corolla papilionaceous (standard, wing & keel petals); aestivation imbricate descending; posterior petal outermost; fruit a legume; gynoecium-all legumes have a single carpel, superior ovary.

Butea monosperma (Lam.) (Flame of the forest)-avenue tree, NTFP.

Dalbergia latifolia Roxb. (Rosewood)-highly valuable and durable timber.

Dalbergia sissoo Roxb. (Shisham)-durable timber.

Pongamia pinnata (L.) Pierre (Karanj)-avenue tree, seed oil is a potential biofuel.

Pterocarpus marsupium Roxb. (Bijasal)-durable timber.

Pterocarpus santalinus L.f. (Red Sanders)-heartwood is a valuable NTFP.

RHIZOPHORACEAE

Mangrove or littoral vegetation; have simple, entire, opposite, coriaceous (leathery) leaves with inter-petiolar stipules and swollen nodes; pinnate venation; ovary mostly inferior; exhibits vivipary; presence of pnuematophores and prop roots; bark contains tannin; fruit is usually a berry.

Bruguiera sexangula (Lour.) Poir. -mangrove species.

Carallia brachiata (Lour.) Merr.- mangrove species.

Rhizophora mucronata Poir.- mangrove species.

COMBRETACEAE

Mostly trees or shrubs; the leaves are simple, alternate or less commonly opposite; stamens 4-10, rarely more; inferior ovary; fruit leathery or drupaceous and contain tannin.

Anogeissus latifolia (Roxb. ex DC.) Wall. ex Guill. & Perr., (Dhawa; Axle wood)-durable timber

Terminalia bellirica (Gaertn.) Roxb.- valuable NTFP

Terminalia paniculata Roth- valuable NTFP

MYRTACEAE

Mostly trees and shrubs; simple, opposite leaves with intra-marginal vein at the sub marginal region; essential oils in the vegetative parts; posses a shiny and scaly periderm; has large and showy stamens; fruit a berry.

Eucalyptus globulus Labill. (Blue gum)-pulpwood; eucalyptus oil of commerce.

Syzygium cumini (L.) Skeels-avenue tree; edible fruit.

LYTHRACEAE

Herbs, shrubs or trees; simple leaves, usually opposite or whorled; young shoots are quadrangular; superior ovary, stamens numerous; petals are clawed and crinkled, margin undulate; fruit-dry dehiscent or indehiscent capsule, sometimes enclosed by a persistent capsule.

Lagerstroemia microcarpa Wight.-deciduous tree.

Lagerstroemia speciosa (L.) Pers. (Queen Crepe Myrtle)-avenue tree.

RUBIACEAE

Leaves simple, entire, opposite decussate; interpetiolar stipules; flowers bisexual; 4-5 connate sepals; inferior ovary; loculicidal or septicidal capsule; drupe, berry or 2 segmented schizocarp.

Haldina cordifolia (Roxb.) (Haldu)-timber

Mitragyna parvifolia (Roxb.) (Kaim)- timber

Morinda citrifolia L. (Noni)-valuable fruit.

SAPOTACEAE

Mostly trees; sap milky (lactiferous ducts); leaves simple, mostly alternate and clustered at the tip of branch; often penni-nerved; fruit a berry, often with a thin leathery to bony outer layer; seed with endosperm usually fleshy or none.

Madhuca longifolia (Koenig) J.F. Macbr. (South Indian mahua)

Manilkara zapota (L.) (Chikku; Sapota)-valuable fruit.

APOCYNACEAE

Sap milky white; leaves simple, entire and opposite decussate; ovary superior; stigma massive and dumbbell shaped; corolla tube often "hairy" or closed with a "corona" (or crown) of scales; fruit a follicle, capsule or berry.

Alstonia scholaris (L.) R. Br. (Scholar wood tree)-avenue tree

Plumeria rubra L. (Frangipani)-avenue tree

Wrightia tinctoria (Roxb.) R. Br. (Pala indigo)-medicinal tree

BIGNONIACEAE

Simple or pinnately compound leaves; opposite decussate, rarely alternate; often glandular (between petiole bases or in leaf vein axils); corolla campanulate or funnelform (oblique lips); fruit usually a capsule, seeds winged.

Jacaranda mimosifolia D.Don (Jacaranda)-avenue tree.

Kigelia africana (Lam.) (Indian sausage tree)-avenue tree.

Millingtonia hortensis L.f. (Indian cork tree)-avenue tree.

Tabebuia rosea (Bertol.) DC. (Trumphet tree)-avenue tree.

VERBENACEAE

Herbaceous or woody plants often with a quadrangular stem; Leaves opposite decussate or whorled, simple or compound; calyx gamo-sepalous and persistent; ovary superior; fruit a drupe or berry.

Citharexylum spinosum L. (Fiddlewood)-avenue tree.

Gmelina arborea Roxb. (White teak: Gamari)-small timber

Tectona grandis L. f. (Teak)-highly valuable timber.

MYRISTICACEAE

Evergreen trees or shrubs; leaves simple, alternate, entire and coriaceous; monoecious or dioecious; aromatic wood or foliage; barks gives a reddish exudation; flowers inconspicuous; monadelphous stamens; ovary superior; fruit a dehiscent drupe; seed partially or fully enveloped by a brightly coloured fleshy aril.

Myristica fragrans (nutmeg of commerce)-valuable fruit.

Knema attenuata (Hook. f. & Thoms.) Warb.-evergreen tree.

LAURACEAE

Mostly evergreen trees; bark and foliage usually aromatic; simple leaves, usually alternate, entire, often pellucid punctuate; small un-differentiated perianth consisting of sepal like segments; petals absent; ovary usually superior; fruit drupe or berry with persistent woody receptacle.

Cinnamomum camphora (L.) (Camphor tree)-valuable NTFP

Persea americana Mill. (Avocado)-edible fruit.

SANTALACEAE

Trees, shrubs or herbs; often hemi-parasites on the roots of other plants; leaves entire, opposite; perianth green or petaloid; ovary inferior or half inferior.

Santalum album L. (East Indian sandal wood)-highly prized timber and wood oil

EUPHORBIACEAE

Extremely diverse family; herbs, shrubs, trees or vines; all forms of leaves are encountered; phyllotaxy- all forms; usually stipulate; sap, often milky white; unisexual flowers; calyx present, mostly apetalous; fruit usually a three segmented schizocarp capsule.

Bischofia javanica Blume, Bijdr. (Bishopwood)-evergreen tree.

Jatropha curcas L. (Jatropha)-biodiesel plant

Mallotus philippensis (Lam.) Muell.-Arg. (Kamala)-natural dye.

Phyllanthus emblica L. (Amla)-fruit is medicinal.

Hevea braziliensis (Willd. ex A. Juss.) Muell.-Arg. (Rubber tree)-latex is commercially valuable.

ULMACEAE

Trees or shrubs; simple, entire to doubly serrate and alternate leaves; base oblique; stipules present; flowers inconspicuous and small; bisexual or unisexual; ovary superior; fruit samara or nutlet; seeds flat, winged or spherical.

Trema orientalis (L.) Blume (Charcoal tree)-colonizer of degraded ecosystems

Ulmus americana L. (American elm)-valuable timber

MORACEAE

Trees, shrubs or lianas; Leaves simple, alternate and deciduous; milky sap; entire to serrate; sometimes lobed; large sheathing stipules; flowers are imperfect and the species are monoecious or dioecious; ovary superior or inferior.

Artocarpus heterophyllus Lam. (Jack)-valuable timber, fruit is edible.

Ficus benghalensis L. (Banyan)-avenue tree, sacred value.

Ficus religiosa L. (Peepal)-avenue tree, sacred value.

BETULACEAE

Leaves simple, alternate, stipulate and deciduous; species monoecious; the individual flowers bloom in clusters of 1-6 in the axils of the bracts; 2-20 stamens; short pistillate flower; fruit a one seeded nut.

Betula papyrifera Marsh. (Paper Birch/White birch)-timber value.

Alnus nepalensis D.Don, (Nepalese Alder)-timber value.

FAGACEAE

Also known as Beech family; deciduous, broad leaved species characteristic of North temperate forests; leaves, simple, alternate, usually penni-nerved and with short petioles; monoecious; staminate flowers has a 4-7 lobed calyx and 4-8 stamens; pistillate flowers has a 4-8 lobed calyx, adnate to 3 (rarely 6) celled ovary; fruit, a one seeded nut with outer cartilaginous coat.

Fagus grandiflora Ehrh. (American beech)-valuable timber

Quercus alba L. (White oak)-durable timber is valuable.

CASUARINACEAE

A monotypic or unigeneric family; trees with noded branches; leaves reduced to many toothed sheaths surrounding the nodes; inflorescence monoecious-male flowers in whorls arising from the branchlets (rust coloured when viewed from a distance); red colored female flowers arise from the branches; calyx and corolla absent; ovary superior; fruit woody globose cone-like with persistent bracts.

Casuarina equisetifolia L. (Beef wood, She-Oak)- pole timber; nitrogen fixing ability, coastal afforestation.

ARECACEAE

A large family of tropical and sub-tropical "woody" palms; also contains woody vines (*Calamus* spp.); leaves (fronds) in a terminal cluster in arborescent species; ovary superior; plants may be monoecious or dioecious.

Areca catechu L. (Betel nut palm)-fruit is an NTFP.

Elaeis guineensis Jacq. (Oil palm)-oil is edible.

Cocos nucifera L. (Coconut tree)-every part is valuable.

POACEAE (GRAMINAE)

Annuals and perennials; leaves flat, long and slender, and form a sheath around the stem; usually parallel veined; stems round, generally hollow; grass inflorescence is known as floret; each small sessile flowers are enclosed between two bracts, palea and lemma; 3 free stamens with versatile anthers; superior ovary; fruit is a caryopsis.

Bambusa bambos (L.) (Thorny green bamboo)-timber value.

Vetiveria zizanioides (L.) (Khus-khus grass; Vetiver)-soil conservation; NTFP.

PINACEAE

Leaves simple; often needle-like; produced in bundles (fascicles) or singly; aromatic foliage and wood due to resin canals; ovulate scales spirally arranged into cones; two seeds produced from each ovulate scale; seeds have a large wing; pollen grains with two bladder like wings.

Pinus radiata D. Don (Radiata pine)-timber.

Pinus roxburghii Sarg. (Chir pine)-commercial resin.

Pinus wallichiana A.B.Jacks. (Blue pine)-timber, resin.

Cedrus deodara (Roxb. ex D.Don) G.Don (Deodar)-durable timber.

CYCADACEAE

Also known as the Sago Palm family; woody, palm–like trees or shrubs; stem thick, mostly tubular and un-branched; leaves compound, alternate in close spirals; leaf vernation erect; leaflets with a single vein; numerous scattered microsporangia; fruiting structure cone like.

Cycas circinalis L. (Queen sago)-NTFP.

FLORAL DIAGRAM AND FLORAL FORMULA

The *floral diagram* represents the numbers and relationships of the parts of a flower using basically four concentric circles to represent the calyx, corolla, androecium and gynoecium. Using internationally accepted symbols, the position of the stamens, type of placentation etc. will be drawn.

A *floral formula* is a representation of the structure of a flower using specific letters, numbers and symbols. Typically, a general formula will be used to represent the flower structure of a plant family rather than a particular species. The following representations are used:

Ca = calyx (sepal whorl; e.g. Ca5 = 5 sepals)

Co = corolla (petal whorl; e.g., Co3(x) = petals some multiple of three)

A = *androecium* (whorl of stamens; e.g., A∞ = many stamens)

G = *gynoecium* (carpel or carpels; e.g., G1 = monocarpous)

G = ovary inferior to insertion point of the other whorls. The floral whorls are epigynous to the gynoecium.

$\underline{\mathbf{G}}$ = ovary superior to insertion point of other floral whorls. The floral whorls are hypogenous to the gynoecium.

X: to represent a "variable number"

∞: to represent "many"

COLLECTION AND SUBMISSION OF HERBARIUM SPECIMENS

WHAT IS HERBARIUM?

A herbarium is a collection of plant specimens that have been pressed, dried and mounted and are arranged in the sequence of an accepted classification and are available for reference or other scientific study. Generations of naturalists have collected plants and mounted them onto sheets of paper, which now serve as historical records. For example, Carl Linnaeus collections numbering approximately 14,300 are still there in the Linnaean Society of London's herbarium.

In olden days, it was common practice to mount each specimen together with a label on a sheet made from good quality paper. Even before that, botanists pasted or stitched plant specimens on notebooks. These collections were housed in herbariums. In 1545, in the University of Padua, Italy, the first herbarium of the world was established. Some of the important herbaria of the world include Royal Botanical Garden (Kew, England), British Natural History Museum, Forest Research Institute (Dehradun, India) and the herbariums of the Botanical Survey of India (Lucknow) and its regional centres.

PRESSING AND MOUNTING TREE SPECIMENS

Press each tree specimen as and when it is collected. The specimen will retain its original features if it pressed and dried without loss of time. Tree specimens should be cut and not torn off from the host plant. A sharp knife or sometimes even a sharp razor blade will be enough for making a neat and clean cut.

While selecting plant specimens, go for a true representative. Avoid specimens which are damaged or suffering from disease or pest attack. Avoid sterile specimens and collect only flowering twigs. This is because, reproductive characters, unlike vegetative characters are more constant and could serve as an easily identifiable trait. The size of the specimen should be restricted to the size of the pressing paper (usually 40 cm x 30 cm made from quality hand made paper). Make sure to leave one or more parts (especially leaves) with its lowerside uppermost.

After collection, the specimen may be kept in between folded newsprint paper and pressed using a steel or wooden herbarium press and left to dry. Brittle stems may be macerated at the point of fold before pressing. Air dry the specimen to a certain degree of crispiness. Ensure that the original colors of all parts are being retained as far as possible.

Properly pressed and dried specimens may be mounted on good quality (usually of hand made paper) herbarium sheets. By applying glue on the lower side of the specimen, the same can be mounted on the herbarium sheet by applying a firm and uniform pressure. Leave it to air dry. Proper entries may be made on the herbarium labels without fail.

PART 3:

SUSTAINABLE TIMBER PRODUCTION

9 PLANTATION FORESTRY

INTRODUCTION

In the history of human development form the time of the earliest agricultural activities, man has cleared the natural forests and wood lands to obtain building materials and fuelwood and to provide lands for domestic animals and crops. The vast clearing of the tropical rain forests in Amazonia was principally to make way for cattle ranches and in Asia, the tropical forests were destroyed for palm oil and rubber plantations. As a result of significant increases in demand for wood, the global wood market is undergoing rapid changes, putting considerable and increasing pressure on the world's remaining natural forests. Without significant investment in promoting sustainable forest management (SFM) efforts and in plantation management, it is expected that increasing demand for wood will lead to further degradation and fragmentation of forests and permanent deforestation in many countries. To successfully change this situation, international wood demand must be met through sustainable wood production from natural forests and plantation management.

Planted forests supply wood, fibre, fuelwood and non-wood forest products for industrial and non-industrial uses. The benefits of wood products over competing products (cement, plastics and metal) are that they are renewable, energy efficient and environmentally friendly. Planted forests, when managed responsibly, can also contribute towards the provision of environmental services (soil and water protection, rehabilitation of degraded lands, restoration of landscapes, habitat development and carbon sequestration) and of social services and livelihood support (regional development, income generation, employment and recreation). They may also offset pressure for wood production from primary forests and valuable forest ecosystems.

At present, forest plantations account for about 7 per cent of global forest area or about 2 per cent of global land area, i.e. slightly less than 300 million hectares. At the same time, they provide more than half the industrial wood produced in the world and their extent and productivity are increasing. Compared with naturally regenerating forests, planted forests represent a higher investment per area unit and normally produce higher values through their products and services. They are also diverse in size and type, ranging from smallholdings to industrial plantations and from primarily protective functions to a wood production orientation. Planted forests stretch from boreal to tropical zones and use native and introduced tree species. Planted forests are also sometimes controversial in achieving a balance among sociocultural, environmental and economic benefits which can be a challenge for its growth.

HISTORY OF FOREST PLANTATIONS

Trees have been planted for thousands of years for food, wood, shelter and religious purposes. The first woody plants to be cultivated were undoubtedly those yielding food or other non-timber products. The first selection and planting of a woody species may have been the olive (*Olea europea*), a thorny shrub growing near Syrian and Anatolian coasts, as early as 4000 BC. Subsequently selected varieties of olive were transported to the eastern Mediterranean e.g. the cultivation of olive has been known in Greece at least since the Minoan ear (3000 BC) for food and olive oil for clay lamps. The olive was exported to Egypt in the same period. The edible fig (*Ficus carica*) was also cultivated in Mediterranean region as early as 4000 BC. The myrrh trees (*Commiphora myrrha*) were introduced to Egypt from Somalia in 1500 BC and planted for sources of perfume. Olives, walnuts and almonds from the Middle East and western Asia were moved into China during 208 BC-265 AD.

The apple (*Malus pumila*) was cultivated for a few centuries BC by the Greeks and Romans and it was spread throughout Europe and Asia as a result of their travels and military conquests. The Romans also introduced the edible, nut-bearing, sweet chestnut (*Castanea sativa*) into Britain. In the fourth century BC, Theophrastus reported trees of frankincence (*Boswellia* spp) and myrrh being planted on private estates in southern Arabia, and Aristotle established what was probably first arboretum in Greece. Cato (234 BC) recorded the planting of willows (*Salix* spp) and poplars (*Populus* spp) and noted that a conifer was planted for ship's timber. There are also several Biblical references to tree planting in the first millennium. Trees were moved around world by the Romans, Greeks, Chinese and by others during military conquests. Voyages of discovery by European navigators to the Americas and Asia provided new opportunities to transfer tree germplasm between distant countries (Turnbull 2002).

Forest plantations are not anything new. In southwestern France, for example, there are records of man-made forests as early as year 1500, as well as evidence of trading in forest products, both wood and non-wood from maritime pine (*Pinus pinaster*) plantations. These early plantations resulted mainly from the initiative of private landlords as well as smallholders and communities, primarily for commercial purposes (i.e., the trade of wood and resin products), but also for direct household needs. In this same area there are records of zones planted with pines and other vegetation to control the advance of sand dunes before 1600 (Sargos 1997). Therefore, there is evidence of commercial, smallholder and community man-made forests, for both production and protection purposes, from as early as 400 to 500 years ago. This particular region of Southwestern France is known as the "Landes de Gascogne" which is nowadays a compact plantation forest of more than one million hectares. While ownership is more than 90 per cent private, with more than 40,000 owners of all sizes, the State plays a limited but effective role in the management of this area and so do local governments (Montane 1994).

Closer to home, teak (*Tectona grandis*) was successfully introduced in Sri Lanka as early as 1680. This same species was successfully planted in Java, India and Bangladesh during early 19th century. Additionally, eucalypts were successfully planted in India in 1790.

Expanding area of planted forests, to a large extent, has resulted from the inclusion of plantation programmes in the national forest policies of many countries in recent years, especially since the mid-sixties (Evans 1992). For the most part, these man-made forests have concentrated on industrial plantations; however, they have also included plantations for protection purposes, as well as rapidly expanding agroforestry and social forestry efforts. This important area of forest plantations reflects a sizable set of investments in which economic agents of several categories have participated, namely, enterprising individuals, private sector enterprises, communities, rural households, governments (both national and local), international funding organizations, and non-governmental organizations (NGOs).

Table 9.1: History of Plantation Forests in the World (Evans 1992; FRA 2010a)

Year	Events
1680	Teak introduced in Srilanka
1823-1825	Introduction of *Eucalyptus* spp to South America
1829	Teak planted in Java, Indonesia
1840	First teak plantation established in Nilambur, India
1843	Eucalypts planted in Nilgiri, India
1871	Teak plantation started in Bangladesh
1877	*Pinus patula* introduced in New Zealand
1880	Teak plantation established in Trinidad
1890	*Eucalyptus globulus* was growing in Addis Ababa, Ethiopia
1900	8,000 ha of Teak was planted under Taungya system in Burma
1900-1920	4,000 ha of *Eucalyptus globulus* established around Addis Ababa, Ethiopia
1910	Taungya system introduced in Kenya
1921	1,15,500 ha of Wattle planted in South Africa
1931	25,500 ha of Wattle planted in Kenya
1945	1,80,000 ha of Pine and Eucalypts plantation established in South Africa
	9,800 ha of *Pinus elliottii* and *Araucaria cunninghamii* planted in Queensland, Australia
	80,000 ha of Teak plantation was established in India
1966-1977	40,000 ha/yr was planted on poor quality Amazon forest in Brazil with *Pinus caribaea*, *Eucalyptus deglupta* and *Gmelina arborea*
1971-1979	Fiji planted about 28,000 ha of *Pinus caribaea*
1965-1980	15,000 ha of Eucalypts established in Congo
	1,80,000 ha of Eucalypts established in Sudan
1966-1979	India established 2.8 million ha of new plantation mainly on Teak
End of 1980	About 11.5 million ha of plantation forestry was established in the tropical countries, where 7.2 million ha and 4.3 million ha were mainly for industrial and non-industrial purposes respectively
1990	Total area of planted forest reported in world: 178.307 million ha
2000	Total area of planted forest reported in world: 214.839 million ha
2005	Total area of planted forest reported in world: 242.965 million ha
2010	Total area of planted forest reported in world: 264.084 million ha

STATUS OF GLOBAL PLANTED FORESTS

Planted forests are composed of trees established through planting and/or through deliberate seeding of native or introduced species. Establishment is either through afforestation on land that until then was not classified as forest, or by reforestation of land classified as forest, after a fire or a storm or following clearfelling. Forests and trees are planted for many purposes and make up an estimated 6.6 per cent of the total forest area, or 264 million ha in 2010. Most of the plantations were established through afforestation i.e. planting of areas not forested in recent times. Three-quarters of all planted forests consist of native species while one-quarter comprises introduced species. At the global level planted forest area has steadily increased since 1990 by an average of 4.3 million ha per year. During 2000-2010, the area of planted forest increased by about 5 million ha per year. Given this trend, a further rise in the planted forest area upto 300 million ha by 2020 can be anticipated. The impact of this development on timber markets should be considered by policy-makers, planners and forest managers and supported by outlook studies that evaluate the future contribution of planted forests to economic, environmental and social services (FRA 2010a).

Table 9.2: Status of Global Planted Forest Development (FRA 2010a)

Region	Planted Forest Area (Million ha)				% to Total Forest Area	Planted Forest Annual Increment (Million ha)	
	1990	2000	2005	2010	2010	1990-2000	2000-2010
Africa	11.663	12.958	14.032	15.409	2.3	0.129	0.245
Asia	76.258	92.928	109.670	122.775	20.8	1.667	2.985
Europe	59.046	65.312	68.502	69.318	6.9	0.627	0.401
North and Central America	20.481	30.261	35.787	38.661	5.5	0.978	0.840
South America	8.276	10.058	11.123	13.821	1.7	0.178	0.376
Oceania	2.583	3.323	3.851	4.101	2.1	0.074	0.078
World	178.307	214.839	242.965	264.084	6.6	3.653	4.925

East Asia, Europe and North America reported the greatest area of planted forests, together accounting for about 75 per cent of global planted forest area. In East Asia, planted forests make up 35 per cent of the total forest area and most of these are found in China. The second largest area of planted forests is found in Europe and the third largest area of planted forests is available in North America. The subregions reporting the smallest area of planted forests are African subregions, the Caribbean, Central America and Western and Central Asia. In total, there are 33 countries with a planted forest area above 1 million ha which together account for 90 per cent of the global planted forest area. Of these, China, USA, Russian Federation, Japan and India together account for more than half the world's planted forests (53 per cent). Top ten countries with greatest annual increase in planted forest area during 1990-2010 are China (1.932 million ha per year), USA (0.805 million ha per year), Canada (0.385 million ha per year), India (0.251 million ha per year), Russian Federation (0.199 million ha per year), Mexico (0.178 million ha per year), Brazil (0.156 million ha per year), Vietnam (0.134 million ha per year), Turkey (0.091 million ha per year) and Finland (0.082 million ha per year).

SIGNIFICANCE OF FOREST PLANTATIONS

Forest plantations are variously defined in the literature and even the standardized definition adapted by FAO has significantly changed over time. The main shared characteristics of these definitions point to forested areas artificially established by planting or seeding. Other features are that the trees usually belong to the same species (native or introduced), have the same age structure and are regularly spaced (FAO 2006; FRA 2010a). Forest plantations are defined as those forest stands established by planting and/ or seeding in the process of afforestation or reforestation. They are either of introduced or indigenous species that meet a minimum area requirement of 0.5 ha, tree crown cover of at least 10 per cent of the land cover and total height of adult trees above 5m (FRA 2000).

Forest plantations embrace a range of forest types with the one common feature that the great majority of the trees present were established on the site by planting and/ or seeding (sowing). There is no internationally agreed definition of forest plantations, although the expression 'planted forest' is now widely used to embrace the continuum of forest types where forest origin is known to be by planting. Mostly plantations contain a single tree species, although plantations of a mixture of species are important in some parts of the world. Plantation forestry is often much like any other agricultural enterprise, aiming to grow highly productive forests on relatively small areas of land, so the land is being used most efficiently. For achieving this, much attention is to be paid to the silviculture of plantations, i.e. to the tending of the trees to achieve some desired objectives.

As the population of the world increases and its economies grow, so does the amount of wood people use, for firewood, for building, for paper and for many other purposes. Plantation forestry will become more important in meeting the growing demand for wood. Today, plantation forests play a significant role for wood production, especially in tropical countries, because of several important characteristics including high yield, short rotations and accessibility. Although some plantations are for protection purposes, most are for production (FAO 2006) and plantation's share in providing roundwood to industry is growing. Plantation forestry is also increasingly changing from large-scale investments in monocultures to small-scale investments in which local households and communities are the principle owners of the means of production. On all the continents, there are large plantation areas grown to produce wood for industrial use. They often cover millions of hectares and are major commercial enterprises. Some are publicly owned and some are privately owned. They are important economically to the countries concerned, often earning export income and providing employement, both in their growing and in the processing of the wood they produce. Many such plantations are grown for only 7-15 year rotations before harvesting. Usually, these produce wood for papermaking. Others are grown on 20-30 year rotations, by which time the trees will be large enough for their stems to be sawn to produce timber.

An emerging type of industrial plantation is being grown to produce bioenergy. The wood from such plantations is used to make energy, either through fermentation to produce ethanol for use as motor fuel or through burning the wood to generate electricity. The primary stimulus for the establishment of bioenergy plantations comes from concerns about the global warming of the atmosphere, which is believed to be caused by the release of greenhouse gases including CO_2. Bioenergy plantations are usually grown on very short rotations (perhaps 3-5 years) and the small size of their trees at harvest is of no importance, because they are simply going to be burnt or used for fermentation. Additionally, plantation forestry in the developing nations of tropical areas is very strongly nested on the political agenda due to the planted forest's carbon storage capacity as noticed in the climate change debate. New incentives such as the Clean Development Mechanism (CDM), afforestation/reforestation methodologies and the debated mechanism of Reducing Emissions form Deforestation and Degradation (REDD), emphasize plantations as a relatively cheap, clean and harmless means of emission compensation.

PARADIGM SHIFTS IN TROPICAL FOREST PLANTATIONS

While forest plantations are a phenomenon known for millennia, in the tropical context they have evolved significantly since the early colonization period. Planting activities, associated predominantly with technical skills and silvicultural knowledge, have been affected by ideational factors, norms, social context and the broader historical and economic developments. Although the humanistic and social scientific research on plantations has been recognized (Evans 2009; Evans and Turnbull 2009), it remains rather marginal and weak in theory-building, macro diagnosis and critical reflections. The theoretical concept of a paradigm is defined as a set of assumptions, values and practices that constitutes the way of perceiving reality shared by the community, especially in intellectual and scientific circles. The tropical forest plantations have undergone many paradigm shifts worldwide in various time periods viz. pre-industrial paradigm, industrial colonial paradigm, industrial post colonial or industrial national paradigm, protective paradigm, social paradigm, neo-liberal paradigm, neo-liberal modified paradigm, global political (Julia Szulecka 2011).

PRE-INDUSTRIAL PARADIGM: Tree planting in ancient times was limited due to the wide availability of natural forests. However, with the migration of people and functions of valued tree species, tree planting was a common practice. Ancient tree planting practices worldwide share many common properties as non-exclusive land use, community management, multipurpose functions, small scale and use of local knowledge during pre-industrial paradigm.

INDUSTRIAL COLONIAL PARADIGM: Forest plantations on a larger scale with a commercial purpose were started in the global south in the 16th and 17th century during industrial colonial paradigm. Firstly, plantations of hardwood species were developed for strategic purposes (teak) e.g. ship building and later for commercial purpose (teak, eucalypts, etc.). This paradigm introduced a restricted exclusive land use for tree planting with specially designed authorities responsible for plantation management. The vertical top-down decision making scheme, introduction of Western scientific forestry knowledge and delegitimizing the local traditional cognizance were important elements of industrial colonial paradigm. The other was the way plantations were justified and explained to the populations as a means to bring progress and economic development (Bryant 1996). The new plantations required much more initial capital investments but were also able to bring economic benefits for the colonial enterprises.

INDUSTRIAL NATIONAL PARADIGM: The important elements of this paradigm are exclusive land ownership, centralization, land accumulation, forest management by national forest departments, economic goals, progress-discourse for justification, vertical top-down decision making scheme, national scientific forestry knowledge from the West, State incentives and plantations of softwood and hardwood.

PROTECTIVE PARADIGM: It was initiated as a counterbalance for industrial productive plantations and due to high deforestation with the goal to use tree planting for soil, water, wind regulation and environmental services. The protective paradigm intensified in the 20th century with soil protection in many parts of Africa (e.g. Ethiopia), combating desert encroachment (e.g. Algeria, Libya, Morocco, Sudan) or flood protection (e.g. China's reforestation programmes). Protective plantations bring new justification for planting and promote generally non-exclusive land use, national priorities and national scientific forestry knowledge with local functions. Investments in plantations are made by State with the objective of no or low production.

SOCIAL PARADIGM: The world forestry congresses in the 1970s with the themes "forests and socioeconomic development" and "forests for the people" clearly illustrated the ideational change from technocratic towards more responsible and socially-engaged forestry (FAO 1978). Social paradigm partly reflected the pre-industrial paradigm by empowering local traditional knowledge, role of the communities in management, bottom-up decision making and horizontal governance. The major elements are non-exclusive land ownership, adapted technology transfer, multiple knowledge, agroforestry, community investments and external organization finance and moderate productivity due to multi-functionality.

NEO-LIBERAL PARADIGM: Global economic re-conceptualization according to neo-liberal thought assured a big change in plantation actors and financing, especially in the developing world. The State, the previous leader in promoting industrial plantations has been widely replaced by the private actors. Due to increased silvicultural technology and high growth rates in the tropics, many private transnational actors started to engage in plantation initiatives. The neo-liberal paradigm in plantations stressing "growth" in monetary terms promoted again exclusive land use, private management, top-down decision making and international forestry knowledge. Huge private capital investments resulted in an extremely high productivity as the increment records from Brazil of 90 m^3 per hectare per year (Evans and Turnbull 2009).

NEO-LIBERAL MODIFIED PARADIGM: Social conflicts caused by neo-liberal plantations lead to a new approach called neo-liberal modified paradigm. It refers to constraints of the private companies and a need to limit land grabbing by establishing partnerships with local farmers, who remain landowners and can be contracted to plant trees (Race and Desmond 2001). The major elements are exclusive land use without ownership, management according to the ownership schemes, profit maximization, constrained top-down outlook, international scientific forestry with possible local inputs, investments reduced by land costs, possible benefit-sharing, high growth and productivity.

GLOBAL POLITICAL PARADIGM: The new emerging paradigm in forest plantations called global-political originates from the ideas of international community present in international agencies and global conventions. It promotes project-management style of plantation projects and global priorities are negotiated between States, sectors and scientific disciplines. Examples include short rotation plantations for the global resources supply (UNFF), plantations to mitigate climate change and carbon storage (UNFCCC), plantations to protect global biodiversity (UNCBD) and plantations to combat desertification (UNCCD).

SUSTAINABILITY OF PLANTATIONS

BROAD AND NARROW SENSE SUSTAINABILITY: Forest plantations are an increasingly important resource worldwide, a trend that is expected to continue strongly. The question of sustainability in plantation forestry has two components. There are the general or broad issues of whether using land and devoting resources to tree plantations is a sustainable activity from the economic, environmental or from the social sense. They can be labelled 'broad sense' sustainability. The second component, 'narrow-sense' sustainability, is largely a biological and silvicultural issue. The question raised is: can tree plantations be grown indefinitely for rotation after rotation on the same site without serious risk to their well being? More specifically, can their long-term productivity be assured, or will it eventually decline over time? These questions are pertinent owing to the increasing reliance on plantation forestry, but are also scientifically challenging since in previous centuries trees and woodlands were seen as soil improvers and not impoverishers.

Today's silvicultural practices are more damaging because of greater intensity and the high timber yields achieved, typically 2-4 times that of natural forest increment. And, of course, are resources such as genetic improvement, targeted fertilizer application, and sophisticated manipulation of stand density, along with rising atmospheric carbon dioxide, likely to lead to crop yield improvement, but such intensive silvicultural practices also lead to genuine site degradation or increasing risk of damage by pests and diseases. The concept of sustainability is central to sound forest management and the subject of much debate. In the case of forest plantations established with the specific purpose of producing wood, whether for industrial purposes or domestic use, it must be assured for the perpetual supply of forest products. Accurate evidence about the sustainability of plantation forestry is needed, since future supplies of forest products will increasingly come from intensively managed forests and plantations (Evans 1997).

SOIL FERTILITY CHANGES: Demonstrating that soil changes may be caused by forestry practices, is usually difficult to establish conclusively both in fact and in scale. An absence of sound baseline data is common and there is a question about reported change actually induced by plantation silviculture. Plantations may have three impacts viz. nutrient removal from soil as trees grow and then are harvested; changes in the chemistry of the soil surface as the litter layer and organic matter are dominated by one species and hence uniform composition and decay characteristics; and site preparation practices such as ploughing, drainage and fertilizing which directly affect soil physical parameters and in turn nutrient and moisture availability.

LITTER: The influence of litter on soil nutrient status may be important since leaves of different species decay at different rates. For example, substantial accumulations may develop under *Pinus patula* on certain sites while this is unusual beneath the more lightly canopied *Pinus elliottii*. In broad-leaved stands, accumulation of litter is uncommon though not unknown. Even under teak and Gmelina, which usually suppress all other vegetation, the large leaves readily decay. Similarly under the light crowns of eucalyptus and the nitrogen rich foliage of leguminous trees such as *Acacia*, *Leucaena* and *Prosopis* spp and non-legume N-fixers such as casuarinas, litter build up is rare owing to rapid decay of the rich organic matter.

SOIL PHYSICAL CONDITIONS: Plantation forestry may impact soil physical conditions and hence sustainability through (1) site preparation and establishment operations, (2) the effects of tree growth itself for example on water uptake, and (3) harvesting practices.

ORGANIC MATTER DYNAMICS: What happens to the litter and organic matter layer at the soil surface is critical to the question of sustainability. The surface litter layer helps prevent soil erosion. Litter and organic matter represent a significant nutrient store, albeit a dynamic one. The 'litter-organic matter-mineral soil interface' is the seat of nutrient cycling and microbial activity. Any activity that disturbs these roles in the ecosystem can have large effects of which perhaps most serious of all, and still practiced in some countries, is regular and frequent litter raking. In commercial plantation forestry, the cost of managing debris and site preparation when restocking plantations is expensive and a high proportion of the establishment costs (Evans 1997).

WEEDS: Establishment of plantations greatly affects ground vegetation with many operations designed directly or indirectly to reduce weed competition to ensure that the planted tree has sufficient access to site resources. A neglected but critical phase is managing the weed problem through crop harvesting and restocking. In subsequent rotations, the weed spectrum often changes. Birds and animals may introduce new weed species, grass seed may be blown into plantations and accumulate over several years, and roads and rides in plantations can become sources of weed seeds. Weed management must be a holistic operation. Failure of proper weed management will always result in yield declines and less sustainability. Often, the significance of weeds has been insufficiently recognized on restocked sites in second or third rotations (Evans 2000).

A great majority of forest plantations are of uniform age and composition (monoculture) and most are managed to optimize the yield of wood from a site. Clear-felling and replanting is the most common silvicultural system although, where appropriate, coppicing is used as a means of restocking. These features of plantation silviculture have raised concern that many of the sites on which trees are planted may be incapable of sustaining their productivity. Models of nutrient export, examination of physical damage of soil structure and claims of greater risk from pests and diseases have all been advanced as hypotheses for the inherent unsustainability of intensive plantation forestry.

PLANTED FOREST MANAGEMENT

Planted forest management means the planning and implementation of all types of regulations, institutional arrangements, research and development activities, policies, monitoring and forest operations related to the planted forest, whether at strategic policy levels or at the operational field level.

EXOTICS IN PLANTATION FORESTRY

Majority of the industrial plantations in most countries are being raised with exotic tree species like *Eucalyptus* spp, *Acacia* spp, *Pinus* spp, etc. The reasons for greater preference for the selection of exotic species include:

- There is generally a lack of adequate knowledge in the propagation and silvicultural management of indigenous species

- There is generally plentiful supply of seeds of the exotic species
- The exotic species are easy for handling
- The exotics are fast growing and high yielding

The selection of indigenous species for raising forest plantations are poorly recognized in many countries. There is a need for consolidated research works and concrete evidences for promoting the identified and priority indigenous tree species in plantation forestry at national level. Many countries have already initiated research works for identifying promising indigenous trees and plantations have been raised at national level also.

GENETIC IMPROVEMENT

Better adapted species, change in seed origin, use of new clones, use of genetically improved seed and, in the future, genetically modified trees all offer the prospect of better yields in plantation forestry. The impact of all these genetic improvements will affect yield and outturn directly and indirectly through better survival and greater suitability to the site which may lead to increased vigour and perhaps greater pest and disease resistance. Some of the world's most productive tree plantations use clonal material, including both eucalyptus and poplars. It is clear that both the potential productivity and the uniformity of product make plantation forestry attractive. Although clonal forestry has a narrow genetic base, careful management of clone numbers and the way they are interplanted can minimize pest and disease problems e.g. use of 30-40 unrelated clones will generally provide security against catastrophic failure.

Through an array of selection, crossing, and propagating techniques traits can be favoured that may improve vigour, stem and wood quality, pest and disease resistance and other parameters such as frost tolerance. There are many examples of successful tree improvement strategies most of which are only beginning to bear fruit owing to long tree rotations and the slow process of tree breeding, particularly in orchard establishment and promotion of flowering, and in field testing of selections and progenies. Genetic tree improvement offers by far the greatest assurance of sustained and improved yields from successive rotations in the medium and long-term. There are no widely planted examples at present where genetic engineering has modified trees. The expectation is that genetically modified trees will be used to develop disease resistance, modified wood properties, cold or drought tolerance rather than increase in vigour (Evans 2000).

SILVICULTURAL TECHNIQUES

Silvicultural knowledge continues to increase through research and field trials and greater understanding of tree and stand physiology. While large yield improvements appear unlikely, incremental gains can be expected. Manipulation of stocking levels will achieve greater output of fibre or a particular product, by fuller site occupancy, less mortality and greater control of individual tree growth. Matching rotation length to optimize yield - the rotation of maximum mean annual increment - offers worthwhile yield gain in many cases. In some localities, prolonging the life of stands subject to windthrow by silvicultural means will increase yield over time. Use of mixed crops on a site may aid tree stability, may lower pest and disease threats, but is unlikely to raise productivity over growing the best suited species (FAO 1992). Silvicultural systems that maintain forest cover at all times - continuous cover forestry - such as shelterwood and selection systems are likely to be neutral to slightly negative in production terms while benefiting tree quality, aesthetics, and probably biodiversity value. Crop rotation, as practiced in farming, appears unlikely. There are examples of forest plantations benefiting from a previous crop of nitrogen fixing species e.g. *Acacia mearnsii* but industry is likely to require a similar not a widely differing species when replanting.

Fertilization: Most forest use of fertilizer is to correct known deficiencies e.g. micronutrients such as boron in much of tropics and macronutrients such as phosphorus on impoverished sites in many parts of both the tropical and the temperate world. In most instances, fertilizer is only required once in a rotation. Fertilizer application is likely to be the principal means of compensating for nutrient losses on those sites where plantation forestry practice does cause net nutrient export to detriment of plant growth.

Site Preparation: Ground preparation to establish the first plantation crop will normally introduce sufficient site modification for good tree growth in the long term. Substantial site manipulation is unlikely for second and subsequent rotations, unless there was failure first time around, except to alleviate soil compaction after harvesting or measures to reduce infections and pest problems.

Weed Control: Weed control strategies may change from one rotation to the next owing to differing weed spectrum and whether weeds are more or less competitive. The issue is crucial to sustainability since all the main examples of yield decline reflect worsening weed environments, especially competition from grasses and bamboos.

Conservation of Organic Matter: It is clear from many investigations that treatment of organic matter both over the rotation and during felling and replanting is as critical to sustainability as coping with the weed environment. While avoidance of whole tree harvesting is probably desirable on nutrition grounds, it is now evident that both prevention of systematic litter raking or gathering during the rotation and conserving organic matter at harvesting are essential.

PEST AND DISEASE INCIDENCE

A serious threat to plantations can arise from a massive build-up of a pest or disease. It has been much disputed whether monoculture itself is more susceptible to devastation from these causes. The broadly accepted ecological principle of stability is that the stability of a community and its constituent species is positively related to its diversity. Following this reasoning, foresters have stressed that substitution of natural forest by even-aged monoculture plantations may remove many of the natural constraints on local tree pest and pathogens and thus increase risk of attack. Increased susceptibility mostly arises from conditions in plantations rather than because only one tree species is present. The relative susceptibility of monocultures to pest and disease attack is complex ecologically.

Plantations of one or two species offer an enormous food source and ideal habitat to any pest and pathogen species adapted to them. Uniformity of species and closeness of trees including branch contact above ground and root lesions in the soil, allow rapid colonization and spread of infection. Narrow genetic base in plantations e.g. one provenance or no genetic variation (e.g. clones) reduces the inherent variability in resistance to attack. Trees grow on a site for many years and permit pest or disease to build up over time. Many plantations are of introduced species and while without the insect pests and pathogens that occur in their native habitat also missing are the many natural agencies controlling pests and diseases. Efficient management of pest and diseases in forest plantations is very important factor for the success of plantations (Evans 2000).

CONSTRAINTS IN PLANTATION FORESTRY

In many countries, while forest plantations are a lucrative option to supply the ever increasing demand for wood on a sustainable basis, the planting of timber trees on a plantation scale is constrained by a number of other factors that are critical. Ecology, land, site and species selection, planting material, labour and mechanization and economics are the major constraints for the success of plantation forestry (Sheikh and Tahir Siddiqui 1999).

ECOLOGY: The establishment of forest plantation involves extensive alteration of the ecosystem particularly when heavy equipment is employed. The complex closed nutrient cycle in tropical rain forests is disrupted for a long time. This can lead to reduction of productivity unless ameliorative measures are undertaken. Furthermore, monocultures further destabilize the system and require heavy use of fertilizers and pesticides. Many slow-growing species grown under fast plantation conditions have poorer quality. There is the problem of species-site matching for the heterogeneous area of large plantations. The danger of fire may also increase in exotic species plantation.

LAND: It is an established fact that land is the world's most valuable resource and public scrutiny of land is becoming more intense with each passing year. With increasing population, the competition on land for agriculture, forests and plantations, pasture and development activities is ever increasing. For a forest plantation investment to be commercially viable, a large area is required. The size of the land required will vary with the objective of the plantation. If the timber is for sawmilling and furniture manufacturing then a smaller area would suffice. On the other hand, if the objective is in establishing a chip or pulp and paper mill then larger areas are required for raising plantations.

It would always be desirable to have a single contiguous piece of land area and if not, the required land area should just be in about two or three nearby parcels only. This is to ensure easy and efficient management of the activities. It is always preferable that the acquired land is close to basic amenities and near a relatively accessible road system and within an economic range to a processing mill or market. As an example, to operate efficiently a pulp or a chip mill the plantation should be located within a 100 km radius. Otherwise, exorbitant cost for the transportation of logs would render the operations uneconomical. Another factor is that land is under the State's jurisdiction. Land being a state prerogative implies that commercial organizations may have to deal with different procedures adopted by each State Governments. Often inquiries on information regarding land can become very difficult. Besides sufficient land, the location with suitable infrastructure and the premium rate for leasing are also crucial factors.

SITE SELECTION: It is essential to have an in-depth knowledge on local site conditions and silviculture, for deciding which species is to be raised under what conditions. It is not advisable to raise some susceptible tree species on unsuitable soil and climatic condition. Susceptible species bears the huge risk of pest and disease attack and failure to produce the desired yield. It is also advisable to have the knowledge of plant and root growth of a particular tree species on the site conditions. Soil and climatic conditions will be the major factors which will decide the success of plantation programme.

SPECIES SELECTION: A review of many reforestation projects in the tropics revealed that about 95 per cent of projects utilized exotics in the reforestation programmes (Evans 1982; Abod 1995). For plantations, although indigenous species are available, a greater preference is given for the selection of exotic species. There is a great challenge ahead to carry out adequate studies on the indigenous species to see their viability for forest plantations.

INADEQUATE SUPPLY OF ELITE PLANTING MATERIAL: In general, high levels of productivity are achieved when genetic and physiological potential of the species are well matched with management practices which promote rapid growth. Valuable improvements can be made in important properties such as stem form and wood density through selection and breeding. One major constraint that is currently perceived is the shortage of good planting material for the various plantation programmes. Quality seeds and plus trees that have been selected and reproduced by vegetative methods are inadequate to meet current and projected needs. While efforts are being stepped up to overcome this problem, middlemen and overnight nurseries are providing planting materials whose genetic sources are unknown.

LABOUR AND MECHANIZATION: Labour supply is another issue of great concern. In many countries, the agricultural sector is experiencing a shortage of labour because of the rural to urban migration of youth in search of better job opportunities. Although the labour requirement in forest plantations is less than in agriculture, it still has to compete for labour in an expanding economy. One option to alleviate the labour shortage is increased mechanization. Machines developed in most advanced countries are environmentally friendly and highly flexible in their operation in forest plantations. But again the mechanized plantation activities are limited by the size of plantation area, higher initial and operation cost.

ECONOMIC FACTORS: The planting of timbers on a plantation scale is constrained by a number of economic factors as well. The first thing will be high initial capital investment to establish the forest plantations. The long period, between initial planting efforts and harvesting, is another important concern and thus the corresponding concern for interest being carried until harvesting period. The high biological and economic risk involved in forest plantations are major limiting factors. Unattractive and inappropriate investment incentives provided by the Government for forest plantation investments result in the lack of interest from private sectors in commercial ventures in forest plantation development.

10 SUSTAINABLE FOREST MANAGEMENT

INTRODUCTION

Forests are areas with rich biodiversity that provide us with services and goods on which our wellbeing depends upon. Forests produce the raw material for renewable and environmentally friendly products and play an important role for the global climate and carbon cycle, biological diversity and water balance. The protecting services of forests are essential for mitigating natural disasters such as floods, droughts and avalanches. Forests are an important resource for economic welfare and rural development in local communities. They provide employment for millions of people in forestry and forest-based enterprises. Forest-derived recreation opportunities are of increasing importance in light of the growing urban population and the interest in nature tourism.

Forests play a critical role in supporting the livelihoods of people worldwide, particularly, in meeting the daily subsistence needs of the world's poor. According to the World Bank, over 90 per cent of the 1.2 billion people living in extreme poverty depend on forests for many parts of their livelihoods. Eradicating poverty is therefore impossible without paying specific attention to the 410 million people (including 60 million indigenous people) who live in or near tropical forest areas and depend on these forests for their subsistence and survival needs. Sustainable forest management can contribute to economic development by providing income, employment, food security and shelter where it is most urgently needed. Finding ways to balance human needs with concerns over the long-term sustainability of forest resources is the very essence of sustainable forest management.

Even despite the enormous importance of forests, thousands of hectares of forest are lost each year. Deforestation is the worst in Indonesia and Africa where tropical rainforests are disappearing at an alarming rate. Forests, especially rainforests, together with oceans are the most important carbon sinkers. Therefore they play a key role in reducing the total climate change impact. The forest loss around the globe leads to release of billions of tons of CO_2 into the atmosphere. Current forest conservation efforts have not been very successful and world needs to find quick solutions to preserve forests for our future generations. Among the proposed solutions are halting the current rate of forest loss, establishing more protected areas such as national and state parks, ensure better land-use planning and improve the current management of protected areas.

Since the Earth Summit of 1992, the need to manage forests sustainably has been well-recognized by the international community. The principal focus of the UN forest-related forums since then has been to implement the aims of the Summit through promoting sustainable forest management (SFM). UN Forum on Forests set four Global Objectives (UNFF 2006) that are central to SFM. These include: addressing the loss of forest cover and forest degradation, forest-based economic, social and environmental benefits and protecting forests, as well as mobilizing financial resources for the implementation of sustainable forest management.

For several reasons, past efforts to achieve sustainable forest management at the global level have not been very successful. One primary reason is the lack of broad recognition of the value of well-managed forests for society in the long term. Another reason is the unique feature of forestry where the same unit of forest may represent a variety of sometimes conflicting values. In such a situation, pursuing one objective implies sacrificing another. However choosing one objective over the other may lead to debate, particularly in regard to public forests, given the very different objectives of the various stakeholders. Other complicating factors are the uncertainty associated with interventions in complex forest ecosystems and the long time dimensions. Different forest management approaches not only result in different ecosystems, but also in different combinations of output of products and services, over a period of time.

WHAT IS FOREST MANAGEMENT?

The term forest management is applied in situations where an integrated, coordinated series of actions, which are directed towards the achievement of specified objectives. Forest management is a process which effectively integrates the biological, social and economic factors which influence the decisions leading towards the implementation of one or more specified objectives.

Forest Management deals with the overall administrative, economic, legal, social, technical and scientific aspects related to natural and planted forests. It implies various degrees of deliberate human intervention to safeguard and maintain the forest ecosystem and its functions, in order to favour specific valuable species or groups of species for the improved production of goods and services. Sustainable forest management will ensure that the values derived from the forest meet present-day needs while at the same time ensuring their continued availability and contribution to long-term development needs (FAO 1993a).

Historically, forest management has mostly considered biological issues with a strong focus upon silviculture for the production of wood. As the forestry profession has grown, an understanding of the term "forest management" has broadened to span wider environmental issues, such as conservation of biological diversity, social and economic matters and, more generally, the concept of sustainability. The Forest Principles, developed at the Earth Summit - the United Nations Conference on Environment and Development (UNCED) - have defined forest management as a part of a statement of principles for a global consensus on the management, conservation and Sustainable development of all types of forests.

Forest resources and forest lands should be sustainably managed to meet the social, economic, cultural and spiritual needs of present and future generations of human beings. These human needs are for forest products and services, such as wood and wood products, water, food, fodder, medicine, fuel, shelter, employment, recreation, habitats for wildlife, landscape diversity, carbon sinks and reservoirs, and for other forest products. Appropriate measures should be taken to protect forests against harmful effects of pollution, including air-borne pollution, fires, pests and diseases in order to maintain their full multiple values (UNCED 1992).

The basic principle of forest management comprises the following elements: national policy and legal framework, security of tenure of forest resources and land, effective forest protection, knowledge on sustainability of forest ecosystems, maintenance of site productivity, forest management planning, goals and objectives for forest management, definition of forest resources, application of appropriate silvicultural systems, minimization of adverse environmental impacts, a regard for the interests of forest-dependent communities, commercial sustainability and business management, monitoring of managerial performance

CONCEPT OF SUSTAINABLE FOREST MANAGEMENT

The perception that human development needed to be carried forward in a sustainable way was formulated during the 1980s and detailed discussions at UNCED constituted the concept of sustainable forest management (SFM). During 1990s there have been a number of processes, launched by different constituencies to define SFM viz. Helsinki Process (the agreements of European governments); Montreal Process (the agreement among other temperate latitude governments); ITTO guidelines (the product of International Tropical Timber Organization representing producers and consumers); Forest Stewardship Council (FSC). Considerable procedural differences still exist among derivations of general standards, termed criteria and indicators (C&I) for sustainable forest management.

Sustainable forest management (SFM) is the management of forests according to the principles of sustainable development. Sustainable forest management uses very broad social, economic and environmental goals. The "Forest Principles" adopted at The United Nations Conference on Environment and Development (UNCED) in Rio de Janeiro in 1992 captured the general international understanding of sustainable forest management at that time. A number of sets of criteria and indicators have since been developed to evaluate the achievement of SFM at both the country and management unit level. These were all attempts to codify and provide for independent assessment of the degree to which the broader objectives of sustainable forest management are being achieved in practice. In 2007, the United Nations General Assembly adopted the Non-Legally Binding Instrument on All Types of Forests. The instrument was the first of its kind and reflected the strong international commitment to promote implementation of sustainable forest management through a new approach that brings all stakeholders together.

The concept of sustainable forest management contains guidelines and criteria to secure the optimal balance of goods and services from forests. It is an efficient tool to help countries and local communities achieve an appropriate equilibrium between the multiple needs of society. A common understanding of what sustainable forest management encompasses has been the basis of the collaboration between various countries to evolve joint definition of sustainable forest management. SFM is the management to maintain and enhance the long-term health of forest ecosystems, while providing ecological, economic, social and cultural opportunities for the benefit of present and future generations.

FOREST EUROPE

The good definition of the present day understanding of the term 'Sustainable Forest Management' was developed by the Ministerial Conference on the Protection of Forests in Europe (MCPEE) and has since been adopted by FAO. It defines sustainable forest management as *the stewardship and use of forests and forest lands in a way and at a rate, that maintains their biodiversity, productivity, regeneration capacity, vitality and their potential to fulfill, now and in the future, relevant ecological, economic and social functions, at local, national, and global levels and that does not cause damage to other ecosystems.*

FAO

Sustainable forest management implies various degrees of deliberate human intervention, ranging from actions aimed at safeguarding and maintaining the forest ecosystem and its functions, to favouring specific socially or economically valuable species or groups of species for the improved production of goods and services (FAO 1999). SFM ensures that the values derived from the forest meet present day needs while at the same time ensuring their continued availability and contribution to long-term development needs (FAO 2008). It has a tremendous potential to serve as a tool in combating climate change, protecting people and livelihoods, and creating a foundation for more sustainable economic and social development.

ITTO (2005)

ITTO defines sustainable forest management (SFM) as the process of managing forest land to achieve one or more clearly specified objectives of management with regard to the production of a continuous flow of desired forest products and services without undue reduction of its inherent values and future productivity and without undue undesirable effects on the physical and social environment.

It means that forest-related activities should not damage the forest to the extent that its capacity to deliver products and services - such as timber, water and biodiversity conservation - is significantly reduced. Forest management should also aim to balance the needs of different forest users so that its benefits and costs are shared equitably.

UNITED NATIONS

The general assembly of the United Nations adopted in December 2007 the most widely, intergovernmentally agreed definition of SFM as a dynamic and evolving concept aims to maintain and enhance the economic, social and environmental value of all types of forests, for the benefit of present and future generations. It is characterized by seven elements viz. extent of forest resources, forest biological diversity, forest health and vitality, productive functions of forest resources, protective functions of forest resources, socio-economic functions of forests, and legal, policy and institutional framework (UN 2008).

CONCEPT OF SUSTAINABILITY

Large areas of forest land, especially in the tropics, will inevitably be converted to agricultural use in the coming decades. Logging and cutting for fuelwood will continue. The challenge is not to prevent these activities but to manage them. The aim must be to ensure that wood and other forest products are harvested sustainably, that forests are cleared only in a planned and controlled way and that the subsequent land uses are productive and sustainable. The World Commission on Environment and Development (1987), commonly called the Brundtland Commission, clearly recognized the necessity for a broad approach to sustainability. It said "sustainable development is a process of change in which the exploitation of resources, the direction of investments, the orientation of technological and institutional change are all in harmony and enhance both current and future potential to meet human needs and aspirations".

The concept of sustainability has deep historical roots in forestry. Much early forestry in World was concerned with the conservation of forests as wildlife reserves for hunting by kings and nobles. Later came the concept of managing forests for a sustainable yield of timber. This was achieved by balancing the volume to be harvested against the growth predicted from regeneration and planting. "The sustainable management of forests for the production of wood is based on a deceptively simple principle. All that needs to be done is to harvest the wood at an average annual rate not greater than the forest in question can grow it" (FAO 1993b).

Management of the forest to provide a sustained yield of timber is still what many foresters have in mind when they talk of sustainable forest management. This definition focuses on the production of wood and does not address the wider issues of the ecological and social functions of forests, with which timber production may only incidentally be compatible or may even lead to conflict. Over the past two decades, management solely for wood production has been a cause of steadily growing concern to those affected by the loss of other benefits. It has led, in an increasing number of areas, to confrontation and even physical conflict between loggers and people living in and around the forest areas being harvested. The concept of sustainable forest management has therefore evolved to encompass these wider issues and values. It is now seen as the multipurpose management of the forest so that its overall capacity to provide goods and services is not diminished. A forest managed in this way will provide timber on a sustainable basis and will continue to provide fuelwood, food and other goods and services for those living in and around it. Its role in the preservation of genetic resources and biological diversity as well as in the protection of the environment will also be maintained.

The application of forest management in this wider sense would represent enormous progress in most of the forest areas of the world. It could be interpreted as confining forest management to those areas where a sustained yield of forest products and services is, at least in principle, achievable. There is a need for explicit attention to those huge areas where the forests are disappearing as a result of encroachment and clearing for agriculture, where excessive grazing is preventing natural regeneration of trees or where cutting for charcoal-making and fuelwood is leading to the degradation or disappearance of forests.

If sustainable forest management is to be achieved such issues must be addressed and the concept needs to be widened further or perhaps better viewed as management for sustainability. This approach of necessity must be a holistic one encompassing land-use planning and the wider questions of rural development. Management for sustainability will therefore first be concerned with securing an improved livelihood for the present generation, while maintaining the potential of the forest heritage for future generations. Second, the forest potential must be seen within the broader context of rural development, in which the allocation of land to different uses is part of a dynamic process but where a balance is maintained between forests and other forms of land use in which trees have a role. Third, responsibilities for forest management must be clearly identified and competing interests must be reconciled through dialogue and partnership. Finally, forestry activities will have to compete for scarce financial resources and both the production and the environmental functions must be shown to be worthwhile to both users and financers (FAO 1993c).

COMPONENTS OF SUSTAINABLE FOREST MANAGEMENT

The following seven thematic elements, acknowledged by UNFF, are based on the criteria of the nine on-going regional/international processes on criteria and indicators for sustainable forest management, and were acknowledged by the FAO Committee on Forestry in 2003. In February 2004, the FAO/ITTO Expert Consultation on Criteria and Indicators recognized that these elements are important for facilitating international communication on forest-related issues.

EXTENT OF FOREST RESOURCES

The theme expresses an overall desire to have significant forest cover and stocking, including trees outside forests (TOF), to support the social, economic and environmental dimensions of forestry. For example, the existence and extent of specific forest types are important as a basis for conservation efforts. The theme encompasses ambitions to reduce deforestation and to restore and rehabilitate degraded forest landscapes. This theme also includes the important function of forests and trees outside forests to store carbon and thereby contribute to moderating the global climate.

BIOLOGICAL DIVERSITY

The theme concerns the conservation and management of biological diversity at the ecosystem (landscape), species and genetic levels. Such conservation, including protecting areas with fragile ecosystems, ensures that diversity of life is maintained and provides opportunities to develop new products, for example medicines, in the future. Genetic improvement is also a means to improve forest productivity, for example to ensure a high wood production in intensively managed forests.

FOREST HEALTH AND VITALITY

Forests need to be managed so that risks and impacts of unwanted disturbances are minimized, including wildfires, airborne pollution, storm felling, invasive species, pests, diseases and insects. Such disturbances may impact social, economic as well as environmental dimensions of forestry.

PRODUCTIVE FUNCTIONS OF FOREST RESOURCES

Forests and trees outside forests provide a wide range of wood and non-wood forest products. The theme expresses the ambition to maintain a high and valuable supply of primary forest products, while at the same time ensuring that production and harvesting are sustainable and do not compromise management options of future generations.

PROTECTIVE FUNCTIONS OF FOREST RESOURCES

The theme addresses the role of forests and trees outside forests to help moderate soil, hydrological and aquatic systems. This includes maintaining clean water including e.g. healthy fish populations, as well as to reduce risks or impacts of floods, avalanches, erosion and droughts. Protective functions of forest resources also contribute to ecosystem conservation efforts. Protective functions of forest resources have strong cross-sectoral aspects, as the benefits to agriculture and rural livelihoods are high.

SOCIO-ECONOMIC FUNCTIONS

The theme addresses the contributions of forest resources to the overall economy, for example through employment, values generated through processing and marketing of forest products and energy, trade and investments in the forest sector. The theme also addresses the important functions of forest to host and protect sites and landscapes that have high cultural, spiritual or recreational values and thus include aspects of land tenure, indigenous and community management systems and traditional knowledge.

LEGAL, POLICY AND INSTITUTIONAL FRAMEWORK

The theme includes the legal, policy and institutional arrangements necessary to support the above six themes, including participatory decision making, governance and law enforcement, monitoring and assessment of progress. The theme also addresses broader social aspects; including fair and equitable use of forest resources, science, research and education, infrastructure arrangements to support the forest sector, transfer of technology and capacity building and public information and communication.

CRITERIA AND INDICATORS OF SUSTAINABLE FOREST MANAGEMENT

During the 1990's, the attention drawn to the forests and their change enhanced the concept of durability and the need for criteria and indicators to describe and measure forest management. Different international conferences have established lists of indicators to be applied essentially on a national level. The Processes of Helsinki and Montreal for boreal or temperate countries, the Proposition of Tarapoto and the Tegucigalpa Meeting for Latin American countries and the UNEP/FAO initiative (Nairobi Meeting) and the Cairo Meeting for the dry African Zone, Near and Middle East, are the realization of these efforts.

As the task of the Inter-Governmental Panel on Forestry (IFF) was to reach, in an open and participatory manner, a consensus and formulation of propositions coordinated for the sustainable management of forests, the role of Criteria and Indicators was emphasized and it was specified that their development must be "a gradual process based on national sustainable forest management policies".

WHAT SHOULD BE UNDERSTOOD BY CRITERIA AND INDICATOR?

Criteria: Criteria define the essential factors of forest management against which forest sustainability may be assessed. Each criterion relates to a key management factor which may be described by one or more qualitative, quantitative or descriptive indicators. Criteria describe the different aspects of sustainability at a conceptual level. A criterion is a characteristic feature or a set of conditions, based on which it is possible to evaluate the various aspects of forestry. A criterion contains a built-in goal to strive for. The evaluation of a criterion is based on evaluation of indicators.

Table 10.1: Features of the ITTO Guidelines Concerning Sustainable Forest Management at National and Forest Levels (Source: ITTO 1992)

NATIONAL LEVEL	
Policy and Legislation	
- Forest policy	- National forest inventory
- Permanent forest estate	- Forest ownership
- National forest service	
FOREST LEVEL	
Planning	
- Static and dynamic inventory	- Setting of management objectives
- Choice of silvicultural concept	- Yield regulation
- Annual allowable cut	- Management inventory
- Mapping	- Preparation of working plans
- Environmental impact assessment	
Harvesting	
- Pre-harvest prescriptions	-Roads
- Extraction	-Post-harvest stand management
Protection	
- Control of access	- Fire protection
- Use of chemicals	
Legal Arrangements	
- Concession agreements	- Salvage permits
- Logging permits on private or customary land	
Monitoring and Research	
- Yield control and silviculture	- Environmental impact studies

Indicators: Through measurement and monitoring of selected indicators, the effects of forest management action, or inaction, can be assessed and evaluated and action adjusted to ensure that forest management objectives are more likely to be achieved. Indicators show or reflect the state of art as well as time related changes. They indicate how well each criterion meets the goals set. Typically an indicator shows a quantitative change. As all aspects of forestry cannot be measured with *quantitative indicators,* some *descriptive indicators* have been formulated to reflect the change regarding the legal, institutional and economic policy framework as well as the informational means to implement the policy.

INTERNATIONAL PROCESSES

Approximately 150 countries are currently participating in one or more of the nine on-going processes on criteria and indicators for sustainable forest management. Over the past years, criteria and indicators processes have helped promote a better understanding of the concept of sustainable forest management. FAO's Forestry Department has collaborated with and supported on-going international processes on Criteria and indicators for sustainable forest management, namely the: African Timber Organization (ATO) process, Dry forest in Asia process, Dry-zone Africa process, International Tropical Timber Organization (ITTO) process, Lepaterique process of Central America, Montreal process, Near East process, Pan-European forest process and Tarapoto proposal for the sustainability of the Amazon forest.

INDIAN INITIATIVES FOR SFM

The importance placed on the development and implementation of criteria and indicators for sustainable forest management by countries has resulted, during the past several years, in the development of nine separate but conceptually linked initiatives. In all these processes, the definition of sustainability remains virtually the same. This holds great promise for convergence or mutual recognition, so that over a period of time, a common approach can be used globally to measure progress in sustainable forest management. It was realized to develop sustainable forest management in India, to accomplish establishment of a benchmark for sustainability according to the policy framework.

SFM process was initiated by IIFM, Bhopal in 1998 and was christened as 'Bhopal-India Process' (B-I). B-I Process identified eight national level criteria and 49 indicators for the sustainable management of dry forests in the region. Participating countries are proceeding with implementation based on a two-year plan of action elaborated during the meeting. The action plan also commits participating countries to seek political and technical support from national forestry authorities for its implementation. The Asia regional initiative was endorsed by the 'National Task Force on Sustainable Forest Management' appointed by the Ministry of Environment and Forests, Government of India in November 1999 and IIFM, Bhopal has been designated as nodal agency for operationalizing SFM in India. In order to coordinate sustainable forest management activities in India, SFM Cell has been set up in the Ministry of Environment and Forests, Government of India. States and Union Territories are in the process of setting up similar cells.

CHALLENGES OF SUSTAINABLE FOREST MANAGEMENT

Forest management and harvesting is important for the national economy and the livelihoods of many individuals. In a world with an increasing population and consequently increasing consumption of wood products, the important question is therefore: "how can these forests be managed even more effectively than today in order to produce industrial roundwood for the benefit of mankind?" The challenge that forestry is currently facing is how to combine this task with demands to preserve a reasonable portion of unaffected forest ecosystems and demands to conserve biodiversity more widely.

Intensive forest management, which may include operations such as site preparation, tree planting (including, the use of genetically improved trees and/or exotic tree species), tending, thinning and fertilizer application, is often criticized as being inconsistent with the goal of achieving an acceptable level of forest biodiversity. However, forest managers argue that a reasonable level of biodiversity can be preserved even if a major portion of the forest is managed intensively as long as these forests are managed in a way that is reasonably environmentally friendly. Furthermore, even under intensive forest management, some forest areas can be left untouched as reserves for biodiversity conservation. These areas should be selected so that they form a continuous web in the landscape and include ecotypes that are particularly valuable for conserving biodiversity. This relatively new management concept is sometimes called "landscape ecosystem management" or "landscape planning".

Another approach to sustainable forest management is to practice low intensity forestry. This has lower reforestation and silvicultural costs than the landscape planning approach. It is often based on natural regeneration or some type of selective harvesting system and results in little interference with the natural development of the forest stand. Biologically, this approach may work well in stands that are easily regenerated or stands that contain trees that regenerate well. If well planned, this approach also has good prospects to satisfy nature conservation objectives and result in minimum damage to valuable ecosystems. However, the long-term consequences of low intensity forest management, particularly concerning future yields of industrial roundwood and wood quality, are uncertain and difficult to predict. The validity of this approach as a viable approach to sustainable forest management is, therefore, somewhat questionable.

CONSTRAINTS TO SUSTAINABLE FOREST MANAGEMENT

Other than difficulties caused by wars and armed conflicts, which are profound, several constraints frequently recur in the country profiles. Probably the most important is that the sustainable management of natural tropical forests is less profitable as a land use than other ways of using the land, especially some forms of agriculture, urban development and mining. As a result, SFM tends to be a low priority for governments and the private sector often lacks incentives to pursue it. In general, tropical timber prices remain relatively low. It is possible that they will increase in the future to better reflect the true cost of production, including the opportunity cost of retaining natural forest, but to date there is no sign of this.

Nevertheless, natural tropical forests are recognized incresingly as a valuable resource at the local, national and global levels, for the ecosystem services they supply. In some countries, payments are being made for such ecosystem services and REDD+ offers a potentially important revenue-generating opportunity for forest owners. In the long run, the extent of payments for the ecosystem services supplied by tropical forests, made at either the national level or the global level, are likely to play a larger part in determining the fate of the remaining tropical forests. In order for such payments to achieve their potential to impact forest management, constraints related to governance also need to be overcome.

Another constraint to SFM is confusion over ownership. Without the security provided by credible, negotiated arrangements on tenure, SFM is unlikely to succeed. In many countries, resolving disputes over land tenure is not an easy task. But it must be tackled, preferably through a transparent and equitable process, if resource management is ever to become sustainable.

FOREST CERTIFICATION

Forest certification is a process which results in a written certification being issued by an independent third party, attesting to the location and management status of a forest which is producing timber (Baharuddin and Simula 1994). The certification is the process of independent third party verification that forest management has reached the level required by a given standard. In some cases, when combined with a chain of custody certificate, certification allows products from a particular certified forest area to carry an ecolabel. Ecolabel helps consumers in selecting environmentally friendly products. Certification is a market-based tool that provides the capacity to customers to select the commodities based on their social and environmental concerns (Manmohan Yadav et al 2007).

Forest certification refers to two separate processes viz. forest management unit certification (FMU) and chain of custody certification (COC). Forest management certification is a process which verifies that an area of forest or plantations from where the wood, fiber and other non-timber forest products is extracted is managed to a defined standard. COC certification is a process of tracking forest products from the certified forest to the point of sale to ensure that product originated from a certified forest.

The innovative idea of forest certification was developed during the Parallel NGO Rio meetings in 1992. The concept was to develop a system for certifying and labeling forest and forest products. As a result, a voluntary non-profit organization Forest Stewardship Council (FSC) was launched in 1993 with the coalition of Worldwide Fund for Nature (WWF) for forest certification. Since then, several forest certification organizations have emerged. Although it is often viewed as an exclusively market-driven tool for sustainable forest management, forest certification actually encourages collaboration, facilitates conflict resolution, builds confidence and trust, promotes partnership and promises a premium price.

11 WORKING PLAN

INTRODUCTION

Forests in India occupies nearly 1/4th of the land area and enjoys various level of protection either under forest or wildlife act and invariably managed under certain approved plans which covers the working of the forest or wildlife area. But the connotation "working plan" is more suited to demarcated forest areas than the protected areas (PAs) which are managed under special plans called management plans. Since it is an age old method to work the forest areas under approved plan, the working plans are conveniently defined as "A written scheme of management aiming at continuity of policy and action and controlling the treatment of forest or otherwise it is an instrument of forest management." Its synonym word is management plan.

According to National Working Plan Code (2011), working plan is a tool for scientific forest management. It is very useful for evaluating the status of forest resource of a division, assessing the impact of past management practices and deciding about suitable management interventions for future. Periodical up-dating and revision of Working Plans is essential to keep pace with the trends emerging out of forest–people interface and to address other national and international obligations (FDA 2007). The working plan deals with the demarcated forests and are formulated with the aim of prudent utilization of forest produce by extracting yield and while doing so sustainability is never compromised at least theoretically to safeguard against any kind of misuse.

Working plans are generally drafted for a forest division once in 10 years and it prescribes treatment in lieu with the management objectives. For the purpose of treatment, the entire working plan area is divided into working circles and further into compartments and sub-compartments. A working circle receives same set of silvicultural treatments.

WORKING PLAN CODE

A working plan code (2011) was formulated by the Forestry Commission and this code lays down detailed guidelines for formulation (procedure & content) of the working plans. This code defines the purpose of working plan as follows:

1- To ensure that National Forest Policy is made effective at all levels of forest management.

2- To ensure the orderly and economic development of the forest within the framework of policy and to prevent uncoordinated, adhoc management decisions being taken without regard to their wider implications.

3- To ensure that the aims of management are clearly defined and the methods of achieving them are realistic and are the result of a careful study of all relevant factors.

4- To ensure that all supervisory staff concerned with forests is aware of its intended future development and thereby achieve continuity of management when there are changes in staff.

5- To bring together in a readily accessible form such information as is necessary for the efficient management of the forest.

ORGANIZATION

All forests are managed under the prescriptions of a working plan/scheme prepared on the basis of principles of sustainable forest management and recognized and innovative silvicultural practices. The authority as designated by the Ministry of Environment and Forests, Government of India, approves the working plan for this purpose. Generally no timber harvesting is done in any forest area without an approved working plan/scheme.

It is the duty of the owner/manager of the forest area to ensure the preparation of the working plan/scheme. While a detailed working plan is prepared for large areas such as forest division, working schemes are prepared for smaller areas for a specific purpose or areas like private, village, municipal, cantonment forests, etc. Even working schemes have all major elements of working plan; and these schemes also need the sanction of competent authority designated by the Ministry of Environment & Forests. At the central level the structure of organization is listed below.

Table 11.1: Structure of Organization at Central Level

Hierarchy
Director General of Forests & Special Secretary to the Government of India (Ministry of Environment & Forests)
Additional Director General of Forests
Inspector General of Forests,
Deputy Inspector General of Forests
Assistant Inspector General of Forests
Regional Chief Conservator of Forests

In the States, there is no uniformity in the constitution of the organization. There are various levels of which Conservator of Forests Working Plan and Working Plan Officer (DCF) are the key functionaries. Overall situation is as under:

Table 11.2: Structure of Organization at State Level

S No	Hierarchy	Position held
1	Policy/Supervisor Level	PCCF/APCCF/CCF
2	Field Supervisor Level	Chief Conservator/Conservator of Forests
3	Working Plan (Field Function Level)	Working Plan Officer (DCF/CF) assisted by ACF, RFO and Foresters

The number of working plan circles and working plan divisions depends upon the workload, i.e., number of territorial divisions for which working plans are to be prepared/ revised. Working plan is generally revised every 10 years; and on an average, one Conservator supervises the work of four working plan divisions. The working plan of forests other than those under the control of Forest Department like municipal, cantonment, private, village, etc., forests can be prepared by Working Plan Officer on the request of the owner by owner themselves or through outside consultants. Microplan of jointly managed forests is prepared by forest staff of territorial division as per MOU and in consultation with the communities involved. The microplan so prepared is integrated with the working plan of that territorial division.

OBJECTIVE AND SCOPE OF WORKING PLAN

OBJECTIVE

- To provide set of guidelines and standards for improved forest management practices
- Provide framework for sustainable management of natural resource and to strike a balance between development Vs conservation
- To make forest practices environmentally responsible, socially beneficial and economically viable

SCOPE

- In forest certification, particularly in meeting the objectives of criteria and indicators of sustainable forest management
- In preparation of microplans for JFM areas
- In assessment of climate mitigating potential of forests
- It may form the basis for EIA and SIA
- Social perspective of natural resource management for poverty alleviation particularly in fringe areas of forest

Working plan document is a means of enforcing systematic, obligatory and mandatory regulations for continuous management of given forest property. Guidelines for drafting of a working plan are primarily derived from the current forest policy of the concerned state.

BRIEF HISTORY

The need for management plan to regulate the felling in Indian forest was felt during beginning of 18th century. But it was materialized only after 1864 when Dietrich Brandis was appointed as the first Inspector General of forest. He initially employed basic tenets of regulation of yield in pegu forest of Burma to harvest teak in a sustainable manner. But in the year 1885, the forest department code (3rd edition) laid down in great detail the procedure of preparing working plans in India (D'arcy 1898).

In the year 1988, the third forest policy of India emphasized the need for working of forest areas of India under approved working plan. Para 4.3.2 of the National Forest Policy clearly states "No forest should be permitted to be worked without the Government having approved the management plan, which should be in a prescribed format and in keeping with the National Forest Policy. The Central Government should issue necessary guidelines to the State Governments in this regard and monitor compliance."

The Hon'ble supreme court of India vide a judgment dated 12 December 1996 in writ petition no. 202/95 have laid down well-defined guidelines regarding working plans. This judgment makes it mandatory for all forest areas (irrespective of ownership and legal status) to be managed according to a scientific working plan. This judgment also makes it mandatory for the working plans to be approved by the Government of India. This is in accordance with the fact that forests are a concurrent subject.

Recent working plan code for the entire county was revised in the year 2011 by Ministry of Environment and Forest to be followed in spirit and for drafting new plans. As of now Clear-felling of forest area is not included in any of these plans. Much higher emphasis has been given to conservation, people's involvement for regeneration of forests, increasing the NTFP outputs, biodiversity conservation, medicinal plants, etc.

CONTENTS OF WORKING PLAN (Source: Working plan of Doda Forest Division (2004-2014))

As a rule a working plan document is divided into two parts part I, II (roman numerals) and each part is further sub-divided into chapters I, II, III

PART I

Part I deals with the summary of facts on which proposals are based and it consists of following chapters

GENERAL DESCRIPTION OF TRACTS

- Name and situation of the area dealt with: Civil District, Forest Division, etc.
- Configuration of the country: Flat, hilly or plateau; altitude; prevailing aspects, etc.
- Rock and soil: Geological formation; underlying rock; general characters of the soil and their effect on forest growth.
- Climate: Temperature, mean and extreme, with its effect on the forest growth; rainfall, periods of drought and rain; prevailing winds, storms; other climatic influences on forest treatment.

- Population: Numbers and condition; agricultural customs or requirements which influence the system of forest management.

THE COMPOSITION AND CONDITION OF THE FORESTS

- Distribution and area: The distribution of the forest; total area and areas under tree-growth, grass, streams, etc., in each class of forest; enclosures within the boundaries; maps available, their degree of accuracy, etc.
- Boundaries and boundary marks: Nature and state of repair of the marks used; enclosures and their boundaries; whether a revision of the boundaries is required or not.
- Legal position of the forests: Brief history of the acquisition by Government and settlement (where there has been one) of the forest; Act and section of the Act under which gazetted, etc.
- Rights: Their origin and general character; statement of the areas burdened with rights and free from them; summary of rights to produce; privileges or other concessions granted at the will and pleasure of Government; adequacy of the forest resources to provide for the exercise of recorded rights or of admitted privileges; actual effect of such exercise upon the crops.

GENERAL DESCRIPTION OF THE FOREST CROP

Composition and condition of the crop; types of forest and areas occupied by each crop; origin of the crop; the principal species and their relative proportion and importance, habitat, mode of reproduction, size attained and rate of growth; density of the crop, blanks and glades; state of the reproduction; principal products yielded; grass and other minor products, etc.

INJURIES TO WHICH THE FOREST IS LIABLE

Causes of injury; fires, grazing, offences of common occurrence; the best means of regulating or combating the causes of injury.

UTILIZATION OF PRODUCTS

It covers the following vital aspects of forest division.

- Marketable products: Yield of timber, fuel, minor produce, etc; quantities consumed in past years in Government works, by purchasers, by right-holders and by free grantees; value of products consumed.
- Extraction of the produce: Extraction roads or rivers leading from forests to centres of consumption; their adequacy and state of repair; manner and cost of extraction; road and river improvements required.
- Markets: Distance, size and importance of principal centres of consumption; produce consumed at each centre and ruling prices obtained.
- Net value of produce: Net value in the forest of each class of produce.
- Demographic statistics: Democratic statistics of the forest division and agricultural customs and needs of the population

STAFF AND LABOUR SUPPLY

It mentions briefly the sanctioned staff strength and shortage if any

THE PAST SYSTEM OF MANAGEMENT

This section elaborates the systems of management in the past and evaluates for any modification or adoption of new silvicultural systems based on the past results.

STATISTICS OF GROWTH AND YIELD

It forms the basis for justification of certain mathematical and statistical procedures used to assess the inventory of the forest. It plays pivotal role due to prevalence of sampling technique in assessment of growing stock.

PART II

In Part II, future management interventions are discussed and prescribed. It consists of following chapters

BASIS OF PROPOSAL

It briefly mentions the chief object of the management with which forest is to be managed, proposed method of treatment (production or protection lines), constitution of working circles, area statement of various working circles, marking of boundaries and period of plan.

WORKING CIRCLES: CONSTITUTION AND PRESCRIPTION

Subsequent chapters cover various working circles and the management prescription as proposed under current plan. It mentions elaborately character of vegetation in a particular working circle, area statement, silviculture system adopted and rotation/ conversion period and allotment of felling series, periodic blocks, annual coupe size and valuation of growing stock. Working plan prescriptions for various working circles are covered under each chapter. Subsequent chapters will be allotted for miscellaneous regulations, establishment and labour, financial forecast and cost of plan, control and summary of prescriptions.

PREPARATION OF WORKING PLAN

The preparation of working plan is a highly technical and time-framed scheduled operation. The preparation of the working plan is based on stock maps which is prepared through ground surveys. Recently, the use of modern tools like remote sensing, GIS & GPS is being done for preparing the forest cover maps of divisions. Every working plan includes the area-specific scientific prescriptions for proper management of forests of a particular forest division. Accordingly, the working circles, felling series and annual coupe area computations are done for the plan period for implementation of the working plan prescriptions.

Working plans of yesteryears are drafted to exploit the commercially viable forest areas to generate revenue. It involves regulation of felling and facilitation of regeneration in a phased manner for 10 years at a time. As of now, no forests areas of India is commercially managed after a Supreme Court's ruling of ban of green felling with effect from 1996.

Keeping this in the backdrop, this chapter tries to explain the methodologies adopted in the past plans as well as with improvisation as few paras are not relevant (yield regulation) to current working of forest areas. Nevertheless it may throw light on systematic management of forest areas in the past. More over, the estimation of growing stock and other relevant information may serve as baseline data to facilitate research in the era of global warming and relevance of forest areas mitigating agents. The preparation of entire working plan can be conveniently divided into following stages:

1. Preparation of preliminary working plan report
2. Field work
3. Collection of statistical data for enumeration or assessment of growing stock
4. Enumeration
5. Regeneration survey
6. Biodiversity survey
7. Calculation of growing stock and yield
8. Prescriptions

PREPARATION OF PRELIMINARY WORKING PLAN REPORT (PWPR)

The purpose of PWPR is to evaluate the past system of management and to propose any modifications there on. This process is initiated by Conservator of Forest (Working Plan) who asks the D.F.O (territorial) through C.F (territorial) to prepare a preliminary reconnaissance report on the operations of current working plan at least six months in advance. This report briefly reviews for each working circle on parameters such as the result of past management, the success and failure of past prescriptions and suggestions if any.

This is followed by a discussion between the C.F (W.P) and C.F (territorial) to arrive at a agreement and C.C.F is consented to if need arise. In the following season, the C.F (W.P) inspects the forests and prepares the first preliminary report. This preliminary report forms the basis for the Working Plan Officers to initiate the immediate work. Working plan officer may be either a D.F.O (territorial) or special duty officer i.e. WPO.

FIELD WORK

After the approval of PWPR filed work is initiated it involves following steps:

(a) Preparation of stock maps
(b) Estimation of site quality
(c) Marking of compartment boundaries
(d) Compartment description
(e) Stump and stem analysis

PREPARATION OF STOCK MAPS: Forest stock maps constitute the basis of working plan for the territorial forest divisions and are prepared in either 1:15,000 or 1: 5,000 scales. They contain detailed spatial information on extent of recorded forest lands including the administrative jurisdiction at various levels, infrastructure and communications facilities, water resources and the status of forest vegetation. The component includes categorization of forest into density classes, species distribution, assessment of growing stock, growth data for various species and age distribution. All these information are extremely vital in presenting the management options like afforestation of blank areas, reforestation of degraded forests, harvesting of mature crop (plantation) followed by regeneration, restoration of biodiversity in monoculture areas and management of Protected Areas for biodiversity conservation.

Stock maps are prepared by tracing the previous management map or topo-sheets with boundaries of administrative units and other vital information like roads, buildings, streams etc. in the field, species-wise vegetation component layered on the map manually by the field staff. Field stock maps are prepared by laying transect lines of 100 meters width (50 meters on each side of the centre line) which are laid at perpendicular to base line. During the cruise along the transect lines details regarding crop composition, crown density (using densiometer), age-class distribution, general quality, presence of soil conservation structures are recorded. The process is repeated to scan the entire compartment. Using this information compartment history is updated for the current plan. Following index are used for the purpose of stock mapping in Sal forests with colour code.

Table 11.3: Index Used for the Purpose of Stock Mapping in Sal Forests

Parameter	Categories	Colour Code
Age	Young	Green
	Middle aged	Blue
	Mature	Red
Crop composition	Pure Sal	Dark blue
	Teak	Light blue
	Mixed	Red
	Barren	Yellow
Site quality	Site quality I	Single lines
	Site quality II	Double lines
	Site quality III	Triple lines

ESTIMATION OF SITE QUALITY: Site quality is a quantitative measure relative productive capacity of soil. The diameter and height of 3-5 trees in each diameter class in a locality is measured and average diameter-height curves are plotted and matched with standard diameter-height curves for different site qualities drawn on the basis of yield table data. The site quality of the compartment is determined from the closeness of the curve to standard curve and indicated in the stock map by drawing lines according to site quality (I, II, III).

MARKING OF COMPARTMENT BOUNDARIES: The layout of each compartment of the division is marked either on trees or on any permanent features on the ground. As a general practice boundary of compartments are marked on border trees by making two rings in coal tar paint and sometimes to differentiate the range boundary with the division boundary. Division boundary is given three prominent rings and range boundary with two prominent rings.

COMPARTMENT DESCRIPTION: It is systematic description of each compartment of division in a concise manner incorporating vital information like: location and name of compartment; topography and soil aspect; general characteristics of growing stock; composition and age of crop; stand density; past history of management; incidence of fire, grazing, disease or pest or any other kinds of damage; plantation areas; allotment to working circle (in the proposed working plan); potential for wildlife, soil and water conservation. It is usually filled up by W.P.O and compiled record is kept for future reference.

STUMP AND STEM ANALYSIS: In so many circumstances, there may not be volume table or yield table for species under consideration. To obviate the drawback, stump and stem analysis is being carried out in selected trees from different site qualities to arrive at table or graphs to determine the growing stock. In rare cases, stump and stem analysis will be carried out to analyze the change in the growth pattern of species under consideration. Stump analysis determines the progress of diameter increment on stumps throughout the life of the tree analyzed over any desired period but its important object is to find out age-diameter relationship to determine the rotation age corresponding to a particular exploitable diameter in absence of yield table. Similarly, in the absence of local volume table for species under observation, the data of the stem analysis can be used for preparing local volume table by correlating diameter volume relations.

COLLECTION OF STATISTICAL DATA FOR ENUMERATION OR ASSESSMENT OF GROWING STOCK

Normally a forest area of a division encompasses vast area with variety of land forms and vegetation types. It is rarely a uniform entity. The total enumeration and assessment of vegetation over the entire area is a herculean task and normally involves intensive labour, investment and training. Except for smaller areas, total enumerations are avoided and sampling techniques are followed. To minimize the cost and time and to improve the accuracy level, a good sampling design combined with suitable sampling techniques may be followed during inventory. A good sampling design may be either random sampling such as simple random sampling, stratified random sampling and multiphase sampling or non-random sampling such as selective sampling, systematic sampling and sequential sampling.

Normally stratified random sampling is followed where the entire forest areas are divided into various strata or sub-strata according to composition and condition of crop and in each stratum sampling is carried out. Sampling technique may be either point sampling or plot sampling.

POINT SAMPLING OR BITTERLITCH METHOD: Point sampling is an age old technique used to assess the growing stock in conifer forests and still being followed in few forest divisions of India in growing stock estimation. Point sampling is usually carried out with the help of simple angle gauge or wedge prism. Nowadays it is used only for quick reconnaissance survey.

PLOT SAMPLING: It is an extensively used technique to assess the growing stock particularly in tropical forests. Square shaped plots of 0.1 ha are used in most of the forest division in India in inventory or other scientific experiments. Though there a considerable debate exists over the two methods, the point sampling estimates basal area more effectively and on the other hand, plot sampling estimates the number of stems/ha effectively.

METHODOLOGY FOR SAMPLING: Sampling in a forest division varies according to crop composition and diversity. After having done the stratification, the number of sample points required to be surveyed and measured in each of the above strata except non-wooded area, to achieve the desired accuracy of 20 per cent at 95 per cent level probability obtained as result of preliminary survey of 60 sample points selected and surveyed at random, over more or less similar area in a forest division. The number of sample points has therefore been calculated by adopting following formula,

$$N = (t^2) \times \frac{(CV)^2}{E^2}$$

where,

N = number of sample points required to achieve the desired accuracy with the probability level implied by the value t

t = constant denoting the reliability of estimate or level of statistical significance or statistical probability

E = stated percentage accuracy desired for average. It is also called as the maximum permissible error acceptable in the sampling design.

Using the formula, the variation for basal area, height and number of trees in each pilot survey points are statistically analyzed for standard deviation, coefficient of variation at 95% probability level. Accordingly number of sample points is estimated by substituting the statistical parameters in the above mentioned formulae. The aim of arriving at sampling intensity is to minimize the cost of enumeration and improve the accuracy in inventory of growing stock. A sampling frame comprising the above number of sample points are distributed at random in the strata by referring to the random table using random numbers. The sample points transferred on the GT sheets concerned have been objectively located on the ground.

ENUMERATION

Partial enumerations are done with laying sample plots of 0.1ha after locating sample points in each sub stratum. Following procedures need to be followed while laying out sample point:

- Objectively sample points are located on ground
- A peg driven on ground to mark middle point of enumeration plot
- From the point, a length of 22.6 is measured towards north and a peg is driven to mark the point
- Similarly length is measured from the middle point which is located and marked on the southern side
- Similarly points are located and marked on eastern and western side
- The enumeration is carried out inside the marked point in 10 cm diameter classes

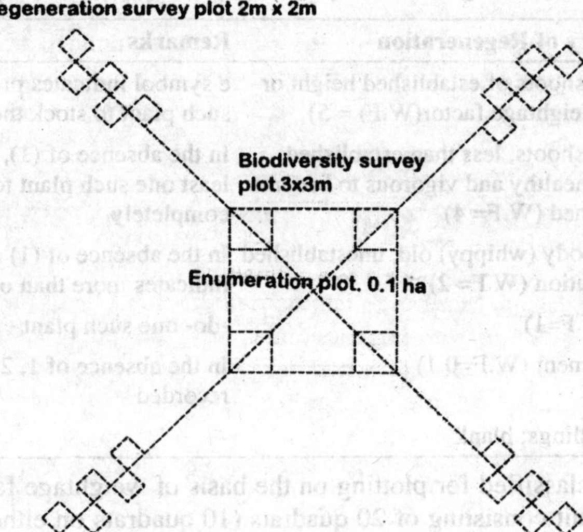

Fig. 11.1: Layout of Sample Plot (0.1 ha) for Enumeration, Regeneration Survey and Biodiversity Survey

CALCULATION OF GROWING STOCK: After having done enumeration in sample plots, the details of diameter class-wise distribution of stems are taken up for calculation of volume of each enumeration plot and later interpolated for the entire compartment. Thus each and every compartment of the given division is covered. The growing stock is calculated in three ways: volume table equation, volume table and yield table (Chaturvedi and Khanna 1994).

From volume table equation (derived from local volume table): Volume table equations are drawn species-wise and it is available only for commercially important trees.

From volume table: The volume of stem timber and small wood for the mean tree in each diameter class is read from volume table. It is converted to volume/tree by dividing it by number of trees and multiplying with the number of trees per ha in that diameter class to get the volume /ha in that diameter class. It is then summed up for all the diameter class and multiplied with the gross area to get the growing stock of the compartment.

Yield table: Yield table gives all the quantitative information regarding development of a crop. Yield table is not applicable to even aged crop.

REGENERATION SURVEY

Regeneration survey is an essential part of field exercise as it provides fair idea about the status of regeneration of forest. The survey is usually combined with partial enumeration by strips. Usually 2-5 per cent sampling intensity is followed to assess the regeneration status. When it is combined with the sample plot enumeration at least 20 sample plots of 2m x 2m are laid in strips. In regeneration surveys, it is sufficient to record the occurrence of most promising categories of plants.

Table 11.4: Method of Recording inside Regeneration Plot

No	Symbol	Category of Regeneration	Remarks
1.	E	Woody shoots of established height or over (Weightage factor(W.F) = 5)	e symbol indicates presence of atleast one such plant to stock the quadrat
2.	W	Woody shoots, less than established height, healthy and vigorous to become established (W.F= 4)	In the absence of (1), symbol w indicates at least one such plant to stock the quadrat completely
3	u +	Non woody (whippy) old, unestablished regeneration (W.F= 2)	In the absence of (1) and (2), symbol u + indicates more than one such plant
4	U	-do- (W.F=1)	-do- one such plant
5	R	Recruitment (W.F=0.1)	In the absence of 1, 2, 3 and 4, R may be recorded
6	O	No seedlings; blank	-

The data may be classified for plotting on the basis of weightage factors (W.F) given in Table 11.4. Taking a unit consisting of 20 quadrats (10 quadrats on either side of cruise line: 4m wide x 20m long), the total weightage is graded as in Table 11.5.

Table 11.5: Grading of Regeneration

Cumulative Weightage value	Regeneration Status	Colour for Regeneration Map
81-100	Excellent	Green
61-80	Good	Blue
41-60	Moderate	Yellow
21-40	Fair	Red
0-20	Deficient	Blank

Note: the calculation of cumulative weightage value does not give information regarding the individual's contribution of different categories of regeneration.

BIODIVERSITY SURVEY

Biodiversity survey for herbs and shrubs are carried out in sample plots of 1m x 1m and 3m x 3m size respectively. The plots for estimation of biodiversity for herbs are laid inside 3m x 3m plots and are located in the four corners of sample plot at 25 meters away from the centre of enumeration plot. Biodiversity Index values are used to determine the Alpha, Beta, Gamma diversity of the respective area.

CALCULATION OF YIELD

Yield is the amount of material which may be removed annually or periodically from a forest over a stated period. Yield is an important aspect of commercially managed forest where matured stock is removed annually or periodically. As stated above calculation and realization of yield is only theoretical importance in India as blanket ban green felling has been in operation since 1996.

However, few relevant aspects of yield regulation as practiced in the past shall be recalled in this section. In this back drop the relevance of silvicultural system in yield regulation shall be given its rightful place. For the purpose of considering the effect of silvilcutural systems in on yield regulation, following three broad types of systems will be recognized. The yield is realized in a systematic manner by laying out coupes and felling series on management map and later it is objectively transferred on ground.

Table 11.6: Yield Regulation for Three Silvicultural Systems (Ram Prakash 2006)

Silvicultural System	Nature of Crop	Systems of Yield Regulation
Clear felling system	Regular even aged forest	Von montels formula/ modifications
Regular shelter wood system	Regular even aged forest	Cottas formula
		Austrian assessment method
Irregular shelter wood system or selection system	Irregular forest	Swiss method
		Brandis method

CONTROL FORMS

Control forms form the basis for evaluation of management prescriptions in the field. It is duly filled by the field staff and kept as a reference to analyze the outcome of relevant prescriptions.

Table 11.7: Working Plan Vs Management Plan

S.No	Particulars	Working plan	Management plan
1.	Object	Drafted for forest area (Reserve Forest, demarcated Forest) as defined in the forests area of the concerned state	Are drafted for Protected areas (National parks, sanctuaries) as declared under wildlife protection act
2.	Intent	Originally meant to regulate yield from forest	Originally meant to protect and conserve the wildlife
3.	Purpose	Normal working plan for a commercially important stand revolves around certain aspect of felling, regeneration , tending of a crop	Management plan of PAs monitors and improve the wildlife and habitats and regulate the biotic pressure
4.	Emphasize	Emphasize laid more on biomass, volume and yield of vegetation. Faunal component finds only a brief mention	Emphasize laid more on population dynamics of target animal including total population, migration and prey predator relation. Floral component is given secondary importance.

ROLE OF SILVICULTURAL SYSTEMS IN THE WORKING PLANS

Silvicultural system is a program of procedures or treatments that include all the management functions which meet the landowners present and long-term objectives. Silvicultural systems are the inseparable component of working plans as long as Indian forests are managed on commercial lines especially for timber(Now generally not in practice). As a rule operational guidelines for management prescriptions for a working circle are derived from appropriate silvicultural system.

12 TIMBER MEASUREMENTS AND TIMBER INVENTORY

CONCEPT AND SCOPE OF FOREST MENSURATION

Mensuration is an essential part of mathematics. Mensuration is derived from the Latin word 'Mensura' meaning 'measure'. It is concerned with the determination of length, area and volume. In ancient times also, there was a system of standard relating to weights and measures with the use of standardized measuring rods and scales. Forest mensuration is the science of measuring the contents of standing or felled timber and estimating growth and yields. Forest Mensuration is that branch of forestry that deals with the determination of dimensions (e, g; diameter, height, volume etc.), form, age and increment of single trees, stands or whole woods, either standing or after felling (Chaturvedi and Khanna 1982). It is also defined as the theory and practice of measuring standing trees and logs to determine yields and other parameters.

Forest mensuration or dasometrics or dendrometrics deals with the quantification of forests, trees and forest products. It can be distinguished with the techniques for direct or indirect measurements, estimation procedures using statistical relationships and methods of prediction where the variable time takes part. Forest mensuration, therefore, concerns with linear, area, volume and weight measurements (Bruce and Schumacher 1930). In forest mensuration, the following parameters can be described quantitatively:

- Age
- Diameter- over and under bark
- Length or height
- Form or shape
- Taper or the rate of change of diameter with length
- Volume over or under bark
- Crown width and height
- Wood density

Forest mensuration includes measurements of felled and standing trees, sawn wood and round logs and Non-wood forest products such as bamboos, bark, fruits, etc. In addition to describing the current status of tree or crop, mensuration also deals with measuring and predicting the growth of trees and crops that is the change in the parameter with time. The main objective of measurement of individual trees is to estimate the quantity of timber, firewood or any other forest produce which can be obtained from them. It covers: diameter measurement, height measurement, volume measurement and measurement of logs and fuel wood. Forest mensuration deals with measurements of trees and logs which are basis for sale, management, research and planning purposes (Husch et al 2003).

Forest mensuration has undergone a number of important changes in recent years. Electronic measuring devices using laser or ultrasound emissions have become commonly used to measure distances. These devices have reduced the time to measure tree heights considerably, and can be used to quickly obtain other tree parameter measures including spatial positions. Ground Positioning Systems (GPS) that triangulate satellites to determine ground positions have become widely used. Aerial photographs continue to be used in navigation, planning, forest inventory and management purposes. However, these are often digital or are converted to digital formats when used in forestry today. Augmenting aerial photographs with other remotely sensed imagery has become more common for modelling crown structure and other features for inventory purposes. Other types of new advanced imagery have been used to estimate net primary productivity (NPP), leaf area index (LAI) and other productivity measures related to forest function and growth. New methods of tree-ring analyses, including physical, chemical and anatomic features, have facilitated reconstructions of growth conditions and investigations of impacts of environmental changes on forest growth. These data have improved studies of growth-environment relationships.

TREE DIAMETER MEASUREMENTS

The size of the tree is best described with the diameter of the tree. The diameter of the tree provides a measure of tree performance and is a useful starting point for estimating tree volume. It is the linear measurement, the main objective of which is to estimate the volume of the trees. The volume of a tree is dependent on diameter or girth at breast-height, total height and form factor. It is not only necessary for calculation of volume of logs, but also necessary for making inventory of growing stock as well as to correlate height, volume, age, increment of trees. In forestry, diameter should always be used for the description of the tree (Avery and Burkhart 1983). In the past, girth has been used but this is no longer recommended.

A carefully defined point should be used for measuring the diameter of the tree. In many countries, especially India, this will be at 1.37 m above the base of the tree where the tree meets the ground. By convention, the diameter of forest trees is measured in centimeters at 1.37 m above the ground and is termed the "Diameter at Breast Height" or DBH. Because trees are measured with the bark on, this is also called the Diameter at Breast Height Over Bark (DBHOB). When measuring live trees most information is presented as over bark dimensions. Diameter at Breast Height (DBH) is the most common parameter used in the measurement of standing timber. Generally DBH is not measured on dead trees or on those of less than 7 cm DBH. Universally adopted standard height for measuring girth, diameters and basal area of standing trees is 1.37 m and being practiced in India, Burma, America, Union of

South Africa and other British Colonies. In Europe, Australia and UK, DBH is taken as 1.3 m which is recommended by FAO as standard. In New Zealand, DBH is measured at 1.4 m.

THE SIGNIFICANCE OF DBH

- Convenient height for taking measurement
- Avoids the fatigue caused unnecessarily
- Saves extra expenditure from not clearing the base
- Abnormalities, e.g. root swell, disappear below breast-height
- Standardizes diameter measurement giving a uniform point of measurement. Diameter measurement at stump height is preferred, but standardization is lost because height of stump depends upon skill of the labour and the commercial value of the tree

STANDARD RULES GOVERNING BREAST HEIGHT MEASUREMENT DBH should be measured using calipers or tapes and the following conventions (Chaturvedi and Khanna 1982; Patric Purser 1999):

- Breast height point should be marked by intersecting vertical and horizontal lines 12 cm long, painted with white paint.
- On flat ground, breast height (BH) should be marked by means of a measuring stick on standing trees at 1.37 m or 4 ft 6 inch above the ground level.
- On slopy ground, BH should be measured from uphill side after removing any dead leaves or needles lodged there.
- In case of leaning trees on flat ground, BH is measured along the tree stem and not vertically on the side of the lean of the trees. On slopy ground, BH on leaning trees should be measured from the uphill side along the leaning side of tree stem but not vertically.
- In abnormal trees, BH mark should be shifted little up or down as little as possible to a more normal position of stem and diameter is measured.
- If tree is forked above BH, it should be considered as a single tree. If tree is forked below the BH, each fork should be treated as separate tree. If forking renders the BH point abnormal, foregoing rule should be applied and the tree is counted as one or two depending on the place of measurement.
- BH of the tree with aerial root/ buttress formation is measured by shifting BH mark a little above the buttress formed.
- Callipers should be held in such a way that the arm of the calipers is at right angles to the stem.
- Tapes should maintain a loop, at right angles to the stem, around the tree. Obstacles such as branches, climbers, loose bark etc. should be removed before measurement so that they do not distort the result.

DBH vs Girth

DBH = Diameter at Breast Height = $2r$

where, r is the radius of stem

GBH = Perimeter of Stem at Breast Height = $2\pi r$

Thus, GBH = DBH × π

UNDER BARK vs OVER BARK

DUB = DOB – $2t$

GUB = GOB – $2\pi t$

where, t is the bark thickness

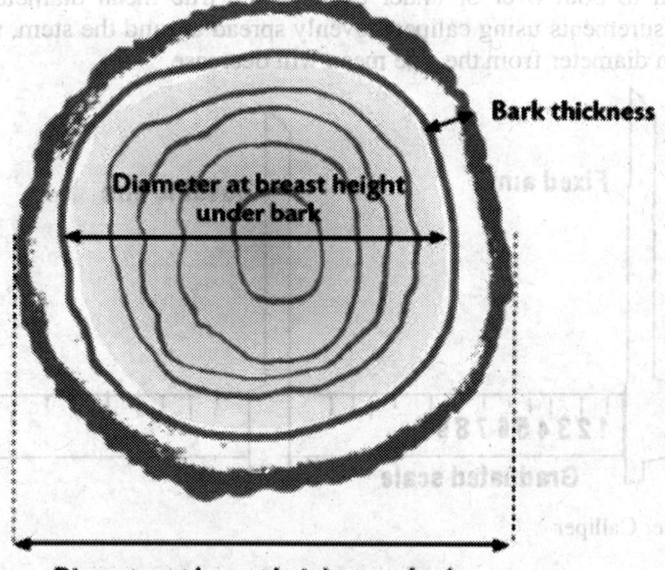

Fig. 12.1: Under Bark and Over Bark DBH Measurements

INSTRUMENTS FOR DIAMETER MEASUREMENT

Instruments such as diameter callipers, tapes and wooden scales are most commonly used and the details of their descriptions are presented below (Chaturvedi and Khanna 1982; Yadav 2012a).

DIAMETER CALLIPERS: It is generally made of wood and consists of a graduated rule and two arms. One is fixed at right angles to one end of the rule and the other movable parallel to the fixed rule. Size more than 120cm in length is rarely used. The use depends on the desired accuracy. Callipers marked are painted to differentiate the reading. In some callipers, the arms are made parallel by screw adjustment. Metal callipers made of aluminum alloy are now in use. They are not heavier than wooden callipers and easy to keep clean and adjustment. The different types of callipers are:

- Flurry Callipers: Arms from aluminum and graduated rule from wooden and bound with brass of sizes available in lengths of 80, 60, 35cms
- Fommes Calipers: All aluminum and same size as Flurry.
- Indian Aluminum Callipers: They are variable in the length of 50 cm, 75cm & 100 cm and they are graduated to show cm and mm.

Callipers are used to measure diameter of standing trees and logs. Diameters can be read directly in cm and mm thus making the instrument applicable for precise scientific work. If just one measurement is taken, an assumption is made that the cross-section at that point on the stem is perfectly circular. More than one measurement may be taken at that point on the stem. The average reading of these measurements is used as an estimate of the mean diameter. The measurement of more than one diameter and the subsequent estimation of mean diameter are made easier with the use of electronic calipers. A single calliper measurement has the potential to both over or under estimate the true mean diameter. With increasing numbers of measurements using calipers, evenly spread around the stem, the deviation of the estimate of mean diameter from the true mean will decrease.

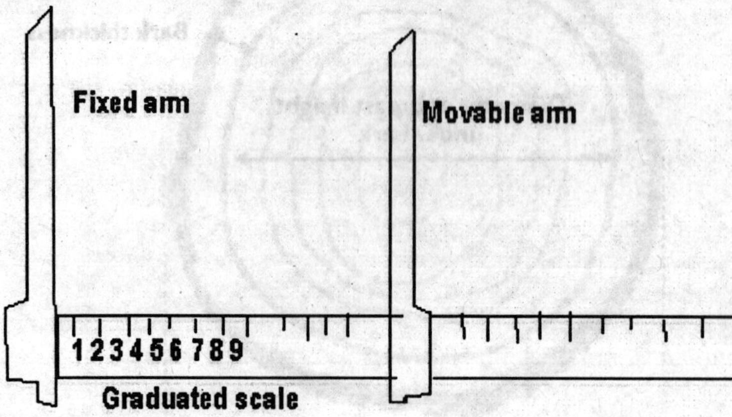

Fig. 12.2: Diameter Calliper

TAPES: It is a band of cloth, reinforced cloth, plastic or steel about 1.5 cm wide and of varying length and is used to measure girth of trees and logs. It is usually graduated one side in cm and mm but sometimes it is graduated on both sides to give measurements in metric system on one side and those in British system on the other. The ends of the tape are always plated some metal to prevent their tearing off but in case of longer tapes which are kept encased in some cover by winding it in or in some other cases too, the beginning of tape has a metal ring to hold it. In western countries, tapes are often provided with hook enables one person to measure large trees with tapes lying flat in correct position on the tree. The different types of tapes are:

- Cloth Tape: The tapes are made of cloth, though they may be painted with some paint on both sides to give better look and correct them from the influence of water. They are also affected by fluctuations in length due to expansion in use.

- Metallic Tape: The better quality of tapes are usually reinforced inside by metal wires and are, therefore called metallic tapes. They are painted with some durable paint. So, they are more durable and reliable.
- Steel Tape: Steel tapes are used for precise work and are mostly used in forest for measurement in sample or research work.
- Nylon and Fiberglass Tapes: In western countries, nylon coated steel tapes and fiberglass tapes are available.

The tree measuring tapes are generally 3 m long or at the most 5 m. But land measurement tapes may be 5 m, 10 m, 30 m or even 50 m long. The tree measuring tapes, which are usually small, do not generally have cases to cover them, though some steel tapes are 2 or 3 m long are kept in cases with some spring device to wind them back when not in use. Diameter is best measured with a diameter tape, which is a tape marked with units that convert the girth to the diameter. It is important to make sure that a correctly calibrated tape is used.

BILTMORE STICK: The Biltmore stick is a rule graduated to indicate the diameter of a tree. The rule is placed against the trunk of the tree at a tangent. With the eye about 25 inches away (average arm length), the diameter may be read by lining up the end of the stick with the line of vision to one side of the trunk and sighting across the rule to the other side. At the point where the line of vision crosses the stick, the graduation will indicate the diameter of the tree (CCC 2009).

WOODEN SCALES: It is a flat wooden piece marked in cm and mm. Wooden scales are used to measure diameter of stumps or end sections of logs exposed as a result of cross-cutting. It is also used in stump and stem analysis for measuring radius at successive decade marks. The diameter should be measured along the passing through pith. In case of eccentric stumps or logs, two diameters, one along major axis and other at right angles to it should be measured. As the end of the scale gets worn off by continuous use, the measurement should be taken from first centimeter and not from zero and one cm is deducted from the reading. The scale should be placed on edge so that the ends of the line to be measured coincide with marks of the scale. While reading measurement, the eye should be just above the mark. If this is not done, some error will creep in the measurement which is called error of parallax. Two types of wooden scales are available provided with folding arrangements at every 15 cm lengh:

- 30 cm long and 3 cm width
- 60 cm long and 1.5 cm width

WHEELER'S PENTAPRISM CALIPER: The pentaprism is a moderately expensive instrument used for measuring the diameter of the stem at heights normally above reach from the ground. The Wheeler's pentaprism is an optical caliper that has two pentaprisms: a fixed one and a movable one. The construction allows generating parallel beams that correspond to the arms of a mechanical caliper. The sighting is situated at the position of the fixed prism and is separated into two parts. Through the upper part, the observer aims directly at the left hand side of the trunk. The second beam goes through the two prisms where the movable prism is moved back and forth in such a manner that the second beam aims at the right hand side of the stem. This is achieved by simultaneously looking through the lower part of the sighting where the picture of the trunk in the prism is mirrored to the left along the left side of the tree. For measuring the diameter, these two images must coincide. That means that the left side of the trunk in the direct and the indirect view is brought to a match. In that special case, the two parallel rays are tangent to the left and right hand side of the trunk and the diameter at the correspondent height can directly be read from the measuring bar.

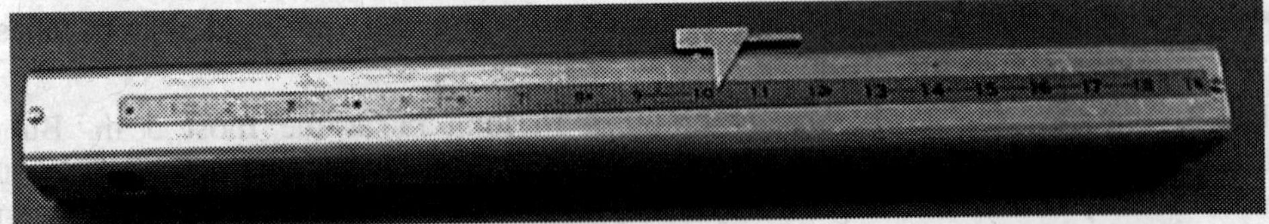

Fig. 12.3: Wheeler's Pentaprism Caliper

SPIEGEL RELASKOP: The Spiegel Relaskop, also known as a Relaskop, is a sophisticated instrument that can be used to measure stand basal area and tree height and diameter at any point up a tree bole. In conjunction with other equipment, the Relaskop can be used in the estimation of distance (range) to an object and the number of trees per ha. The Relaskop has a peep-hole at the rear and a clear window at the front. Three additional windows in the lower half of the instrument allow light to enter and illuminate the scale. A brake button, bottom half at the front of the instrument, allows a weighted wheel within the Relaskop to rotate. When looking through the peephole, a circular field of view is seen. The scales are seen in the bottom half of this field of view and scale readings are taken where the scale touches the line halfway up the field of view.

Determination of the stem diameter for a known height on the tree: First depress the brake button and move back until the tree stem appears to cover all of band 1 and the 4 quarter bands, i.e. the left side of the tree is aligned with the left of band 1 and the right side of the tree is aligned with the right side of the right most quarter band. Measure the horizontal distance (D in m) to the tree (directly beneath the point of interest). This method provides a maximum precision for determining diameter with the Relaskop.

Calculate the stem diameter (d in cm) as $\mathbf{d = 4 \times D}$

Measurement of both height and diameter at same time at one or more points up the stem:

1. Measure out a horizontal distance (D in m) appropriate for the height of the tree.
2. Sight to the base of the tree and the point of interest and determine height using the appropriate scales and calculations.
3. With the brake button depressed, align the right side of the tree with the right side of band 1 and ensure that the left side of the tree falls within the quarter bands. If the left side of the tree goes beyond the quarter bands, you will need to move further away from the tree - you will need to be 5 m away for every 20 cm of diameter. If the left side does not make it to the quarter bands, i.e. it falls within band 1, align the right side of the tree with the left side of band 1.
4. Estimate the number of quarter bands that are covered by the tree bole (between 0 and 4). Experienced operators can read down to fifths of a quarter band:

1/5 just into the quarter band

2/5 almost half a quarter band

3/5 at least half a quarter band

4/5 almost the whole quarter band

5. Calculate diameter (d in cm) as:

$$d = 2 \times D \times [b + (q/4)]$$

where

b equals 1 if band 1 is covered and 0 otherwise,

q denotes the number of quarter bands (read down to one fifth of a quarter band).

6. The precision (p) in per cent of the estimate depends on the distance from the tree (D in m) and the estimated diameter (d in cm) :

$$p = 10 \times D/d$$

TREE HEIGHT MEASUREMENTS

Total height of the tree is the straight line distance from the tip of the leading shoot (or from the highest point of the crown where there is no leader) to the ground level, usually measured on slopes from the uphill side of the tree. It is the straight-line distance from the highest point of the crown to the ground level. It is used to find out tree volume. It is also essential to read volume tables, form factor table, yield tables, etc. Tree height is normally used to find out productive capacity of the site and site quality of a locality. There are different measurements of height on trees:

- *Bole Height*: It is the distance from the ground level to the position of the first crown forming living or dead branch
- *Commercial Bole Height*: The height of bole that is usually fit for utilization as timber is called commercial bole height
- *Height of Standard Timber Bole*: It is the height of the bole from the ground level up to the point where average diameter over bark is 20cm
- *Timber Height*: It is the vertical distance from the base of the tree to the point on the main stem where the diameter is 7 cm for conifers. For hardwoods, timber height is the vertical distance from the base of the tree to the point on the main stem where the diameter is 7cm or where the main stem becomes the crown, whichever is the lower
- *Stump Height*: It is the height of the top of the stump above ground and it is normally 20-30 cm
- *Crown length*: It is the vertical measurement of the crown of a tree from the tip to the point half way between the lowest green branches forming green crown all round and the lowest green branch on the bole
- *Crown Height*: It is height from the ground level to the point half way between the lowest green branches forming green crown all round and the lowest green branch on the bole

- *Breast Height*: If refers to the usual point of measurement of standing tree or stem diameter (1.37 m or 1.3 m) above ground on the uphill side of the tree
- *Mean Crop Height*: The mean height of all the trees growing in forest
- *Top height*: Mean height of the trees with the largest dbh in a stand

HEIGHT MEASUREMENT PRINCIPLES (Chaturvedi and Khanna 1982; Yadav 2012a)

TRIGONOMETRIC PRINCIPLE: Trigonometry is a branch of mathematics dealing with measurements of the angles and sides of triangles and functions based on these measurements. *Tangent Law*: The height of the tree is calculated with the help of the tangents of the angles to the top and base of the tree and the distance of the observer from the tree. Simple example is of standing tree in plane ground.

Tree Height (BD) = AE + ED Tan α

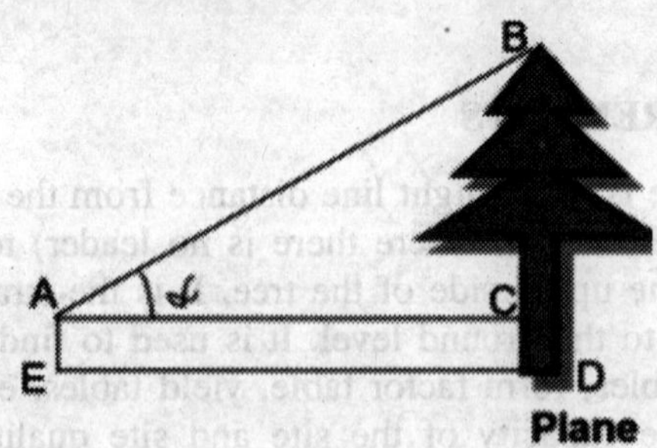

Sine Law: In trigonometry, in any triangle, sines of angles are proportional to the opposite sides. The three basic trigonometric functions that we are concerned with here (sine, cosine and tangent) are ratios of the lengths of two sides of a triangle. These ratios are the trigonometric functions of an angle, theta, such that here, opposite side refers to tree height.

PRINCIPLE OF SIMILAR TRIANGLES: Two triangles are said to be similar when the corresponding angles are equal and the corresponding sides are proportional. ADE and ABC are similar triangles.

Thus, BC:DE = AC:AE

Tree Height (BC) =DE × (AC/AE)

where, DE and AE are known and AC can be measured in ground

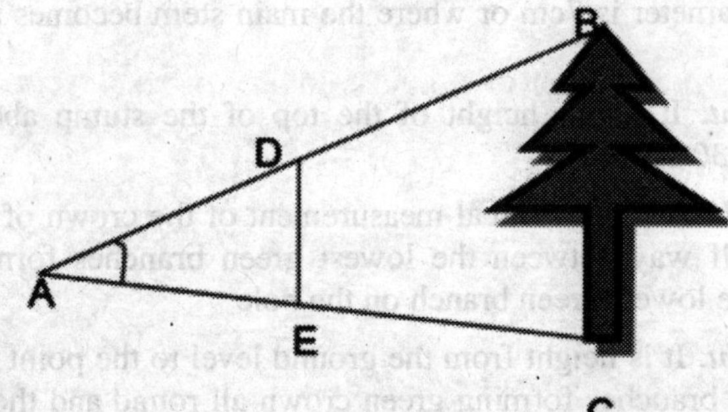

INSTRUMENTS FOR HEIGHT MEASUREMENT

Height sticks are used to measure the height of small trees. Hypsometers, altimeters and clinometers are used to measure height of tall trees (CCC 2009).

- *Hypsometers*: Used for determining the height of standing tree from observations taken at some distance from the tree.
- *Altimeters*: Generally altitude measuring instruments, which can be devised to determine heights of tree.
- *Clinometers*: It measures angle of slope. Any instrument which measures angles of slope can be used for height of tree by trigonometrical methods.
- *Laser Dendrometers*: They basically rely on the same geometric principles as the hypsometers.

Some clinometers designed for this purpose called hypsometers are based on geometrical principles of similar triangles or based on relations between the sides of right angled triangles. Christen Hypsometer, Improved Callipers, Smythies' Hypsometer, Improved Smythies' Hypsometer are the instruments based on similar triangles. They are very slow, fatigue, heavy and rough information. These instruments are very easy to use manually in the field even by unskilled labour.

Abney's Level, Brandis Hypsometer, Relaskop, Topographical Abney's Level, Haga Altimeter, Blume-leiss Hypsometer are the height measuring instruments based on trigonometrical principles. They are manufactured scientifically and repair is difficult on spot. They are used only by skilled labour and limited use and expensive. It is fast, easy to carry and accuracy is maintained. They are easily available in markets and adopted by many countries.

CHRISTEN HYPSOMETER

The Christen hypsometer is a simple instrument consisting of a rule or scale about 10 inches long which may be folded and carried in the pocket. It is made of metal, thin wood or even card board of about 2.5 cm thickness having two flanges to be used with a staff of known length. Upper flange will suspend by thread and next one to be suspended with weight to prevent swinging. It is based on similar triangles and usual length of instrument is 33cm (1.3 feet) used with a staff of 3.6 m (12'). The estimator, facing the tree at sufficient distance to permit him to see its top and base, holds the instrument vertically before him. An assistant holds a 12-feet pole upright at the base of the tree. The estimator moves the scale nearer or away from the eye until the whole length of the scale just covers the entire view (in height) of the tree. The marking on the scale which is on the line of sight with the top of the upright pole indicates the height of the tree.

ABNEY'S LEVEL

It is commonly used tree height measuring instrument based on trigonometrically principles. It gives accurate angle of elevation and depression. Readings can be taken after sighting the tree without disturbing the index arm. The instrument is small and light and can be used even in hills without difficulty. It has eye piece (hollow telescopic piece), magnifier glass, protector, wire and screw. It is an instrument with hollow tube with an eye piece at one end and a short sighting tube fitted at the other. Eye piece consists of 2 or 3 telescopic hollow tubes and a sighting tube is a small detachable tube fitted with a horizontal wire at the centre. A mirror behind the horizontal wire covering only half of the tube so fitted that it makes 45^0. A spirit level is fitted to the main tube, which can be rotated by one screw. Wheel is for quicker movement and screw is for final adjustments. An index arm is also attached to the spirit level. As the spirit level rotated, index arm moves on a graduated semi-circular arc. The angles of elevation and depression are noted in graduated arc in degrees up to 90^0. On either side of the zero mark at middle, each division gives the reading of 10' (Chris Brack 2007a).

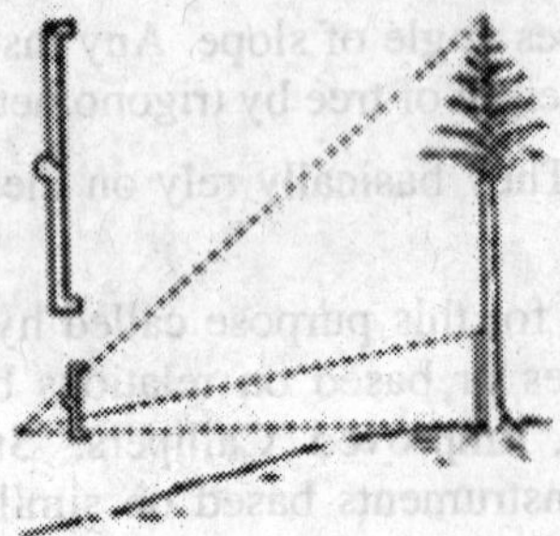

Fig. 12.4: Height Measurement using Christen Hypsometer

The horizontal distance from the base of the tree is measured to a location where the required point on the tree (e.g. tree tip) can be seen. The tree tip is sighted through the eye piece and this makes the instrument inclined and the bubble is not seen in the mirror. Therefore, while sighting the top, the screw is rotated to bring the spirit level in a horizontal position. The spirit level is continued to be moved slowly to the position when the bubble image is bisected by the line of horizontal wire on the mirror and in the other half the tree top is seen touching the horizontal wire. The spirit should be balanced by tuning with screw as well as moving to and back ward. The degrees and minutes for angle of elevation (α) are read. Again the base of the tree is sighted through eye piece and angle of depression (β) is measured. If the base of the tree is level with the observer or below observer (i.e. observer on the upward slope), total tree height is calculated using following formula.

$$\text{Tree Height} = \text{Horizontal Distance} \times [Tan\ (\alpha) + Tan\ (\beta)]$$

If the base of the tree is above the observer (i.e. observer on the downward slope), the following formula is used to get a total tree height.

$$\text{Tree Height} = \text{Horizontal Distance} \times [Tan\ (\alpha) - Tan\ (\beta)]$$

SUUNTO CLINOMETER

The Suunto Clinometer (clino) is a tool commonly used by foresters to measure tree heights and also slope angles. Tree height is measured using the principle of triangulation with a clinometer. Of all the forestry tools, the clinometer requires the most practice and skill. It is assumed that the tree grows at a right angle to the ground (even on a slope). At the rear of the clino is a peephole, which shows a percentage scale and a horizontal line. First, the horizontal distance between the base of the tree and the operator is measured. Looking through the peephole, the horizontal line is lined up with the top of the tree and the corresponding number from the percentage scale, which is on the right hand side is read off (top reading). The scale on the left is in degrees and should not be used. Similarly, the corresponding percentage scale for base of the tree (bottom reading) is measured. If the base of the tree is level with the observer or below observer (i.e. observer on the upward slope), total tree height is calculated using following formula (Chris Brack 2007a).

Tree Height = Horizontal Distance × [Top Reading + Bottom Reading]

If the base of the tree is above the observer (i.e. observer on the downward slope), the following formula is used to get a total tree height.

Tree Height = Horizontal Distance × [Top Reading - Bottom Reading]

HAGA ALTIMETER

Haga altimeter is based on the trigonometrically ratio. It is very easy to use and read height directly. This instrument is based on the principle of 'sight the object and shoot'. It has eye piece, pronges, scale, trigger button, release trigger button and tuning knob. It has several percent of usual scales: 15, 20, 25, 30 for the corresponding horizontal distance from the tree: 15 m, 20 m, 25 m, 30 m. For example, if the observer is standing at a distance of 20 m from the base of the tree, then he must use the percentage scale of 20 for measuring the reading. First, the observer stands at a distance of 20m from the tree and looks the top of the tree through eye piece. He sees the tip of the tree through eye piece and coincides it with the pronge of the instrument. At this position, trigger is pressed to arrest the pointer on the scale and now the reading on percent scale 20 is noted (top reading). Now the pointer is released by pressing trigger release button. Again the base of the tree is viewed through eye piece and corresponding reading on 20 percent scale is noted (bottom reading). The total tree height is calculated using the formula.

Tree Height = Top Reading + Bottom Reading

Ravi altimeter/multimeter and Blume Leiss altimeter also work with the above same principles and readings are taken in the same manner of Haga Altimeter.

VERTEX IV AND TRANSPONDER T3

The vertex IV is primarily designed to measure the height of standing objects and most often trees. The instrument can also be used to measure distance, horizontal distance, angle and inclination. The vertex instrument with its ultrasonic measuring technique proved to be specially useful in dense terrain with thick undergrowth, where conventional methods such measuring tapes, laser instrument and mechanical height measures are difficult to use. To define a reference point in a secure and reliable way, the vertex IV works with the transponder T3. The vertex IV communicates with the transponder. This communication eliminates any mix up of signal from other instruments or places (echos) in an efficient way. The measuring operation will not be disturbed by objects in between the vertex IV and the transponder T3 in any significant way. The reference point i.e. the T3 is used as a sight mark for height measuring and can be placed at optional height. The reference point height is set in a special menu in the vertex instrument and automatically added to the measured height.

The vertex IV uses ultrasound to measure distance. Unlike measuring tapes and laser instrument, ultrasound can be used also when there is no free aim to the reference point. The ultrasound will not pass through an obstacle, but looks for the shortest way. Heights are calculated trigonometrically using the variables contained when measuring angle and distance. The vertex IV automatically assumes that the measuring object is perpendicularly positioned to the ground. With the vertex IV, an unlimited number of heights or objects can be measured. The instrument display can show the four lastly measured heights per object at a time. When using a telescopic method to measure, an in built BAF function can be used for the vertex IV instrument to control the minimum diameter for trees. The function is useful when some trees in an area are covered by other, making the decision whether to include the tree or to exclude it from the area, difficult. By simply measuring the distance between the tree and plot centre, the vertex IV can calculate the minimum diameter the tree should have in order to be included into the counting. Data can be sent through IR or bluetooth and results can be stored and processed in the digitech professional calliper, other PC or handheld computer (Chris Brack 2007a).

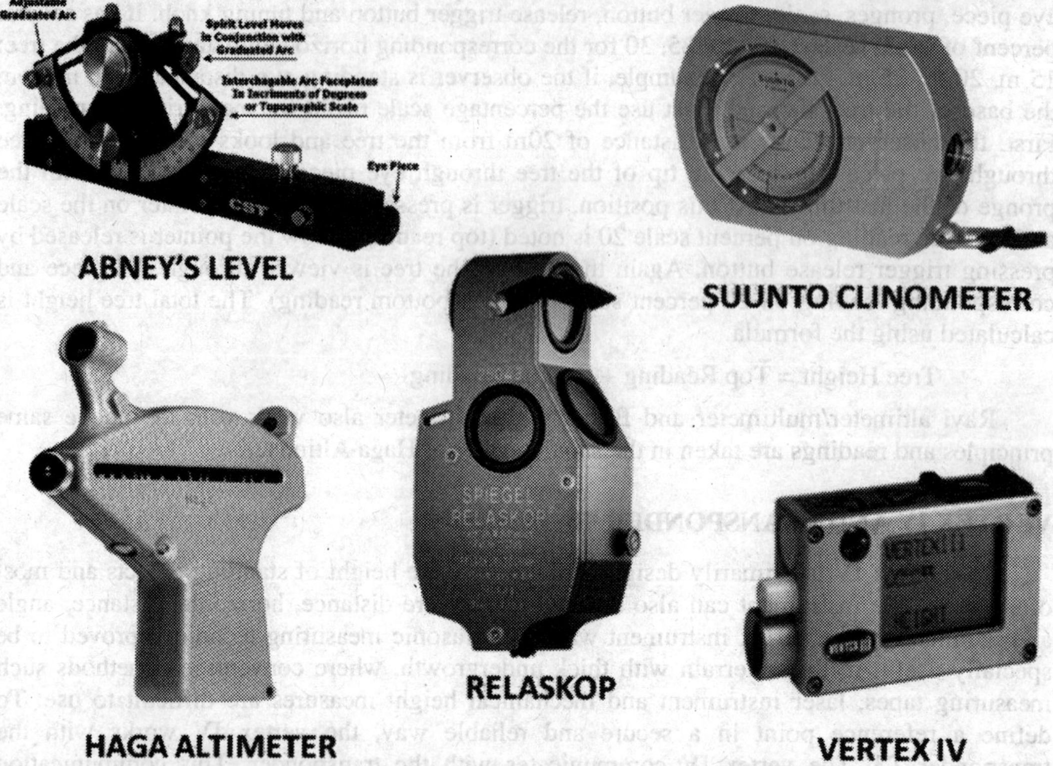

Fig. 12.5: Height Measuring Instruments

SPIEGEL RELASKOP

The standard metric Relaskop has three scales for measuring (vertical) height. The appropriate scale will depend on the horizontal distance from the tree.

- left-most scale - 20 m from the tree.

- middle-left - 25 m from the tree.
- middle-right - 30 m from the tree.

After depressing brake button, one should look straight up or down and the appropriate distance values can be seen alongside their scales. A point must be selected that is 20, 25 or 30 m (horizontal) from the tree of interest. Other horizontal distance can be selected for which the final scale reading must be doubled or halved or multiplied with respective conversion factors. The base and tip (or any other points of interest) must be clearly visible from the selected point. If the tree is leaning, the point where you observe the tree should be at 90 degrees to the plane of the lean. If the lean is severe (i.e. more than 10 degrees from vertical), the point directly beneath the tip of the tree should be located (e.g. using a plumb-bob) and the horizontal distance taken from this point. The actual distance chosen is related to an initial estimate of the tree height. Once the observation point is found and the appropriate scale selected, sight through the peephole to the base of the tree. Tap the brake button several times until the scale settles then read the height directly from the appropriate scale. If you are looking down towards the base of the tree, this reading is the vertical height that the base of the tree is beneath your eye. Sight to the top of the tree (or other point) and again tap the brake button until the scale settles. The scale is a direct reading of the height above your eye to this new point. Add the two heights together if the base of the tree was below your eye to determine total (vertical) height. If the base of the tree was initially above your eye (i.e. the base if above you on sloping ground) subtract the initial height from the upper height to determine (vertical) height (Chris Brack 2007b).

TREE BASAL AREA MEASUREMENTS

Tree basal area (TBA) is the cross-sectional area (over the bark) at breast height measured in square meter (m^2). TBA can be used to estimate tree volumes and stand competition. To determine tree basal area simply measure the diameter at breast height in centimeters (DBHOB) and calculate the basal area (m^2) using an equation based on the formula for the area of the circle (area = πr^2 where r = radius). The formula below also converts the diameter in centimeters to the basal area in m^2. The same technique can be used to calculate the cross sectional area of the tree at any point along the stem.

Tree Basal Area (TBA) in $m^2 = (DBH/200)^2 \times 3.142$

Where, DBH is the Diameter at Breast Height in centimeters and 3.142 is π

TREE VOLUME MEASUREMENTS

Volume is the most widely used measure of wood quantity and is usually estimated for the assessment of economic value or commercial utilization potential. The wood volume of a tree includes stem, branches, stump and roots. For standing trees, aboveground volume production is generally based on stem wood volume for conifers, but may include branch volume for broad-leaved tree species. Depending on measurement objective and local traditions, measurements or predictions of wood cubic volume may refer to, for example, total stem volume, total tree volume (stem and branches), or volume above a certain merchantable limit. Volume estimates may include or exclude bark and, for aboveground estimates, include or exclude the stump. Volume is always a cubic measure and usually expressed in cubic meters. Merchantable volume, however, is sometimes expressed in other units related to commercial use.

Volume is usually estimated for standing trees from such measurements as diameter, or diameter plus merchantable height, using a volume equation or a log rule. Volume may be measured directly on felled trees or logs, but is often estimated from dimensions such as minimum diameter or piece length. Direct measurement of volume is usually done by sectioning the tree into smaller pieces assumed to be cylinders. The general formulas for calculating the volume of standing trees are:

- Volume = BA × H × Form Factor
- Volume = $\pi r^2 H$ × Form Factor
- Volume = D^2H × Form Factor

where,

BA = Basal area of the tree

r = radius of the tree

H = Height of the tree

D = Dbh of the tree

STEM/LOG VOLUME MEASUREMENTS

It is possible to utilize geometric relationships to approximate volume. The volume of a cylinder is simply the area of the base times the height, and the volume of a cone is one-third of the volume of a cylinder with the same area of the base and height. Trees are neither cones nor cylinders, but empirical analyses often indicate that the volume of a single-stemmed tree is between that of a cone and a cylinder, with tree volume often lying between 0.40 and 0.45 times that of an equivalent cylinder.

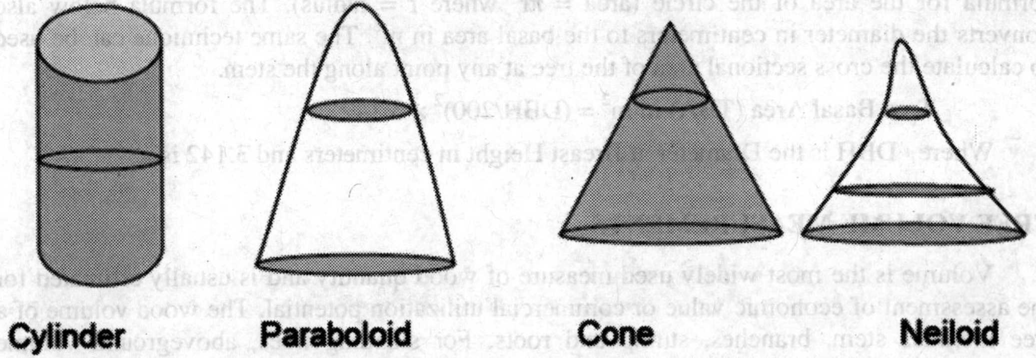

Fig. 12.6: Geometric Shape of Tree Stems/Logs

Volume of Cylindrical Stem = S × L

Volume of Conical Stem = 1/3 × S × L

Volume of Paraboloid Stem (Huber's Formula) = S_m × L

Volume of Paraboloid Stem (Smalian's Formula) = $(S_1 + S_2) \times L/2$

Volume of Neiloid Stem (Newton's Formula) = $(S_1 + 4S_m + S_2) \times L/6$

Quarter Girth Formula or Hoppus Volume = $\left(\frac{G}{4}\right)^2 \times \frac{L \text{(in ft)}}{144}$

where,

V = Volume of log

L = Length of log

S = Cross-sectional area of cylinder or cone

S_m = Mid cross-sectional area of log

S_1 = Cross-sectional area at the smaller end

S_2 = Cross-sectional area at the larger end

G = Mid Girth of log in inches

Volume Calculation of Sawn Timber

V = L x B x H (cubic meter or cubic feet)

Volume Calculation of Firewood/Stacked Wood: Volume may be estimated for stacks of logs or processed products by measuring dimensions. In these cases, local knowledge is often needed for appropriate estimation of volume. Firewood is stacked in the form of rectangular parallel piles and the volume of the stacked firewood is calculated by:

Volume = Length (L) × Height (H) × Breadth (B) of the stack expressed in cubic meter or cubic feet

STEM FORM

Form is defined as the rate of taper of a log or stem. Taper is the decrease in diameter of a stem of a tree or of a log from base upwards. The taper varies not only with species, age, site and crop density but also in the different parts of the same tree.

FORM FACTOR

Form factor is defined as the ratio of the volume of a tree or its part to the volume of a cylinder having the same length and cross-section as the tree. The form factors are of the following types (Chaturvedi and Khanna 2000):

- **Artificial Form Factor:** This is also known as the breast height form factor. Here, the basal area is measured at breast height and the volume refers to whole tree both above and below the point of measurement.
- **Absolute Form Factor:** For this form factor, basal area is measured at any convenient height and the volume refers only to that part of the tree above the point of measurement.
- **Normal Form Factor:** In this form factor, basal area is measured at a constant proportion of the total height of the tree.

FORM QUOTIENT

It is the ratio of diameter to diameter at breast height. Form quotients are of two types:

- **Normal Form Quotient:** It is the ratio of mid-diameter of a tree to its diameter at breast height.
- **Absolute Form Quotient:** It is the ratio of a stem diameter at one half its height above the breast height to the diameter at breast height.

FORM HEIGHT: It is the product of form factor and the total height of the tree.

Form Height = Volume of tree/Basal area

FORM CLASS: It is one of the intervals in which the range of form quotient of trees is divided for classification and use.

FORM POINT RATIO: Form point is the point in the crown where wind pressure is estimated to be centered. Form point ratio is the relationship of height of the form point above ground level to the total height of the tree, usually expressed as percentage.

TIMBER INVENTORY

Timber inventory is defined as the procedure for obtaining information on the quantity and quality of timber resources available in an area on which the trees are growing. The timber inventory is an extensive survey of the timber resources on a particular forest area. Timber inventory is an accounting of timber and their related characteristics of interest over a well defined land area. It is the enumeration of trees in a particular area for assessing the timber availability. It may be compared to census methods of human population. Similarly timber inventories seek to enumerate the population of trees within a forest and ascertain other information, such as species composition, timber volume, value and growth. Complete enumeration is carried out only in exceptional cases where very high value timber species are present in smallest tracts of land. Complete enumeration of individual trees is usually infeasible and survey sampling techniques are required. A timber cruise is a sample measurement of a stand used to estimate the amount of standing timber that the forest contains.

PURPOSE

The prevailing reason for conducting a timber inventory is to make informed decisions about forest management. The primary need has been the quantification of volumetric product yield and structural composition of the forest. The volume of the timber resource is typically categorized by species, product and size. In addition, quantities like the number of trees and basal area per unit area are often desired. While collection of data on other components of ecosystem is increasing, the timber resource still remains the main focus for most timber inventories. Both spatial and temporal scales are normally addressed in the planning of an inventory. The most obvious dichotomy is in time: a timber inventory may be initiated for an assessment of current conditions (growing stock), or it may be repeated at future periods in time, yielding estimates of change in volume or basal area and possibly other components of growth, mortality and removals (due to harvest). Spatially, the size of the area in question may range from a small woodlot or individual stand to a national inventory. Different methods are required for differing spatial scales. The objectives for conducting a timber inventory include:

- To examine stands that focus specifically on making stand-level decisions over a short planning horizon with regard to the prescription of silvicultural treatments e.g. thinning
- To prepare short term operational plans for harvesting
- To estimate the volume and value of an area and to determine the total availability of timber for sale prior to a commercial harvest
- To assess the damage on residual trees through postharvest inventories
- To assess the adequacy of regeneration stocking following a regeneration treatment and such efforts may be repeated to estimate survival rates
- To develop long term strategic plan through large-scale inventories for determining allowable harvest rates or optimal harvest scheduling plans
- To determine the value of the forest for accounting purposes
- To conduct survey for the valuation of land and timber to be purchased, sold or exchanged
- To conduct countrywide surveys often mandated by law for making high-level policy decisions and broad-scale resource monitoring

ESTIMATING THE AREA AND MAPPING THE SITE

The starting point for forest measurement involves developing a good map of the area that shows the type of the forest, age (if planted) and past management. If tree management, performance, provenance, age, soil types or environment are quite clearly different and have effected the growth in parts of forest, these should be treated as separate areas and marked on the map. A common method of measuring the area of a stand is to use a large-scale aerial photograph (1:10,000) and calculate the area using a dot grid or planimeter. Topographic maps may be suitable where a person can clearly delineate the boundaries of different forest areas (e.g. 1:25,000 map). Many computer programmes can determine areas of irregularly shaped polygons drawn over scanned photographs or maps, thereby providing an easy estimate of ground areas.

SAMPLING DESIGN

Timber inventory field work is extremely costly and so it is generally well worth the effort to spend time ensuring that the design of the inventory is as efficient as possible and so gain maximum accuracy at minimal cost. Sample points should be located on a map and the same map is used in the field to locate the sampling points. It is desirable to ensure that the samples cover the complete area of interest.

Sampling is a process in which inventory is carried out only in a representative portion of the whole. For example, a forest may be of 1,000 ha out of which only 100 ha has been selected for inventory and estimate of the whole population of 1,000 ha is made, it is called sampling. Sample is defined as a part of population consisting of one or more sampling units, selected and examined as a representative of the whole. The advantages of sampling over total inventory are reduced cost and saving of time, relative accuracy, knowledge of error and greater scope. The important sampling designs used in timber inventory are simple random sampling, stratified random sampling, systematic sampling, multi-phase sampling and multi-stage sampling.

Simple Random Sampling: It is a selection process in which every possible combination of sample units has an equal and independent chance of being selected. Sampling units are chosen completely at random. Samples are either chosen with replacement or without replacement, the latter means that once a sampling unit is chosen it may not be chosen again.

Stratified Random Design: The most commonly used sampling design is the stratified random design where plots or points are located at random within strata. The stratification into approximately homogenous strata reduces the number of samples that are required to meet an overall precision target for sampling precision. A population is divided into sub-population of known size and a simple random sample of at least two units is selected in each subpopulation. In forests, stratification can be done on the basis of stand type, stand age, site quality, climate region or ecosystem.

In principle, variable probability is more efficient as the strata with higher timber yields can be sampled at greater intensity and this will improve the overall precision of the inventory. It is efficient both statistically and economically. If the forest is very heterogeneous, then it is possible to use variable probability sampling and to concentrate the samples in the strata with greatest volume or basal area. This approach may also assist future practical management of the forest as it may be sensible to treat each stratum, or perhaps a group of strata, separately, in order to best meet harvesting or other silvicultural objectives. If strata differ widely in terms of a variable of interest (basal area or volume) then it is possible to adjust the sampling level accordingly and so reduce the number of samples required to achieve a desired precision. Some inventory specialist prefer to use equal probability sampling between strata even though it is more expensive as it makes computation easier and it is easier to explain what is happening to managers (Jerry Leech 2007).

Systematic Sampling: The initial sample unit (i.e. first plot) is randomly selected. All other units (plots) are spaced at uniform intervals throughout the forest. Sampling units are easy to locate and appear to be representative. Generally acceptable estimates for the population mean are obtained. Systematic sampling is generally satisfactory for estimating means in typical forest conditions. When an objective numerical statement of precision need not be appended to inventory estimates, systematic sampling may provide more information for the time expended than simple random sampling. In cases in which estimates of variance are important or little is known about the basic characteristics of the population being sampled, it is good to use simple random sampling than systematic sampling. If adequate care is taken not to introduce a systematic bias then systematic sampling may be quite appropriate in some inventories.

Multi-phase and Multi-stage Sampling: The differentiation between multi-phase and multi-stage sampling is often clouded. These sampling procedures occur more generally in large national timber inventories than in timber inventories generally used for valuing smaller forests. The idea is to start from a complete area and to stratify that, selecting sample areas, such as forest reserves, watersheds, or bio-geographic regions. This provides the first stage samples and a number of these are selected, but not all. Then for each first stage sample the area is stratified in more detail and second stage samples are selected from this level. This procedure can be repeated to any depth. This is the multi part of the names.

The separation into multi-phase or multi-stage is generally based on the methods of stratification. If the stratification of each level is basically the same, only increasing in depth, then it is called multi-stage sampling. If the stratification is based on quite different bases, for example the first stage being administrative regions, the second remote sensing imagery and the third on aerial photography or more intensive, then the procedure is generally called multi-phase. For most forest valuation work, it is often considered more appropriate to ensure that all first stage and generally all second stage strata are sampled. In this case, the sampling is just stratified random sampling.

POINT SAMPLING

Point sampling or angle count sampling use angle gauges such as spiegel relaskop or the simple optical wedge prism. Point samples have some definite advantages in that they give a higher priority to the larger trees which are generally the more important, and there are generally less of them in the forest. On the other hand, it is more difficult to address problems near the boundary of the forest or the stratum and quite sophisticated procedures have been developed to account for this. If point samples are used to determine growth then there is generally an increase in sampling error as trees that were considered "out" during the first measurement may now be considered "in". A commonly adopted rule of thumb is to use point samples if re-measurement is not expected.

OPTICAL PRISMS (*WEDGE PRISMS*): The optical properties of wedge shaped prisms are particularly suited to angle count sampling, and since the early 1960s, the optical wedge or wedge prism has been used extensively as an angle gauge for basal area estimates. A wedge prism can be used to estimate quickly the Basal Area per hectare and one costs only 2 per cent the price of a Spiegel relascope. It is a simple wedge-shaped prism of glass or see-through plastic, typically 5 x 2 cm. It distorts the light and shifts the position of a tree stem when looked at through the prism. Different factors of prism relascopes are available, with common Basal Area Factors being 5, 8, or 10 (Chris Brack 2007a).

The technique with the wedge prism is to stand at one point among the trees and to complete a 360 degree sweep around, counting all the trees that are 'IN'. Those that are 'borderline' should be counted every other time considering it as "Half In", and those that are 'Out' are not counted. To estimate the basal area simply multiply the number of counted trees by the basal area factor (e.g. 5, 8 or 10). One should conduct as many sweeps around the stand of trees, as this will provide a more accurate estimate when averaged over the stand. Thus, the wedge prism is sometimes claimed to be faster and more accurate than the spiegel relaskop.

BASAL AREA MEASUREMENT WITH SPIEGEL RELASKOP: The relaskop can generate nine basal area factors (BAFs) using bands 1, 2 and combinations of the quarter bands. Choose the appropriate BAF and combination of bands. Then, standing over the sample point, hold the brake button down and make a sweep of 360 degrees while comparing the tree diameters at breast height with the selected bands. Count the number of trees whose diameter at breast height appears greater (wider) than the selected bands (IN trees). Ensure trees are not hidden behind closer trees by stepping to one side and checking before returning to the sample point. Where trees appear to be the same width as the selected Relaskop bands, it refers to the borderline calculations and considered as "Half In". Once the sweep has been completed, multiply the count of IN trees by the appropriate BAF to get stand basal area (m^2 per ha).

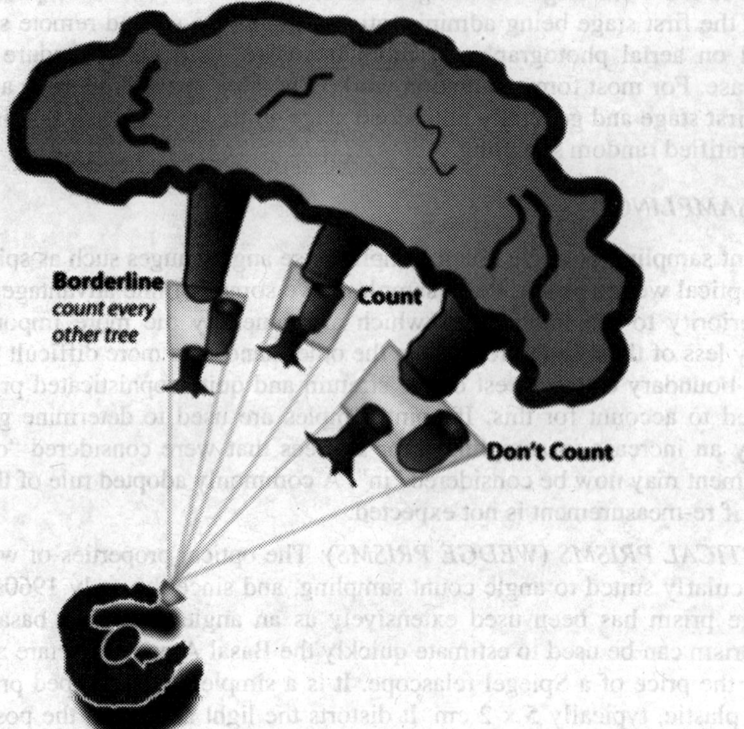

Fig.12.7: Wedge Prism Usage

Table 12.1: Spiegel Relaskop Bands and BAFs

Bands	BAF	K
1 quarter (narrow band)	0.0625	2
2 quarter (narrow band)	0.25	1
3 quarter (narrow band)	0.5625	0.6
Band 1	1.0	0.5
Band 1 + 1 quarter	1.5625	0.4
Band 1 + 2 quarter	2.25	0.3
Band 1 + 3 quarter	3.0625	0.2857
Band 1 + 4 quarter (also called Band 4)	4.0	0.25
Band 2	2.0	0.3536

PLOT SAMPLING

Because measuring trees can be a time consuming and costly operation, most foresters only measure a sample of trees as an estimate to the growth of different parts or blocks of the forests. An absolute minimum of 3 plots should be established in any uniform section, unless one or two plots cover most of the site, in which case all trees should be measured. For very large uniform forests, the total area of all the plots should be an absolute minimum of 2 per cent of the total forest area. For example, in a forest of 10 hectares, a total area of 0.2 hectares should be measured. If the plots are to be 0.04 hectares in size then at least 5 plots would be required for a sufficient sample. Plots can be kept within the stratum or forest boundary more easily and do provide a consistent area base for determining growth. However because they sample a fixed area, they may not provide as good an estimate for the largest tree diameter classes and commonly over sample the smaller diameter classes. A commonly adopted rule of thumb is to use plot sampling if re-measurement is expected.

Circular or Rectangular Plots: Plots are generally circular or rectangular. Rectangular plots are useful and preferred in stands where planting rows are well defined, as in young stands or in heavily stocked forests. Circular plots, on the other hand, are easier to lay out in mature or irregularly spaced stands with low stocking rates or where the rows are poorly defined or not present, such as in native forests.

Plot Size: In order to obtain a representative sample, between 15 and 30 trees per plot is required (12 being the absolute minimum). Plot size will therefore depend on the stocking rate. The more sparse the trees, the larger the plot will need to be to include sufficient trees. Tree stocking can be quickly estimated from the average spacing or, alternatively the size of the plot can be gradually increased until it includes sufficient trees. Plot area for each shape is calculated as follows:

Rectangular Plots: Plot Area (Ha) = Length (m) × Width (m)/10000

Circular Plots: Plot Area (Ha) = $[\text{Radius (m)}]^2 \times 3.142/10000$

Table 12.2: Sample Plots Size for Different Stocking Rates that will include about 20 trees

Stocking Rate (Trees/hectare)	Sample Plot Size (hectares)	Dimension of Rectangular Plot	Radius for Circular Plot
100	0.2	50m x 40m	25.2m
200	0.1	40m x 25m	17.8m
400	0.05	25m x 20m	12.6m
600	0.033	20m x 16m	10.2m
800	0.025	20m x 12.5m	8.9m
1000	0.02	16m x 12.5m	8.0m

Highly Irregular Forests: Where the forest is very irregular, rather than establishing fixed area plots, it may be preferable to identify a number of individual "plot trees" scattered through the forest that are measured individually.

Farm Forestry Trees: In farm forestry, trees are planted in small blocks, belts or strips with many, perhaps most, of the trees growing on the edge. In this case, it is better to measure the trees in a length of belt to assess yield per 100 meters of belt (rather than yield per ha).

STAND MEASUREMENTS

STOCKING RATE

The tree density or stocking rate of a forest is described as the number of trees per hectare

Stocking Rate (stems/ha) = Trees in Plot/ Plot area (ha)

STAND DIAMETER

The diameter of every tree in the sample plot is measured as per the methods described earlier. While measuring diameter, the form of the tree must be inspected and form factor should be recorded.

STAND HEIGHT

Measuring the heights of trees can be difficult and time consuming. The heights of the tallest trees in a plantation or native forest are usually quite uniform and therefore, rather than measure the height of all trees in the sample plot, it is common to select a sub-sample. In most cases, a number of the fattest trees (largest DBH) of good form are measured for height and this is called the "Mean Dominant Height" (MDH).

Mean Top Height (MTH) is most commonly used in some countries. It is the height predicted by reading from a height - diameter curve (commonly the Pettersen Curve) for a DBHOB corresponding to the mean top dbh (MTDBH). Mean top dbh (MTDBH) is defined as the quadratic mean of the largest 100 trees (by diameter or dbhob) per hectare. If the stocking is less than 100 trees per ha then the mean is across all the trees.

EMPIRICAL HEIGHT EQUATIONS

Because tree height measurement is costly, it is common to measure all trees for diameter but to measure only a sample of trees for height and then to use a model based on these data to predict tree height for all trees. Tree height is necessary as tree volume (V) is commonly derived from a function of diameter (DBH) or (D) and tree height (H), i.e. (V = f[D,H]). Tree height generally increases as tree diameter increases with the rate of increase decreasing with increasing diameter. Tree height often approaches an asymptote. A wide range of height equations are used to estimate the height of trees using diameter values:

$$H = b_0 + b_1 D \quad \text{(Linear)}$$

$$H = 1.3 + b_1 D + b_2 D^2 \quad \text{(Quadratic)}$$

$$H = b_0 + b_1 \ln(D) \quad \text{(Log linear)}$$

$$H = b_0 + b_1 / D^{b_2} \quad \text{(Inverse linear)}$$

$$H = b_0 D^{b_1} \quad \text{(Power)}$$

$$H = b_0 \exp(b_1/D) \quad \textit{(Exponential)}$$

$$\ln(H) = b_0 + b_1 \ln(D) \quad \text{(Log - log)}$$

$$\ln(H) = b_0 + b_1/D \quad \text{(Log inverse linear)}$$

$$1/\{(H - 1.3)^{0.4}\} = b_0 + b_1/D \quad \text{(Pettersen Curve)}$$

$$D/\{(H - 1.3)^{0.4}\} = b_0 + b_1 D \quad \textit{(Alternative Pettersen Curve)}$$

It can be argued that it is important to see which function best suits a particular species and region and that the best curve may also change during the life of a plantation. Of the models presented, four latter models will produce biased predictors as the model is fitted to a transformed dependent variable and so when the predictor is transformed back to predict height a small bias is introduced. This is generally ignored as it is small relative to the errors in the basic measurements.

STAND BASAL AREA

Stand Basal Area (SBA) is simply the cross sectional area of all the trees at breast height per hectare of forest or plantation (m²/ha). Stand basal area can be used to estimate stand volume and is a useful measure of the degree of competition in the stand. SBA is often quoted when planning thinning prescriptions. The basal area of stand or plot can be measured in two ways: sum of individual tree basal areas and optical method using spiegel relaskop or wedge prism.

Sum of Individual Tree Basal Areas: The most accurate method of assessing the basal area of a stand of trees is to measure all tree diameters at breast height in a plot, calculate individual tree basal areas and add these up. A quicker method is also available to calculate the stand basal area using the average tree diameter.

$$\text{Tree Basal Area } (m^2) = (\text{DBH}/200)^2 \times 3.142$$

$$\text{Stand Basal Area } (m^2/\text{ha}) = \frac{\text{(Sum of the basal area of each tree in plot)}}{\text{Area of the plot (ha)}}$$

$$\text{Stand Basal Area } (m^2/\text{ha}) = \text{(Basal area of the average tree diameter)} \times \text{Stocking (trees/ha)}$$

STAND VOLUME/ FOREST VOLUME

Total forest volume or stand volume can be calculated from the plot measurements:

$$\text{Standing Forest Volume } (m^3/\text{ha}) = \frac{\text{Plot Volume } (m^3)}{\text{Plot Area (ha)}}$$

Forest basal area measurements can be used to calculate tree and butt log volumes in the same way that tree volumes are calculated. In plantations where we might assume all the trees are "quite uniform, a quick estimate of total volume can be made from the Stand Basal Area and Dominant Tree Height using form factor.

$$\text{Standing Forest Volume (m}^3\text{/ha)} = \text{SBA} \times \text{MDH} \times \text{Form Factor}$$

where, SBA = Stand Basal Area, MDH = Mean Dominant Height

MEAN ANNUAL INCREMENT (MAI) AND CURRENT ANNUAL INCREMENT (CAI)

Mean annual increment is simply the average volume production per year for a forest of known age.

$$\text{Mean Annual Increment (m}^3\text{/ha/yr)} = \frac{\text{Stand Volume (m}^3\text{/ha)}}{\text{Age of Stand (yrs)}}$$

Current annual increment (CAI) is the increase in volume at a particular age and is determined by annual measurements of standing volume. It is simply defined as the volume increment of a stand in a single year.

CAI (m^3/ha/yr) = Stand Volume at current year – Stand Volume of previous or last year

VOLUME EQUATIONS

There is need to predict tree volume or stand volume and one way is to develop tree volume equations that just require tree diameter and either tree height or upper stand height. The most commonly used model form is:

$$Y = b_0 + b_1 D^2 H$$

where

Y is the tree volume

D is tree DBHOB (diameter at breast height over bark)

H is the tree height

It is often the case that the residuals suggest a σ proportional to the independent variable D2H. Better estimates for the parameters b_0 and b_1 are therefore obtained by fitting the equation

$$\frac{Y}{D^2H} = b_0 \frac{1}{D^2H} + b_1$$

Commonly used typical local volume equation forms (based on DBH)

$$Y = b_0 + b_1 D^2$$

$$Y = b_0 D^{b_1}$$

$$Y = b_0 D + b_1 D^2$$

$$\ln(Y) = \ln(b_0) + b_1 \ln(D)$$

$$\frac{Y}{D} = b_0 + b_1 D$$

The parameters b_0 and b_1 may be estimated by simple linear regression.

Commonly used general volume equation (based on DBH and height)

$$Y = b_0 + b_1D^2H + b_2D^2 + b_3H$$

$$Y = b_0 + b_1D^2 + b_2H + b_3D^2H$$

$$Y = b_0 + b_1 (D - b_2)^2 (H - b_3)$$

$$Y = b_0 + b_1 (D^2 - b_2)(H - b_3)$$

and even

$$Y = b_1 D^{b_2}H^{b_3})$$

which can be transformed by taking logarithms into an intrinsically linear model form

$$\ln(Y) = b_1 + b_2 \ln(D) + b_3\ln(H)$$

Using the concept of dimensional analysis the sum $(b_2 + b_3)$ in the last equation should equal 3.0 with $(b_2 = 2.0, b_3 = 1.0)$. However as trees are not cylinders but can generally be approximated by a neiloid or second degree paraboloid and then the parameter values generally fall in the ranges $(b_2 = 2.2 - 2.5, b_3 = 0.5 - 0.8)$ with the $(b_2 + b_3 = 3)$.

Choice of model may depend on modeling objective, data used in the estimation of coefficients and error structure. These basic equations implicitly assume a single-stemmed form and may require modification or replacement for species with a more complex form. In addition, when equations are used to estimate the logarithm of a variable, a negative bias is introduced when the predicted logarithm is converted back to arithmetic units. This bias is approximately the order of magnitude of one-half of the residual variance of the equation.

VOLUME TABLE

Volume table is defined as a table showing for a given species the average contents of trees, logs or sawn timber for one or more given dimensions. The main objective of the volume table is to estimate the volume of average standing trees of known dimensions.

The given dimensions may be:

- dbh alone
- dbh and height
- dbh, height and some measure of form or taper

On the basis of application, volume tables are classified as follows:

GENERAL VOLUME TABLE **(GVT):** These tables are usually based on two variables named diameter at breast height (dbh) and total tree height. These are based on the average volume of trees growing over a large geographical area. These are used to derive local volume table and to estimate the volume of trees in the form of standard timber.

LOCAL VOLUME TABLE **(LVT):** These tables are generally based on one independent variable i.e. dbh. These tables are derived only from restricted locality (local area). These are used to estimate the volume of growing stand and to make confidential volume of coupe or compartment.

13 FOREST PEST MANAGEMENT

INTRODUCTION

Forest ecosystem consists of a variety of components, all of which are critical to the forests long term sustainable health. All parts of a tree roots, stems, foliage, shoots and terminal leaders are vulnerable to attack by pests. Pest damage can range from slight damage that has no effect on the value of the harvested products to severe damage that stunts or kills the trees or reduces their market value. Tree pests include insects and mites, diseases, weeds, vertebrates and nematodes. Among these insects, fungi and other microorganisms play an important role in forest stability, succession and productivity by contributing to the degradation of dead and dying trees, nutrient recycling of decaying organic matter, creating opening for regeneration, providing a food source for other organisms, pollination and predation or parasitism of potentially harmful pests. Thus it is important for forest managers to recognize that forest management practices can have an impact on the other components of the ecosystem that could be detrimental to long term forest stability. Interest in protecting forests from insect, disease, weed and vertebrate pests has increased in recent years. This has come about largely because of, increased awareness of the destructive capacities of pests, heavy toll they take on supplies of commercial and recreational timber, environmental concerns, effects on threatened and endangered species and availability of new specific pesticides.

Destructive insects are an integral part of forest ecosystems, that they have significant economic and social impacts as well as ecological effects on forest productivity and values. The adverse impacts should be prevented or kept at tolerable levels by ecologically sound means, which are compatible with forest management practices. Insects causing damage to forests are more explicitly defined in terms of its economic effects and disruptive influence on forest pest management schedules and operations. Augmentation of natural controls and the manipulation of forest conditions are necessary to make them less favorable for outbreaks and to ameliorate losses on a long term, continuing basis. The silvicultural methods and management practices are considered to have some influence on the probabilities of outbreaks, reduction in damage, or reclamation of damaged stands with respect to specific kinds of pests. Forest pest management, as now conceived, is firmly rooted in quantitative methodology related to both the measurement and analysis of forest growth and development and the dynamics of forest insect populations. Forest ecosystems are diverse and complex.

Modern concepts, methods of mathematical modeling and systems analysis are drawn upon at an early stage to gain a better understanding of their basic functional processes and relationships.

The goal of forest management should be to maintain all components of the system in a balance that enhances the likelihood of long term sustainability. Managing tree pest effectively should be based on the thorough consideration of ecological and economic factors. An understanding of pest identification, its biology and type of damage are some of the factors that determine which control strategies and methods, if any, should be used. Ultimately, pest management decisions represent a compromise between the value of the product, the extent of the pest damage, the relative effectiveness and cost of the control measures, and the impact on the environment. Thus, forest pest management has a sound conceptual, methodological, and technological foundation for implementation as an integral part of forest resource management planning and operations. For pest management to be an integral part of forest resource management, a "look-ahead" policy must displace the "wait and see" policy. Pest monitoring and treatment strategies must be fully incorporated into the overall planning and decision process.

The conventional control measures continue to be the major tool for most pest situations as well as basis of integrated pest management systems. It is therefore, essential that a forest manager and the upcoming forest entomologists become fully conversant with various pest control procedures and options to employ them either singly or in combinations as a component of forest pest management. Forest pest control can be classified into natural control and artificial control. Natural control includes climatic condition, nutrition, host resistance and biotic factors, whereas artificial control includes methods such as silvicultural control, biological control, microbial control and chemical control.

NATURAL FOREST PEST CONTROL

CLIMATIC CONDITION

Climatic condition is expressed in terms of average physical conditions occurring throughout the year in a locality experienced over a long period of time. It also determines the length of the life cycle, successive stages and number of generation of insect species in a year. The more important climatic factors are meteorological factors, temperature, rainfall, atmospheric humidity, wind, sunshine, moon light, topographic features, mountains, plains and zoogeographical distribution. Each factor affects the insect activities in one way or the other, but in nature they seldom operate independently.

TEMPERATURE: It is the single most important physical factor which regulates the insect activity in the environment. Each species of insect has a definite range of temperature, within which the insect is active and its metabolism or growth is apparent. Any deviation from this point on either side retards metabolism and which in turn influences its activities. Near the upper and the lower limits of effective temperature zone lies the dormant zones in which the metabolism is greatly retarded and the insect becomes completely inactive and its movement ceases. The fatal temperature at which practically all the species of insects are killed instantly is 60ºC.

ATMOSPHERIC HUMIDITY (MOISTURE): Like temperature, atmospheric humidity (moisture) is an important factor of environment, which profoundly influences the development, behaviour and distribution of insects. Each species has an optimum favourable zone and any deviation from this zone leads to a state of dormancy and beyond the effective range, the insect dies. In other words, "with increase in temperature, there is a corresponding decrease in relative humidity". Under normal conditions, both excess and deficient moisture conditions do not result in immediate death of the insect, though its activities are definitely disturbed. However, certain wood boring forest insects are profoundly influenced by variations in the moisture contents within the microclimatic conditions of its host plant. This is best explained by studying the ecological succession of wood borer fauna of a felled wood. While the coleopterous borers of families Buprestidae and most Cerambycidae require higher moisture conditions, the powder-post beetles families Bostrychidae and Lyctidae and Hymenopterans wood borers (*Xylocopa* spp., family: Xylocopidae) require dry and seasoned wood and cannot survive in moist wood.

HEAVY PRECIPITATION: Heavy precipitation (snowfall, hailstorm, rainfall or frost, etc.) may also sometimes cause direct mortality of different stages of forest insects. While the effect of snowfall is like that of low temperature, hailstorm and heavy rainfall act mechanically. Lepidopterous larvae, beetles and their larvae, which are open feeders, feeding exposed on the foliage, are beaten and washed off by sudden heavy hailstorm and rain or killed by frost. They are either drowned or injured, thus preventing their ascendance to the crown of the trees. Rainfall also increases atmospheric humidity which influences the emergence period of insects, such as soil insects and wood borers. For example, winged form of termites and ants, wood borer adults of many species of Cerambycidae and Curculionidae, start emerging with the onset of monsoon rains.

WIND: Wind is yet another physical factor and its discernible effects are both mechanical and eco-physiological in nature. Strong wind shakes the crown of trees violently, blowing off and dislocating the insects, especially the more delicate forms (larvae) from their food plant and subsequently prevent them from gaining access to the food source and shelter.

NUTRITION

Nutrition (food) is the second important natural control factor, which, unlike physical factors of environment, is amenable and can be controlled with certain degree of accuracy. It constitutes the primary driving force for seeking not only the right type of food for its existence and development but also to see that it is available in sufficient quantities. Thus food most frequently act as a controlling factor on insects and influence the structure and activities of the insect communities (for that matter all animal communities). Thus, if food conditions are controlled in a way that even if the weather conditions are favourable, it checks the pest epidemics. Nutritional control may be considered under the following aspects like, food abundance, competition and quality of food.

Food abundance plays an important role in all the living organisms which are capable of multiplying to the extent of food available in an area. Forest monocultures offer ideal conditions for population build up of insect species feeding on it. The competition may be intra-specific or inter-specific, the outbreak of secondary species under certain conditions competes with the major species for feeding space. In intra-specific competition, all the individuals of the same species compete for all factors in the habitat i.e., space and food. In case of outbreaks, food may become the limiting factor for their survival. The quality of food has a definite effect on the abundance of forest insects and their development. The principal factors affecting the quality of food are moisture content, alternate food plants and host resistance.

HOST RESISTANCE

All plant species possess certain physiological and genetic traits which impart some degree of pest resistance by making environmental inherent conditions of less favourable for the pest species. The resistance may be due to inherent conditions of unpalatability, such as hardness of leaf lamina, shoot, sap density and presence of certain chemical compounds, such as essential oils, acids, gums, resins, latex, etc. There are three mechanism of host resistance to pest recognized viz. non-preference, antibiosis and tolerance.

BIOTIC FACTORS

Biotic control comprises all such factors which tend to control the population of an organism (pest) through the activities of other organisms. The biotic factors, which play an important role in limiting the pest populations, are competition, parasitoids, predators and pathogens.

ARTIFICIAL CONTROL OR APPLIED CONTROL

It is defined as the introduction of external elements, such as introduction of biocontrol agents, chemical control, etc. into the forest environment and manipulation in sylvan ecosystem (silvicultural control), which though temporarily, is under siege of a pest. Most common control procedures envisaged are monitoring, mechanical control, cultural control, legislative control, chemical (insecticidal) control and integrated pest management.

SILVICULTURAL CONTROL

The regulation of insect pest population by modifying forest stands or methods of growing and harvesting trees to suppress insect pests is called silvicultural control. It is a type of artificial control as it is achieved by augmenting natural controlling factors of the environment by deliberate human intervention. The silviculture control requires detailed knowledge about the prevailing ecological conditions of a forest that are favourable to the pests and then manipulate these conditions to the disadvantage of insect pests. Before taking-up any silvicultural pest management measures, it is necessary to have an idea of forest type and their relationship with insects.

INTER-RELATIONSHIP OF FOREST TYPES AND INSECTS

It has long been known to foresters that a forest growing under natural conditions is less vulnerable to insect attack in contrast to those that grows in disturbed ecological conditions. It shows that a forest is much safer in an equilibrium state than a forest in broken biotic conditions from the point of view of pest attack. If a forest is grown under such conditions that are, conducive for maintaining the biotic balance throughout the rotation period, costly measures to restore the disturbed balance are not needed. To achieve a successful silvicultural control, it is of paramount importance to have an accurate knowledge of the environmental factors involved. A forest represents complex biotic communities which may vary in a virgin or natural and in an artificial forest, particularly with reference to insect attack. A climax forest, undisturbed by anthromorphic intervention is a virgin forest where a biotic balance is always maintained. Generally, no pest epidemic occurs in such type of forest. However, in pure coniferous temperate forest, outbreaks of defoliator and bark beetles are known to occur. A virgin forest in the true sense does not really occur in India.

The biotic complex of these forests remains in somewhat equilibrium state and that insects epidemics are rare. Epidemic occurs only under severe stressed conditions like prolonged drought or other natural calamities like fire or frost. Most of natural forests of India where biotic and abiotic pressures are minimum, exhibit no insect problem but whenever these pressures increase, the insect problems crop up. For example, epidemics of deodar defoliator *Ectropis deodarae* epidemic in outer Himalaya, bark beetle *Scolytus major* epidemic in deodar forest, chirpine defoliator *Cryptothelia crameri* epidemic in chir forest in outer Himalayas are instances supporting this contention. In tropical and subtropical natural forests, varied types of insect problems have arisen owing to biotic and abiotic pressures.

KIND OF LOSSES BY INSECTS IN FORESTS

Losses to insects in a forest are of varied types and range from tree mortality to reduced vigour. Losses due to insects may be of following types (a) Mortality of trees particularly young ones may primarily be caused by defoliators, bark beetles and termites. (b) Wood boring insects may damage the solid wood of a living or felled tree. (c) Insects may cause great losses to growth increment. (d) Natural regeneration is hampered by insect damage. (e) Deterioration of site and degeneration of species composition may be the result of insect activity in a forest. (f) Insect attack may cause complete failure of planting a species in an area which may be the cause for abandonment of the plantation. It has happened in case of *Acacia nilotica* plantation in Madhya Pradesh which was abandoned because of severe damage by root and shoot borer *Celosterna scabrator.*

SILVICULTURAL PRACTICES TO REDUCE INSECT ACTIVITY

Various silvicultural practices have been recommended and practiced for effective control of insect pests.

HOST SPECIES AND STAND MANIPULATION

Regulation of forest composition by mixed stand is known to foresters since long that more diverse type of forest has least insect problems. Increased ecological stability results from an increased interaction between insects and other biotic and abiotic factors in the forests. In general, outbreaks of insects are less common in a forest with a diverse species than in a forest with less diverse flora or consist of single species (monoculture). Monoculture should be avoided as far as practicable especially in case of species which are not gregarious. Monoculture not only increases the availabilities of quality and quantity of food for the pests but also discourages natural enemy complex. In general, clear felling system is more dangerous than any other silvicultural system like shelter wood or selection system. A large clear-filled area is more susceptible to insect attack than other areas. From the silvicultural point of view, it will be beneficial to distribute the area to be regenerated annually by means of small separate coupes in several felling series with variation in their age classes.

DENSITY AND TREE GROWTH MANIPULATION

REGULATION OF DENSITY BY THINNING: Stand density may influence the physical conditions of a forest. Degree of density is also important in determining the ability of forest tree to grow vigorously. Stand density is regulated by thinning which depends on a number of factors like pest and tree species, site conditions, tree age, area and cost-benefit ratio. Thinning must incorporate removal of dead and dying trees coupled with removal of refuse resulting from thinning of overstock stand. Dying dead trees and refuse are most susceptible to insect attack like defoliators, sap suckers and bark beetles.

Badly attacked plantations of *Toona ciliata* and *Bombax ceiba* have been recovered by regular thinning in young stands after the attack by shoot borer *Hypsipyla robusta* and *Tonica niviferana,* respectively (Beeson 1941). An early thinning of *Dalbergia sissoo* produces a more favourable ratio between the pest population and the quantity of foliage per tree. Attack of *Cyrtotrachelus dux* weevil on bamboo can be reduced by thinning the dense area (Singh and Bhandari 1988) because split clumps provide shelter to beetle in hot dry season. Thinning of average dense sal forest is essential to reduce the effect of *Hoplocerambyx spinicornis* beetles.

PURE STAND: It seems that some trees reach a certain "entomological rotation" age at which they are more a *pure stand.* The monoculture plantation is more susceptible to insect attack because of even aged structure, food abundance and lack of diverse enemy complex. All these congenial conditions will provide optimum environment for multiplication of the pest species. The annual epidemics of teak defoliator (*Hyblaea puera*) and teak leaf skeletonizer (*Eutectona machaeralis*) in pure stand of teak are the most common examples of adverse effect of monoculture plantations in India (Nair et al 1985). Poplar defoliators *Pygaera cupreata* and *P. fulgurita* and stem borer *Apriona cinerea* (Chatterjee and Singh 1968) provide other examples of insect epidemics in monoculture plantations. Many factors like water supply, soil, soil cover etc., affect the severity of economic damage in pure stands. Water supply affects the sprouting of *Dalbergia sissoo* and therefore, this knowledge can be manipulated to the disadvantage of defoliator *Plecoptera reflexa* by adjusting irrigation dates.

Monoculture of *Ailanthus* plantation suffers from two notorious pests namely *Atteva fabriciella* and *Eligma narcissus* (Varma 1986; Roonwal 1990). Termite species *Odontotermes obesus, O. distans* and *Microtermes unicolor* are major pests of *Eucalyptus* plantations in India (Thakur et al 1989). An even-aged monoculture plantation can be managed by developing a mosaic of age classes in small blocks. This also helps to reduce easy access of the pest to the food. A mixture of tree species can be achieved by cutting in groups or strips or by creating gap to enhance light penetration so that a light demander species could be established. Similarly, mixed age stands can be produced by cutting adjacent areas at intervals of about 15 meters.

MIXED STAND: In perspective of pest management, mixed stand is an association of several species of plants in which auxiliary flora assumes importance in proportion to their effect on the abundance of insect pests of timber trees. Generally, there are good and bad mixer. In a mixed stand, generally oligophagous or poliphagous insect species predominate over monophagous species and a polyphagous species rarely become a pest unless abnormal conditions in the environment exist. A mixed stand protects the main timber species from the attack of monophagous insects by reducing food supply and encouraging growth and survival of predators and parasites by providing them appropriate niches including food nectar of the adult parasites.

Different tree species of different age in a mixed forest often differ in their susceptibility to insect attack. When any insect species in a mixed stand exhibit a strong preference for a certain host species, selective elimination of that species and encouragement of a variety of non-host species may reduce the potential for injury to the primary host species.

ALTERNATE HOST AND GROUND COVER

Population of pests can be regulated to some extent by manipulation of weak links in their life cycle. One of those is the need for alternate host. A horizontal mixture of the main tree species and other over wood species of the same crown size has salubrious effect. Mixture of two different tree species at width of 15 to 30 meter can hardly be able to check the inter-exchange of winged insects and subsequently egg laying. An advantage may be obtained by such mixture to check the spread of larval stages that may spread from crown to crown. A more resistant mixture of block against insect attack can be created by choosing most suitable species to plant on a suitable soil in patches than the mixture arranged in lines or furrows. The abundance of certain insect species is often influenced by the occurrence and abundance of alternate hosts. Alternate hosts form an integrated part of the life cycle of various best species. The presence of alternate hosts in a plantation increases the chances of an infestation. For example, incidence of attack of collar weevil, *Pagiophleous longiclaris* in plantation of mahagony (*Swietenia macrophylla*), is primarily determined by the incidence of attack on *Toona ciliata* (Stebbing 1914). Teak canker grub *Dihammus cervinus* depends on the presence of *Clerodendron* and gamhar (*Gmelina arborea*) (Beeson 1941) and babool borer *Celosterna scabrator* breeds in *Acacia* species, *Eucalyptus* spp. and *Casuarina equistifolia* (Beeson 1941; Sen-Sarma and Thakur 1985; Chatterjee and Singh 1968).

FIRE AND LOGGING

There are number of other interesting examples of heavy infestation where forest composition has been changed by logging, fire or due to other causes. For example, heaviest defoliator attack was seen in Nilambur teak plantation when undergrowth in 2,600 acre was burnt (Beeson 1941). Ground fire was used to clear an area from refuse for regeneration purpose but it caused extensive damage to parasites and predators along with the pests. Soil inhabiting insects like termites, white-grubs and wood borer largely escape unhurt from ground fire. Ground fire is not advocated to control insect epidemic. The prevention of fire in a forest provides protection to natural enemies of forest pests.

INCREASING HOST VIGOUR

Host vigour can be improved by reducing the number of trees from the dense stands where competition has subsequently slowed down the growth of even the dominant trees. Reduction in basal area of an overstocked stand either by thinning or partial branch cutting will enhance the tree growth and vigour which in turn will lower the incidence of an insect outbreak. Vigorously growing trees are not liable to insect injury in comparison to weak trees.

For example, vigorously growing trees of conifer forests are able to withstand bark beetle attack, while beetles and borers establish themselves in weakened trees. The successive insect attack occurs in forests as these developed from seedlings to pole and pole to mature trees. Some insects may thrive only on a certain stage of a tree. If a vigorous tree is attacked, it may soon recover from the insect attack. Tree vigour can be estimated by a variety of quantitative and qualitative methods. In general, when vigour is reduced, trees become stressed and supply low quantity of carbohydrate from their reserve for defensive functions. Thus, nutritional factors are greatly responsible for regulating insect activities.

SILVICULTURAL MANAGEMENT FACTORS

Over all, chief sound silvicultural management factors which ensure healthy and vigorous growth of plantation species are (i) site selection (ii) choice of forest tree species (iii) conditions in nursery (iv) transplanting of nursery stock (v) eradication of collateral hosts and (vi) encouragement of desirable and beneficial plant species. However, in practice, such ideal silvicultural conditions are seldom available to forest managers, who are often called upon to carry our plantation schemes on unsuitable sites. Most of these short comings can be partially mitigated through soil improvement practices and by using cover or nurse crops. Many of these practices are considered as cultural practices.

SITE SELECTION: Growing of a tree species at a suitable site is one of the most fundamental methods to reduce pest hazards through silvicultural practices. In fact, often this important factor is overlooked by foresters which results in poor development, reduced vigour and increased susceptibility of trees to insect attack. The most suitable example of this type in the history of Indian forest entomology may be the failure of *Acacia nilotica* plantation in Berar area of Madhya Pradesh (Beeson 1941). Unsuitable soil and deficient water supply were responsible for creating permanent breeding centers for the multiplication and spread of shoot and root borer *Celosterna scabrator*. The only practical remedy is to clear the infested areas and replant the area with a species which is not attacked by this pest. Therefore, choice of suitable site for planting is also an important consideration in silvicultural control. Under suitable site conditions, trees will be able to tolerate greater tree injury than growing under less favourable conditions.

CHOICE OF RIGHT SPECIES AT RIGHT PLACE: Sometimes, a site is poor because of acidic soil which inhibits tree growth. Such type of soil can be reclaimed by proper water drainage. Poor or problematic soils can also be amended with the help of fertilizer and gypsum to make suitable for tree planting with right species in a given area. Inadequate or poor forest site conditions have frequently been associated with increased injury and pest epidemics. For example, planting of *Dalbergia sissoo* should be restricted to suitable and well irrigated site and for planting of teak it should be insured that teak is not planted on the site previously occupied by *Clerodendron infortunatum* or adjacent to *Gmelina arborea* plantation as both the species act as collateral host for canker grub *Dihammus cervinus.*

BETTER SITE CHARACTERS: Establishment of new plantations, nurseries or reforesting the area after clear felling often cause precarious situation. In forests, the establishment of new plantings to create stand diversification often requires specific preparation of sites to ensure tree establishment and survival. Researches on bark beetles in European countries have shown clear correlations between site characteristics of poor quality soils and low moisture and outbreaks of pest insects.

MONITORING: When formulating any management strategies in forestry, monitoring is an important step. Monitoring can take on many forms: scouting and sampling, pheromone trapping, tree baits and aerial surveillance. Sometimes, combinations of methods are used such as aerial surveillance to detect dead trees and ground checking to determine the cause of mortality. In some cases, more than one life stages of a pest is required before control measures are taken. For example, pheromone trapping may detect a large concentration of adult male moths in a given location. This is followed by a ground survey for the egg stage to verify that mating and reproduction has taken place. Regular observations, good record keeping and a thorough knowledge of the pest's life cycle are the foundation for an IPM program. Insect pheromones are chemicals that members of a species use to communicate with each other. Females of many species produce sex pheromones that attract males for mating. Traps baited with pheromones lures can be used to survey for invasive insects, or in some cases, to control limited infestations.

The chief advantage of silviculture control is cheaper as compared to direct control (chemical control) measures and this procedure is more in tune with the ecosystem, environmentally friendly, with inherent capacity to regulate the pest populations on long term basis. It aims at developing preventive control measures and encouraging forces of natural enemies of the pests. But the disadvantages of silvicultural control includes, a long term control measure and therefore, proves failure during emergencies, particularly when the pest has assumed epidemic status and its success depends on in-depth knowledge of the pest and its host. Unfortunately, precise information in most cases is lacking and unsound and ill planned silvicultural procedures may even prove counterproductive, with undesirable side effects.

MECHANICAL AND PHYSICAL CONTROL

Though non-conventional and highly specified, the mechanical and physical methods of control play an important role in pest management. These methods aim in reducing the pest population by employing certain mechanical devices or manual operations which ensure killing of insects by trapping or mechanical means. It act as barriers and prevent insects or its life stages from gaining access to host plants (mechanical control) or by manipulating certain physical conditions of environment more particularly the temperature and humidity, in a way that they prove destructive to the pest species (physical control) and thus preventing or minimizing the insect attack. These measures also involve the use of radiant energy in various ways, which could destroy the insect pests. In broader sense, the physical control measures comprise one or more of the following measures, viz. heat treatment, refrigeration, radiant treatment, trapping, sound percussion devices and self destruction or autocidal treatment.

Many insect pests of forest tree species are amenable to control by mechanical and physical methods, however, these methods rarely provide commercial control. These methods hardly differ from cultural methods except that they involve the use of special equipment or operation and are employed in situations where insects are to be eliminated from an area without any deleterious effect on the environment.

Hand picking of noxious insects or their life stages is the simple perhaps the most ancient effective method of insect control practiced by man and which continues to be one of the most profitable methods for some insect pests (e.g. white-grubs and crickets in seed beds; beetles and larvae of some pests in forest nurseries). Use of hand nets or dragnets for catching small and noxious insects is an extension of hand-picking procedures (e.g. beetle defoliators, grasshoppers, plant hoppers, etc.). Similarly, wire probes or other flexible materials are also employed for poking tunnels of some wood borer species (e.g. *Batocera rufomaculata, Tonica niviferana, Zeuzera* spp., etc.). Jarring of branches of trees is also one of the mechanical methods employed in shaking off adult or larval population of species from the host plants. The purpose of control should be to maintain a balance between pests and hosts at such a level that economically important injury will not occur. Economic consideration is one of the most important factors to control a pest in a forest. In case of direct pests, even moderate levels of pest abundance cannot usually be tolerated. The appropriate level of control is determined by economic value of the forest stand.

Over-all, mechanical control in forestry generally means salvage or the removal of infested trees to prevent the spread of a pest. If the cut trees can be utilized, the control operations may be cost effective. In some instances, the infested trees may be too few or too remote for a cost effective operation. Cut-and-leave or cut-and-treat are options that are frequently employed for bark beetle control. The infested trees, along with a buffer, are simply cut down and left or treated by debarking, applying a pesticide, or burning. It is assumed that any surviving beetles will disperse and not create additional infestations. On high value trees, mechanical control could include simple hand picking, erecting insect or pathogen barriers or sticky traps. Non-toxic control strategies such as insect lure method (use of attractant), insect bait station, cellulose baits, pheromone traps, sealed baits, electronic bait etc., are also found useful in control of certain forest insect pests.

Table 13.1: Predators of Insect Pests

S No	Predator	Insect Pests
1	Asian lady beetle (*Harmonia axiridis*)	Aphids
2	12-Spotted Lady Beetle (*Coleomegilla maculata*)	Aphids, Eggs of beetles, Egg masses of caterpillars
3	Tropical Lady Beetle (*Cryptolaemus montrouzieri*)	Mealybug
4	*Lindorus lopanthae* Beetle	Scales
5	Green lacewings (*Chrysoperla rufilabris*)	Aphids, Soft bodied insect larvae
6	Gall midge (*Feltiella acarisuga*)	Spider mites
7	Six spotted thrips (*Scolothrips sexmaculatus*)	Spider mites
8	Damsel Bug	Caterpillars, mites, aphids, plant bugs
9	Ground Carabid Beetles	Root worms, Caterpillars, soft bodies insects, Aphids
10	Birds, spiders and insects such as shield bugs, robberflies, lacewings, hoverflies, ladybirds, predatory beetles, dragonflies and praying mantids	Autumn gum moth (*Mnesampela privata*), Sawflies (*Perga* spp.), Chrysomelid beetles (various species), Gum tree scale (*Eriococcus coriaceus*), Wingless grasshopper (*Phaulacridium vittatum*) and various beetles: e.g. *Heteronyx* spp., Spring beetles, *Cadmus* spp.

BIOLOGICAL CONTROL

Biological control is also referred as bioremediation and is defined as ingenious and resourceful employment of certain promising biotic elements (predators, parasitoids and pathogens) for suppressing pest species. Biological control refers to using natural insect enemies to control pest populations. Enemies can include vertebrate predators such as birds or mice, arthropod parasitoids such as the tachinid fly, arthropod predators such as spiders and mites. The phenomenon involves certain degree of deliberate human interference, which, unlike biotic natural control, is manipulated by man for regulating pest populations in an ecosystem. Simply, biological control is the reliance of natural enemies to keep a pest in check. Three principles or approaches are involved in biological control viz., importation, conservation and augmentation.

Normally importation is utilized against exotic pests where natural enemies from the pest's home land are introduced into the infested region. It is assumed that once the natural enemy is established it will be self-maintained. Conservation includes the identification and remediation of negative influences that suppress natural enemies already in the system, and the enhancement of habitat for natural enemies. For example, providing alternative food sources for natural enemies, such as pollen and nectar, by encouraging selected flowering plants. Using more selective pesticides that have a minimum impact on natural enemies is another approach.

Augmentation is the release of enemy populations, which were artificially reared, for pest suppression. It is used when natural enemies are missing, too scarce, or slow to arrive. There are two strategies for augmentation releases. The first is "inoculative" where the released organisms and the progeny of the released organisms accomplish control of the pest. The second is "inundative", where the released organisms only accomplish control of the pest. Forestry has a great potential for utilizing biological control strategies because of the perennial nature of the system and the potential for alternative food sources. The fact that pesticides are generally not cost effective is another consideration.

Table 13.2: Parasitoids of Insect Pests

S No	Parasitoid	Insect Pests
1	Ichneumon Wasp	Larvae of Moths and Butterflies, Saw flies
2	Braconid Wasp	Larvae of Moths and Butterflies, Beetles, Flies, Aphids
3	Trichogramma Wasp	Eggs of Moths and Butterflies
4	Parasitic Tachinid Flies	Larvae of Moths and Butterflies, Beetles, Saw Flies
5	*Encarsia formosa*	Whitefly nymph
6	*Psyllaepgagus pilosus*	*Ctenarytaina eucalypti* (Blue gum psyllid)
7	*Enoggerra nassaui*	Gold-ring Eucalyptus leaf beetle (*Paropsisterna gloriosa*)

MICROBIAL CONTROL

Microbial control is another important component of biological control. Apart from parasitoids and predators, certain entomogenous microbial pathogens, comprising chiefly the bacteria, fungi and viruses are the well known biocontrol agents. These microorganisms, as the name suggests, constitute a specialized group of biocontrol agents, having extremely smaller size and eco physiological characteristics, which has imparted these organisms a distinct identity, quite different from insect parasitoids and predators. Nevertheless, the principles involved for profitable utilization (inoculum, augumentation and conservation) more or less remain the same.

BACTERIAL PATHOGENS: Among the bacterial pathogens, particularly the sporulating (sporeforming) bacterial diseases constitute an important group of microorganisms as they are highly resistant to drying and high temperatures. *Bacillus thuringiensis*, commercially named as Thuricide, is the best known bacterial pesticide with worldwide recognition. Though broad spectrum, this biocontrol agent is selective and ecologically safe in most situations. In India, it has been successfully tested against over two dozen insect pests of forestry importance, achieving high mortality rate ranging from 30 per cent (*Lymantria mathura, Dioryctria abietella*) to 90 per cent (in *Ascotis selenaria, Eutectona machaeralis, Clostera fulgurita, Plecoptera reflexa* and *Selepa celtis*).

ENTOMOGENOUS FUNGI: The role of entomogenous fungi as effective biocontrol agent has been recognized throughout world. In India, a few fungal pathogens have been recorded infesting and killing larvae and pupae of some important forest insect pests. *Beauveria bassiana,* a well known insect pathogen has been found infecting and killing larvae of *Atteva fabriciella, Eligma narcissus, Hypsipyla robusta, Philosamia cynthia* and Scarabaeid beetles, causing large scale larval mortality.

VIRUSES: Of the numerous viral pathogens, which are known to cause diseases in insects, only those (viruses) belonging to families Baculoviridae and Reoviridae reportedly hold promise as biocontrol agents against insect pests. In India, nuclear polyhedrosis viruses (NPV), isolated from the diseased larvae of poplar defoliator *Clostera fulgurita,* have been successfully tested in the field and found very effective in controlling the pest population. More recently another NPV isolated from infected teak defoliator *Hyblaea puera*, has been tested in suppressing the defoliators population. Elsewhere, NPV of *Helicoverpa, Baculovirus heliothis, Pieris brassicae* and *Pieris rapae* are successful microbial pesticides.

CHEMICAL CONTROL

A key part of pest management is to use a pesticide only when it is needed to prevent an unacceptable amount of damage. Use of pesticides may not be justified if the cost of control or potential harm to the environment is greater than the estimated damage or loss. Pesticides are a very important tool, when large pest population threatens high value trees. Knowledge of the pest's life cycle, selection of an appropriate pesticide, proper timing of the application and use of the right application equipment will improve coverage and effectiveness. A pesticide application may be needed to control a pest outbreak or to eradicate limited infestations of an invasive species. Pesticides should be selected and used in a manner that will cause the least harm to non-target organisms in forests, seed orchards and nurseries, while still achieving the desired management goal.

Pesticides are labeled for specific pests, crops and land-use situations. Use of insecticides, fungicides and herbicides is common in managed seed orchards, forest nurseries, intensive short-rotation plantations and in Christmas tree production. In general, the most commonly used forest pesticides are herbicides used for site preparation, herbaceous weed control, etc. Insecticide applications are seldom used in general forest management because of high treatment costs and potential effects on non-target organisms. Situations justifying the widespread use of fungicides also are rare. In some cases, vertebrate animals must be controlled through trapping or hunting and similarly use of repellents and poison baits may be employed.

INSECTICIDES FOR FOREST PEST CONTROL

INSECTICIDES: Insecticide applications are rarely practical against forest insect pests. However, they may be useful in specific situations (For e.g., Chlorpyriphos 20EC against borers and termites in nursery, Methyl parathion 50EC against defoliators, Dimethoate 30EC, Imidachloprid and Thiometaxm 25G against sucking pests) such as limited infestations of an invasive species. Insecticidal soaps, horticultural oils, Bt-insecticides, and systemics are common choices because they have very low potential to harm the environment, non-target species and applicators.

INSECTICIDAL SOAPS: They are made from of salts of fatty acids. When sprayed directly on vulnerable stages of soft bodied insects, such as aphids and adelgids, they kill by damaging individual cells. However, they are not very effective against stages of the life cycle that are inactive or hidden, or against larger insects such as caterpillars and beetles. Insecticidal soap sprays must come into direct contact with the target pest and often results are best against specific life stages. Timing and thorough spray coverage is essential for best results. If the vulnerable stage is active over a long period of time, several applications are needed to control most insects. Insecticidal soaps should not be applied directly to water or near a water source. Hard water is not effective for mixing soap sprays, so softened or distilled water is used for best results. Tender young growth of evergreens and shrubs in the spring can be sensitive to insecticidal soaps.

Bt INSECTICIDES: Bt Insecticides (*Bacillus thuringiensis*) are protein toxins that are produced by a common soil bacterium. Many provide specific control of caterpillars without affecting other types of insects (beetles, sawflies, etc.). Bt insecticides disrupt the gut wall cells in the caterpillar digestive tract so a formulation of the insecticide must be sprayed on foliage that will be eaten by the caterpillars. Caterpillars stop feeding soon after eating the insecticide but usually do not die for several days. Bt insecticides work best against young caterpillars that are less than half-grown. These insecticides are relatively non-toxic to mammals and other animals.

SYSTEMIC INSECTICIDES: Imidacloprid is the common name of a systemic insecticide that is used to control sap-feeding insects, such as the hemlock woolly adelgid. It is a nerve poison but it is much more toxic to insects than to warm-blooded animals. Imidacloprid can be diluted in water and applied as a drench poured around the base of a tree, or injected into the soil. The insecticide is taken up by the roots and moved throughout the tree.

CLASSIFICATION OF INSECT PESTS

There are three main feeding styles of insect pests viz. woody tissue feeders (which have cutting and chewing mouth parts and feed on the outer sapwood of the main stem, branches or roots), defoliators (which feed on foliage and buds by means of cutting and chewing mouth parts), sucking insects (which have piercing or sucking mouth parts for removal of sap from stems, needles or branches). General information on major groups of insect pests and strategies to be followed for taking up control is presented below.

WOOD BORING INSECTS

BARK BEETLES: Bark beetles use trees as breeding sites and have an important natural role in killing weak or old trees or aiding the decomposition of dead wood. Odours from damaged trees attract bark beetles so initial attacks in an area often occur on stressed or injured trees. Beetles that develop in these trees emerge through small round holes in the trunk and move to other trees in the area. Adults enter trees to lay their eggs, often creating distinctive tunnels or brood chambers. The grub-like larvae bore extensively into the wood feeding on tree tissue or fungi brought to the tree by colonizing individuals.

Most bark beetle species attack trees that have been significantly weakened by disease, smog, competition, or physical damage. Extensive tunneling by the larvae can girdle and weaken or eventually kill the tree. Healthy trees have defensive compounds that can kill or injure attacking insects, or simply immobilize and suffocate them with the sticky fluid. However, under outbreak conditions, large numbers of beetles can successfully attack healthy trees with disastrous results for the timber industry. Some bark beetles (Ambrosia beetles) carry a fungus with them that grows within their galleries in the tree. These bark beetle larvae bore in the tree but feed on the fungal growth. The fungal growth clogs the vascular system of the tree and causes death.

Good silvicultural practices reduce the potential for bark beetle attack by reducing stress and keeping trees actively growing. When feasible, prompt removal of damaged trees significantly reduces the likelihood of successful bark beetle attacks. Insecticides used in bark beetles control do not penetrate the tree and kill the developing larvae, so trees that have been successfully attacked by bark beetles cannot be saved by insecticide applications. However, uninfested high value trees, judged to be at high risk, can be sprayed with an insecticide as a preventive measure against attack. The area of the tree requiring insecticide treatment depends upon the insect species for which the application is being made. The appropriate area of the tree should be thoroughly wetted with the insecticide spray mixture.

ROUND HEADED AND FLAT HEADED WOOD BORERS: This group includes mostly beetles, and a few caterpillars, that infest terminals, shoots, twigs and roots of living trees. Terminal and shoot insects are of particular importance in the initial stages of forest regeneration and early stand growth. These insects are very important in forest nurseries and ornamental trees. Other species in this category damage or destroy trees that would otherwise produce quality timber or other wood products. Most insects that cause this damage are borers, either adult or larval stages or both.

As like bark beetles, most borers are secondary invaders that attack bark and wood of trees that are seriously weakened, dying or recently cut. Examples include carpenter worms, ambrosia beetles, oak clearwing borers, metallic wood borers, long horned beetles and pine sawyers. Trees attacked by these pests are usually scattered so that most control measures are difficult and not economically feasible. However, few silvicultural and insecticidal treatments can be adopted to save the high value timber trees.

LEAF FEEDERS OR DEFOLIATORS

This diverse group of insects, which includes many species of caterpillars, sawflies and beetles, eats leaves and needles. Trees attacked by defoliators can be recognized by missing foliage and uneaten leaf parts such as veins and petioles. Some members of this group feed within a leaf, mining between the upper and lower epidermis. Correct identification of the leaf feeders gives information needed to assess the problem. Defoliation reduces photosynthesis, interferes with transpiration and translocation within the tree. Light defoliation normally has little effect on the tree but moderate-to-heavy or repeated defoliation can reduce tree vigor. The impact on a tree varies with time of attack, tree species and health and single or repeated defoliations.

SAWFLIES: Several species feed on conifers or deciduous trees in forest and plantation stands. The adults are small broad-waisted wasps. Larvae resemble caterpillars but are usually without hairs and have pairs of fleshy prolegs on the underside of every segment on their abdomen (caterpillars normally have four or fewer pairs). Larvae of the more commonly found sawflies vary from 2/3 to 1/4 inch long, are usually greenish to dusky gray, and have conspicuous stripes or spots. Outbreaks occur periodically, sometimes over large areas, and can result in loss of tree growth and sometimes tree mortality.

RED-HEADED PINE SAWFLY: It has a red head and a yellow-white body marked with six rows of black spots. The larvae are usually found on trees from 1-15 feet tall, where they feed gregariously on old and new needles and on tender shoots of the young trees.

INTRODUCED PINE SAWFLY: It has a black head and black body covered with yellow and white spots. The larvae prefer to feed on the needles of eastern white pine but also will eat scotch, red, Austrian, jack, and swiss mountain pine. Short leaf and Virginia pines have been attacked but usually are not heavily damaged. Defoliation is most severe in the crown to upper half of the tree but heavily infested trees can be completely defoliated. If this occurs after the winter buds have formed, many branches or even the entire tree can be killed. There are two generations each year. Larvae of the first generation feed on needles from the previous year. Young sawflies eat the more tender outer parts of the needles while older larvae consume them entirely. They are full-grown (about inch long) in July. The second generation of this sawfly feeds on both old and new needles during August and September.

EUROPEAN PINE SAWFLY: It is a green and black striped larva with a black head. A full grown larva is about 1 inch long. This species can feed on many hosts including scotch, Eastern white and Austrian pine. It feeds on the previous year's needles and does not damage new needles.

CATERPILLARS: Many species of caterpillars (such as teak defoliator and sketetonizer, *Ailanthus* webworm, hairy caterpillars) feed on deciduous trees in forest and a few feed on conifers. Adults are usually moths but a few are butterflies. Most caterpillars have five pairs of fleshy prolegs - four along the abdomen and one pair at the end. Full grown larvae of the more common caterpillars range from 3/4 inch long to more than three inches. Colour and markings are variable. Outbreaks occur periodically, sometimes over large areas, and can result in loss of tree growth and sometimes tree mortality.

TEAK DEFOLIATOR CATERPILLAR: It is darker with variable color patterns; wholly dark, almost black or grayish green with faint longitudinal middle line or black with a broad dorsal orange band from pronotum to the anal segment. Larva feeds on the leaves and cause coarse skeletonization and turns over flaps in the early larval stages and complete stripping between the veins of a teak leaf by the ultimate and penultimate larval stages is a typical characteristics of *H. puera.*

TEAK SKELETONIZER CATERPILLAR: It is green to brown or purplish in colour with four black dots and inconspicuous hairs, framed in a white or yellow rectangle on the dorsum of each abdominal segment. The grown up larva scrapes green tissues and causes uniform skeletonisation.

AILANTHUS WEBWORM AND DEFOLIATOR: They are the caterpillars of different colours, which also feed on leaves and cause defoliation

COMMON OAK MOTH CATERPILLAR: It is brown with tan to black blotches on the sides, there are diamond shaped markings and slanted lines on its back. This caterpillar moves in a looping manner and is about 1-1/4 inches long when mature. There is one generation each year with the caterpillars active from May to June. Common oak caterpillars seem to be able to feed on many kinds of oaks but prefer white oaks. In many cases, trees can be severely or completely defoliated. While a single defoliation do not adversely affect established and healthy trees but previous droughts or other stresses can increase the impact of this damage.

EASTERN TENT CATERPILLAR: It feeds on trees in the genus *Prunus*, black cherry is the preferred host. The hairy larvae are black with a white stripe down the center of the back. A row of pale blue spots along each side is bordered by yellowish orange lines. Full-grown larvae are about 2 inches long. Defoliated trees normally suffer only minor growth loss.

FOREST TENT CATERPILLARS: They are very similar to eastern tent caterpillars but have a row of light keyhole-shaped spots down the center of the back rather than a stripe. They feed on a wide range of trees including sweet gum, oak, birch, ash, maple, elm and basswood. Like the eastern tent caterpillar, there is one generation in the spring.

FALL WEBWORM: It is a hairy pale green to yellow caterpillar that is about one inch long when full grown. There are two or more generations per year. Webworms enclose leaves and small branches in light gray, silken webs. They feed on more than100 tree species.

ORANGE STRIPED OAK WORM: It is black with eight narrow yellow stripes along the length of the body. They have a distinctive pair of long, curved horns behind the head. They can rapidly strip leaves from small trees but the defoliation usually occurs late in the summer or into the fall, their economic impact is relatively minor.

SAP FEEDING/ SUCKING INSECTS

This diverse group of insects includes many species of bugs, aphids, adelgids, whiteflies, scales, mealy bugs or wooly aphids, cicadas and thrips, which cause damage by way of draining out the sap followed by yellowing, sooty-mould growth and sickly appearance. Further, many of them act as vectors for the transmission of diseases. All these pests occur throughout the tree growth period. All of them can be managed by adopting silvicutual practices, conserving natural enemies, developing tolerant/resistant clones and need based application of systemic insecticides and botanicals. Some common sap suckers are:

(1) Balsam woolly adelgid (*Adelges piceae*), cooley spruce gall adelgid (Adelges cooleyi), pine bark adelgid (*Pineus strobi*)

(2) Aphids (*Cindara* spp, *Essigella* spp, *Eulachnus* spp, *Chaitophorus* spp, *Pterocomma* spp, *Caveriella aegopodii*,),

(3) Scales (Pine Needle Scale - *Chionaspis pinifoliae*, Spruce Bud Scale - *Physokermes piceae*, Oystershell Scale - *Lepidosaphes ulmi*)

(4) Cicadas (*Magicicada* spp)

CONCLUSION

Before choosing a control method(s)/IPM, one should consider following points:

1. **Correctly identify the organism/pest.**
2. **Assess the infestation and determine the potential economic damage.**
3. **Determine the available control methods.**
4. **Evaluate the benefits and risks of each method or combination of methods.**
5. **Are there threatened or endangered species or sensitive sites in the area to be treated?**
6. **Choose effective method(s) that will be least harm to humans and the environment.**
7. **Follow applicable local, state and federal regulations.**
8. **Correctly carry out the control practices and keep accurate records so results can be evaluated.**

14 FOREST DISEASE MANAGEMENT

INTROUDUCTION

Forests provide multipurpose benefits as sources for timber, fuel, fodder and minor forest produce, for conserving soil and water, moderating climate, offering food and shelter for wild life and adding to the aesthetic value and recreational needs of man. The prime objectives of forestry are to develop and protect forests which are one of the world's greatest renewable natural resources for maximum productive and protective functions. Protection of forests from diseases constitutes one such objective.

In the natural forest, the productivity is generally low due to inherent slow growth of the species and mixed composition where all species may not be valuable. In India, the growing stock of the forests is estimated at 2400 million m^3. The annual increment is only 0.53 m^3 per ha as against world average of 2.1 m^3 per ha (Anon. 1961). India with just 2.5 per cent of the global land has to support nearly 15 percent of the world's population and 16 percent of world's livestock. Due to this, forests are under intense biotic pressures leading to degradation of forest resources. Forests in India have low growing stock with 74 m^3 per ha compared to the world's average 110 m^3 per ha. Likewise, average mean annual increment of forests in India is very low at less than one m3 per ha per year compared to the world's average of 2.1 m^3 per ha per year (Lal 2000). Indian forestry has reached the threshhold of the new millennium additional burden of increased demand for food and fodder. There is huge domestic demand for timber and wood-based products. Demand for timber during the year 2006 was 82 million m^3 as against supply of 27 million m^3. The gap between demand and likely domestic production is projected to grow to 51 lakh tons by 2015 unless immediate steps are taken to promote pulp wood plantations for securing the future raw material supplies for the industries. Insect pest and diseases bring about losses making the productivity and production still low (Lal 2000).

PLANT DISEASE

Disease refers to the abnormal state of the individual. Disease in plants may be defined as sustained physiological and resulting structural disturbances of living tissues and organs ending sometimes in death (Bakshi 1973). A characteristic of the disease is that it is a continuous process and not transient and temporary one, such as caused by injury during lopping of branches, thinning and so on (Bakshi 1973). Forest pathology is the study of diseases in forests mainly due to pathogens. Diseases in trees are broadly classified into two categories viz. pathogenic and non-pathogenic.

- Pathogenic, when they are caused by pathogens like fungi, bacteria, virus, nematode, phytoplasma, etc. and diseases occur in trees at various diseases (Naik 1998).
- Non-pathogenic or physiological, when they are caused by adverse environmental factors. They are often called physiological disorders. High temperature and drought, deficiency of moisture, waterlogging conditions, forest fire and frost cause physiological disorders in plants and trees. Once the normal condition returns, the plant will change to normal healthy condition.

SYMPTOMS OF PLANT DISEASES

Symptoms of disease include death and destruction of host tissue, wilting, abnormal growth and differentiation and discolouration of host tissue. Plant disease symptoms are generally categorized into following groups,

Underdevelopment of Tissues or Organs: Examples include such symptoms as stunting of plants, shortened internodes, inadequate development of roots, malformation of leaves, inadequate production of chlorophyll and other pigments and failure of fruits and flowers to develop.

Overdevelopment of Tissues or Organs: Examples include galls on roots, stems, or leaves, witches brooms and profuse flowering.

Necrosis or Death of Plant Parts: These may be some of the most noticeable symptoms, especially when they affect the entire plant, such as wilts or diebacks. Other examples include shoot or leaf blights, leaf spots and fruit rots.

Alteration of Normal Appearance: Examples include mosaic patterns of light and dark green on leaves and altered colouration in leaves and flowers.

NURSERY DISEASES

DAMPING-OFF

Damping-off is probably the most serious of all the nursery diseases and is prevalent all over the world. The damage by damping off is so serious that the entire nursery stock may be destroyed. It causes large scale mortality in nursery upsetting all plantation activity.

SYMPTOMS, CAUSES AND PREDISPOSING FACTORS: Mortality in seedlings may occur at three stages during their growth due to pre-emergence blight, post-emergence mortality soon after germination and root rot in older seedlings resulting in poor growth and in extreme cases even death. The most important fungi causing damping-off belong to *Pythium, Rhizoctonia, Phytophthora* and *Fusarium*. Damping-off is favoured in clayey and wet soils, because of anaerobic condition in the rooting zone and relatively high temperatures. Because of the effect of warmth damping-off tends to develop more during summer on late sowings than in the seeds sown early in spring or in autumn. Soil pH influences development of the disease, which is more serious in alkaline or neutral soils than in acid soils (Griffin

1958). Low light intensity acting both by reducing the metabolism and increasing the soil moisture favours development of damping-off. In addition, dense sowings, application of nitrogen in high dosages during early stages of growth and addition of raw undecomposed humus favour disease development (Vaartaja 1952).

CONTROL: Light textured soil and good drainage will prevent damping-off. Damping-off and other soil-borne diseases may be controlled by seed and soil treatment with chemicals and bioagents. Soil acidification, steaming, fumiagation and fungicidal seed and soil treatments will control damping-off.

ROOT DISEASES

The underground part of the plant, the root system, may suffer from diseases which generally remain unnoticed until manifested by symptoms on the above-ground parts. Diseases may be due to fungi, bacteria, nematodes etc. Fungi parasitic on plant roots and root infecting fungi constitute an important group of organisms causing root diseases of forest trees. Root diseases are of two types: vascular wilt and root rot (Negi 1996).

VASCULAR WILTS

Wilt is a systemic disease where the entire individual or its part exhibits wilting of the foliage in acropetal succession up the shoot. The leaves become yellow, lose turgidity and fall off. The affected plant or the branch soon dies. A growing imbalance of water economy in affected plants is common to all wilt diseases. A characteristic symptom of several wilts is the vascular discoloration in the outer layers of the sapwood, green streaks in v*erlicillium* wilt in many hardwoods and pink stain in fusaral wilt. The vascular browning is ascribed as due to enzymatic action of the wilt fungus. Such discolouration may not be, however, diagnostic and cultural studies are necessary to diagnose the cause of wilt.

CONTROL: Fungicides like mancozeb, carbendazim and Thiram are good seed dressing chemicals. Bioagents like *Trichoderma* and *Psedomonaus fluorescens* may also be used in nurseries. Application of FYM, neem cake encourages growth of bioagents present in the soil (Naik 1998). Two theories are advanced to explain the mechanism of wilt: (i) *occlusion theory*, which assumes plugging of vessels by the mycelium of the wilt fungus interfering with translocation of sap. (ii) *toxin theory*, which postulates production of plasma poison by the wilt fungus and these toxins destroy the osmotic function of the living cell (Dimond 1955).

ROOT ROTS

Root rot fungi generally possess a wide host range. Once the infection has established in the roots, its further spread depends on the host resistance. In the susceptible hosts, the pathogen may spread quickly into roots and causes disease. The symptoms of root rot attack, however, do not manifest till a large percentage of roots is attacked. In the less susceptible hosts, the fungi may colonise roots, but then the progress inside the roots is slow because a large mass of healthy roots react strongly against spread of infection. Such hosts, therefore, generally do not show visible symptoms of infection till most of the roots become attacked. In the natural forests, therefore, where resistant and susceptible species are in mixture, the fungi remain in an endemic state. When natural forests are clear felled for raising plantations, the host resistance is lost and the fungi on roots of susceptible hosts are able to spread quickly (Negi 1996).

When trees are felled, the stumps and their roots remain alive for a considerable period. Gradual lowering of host resistance paves the way for colonization of roots first for specialized parasites like root rot fungi followed by weak secondary parasites and finally for saprophytes. *Armillaria mellea* which is able to spread quickly on roots of felled trees become greatly restricted or is absent if the roots are killed more quickly by ring barking the trees when green (Bakshi 1973). While the root rot fungi are considered to be specialized parasites, two groups may be distinguished: in one, the fungi cause typical root rot and do not attack the heartwood as in *Ganoderma lucidum*; in the other, the fungi may cause root and butt rot as in *Heterobafrehi annosum, Armillaria mellea* and *Polyporus* spp.

CONTROL: The control of root diseases generally is more difficult because of the difficulty of combating the disease on underground plant parts which are not easily accessible as compared to diseases attacking aerial plant parts. Chemical control of root diseases with fungicides applied to soil may not be effective as these chemicals may not reach the affected parts and also they may soon become inactivated due to reaction with other chemicals present in the soil or through leaching. Difficulty in the control of root diseases of forest crop arises because of the long rotation age of such crops. In mixed plantations of susceptible and resistant species, root disease remains very much in check. However, in pure plantations, once the root disease breaks out, the spread is very rapid from tree to tree and therefore, within a short time disease becomes alarming.

MEASURES FOR ROOT DISEASE CONTROL

The important effective methods for controlling root diseases are decribed below,

Suitable Site: For devising successful control measures for root diseases, a through knowledge of silviculture of the species is essential. Tree vigour is important in determining the susceptibility. It is also advisable to have the knowledge of root growth of a particular tree species on the site conditions and species resistance to root diseases. Raising forest species on sites favourable for their healthy growth will prevent root diseases.

Control Burning: Control burning, a necessary silvicultural measure keeps the root rot in check (Bakshi and Boyce 1959). Burning results in the increase of *Trichoderma* in the soil.

Resistant Species and Mixed Stands: In case, a species susceptible to root rot is raised in the first rotation, it should be planted in intimate mixture with resistant species which will serve as barriers to rapid spread of the disease.

Isolation Trenches: Root diseases generally spread in plantations from root to root by contact or root graft resulting in diseased plants occurring in groups. Isolation trenches may prove effective in containing the disease in between the trenches thus preventing its spread beyond the trenches.

Chemical Control: An ideal stump treatment should be effective in preventing infection at the stump. Among chemicals tested and recommended for use are creosote, ammonium fluoride, sodium nitrite, borax, urea and ammonium sulphamate. Root feeding with Tridemorph @ 0.1% is also effective.

Biological Control: Root rot due to *Fomes annosus* is severe in plantations soon after thinning because the pathogen is able to colonise freshly cut stumps. *Peniophora gigantea* is a vigorous competitor of *Fomes annosus* for initial colonisation of stumps and may also replace *Fomes annosus* in stumps (Rishbeth 1963).

HEART ROTS

In the tree, decay does not set in till the heartwood is formed and this usually occurs between 15 to 30 years of age depending on the species. When the heartwood becomes exposed through openings in the bark and sapwood, decay may establish in the heartwood as heart rot which becomes progressive with age (Bakshi and Sujan Singh 1970).

INFECTION OF DECAY FUNGI

Tree wounds due to various causes are the principle infection courts for trunk and basal rots. When the wood of a living tree is freshly exposed following an injury, chemical and biological changes occur. Toxic natural extractives which impart resistance to decay are formed during development of the heart wood and these substances are mostly deposited in the cell wall.

DETECTION OF HEART ROTS

Heart rot causes progressive decay and therefore, the volume of decay increases in the tree and in the stand with age. Heart rots in trees may be detected either from external symptoms manifested by diseased trees or through direct probing into the heart wood.

EXTERNAL INDICATORS (Boyce 1961)

Sporophores: The presence of sporophores of heart rot fungi on the tree is a sure indication of decay.

Punk Knots: Punk knots are cylindrical cores of soft punky tissue of the wood decayed by the fungus, embedded in the sapwood up to different depths and opening upon the bark as round or oval hard structures.

Swollen Knots: Swollen knots are formed when living wood tries to grow over a knot in which a sporophore is forming.

Swollen Boles: A swollen bole is a reliable indicator of decay.

DIRECT PROBING

Heart rot may be detected by taking increment cores from trees suspected to contain decay and isolate the fungus in culture from the decayed wood. Advanced decay trees may give a hollow sound if struck on the bark with a hammer or back of an axe at the region of decay. Zycha and Dimitri (1962) used device for determining hidden decay in trees by forcing a steel needle into the stem by means of spring. Eslyn (1959) reviewed the use of X-ray photography to determine the extent of decay and successfully used a thulium X-ray unit to detect the presence of cavities and advanced decay in trees.

CONTROL

IMPROVEMENT IN STAND MANAGEMENT: Any recommendation concerning management of forests in order to minimize damage due to heart rot must take into consideration the silvicultural characteristics of the tree species. Fire is the most important cause of tree wounds and associated infection courts for decay fungi and should, therefore, be kept out of forests unless necessitated in the management of a species. Thinnings and improvement fellings remove diseased, forked, malformed and injured trees and also those bearing sporophores of decay fungi thus leading to stand improvement.

DECAY IN RELATION TO SILVICULTURAL SYSTEMS: Even-aged stands are more susceptible to damage by snow, frost and high winds than uneven-aged stands. If there is a high incidence of heart rots, it is inadvisable to allow the stand to complete its normal rotation age. Instead, a pathological rotation, as indicated by net mean annual increment, should be prescribed when the maximum volume of sound wood may be salvaged profitably. Management of coppice of forests where there is a high incidence of decay in the stools may be practiced by gradual change over to seedling crop and preventing further coppicings of stumps after the seedling crop is established.

Table 14.1: Diseases along with Causal Pathogens occurring in Important Tree Species

S.No	Host	Name of the Disease	Causal Pathogen
1	Sal (*Shorea robusta*)	Root rot	*Polyporus shoreae*
		Heart rots in sal	*Hymenochaete rubiginosa*
			Fomes caryophylli
			Fomes fastuosus
			Trametes cubensis
			Trametes incerta
2	Shisham *(Dalbergia Sissoo)* and Rosewood (*Dalbergia latifolia*)	Wilt disease	*Fusarium solani*
		Ganoderma root rot	*Ganoderma lucidum*
		Polyporus gilvus root and butt rot	*Polyporus gilvus*
		Leaf spot	*Cercospora sissoo*
			Colletogloeum sissoo
			Phyllachora dalbergiae
			Phyllachora spissa
			Phyllosticta sissoo
			Mycosphaerella dalbergiae
		Powdery mildew	*Phyllactinia dalbergiae*
		Leaf rusts	*Maravalia achroa*
			Uredo sissoo
		Phanerogamic parasites	*Tolypanthus involucratus*
			Tapinanthus dodoneifolius
		Pod fungi	*Glomerella cingulata*

3	Khair (*Acacia catechu*)	Ganoderma root rot	*Ganoderma lucidum*
		Polyporus gilvus root rot	*Polyporus gilvus*
			Fomes badius
		Powdery mildews	*Erysiphe acaciae*
			Phyllactinia acaciae
		Powdery leaf spot	*Microstroma acaciae*
		Leaf rust	*Ravenelia tandanii*
		Phanerogamic parasites	*Macrosolen cochinchinensis*
4	Teak (*Tectona grandis*)	Bacterial wilt	*Psedomonas solanacearum*
			Psedomonas tectonae
		Armillaria root rot	*Armillaria mellea*
			Armillaria tabescens
		Fomes lamaoensis root rot	*Fomes lamaoesis*
		Fomes noxius basal rot	*Fomes noxius*
		Helicobasidium compactum root rot	*Helicobasidium compactum*
		Peniophora root rot	*Peniophora rhizomorpho-sulphurea*
			Polyporus zonalis
		Polyporus zonalis root rot	*Polyporus zonalis*
		Pink disease	*Corticium salmonicolor*
		Nectria stem canker	*Nectria haematococca*
		Decay in dry coppice teak	*Fomes lividus*
			Polyporus zonalis
		Leaf spot	*Cercospora tectonae*
			Phyllosticta tectonae
			Sphaceloma tectonae
		Powdery mildews	*Phyllactinia corylea*
			Phyllactinia guttata
			Uncinula tectonae
		Leaf rust	*Olivea tectonae*
		Thread blight	*Marasmius* spp
		Phanerogamic parasites	*Dendrophthoe falcata*
			Macrosolen cochinchinensis
			Phoradendron piperoides
5	Casuarina	Algal parasites	*Stomatochroon* spp
			Cephaleuros spp
		Bacterial wilt of seedlings	*Pseudomonas solanacearum*
		Seedling blight	*Pellicularia rolfsii*

		Black root rot of seedlings	*Macrophomina phaseolina*
		Ganoderma root rot	*Ganoderma lucidum*
		Phytophthora root rot	*Phytophthora cambivora*
		Armillaria root rot	*Armillaria mellea*
		Stem wilt	*Trichosporium vesiculosum*
		Pink disease	*Coticium salmonicolor*
		Heart rot	*Polyporus glomaratus*
			Fomes fastuosus
			Fomes senex
		Leaf spot	*Phoma casuaranae*
		Phanerogamic parasites	*Cuscuta campestris*
			Dendrophthoe falcata
			Dendrophthoe lanosa
6	Sandal	Spike disease	*Phytoplasma*
7	Semul (*Bombax ceiba*)	Root rots	*Ganoderma lucidum*
		Phomopsis die-back	*Phomopsis salmalica*
		Pink disease	*Corticium salmonicolor*
		Leaf blight	*Dicellomyces bombacis*
		Leaf spots	*Myrothecium roridum*
			Phyllosticta bombacis
			Sphaceloma bombacis
			Cercospora bombacina
		Leaf rusts	*Chrysomyxa bombacis*
8	Albizia	Root rots	*Clitocybe tabescens*
			Ganoderma lucidum
		Pit canker	*Fusarium solani*
		Nectria canker	*Nectria ditissima*
		Gibberellia die-back	*Gibberellia pulicaris*
		Heart rots	*Fomes fatuosus*
			Fomes robiniae
			Ganoderma applanatum
			Polyporus gilvus
		Leaf spots	*Cercospora albizziae*
			Endodothella albizziae
			Phyllosticta albizzinae
		Leaf rusts	*Revenelia japonica*
			Revenelia sessilis
			Revenelia clemensiae
			Revenelia indica

		Sooty mold	*Meliola albizziae*
9	Mulberry	Crown gall	*Agrobacterium tumefaciens*
		Rosellinia root rots	*Rosellinia necatrix*
		Helicobasidium root rots	*Helicobasidium mompa*
			Helicobasidium tanakae
			Helicobasidium purpureum
		Diplodia root disease	*Diplodia morina*
			Armillaria mellea
			Botrytis cinerea
			Ganoderma lucidum
		Hendersonula die-back	*Hendersonula toruloidea*
		Gibberella twig canker	*Fusarium lateritium*
			Fusarium mori
			Gibberella lateritia
		Nectria cankers	*Nectria cinnabarina*
			Nectria coccinea
		Sphaeropsis die-back	*Sphaeropsis mori*
		Thyrostroma twig blight	*Thyrostroma mori*
		Heart rots	*Fomes fomentarius*
			Fomes robiniae
			Fomes senex
		Leaf spot	*Ganoderma applanatum*
		Pink disease	*Polyporus hispidus*
			Polyporus squamosus
			Colletotrichum moricola
			Corticium salmonicolor
		Popcorn disease	*Ciboria carunuloides*
		Leaf spots	*Cercospora missouriensis*
			Cercospora mori
			Cercospora moricola
			Cylindrosporium maculans
			Mycosphaerella mori
			Mycosphaerella indica
			Pseudomonas mori
			Clasterosporium mori
			Phyllosticta kuwacola
			Phyllosticta morifolia
		Powdery mildews	*Phyllactinia corylea*
			Phyllactinia guttata
			Phyllactinia moricola
			Phyllactinia suffulta

			Uncinula mori
		Leaf rusts	*Aecidium mori*
			Cerotelium fici
		Virus disease	Mulberry dwarf, curly little leaf virus and mulberry small leaf curl
		Phanerogamic parasites	*Dendrophthoe falcata*
			Viscum album
10	Wattles (*Acacia* spp.)	Seedling blights	*Glomerella cingulata*
		Black butt disease	*Phytophthora nicotianae*
			Corticium salmonicolor
			Hypoxylon hypomiltum
			H. rubiginosum
		Heart rots	*Fomes mastoporus*
			Ganoderma applanatum
		Leaf sports	*Calonectria theae*
			Camptomeris albizziae
			C. verruculosa
		Leaf and twig rusts	*Uromyces phyllodiorum*
11	Eucalyptus	Cylindrocladium seedling blight	*Cylindrocladium scoparium*
			C. quinqueseptatum
		Other seedling diseases	*Cercospora eucalyptii*
			Botrytis cinerea
		Phytophthora cinnamomi root rot	*Phytophthora cinnamomi*
		Ganoderma root rot	*Ganoderma lucidum*
		Crown gall	*Agrobacterium tumefaciens*
		Pink disease	*Corticium salmonicolor*
		Diaporthe canker	*Diaporthe cubensis*
		Other canker diseases	*Hypoxylon annulatum*
12	Deodar (*Cedrus deodara*)	Fomes annosus root and butt rot	*Fomes annosus*
		Armillaria mellea root rot	*Armillaria Mellea*
		Peniophora butt rot	*Peniophora luna*
		Polyporus tomentosus root rot	*Polyporus tomentosus* Var. *cricinatus*
		Peridermium witches-broom	*Peridermium cedri*
		Other diseases	*Phytophthora cactorum*
			Phytophthora cinnamomi
			Fusarium fuliginosporum

			Ganoderma lucidum
			Fomes pini
			Fomes pinicola
			Phytophthora schweinitzii
13	Fir (*Abies*) and spruce (*Picea*)	Root rots	*Armillaria mellea*
			Polyporus tomentosus
			Polyporus schweinitzii
			Ustulina deusta
		Heart rots	*Fomes fomentarius*
			Fomes robustus
			Ganoderma applanatum
			Polyporus sulphureus
			Fomes pini
			Fomes pinicola
			Fomes igniarius
			Polyporus cuticularis
		Spruce needle rusts	*Peridermium thomsoni*
			Chrysomyxa deformans
			Cercospora piceae
			Peridermium piceae
		Spruce cone rust	*Pucciniastrum areolatum*
		Other needle diseases	*Rehmiellopsis abietis*
			Lophodermium piceae
			L. pinastri
			Rhytisma piceum
14	Chirpine (*Pinus roxburghii*)	Root rots	*Armillaria mellea*
			Fomes annosus
			Ganoderma lucidum
			Helicobasidium compactum
		Pine-swertia felt rust	*Cronartium himalayense*
		Heart rots	*Fomes pini*
			Ganoderma applanatum
			Poria monticola
		Coleosporium needle rust	*Coleosporium campanulae*
			Dothistroma pini
			Lophodermium pinastri
			Capnodium pini
			Cersospora pinidensiflorae
			Pestalotia funereal
			Dendrophthoe falcate
15	Blue pine (*Pinus wallichiana*)	Root and butt rots	*Armillaria mellea*
			Polyporus tomentosus

			Polyporus schweinitzii
		Fomes pini heart rot	*Fomes pini*
		Other heart rots	*Fomes pinicola*
			Polyporus gilvus
			Stereum purpureum
			F. roseus
			F. fomentarius
		Stem blister rust	*Cronartium ribicola*
		Stem cankers	*Cenangium ferruginosum*
			Scleroderris lagerbergii
			Leucostoma kunzei
			Trichoscyphella resinaria
			Valsa superficialis
		Coleosporium needle rust	*Coleosporium barclayense*
		Melamspora needle rust	*Melamspora oblonga*
		Pucciniastrum needle rusts	*Pucciniastrum agrimoniae*
			P. aceris
		Needle casts	*Dothistroma pini*
			Hypoderma desmazierii
			Leptostroma ahmadi
			Lophodermium nitens
			L. pinastri
			L. piniexcelsa
			Botrysphaeria obtuse
			Capnodium pini
			Sclerotinia fuckeliana
		Flowering plant parasite	*Arceuthobium minutissimum*
16	*Cupressus* and *Juniperus*	Monochaeta canker	*Monochaeta unicornis*
17	*Cryptomeria*	Armillaria root rot	*Armillaria mellea*
18	Poplar	Ganoderma root rot	*Ganoderma lucidum*
		Stem cankers	*Corticium salmonicolor*
			Botryodiplodia palmarum
		Rust, mildew and leaf spots	*Melampsora rostrupii*
			Uncinula salicis
			Myrothecium roridum
			Septoria populi
			Phyllosticta adjunct
		Leaf spot	*Botryodiplodia palmarum*
19	Salix	Stem cankers	*Vacla salicina*
			Aposphaeria Salicis

		Melampsora leaf rust	*Melampsora salicis-albae*
20	Silver oak (*Grevillea robusta*)	Root rots	*Rosellinia bunodes*
		Pink disease	*Corticium salmonicolor*
		Leaf spots	*Botryodiplodia theobromae*
			Mycospharella grevilleae
			Cytospora grevilleae
			Seimatosporium grevilleae
			Phomopsis bakeri
		Phanerogamic parasites	*Dendrophthoe falcata*
			D. neelgherensis
21	*Acacia leucophloea*	Gall rust	*Catacauma acaciea*
			Hapalophragmium pondersberosum
			Hapalophragmium tandonii
22	*Acacia nilotica*	Heart rot	*Fomes badius*
			Ravenelia accaiae-arabicae
			Cylindrosporium accaiae
23	*Acacia senegal*	Trunk rot	*Polyporushispidus*
			Ravenelia accaiae- senegalae
24	*Acer oblongum*	Leaf diseases	*Schizothyrium annuliforme*
			Uncinula aceris
25	*Aegle marmelos*	Shoot blight	*Fusarium semitectum*
		Leaf spot	*Chaetothyrium guananiticum*
			Cercospora aeglicola
26	*Ailanthus excelsa*	Leaf spot	*Cercospora glandulosa*
27	*Anogeissus latifolia*	Leaf blight	*Pestalotiopsis versicolor*
		Black spot	*Monochaetia jabalturensis*
			Marssonina poonensis
		Pink disease	*Corticium repens*
		Heart rot	*Fomes caryophylli*
			F. rabiniae
28	*Azadirachta indica*	Root rot	*Ganoderma lucidum*
		Pink disease	*Corticium salmonicolor*
		Leaf spot	*Cercospora subsessilis*
			Cercospora azadirachtae
		Bacterial leaf spot	*Xanthomonas azadirachtii*
			Pseudomonas azadirachtae
		Phanerogamic parasite	*Dendrophthoe falcata*
			Tapinanthus spp

29	*Bambusa* spp.	Decay fungi	*Polyporus durus*
			P. friabilis
			Ganoderma lucidum
			Trametes corrugate
			Merulius similis
		Rust	*Puccinia gracilenta*
			Puccinia xanthosperma
			Puccinia longicornis
30	*Dendrocalamus strictus*	Rust	*Dasturella divina*
			Tunicopsora bagchii
31	*Emblica officinalis*	Rust	*Ravenelia emblicae*
			Phakopsora phyllanthi
32	*Gmelina arborea*	Heart rot	*Poria rhizomorpha*
			Phyllactinia suffulta
		Powdery mildew	*Cercospora ranjita*
			Armillaria mellea
		Leaf spot	*Sclerotinia rolfsii*
		Root rot	*Fomes roseus*
			Polyporous baudoni
			Trametes straminea
			Tapinanthus spp
33	*Hopea parviflora*	Heart rot	*Fomes lamaoensis*
			Trametes spongipellis
34	*Madhuca latifolia*	Heart rot	*Fomes caryophylli*
		Rust	*Scopella echinulata*
		Phanerogarmic parasite	*Dendroththoe falcata*
35	*Pongamia pinnata*	Root rot	*Ganodema lucidum*
		Heart rot	*Fomes merilli*
		Rust	*Ravenelia hobsoni*
			R. stictica
		Leaf spot	*Microstroma pongamiae*
			Sphaceloma pongamiae
			Fusicladium pongamiae
		Tar spot	*Phyllachora pongamiae*
			P. henn
		Phanerogarmic parasite	*Cuscuta reflexa*
36	Pterocarpus	Rust	*Maravalia pterocarpi*
		Root rot	*Ganoderma lucidum*
			Fomes fastuosus
			Polyporus gilvus

		Leaf spot	***Cercospora vanescens***
		Phanerogarmic parasite	***Loranthus scirrula***
37	*Syzygium cuminii*	Leaf spots	***Elsinoe kamatii***
			Kamatella indica
			Maublamcia indica
			Corylespora cassiicola
			Dodymosporella spp
			Neobarclaya congesta
			Pestalotiopsis versicolor
		Tar spot	***Phyllachora eugeniae***
			P. ambigua
		Heart rot	***Fomes caryophilli***
		Leaf spot	***Meliolina cladotricha***
		Phanerogarmic parasite	***Asterina jambolanae***
			Dendrophthoe falcate
38	*Terminalia arjuna*	Root rot	***Armillaria mellea***
		Leaf spot	***Colletotrichum arjunae***
			Sphaceloma terminaliae
39	*Toona ciliata*	Root rot	***Armillaria mellea***
			Ganoderma lucidum
			Polyporus zonalis
			Polyporus sulphureus
			Fomes fastuosus
			Ganoderma applanatum
			Tramelles straminea
		Powdery mildew	***Discella cedrelae ramakr***
			Phyllactinia suffulta
		Phanerogarmic parasite	***Cuscuta reflexa***
			Loranthus

FOREST DISEASE CONTROL: GENERAL PRINCIPLES

One of the prime objectives of forestry is to grow forest crops free from diseases and to obtain a profitable harvest. In the event of an outbreak of a disease, control measures that have to be adopted should be economically justifiable and ecologically sustainable.

DIRECT MEASURES

SANITATION: Sanitation is the process that reduces or eliminates the initial inoculum from which the disease develops. Such inoculum may be present either on principal hosts or on collateral or alternate hosts. *Corlicium salmonicolor*, the cause of pink disease, occurs on rubber and many other hosts and attacks *Eucalyptus* planted in sites under suitable conditions of development of the disease. Control by sanitation requires an early detection of the disease in the forest and taking suitable steps to remove the disease inoculum before it can spread the disease.

ERADICATION: Eradication may be done mechanically or through use of weedicides, thereby keeping the other host plant of economic importance free from the disease. In natural forests, fungi normally remain endemic. Root and butt rot fungi have restricted growth on roots but quickly colonise residual roots and stumps after clearfelling. The large inoculum potential thus builds up and becomes a potential threat to susceptible hosts later planted in such areas. Control of root rot may be obtained by removal of residual stumps and roots prior to planting either by raising mechanised plantations or through extraction by physical means.

ISOLATION TRENCHES: Once a root disease establishes in forest, it spreads centrifugally from the infection centre to adjacent trees. For pathogens which do not have a free spread through the soil, the disease is mainly communicated to adjacent healthy plants through root contact or root graft. In such cases, the spread of the disease can be checked by isolating diseased plants by trenches.

CHEMICAL CONTROL: Chemical control of forest diseases is generally prohibitively expensive and, therefore, not practicable except probably in nurseries and plantations where the incidence of the disease and the damage are high. Fungicidal chemical may act as *prophylactics* in preventing infection. A systemic fungicide may be used and the chemical is distributed throughout the system of the host. A fungicide will have its desirable effect if it attacks the pathogen at the weakest point in its life cycle and at the same time proves non-toxic to the host at the dosage used. The fungicide must adhere well to produce a lasting effect. For this, stickers as well as spreaders are added to the fungicide. To control a disease of aerial plant parts by a fungicide, the timing of application is important. Mancozeb, Copper oxychloride, Carbendazim, Tridemorph and Triazoles are commonly used.

INDIRECT MEASURES

CHOICE AND IMPROVEMENT OF SITE: Natural stands are adapted to the site they occupy. Such forests, therefore, generally grow healthy as long as the stands are properly managed. Suitable selection of site is important for plantations, particularly of exotics, which are far remote from their natural range of distribution. Good site is necessary for maintaining stand vigour to resist attack due to diseases. An extreme case of species conversion may be seen in the high rainfall areas in Karnataka and Kerala States in South India where semi-evergreen and evergreen natural forests are clearfelled and planted with *Eucalyptus*. The situation becomes aggravated with large scale mortality in *Eucalyptus* due to pink disease resulting in large gaps in plantations.

CHOICE OF SPECIES: Natural forests have inherent slow growth and low productivity. Also, such forests are of mixed composition where only some species are economically valuable. Indigenous species are often not capable of making full economic use of the sites. Exotics should possess the ability to thrive in their new environment for which a great deal of care is needed in selection of suitable species. Plantations may be raised either pure or in mixer. A pure stand is, however, ideal for a pathogen to build up to epidemic proportions since infection is direct and rapid from tree to tree. Mixed stands may be raised in plantations where the species can be in intimate, strip, group or block mixers. In using clonal stocks in planting, it is advisable to plant a number of clones in mixture and limit the area occupied by each clone (Hartley 1939).

SILVICULTURE AND MANAGEMENT PRACTICES: Natural forests which are worked on sound silvicultural practices and properly managed, do not normally suffer from any serious disease. Fire, which is a regular feature in forests, is the principal cause of injury to trees through which decay fungi establish in the tree. In an uneven-aged stand, suppression may occur to a certain degree to the understorey crop under the shade of dominants. In forest raised for timber, a high incidence of decay in the heartwood causes serious loss in yield of timber. In stand with high incidence of decay, it is, therefore, advisable to lower the rotation age. Infestation by dwarf mistletoe is favoured by selective cutting while in cleared cutting, the infestation is less. A pure even-aged stand may suffer from injury due to frost and from excessive insolation resulting in sun scorch. An even-aged crop is also liable to wind damage which is accentuated by thinning and pruning.

CULTURAL PRACTICES: Timely thinning with a view to provide spacing is therefore important as it increases plant vigour which in turn reduces damage from diseases. Diseased trees are likely to be randomly distributed for which reason thinning has to be irregular. On the contrary, stumps left after thinning in conifer stand may be colonised by *Fomes annosus*. Such stumps serve as infection centres for spread of the disease in plantations. It is also done for trees in parks and avenues to remove dead and diseased limbs, create openings for sun light or make clearance for overhead communication lines.

BIOLOGICAL CONTROL: Biological control involves the use of an innocuous organism to prevent attack by a pathogenic organism. Control of *Armillaria mellea* root rot in tea bushes is possible. Likewise, colonisation of fresh conifer stumps soon after thinning by *Peniophora gigantea,* an innocuous fungus, prevents stumps colonization by the pathogen *Fomes annosus*.

RESISTANCE BREEDING: Resistance is the ability of an organism to hinder the development of a disease. There may be active resistance where the host reacts to the attack by a pathogen or passive resistance due to qualities inherent in the host. A *pathodeme* is a host population in which all individuals have a given resistance in common, while a *pathotype* is a pathogen population in which all individuals have a given pathogenicity in common. If in a seies of inoculations, the amount of disease shows a differential interaction between pathodemes and pathotyes, the resistance and pathogenicity are called vertical, which involves mechanisms which are within pathogen's capacity for change. In the absence of any such interaction between pathodems and pathotypes, the resistance is termed horizontal, where the mechanisms involved are beyond pathogen's capacity for change (Van der Plank 1963). In vertical resistance, the inheritance is governed by single gene and, therefore, easy to manipulate in a breeding programme. Horizontal resistance is controlled by ploygenes and, therefore, difficult to manipulate genetically. The effectiveness of horizontal resistance does not break down. Vertical resistance, however, is liable to breakdown (Robinson 1973).

The objective of disease resistance improvement is to select or breed trees to establish a host-parasite relationship to yield an economic harvest. The freedom from infection may either be due to escape or to inherent character in the presence of genes for resistance in the host or absence of gene for infectivity in the parasite. The procedures for resistance breeding include selection of such disease-free individuals and testing their progenies by raising them in a heavily diseased area or through inoculations which will eliminate the escapers and ensure the heritability and selection of the resistance factor. Artificial inoculations, where massive spores or mycelium are used as inocula, can be so severe that they assess only immunity and not relative resistance. This should, therefore, not be relied upon fully. During selection, due consideration should be given to other desirable characters in the tree necessary for tree improvement. Parents and progenies should be tested against all known biotypes of the pathogen by exposing to natural infection and to inoculation. The improved stock may be outplanted providing means of testing at the same time resistance under different environmental conditions favourable for disease development. Individuals that have not developed disease resistance can be salvaged during improvement cuttings. Further propagation of these trees may be done by establishment of grafts in seed orchards. Intra-specific crosses between the selected individuals may increase the degree of resistance in the progenies. Seed orchards yielding such progenies will thus form valuable basis for further propagation of a resistant stock.

Though genetic resistance has been recognized in a large number of forest tree species, remarkably little is known about the nature of host resistance. Hairiness in leaves can provide complete protection against some virus vectors such as jassids and whiteflies (Robinson 1973). In *Castanea*, high concentration of pyrogallic tannins in the bark of genotypes resistance to *Endothia parasitica* may be responsible for retarding the growth of the fungus (Nienstaedt 1953). Enzymatic activity is responsible for resistance of oaks against *Microsphaera alphiloides*. A close correlation between bark moisture and disease development exists in canker diseases in trees, as in willow due to *Cryptodiaporthe* (Bier 1959a); in poplar due *Fusarium* (Bier 1959b) and in western hemlock due to *Cephalosporium*. In absence of much data on the nature of host resistance, however, breeding forest trees for producing disease resistance clones is a potential way for disease control. In *Populus,* significant progress has been possible because of the variation in disease resistance between and within the natural species, wide range of genotypes and ease of vegetative propagation for purpose of clonal multiplication (Schreiner 1960).

DISEASES OF TEAK

BACTERIAL WILT

Bacterial wilt in teak is caused by the pathogen *Psedomonas solanacearum* which attacks species belonging to as many as 17 families in the tropical and sub-tropical parts of the world. In the initial stages, small patches of brown tissue appear between the veins of the leaves. They may form a network to give a scorched appearance to the leaf. Wilting of the leaves occurs and even young plants may wilt. As a result of this wilting, the plant upto pole stage becomes yellow. Discouloration occurs in vascular tissues of the roots. The main root and lower part of stem become soft, turn black and may decay in advanced stages.

Nurseries must have good drainage and they should not be located in waterlogged sites. Plants that have previously been affected by this wilt must be removed and burnt. Weeding must be done carefully to avoid root injuries which form the main avenue for infection. It is advisable to sterilize the soil of nurseries where this disease is prevalent. In extreme cases, the site may be abandoned and a new site selected.

LEAF RUST

Leaf rust in teak is caused by *Olicea tectonae* which occurs all over its natural range in India, particularly in teak nurseries. It attacks the leaves generally between September and February. Uredia develop profusely and rarely produce Telia. They cover the lower surface of the leaves with a rusty coat while the upper surface has a grayish appearance. Survey conducted for rust disease in teak indicated that plantations under high humid conditions suffer more damage. Rust severity, rate of disease development and maximum defoliation is higher in 3-6 year old teak plants. In forests, opening of the canopy helps to prevent rust disease spread. Tridemorph @ 1ml/lit is effective and two sprays at an interval of 15 days may be followed.

POWDERY MILDEW

In India, powdery mildew in teak is caused primarily by *Uncinula tectonae. Phyllactinia corylea* and *Phyllactinia guttata* also cause damage. Pathogen attacks the upper surface of the leaf and adheres to it by serrated appressoria. Haustoria penetrates upto only the upper epidermal cells. Conidia are dispersed by wind, particularly on hot days and cool nights. The infected leaves are coated with a dull white mycelium some time dotted with black fruiting bodies. Soon after attack, most leaves become infected and they may fall down. However, no serious damage is caused though the process of photosynthesis is retarded due to the coating on the leaves. The affected leaves are shed during the normal period of leaf fall.

PHANEROGAMIC PARASITES

In India, the mistletoe *Loranthus longiflours, Dendrophthoe falcate* are common on teak trees in different parts of the country. Initially, the infection causes mutilation or death of branches or tops of the trees. In case of heavy infection, the trees, particularly upto pole crop may even die or growth is severely retarded. Sanitary cutting of the affected branches below the point of infection is useful. Heavily infested trees may be felled. Spraying of chemical herbicides in early summer when teak is leafless is recommended.

DISEASES OF SAL

POLYPORUS ROOT ROT

In sal, polyporus root rot occurs almost all over the range of the host species. Though its occurrence is less common under relatively drier conditions and fairly common in localities with heavy rainfall, it has been recorded from all states bearing natural sal forests. *Polyporus shoreae* has an immense capacity for infecting both healthy and injured or damaged roots. Due to the infection, the tree may die back; its top may become broken or it may even be uprooted in severe windstorms. In later stage, it also causes white pocket rot in the bark and sapwood though the heartwood largely remains unaffected. Large trees show symptoms of top dying as their remaining healthy roots are unable to meet the water requirement of the crown. This dying of branches gradually extends downwards leading the death.

Control burning reduces soil moisture and checks the growth of weeds which ultimately helps in lowering the incidence of this disease. All dead, dry, diseased and dying trees must be removed from sal areas where the incidence is high (Bakshi and Boyce 1959).

HEART ROTS

In high sal forests, heart rots establish in trees from the trunk and as butt infections. In Coppice sal forests, infection usually establishes on the stool and progresses up the tree when the heart rot develops in coppice shoots. Heart rot in sal is caused by *Fomes caryophylli, Fomes fastuosus, Hymenochaete rubiginosa, Trametes cubensis* and *Trametes incerta.* Punk knots, swollen boles and presence of sporophores, wounds, branch stubs, dead branches etc. are the symptoms of heart rot disease.

Frost Injury to trees may be lessened by suitable adjustment of the canopy so that large gaps are not created. In uneven aged stands, suppression is prevented by canopy adjustment. Fire control measures help in preventing fire injury, thus lowering the incidence of heartwood rot. All trees injured by fire and frost and those with significant symptoms of decay must be removed in cultural operations. Trees to be felled must be lopped so that while falling their branches not damage the standing trees. Felling debris consisting of lops and tops must not be left in the forest. Severely affected stands may be worked on basis of the pathological rotation age. Pathological rotation is the age at which the progress of decay is equal to the rate of formation of new wood and this may differ from the rotation age. Coppice forests of sal are severely affected by heartwood rot decay due to the vulnerable stools. Prevention and control of this disease in coppice stands include: lowering the rotation age; dressing the stool with chemical; and girdling old abandoned stumps so as to kill them.

DISEASES OF SHISHAM

WILT DISEASE

This disease is caused by the soil inhabiting fungus *Fusarium solani.* The adverse effects are yellowing and death of leaves and subsequent defoliation. Young affected trees die off after an intense attack. Waterlogging favours the incidence of this disease. Silty and clayey soils favour growth and infection by this pathogen. Selection of the proper site for raising shisham plantations is the most important preventive measure. The soil should be light textured, with adequate soil moisture and good drainage. Soil drenching with carbendazim, around the infected trees is recommended.

GANODERAMA ROOT ROT

Disease is caused by *Ganoderma lucidum* both in natural forests and plantations. The pathogen spreads rapidly in the system of trees growing on light textured soils. The trees may die within a short span. On the other hand, the spread is relatively slow in heavy textured soils. It spreads through root contact and hence it damages pure plantations very quickly, resulting in the death of trees. The diseased trees present a *stag headed* appearance and die after a few years. Selection of suitable site, removal of the affected individuals and sanitations are the recommended measures. This disease can be prevented by removing residual roots and stumps before raising a new crop. Isolation trenches also help to prevent the spread of this pathogen.

Ganodrma lucidum also causes root rot in *Dalbergia latifolia* natural forests as well as plantations. The symptoms and control measures are same as that of *Dalbergia sissoo.*

DISEASES OF KHAIR

GANODERMA ROOT ROT

Root rot caused by *Ganoderma lucidum* is serious disease in many khair plantations and coppices. This pathogen is normally endemic in natural forests though it does not cause any serious adverse effects there as a certain level of resistance is noticed. There is a poor development of lateral roots in the early stages. The infected plants are killed in the first few years of a plantation and their small mass of roots disintegrates quickly with no effect in the further spread of the disease. With increasing age, the chances of lateral root contact become more. Infection is through root contact and injured surface in plantation.

Mechanized soil working and removal of residual roots and stumps are considered major prevention and control methods. Isolation trenches are dug for preventing root contact between the healthy and diseased individuals. These trenches are about 1 m deep, 1.5 m long (across the line) and upto 0.5 m wide.

DISEASES OF NEEM

NURSERY DISEASES

Rhizoctonia Leaf Web Blight: This disease affects the foliage of young plants. It is caused by *Rhizoctonia solani.* Development of greyish brown blotches that increase in size with the advancing fungal hyphae and then engulf the entire leaf blade. The fungal hyphae joins the infected leaves giving the appearance of spider's web and hence the name of the disease. The leaflets or the entire pinnae become detached prematurely. However, the detached pinnae remain clinging to the stem for quite some time, being bound together with the fungal hyphae and presenting a web by appearance.

Colletotrichum Leaf Spot and Blight: This disease is caused by *Colletotrichum gloeosporioides* which initiates leaf spots that increase rapidly in size, covering large areas of the leaf surface. The affected leaves present a blighted appearance and there is premature defoliation. The growth is retarded due to the premature defoliation. Under nursery conditions treatment with Bavistin @ 1% is highly effective.

Alternaria Leaf Spot and Blight: This disease is caused by *Alternaria alternate* which affects the foliage of seedlings.

Pseudocerocopora Leaf Spot: This disease is caused by *Pseudocerocopora* spp. It produces brownish infection spots interspersed with white patches. The fungus sporulates on the upper leaf surface and produces conidia which appear grayish in mass. The heavily infected leaves turn pale and are shed prematurely.

Bacterial Leaf Spot: Disease is caused by *pseudomonas viticola.*

Other Diseases: Leaf spot (*collectotrichum capsic*), blight and stem rot (*Sclerotium rolfsii)* and leaf wilt (*Fusarium solani)*.

DISEASE OF SANDALWOOD

SPIKE DISEASE

The spike disease occurs only in sandal trees growing in south India. It has not been reported from other regions or other species of the genus *Santalum.* The disease is also prevalent in Tamil Nadu. There is a drastic reduction in the size of the leaves, both in length and breadth. The internodes are shortened so that the leaves become crowded together. The leaves show a tendency to stand out stiffly from the branches. This gives a rosette appearance to the affected branches. The leaves appear like spikes on the branches and hence the term spike disease. At times, the infected shoots tend to show continuous apical growth, becoming drooping and giving a pendulous appearance to the spike. The affected trees may perish after a period of 12 to 24 months. A branch or tree that is fully affected by the spike disease does not bear flowers or fruits. Even if flowers appear before the branch is affected by this disease, they may develop into fruit but seeds may abort and not form fully. Flowers exhibit phyllody in infected branches. There is a significant accumulation of starch in the branches and leaves of diseased sandalwood trees.

Recent researches suggest that spike disease is caused by Phytoplasma like organism. This disease belong to the yellow group of plant diseases in which there is reduction in leaf size, development of witches broom and phyllody. Spread is also through weak-flying insect vectors that may be carried over long distances by strong winds. These insect vectors include leaf hoppers *Jassus indicus* and *Moonia albimaculata.* The following measures should be adopted for prevention and control,

- *Sanitation*: The affected individuals or their parts may be removed and burnt.
- *Host Selection*: A number of host species impart resistance to spike disease in sandalwood. It may be raised in association with such hosts (Redgram or Neem).
- Insecticides may be sprayed to control.
- *Genetic Resistance:* Spike-resistant sandal may be evolved through selection of apparently resistant sandal trees in a heavily spiked area. There are two types of resistance to spike disease in sandal: *Autogenic resistance* which is largely independent of the nature of the associated host plant. *Acquired resistance* is built up by sandal if it parasites certain species of hosts and certain environmental conditions are provided.

DISEASE OF CASUARINA

STEM WILT

Trichosporium vesiculosum causes stem wilt in casuarina. Numerous spores are produced by the pathogen which is dispersed by wind. Wounds caused to the stem or branches serve as channels for infection. The spores carried by wind attack the trees through these wounds or injuries. The pathogen spreads to the shoot and root system of the tree after infection. After this, there is secondary spread of the pathogen in the plantation through root contact and root grafting. *Casuarina* trees have a well developed network of roots and thus the disease spreads very rapidly in the plantation.

Symptoms of wilting like paling of needles which dry up completely after some time. Formation of spore mass beneath the bark results in the development of blisters on the bark. After some time, these blisters burst or rupture, exposing the black powdery mass of spores on the upper surface of the bark and sporulation also occurs on the roots. Diseased trees may be found in groups as the secondary spread of this pathogen is by root contact. Prevention of primary infection by minimising all forms of injuries to the trees is important. Isolating the diseased trees by means of trenches and groups of affected trees may thus be isolated. All the infected trees must be removed as early as possible.

DISEASES OF EUCALYPTUS

PINK DISEASE

Eucalyptus pink disease is caused by the pathogen *Corticium salmonicolor*. Repeated die back in young plants occurs which lead to their death. Infection occurs, in trees which may have escaped being affected by this pathogen initially, at a later stage in the form of cankers. There occurs girdling of branches and the main stem of young plants towards the later part of the rainy season. The parts lying above girdled portion may die, causing many epicormic branches to develop. One of these becomes the leading shoot which is again infected and killed in the next season. High summer rainfall (over 250cm) and warm tropical conditions may cause epidemics.

The pathogen *Corticium salmonicolor* occurs in four growth forms on the host tree viz. (1) *Pustule*: These appear as pinkish cellular sterile bodies having a diameter of upto 0.1 cm on the branches and main stem soon after infection at the commencement of the monsoon season. (2) *Cobweb mycelia*: These are thin white arachnoid mycelia produced on the bark and main stem by the infected parts. They originate either from the pustule or from the edges of the pink incrustation. (3) *Necator*: These are orange to reddish fruit bodies having a diameter of about 0.2 cms. They develop on the side of the attacked branch which is exposed to the sun. (4) *Pink incrustation*: These are light pink, thin incrustations on the bark occurring as the perfect form of the pathogen. This develops in autumn on the shady portion of the stem and branches. Later on the incrustation spreads all around the stem.

Individuals showing a fair degree of resistance to this disease in an infected plantation could be identified and selected for breeding programme for introduction on a large scale. *Eucalyptus torelliana* is found resistant to this disease and may be used in crossing programme. Spraying of systemic fungicides like Trdemorph @ 0.1% may be done on young plantations before and during the rainy season. Diseased individuals must be removed from the plantation as early as possible.

SEEDLING DISEASES

Seedling Blight: Seedling blight is caused by *Cylindrocladium clavatum, C. quinqueseptatum* and C. *scoparium*. Damage to the leaves and in extreme cases there may be complete defoliation. Premature defoliation slows down the growth of the seedlings.

Cercospora eucalypti: It causes damage to the leaves of Eucalyptus seedlings in the nursery.

Puccinia psidii: It attacks the foliage and young twigs of seedlings upto 1 year in age. It may destroy the whole nursery crop.

Botrytis cinera: It causes damping off and leaf spots in nursery

Oidium eucalypti: Powdery mildews on Eucalyptus seedlings growing in the nursery are caused by this pathogen.

GANODERMA ROOT ROT

It is caused by *Ganoderma lucidum* in different species of Eucalyptus. It is prevalent in areas where the previous crop is affected by *Ganoderma.* The residual stumps and roots serve as infection centers. It then spreads by root contact. Removal of residual stumps and roots of the previous crop before planting prevents disease. However this is possible only in the case of mechanized plantations. Eucalyptus may be raised in mixture with resistant species so as to prevent the spread of the disease through root contact. Affected trees may be isolated by digging isolation trenches around them.

PHYTOPHTHORA ROOT ROT

Root rot in many tree species including Eucalyptus is caused by *Phytophthora cinnamomi*. It spreads through roots or even soil particles and results in decay of the root system.

CANKER

Diaporthe cubensis causes canker in various species of Eucalyptus, though some species are resistant to this pathogen. Young plants are attacked at the root collar causing discolouration, girdling and even death. There is development of attacks around the base of the dead branch stubs. Initially there develops longitudinal cracks in the bark, while open cankers form later on having a length of more than 1 m. There is exudation of gum from these cankers. In tracts prone to this pathogen, only resistant species should be planted. The diseased individuals must be identified, isolated and removed from the plantation as early as possible.

GUMMOSIS

Gummosis or the exudation of gum kino from Eucalyptus trees may be due to severe frost which causes injury to the cambium, resulting in gummosis. Fires too have been attributed to injure the cambium, thereby causing the exudation of gum kino. Due to this disease, there occurs swelling and splitting of the bark at various parts. This is accompanied by the exudation of gum kino which is a shining, golden brown coloured viscous to semi-viscous liquid. It hardens as brownish or red lumps soon after exposure to air. There is lateral displacement of the medullary rays and the ray cells are broadened at the region of the injury. The right species of Eucalyptus must be planted on a particular site. Control burning helps to minimize injury to trees by fire.

DISEASES OF BAMBOOS

LEAF DISEASES

- Leaf blight: It is caused by *Rhizoctonia solani* and the infected leaves have profuse mycelial growth of the fungus. It causes premature defoliation.

- Leaf rust: It is caused by a number of pathogens such as *Dasturella divina, Dasturella bambusina, Puccinia xanthosperma* and *Tunicospora bagchii.*
- Leaf spots: Leaf spots are caused by a number of pathogens such as *Ascochyta bambusina, Balladyna butleri, Corticium koleroga, Drechlsera rostrata, Phyllachora bambusae, Phyllachora graminis, Phyllachora malabarensis* and *Phyllachora shiraina.*

BAMBOO BLIGHT

This is one of the most destructive bamboo diseases in India caused by the pathogen *Sarocladium oryzae*. The pathogen also causes sheath blight in rice and bamboo acts as collateral host. Burning of debris in clumps in April and May before the onset of the rains, application of certain chemicals as soil drench and removal of blighted culms are reommneded.

RHIZOME AND CULM ROT

- *Rhizome Bud Rot*: This disease is caused by *Fusarium* and *Pythium* spp.
- *Basal Culm Rot*: This disease is caused by the pathogen *Fusarium moniliforme* and occurs in culms that have just emerged from the soil.
- *Culm Rot*: This disease is caused by two species of *Fusarium* and develops on wounds of culms caused by the sap sucking insects.

RHIZOME AND ROOT ROTS

- *Ganoderma lucidum*: This is a serious root rot pathogen affecting bamboos and other trees. It is wide spread in *Bambusa bambos* in which there occurs lateral spread of the disease through root contact.
- *Merulius eurocephalus*: This pathogen causes severe losses in various species of *Bambusa* and *Thyrsostachys*. Sporocarps develop at the base of the culm and also on humus around the infected clumps.
- *Poria rhizomorpha*: It occurs as a saprophyte in the soil, forming abundant rhizomorphic stands on decaying roots and other vegetable matter. It becomes a parasite on poorly drained soils, spreading through rhizomorphs.

DECAY IN BAMBOO

A number of pathogens cause decay in bamboos both in standing culms and also in bamboos in storage and use. They also affect the rhizomes. Important decay causing pathogens in bamboos are *Amauroderma rugous, Daedala Flavida, Fomes* spp., *Ganoderma lucidum, Guepinia ramose, Polystictus* spp., *Sphaerostilbe* spp., *Trametes corrugata* and *Tremellodon gelatinosum.*

15 FOREST FIRE MANAGEMENT

INTRODUCTION

In India, about 55 per cent of the forest area, which is predominantly covered by deciduous forests, is prone to fires every year (Gubbi 2003) causing loss of about Rs. 440 crores (approximately 104 million dollars) (Bahuguna and Satyendra Singh 2002). It is estimated that proportion of forest areas, prone to forest fires annually ranges from 33 per cent in some of the states to over 90 per cent in other (FSI 1987). Despite the natural fires, the major sources of forest fires in India are anthropogenic, which include shifting cultivation practices, controlled burning, deforestation, fire wood burning and others. Besides conventional methods of fire control systems like fire line, fire watch tower, block lines and other manual fire control systems, application of remote sensing with GIS is also used to address the problem with good scientific and technical strength in a time effective and cost effective way (Murthy et al 2006).

Forest fires in the country area mostly experienced during summer months from April to June. The extent and type of fire varies from state to state based on type of forest as well as climatic conditions like prolonged spell of dry conditions or delay in arrival of monsoon, etc. (Chetan Kumar 2002). Forest fires not only causes loss of biodiversity, loss of valuable timber, degradation of natural forests and water catchment areas but also result in depletion of carbon sinks, reduction in water table level, global warming, ozone layer depletion and also loss of agricultural production (Bahuguna 1999).

CLASSIFICATION OF FOREST FIRES

Forest fires can be classified into two categories viz. on the basis of causative factors and on the basis of the place of their action.

ON THE BASIS OF CAUSATIVE FACTORS

NATURAL CAUSES: The natural causes may be lightning, rolling stones or rubbing of dry bamboos with each other. It comprises about 5 per cent of total occurrence of all fires.

ANTHROPOGENIC (MAN-MADE) CAUSES: About 95% of the fires in the country are caused by man. Man may cause these fires either unintentionally i.e. by his carelessness or deliberately and intentionally.

ACCIDENTAL FIRES

The fires that are caused unintentionally by man are called "*accidental fires*" as they are caused accidentally by his carelessness. The accidental fires may be due to the following causes,

- Leaving fires burning by cartmen or travelers, lighted by them for cooking their mid-day meal or spreading of fire from labour camps
- Throwing of burning match stick, or *bidi* or cigarette stumps in the forest by grazers or other travelers
- Throwing of glowing coal pieces by trains speeding through the forests
- Throwing of torch wood by villagers while passing through the forest in the night
- Falling of dry needles or leaves on electric poles giving out sparks
- Burning of fields or grass lands in the villages adjacent to the forest and leave such fires to spread in the forest area and
- Accidental spread of fires in the forest while burning firelines departmentally.

DELIBERATE OR INTENTIONAL FIRES

Often forest fires are caused by man deliberately and intentionally. These fires are therefore called *deliberate* or *intentional fires*. The main causes of such fires may be any of the following.

- Burning the undergrowth and grass to collect Non-Timber Forest Produce (NTFP) such as horns, which otherwise is not easily visible
- Inducing new shoots of grass in summer by burning the dry grass
- Burning undergrowth and grass in the forest to search for injured wild animals or even to trap wild animals
- Scaring away wild animals from nearby villages
- To drive bees/wasps from trees
- Burning forest in the charge of a particular official out of enmity with him and
- Destroying or at least charring the stumps of illicitly felled trees in a forest

ON THE BASIS OF THE PLACE OF THEIR ACTION

On the basis of the place of their action forest fires are classified into four categories: creeping fire, ground fire, surface fire, crown fire.

CREEPING FIRE: It's defined as *a forest fire spreading slowly over the ground with low flame. It usually occurs in forest floor covered with a layer of dry leaves which burns slowly in absence of strong wind.*

GROUND FIRE: It's defined as *a forest fire that burns the ground cover only, i.e. the carpet of herbaceous plants and low shrubs, which covers the soil.*

SURFACE FIRE: It's defined as ***a forest fire which burns not merely the ground cover but also undergrowth.***

CROWN FIRE or TOP FIRE: It's defined as ***a forest fire which spreads through the crowns of trees and consumes all or part of the upper branches and foliage.*** This usually occurs in coniferous forests.

The above classes of forest fires are not exclusive and independent of each other. Usually one kind of fire may start and may develop into some other class depending upon the circumstances such as conditions of wind, presence of under growth, lichens and dry climbers and other factors forming fire environment (Khanna 1982).

FIRE ENVIRONMENT

Fire occurs as a result of certain circumstances which constitute its environment. The following factors make up fire environment.

WEATHER

High temperature, low humidity and hot strong wind together prepare an environment in which the fire risk is maximum.

INFLAMMABLE MATERIAL

Even in a favourable fire environment, fire occurs only when there is inflammable material/ fire hazard on the forest floor. Grass, dry leaves and dry fallen wood constitute inflammable material in any forest. In *chir* forests when resin tapping is in progress, it adds to the inflammable material. If the grass and shrubs, fallen leaves, fallen wood, etc., have all been control burnt in the forest floor before onset of summer by early burning, even high temperature and dry hot winds, cannot produce fire in the area.

TOPOGRAPHY

Topographic features of hilly terrain vary from place to place and affect not only the fire environment but also the behaviour of fire when it occurs. Altitude and aspect both have a great effect on temperature, rainfall, etc and thus affect the formation of fire environment. As the southern or south western aspects are hotter and drier, there is greater fire risk on these aspects when compared to northern or north eastern aspects. As temperature falls with increasing altitude, lower altitudes have a greater risk of fire. Also, fire spreads very fast while travelling up a hill slope, but its speed slows down considerably when it travels downhill. Similarly aspect and elevation influence fire behaviour through their combined effects on fuel moisture and on ambient weather conditions.

OCCURRENCE OF FOREST FIRE, ITS BEHAVIOUR AND DYNAMICS

From the place of ignition, the direction of spread, speed, extent and shape of fire depend upon wind, inflammable material and topography.

DIRECTION OF SPREAD

Fire spreads fast in the direction of wind and in the direction where more inflammable material is present. Fire also spreads fast uphill and slowly downhill. Normally fire spreads along the ground in horizontal plane but presence of dry climbers on trees, resin channels in chir forests or lichens in deodar forest, induce it to spread at planes at right angles to the normal plane i.e. along the crowns of trees.

SPEED OF FIRE

Spread of fire is proportionately influenced by speed of wind and also the quantum of dry inflammable material present. The spread of fire is greater during the day when temperature is high and strong winds blow and is comparatively lesser during night when the temperature falls and the wind slows down.

EXTENT

The extent of fire depends not only on wind, inflammable material and topography but also on the promptness with which the fires is extinguished. The greater the delay in detecting the start of a forest fire and to control it, the greater will be the area affected by it.

SHAPE OF BURNT AREA

The shape of burnt area is generally irregular. However, atleast some of the sides of burnt area can be straight lines because counter-firing is done from artificially cut and cleared fire traces in the form of straight lines.

FOREST FIRE MONITORING IN INDIA

Forest Survey of India has started the monitoring of forest fires across the country since the year 2005 using data from web fires mapper (using inputs received from MODIS satellite system, a joint collaboration of NASA and Geographic Department of University of Maryland). The coordinates of active fire location from this site are projected on the forest cover map of India to select active forest fire locations lying within the forest cover. The information is then disseminated to the State Forest Departments. From March 2010 onwards the information is being sent through e-mail/ SMS to the registered users through its website www.fsi.nic.in.

Any user can register for the next system by providing his/ her mobile number and e-mail address and the names of district/ state/ UT for which the information is sought. Every day, between 11:00-12:00 hours e-mails/ SMS alerts reach the registered users giving a summary of total number of forest fires detected in their chosen areas. Presently, there is a time lag of 12 to 24 hours in the reporting of these fires owing to late availability of this data. Efforts are on to reduce this time lag so as to provide information on near real-time basis. A total of 13,898 fire incidences have been reported by FSI to different states in the year 2010-11 (FSI 2011).

The on-going exercise has helped in identifying the forest fire prone areas in the country and also the critical time period of fires occurrence for each state and UT. The identification of fire sensitive zones as well as the fire seasons is likely to help in formulation of effective forest fire control strategy in term of prevention, alertness, mitigation, fund allocation and deployment of personnel and equipment. The work has generated basic data on the pattern of forest fire in the country which can be used for preparing national level strategy for early warning and burnt area assessment (FSI 2011). The detailed geo-coordinates of the forest fire point locations are also made available on the FSI website. All the archival forest fire data is also available on the website. The service was widely publicized amongst the SFDs and the feedback received from them indicated that the forest fires detection has accuracy over 95 per cent.

DAMAGES CAUSED BY FOREST FIRES

DAMAGE TO THE TREES

Damage to trees varies with species, age of trees, their condition and the season. The species with thick bark when compared to the species with thin bark; the broad leaved species when compared to conifers are generally more fire resistant. Younger trees when compared to the older trees of the same species are more fire sensitive due to their relatively smaller poles, thinner bark, lower height and lesser diameter.

The green trees are usually less damaged. However, burning of the bark and cambium layer causes serious injury to the tree stem and reduces their growth and vitality. Consequently, the chances of their getting attacked by insects and fungi are increased. Deciduous forests particularly during summer season are more vulnerable to fire. The condition of the grass i.e. whether it is dry or green, also affects the damage caused by fire. Fires in summer season are not only common but also destructive because of high temperature, strong wind, dry undergrowth and ground cover and thick layer of dry fallen leaves.

DAMAGE TO REGENERATION

Even a surface/ ground fire is sufficient to wipe out the young regeneration completely. Repeated fires in natural regeneration areas make it difficult to satisfactorily complete the regeneration within the *regeneration period*. If the species possesses coppicing power and in case the roots of seedlings have not got damaged due to fire, regeneration in the form of seedling coppice appears. If, however, the species does not possess coppicing power, the regeneration is completely destroyed.

DAMAGE TO SOIL

Destruction of organic matter affects the structure of the soil adversely and soil is laid bare to the action of elements viz., sand, wind and rain. Fire also makes soil compact and impervious. Consequently, soil erosion starts resulting in loss of top fertile soil.

DAMAGE TO PRODUCTIVE POWER OF THE FOREST

Repeated fires degenerate an evergreen forest into inferior deciduous forest or even grass lands in extreme cases. Valuable species disappear and are replaced by inferior fire hardy species. Density and increment of forest is reduced and it affects the yield. As timber becomes defective, it fetches lesser price in the market.

DAMAGE TO PROTECTIVE POWER OF THE FOREST

Forest can discharge its protective functions when it is a forest in the real sense, i.e., it has not only trees but also grasses, shrubs, small trees, etc. As already stated, even an ordinary fire burns down the ground cover and undergrowth completely and therefore, it affects the protective power of forest. When protective power of forest is reduced, the result is increase in soil erosion and run-off. Fire increases flood havoc as it destroys the protective cover of the watershed. Heavy rains on newly denuded slopes result in devastating floods.

DAMAGE TO WILD ANIMALS

Forest fires result in enormous loss to wild animals, birds, insects and other fauna. It burns not only the eggs or young ones but sometimes even the bigger animals also. As the destruction of fauna destroys a valuable part of the eco-system, natural equilibrium is seriously affected with consequent adverse effect on vegetation itself.

DAMAGE TO THE RECREATIONAL AND SCENIC VALUE

As fire destroys the greenery of the forest, it destroys its recreational and scenic value. The forest, no longer remains a fit place for recreation as the ground is littered with ash and the blackened stem of the shrubs and lower portions of the boles of bigger trees make the entire place desolate.

METHODS OF EXTINGUISHING FOREST FIRE

In India, fire is usually extinguished by one of the following methods: by water, by earth, by beating and by counterfiring.

BY WATER

This method cannot be applied on a large scale in our forests because of dearth of water. Water is, however, used, after the fire has been controlled, in extinguishing burning stumps near the line of control as sparks from such stumps, which may keep smouldering for weeks, can start fresh fires.

BY EARTH

Though earth can be dug out at site, it is a time-consuming operation. Therefore earth is also not used to extinguish fires. Like water, it is only used to extinguish smouldering stumps near the line of control after the fire has been extinguished.

BY BEATING

The best way to extinguish all mild surface fires is to beat them out. For beating out fires, brooms are made by cutting branches of shrubs and then men standing near the fire beat it back with the brooms. Brooms should not be struck vertically as this often results in spreading sparks in unburnt area. They should be struck in a slanting manner so that the fire is pushed inside the burnt area. As already stated this method can be used in case of fire which is mild and therefore fire extinguishing men can stand close to it.

BY COUNTER-FIRING

When the fire is so fierce that men cannot stand close to the burning front to beat it out, it is extinguished by *counter-firing* which is defined as *an attempt to extinguish an advancing fire by deliberately burning the forest from the opposite direction.* When a fire is started some distance away from an advancing fire and is made to proceed towards it, the two fires meet and get extinguished. This happens because the advancing fire finds before it a burnt strip and therefore gets extinguished. In order to counter-fire, a fire trace is made. A fire trace is defined as a cleared (often burnt) line used as a base from which to counter-fire.

METHOD OF PROTECTION AGAINST DAMAGE BY FIRE

Methods of protection against damage by fire can be broadly grouped into preventive and remedial measures.

PREVENTIVE MEASURES

Those measures which prevent the occurrence or at least the chances of occurrence of fire are called preventive measures. They can be further grouped into indirect and direct preventive measures.

INDIRECT PREVENTIVE MEASURES

Indirect preventive measures are those measures which, though not directly connected with the causes of fire or factors affecting fire environment, reduce the fire risk. The following are some of the indirect preventive measures.

GOODWILL OF THE LOCAL PEOPLE: Goodwill of the local people will effectively stop cases of deliberate fires. The goodwill of the local people can be earned and they can be prevented also from making up their minds to set fire to the forest, by granting them rights and concessions in collecting NTFP, etc, by providing alternative livelihood, by giving them as much employment in the forest works as possible, etc. By earning the goodwill of the local people, their spontaneous help in extinguishing fires when they occur accidentally will also be available.

EDUCATION OF PUBLIC OPINION: Accidental fires can be avoided by educating public opinion through press, radio and television, posters, film shows and lectures in local languages, about the causes of fires and their effect.

FORBIDDING COLLECTION OF CERTAIN ITEMS OF NTFPS DURING SUMMER: Permission for collection of horns and bones of wild animals, collection of honey by smoking bees away during summer season increases the chances of accidental as well as deliberate fires.

DENIAL OF BENEFITS WHICH ACCRUE FROM FOREST FIRES: Villagers sometimes deliberately burn the forest so that new flush of green grass may appear on the forest floor for their cattle to graze. The easiest way to prevent such fires is to deny the benefit of grazing to villagers in fire-burnt areas. The provisions of section 26(3) and 33(2) of Indian forest Act, 1927 should be invoked to close burnt areas in reserved and protected forests respectively for the exercise of all rights of pasture and other forest produce for some years. This will have a deterrent effect on the villagers.

ACTION AGAINST DERELICTION OF DUTY AND REWARD FOR CONSCIENTIOUS WORK IN PREVENTING FIRES: During the fire season, the action taken by staff and other people should be properly supervised and reviews so that those found guilty of dereliction of duty could be punished and those found working conscientiously should be rewarded. Efforts should also be made to involve local people and JFMC members in fire protection work.

PUTTING UP NOTICES PROHIBITING KINDLING, KEEPING AND CARRYING OF FIRE IN FOREST AREAS IN FIRE SEASON: It creates public awareness and may be useful to some extent.

DIRECT PREVENTIVE MEASURES

Direct preventive measures are those measures which directly affect the factors of fire environment and causes of fire and thus help in prevention of fire. These measures are called pre-suppression measures and cover all activities carried out in advance of fire occurrence. The following are the important direct preventive measures or pre-suppression operations:

- Forecasting of dangerous burning days
- Organization, training and detailing of staff for fire control activity
- Hazard reduction

Hazard reduction comprises the following activities:

CLEARING: Clearing camping sites and areas along paths and roads particularly at the beginning of summer season and keeping hazard free till the onset of monsoon.

EARLY BURNING: Early burning is defined as *controlled burning early in the dry season before the leaves and undergrowth are completely dry or before the leaves are shed, as an insurance against later fire damage*. The object of early burning is to burn down all inflammable material such as grass shrubs, fallen leaves and wood before the commencement of hot weather so that occurrence of fire may be prevented.

BURNING OF GRASSY BLANKS: Grassy blanks inside the forests are burnt well before the fire season as a measure of hazard reduction and this has proved fairly efficacious in reducing fire incidence in many parts of the country.

SLASH DISPOSAL IN HILL FOREST AND CONTROLLED BURNING IN RESIN TAPPING AREAS: Slash is the unusable residue after logging viz., branches, tops, bark, un-utilizable logs, uprooted stumps, broken or uprooted trees left on the area and also any large accumulation of debris after wind or fire. Slash disposal is defined *as the treatment or handling of slash for reducing hazards from fire, insects or fungi and for providing the seeds with access to soil.*

BURNING A BELT AROUND PLANTATIONS/ NATURAL REGENERATION AREAS: In order to protect some valuable forests, plantations or natural regeneration areas, it is usual to burn a belt of sufficient width around such areas so that in case of accidental fire, it may not enter the plantation/ natural regeneration area or some other valuable forest proposed to be fire protected. But, such burning should be done carefully so that it may not start an accidental fire.

RAISING OF FIRE-BREAKING GREEN BELTS: Evergreen species such as *Wendlandia*, *Eugenia*, etc., are sometimes grown to form fire-breaking belt around Shola forests in Nilgris, Tamil Nadu. *Strobilanthes* with its juicy stems often acts as a natural fire break in evergreen forests.

CLEARANCE OF FIRELINES: Fireline is defined as a cleared permanent fire break intended to prevent fires from crossing from one area to another. Fire break is an existing barrier, natural or otherwise, or one prepared before a fire occurs from which all or most of the inflammable materials have been removed, designed to stop light ground or surface fires and to serve as a line from which to work and counter-fire, if necessary; also to facilitate the movement of men and equipment in fire fighting. Firelines are of two kinds: internal firelines and external firelines.

Internal firelines are those *firelines which are situated inside the forest to prevent fires raging inside the forest to spread to that part of the forest which is beyond the fireline*. Thus they localize the fire burning inside the forest. *External firelines* are those *firelines which are located along the outer boundary of the forest to prevent any fire raging outside the forest from entering the forest.*

REMEDIAL MEASURES

Remedial measures refer to those measures which are taken to extinguish fires when they break out in spite of the preventive measures. As the greater the delay in starting fire control measures after its occurrence, the greater are the difficulties because of large area and fierceness of fire and the greater are the losses. The delay can be avoided by quick detection, quick communication of occurrence of fire to range or other sub-headquarter and quick action to suppress fire.

DETECTION

For quick detection of forest fires, fire watchers are engaged during fire season for ground patrolling and fire water towers are erected in fire prone areas. On the hills, however, the fire watcher is supposed to go to high ridges in his beat and locate the occurrence of fires. FSI (Forest Survey of India) has also started the monitoring of forest fire across the country and the information on active fire locations are disseminated by SMS/ e-mail to the State Forest Departments to enable them to quickly act upon to extinguish the fire.

QUICK COMMUNICATION OF THE OCCURRENCE OF FIRE TO RANGE AND OTHER SUB-HEADQUARTERS

Communication has now become a possibility with the arrival of mobile phones. Wireless sets also serve the purpose efficiently particularly in remote localities.

QUICK ACTION FOR SUPPRESSION OF FIRE

On receipt of the report of the occurrence of fire, the range officer has to take quick action to arrange for the suppression of fire. This requires arrangement for labour (from adjacent villages) / JFMC members and tools, and for food, water and lighting and quick transport for all these items / personnel to the place of fire occurrence.

PARTICIPATORY APPROACH IN FIRE PREVENTION AND CONTROL

The community based fires management has to rely extremely on the positive relationship between the people in the rural space and their forest. Mutual confidence and public support has to be created by participatory approached e.g., incentives, income generation activities, involvement in production enterprises, etc for involvement of communities in fire prevention and control (Godammer 2000).

COMMUNITY INVOLVEMENT IN FOREST FIRE PREVENTION AND CONTROL

Over the years, there has been a significant decline in the prioritisation of fire management in the forest management objectives. With various social sectors competing for funds, the funding for the fire prevention and control has also gone down or has been diverted to schemes like 'employment generation' or even the establishment expenses of the forest department. In fact at present most of the states do not have any regular schemes/funds for prevention and control of forest fires (Singh 1997). With meagre human resource at its disposal (e.g. as per one of the estimates on an average 500 ha of forests have to be patrolled by one forest guard), the forest departments in most of the states are poorly equipped to prevent or control the spread of forest fires. This situation and the fact that forests are under tremendous pressure, due to increasing population pressure and hence commensurate demand of land, forest products etc necessitates exploration of alternatives to arrest this phenomenon. Attempts to elicit people's participation in fire control offers hope of minimising the damage caused by fires. In this context Joint Forest Management (JFM) assumes an important role in fire prevention and control. JFM has been a significant development in the context of institutional arrangements pertaining to forest management in India. The effective involvement of local communities in evolving sustainable forest management systems was looked upon as an important approach to address the long-standing problems of deforestation and land degradation in India.

The National Forest Policy (1988) and Joint Forest Management (JFM) Guidelines (1990) of the Government of India acknowledged and endorsed this system of management, which supports the involvement of village communities and non-governmental organizations (NGOs) in the regeneration, management and protection of degraded forests. The conducive environment created by these enabling legal and administrative measures is manifested in the fact that as many as 22 State Governments have issued directions to the respective State Forest Departments for adoption of JFM. In India about 36,130 forest protection committees are protecting about 10.25 million ha of forest area in the country (MoEF 1999). These committees operational in various states are assisting the forest department in forest protection (including fire prevention and control) and management, though the extent of participation and contribution to efforts varies.

A very definitive lesson and pre-requisite for community based approach to fire management, which emerges out of the JFM experience in forest protection, is that the forest dependent communities with stake in forests would be sufficiently motivated to prevent and control forest fires if their livelihood and subsistence needs are met. The JFM program is an example of a participatory approach in which people co-operate with forest department in forest protection in return for economic benefits (Chetan Kumar 2002).

AWARENESS AND TRAINING ON FOREST FIRE PREVENTION AND LOSSES DUE TO FIRES TO THE COMMUNITY

Most important reason for failure of prevention of forest fires is related to the fact that communities do not realize the economic and ecological losses due to forest fires. Therefore, an efficient motivation strategy for fire prevention requires an initial understanding of the cultural, socio-economic and psychological background of community perception of fires losses. The level of community awareness of the potential losses that could result from forest fires and the assurance of economic incentives in the form of fuel wood, fodder or NTFP, etc to JFMC/ VFC members are the important factors that can motivate the communities to protect the forests from fire. Formal training in forest fire prevention and control is invaluable for preparing a nucleus of JFMC members for leading fire prevention and control programs.

STRATEGY TO PREVENT AND CONTROL FOREST FIRE THROUGH JFM

Proper planning is imperative for fire prevention. This calls for the three general approaches to work in tandem i.e. Education, Engineering and Enforcement (Nair 1992). Motivation of community to participate in fire prevention and control should follow education to underscore its importance. An important strategy in fire prevention is to educate the villagers in the forest area and along its fringe regarding the care required to keep fires well under control if lit for legitimate purposes like for example, subjecting agricultural plots to a light burn as a pre-monsoon preparation. An education strategy must appreciate that a series of edicts will not work unless the villagers are convinced about the harmful effects of fire in context of their dependence on the forest resource. Also villagers believe most in what they see than what they hear. Taking groups of villagers to burnt areas and explaining the fire effects will be useful.

Hazard reduction or limiting the exposure of forests to fire risks constitutes mainly the engineering aspect (Brown and Davis 1973). This also included clearing along paths, early and control burning of vulnerable areas, fire lines etc. Ensuring that the public abides by the rules and regulation set out for prevention of fires calls for effective enforcement of regulations.

Finally, while community participation is important, it needs to be further augmented with appropriate (a) pre-fire planning and fire prevention strategy like developing fire plans, fire maps, capacity building through training, pilot demonstration, (b) fire suppression mechanism, and (c) and if necessary post-fire rehabilitation and management (Chetan Kumar 2002).

LEGAL PROVISIONS

The Indian Forest Act 1927 has two types of provisions, viz

(i) for those who are bound to extinguish fire but do not discharge their duty in putting out that fire and

(ii) for those who set fire to or kindle or carry fire in reserved and protected forests in certain seasons.

(i) According to section 79 (1), a person who exercises any right in a reserved or protected forest, or is permitted to take any forest-produce from, or to cut and remove timber or to pasture cattle in, every person in any village contiguous to forest who is employed by the government, or who receives emoluments from the government for services to be performed to the community, is bound to take steps, whether required by any forest officer or police officer or not, to extinguish any forest fire in such forest of which he has information. If such a person does not discharge his obligation and does not take steps to extinguish any forest fire in reserved or protected forest he shall, according to section 79 (2), be punishable with imprisonment for a term which may extend to one month or with fine which may extend to two hundred rupees or with both. In order to get a man convicted under this section, it is necessary to prove that he belongs to one of the categories of persons mentioned under section 79 (1) and that he has not done one of the things he was required to do by that section.

(ii)(a) According to section 26(1)(b), any persons who sets fire to a reserved forest or in contravention of any rules, made by the State Government in this behalf, kindles fire or leaves any fire burning in such manner as to endanger such a forest is punishable. This section provides for setting fire to a reserved forest from outside and for kindling fire and leaving it burning outside the reserved forest against rules made by State Government, so as to endanger such forest.

(ii) (b) According to section 26 (1) (c), any person who, in a reserved forest kindles, keeps or carries any fire except at such seasons as the forest officer may notify in this behalf shall be punishable. This section makes kindling, keeping or carrying of fire inside the reserved forest an offence if it had been done against the order of the forest officer. In such cases it has to be proved that the Divisional Forest Officer had prohibited these acts and had notified the season for information of general public.

The persons convicted under section 26(1) (b) as well as section 26(1) (c) of I.F.A. are punishable with imprisonment for a term which may extend to six months or with fine which may extend to five hundred rupees or both in addition to such composition for damage done to the forest as the convicting court may direct to be paid. The M.P. Government has, by an amendment, made the maximum term of imprisonment as one year and the maximum amount of fine as one thousand rupees.

(ii) (c) According to section 33 (1) (d) any person who sets fire to a protected forest or kindles a fire without taking all reasonable precaution to prevent its spreading to any tree reserved under section 30, whether standing, fallen or felled or any closed portion of such forest shall be punishable with imprisonment which may extend to six months or with fine which may extend to five hundred rupees or with both.

FIRE TERMINOLOGIES

(A) *HAZARD*: It is a measure of that part of the fire danger which is dependent on the materials available for burning and its relative amount, class, character, moisture content and condition. Hazard is also concerned with inflammable material constituting a threat of special suppression difficulty or of likelihood of ignition because of location. In forest protection, the following terms have each the limited meaning given against it:

- *Hazard*: It concerns inflammable material only.
- *Inflammability*: It is the susceptibility of the material to ignition.
- *Risk*: It refers only to the agencies that cause fire.

- *Danger*: It is the sum of them all plus damage and other factors (BCFT Modified).

(B) *EARLY BURNING*: Controlled burning early in the dry season, before the leaves and undergrowth are completely dry or before the leaves are shed, as an insurance against later fire damages (BCFT).

- *Controlled Burning*: Any deliberate use of fire where by burning is restricted to a predetermined area and intensity (BCFT).
- *Slash*: The unusable residue after logging, viz. branches, tops, bark, unutilizable logs, uprooted stumps, broken or uprooted trees left on the area. Also any large accumulation of debris after wind or fire (BCFT Modified).
- *Fire break*: An existing barrier, natural or otherwise, or one prepared before a fire occurs from which all or most of the inflammable materials have been removed, designed to stop light ground or surface fires and to serve as a line from which to work and counterfire, if necessary; also to facilitate the movement of men and equipment in fire fighting (BCFT).

16 PROBLEM SOILS AND ITS MANAGEMENT

INTRODUCTION

Soil fertility is defined as the capacity or ability of the soil to supply the plant nutrients required by the plants in available and balanced forms. It is the capacity of the soil to produce crops of economic value to human and maintain the health of the soil for future use. The soil is said to be fertile when it contains all the required nutrients in the right proportion for luxuriant plant growth. This principle is explained by the "Law of the Minimum" proposed by Justus von Liebig in the mid of 19th Century. The law states that if one of the nutritive elements is deficient or lacking, plant growth will be poor even when all the other elements are abundant. Any deficiency of a nutrient, no matter how small an amount is needed, will hold back plant development. If the deficient element is supplied, growth will be increased up to the point where the supply of that element is no longer the limiting factor. Increasing the supply beyond this point is not helpful, as some other element would then be in a minimum supply and become the limiting factor. Problem soils normally hinder the availability of certain nutrients or have certain elements in excess concentration which pose toxicity to the plants. This results in less or even no plant growth in the problem soils. Once problem soils are reclaimed successfully, the vegetation growth appears again.

Soil fertility management (SFM) is the basis for achieving sustainability of plant growth and crop production system. Many different practices influence soil fertility viz. application of organic and mineral fertilizer, green manuring, mulching, burning, soil tillage, soil and water conservation, crop rotation, animal and tree integration. Judicious application of combination of these factors will ensure the soil fertility on sustainable manner. Similarly application of appropriate measures to reverse the problem soils back to the normal condition will result in green development and soil fertility improvement. The problem soils are categorized into three classes: salt affected soils, acidic soils and waterlogged soils.

SALT AFFECTED/SALINE SOILS

Soils of arid and semi-arid regions contain high concentration of either soluble salts or exchangeable Na or both due to poor leaching of base forming cations. The soil pH is generally >7 and soil colloidal complex dominated by Ca^{++}, Mg^{++}, K^{+} and Na^{+} ions. On the basis of soil pH, SAR, ESP and EC, salt affected soils are classified into four groups: saline soil, non-saline alkali soil, saline-alkali soil, degraded alkali soil

Table 16.1: Characteristics of Salt Affected Soils

Particulars	Saline Soil	Non-Saline Alkali Soil	Saline-Alkali Soil	Degraded Alkali Soil
Other Names	White Alkali, Solanchak, Szek and in India, their local names are Reh, Kari, Kallar, Khar, Pokkati, Shora, Soulu, Loni, Thur (Punjab), Brown alkali soil	Black alkali, Solonetz, Szik and in India, their local names are Black alkali, Typical usar, Alkali soil, Sodic soil, Chaupansoil (Bombay), Chaundusoil, Kshar soil, Rakkar, Bara-bari soil (Punjab)	Usar soil	Solod soil, Soloth soil
Soil pH	< 8.5	8.5-10	> 8.5	6 (surface soil), 8.5 (sub surface)
SAR	< 13	> 13	> 13	<13 (surface), >13 (sub surface)
ESP	< 15	> 15	> 15	> 15
EC of Saturated Extract (dSm^{-1})	> 4	< 4	> 4	< 4
Soluble Salts (%)	> 1%	< 1%	> 1%	< 1%
Dominant Salts	SO_4^-, Cl^-, NO_3^- of Na, Ca, Mg and K $CO3^-$ and HCO_3^- are also found in small amount	Na_2CO_3	–	Na_2CO_3 in lower layer
Content in Soil	High soluble salts of Na	High exchangeable Na on the soil colloidal complex	Na clay and soluble salts	H^+ in upper layer and Na^+ in lower layer
Soil Colour	White due to SO_4^- and Cl^- of Na, brown due to $NaNO_3$	Black colour due to Na_2CO_3 content	-	Black colour in lower layer due to Na_2CO_3
OM Content	Slightly less than normal soil	Very less amount due to dispersion of OM	Low	Low
Climate	Arid and semi-arid regions, Rainfall < 550 mm annually	Semi-arid & Sub humid region , Rainfall > 550 mm annually	Semi-arid & Sub humid region	High rainfall areas
Formation Process	Salinization	Alkalinization	Salinization, Alkalinization	Solodification or solodization
Physical Condition of Soil	1. Flocculated or granulated condition 2. Good permeable to air and water 3. Soil structure is optimum	1. Deflocculated or un-granulated condition 2. Poor permeable to air and water 3. Soil structure is very poor	Flocculated and Deflocculated condition	1. Low rate of infiltration and permeability 2. It develops columnar or prismatic structure 3. Compacted soil

Recommended Fertilizer	$(NH_4)_2SO_4$, CAN, SSP, TSP	$(NH_4)_2SO_4$, NH_4NO_3, DAP	NH_4NO_3, DAP, TSP	Basic fertilizer i.e. $Ca(NO_3)_2$, $NaNO_3$
Harmful Effect	1. Available moisture is less 2. Poor seed germination due to poor aeration 3. Soils are generally barren 4. Wilting coefficient is very high 5. Soil mass is structureless and dusty 6. Conc. of K is low 7. Conc. of B, F, NO_3 are high which affect the plant growth 8. Toxic effect of CO_3, Cl, SO_4 of Na 9. Root injury 10. Leaf tip burnt / chlorosis	1. Dispersion of soil particles 2. Poor soil physical condition 3. Na generated Ca and Mg deficiency 4. N, Zn, Fe, Cu are low 5. P, K, B are high 6. Toxicity due to high pH, CO_3 or HCO_3 of Na, and B 7. Crust formation during dry 8. Sticky during wet	1. Poor internal drainage due to hard clay pan formed by dispersed Na-clay 2. Poor plant growth	Very low P and due to leaching of Phosphorous

OCCURRENCE OF SALT AFFECTED SOILS

Salt affected soils are grouped under Aridisols order according to 7th Approximation. Some soil scientists, such as Thorp and Smith, group salt affected soils under Intrazonal order or Halomorphic sub-order (Genetic system of soil classification). It is reported that 957 M ha area of the land in world are categorized under salt affected. In India, the available literatures report the area of salt affected land in the range of 7-13 M ha.

Sodic soil (3.74 M ha): Haryana, Gujarat, Maharashtra, Coastal areas

Saline soils (3.0 M ha): Rajasthan, Punjab, U.P., Bihar

FORMATION OF SALINE AND ALKALI SOILS

SALINIZATION: Accumulation of soluble salts in soil results in the formation of saline soil . The sources of soluble salts are: arid and semi arid regions where low rainfall and high evaporation occurs; Irrigation by more salty water upto long period; poor water drainage or restricted water drainage; very high water table; presence of hard layer in soil profile; nature of parent material (basic and sub-basic rocks); flooding by sea water; ploughing in same depth upto long period; slow rate of permeability of soil; basin shaped topography; more use of basic fertilizers up to long period.

ALKALINIZATION: The accumulation of Na^+ ions on soil colloidal complex results in the formation of alkali or sodic soil. Na^+ replaces adsorbed exchangeable cations i.e. Ca^{++}, Mg^{++}, and K^+ from soil colloidal complex by base exchange process and then Na^+ adsorb in place of replacing cations, resulting Na^+ increases on soil colloidal complex.

DESALINIZATION: Soluble salts goes down to lower horizons by leaching from the upper soil surface but adsorbed Na^+ remain on soil colloidal complex.

DEGRADATION OR DEALKALIZATION/ SOLODIZATION

In this stage, the exchangeable Na^+ ions are hydrolyzed and form NaOH and H^+. NaOH reacts with CO_2 present in soil and form Na_2CO_3. Na_2CO_3 is more soluble in water and goes down in lower horizons by leaching. H^+ replaces adsorbed Na^+ from soil colloidal complex, resulting H^+ increases on soil colloidal complex and finally surface soil pH is about 6 and result in the formation of degraded alkali soil.

REGRADATION: The soil in which the process of degradation has occurred and bases are again re-accumulated. The amount of Na salts increases in the surface soil and the process is known as regradation.

EFFECT OF SALINITY AND ALKALINITY ON SOIL

- Poor soil structure
- Poor aeration
- Slow permeability rate
- Slow bacterial activity
- Pudding of soil
- Difficulty in agricultural works
- Nutrient imbalance
- Poor physical condition of soil
- Deficiency of Fe, Mn, Zn, Cu
- Oxygen deficiency

EFFECT OF SALINITY AND ALKALINITY ON PLANTS

- Hindrance in water absorption
- Poor uptake of Ca^{++}, Mg^{++} by plants
- Toxic effects of individual ions

Table 16.2: Properties of Saline and Sodic Soils

Particulars	Saline Soil	Sodic Soil
Surface soil under dry condition	Friable soil surface	Hard surface
Sub surface hard pan	Sometimes thin hard pan	Generally thick hard pan
Collected water in surface	Clean	Muddy
Collected water in sub-surface	Bad	Good
Absorption of rain water or irrigation water	Soon	Waterlogged on surface soil due to presence of hard pan

PHYSICAL MEASURES OR MECHANICAL MEASURES FOR RECLAMATION

Scrapping: It includes scrapping of surface soil to 15-30 cm depth suitable for patchy field

Leaching: leaching of soluble salts below the plant root zone, Leaching process is more useful in salt affected soil where water table is down. Summer season is more suitable for leaching.

Flushing: Removal of soluble salts from soil surface by runoff from upper portion to lower portion is called flushing. This process is more useful for salt affected areas where water table is high and soils having gentle slope.

Trenching: Using 50 cm x 50 cm x 50 cm trenches with 2m distance found to be effective. This is suitable in places where sub-soil is good.

Washing: Washing out of soluble salts from surface layer. This is more suitable where water is cheap and available in excess amount

Drainage: Facilitating proper water drainage in the salt affected areas.

AGRONOMICAL PRACTICES OR CULTURAL PRACTICES FOR RECLAMATION

It includes deep ploughing with the help of soil turning plough i.e. victory plough to open hard layer, growing of salt tolerant crops, suitable crop rotation, increased plant population / unit area, sowing of seeds in furrow, use of acidic nature fertilizers, salt free water for irrigation purpose, using light irrigation than heavy irrigation, sprinkler irrigation and application of organic residue mulch @ 4 tons per ha on surface soil to reduce the evaporation rate.

BIOLOGICAL MEASURES FOR RECLAMATION

It includes application of FYM @ 50 tons per ha (0.5% N + 0.25% P_2O_5 + 0.5% K_2O), growing of green manuring crops i.e. Dhaincha (*Sesbania acculeata*), use of molasses i.e. sira, pressmud, use of 2.5 tons per ha of dry powder of Satyanasi weed (*Argemone maxicana*), use of Usartor Masala which is an organic amendment.

CHEMICAL RECLAMATION

GYPSUM: Main aim of reclamation of sodic soil is to replace exchangeable Na with Ca. Of all Ca compounds, gypsum is considered best and cheapest for reclamation of sodic soil because it has high soluble Ca salt and easily available in the market. The Chemical formula for Gypsum is $CaSO_4.2H_2O$. It has the composition of 23.2 per cent Ca, 18.6-22.5 per cent S and 20.9 per cent water. The amount of gypsum required to be added in a sodic soil to lower the ESP to a desired value is known as gypsum requirement (GR). It is expressed as me of Ca /100 g soil. GR is governed by degree of sodicity, crops to be grown, soil texture, drainage condition and quality of irrigation water.

$$GR = \frac{CEC \times [ESP\ (initial) - ESP\ (final)]}{100}$$

where,

ESP (initial)= ESP value before application of gypsum

ESP (final) = It is kept at 10

CEC = Cation exchange capacity as Cmol(p^+)/kg soil

GYPSUM APPLICATION METHOD: Agricultural grade gypsum (70% pure) should be powdered and sieved (sieving with 2 mm size sieve). It is applied and mixed thoroughly in 10-12cm soil depth with shallow tillage followed by planking. About 25 tons per ha of FYM should be applied. The field should be irrigated and kept ponded for 15 days with 6-8 cm standing water.

For planting tree seedlings, the pit should be dug up upto 1 m depth and excavated soil should be discarded. Fresh soil mixed with gypsum should be filled into the pit and FYM should be added. Finally salt resistant tree seedlings should be planted in the pit.

PYRITE: The chemical formula for pyrite is FeS_2 (Iron sulfide). It has the composition of 20-22 per cent Fe, 22-24 per cent S, 35-40 per cent Si, 6-8 per cent Al, 2-3 per cent C and other elements such as MgO, CaO, Zn, Cu and Mn. Pyrite requirement (PR) for reclamation of sodic soil is calculated using the formula, PR=GR x 0.727 (GR is the Gypsum Requirement).

Other methods include application of sulfur, application of H_2SO_4 (suitable where alkalinity is very high but costly and risky method) and application of ferrous sulfate.

ACIDIC SOILS

Soils in which acidity of adsorbed H^+ ions on the colloidal complex is more than bases and pH is < 7 are called acid soils. In acid soils, most of the plant nutrients are unavailable and resulting plant growth is affected. Area of acid soils in India is about 49 M ha, out of which 26 M ha of land having the pH < 5.6 and 23 M ha of land having pH ranging from 5.6 - 6.5.

Acid soils occur generally in all major soil groups except black cotton soil. In India , acid soils are found in North-Eastern and peninsular regions (including Bihar plateau and sub-plateau of West Bengal and major part of the Odisha) and Himalayan regions (comprising Jammu and Kashmir). There are large area of acidic soils in Assam, Manipur and Tripura.

CAUSES OF FORMATION OF ACID SOILS OR DEVELOPMENT OF ACID SOILS

LEACHING OF ADSORBED SOLUBLE BASES FROM SOIL COLLOIDAL COMPLEX: In heavy rainfall areas, adsorbed soluble bases i.e. Ca^{++}, Mg^{++}, Na^+, and K^+ on soil colloidal complex becoming soluble and goes downward in lower horizon by leaching and H^+ ions adsorbed in place of replacing bases, as a result, soils become acidic. The regions having annual rainfall 100 cm or more have high percentage of acid soils.

ACIDIC PARENT MATERIAL: Soil acidity is produced by following rocks:

Acidic igneous rocks (65-75% silica): Granite, Rhiolite, pitchstone, obsidian rocks

Sub-acidic igneous rocks (60-65% silica): Syenite, Trachyte rocks

ACID FORMINING CHEMICAL FERTILIZERS: Soil acidity is produced by fertilizers. Nitrogen source fertilizers through nitrification reaction produce acid residues (H^+ ions) in soil, as a result soil becomes acidic. E.g. Anhydrous NH_3, Urea, NH_4NO_3, $(NH_4)_2SO_4$

ORGANIC MATTER: CO_2 is formed by the decomposition of organic matter, then CO_2 combining with water and forms carbonic acid (H_2CO_3). H^+ ions are formed after decomposition of H_2CO_3 by replacing adsorbed cations on soil colloidal complex which goes downward in lower horizons by leaching and as a result soils become acidic.

CO_2: If the soil is containing high concentration of CO_2, the pH of such soil is low i.e. soils become acidic.

MICROORGANISMS: The decomposition of organic matter, nitrification etc. in soil is caused by microorganisms. As a result, organic and inorganic acids are produced in the soil causing the acidity in soil.

CHARACTERISTICS OF ACID SOILS

Acid soils are low in soil pH and high in exchangeable hydrogen and aluminum ions. These soils have low CEC and high base unsaturation. They are low in available P and have high phosphate fixation capacity. The status of available micronutrients except Mo is generally enough in these soils. Soil acidity checks biological N-fixation in the soil. The population of bacteria and actinomycetes are lower and fungi are higher in acidic soil than normal soil. Soil acidity changes number and activity of microorganisms. Nutrient imbalance and deficiency of Ca and Mg are prevalent in these soils.

CLASSIFICATION OF ACID SOILS

Strongly acidic soils : pH < 4

Moderately acidic soils : pH 4 - 6

Slightly acidic soils : pH 6 - 7

ACID SOILS IN INDIA

- Laterite soil
- Lateritic soil
- Peat soil
- Muck soil
- Foot hill soil
- Podzolic brown soil
- Mixed red, black and yellow soil

EFFECT OF SOIL ACIDITY OR PROBLEMS OF SOIL ACIDITY

Soil acidity increases the concentration of Fe and Al in soil solution resulting toxicity to plants. Availability of micronutrients i.e. Fe, Mn, Zn, Cu, Cl, B increases in acidic soils resulting in poor plant growth. Availability of Ca and Mg reduces in soil. P fixation occurs in acid soils resulting deficiency of available P which affect the roots growth. Beneficial activities of soil microorganisms are adversely affected. Soil acidity has bad effect on plant phloem and activity of enzymes related to root and stem. H^+ ions exert toxic effect of on root tissues. Various diseases are reported to occur in plants due to soil acidity.

RECLAMATION AND MANAGEMENT OF ACID SOILS

The methods for reclamation and management of acidic soils include addition of liming material, growing of acid tolerant crops, use of basic fertilizers, soil and water management

LIMING MATERIALS

Various liming materials are added to the soil in relation to the acidity of the soil to neutralize the pH.

Table 16.3: Liming Materials

S No	Liming materials	Chemical Composition	Remark
1.	Calcic lime stone or Calcite or calcitic limestone	$CaCO_3$ (36 % Ca)	It is ground limestone, fine power, slow solubility and slow reacting
2.	Quick lime, Un-slaked lime, caustic lime, lump lime or burnt lime	CaO 70% Ca and purity between 85-98%)	It is burnt lime stone with white power and fast reacting
3.	Hydrated lime, slaked lime or builder lime	$Ca(OH)_2$ (50% Ca)	Coming from quick lime which has changed to a hydroxide form after water reaction. More causting and fast reacting
4.	Dolomite or Dolomite limestone	$CaCO_3.MgCO_3$ or Ca Mg $(CO_3)_2$ (Ca =20%, Mg =13.5%)	It is ground limestone and contains high Mg. Fine powder and slow reacting than Calcic limestone. Dolomite is calcium magnesium carbonate.
5.	Blast furnage slage	$CaSiO_3$ or Ca_2SiO_4	It is by product of Iron industry. It contains 1-7% P and 31% Silica
6.	Chalk	$CaCO_3$	Resulting from soft limestone
7.	Marl or vog lime	$CaCo_3$	Soft and unconsolidated $CaCO_3$ generally mixed with clay is called Marl /or Bottom of small fresh water ponds in areas where soils are high in lime is called marl.

EFFECTIVE CALCIUM CARBONATE (ECC) *OR NEUTRALIZING INDEX* (NI): It is defined as the product purity of $CaCO_3$ equivalent and fineness factor. It is expressed as percentage.

CALCIUM CARBONATE EQUIVALENT (CCE) *OR NEUTRALIZING VALUE* (NV) *OR TOTAL NEUTRALIZING POWER* (TNP): CCE is defined as the acid neutralizing capacity of the liming material and expressed as the weight percentage of $CaCO_3$.

$$CCE = \frac{\text{Molecular Weight of Pure } CaCO_3}{\text{Molecular Weight of a Liming Material Whose CCE is to be determined}} \times 100$$

CCE of different liming materials are $CaCO_3$ (100%), CaO (179%), $Ca(OH)_2$ (135%), $CaCO_3.MgCO_3$ (108.7%), $CaSiO_3$ (86%).

LIME REQUIREMENT: To a chemist, lime is a CaO but to a farmer, agronomist and soil scientist, lime means $CaCO_3$ or $CaCO_3$ equivalent. Amount of liming material required to be added to an acid soil to increase the soil pH to a desired value is known as lime requirement. It depends on pH of soil, soil texture, organic matter, type of clay minerals, buffering capacity (BC)

pH of Soil: More lime is needed for strongly acidic soil than slightly acidic soil.

Soil Texture: More lime is needed for heavy textured soil than light textured soil.

Organic Matter: More lime is needed for organic soil due to more CEC and buffering capacity than mineral soil.

Type of Clay Minerals: More lime is needed for Montmorillonite type of clay mineral due to high CEC and buffering capacity than Kaolinite and Illite clay mineral present in soil.

Buffering capacity (BC): Higher BC of soil, the greater amount of lime applied to attain satisfactory change in soil pH.

LIME APPLICATION: The most efficient way to use lime is to apply small amounts in each year or alternate year. Generally, it is applied one or two months before the sowing of crop or planting of tree seedlings. Lime should be powdered (sieved with mesh sieve) and mixed thoroughly in 10 cm soil depth with shallow tillage followed by planking.

EFFECT OF LIME ON SOIL: The physical effects include improvement of soil structure, reduction in bulk density, increase in infiltration and permeability rate, indirect help in soil erosion control. The chemical effects include increase in soil pH, decrease in H^+ ions concentration and increase in OH^- concentration, reduction in toxic effect of heavy metal of Fe, Al, Mn, supply of Ca and Mg for plant growth, increasing the availability of N by the decomposition of OM, making P and K more available. Its biological effects include fast decomposition of OM and increasing the activity of aerobic microorganisms.

USE OF BASIC FERTILIZERS

Use of basic fertilizers such as Sodium nitrate ($NaNO_3$), Calcium nitrate [$Ca(NO_3)_2$], Calcium cynamide ($CaCN_2$), CAN, Basic slage [$(CaO)_5.P_2O_5.SiO_2$], and Rock phosphate [$Ca_3(PO_4)_2]_3.CaCO_3$ will help to neutralize the soil acidity.

WATERLOGGED SOILS

Waterlogged or wet land can be defined as the soils whose development and properties are strongly influenced by temporary or permanent saturation in the upper part of the soils. The soil is intermittently saturated with water and oxidation of OM is slow and it accumulates in the 'A' horizon. In the second horizon, if the diffusion of O_2 into the soil is low, Fe and Mn are deposited as rusty mottles. If the diffusion is rapid, they are deposited as concretions.

Certain areas in high rainfall zones where drainage is poor become waterlogged, generally in depressions and along rivers and streams. In India, nearly 6 M ha area of land is under waterlogged condition.

TYPES OF WATERLOGGING

Waterlogging is of following types:

Riverine Flood Waterlogging: In the rainy season, flood may come to the nearby land from the river having excess flood water.

Oceanic Flood Waterlogging: Ocean water spreads in the nearby land causing waterlogging.

Seasonal Waterlogging: Runoff water accumulates in the low lands in the rainy seasons.

Perennial Waterlogging: Deep water, swamp etc. get rain water, runoff water and seepage water from canal causing perennial waterlogging.

Sub-Soil Waterlogging: High water table in the rainy season causes sub-soil waterlogging which is normally unsuitable for root growth.

FACTORS AFFECTING FORMATION OF WATERLOGGED SOILS

Climatologic Factor: Due to high rainfall, water accumulates on the soil surface.

Flood: Flood water is generally causing waterlogged situation in the field.

Seepage from Canal: Ground water levels are closer to the surface due to seepage from canal.

Land Shape: Saucer shaped land gets water from highland which results in waterlogging.

Uncontrolled and Unwanted Irrigation: Excess irrigation may cause accumulation of water on the soil surface.

Drainage: Poor drainage system is responsible for waterlogging in the field.

CHARACTERIZATION OF WATERLOGGED SOILS

The following are the classes of depth to the wet state:

Class 1: Not wet above a depth of 400 cm

Class 2: Not wet above a depth of 100 cm

Class 3: Not wet above a depth of 50 cm

Class 4: Not wet above a depth of 25 cm

EFFECT OF WATERLOGGING

Waterlogged soils have poor aeration and poor structure. It makes compact soil and lower down soil temperature. pH tends to increase in a acidic soil and decrease in sodic soil when they are waterlogged. Many trees and crops cannot survive in waterlogging except few species. Nitrogen deficiency is extremely common in waterlogged soils. Due to low temperature and reduced condition, mineralization of organic N is affected. The inorganic form of P is generally present at higher levels in flooded soils than in upland soils. Soil flooding decreases P absorption by 28-70 per cent. Waterlogged soil may increase the soil solution concentration of K because of displacement of exchangeable K by the large amount of Fe^{++} and Mn^{++} in the soil solution.

Sulfur deficiency has been reported from many low land areas. The reduction of sulfates in flooded soils has three complications, the S supply may become insufficient, Zn and Cu may be motionless and H_2S toxicity may arise particularly in soils low in Fe. Widespread deficiency of Zn in wetland conditions is reported. Fe^{++} and Mn^{++} are available in excess causing toxicity to the plants.

MANAGEMENT OF WATERLOGGED SOILS

Leveling of Land: Leveling of land in many wetlands removes water by runoff.

Water Drainage: Drainage removes excess water from root zone that is harmful for plant growth. Land can be drained by surface drainage, sub-surface drainage and drainage well method.

Controlling Seepage: Controlling the seepage from the canals and irrigation channels will improve the land.

Flood Control Measures: Construction of bund may check water flow from the rivers to the cultivated lands.

Plantation of Trees having High Transpiration Rate: Transpiration rate in certain trees like *Eucalyptus* spp., *Acacia* spp., *Zizyphus* spp. and *Salix* spp. are very high. In transpiration process, the underground water is consumed by trees, thus lowering the ground water table.

Selection of Crops and their Proper Varieties: Waterlogging tolerant crops and their varieties should be introduced

Sowing and Planting on the Bunds or Ridges: In the waterlogged area, sowing and planting of tree seedlings should be done on bunds or ridges. In this method, there is scope of good aeration near the root zone of the seedlings.

Nutrient Management: Low N fertility is an important constraint in the waterlogged area. The predominant form of inorganic N in low land soils is NH_4^+. The following concepts may increase N use efficiency viz. deep placement of fertilizers, use of slow and controlled release fertilizers, use of nitrification inhibitors (e.g. *Nitrosomonas* bacteria), use of urease inhibitors (e.g. Phenyl phosphorodiamidate)

Table 16.4: Tree Species Suitable for various Problematic Soils

Salt Affected Soils	Acidic Soils	Waterlogged Soils	Marshy Areas
Acacia auriculiformis	*Acacia nilotica*	*Acacia nilotica*	*Eucalyptus robusta*
Acacia nilotica	*Albizia procera*	*Albizia procera*	*E. saligna*
Albizia procera	*Albizza lebbeck*	*Anthocephalus indicus*	*E. rudis*
Casuarina equisetifolia	*Terminahia arjuna*	*Butea monosperma*	*Baringtonia spp.*
Emblica officinalis	*Prosopis julifora*	*Dalbergia sissoo*	
Pongamia pinnata	*Tectona grandis*	*Eucalyptus pinnata*	**Coastal Areas**
Prosopis cineraria	*Eucalyptus tereticornis*	*Salix tetrasperma*	*Pongamia glabra*

Zizyphus nummularia	***Eucalyptus camaldulensis***	***Avecenia officinalis***
Albizia lebbeck	***Syzygium cumini***	***Eucalyptus tereticornis***
Cassia fistula	***Terminalia arjuna***	
Prosopis juliflora		
Butea monosperma		
Melia azadirach		
Salvadora oleoides		
Acacia farnesiana		
Azadirachta indica		
Eucalyptus hybrid		
Parkinsonia acculeata		
Terminalia arjuna		

PART 4:

FORESTRY CONCEPTS AND TECHNOLOGIES

17 TREE SEED TECHNOLOGY

INTRODUCTION

Seeds constitute a unique genetic composition resulting from the mixing up of parental genetic material. They produce offspring of genetic variation, which in turn enhances ecological adaptability. Seeds are produced in large numbers and are readily available in each year or at longer intervals. They are much more resistant to damage and environmental stress than vegetative propagules. In essence, seed is a miniature tree because it is responsible for its regeneration and ultimately for its reproductive success. Following fertilization, growth sets in various parts of the ovule resulting into a seed; the zygote develops into the embryo, the primary endosperm nucleus gives rise to endosperm and the integuments form the protective seed coat, which is the seed's primary defense against adverse environmental conditions. Seed quality is the first and foremost factor, which decides the success and viability of a tree plantation in an afforestation programme or genetic improvement programme. Management of seed quality starts from the selection of plus trees and continues through seed collection, extraction, processing, treating and storage. Each and every step has its own significance and each process should be done with a great care because they decide the basic quality of the seed while rest of the practices only helps to improve the quality attributes.

Collection of seeds in large quantities poses several problems as a result of seed production behaviour of the species. Some species bear fruit or seed in good quantities almost every year, others do so in intervals of one to several years. The maturation pattern of fruits presents enormous problems. Simultaneous type of fruit which mature at the same time, are easy to harvest. Intermittent type matures over a long interval making harvesting operations difficult and expensive. Important aspects of trees seed technology are presented in this chapter on seed collection, extraction, processing, dormancy, presowing treatments, storage and testing of seeds to guide the foresters and seed managers to obtain seeds of high quality.

SEED COLLECTION

SELECTION OF TREE SEEDS

Basic material is the plant material from which the forest reproductive material (FRM) is derived and includes seed stands, seed orchards, parent material held by tree breeders in archives, individual clones and mixtures of clones. Seedlings inherit the qualities of their parents, therefore seeds should be collected from trees with characteristics that are preferred by the eventual tree planters. Useful guidelines for selection of suitable trees for seed collection (Mapier and Robbins 1989) are listed below:

- All seed trees should be healthy, vigourous and should have good form.
- Trees intended for fodder should have abundant foliage and a bushy crown.
- Trees intended for fuel wood can have plenty of branches.
- Trees intended for timber should be straight and without too many thick branches.
- If a species has multipurpose or more than one use, it will be necessary to compromise on these characters.
- Collect seeds from middle aged trees to matured trees, over matured trees should be avoided, since seeds from them may be of low viability.
- Seeds should be collected from dominant and co-dominant trees of the stand.
- Avoid collecting seeds from isolated trees.
- Tree stands containing numerous poorly formed individuals should not be considered for seed collection. Such trees are generally abnormal or diseased.

PROVENANCE

Provenance is defined as the geographic source or location to which plants are native and within which their genetic characteristics have been developed through natural selection. The term provenance is used to describe the location of the source from which the reproductive material is collected. Seeds from unknown origin are risky and those from inferior origin are disastrous. Thus, use of seeds from known and genetically superior origin is desirable since while raising plantation, millions of inputs and even more millions of outputs are involved. Due to the considerable variation in the characteristics both between seed lots from different sources and within seed lots from the same source, many pretreatments are required for different tree species. It is better to use locally grown seed where possible, as this is usually cheaper, and trees growing in a particular locality tend to be better adapted to local conditions than those from distant areas, often with very different conditions of climate and soil (Jackson 1987).

Locally collected seed is performing better than those from distant areas in broad-leaved species like *Eucalyptus* spp., *Grevillea robusta, Casuarina* spp. etc. (Thapliyal and Rawat 1982). Teak drupes collected from moist tropical forest of Top Slip in Tamil Nadu showed superiority and gave maximum germination and seedling vigour than the drupes from far areas like Karnataka, Andhra Pradesh and Maharashtra (Masilamani et al 1999a). Dharmalingam and Masilamani (1997) reported that teak drupes from two provenances viz. Top Slip in Tamil Nadu, India and Lampang provenance of Thailand differed in size and weight within and between seed lots.

SEED SOURCE

Seed source covers all material from a single tree or collection of trees within a region of provenance or seed zone. In fact, different seed sources differ in germination and seedling vigour, they may require different type of pretreatments to enhance or activate the germination. Seed source variations are reported in several tropical tree species viz. Sivasamy (1991) in *Azadirachta indica, Ailanthus excelsa* and *Pongamia pinnata*; Masilamani (1992) in *Cassia siamea, Hardwickia binata* and *Prosopis juliflora*; Bahuguna et al (1989) in *Albizia falcataria*; Jambulingam (1990) in *Casuarina equisetifolia* and Masilamnai et al (1999a) in *Tectona grandis.* Location effects on seed quality of tree species are well studied and clearly indicate the influence of seed source on physical and physiological quality of seeds.

AGE OF MOTHER TREE

Seed production depends mainly on climatic conditions, but is also affected by the age of the tree and manner of growth. Few trees seed after two to three years, others may take 20-30 years. Seeds are borne by trees that have sufficient maturity and it is believed that, in most species, there is an optimum age of seed bearing. Seeds from trees just beginning to bear are often of poor quality. Tree scientists therefore should know the optimum age of the tree for seed collection and effective seedling production. Swaminathan et al (1994) reported that in *Ceiba pentandra,* seeds collected from 7 year old stand showed superior physical and physiological parameters and gave maximum germination and seedling vigour compared to seeds collected from 4, 5 and 6 year old stand. Similarly, an increase in germination percentage, seedling dry matter and seedling vigour of neem seeds from 20 year old trees compared to neem seeds from lower ages (5, 10 and 15 years) and higher ages (25 and 30 years) was reported by Masilamani et al (2000).

Suresh (1994) reported maximum germination (90 per cent) in the seeds of *Acacia nilotica* collected from 9 and 11 year old trees, which was drastically reduced viz. 56 and 34 per cent in 7 and 3 year old trees. Prasad and Kandya (1992) recommended that the seeds should be collected from mature or nearly mature trees. Over mature or immature trees should be avoided, since seeds from them may be of low viability. Various research results suggest that tree age may significantly influence seed quality characteristics.

SEED MATURATION

The maturation of a seed is a positive growth process, which includes the following physical changes: increase in size and weight; accumulation of dry matter; development of essential structures of the embryonic axis; an increase in viability and vigour and finally, a loss of moisture.

Changes that occur in seeds as they mature are summarized below,

1. Moisture content decreases rather uniformly from 70-80 to 15-20 per cent.
2. Seed size increases to a maximum then decreases some what as the seed dry.
3. Dry weight increases to a maximum then may decrease slightly.
4. A few seeds become capable of germinating within a few days after fertilization, maximum germination is reached at a somewhat later date.
5. Some seeds may become dormant as they ripen, and require special treatments to obtain maximum germination per cent.
6. Seedling vigour increases as seed dry weight increases and reaches a maximum at the time of maximum dry weight.

Maturity is defined as the point of maximum dry weight of seed. This point is reached when the seeds are relatively high in moisture content (35 to 45 per cent). The sooner the seeds are harvested after maturity, the higher the quality. Maturity stage is differentiated into physiological maturity and harvestable maturity.

PHYSIOLOGICAL MATURITY: This is the stage at which the, maximum dry matter accumulates, where the moisture content of the seed will be 26 to 50 per cent. If the seed is devoid of dormancy, it registers highest seed germinability and vigour index at that stage. The seeds have to be effectively sandy dried to safe moisture contents.

HARVESTABLE MATURITY: This is the stage at which the seed first time dries to a moisture level of 14 per cent or less, which allow commercial harvesting and the seeds can be stored without supplemental drying. Characteristics features are,

- Loss of the fruit chlorophyll content and photosynthetic activity as they ripen. Loss or gain of a distinctive colour.
- Softening in fleshy fruits or achievement of brittleness as in many conifer seeds.
- A change in flavour or odour.
- In most seeds, complex carbohydrates, fats, oils and proteins usually accumulate and lipids form the major food reserve in many tree species. In general, nitrogenous constituents and organic acids are found in increased quantity.
- Respiration reaches a peak as maturation is attained and then declines as senescence begins.
- Developing seeds of both angiosperms and gymnosperms are rich source of growth regulating substances and changes in the levels of these compounds have also been associated with ripening. Normal levels of growth regulating substances are mostly associated with meristamatic activity and the highest levels are found in immature seeds.

The planting value of seed and its storability is directly related to the level of maturation of the seed at the time of collection. Seed collection in bulk quantities in the field can only be economical and feasible, if exact stage and time of seed maturation are known. Seed collectors must be aware of the physical (length, breadth and thickness of pod, number of seeds per pod, moisture content of pod and seed, specific gravity of pod and seed and seed colour), physiological (germination, seedling length, drymatter accumulation and seedling vigour) and bio-chemical indices (crude fat, reducing and non reducing sugars, starch, soluble nitrogen, protein nitrogen and isoenzymes) of the developing fruit and seed. Practical maturity indices for different forest fruits are given in Table 17.1.

Table17.1: Practical Maturity Indices for Forest Tree Fruits (Source: Schmidt 2000)

Maturity event	Method of Examination
Colour change	Visual
Dry Fruits: green to yellow, brown, black etc.	
Fleshy Fruits: green to conspicuous	
Dehydration (dry fruits)	Visual, touching or 'weighing' in the hand Measurement of specific gravity

Dehiscence and abscission	Observation of fruits fall or opening of dehiscent fruits, Shaking or beating fruit-bearing branches, Beating or manual splitting of dehiscent fruits, Examination of opening structures in dehiscent fruits, e.g. valves. scales and margin, Breaking off fruit stalks
Hardening of fruit or seed coat	Cutting, pricking, breaking of seed or fruit coat
Hydration (fleshy fruits): Softening of fruit flesh	Squeezing
Loosening of fruit pulp (fleshy fruits)	Squeezing, rubbing or other separation of fleshy part from seed or endocarp
Accumulation of sugar substances (fleshy)	Taste (careful as some fleshy fruits are poisonous to humans)
Endosperm and embryo development of seed	Cutting of seed

Change in fruit colour (physical indices) and germination (physiological indices) of developing seed is by far the best available criteria for judging the maturation of different tropical forest species as shown in Table 17.2.

Table 17.2: Criteria for Judging the Maturation of different Tropical Forest Species

S.No.	Species	Weeks after Anthesis	Change of Fruit / Pod / Samara Colour	Germination Percentage	Authors
1	*Acacia nilotica*	13	Green to greenish white	61.0	Rajasingh 1987
2	*Ailanthus excelsa*	13	Yellow to brown	52.0	Ramakrishnan et al 1990
3	*Azadirachta indica*	12	Green to pale green	85.0	Maithani et al 1989
4	*Cassia siamea*	19	Green to brown	90.0	Masilamani and Vadivelu 1997
5	*Casuarina equisetifolia*	18	Green to brown	49.5	Vinaya Rai 1990
6	*Ceiba pentandra*	11	Green to greenish yellow	57.5	Gawande 1985
7	*Delonix regia*	12	Green to brown	0	Sivasamy 1987
8	*Eucalyptus camaldulensis*	26	Green to greenish brown	100	Anon 1996
9	*Eucalyptus tereticornis*	25	Green to greenish brown	100	Anon 1996
10	*Hardwickia binata*	17	Green to brownish yellow	96.0	Masilamani 1992
11	*Pongamia pinnata*	19	Green to light straw	87.0	Sivasamy 1991
12	*Prosopis juliflora*	13	Green to light yellow	90.0	Masilamani and Vadivelu 1997
13	*Tamarindus indica*	41	Green to light brown	100	Parameswari 1999
14	*Tectona grandis*	17	Green to light brown	0	Hedegart 1973

It is important to know the time requirement between anthesis (flowering) and physiological maturity for each species. Seed harvested before reaching the physiological maturity, being high in moisture, will shrivel when dry and probably will not germinate. On the other hand, delaying harvest excessively beyond physiological maturity generally will result in increase in field deterioration attack by insects and disease causing organisms and harvest loss. Most of the leguminous species seed must be collected at physiological maturity stage to achieve higher germination and seedling vigour. Beyond this stage, germination and seedling vigour of these species start declining due to the formation of hard seed coat.

TIME AND SEASON OF SEED COLLECTION

The time of seed collection depends on the local weather conditions through the natural ranges of each species. Progress of ripening and dispersal time of seeds need to be closely monitored. In a number of species, for example teak, lagerstroemia and most legumes, seeds ripen in a fairly definite time. In such cases, dry seed or fruits must be collected as soon as possible. Bhardwaj and Chakraborty (1994) reported that the *Terminalia bellirica* and *T. chebula,* when the seeds were collected during first fortnight of January and sowing was done at last week and third week of March gave better germination of 52.2 percent and 44.4 per cent, respectively. *Dalbergia sissoo* is profusely seeding almost every year and collection time of seed is different from place to place (Table 17.3).

Table 17.3: Seed Collection Months of *Dalbergia sissoo* in different States (Tewari 1994)

State	Seed Collection Time for Shisham
Assam	December – January
Bihar	November – March
Odisha	November – March
Punjab	December – February
Uttar Pradesh	January – February
West Bengal	December – Mid-February

The fruiting season of teak in India is between November and March or April. Collection of fruits during the later part of the season gives better results. Studies conducted on the effect of time of collection on germination of teak fruits from Top slip source in Tamil Nadu revealed that fruits collected in March-April gave better germination per cent than those collected in other months. The natural fall of teak fruits take place in March-April (Gurudev Singh et al 2001). Seeds with the highest germination capacity and with the least damage to the seed are found at the end of the season (Hedegart 1975). In Cauvery delta zone of Tamil Nadu, neem seeds collected during June-July (normal bearing season) gave maximum germination and seedling vigour as compared to seeds collected during off season (December-January) (Masilamani et al 1998).

METHODS OF SEED COLLECTION

There are two general methods of collecting the fruits and seeds of forest trees in India. First method is the collection of naturally fallen fruits or seeds from the ground surface, as in the case of *Tectona grandis, Gmelina arborea* and several *Dipterocarpus* and most other heavy seeds or fruits. Second is the direct collection of fruits from felled or standing trees, as in the case of *Cassia fistula, Casuarina equisetifolia, Simarouba glauca, Emblica officinalis, Shorea robusta, Eucalyptus* spp., *Cassia siamea* and many other conifers. There are certain disadvantages in ground collection viz.

- Seed if not collected immediately after shedding from the trees, may be attacked by insects, fungi and animals e.g. *Acacia* spp.
- Seeds having short viability lose germinability quickly. Once the seeds have been shed, any delay in collection may lead to collection of non-viable seed e.g. *Bassia latifolia, Pongamia pinnata* etc.

Collection of seeds from standing trees also poses some practical problems. Specially designed extention ladders, tree cycles etc., have been designed for use in collecting seed from difficult trees. Where as the ladders cannot be safely used in inhospitable terrain, the use of tree cycle needs exceptional energy, skill and training.

Acacia catechu pods must be picked by hand directly from the trees before the pods open and scatter their seeds, pods laying on the ground might already damaged by *Bruchus bilincatopygus* or other insects. Masilamani et al (1995) reported that in *Prosopis juliflora*, pods collected from natural forest floor showed heavy microbial population (fungi, bacteria and actinomycetes) and seeds extracted from infected pods exhibited lesser germination and vigour than seeds of uninfected pods. Masilamani et al (1999b) observed that teak seeds collected from the crown of the tree (straight from the tree) gave maximum germination compared to seeds collected on the ground.

EFFECT OF PEST AND DISEASE ON SEED PRODUCTION

Damages done by insect pest and diseases, are one of the major constrains for seed productivity. Regeneration failures in most of the tropical broad-leaved species are attributed to the heavy insect attack in the natural stands, at the time of inflorescence, seed formation and seed shedding. The problem of seed destruction is increasing due to our increasing reliance on seed sources for the production of seedlings or for trees of known genetic characteristics in reforestation and reclamation programs. The flowers, fruits, seeds and cones, being rich food sources, are vulnerable to insect pests and diseases. Insects that feed on them often cause seed crop failure and thus, exert an adverse impact on natural and artificial reforestation.

Regeneration failure, especially in teak and Sal in India and conifers in other countries can be attributed to two factors, firstly, the heavy insect attack in the natural stands at the time of inflorescence and seed formation, and secondly, insect attack on the ground after seed shedding (Kulkarni and Joshi 1998). Some seeds, such as those of *Acacia catechu*, *Albizzia lebbeck* and *Cassia* species may be attacked or destroyed by insects or disease, if left on the parent trees. Other tree species such as *Cassia siamea* and *Santalum album*, do not have a single, well defined period at which their fruit ripen and they may have two or more seed crop each year (Prasad 1999).

At the time of seed collection, care must be taken to avoid collection of insect or disease infested seeds. It may cause impairing germination and seedling vigour. Mittal and Sharma (1981) have reported seed borne mycoflora of *Cassia fistula* which includes common fungi *Rhizopus nigricans, Aspergillus flavus, A. niger* and *Penicillum* species. These fungal species were also frequently associated with abnormal seedlings, in which either roots were altogether suppressed or their growth was retarded. Randhawa et al (1986) reported that *Rhizopus nigerians* was associated with *Cassia fistula* seed, it causes inhibition of root and collar rot and also caused partial or complete withering of one or both the cotyledons during germination. Conway (1975) observed that bruchid attack on the pods of *Acacia tortilis* takes place between 30-100 days after pod initiation. Pods and seeds of *Acacia nilotica* were destroyed by *Caryedon serratus* and in *Callosobruchus chinensis* and the incidence of attack was 100 per cent. Insect attacked fruit or seed should be avoided during seed collection, otherwise those insect will continue their activity during seed storage.

Table 17.4: List of Major Insect Pests and Damage to Fruits/Cones /Seeds on Standing Trees (Source: Kulkarni and Joshi 1998)

Tree species	Insect Pest	Damaged Parts
Acacia spp.	*Argyoplace* (*Cryptophdbia*) *illepida* Butler (Lepidoptera: Tortricidae)	Pods and seeds
Bauhinia spp.	*Laspeyresia palamedes* Mey. (Lepidoptera: Tortricidae)	Flowers and seeds
Cassia fistula and *C. occidentalis*	*Brachyacma palpigera* walk (Lepidoptera: Gelechiidae)	Pods and seeds
Albizia lebbeck	*Bruchus sparsemaculatus* Pic. (Coleoptera: Bruchidae)	Pods and seeds
Dalbergia sissoo	*Aspidiotus orientalis* News. (Hemiptera: Coccidae)	Pods and seeds
Pongamia pinnata	*Laspeyresia perfricata* Mey. (Lepidoptera: Toruicidae)	Pods and seeds
Prosopis juliflora and *P. velutina*	*Apion subornatus, A. ventricosum* & *Microlychius spp.* (Coleoptera: Curculionidae)	Seeds
Ailanthus excelsa (Samaras)	*Alleva fabriciella* swed. (Lepidoptera: Yponomeutidae)	One generation on pods
Bamboo	*Sitotroga cerealella* Oliv. (Lepidoptera: Gelechiidae)	Seeds
Bombax ceiba (Semul)	*Dysdercus cngulatus* F. (Hemiptera: Pyrrhocoridae)	Green fruits
Casuarina spp.	*Caryedon gonagra* Fab. (Coleoptera: Bruchidae)	Seeds
Santalum album	*Callosobruchus theobromae* Linn. (Coleoptera: Bruchidae)	Seeds
Shorea robusta (Sal)	*Blastobasis crassifica* Mey. (*B. molinda, B. ochromorpha*) (Lepidoptera: Blastobasidae)	Seeds
Terminalia bellirica (Bahera)	*Mecobaris terminaliae* Marsh. (Coleoptera: Curculionidae)	Fruits
Terminalia catappa (Badam)	*Araecerus fasciculatus* De Geer (Coleoptera: Anthribidae)	Fruits and seeds
Terminalia myriocarpa	*Allagenus glariosae* F. & *A attagenus* (Coleoptera: Dermestidae)	Fallen fruits

	Meliaceae shoot borer	Flowers and fruits
Toona ciliata	*Hypsipyla robusta* Moore (Lepidoptera: Pyralidae)	
Tectona grandis (Teak)	*Pagyda salvalia* walk. (Lepidoptera: Pyralidae)	Inflorescence, fruits and seeds

SEED EXTRACTION

Extraction denotes the procedure of physically releasing and separating the seeds from the enclosing fruit structure. Extraction of seeds in tropical species is generally done close to the site of collection. After the collection of fruit, seeds should be extracted in the shortest possible time to avoid damage to seeds, as well as to facilitate storage and sowing. The main rationale of extraction is to reduce bulk, ease handling, enhance germination and improve storability of seeds and lower the risk of transfer of diseases or pests between countries and this is achieved by removing material that may harbour disease and pests. Different methods of extraction are available for both dry fruits and fleshy fruits (Table 17.5).

Table 17.5: Summary of Extraction Methods for various Fruit Types (Source: Schmidt 2000)

Fruit Type	Extraction Procedure
Dry dehiscent fruits: dehiscent pods, follicles and capsules and cones, e.g. *Pines, Eucalyptus* and most Leguminosae	Drying → shaking or tumbling
Dry indehiscent fruits, e.g. indehiscent pods of *Acacia nilotica* and *A. siberiana*	Drying → threshing
Serotinous fruits, e.g. cones, capsules plus some dry compound fruits	Kiln heating → tumbling Scorching → tumbling
Fleshy fruits with very thin pulp, e.g. *Vitex* spp. and *Ziziphus* spp.	Drying Soaking → maceration → washing
Fleshy fruit with soft pulp. e.g. *Prunus, Olea, Ficus*	Soaking → fermentation → washing Soaking → maceration → washing
Fleshy fruits with soft, fibrous pulp, e.g. *Gmelina*	Soaking → maceration → washing → Abrasion or polishing
Fleshy fruits with felty pulp, e.g. *Tectona grandis, Sclerocarya* spp. and *Vitex* spp.	Soaking → abrasion or polishing

DRY FRUITS

This is very diverse group and examples include follicles that split open down one side, pods from legumes (e.g. *Acacia* spp., *Albizia* spp.) that spilt down two sides and capsules from Eucalyptus that splitdown three (or) more sides. From extraction point of view, dry fruits can be divided into two groups: truly dehiscent fruit that will split open and expel the seed when dried e.g. *Leucaena leucocephala* and *Cassia siamea* etc. and indehiscent fruit that will remain closed after drying e.g. *Albizia lebbeck*, *Cassia fistula*, *Cassia grandis*, *Delonix regia*, *Peltophorum pterocarpum* and *Prosopis juliflora* etc. In case of dehiscent fruits, the fruit should be picked before split open and spread on the cement floor or on trays and exposed to the sun (or) artificial drier. These fruits dry too fast, and they will have to be extracted or threshed after drying to extract seed e.g. *Albizia lebbeck*. The most common seed extraction methods for dry fruits include beating with stick, splitting by hand, using paddy thresher or dehuller machines.

The major advantages for threshing of leguminous fruits or pod through thresher or dehuller are,

1. Complete removal of the seeds from the fruit or pod is achieved.
2. Minimum labour requirement
3. Increase the efficiency of operation
4. Maximum seed extraction with minimum cost
5. A beneficial scarification of seeds may be made due to rubbing action.
6. Since the thresher winnows the threshed seeds, the chaff and other lighter impurities are separated. Hence the quality of the seeds is improved.

Disadvantages are:

1. It may damage the seeds causing loss of viability and vigour, if the machine is not adjusted properly
2. Effective and economical only if large quantity of seeds is to be threshed.

Some seeds may be extracted by light rubbing or tumbling, where as others may need a series of hard mechanical, chemical or acid treatments e.g. *Prosopis juliflora* seed extraction is a challenging task. The pods have tough pericarp, waxy exocarp, spongy mesocarp and a cartilaginous endocarp, enclosing the seeds in tomentaceous segments. As the cartilaginous inner lining of the pod does not allow the seed to escape easily, due to that the natural regeneration through seeds in the normal way is time consuming. The different treatments for seed extraction are passing through steam, hot water treatment (90ºC for 10-20 minutes), cold water soaking (24 hrs), treatment with commercial H_2SO_4 (20-30 minutes), termite feeding, scarification while passing goat or other animal digestive system, electric scarification etc. Soaking of pods in diluted NaOH for 30 minutes followed by washing in water and rubbing with a cloth is the best treatment for removing seed from the septa.

Casuarina equisetifolia seeds are normally extracted by spreading the cones on canvas or plastic sheets in the sun or oven drying in 40ºC until the bracteoles are open. Shaking and sieving can then separate the seeds. In some cases, the only option may be extracting each seed by hand e.g. *Hardwickia binata* and *Pongamia pinnata*.

In case of *Tectona grandis*, artificial regeneration are achieved only through sowing of fruits. The hard fruit is irregularly round consisting of an outer thin epicarp, the middle thick felty mesocarp and the inner strong impervious endocarp encircling the tetracarpellary ovary. The true seed inside the cavity is oval in shape, having a thin seed coat. The delayed germination of teak fruits has been ascribed to the strong endocarp acting as a physical barrier to imbibition or imposing mechanical resistance to embryo enlargement, the presence of water soluble germination inhibitors in the felty mesocarp has also been thought to be one of the reasons for poor germination. Besides the poor germination of fruit, it occupies more space for storage (31 to 50 gram for 100 fruit weight). Methodologies to be developed to extract the seeds from fruit, are really challenging task to early and enhanced germination with lesser space for storing of true teak seed.

Maharashtra Van Sanshodhan Sanstha has developed a prototype machine for teak fruit scarification for removing the mesocarp from the fruits. The scarified fruit were 100 per cent devoid of mesocarp under 2-3 hours of scarification at 50-55 rpm. Teak fruit treatment mainline prototype – II machine completely removes the mesocarp besides exposing the seeds, when the drupes are subjected to scarification. Similarly mechanical scarification of trees seeds in a mechanical scarifier (Westrup LA-H type) is also practiced. The role of some fungi on weathering teak fruits is also found useful e.g. *Scytalidium* spp. decomposes the epicarp and mesocarps of teak seeds and seeds extracted from such decomposed drupes exhibit early and high germination.

FLESHY FRUITS

Species with juicy and pulpy fruits are potentially troublesome. Soon after collection, the pulp begins to soften and various fermentation processes may start. After soaking or decomposition the fruit pulp and fruit skin are separated from the seed. Several methods, often used in combination, are applicable. Selection of the most appropriate method depends on fruit type, quantity of fruits to be processed and equipment available (Table 17.6).

Table 17.6: Seed Extraction Methods of Fleshy Fruits

Extraction Method	Type of Fruit / Seed and Extraction Procedure
1.Individual manual extraction	Very small seed lots, seeds with extremely fragile seed coats or large fruits. Individually extracted seeds may be washed and cleaned e.g. *Syzygium cumini*
2.Washing in deep bowls or drums	Small seed lots and small seeds are extracted effectively. Separation of pulp and skin of seeds is done by flotation in excess water. Seeds or stones remain at the bottom while pulp and skin tend to ascend to the surface where they may be skimmed off e.g. *Ficus* species.
3.Washing on wiremesh screens	Very small seeds or seeds with fragile seed coats. Fruits are rubbed manually on the screen, the mechanical impact can easily be adjusted when extracting fragile seeds e.g. *Anthocephalus cadamba*
4.Cement mixture with abrading material	Fruits with sticky pulp. Abrading material (gravel) rubs against the fruits. The pulp is gradually torn off. The mixture must be separated into three fractions after extraction viz. seed, abrading material and fruit pulp. e.g. *Vitex keniensis* and *Cordia* spp.
5.Mechanical depulping	Applicable for large quantities of fruits or where the flesh tends to remain firm and fruits with relatively hard seed coats or stones. Depulper mechanically abrades the fruit pulp by rupturing and squeezing against or between its mechanical parts and the pure seeds are separated e.g. *Azadirachta indica* and *Gmelina arborea*

The fruit of many species such as *Azadirachta indica, Cinnamomum camphora* and *Gmelina arborea* are fleshy and pulpy. It is necessary to remove the pulp immediately at maturity to prevent loss of germination capacity. In neem, several methods are available for extracting the seeds from the fruits viz., the fruits were depulped in a coffee depulper, fruits soaked in water for six hours, then depulped by rubbing them over a soft wire mesh and the pure seeds were separated. All the afore said methods are laborious and time consuming. The seeds soaked in concentrated sulphuric acid (50 ml /kg) for 3 minutes are found to have higher percentage of mesocarp removal influencing germination and initial seedling vigour attributes of neem, with lesser time and cost. Similarly pulping machines are found to be effective in removing mesocarp.

In amla (*Emblica officinalis*), alternate soaking and drying of fruits in 20 per cent sodium chloride solution upto three cycles could extract seeds. The fruit pulp after treatment is easily extractable and the nuts on sundrying expel seeds. In jamun (*Syzygium cumini*), heaping fruits for a day make the separation of seed easier while longer duration of heaping damages the seed. In ber (*Ziziphus mauritiana*), removal of flesh with a sharp knife and thorough washing with water followed by soaking in lime solution for 30 minutes, make the extraction easier than drying under sun or water washing of fruits. These extraction techniques not only make the extraction easier but also make the edible part economically useful (amla and ber). The extracted seeds have higher seed quality in terms of germination, seedling vigour and storability. In *Tamarindus indica,* seed extraction from fruit is quickest using the thresher while knife is also used which is slower. The thresher results in the highest percentage of damaged seeds and manual maceration gives the lowest percentage of damaged seeds. Fleshy fruits are also adopted to being ingested by animals. Both ants and termites may readily and efficiently clean fruits and seeds of sweet material, both moist and dry.

SEED DRYING

After extracting the seeds from the fruit or pod, drying is essential before the seed can be safely stored until use in the nursery or packing for dispatch. The moisture content of most of the seeds for storage ranges between 10-12 per cent. Recalcitrant seeds lose viability seriously by drying, such seeds have to be stored in moist conditions. Development of all type of storage fungi *(Aspergillus* spp. and *Penicillium* spp.) and storage insects (Internal feeders: *Sitophilus oryzae, Rhizopertha dominica, Callosobruchus maculates, C. chinensis, C. pisorum, Sitotroga cerealella, Caryedon serratus, Trogoderma granarium* and External feeders: *Tribolium castaneum, Oryzaephilus surinamensis, Cryptolestes ferrugineus, Plodia interpunctella, Ephestia cautella, Corcyra cephalonica, Liposcelis divintorius, L. transvallensiss*) are common in seeds stored in higher moisture content. Increase in the moisture content from 15 to 24 per cent will increase in storage fungi and they cause reduction in germination and seedling vigour; discoloration, shrinkage and shriveling of seeds; production of heat and mycotoxin; reduction in nutritional value and bio-chemical constituents of seed; development of mustiness and caking and total decay of seeds.

The seed quality is affected on the basis of insect damage to endosperm or embryo during storage. If embryo portion is damaged, seed fails to germinate. Some seeds may germinate and develop into a seedling but the vigour of the seedling will depend up on the extent and intensity of the damage. The degree and type of insect infestation and insect species involved depend on the type of tree seed and prevailing ecological conditions. To avoid the above said storage fungi and insects, proper precautions are necessary at harvest and during post harvest operations. Removing the inert matter such as dried stick, leaves, soil clod, broken seeds, other tree seeds and weed seeds will reduce the attack from storage fungi and pests. Dried, sound, clean and undamaged seeds should be treated with insecticide and fungicide for better results. The insecticides like Malathian 5 D @250g per quintal (or) carbaryl @200g per quintal) and fungicides captan (or) thiram @250 to 300g per quintal (or) Bavistin @200g per quintal are found effective. The treated seeds should be packed in moisture vapour pervious or moisture vapour resistant or moisture vapour proof containers and labeled properly. Extreme care is required to ensure that treated seeds are never consumed for human or animal food. Treated seeds should be clearly labeled as being dangerous, if consumed.

SEED PROCESSING

Seed processing encompasses all the steps involved in the preparation of harvested seed for marketing. It is done to narrow down the level of heterogeneity of the harvested produce with the sole objective of improving the quality of the seed that is produced carefully in the field, using suitable processing methods. The harvested seed lot can be heterogeneous due to the following reasons,

- Heterogeneous soil, which vary in the fertility due to the availability of nutrients, physical, chemical and biological properties.
- Due to time of pollination and fertilization over a period of time.
- Due to the position of seed on the plant or fruit.
- Due the interference of biotic factors and their variability in infestation.
- Due to the management practices like water, land preparation, leveling, staggered sowing, uneven distribution of fertilizer and irrigation water, uneven plant protection sprays and uneven maturity at harvest.

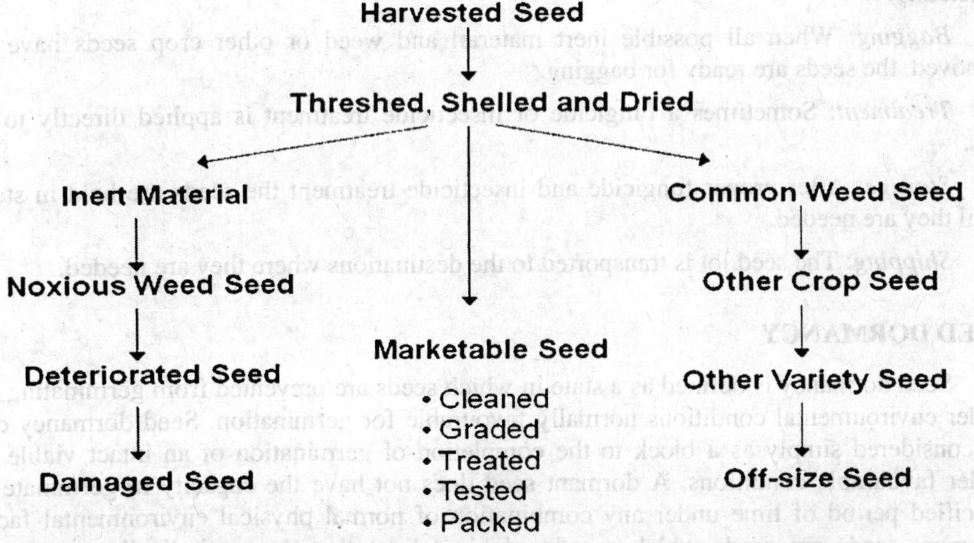

Fig. 17.1: Types of Materials Removed from Harvested Produce during Processing

This heterogeneity can be narrowed down in the processing of seeds by eliminating the undersized, shriveled, immature, ill filled seeds using appropriate sieve size and the physical purity of the seed lot can be improved. The germinability and vigour of the seed lot can be improved by grading the seeds based on size, specific gravity, length and density of the seeds. Researches in seed processing reveal that the inherent qualities such as germinability and vigour are exemplified by certain physical characteristics of the seed i.e., large size, denser seed, optimum length etc. Hence on managing of seed with a particular range of size, shape, length and density, the quality of the total seed lot can be improved.

SEQUENCE IN SEED PROCESSING

Seed processing steps can be broken down into several definite steps that follow in a specific sequence.

Receiving: This is the first step in seed processing. Seed arrive at the processing plant in bags, pallet boxes or in bulk.

Bulk Storage: From the receiving station the seed go into bulk storage to be held for later processing, or directly into the processing line for cleaning.

Conditioning and Pre-cleaning: The air-screen machine is probably the most common basic cleaner. It makes size separations and seperates the seeds. Seed lots may come from the field in good condition with few contaminated seeds and require cleaning only on the air-screen machine.

Separating or Upgrading: Most often, however, it is necessary to send the seed through one or more special separating or upgrading machines to remove a specific contaminant. These special machines separate crop and weed seeds by differences in their specific physical characteristics. For effective and efficient separation, the crop seeds must differ from the weed seed sufficiently in some characteristics so that the machine can differentiate them efficiently.

Bagging: When all possible inert material and weed or other crop seeds have been removed, the seeds are ready for bagging.

Treatment: Sometimes a fungicide or insecticide treatment is applied directly to seed lots.

Storage: After proper fungicide and insecticide treatment the seeds are held in storage until they are needed.

Shipping: The seed lot is transported to the destinations where they are needed.

SEED DORMANCY

Seed dormancy is defined as a state in which seeds are prevented from germinating even under environmental conditions normally favourable for germination. Seed dormancy could be considered simply as a block to the completion of germination of an intact viable seed under favourable conditions. A dormant seed does not have the capacity to germinate in a specified period of time under any combination of normal physical environmental factors. Dormant seeds are seeds which remain viable (alive) though metabolically inactive and incapable of germination under the conditions normally favourable for the species. Seed dormancy introduces a temporal delay in the germination process that provides additional time for seed dispersal over greater geographical distances. Seed dormancy is a condition of delayed germination and can be categorized into induced, innate and enforced types.

Seed dormancy is particularly associated with environment characterized of variable rainfall trends and with extended dry periods within the annual cycle. Within the tropical forest biome, the proportion of species producing non-dormant seeds declines with a decrease in the predictability and total amount of rainfall and an increase in the length of the dry season. While as many as 63 per cent of the tropical rainforest species produce non-dormant seeds, only 24 per cent of the dry tropical tree species produce seeds which are not dormant .

Seed dormancy is a mechanism that insures the survival of the species. In nature, dormancy is a blessing that insures the continuation of the species over time and through periods of environmental stress. Seed dormancy aids species survival by delaying or retarding germination so that overall germination is dispersed over time. This means that if the environmental conditions are not right during the normal germination period, a percentage of seeds will germinate at some later time when the conditions are right. Dormancy pre-disposes the seed to respond to specific environmental conditions that will favour seedling survival and maximize species proliferation. Seed dormancy has been categorized into various types, by different authors based on the manner in which it originates.

SEEDCOAT-IMPOSED AND EMBRYO DORMANCY

SEEDCOAT-IMPOSED OR PHYSICAL DORMANCY: It is dormancy imposed on the embryo by the seed coat and other enclosing tissues, such as endosperm, pericarp, or extrafloral organs. The embryos of such seeds will germinate readily in the presence of water and oxygen, once the seed coat and other surrounding tissues are either removed or damaged. Physical dormancy is more prevalent in dry tropical trees (51 per cent) compared to wet tropical species (5 per cent). Acid, hot water or mechanical scarifications are common treatments employed to overcome the seed coat dormancy. There are five basic mechanisms of physical or seed coat-imposed dormancy:

Prevention of water uptake: Prevention of water uptake by the seed coat is a common cause of seed dormancy in families found in arid and semiarid regions, especially among legumes. Waxy cuticles, suberized layers and lignified sclereids all combine to restrict the penetration of water into the seed.

Mechanical constraint and hardseededness: The first visible sign of germination is typically the radicle breaking through the seed coat. In some cases, however, the seed coat may be too rigid for the radicle to penetrate. Nuts with hard, lignified shells are examples of dormancy caused by mechanical constraint. Such shells must be broken by biotic or environmental forces for the seed to germinate. Even non-lignified tissues, such as the endosperm of some tree seeds, can suppress expansion of the embryo. For the seeds to germinate, the endosperm cell walls must be weakened by the production of cell wall degrading enzymes. Hardseededness is where the seedcoat or pericarp surrounding the seed presents a physical barrier. This type of dormancy is common in the Fabaceae, Convolvulaceae, Geraniaceae, Malvaceae, lamiaceae, and the Poaceae plant family members. Physical abrasion (scarification), or freezing and thawing may be needed to allow water uptake and germination to proceed.

Interference with gas exchange: Some seed coats are considerably less permeable to oxygen. This lowered permeability to oxygen suggests that the seed coat inhibits germination by limiting the oxygen supply to the embryo. The dormancy of such seeds either by making a small hole in the coat with a pin (without weakening the coat mechanically).

Retention of inhibitors: The seed coat may prevent the escape of inhibitors from the seed.

Inhibitor production: Seed coats and pericarps may contain relatively high concentrations of growth inhibitors that can suppress germination of the embryo. Abscisic acid (ABA) is a common germination inhibitor present in these maternal tissues. In certain cases where repeated washing (leaching) removes dormancy, the effect is thought to be due to the loss of such inhibitory compounds.

EMBRYO DORMANCY: It refers to a dormancy that is inherent in the embryo and is not due to any influence of the seed coat or other surrounding tissues. Embryo dormancy is caused by the physiological deficiency in the embryonic axis or due to the existence of metabolic blocks within the cotyledons. Embryo dormancy can be relieved by amputation of the cotyledons in some species which exert an inhibitory effect. A fascinating demonstration of the cotyledon's ability to inhibit growth is found in some species in which the isolated dormant embryos germinate but grow extremely slowly to form a dwarf plant. If the cotyledons are removed at an early stage of development, however, the plant abruptly shifts to normal growth. Embryo dormancy is due to the presence of inhibitors, especially ABA, as well as the absence of growth promoters, such as GA (Gibberellic Acid). The loss of embryo dormancy is often associated with a sharp drop in the ratio of ABA to GA.

PRIMARY, SECONDARY AND ENFORCED SEED DORMANCY

Different types of seed dormancy also can be distinguished on the basis of the timing of dormancy onset rather than the cause of the dormancy.

PRIMARY OR INNATE DORMANCY: Seeds that are released from the plant in a dormant state are said to exhibit primary dormancy. Primary dormancy (also called innate dormancy) prevents germination during development and maturation of the seed on the mother plant and usually also for sometime after shedding and harvesting. Seeds with impermeable coats and those requiring prechilling treatment represent innate dormancy.

SECONDARY OR INDUCED DORMANCY: Seeds that are released from the plant in a non-dormant state but which become dormant if the conditions for germination are unfavorable exhibit secondary dormancy. For example, seeds can become dormant in the presence of temperatures higher than the maximum for germination or at temperatures below the minimum for germination. This type of dormancy is called secondary or induced dormancy.

ENFORCED DORMANCY: When a light-requiring seed is buried deep in soil and fails to germinate, the dormancy is called 'enforced' dormancy.

SEED DORMANCY: AFTER-RIPENING

In a broad sense, after-ripening describes the loss of the dormant state in a seed over some period of time. In the strictest sense, after-ripening refers to the loss of dormancy mechanisms imposed by the mother plant. Seeds maintained in dry storage or imbibed in the soil seed bank tend to lose this maternal control over dormancy and germination without any applied dormancy breaking methods over a period of days to years depending upon the species. After-ripening is a period of quiescence that the seed must go through to finalize the separation from the mother plant and become autonomous and on its own.

Another type of after-ripening involves seeds with rudimentary embryos. With this after-ripening, the rudimentary embryo must develop into a full embryonic axis before germination can occur. This is a physiological process that may occur over a period of time or may have to be stimulated by certain environmental factors to proceed. In some species, there is a dependence upon where the seed is positioned with the inflorescence as to how rudimentary or developed the embryo is at physiological maturity. Those seeds with a well developed embryo germinate readily while those with rudimentary embryos must develop fully over time to germinate.

INTERNAL AND DOUBLE DORMANCY

INTERNAL DORMANCY: Dormancy factors are located in the inner tissues of seed. Based on the intensity of dormancy, this group is sub-divided into three sub-classes physiologically shallow dormancy, physiologically intermediate dormancy and physiologically deep dormancy.

DOUBLE DORMANCY: This group includes seeds in which dormancy is caused both by factors related to seed coat and to the embryo.

PRE-SOWING SEED TREATMENT

The seeds of many native plants have built-in dormancy mechanisms, which protect them from germinating in times of droughts. In the wild, seeds will lie dormant until the proper conditions for growth occur. But in cultivation, the grower must become familiar with several simple pre-sowing seed treatment methods, which will unlock the dormancy mechanisms and stimulate quicker, more consistent germination.

Physical Abrasion: Physical breaking of seed coat will improve the seed germination of certain tree species e.g. cracking the endocarp, making hole in the seed coat and rubbing on the hard floor. For treatment of large quantities of seed to overcome seed coat dormancy, they are tumbled or churned in a concrete mixer with sharp gravel mixer or sand or in a special drum with abrasive material.

Cold Water Treatment: The soaking of seeds in water for 24 hrs helps to improve the seed germination by softening the seed coat.

Hot Water Treatment: Soaking the seeds in boiled water for few minutes and subsequent cooling also improves germination. The seeds, which show rapid imbibitions, germinate rapidly, a gentle but sure indication on the quality of seed being used.

Acid Scarification: In some tree species, poor germination in the nursery is a problem due to seed coat dormancy. To overcome this, seeds are treated with concentrated H_2SO_4 for different durations ranging from 5 to 30 min and are washed in running water after treatment. This is termed as acid scarification. Improvement in germination and seedling vigour has been obtained through acid scarification e.g. *Acacial nilotica, Albizia lebbeck.*

Growth Regulators and Chemical Treatment: Soaking of seeds in growth promoters (IAA, IBA, GA_3 etc.) and aqueous solutions of thiourea or KNO_3 or KCl or sucrose solution enhances the germination percentage and seedling vigour in selected tree species. The enhancement in germination may be attributed to an increase in cytochrome oxidase activity or cytokinin activity, which could break the post harvest dormancy.

Passing through Animal Digestive Track: Certain animals and microorganisms are also known to break seed dormancy. Seeds of *Acacia senegal* and *Ceratonia siliqua* germinates readily when passed through digestive tracts of goat. In *Gmelina arborea* and *Acacia nilotica*, partial rumination of seeds by sheep and goats enhances germination. This is due to the action of strong digestive chemicals present in the gut. Certain rainforest species such as *Cecropia obtusifolia* shows enhanced germination when passed through bat intestines.

Termite Treatment: Termites are also known to break dormancy of *Tectona grandis* fruits and other tree seeds by partially damaging the hard seed coat.

Fire Treatment: Burning with light fire, applied on the surface of fruits of *Tectona grandis* and other tree seeds with hard seed coat, has improved the germination by breaking the dormancy.

In Ber (*Zizyphus mauritiana*), the nut or stone is used as the seed due to the reason that the true seed cannot be easily extracted from the hard endocarp. The germination of ber seed is slightly difficult due to the hard endocarp and requires about 3-4 weeks to germinate. The germination time could also be reduced to one week by cracking the endocarp carefully. However, for easier and improved germination the various treatments are applied viz. soaking the seeds in GA_3 or sucrose solution or thiourea, simple water soaking for 24 hrs, acid scarification with H_2SO_4.

With regard to *Hardwickia binata*, the seeds are readily germinating but could also be improved by several seed management practices such as soaking of seed in cold water for 24 hrs, soaking in ascorbic acid, pre-sprouting of seeds by placing in wet gunny bags for 48 hrs, treatment with IAA at lower concentrations. For improving the germination of seeds of *Emblica officinalis*, the following treatments are adopted to break morpho-physiological dormancy viz. soaking in GA_3, stratification treatment with sand moistened with KNO_3 solution. In *Simarouba glauca*, the seeds or nuts are non-dormant which initiates its germination on 12-15 days after sowing and soaking the seeds in cold water for 24 hours can break hard endocarp and facilitate quicker and enhanced germination.

In *Cassia siamea*, the seed treatments such as mechanical scarification, acid scarification with H_2SO_4, 1-2 minutes soaking in 85°C warm water, or submersion at 85°C for 1-2 minutes with subsequent cooling in the water for 12-36 hours are found effective to increase the germination.

SEED STORAGE

Seed storage is the preservation of viable seed until their sowing or requirement. It is essential to offset the uncertainty of seed production/availability during bad seed years. It delays deterioration, maintains viability and protects seed from rodent and insect damage. The longevity of seeds is a species specific characteristic. The seed of most of the species can be stored at low temperature and low moisture content in sealed containers. It is important to dry the seed uniformly to prevent fluctuation in moisture content during storage. The moisture content of most of the seeds for storage ranges between 10 to 12 per cent. The respiration continues at low temperature, which is necessary to keep the embryo alive. Polythene bags make good containers because they are impermeable to water but less so to oxygen and carbon dioxide. However, many species of moist tropical forests are so thoroughly adapted for germination that their seeds are almost impossible to store or even to transport. On the basis of storage behaviour seeds can be divided into two broad categories: orthodox seed and recalcitrant seed.

ORTHODOX SEED

It is a generally conceived notion that seeds endowed with natural longevity are hard-coated. They include a number of tropical leguminous species. Orthodox seeds acquire desiccation tolerance during development and may be stored in the dry state for predictable periods under defined conditions. Orthodox seeds maintain high vigour and viability at least from harvest until the next growing season or for many decades at -18^0 C. Generally such seeds undergo a period of drying during their maturation and are shed at low water content which is equilibrium with the prevailing relative humidity. The equilibrium water content at any particular relative humidity is determined by seed composition, but all orthodox seeds can withstand dehydration to around 5 per cent (0.053 g H_2O/ g dry material), even when maturation drying is not completed prior to shedding.

Some of the most outstanding examples of maintaining viability, after lengthy periods of storage, are mentioned in a work emanating from FRI as far back as in the year 1948, by T.V. Dent. Recent research has provided more precise information on the conditions of storage, such as low moisture content, low temperature and low oxygen pressure, as the three most important constituents of the storage conditions which must be provided to maintain viability of orthodox seeds in long term storage. The orthodox seeds can further be subdivided into two classes viz. orthodox seeds with hard coats and orthodox seeds without hard coats.

ORTHODOX SEED WITH HARD SEED COATS: Most of the seeds which have been recorded to fall in this category are hard coated. More specifically, this category refers to those in which the seed coat is impermeable to water, and do not swell even after keeping in a moist medium for weeks or even months. They include a number of tropical leguminous species. Since most seeds have pretty long natural viability span, therefore, little well-planned work has been done on their storage in controlled environmental conditions. However, for the purpose of conservation of genetic resources, where the objective is to store seed for a couple of decades, a long term storage will have to be planned, e.g., for species like *Acacia, Prosopis, Albizia, Cassia, Bauhinia* etc., which are being planted on a massive scale in social forestry plantations.

ORTHODOX SEEDS WITHOUT HARD SEED COATS: Many species belonging to important genera of forest trees, fall into this group, e.g. *Pinus, Picea, Eucalyptus* and several other genera. Kandya (1987) listed seeds of some tree species according to their probable period of longevity. The list includes mostly the probable orthodox seeds without hard seed coats.

Two Weeks: *Michelia champaca, Syzygium cumini, Soymida febrifuga, Aegle marmelos, Anogeissus latifolia, A. pendula,* etc.

Two Months: *Chloroxylon, Chukrassia, Podocarpus*

One Year: *Artocarpus heterophyllus, Toona ciliata* (Syn. *Cedrella toona*), *Mesua ferrea, Terminalia tomentosa, T. bellerica, Bambusa arundinacea, Semecarpus anacardium, Pongamia pinnata, Pterocarpus marsupium, Lagerstroemia parviflora, Lannea coromandelica, Diospyros melanoxylon, Garuga pinnata, B. lanzan* (Syn. *Buchanania latifolia*).

Two Years or More: *Dalbergia sissoo, Eucalyptus tereticornis, Anthocephalus cadamba, Tectona grandis*

Some of the short viability seeds include *Dendrocalamus strictus, Holoptelia integrifolia, Terminalia myriocarpa, Pongamia pinnata, Alnus nepalensis* and *Alnus nitida, Bambusa tulda* etc.

RECALCITRANT SEEDS

These seeds are characterized mostly by large size and high moisture content, which cannot be dried without causing injury. Recalcitrant seeds cannot be stored for a longer period. Recalcitrant seeds are those that undergo little, or no maturation drying and remain desiccation sensitive both during development and after they are shed. Wide range of variability is observed among recalcitrant seeds of different species and of individual species under different conditions. Recalcitrant seeds are shed from the tree in hydrated condition and water content can generally be in the range from 0.43 to 4.0 gram per gram of dry material which is 30 to 80 per cent on a wet mass basis. Most of the recalcitrant seeds belong to woody species. Most of the species belong to the family Dipterocarpaceae and Lauraceae. *Quercus, Castanea, Aesculus indica, Shorea robusta, Azadirachta indica, Bassia latifolia* (Syn. *Madhuca indica*), *Mangifera indica, Dipterocarpus, Hopea,* etc. fall into this category.

SEED TESTING

Seed testing is essential to assess the physical and biological aspects of seed. Seed tests are commonly done immediately after extraction and shortly before actual sowing. It is also done periodically on seed lots kept in long storage. For small nurseries, common sense, clean hands, a clean working table and one good knife are sufficient for most seed testing tasks. Some of the common terms and methods have been described below.

SEED LOT

A seed lot is defined as a specified quantity of seeds of reasonably uniform quality from a particular geographic source.

PURITY TEST

It determines what proportion of the seed sample by weight has pure seed and what proportion is other material. The four recognized components of a seed lot are pure seeds, other seeds, damaged seeds and inert matter such as seed wings, twigs, stone, soil or other non-seed materials. The separation is done manually by placing seeds on a working table. The immature, shriveled, cracked, and damaged seeds larger than one-half of the original seed-size, including those with internal insect damage and those starting to germinate, are designated as pure seeds.

$$\text{Purity \%} = \frac{\text{Weight of Pure Seeds}}{\text{Total Weight of Seed Sample}} \times 100$$

Thus, if the initial weight of a seed sample is 50 gm and the pure seed portion weighs 40 gm, then the purity of the lot is:

$$\text{Purity \%} = \frac{40}{50} \times 100 = 80\%$$

SEED WEIGHT

It is normally expressed for 1000 pure and full seeds. Factors affecting seed weight are size, moisture content and proportion of full seeds in the lot. It is generally calculated by taking 10 random samples of 100 seeds from a pure lot. If the difference between any two replicates exceeds 10% of the mean weight, additional replicates should be drawn. To convert number of seeds per kilogram following formula is applied.

$$\text{Number of Seeds per Kg} = \frac{10{,}00{,}000}{\text{1000 Seed Weight in gm.}}$$

SEED MOISTURE CONTENT

Knowledge of seed moisture content is essential to determine the viability and storage conditions. Seeds of high moisture content cannot be stored and over drying can make them non-viable. It can be determined by drying of 10g sample in oven at 103°C for 17 hrs (or at 130°C for 1 to 4 hrs). Seed moisture content is calculated through the following formula.

$$\text{Moisture Content (\%)} = \frac{\text{Original Wt. of Seed - Oven Dry Wt. of Seed}}{\text{Original Wt. of Seed}} \times 100$$

GERMINATION TEST

Seed germination is defined as the emergence and development from the seed embryo of those essential structures which, for the kind of seed under test, are indicative of the ability to produce a useful, mature plant under favourable field conditions. In a laboratory germination test, the plant producing potential of a sample of seed is evaluated. So germination is a reliable, repeatable test which is a great way to compare seed lots based on their highest potential. The most reliable test of seed viability is to germinate a representative sample (four replicates of 100 seeds each) under laboratory conditions. Under field condition, cutting the seed into two equal parts can test viability. Seeds having fully grown, firm and undamaged embryo can be presumed to be good. However, this is not a reliable test for stored seeds because loss of viability in storage may not produce immediate visible changes.

LABORATORY GERMINATION COUNTS (LGC)

Seeds (100 nos.) are placed on moist blotting paper or cotton wool in a petridish after giving the necessary pre-treatment, in case of very small seeds, one-gram seed is taken. The petridishes are placed in a warm (not hot) place and kept moist regularly. The number of seeds which germinate, is counted every day and after 4 weeks or more, LGC is calculated as:

$$\text{LGC}(\%) = \frac{\text{No. of Seeds germinated}}{\text{No. of Seeds sown in petridish}} \times 100$$

It is expressed as the number of seeds germinate per kilogram.

TTZ TEST

Another simple test is tetrazolium (TTZ) staining test, which indicates the presence of live tissue. 1 per cent solution of TTZ (2,3,5 - triphenyl-tetrazolium chloride) is applied to fully imbibed seeds, which have been cut opened length-wise without damaging the embryo. The seeds are left overnight (18 to 24 hrs) in the dark at 30ºC). The live embryo, cotyledons and other tissue stain pink to red indicating that the seeds are viable. Comparatively larger seeds like *Albizia, Bauhinia, Pheobe goalparensis,* etc. can be conveniently tested in this way.

VIGOUR TEST

Seed vigour refers to the ability of seed to germinate, emerge and produce a good crop under a wide range of environmental conditions. Vigour shows how do lots of similar germination compare when conditions are not favourable, which lots will retain their potential after storage, which lots will come out of the ground uniformly, and which one will establish a good stand of plants in the field. Seed vigour is defined by International Seed Testing Association (ISTA) as the sum total of those properties of the seed which determine the level of activity and performance of the seed or seed lot during germination and seedling emergence. Seeds which perform well are termed high vigour seeds. It is defined by Association of Official Seed Analysts (AOSA) as Seed vigour comprises those seed properties which determine the potential for rapid uniform emergence, and development of normal seedlings under a wide range of field conditions. The aspects of performance associated with the seed vigour include:

- Rate and uniformity of seed germination and seedling growth
- Field performance, including extent, rate and uniformity of seedling emergence
- Performance after storage and transport, particularly the retention of germination capacity

Objectives for Vigour Tests (Source: ISTA)

- To provide a more sensitive index of seed quality than the germination test.
- To provide a consistent ranking of seed lots in terms of their potential performance.
- To be objective, rapid, simple, and economically practical.
- To be reproducible and repeatable.

Due to varied soil, seed bank, and environmental conditions vigour cannot be an absolute value representing field emergence but can provide information about field emergence, storage potential, and consistently ranking these features, which are not offered in the germination test.

FACTORS AFFECTING SEED VIGOUR

Seed quality and vigour is greatly influenced by environmental factors, genetic constitution, seed size, physiological maturity, production and storage. The major cause of the loss in vigour is attributed to deterioration and ageing which commences at the time the seed becomes physiologically mature. It is therefore imperative that the seed is handled carefully to prevent an accelerated reduction in performance. Physical damage to cell membranes, particularly in large seeds, is usually the primary cause of seed deterioration. Enzyme activity in immature and dormant seeds, respiration during harvest and storage, impaired protein and the use of protein and RNA synthesis during periods of low temperature stress, genetic damage and the accumulation of toxic metabolites are all associated detrimental factors.

The speed and uniformity of germination, inferior seedling emergence, seedling growth rate, intolerance to environmental stress and other similar reductions in performance capabilities can be detected. Because the loss of seed vigour precedes the loss of the seeds ability to germinate, high germinating seed lots may mask the extent of the damage and consequently have poor vigour. The demand for seed vigour testing continues to increase annually and has now grown to include a wide variety of tree species and subsequently, a variety of vigour tests. Foresters use vigour information as an additional quality assurance tool to rank seed lots for field and/or storage potential.

TYPES OF VIGOUR TESTS

Stress Tests: Tests which subject the seed to adverse conditions and include cool tests, accelerated ageing tests, and controlled deterioration tests.

Seedling Growth and Evaluation Tests: Tests where seedlings are measured at some point under uniform laboratory conditions and include seedling growth tests, seedling weight tests and first counts in germination tests.

Biochemical Tests: They measure metabolic activity as it pertains to the vigour of the seed and include conductivity and tetrazolium tests.

Vigour tests in conjunction with a germination test can give seed sellers and buyers more confidence when choosing seed lots. Precision is an extremely important factor in testing seed for vigour. The ability of the seed testing equipment used should be constantly monitored and verified. The analyst performing the test should be competent, given training updates and should be comfortable and experienced with the test being performed.

CONCLUSIONS

The care taken to address maximum precautionary points during seed collection, extraction, processing, pre-sowing treatments, storage and testing of seeds will pay through the vigorous offsprings and remunerative plantations. Nevertheless the prudent and judicial collection, extraction, processing, storage and testing of seeds go a long way in guarding the physical, physiological, genetic and health quality of seeds. Good seed handling practices are an essential complement to genetic improvement. Therefore, sequential steps in seed

handling operations viz. collection, extraction, processing, seed treatment, packing, storage and seed sowing operations should be meticulously planned and synchronized to improve or maintain seed quality towards sustainable yield achievement.

18 FOREST TREE IMPROVEMENT

INTRODUCTION

The forest plays a dominant role in the progress of our civilization. Apart from being a direct source of timber, fuel, fodder, food, medicine and recreation, they contribute immensely on preserving the ecosystem, controlling air and water regimes ultimately leading to social, cultural and economic prosperity of the nation. Forests are the single most important repositories of terrestrial biological diversity. Forest trees and other woody plants help support many other organisms and have developed complex mechanism to maintain high level of genetic diversity. The genetic variations both inter and intra specific of trees and shrubs serve a number of fundamentally important purposes. It allows them to react against change in environment, including those brought by pests, diseases and climatic change. It provides building block for future evolution, selection and human use in breeding for wide range of sites and uses.

As the world's population grows, so does the need for forest products. However the land available for growing them is shrinking because of agriculture demands, city expansion and road development. At present, there is wide gap between production and demand of forest products. The annual production of wood in India is 0.7 m^3 per ha which compares poorly with the worlds average of 2.1 m^3 per ha. The average growing stock is 65 m^3 per ha compared with World's average of 110 m^3 per ha. The annual cut does not exceed 32 million m^3 per year, which is far short of current requirement of the country. The forester's activities are mainly confined in raising and harvesting of forest crops with little concern to genetic quality of planting stock. Most of the seed collectors are non technical with tendency of collecting seed from easily approachable trees. This result in forest crops having low productivity. The trend can be reversed and productivity of our forest land can be increased by deploying genetically superior planting stock of indigenous and exotic species.

Forest tree improvement means improvement of overall yield and quality of forest products by combining the forest management activities with control of parentage. Main objectives of the tree improvement are to maximize the gain per unit time and space via selection and development of improved genotypes and to maintain large and diverse genetic base for continuing successful tree improvement programme.

ELEMENTS OF TREE IMPROVEMENT PROGRAMME

According to Wellendorf (1991), the main elements of tree improvement programme are tree improvement strategy, forest tree population, management, research and development.

Tree Improvement Strategy: The first step is the identification of species on the basis of its value which can be assessed on the basis of its use in plantation programme, market value and consumption pattern and users preference. After identification of the species possible gains and risks are assessed. Possible gains and risks will depend upon end use characters of interest, their genetic variation, their heritabilities, rotation age and type of value found in value production i.e. commercial subsistence or environmental. The risk of tree improvement is associated with the situation where non local seed sources are used without testing or based on short term testing. Risks also increase when breeding and seed production is based on few individuals (Graudal and Kajaer 2000).

The tree improvement strategy depends upon availability of investment capital. The traditional area of tree improvement is industrial plantations. In developing countries like India, private companies are investing in short rotation forestry. In case of social forestry sector, investment to finance the tree improvement is not generally available. Here low cost tree improvement options can be explored. The tree improvement options like low level improvement programme (seed collection from identified sources/ seedling and clonal seed orchards) or high level improvement programme (advance generation tree improvement) will depend upon ease of breeding of the identified species, its value and availability of investment capital.

Forest Tree Population: Forest tree population are the basic resources of tree improvement. Their structure and handling is the core of tree improvement.

Management: Management aspects which include human and organizational base for tree improvement activities.

Research and Development: Research and development which are necessary for solving key problems in tree improvement.

ADVANTAGES OF TREE IMPROVEMENT

The major advantages of tree improvement strategies are:

- Change in genetic make up of trees is permanent
- By means of vegetative propagation, developed material can be kept for indefinite time.
- Greatest genetic variability is present in most tree stands, which is not changed by action of people.
- The wide genetic variation enables wide adaptability.

LIMITATION OF TREE IMPROVEMENT PROGRAMME

Though there are great advantages of tree improvement programme, it has some limitations also.

- Requirement of large area for storage and testing of desirable genetic material.

- Large size of trees create problems to carry out different breeding activities e.g. measurement of length, diameter, crossing and seed collection.
- The cost and effort of producing new seed crop is large.
- Different trees possess different growth curves. Therefore juvenile mature correlations are not satisfactory for growth characteristics
- Lack of availability of seed with known or desired genetic background.
- Requirement of good permanent record because tree improvement is a long term process.

PECULIAR PROBLEMS IN TREE IMPROVEMENT

As mentioned by Wright (1976), although genetic principles are same for trees, men and fruit flies, the inheritance pattern, method of experimentation vary considerably among these groups. Due to long life of forest trees, the generation interval is quite large giving rise to peculiar problems in the improvement of forest trees. He has outlined some peculiar problems related to forest genetics:

INDIRECT EVIDENCE: Since most of the characters are under control of genes and environment, the genetic control of these characters is identified by progeny testing i.e. growing and measuring the off springs of trees under similar conditions in replicated environment.

UNCERTAINTY AND NEED FOR CONTINUOUS EXPERIMENTATION: Generalization that taller trees will produce taller offspring does not always hold true. Offspring of taller trees may grow 1 per cent taller than the offspring of average trees; however the offspring of some trees may grow 5 per cent faster than average. Thus one can obtain 1 per cent gain by using general rule, but he must test the offspring of each parent to obtain full 5 per cent gain.

TIME ELEMENT: Trees are long lived species which require several years to produce the seed. Species such as Eucalyptus take 3-4 years to produce seed whereas some pines may take 10-30 years to produce seed. However, early flowering can be induced by grafting and nutrient management in some species. Also juvenile mature correlations can help in reducing generation interval.

NECESSITY FOR SEED PRODUCTION: Though most of the trees are not grown for seed yet a tree breeder must dilute some part of his work with experiments on simulation of flowering and fruiting.

SCARCITY ABOUT BASIC INFORMATION: Field crops like wheat, maize, rice etc. have been extensively studied crops. In these crops, genes controlling different characters have been identified on particular chromosomes. Transgenic plants have been produced in number of plant species. There is lack of basic information on most of the forest tree species.

STEPS IN TREE IMPROVEMENT PROGRAMME (Zobel and Talbert 1984)

- Determination of species, or geographic source within species, that should be used within a given area.

- Determination of amount kind and causes of variability within a species.
- Packaging of desirable qualities into improved individuals so as to develop trees with combination of desired characteristics.
- Mass producing improved individuals for large scale plantation purposes.
- Developing and maintaining genetic base populations broad enough for needs in advance generation.

Tree improvement is carried out in stages. Each of these stages provides opportunities for improving quantity and quality of forest reproductive material. The challenge is to make sure each stage is adequately carried out and gains maximized, before advancing to the next. In some cases two stages may run concurrently. The sequence of stages is as follows:

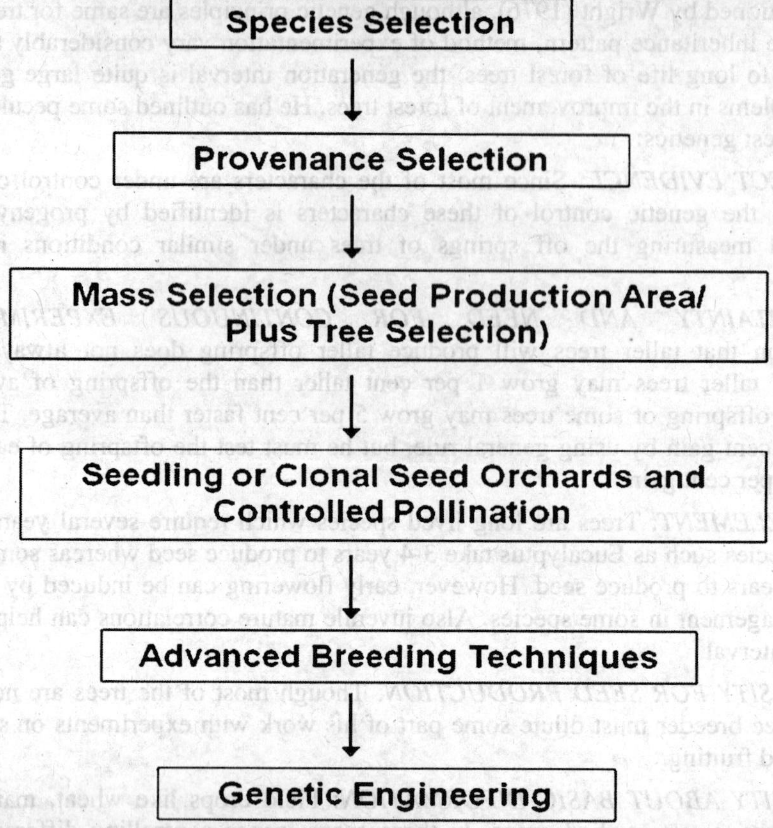

Fig. 18.1: Sequence of Stages in Tree Improvement

SPECIES AND PROVENANCE TRIALS (S & P TRIALS)

The foremost step in any tree improvement programme is the right choice of species followed by selection of suitable provenance within a species. It is necessary to conduct species and provenance test before making right choice of species and provenance for specific site. Species and provenance trial should be conducted according to written plans, which will usually form part of research programme of national forest authority. Plan of this sort must state clearly precise objectives of the particular species or provenance trial project. These should be derived from the objectives of national afforestation schemes and constraints upon them (Burley and Woods 1976). Species and provenance trials must be aimed at reducing a large number of entries to relatively few proven species or provenances suitable for production of desired forest products.

THE BASIC FEATURES OF SPECIES AND PROVENANCE TRIALS (Hughes and Willan 1976)

- Review of literature, correspondence and personal knowledge of distribution and variation of species
- Selection of parent stands of a species in natural forest or seed stand in plantation. Procurement of seed and storage.
- Assessment and selection of site, planning and designs of experiments.
- Design and conduct of nursery experiments, nursery assessment and early testing like juvenile mature correlations
- Techniques and assessment in forest stage. The main activities are the duration of the trials (short, medium and long), designing and conduct of experiments and assessment.

PHASING OF SPECIES AND PROVENANCE TRIALS

Before starting large scale plantation programme, it is essential that adequate information about the requirement of the species and characteristics of site are obtained. Often this information is lacking at initial stages. Therefore the best way to get this information on suitable species/ provenance is to lay trials of number of species in small plots on localities which have been carefully selected. Species or provenance should represent geographic variation of the species. Sites selected for experimentation should be done on ecological requirement of the species (Burley and Woods 1976). The best way is to obtain gradual information by phasing the S & P trials as given below:

SPECIES TRIALS: Initially large number of species is tested. Ultimately, one or two suitable species are recommended for plantation. Species selection passes through number of stages. Species are selected on the basis of their suitability for plantation sites. Initially trials are preliminary and are conducted on small sized plots for short period followed by trials with large plots and larger period. An idealized sequence for species trials will be:

The Arboretum Phase: It involves testing of large number of species. The number of individuals of species is very low. The species are kept for longer duration, probably for their full life.

The Species Elimination Phase: The objective is to compare performance of large number of different species on one or more sites and to select a smaller number for more intensive trials. The number of species to be included in the trial depends upon availability of funds, seed, infrastructure, suitability of species to site and desired end product.

The Species Testing Phase: The objective is to compare restricted number of promising species based on previous experiments within a broad climatic region. Properly statistically designed layouts are important in this phase. Plots must be large enough to enable reliable assessment to be made up to at least first thinning stage. Sites should be stratified in order to evaluate the interaction between site variation and species difference. Again, the number of species tested depends upon availability of funds, infrastructure and adequate supply of seed from natural forests or local seed stands and number of suitable species from previous phase.

The Species Proving Phase: The objective is to confirm under normal plantation conditions, the results shown by the small number of species that have shown themselves superior in early phase. Trials should large enough to provide data on growth and yield for full rotation. Surrounds should be large enough to minimize edge effects. The trials also include different types of nursery methods, different types of planting methods, spacing and thinning trials etc.

Table 18.1: Phases of Species Trial

Phases	Number of Species	Duration of Trial	Plot Size	Spacing
Arboretum	Large	Longer	-	-
Species Elimination	20-40	0.1 to 0.2 x rotation age	5 tree line plot, maximum 25 trees (5x5)	2 x 2 m or 3x 3 m
Species Testing	5-10	0.5 x rotation age	16-25 trees, 1 to 2 row surround	2.5 x 2.5 or 3 x 3 m
Species Proving	2-3	Full rotation	100 trees, 2 row surround	2.5 x 2.5 or 3 x 3 m

PROVENANCE TRIALS: After selection of suitable species it is important that the most suitable seed source for planting site is selected. This is accomplished by selecting suitable provenances. Provenance is the area on which any stand of tree is growing. The stand may be indigenous or non- indigenous. For an indigenous stand of trees the origin is the place in which the trees are growing; for a non-indigenous stand the origin is the place from which the seeds or plants were originally introduced. Provenance trial is also conducted in phases.

Range wide Provenance Phase: The objective is to determine the content and pattern of variation between provenances of promising species. The number of provenances included in this phase depends upon the geographical distribution and variation in the species. This phase is run concurrently with species elimination phase. It is desirable to sample the entire range of the species. This provides insurance against the possibility that some overlooked region is capable of providing the best trees. Seeds should be collected from several trees per locality. Parent trees could be average or selected and then bulked. A range wide provenance test usually indicates total range of genetic variation within a species

Restricted Provenance Phase: The objective is to find sub regions and ultimately provenances most suited to site under test. It detects minor trends and gives location of specific stands or small areas having the best germplasm. The test provides information on comparative amount of geographic and individual tree variability to function ultimately as seed orchard.

Provenance Proving Phase: The objective is to confirm under normal plantation conditions, the results shown by small number of provenances that have shown themselves superior in earlier phase. Here also the plots should be large enough to provide data on growth and yield for full rotation and surrounds should be large enough to minimize edge effects. In addition to normal plantation conditions a range of management techniques need to be tested.

Table 18.2: Phases of Provenance Trial

Phases	Number of species	Duration of Trial	Plot size	Spacing
Range wide Provenance	10-30	0.5 x rotation age	5 tree line plot, maximum 25 trees (5x5), no surrounds	2 x 2 m or 3x 3 m
Provenance Testing	3-5	0.5 x rotation age	25- 49 trees , 1 to 2 row surround	2.5 x 2.5 or 3 x 3 m
Provenance Proving	1-2	Full rotation	100 trees, 2 row surround	2.5 x 2.5 or 3 x 3 m

MASS SELECTION

Once the species has been selected and promising provenance has been delineated, the next step in tree improvement is selection. Simple form of selection is mass selection (Seed Production area etc.) which is an extension of natural selection with choice of trees left to man instead of nature.

SEED PRODUCTION AREA (SPA)

It is a seed-producing stand which has been developed either from existing planted stand or from stand specially planted and generally upgraded and opened by removal of undesirable trees and then cultured for early and abundant seed production. SPA constitutes a reliable source of seed of certain genetic quality, until such time the seed orchards come into production. They are a very effective way of making available seed supply of an inexpensive but somewhat improved genetic quality. Although the amount of genetic improvement from a seed production area is expected to be small, the seed obtained from seed production areas still has better genetic qualities than seed from commercial collections in routine plantations, especially in adaptability, form and pest resistance. SPA is a form of mass selection in which gains can be obtained by selecting or characters showing high heritabilities. A gain of 5-10 per cent may be possible but real genetic value is not known since the trees in the stand are not subjected to progeny trials.

DEVELOPMENT OF SEED PRODUCTION AREAS FROM EXISTING PLANTED STANDS (Khongsak 2005)

The quickest way to make a seed production area is to convert existing mature stands of good quality trees for this purpose. A seed production area can also be developed from a progeny or provenance trial. There are no specific age limitations but the stand should be old enough to produce seed. A planted stand suitable for conversion to a seed production area should have the following specifications.

Details of the seed source used to establish the stand should be known: Knowledge of the geographic origin and genetic base, which ideally will be broad, is important in determining the suitability of the stand for development into a seed production area.

The trees should be mature for seed production, but not too old: There are no age limitations, other than that the stand must be old enough for reliable selection and be sexually mature to produce seed. Individual trees must have sufficient crown density to potentially produce large seed crops. Seeds from very young or very old trees often are of inferior quality compared to those from middle age trees. In addition, old trees tend not to respond to thinning by further crown development.

The stand should be near full stocking and contain a large number of trees of good phenotype: A suitable stand before conversion should be near full stocking to increase the selection intensity. The final stocking of a seed production is usually 150-200 stems per ha. An initial survey should reveal that the number of trees of good phenotypes is equivalent to that.

Free from pests and diseases: Evidence of pests or diseases may be an indication of inferior adaptability of the seed source. If the stand shows widespread symptoms of attack by insects or diseases it should not be used for conversion.

The trees should have proven capacity to produce flowers and seeds in the area: To avoid failure, a detailed survey of the candidate species' ability to flower and seed in a particular environment should be a prerequisite to establishing seed production area. This is especially important for exotic species where flowering may fail or seeds are not produced due to incompatibility with the site. Non-favourable environments, such as drought prone areas, may be unsuitable for seed production area. Availability of pollinators can also be important.

The area should be easily accessible: The conversion of a stand into a seed production area and subsequent management operations, such as harvest of seeds, requires that the area is accessible throughout the year. This implies that it should be relatively flat but well drained. Ideally it should not be in a very remote area or too far from available labour for ease of maintenance and management.

The selected stand should not be subject to commercial harvesting: Seed production areas will be maintained for many years. It is necessary, therefore, to ensure that the stand in question is safe from commercial harvesting operations. Good communication with the forest managers will help avoid the accidental loss of valuable seed production areas.

Size of seed production areas: In general a minimum of 4 ha is recommended for practical management of seed production areas for most tree species. Managing small stands is inefficient and there is a great danger of contamination from outside pollen. However, for some species like eucalypts, an area as small as 0.5 ha may be sufficient due to prolific seed production. The area should be as close to square as possible, rather than a long linear block to facilitate more cross pollination among the trees.

Isolation: Poor isolation will be a common disadvantage of seed production area developed from existing stands. Usually such stands are selected from mature plantations of the same species. It is virtually impossible to eliminate completely, contamination by stray pollen. However, an isolation zone or pollen dilution zone surrounding the seed production area can reduce it. The dilution zone may be an open area of some 200 m. If trees are to be grown in the dilution zone they must be of a species that does not hybridize with the species of interest for seed production.

SELECTION OF TREES FOR A SEED PRODUCTION AREA

Desired attributes of the trees left in a seed production area are similar to, but less rigorous than, the qualifications required for selecting tree to be used in an intensive tree improvement program. Only trees in the dominant and co-dominant crown classes are considered for retention because of their growth and seed-producing potential. For timber species, fast growth, straight bole, good crown development, self pruning etc. are the criteria for selection of trees. Inferior trees should not be kept even if there is a gap. It is essential that the crowns of the trees be exposed to full sunlight of at least three sides if good seed production is to be realized.

Selection of trees should be done in small groups for ease of comparison. By dividing the stand into small plots, a more even spread of seed trees will be retained after final thinning. Trees that do not meet the desirable standard are marked for roguing. Some species such as Casuarinas are dioecious, i.e. trees are unisexual with separate male and female trees. Care must be taken to ensure that there are sufficient male trees left for pollination. A final stocking of 150-200 trees per ha is considered suitable for cross pollination and subsequent seed production in a wide range of commercial tree species.

PLUS TREE SELECTION

Plus trees are individual trees of outstanding merit, initially selected on the basis of superior phenotypic characters like height, diameter, clear bole etc. Phenotypic selection of superior individuals forms the basic part of the most tree improvement programmes. Selection of superior phenotypes in forestry is most effective, as even a small genetic gain can be of enormous economic benefit. Selected individuals are used as a base population for future breeding programme. The selection of superior parents from stands having uniform environment is of great importance. Selection from plantations gives more gain than from the natural stands. Two factors are of paramount importance in developing a grading scheme for selection of superior trees. First the trait under consideration should be under at least moderate to strong genetic control. Second, the trait must have considerable economic value. Growth rate is nearly always key characteristics in a selection programme.

STAND SELECTION: Within a provenance/ seed zone, selection starts with designation of suitable stands. Stands should be chosen on several site classes to avoid selection of trees adapted to narrow range of environmental conditions. On the contrary selection of stands on better sites is also justified because these are the sites that will have priority in regeneration programme. Select middle aged stands, close to expected future rotation as selection in stands of present rotation age will not lead to identification of genotypes that perform best under management conditions. If choice is to be made between even aged and uneven aged stands, the former should be preferred because they allow more accurate tree comparison.

NUMBER OF PLUS TREES: The number of plus trees selected depends upon overall breeding plan for the selected species. A large number of plus trees provide a broad genetic and avoid inbreeding. The trees can be selected within few years. A breeding/ production population should be at least consisting of 20-30 clones/ genotypes.

CHOICE OF TRAITS: The characters to be improved may vary from species to species and objectives of the programme. Selection should be restricted to few traits as too many traits will increase the cost. Traits with long term economic value should be chosen. Improve only those traits which cannot be improved through silviculture or improved industrial technology. The general characteristics of tree for timber purpose should be: (i) rapid growth, (ii) straight bole, (iii) no forking, (iv) branches: thin, small, slow growing, (v) narrow crown, (vi) self pruning habit, (vii) quality timber, and (viii) tolerant to diseases and pest.

METHODS OF PLUS TREE SELECTION

The procedure used for selection of plus tree in the selected stands will depend upon number of factors like type and number of traits considered, heritabilities, selection intensity, stand structure etc. Based on Van Buijtenen (1969), following are the methods of selection of plus trees.

OCULAR SELECTION: Vigorous, healthy trees of good form are rapidly identified without measurement or rating of individual traits. This procedure is adopted in both the situations where heritability of character is either low or high. It is applicable when selection is restricted to very few traits.

COMPARISON TREE: The method is suitable in plantations and natural stands of uniform age. A candidate tree which is phenotypically superior in chosen traits is compared with its nearest neighbours, or average or best trees in the stand. The comparison trees should be selected from dominant or co-dominant crown with similar age and site conditions approximately with 100 meter range from the candidate tree. The superiority of candidate tree over the average of comparison trees is worked out for each trait. The candidate tree is designated as plus tree if it proves superior to comparison trees; otherwise it is rejected.

REGRESSION METHOD: This method is applied for uneven aged or mixed species forests. This requires the development of tables relating the desired character with age. Regression line is developed for a stand or site, relating the expression of characters to tree age by sampling a number of trees. The sampling requires the trees of different ages. Different regressions are developed for different sites. Once the regression curve has been developed, the candidate values are plotted against the regression line and the trees that meet the minimum selection standards or are above the regression line are selected as plus tree. This method cannot be used for those species whose age cannot be determined. .

Table 18.3: Plus Tree Record Form (Source: Paramathma 2002)

Species							
Location in detail							**Candidate Tree No.**
							Longitude
Description of stand							
							Latitude
Soil							
Flora							
Characteristics	**Comparison Trees**						
	1	**2**	**3**	**4**	**5**	**Average**	**Candidate Tree**
Age							
Height							
Total							
Clear bole							
DBH							
Form factor							
Forking							
Height							
Diameter							
Angle							
Crown							
Width							
Length							
Branches							
Angle							
Diameter (L, M, U)							
Bark thickness							
Wood							
Specific Gravity							
Fibre Dimension							
Volume							
Seed bearing (H,M,L)							
Disease/insect pest attack (A, M,N)							

Scoring Measurements (L,M, U: Lower, Medium, Upper; H,M,L: High, Medium, Low; A,M,N: Abundant, Medium, Low)

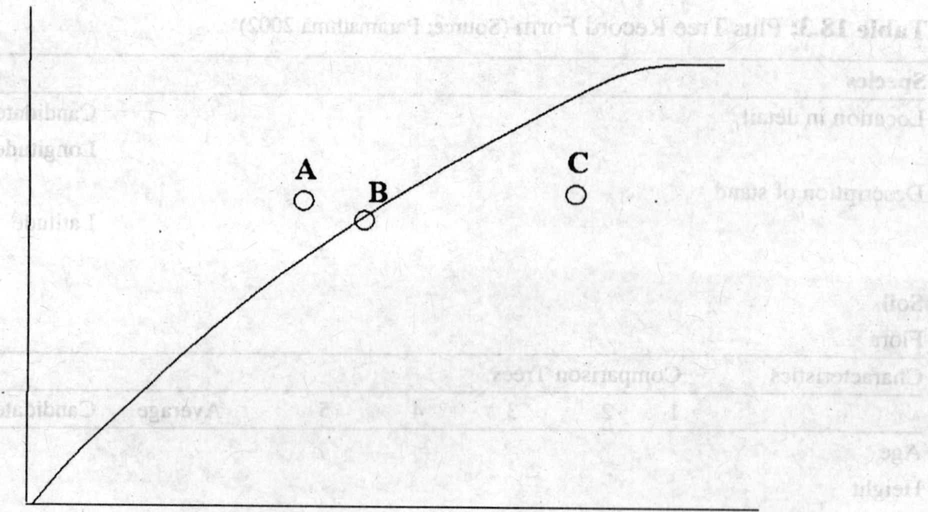

Fig. 18.2: Regression Line for Plus Tree Selection

The regression system of selection is particularly suitable for all aged or mixed species stands. It consists of developing a curve of production (growth is illustrated) for different ages of trees on a given site. Candidate *Tree A* falls above the curve, therefore, it has the desired growth for its age. *Tree B* is average, therefore, its use depends on other characteristics, whereas *Tree C* has inferior growth for its age and should not be used. The regression line should be based on at least 50 trees if the age spread is considerable.

THE MOTHER TREE SYSTEM: It consists of locating good trees. Seed is obtained from these trees to lay out progeny trials. On the basis of progeny trials, either the best parent trees or best trees of best families can be used in a vegetative seed orchard. The progeny test may also be converted to seedling seed orchard. This type of selection is used in situation where comparison tree or regression method cannot be used. The main problem with this method is that time is lost in progeny testing before establishment of seed orchard for commercial seed production. However, this method is highly suitable for hardwood where planting programmes are small and seed is not immediately needed.

SEED ORCHARDS

A seed orchard is a plantation of genetically superior trees, isolated to reduce pollination from genetically inferior outside sources, and intensively managed to produce frequent, abundant, easily harvested seed crops. It is established by setting out clones (as grafts or cuttings) or seedling progeny of trees selected for desired characteristics.

ADVANTAGES OF SEED ORCHARDS (Morgenstern et al *1975)*

- Isolation from surrounding wild populations to avoid pollination by outside pollen.
- Selection of locally favourable site with respect to climate, soil and application of management.
- Promotion of early flowering by grafting of mature scions.

- Avoidance of inbreeding by randomizations of clones or genotypes.
- Maximizing seed yield through cultural practices such as fertilization, maintaining wide spacing, bending or girdling of stems, root pruning, irrigation and control of weeds and animals.

TYPES OF SEED ORCHARDS

Seed orchards are of two types based on propagation material: clonal seed orchards and seedling seed orchards.

THE CLONAL SEED ORCHARDS (*CSO*): Vegetative material (grafts, cuttings or plantlets derived from tissue culture) of selected phenotypes (plus trees) are planted in areas with good isolation, under conditions favouring flowering and fertilization and managed for production of maximum amount of seed. The identity of each ramet (member of a clone) is carefully maintained by tags and maps. It is a characteristic of the clonal orchard that main emphasis is given to crown development for maximum seed production by use of wide spacing. Therefore silvicultural characteristics such as fine stem form, clean bole and fine branches, which are important in production forests or plantations and are valid criteria for selecting clones for inclusion in the seed orchard and are not valid criteria for seed orchard management. The establishment of a clonal seed orchard should be preceded, or at least closely followed, by progeny tests. CSOs established by grafting are most commonly used. The greatest disadvantage of grafted clonal orchards is the laborious work involved with scion collection and grafting, maintaining healthy ramets in nursery before planting and last but not least, incompatibility between stock and scion, which may occur even several years after establishment. To avoid narrowing of genetic base, at least 25-30 clones should be planted. The ramets of clones should be distributed in such a way that inbreeding is avoided. Adequate measures should be taken to keep the contamination of outside pollen below 20 per cent.

Table 18.4: Comparison between Seedling Seed Orchard and Clonal Seed Orchard (Paramathma 2002)

Seedling Seed Orchard	Clonal Seed Orchard
Established from seedling, i.e. by seed propagated from selected parents	Establishment by the use of grafts, cuttings, air layered plants, tissue cultured plantlets or other methods of vegetative propagation
Preferred if genetic tests can be converted in SSO i.e. fulfilling testing and seed production function at one time	The seed orchard serves as production area only for seed or vegetative propagules
Necessary when incompatibility stock or scion or other constraints make vegetative propagation difficult	Generally preferred when vegetative propagation is possible.
Broader genetic base than CSO but less selection differential	Narrower selection base than SSO but higher selection differential.
An outstanding genotype will only appear once. Testing is at family level	An outstanding genotype will appear many times. Testing is on individual level.

First flowering and fruiting occurs later than CSO, applicable in species with early physiological flowering	Generally flowering earlier than SSO
Harvest of seed or fruit more difficult than CSO	Harvest of fruit and seed easier due to low crown branching.

THE SEEDLING SEED ORCHARDS (*SSO*): Progenies from open or controlled pollination of selected phenotypes are planted at normal plantation spacing. Isolation and other conditions should be as for clonal orchards. The identity of families is maintained in order to allow for thinning among families and individuals within families based on their phenotypic performance. This thinning must be done before abundant seed production starts. The design of the orchard is of great importance for the later thinnings. Seedling seed orchards permit selection and rouging within as well as between families. One disadvantage of SSO is that their formation and future quality is based on early selection, at a time when the juvenile/ mature correlation may still be low. SSO also generally take longer to start flowering and to reach seed productivity.

SEED ORCHARD DESIGN

The general requirement in the orchard design is:

- Minimize selfing
- Maximize outcrossing and mating of all genotypes
- Simple and easy establishment and management
- Allowing any number of clones or families to fit into the design

CSO: Planting distance of between 2 meters and 16 meters in both directions have been reported, however, an initial spacing of five to six meters is used. There is also trend towards wider initial spacing since it is now generally accepted as being more economic in the longer term to accept lower early seed yields than to systematically thin expensively raised grafts.

SSO: Planting distances are less than clonal seed orchards. They range between 0.6 to 6 m in both directions with majority at 0.5 to 2.0 m. It is a common practice to use large number of individuals/ family plot thus providing better opportunity for within plot rouging at later stages.

PLANTING PATTERN: Either square or rectangular arrangement of plots is used. A rectangular arrangement provides more access to machinery and full advantage of sun.

NUMBER OF CLONES /FAMILY: Most orchard designs are based on 20-50 clones keeping in view broad genetic base.

SEED ORCHARD DESIGNS: Number of seed orchard designs are in vogue: pure row, chess board, completely random, randomized complete block, fixed block, rotating block, unbalanced and balanced incomplete block, cyclic balanced incomplete block, balanced lattice, permuted neighborhood, alpha design etc. Each design has its advantage and disadvantages. The two most frequently used designs are complete randomized design and randomized complete block design:

Randomized Complete Block Design: The area is first divided into equal blocks, each sufficient in size to contain one ramet of each clone or one member of each family. The ramet position in each block is completely randomized and then manipulated to avoid similar ramets in adjacent planting positions. These design help in accounting for variation across different blocks in the experiment site. They are however, not very effective in accommodating large number of genotypes as their will be much heterogeneity within the large blocks.

Incomplete Block Design: To accommodate large number of genotypes (200 or more), incomplete block designs are increasingly used. Here the blocks are smaller than a complete replication and the genotypes are planted in plots within a block and groups of incomplete blocks constitute a replication.

LOCATION, ESTABLISHMENT AND MANAGEMENT OF SEED ORCHARDS

LOCAL CLIMATE AND SITE REQUIREMENT: Above average soil fertility is essential for good orchard tree development. Shelter from strong winds and good air and soil drainage are also basic requirements. Whereever basic information on site selection is lacking, it should be possible to identify potentially good sites from observation of flowering and seed production in local stands.

ISOLATION FROM UNDESIRABLE POLLEN SOURCES: Good isolation from contaminating pollen sources is a necessity. Complete isolation in wind pollinated species is however very difficult to obtain, unless orchards are established outside the region in which they are found in plantations or natural stands. Ideally seed orchard should be established on sites where contamination of pollen from same or related species is negligible. If this is not possible then pollen dilution zone should be maintained to avoid outside pollen. There should be a gap of 300 to 500 m between plantation and seed orchard and orchard should be located on windward side of plantation (Faulkner 1975).

ESTABLISHMENT OF SEED ORCHARD: The cost of producing planting stock from seed orchard is quite high; therefore it is essential to achieve high standard of general site preparation and planting. Site preparation should start well in advance. Need for fertilization should be determined on the bases of soil analysis. Protection against wild animals should be provided by adequate fencing before planting. Take utmost precautions while marking plant positions on the ground. The position should be marked with stakes bearing an appropriate label for identification. Labels of plants from nursery should be checked against those on stakes before planting.

MANAGEMENT OF SEED ORCHARD: Weeds should be kept under strict control within half meter radius of the plant until it is well established and growing vigorously. Subsequently entire orchard area should be kept weed free. For proper growth and flowering, fertilizers containing nitrogen, phosphorous and potassium should be used.

PROGENY TESTING

Once plus trees are selected on the basis of their physical appearance and planted in a seed orchard, their genetic worth can be known only on the basis of performance of their progeny. The relative contribution of genotype and environment towards phenotype of the selected tree can be determined through progeny test. The objectives of progeny testing are:

1) to evaluate family lines for the purpose of rouging out seed orchards; 2) to establish plantings from which to make second generation selections. Progeny testing is advantageous in case where most of the variation is environmental. One can make substantial gains by separating the total variation in environmental and genetic components through progeny testing. Progeny trials are either half sibs or full sibs. In half sib progeny trials, open pollinated seeds are collected from selected parents to lay out trials for assessment of general combining ability (GCA). In full sib progeny trials, controlled cross pollinated seeds are collected to lay out trials for the assessment of specific combining ability (SCA). A summary of advantages and disadvantages (Van Buijtenen et al 1981) of different progeny testing method is given in Table 18.5.

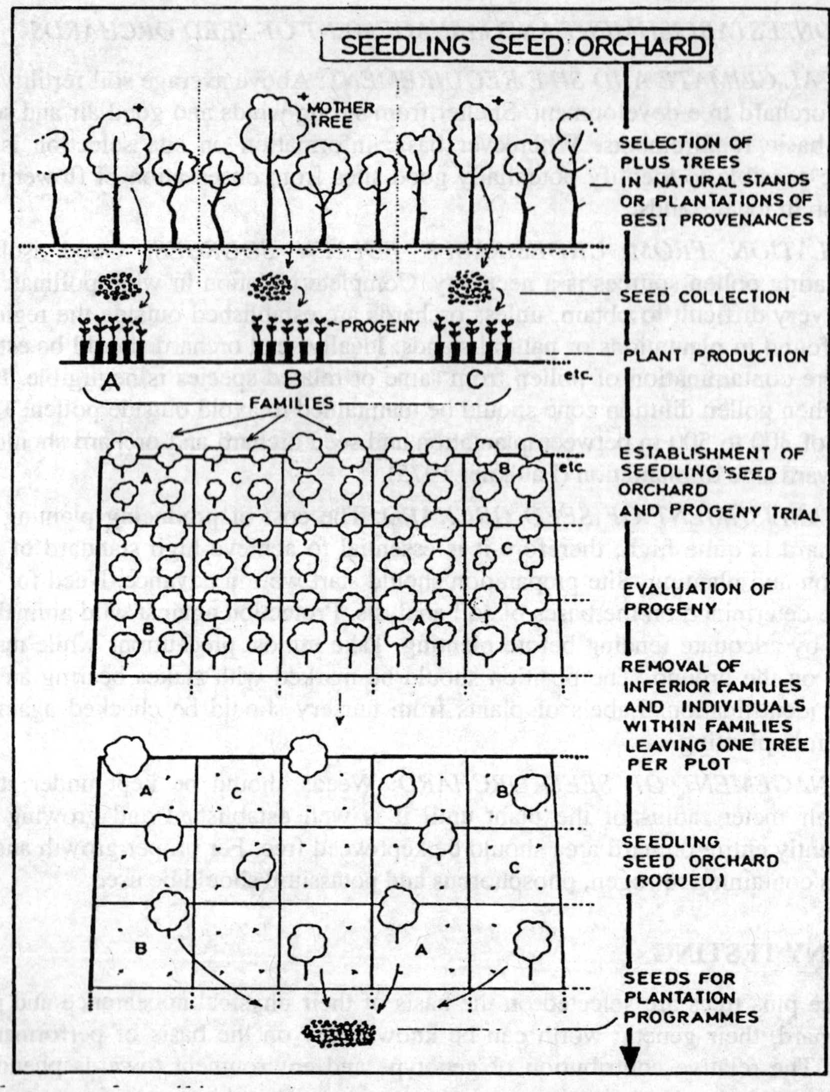

Fig. 18.3: Establishment of Seedling Seed Orchard (Ditlevsen et al 1988)

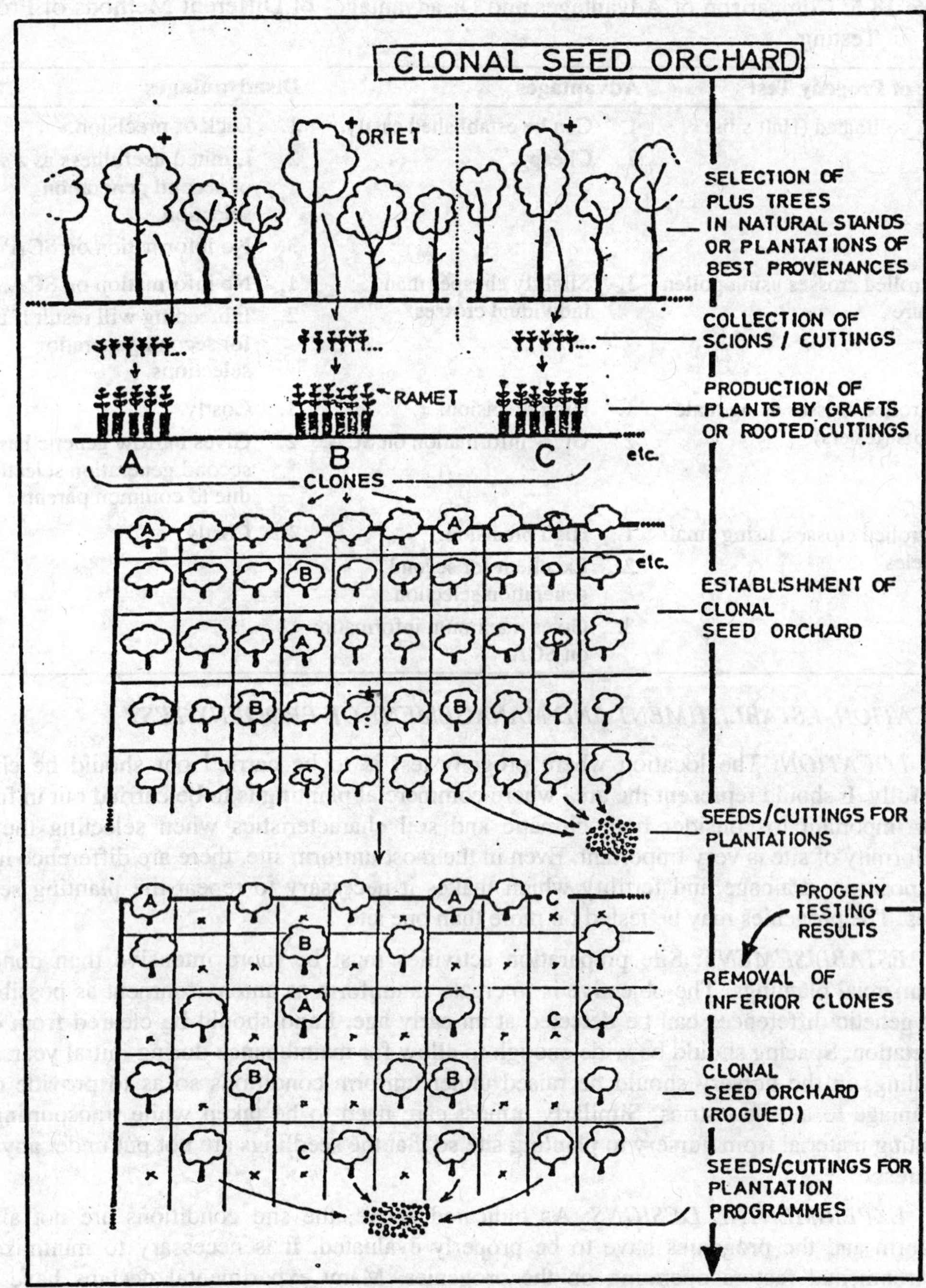

Fig. 18.4: Establishment of Clonal Seed Orchard (Ditlevsen et al 1988)

Table 18.5: Comparison of Advantages and Disadvantages of Different Methods of Progeny Testing

Type of Progeny Test	Advantages	Disadvantages
Open pollinated (Half sibs)	1. Can be established easily. 2. Cheap	1. Lack of precision. 2. Limited usefulness as a source of second generation selection. 3. No information on SCA
Controlled crosses using pollen mixture	1. Slightly cheaper than individual crosses.	1. No information on SCA. 2. Inbreeding will result if used for second generation selections.
Controlled crosses using male parents (testers)	1. High precision. 2. Gives information on SCA.	1. Costly. 2. Gives narrow genetic base for second generation selections due to common parents.
Controlled crosses, using small dialleles	1. High precision. 2. Excellent for second generation selection. 3. Gives maximum information on SCA.	1. Costly

LOCATION, ESTABLISHMENT AND MANAGEMENT OF PROGENY TEST

LOCATION: The location where progeny test is to be carried out should be chosen carefully. It should represent the area where commercial planting is to be carried out in future. It is important to consider both climatic and soil characteristics when selecting the site. Uniformity of site is very important. Even in the most uniform site, there are difference in soil composition, drainage and fertility which makes it necessary to repeat the planting several times. The progenies may be tested on more than one site.

ESTABLISHMENT: Site preparation activities must be more intensive than done for commercial plantings. The objective is to create as uniform as an environment as possible so that genetic differences can be detected at an early age. Land should be cleared from other vegetation. Spacing should be wide enough to allow for maintenance during initial years. The seedlings in the nursery should be raised under uniform conditions so as to provide equal advantage to all the entries. Similarly utmost care need to be taken while transporting the planting material from nursery to planting site so that the seedlings are not put under any kind of stress.

EXPERIMENTAL DESIGNS: As indicated above, the site conditions are not always uniform and the progenies have to be properly evaluated. It is necessary to minimize the environmental factors operating on the progenies. Many experimental designs have been developed to control the environmental effects in the experiments. However, the most commonly used design is the randomized complete block design or incomplete block designs (see seed orchard designs). Replication should be done in both space and time. A poorly replicated plot on a highly variable site will yield little reliable information. Progenies within a replication should be properly randomized. Border rows should be planted to remove the edge effects.

PLOT SIZE: In the past, progeny tests were conducted using the plot size as in the provenance trials that are square plots containing 25-100 trees. This will often result in the large size of the plots thereby increasing variation among the plots. To overcome this problem, in many programmes, progeny tests are conducted using row plots. Use of small plots ensures that the genetic differences among the progenies are accurately evaluated. Row plots are easier to plant and measure than traditional square or rectangular plots. It is often desirable to use 10 tree row plots and six replications at the beginning of the testing programme (Otegbeye 2001).

MANAGEMENT: Usually progeny trials run for half or full rotation age. Therefore, it is imperative that proper labeling and documentation of the trials in the field as well as in the office is done. The maps of the site, layout of the trial should be properly maintained. The trial should be protected from fire by establishing fire lines around the test. Diseases and insects should be controlled as per standard protocols. Thinnings may or may not be done in the progeny test. If thinning is carried out, proper records should be maintained. The measurement of height, diameter etc. in the trial is done periodically. Data should be recorded accurately.

ADVANCE GENERATION BREEDING

Mass selection (Seed production area/ plus tree selection) is carried out in the natural population or plantations which are the source populations. The selected individuals are used to establish seed orchards. The seed orchards are rouged on the basis of progeny test. Information from progeny test is used to selected individuals which are intercrossed using different mating designs to develop populations which are source of second and third generation seed orchards. Advance generation populations also provide estimates of genetic parameters such as additive and non additive genetic variation, heritability, genetic gain and production of commercial end product. In improvement strategy, for every generation, an interim seed orchard has been created, which is improved through grafting of selected material following intermating of good combiners. Broader genetic base is maintained through infusion of new entries from the unimproved populations in later generations. Tree improvement programs consist of two main lines of endeavor, the operational phase and developmental phase, as indicated. The quality of the breeding clone bank will determine long-term success and gains from the production seed orchards determine immediate success.

GENETIC ENGINEERING

Traditional approaches to tree improvement have involved the identification of mature trees with desirable phenotypes, followed by their incorporation into breeding programs. The length of time needed for trees to reach reproductive maturity, often in excess of 20 years, before controlled crosses can be made is the limiting factor for tree improvement (Martin-Trillo and Martinez-Zapater 2002). Furthermore, there is no guarantee that the desired progeny phenotypes will be identified. The application of biotechnology can overcome many of the drawbacks associated with conventional breeding strategies. This may be achieved by the use of genetic modification technologies. A genetically modified (GM) tree (sometimes referred to as a genetically engineered tree or a transgenic tree), can be defined as a tree that "through human intervention in a laboratory, has had its genome, or genetic code, deliberately altered through the mechanical insertion of a specific identified sequence of genetic coding material (generally DNA) that has been either manufactured or physically excised from the genome of another tree. The new genes are passed on to future generations. Genetic modification may be used to alter or introduce a wide range of traits, including insect and disease resistance, herbicide tolerance, tissue composition and growth rate.

The first GM tree trial was carried out in Belgium in 1988. Since that time many tree species, both for fruit and timber production, have been genetically modified and tested in field trials. Trials have been conducted in Europe, North and South America and New Zealand, principally for facilitating pulp production and for increasing productivity. In China the State Forestry Administration has approved GM poplar trees for commercial planting and one million insect resistant GM poplars have been planted. There is enormous potential for speeding up tree breeding cycles by the use of genetic modification (Humphreys 2005).

Systems have been developed for gene transfer, selection of novel gene-containing shoots and stimulating regeneration for both broadleaved and coniferous trees. The systems use either the soil-based microorganism *Agrobacterium* or DNA-coated particle bombardment, known as biolistics. Two species from the genus *Agrobacterium* (*A. tumefaciens* and *A. rhizogenes*) have been used for gene transfer into trees. A small section of the bacterial DNA, known as T-DNA, is transferred into and expressed within the plant nuclear genome. *Agrobacterium tumefaciens*, the causative agent for crown gall disease, is widely used after the functional deletion of the auxin and cytokinin biosynthetic genes from the Ti, or tumor-inducing, plasmid. A separately replicating plasmid, known as a binary vector, can be introduced and used to deliver desired genes into the plant cell, from which phenotypically normal transgenic trees can be regenerated (Gartland and Davey 1995). *Agrobacterium rhizogenes*, responsible for hairy root disease, transfers genes from root-inducing, plasmid. Applications of genetic modifications to trees include altered wood properties, speeding up breeding cycles, forests as pharmaceutical factories, dendroremediation and improved resistance to pests and diseases, as well as the restoration of sensitive landscapes (Gartland et al 2003).

But the application of GM technologies to trees has raised a number of potential public concerns. Many of these concerns are the same as those raised for GM annual crop plants. This includes the potential for spread of antibiotic or herbicide resistance genes to other non target species from GM trees; the potential for long-distance pollen spread over many years from long-lived trees; the potential for adverse effects on biodiversity from forests of GM trees etc. Although so far adverse effect of GM trees on environment has not been found, still the issue is debatable and needs much more research on safety of GM trees.

MOLECULAR TECHNIQUES: Traditionally genetic variation was assessed through morphological, growth and cytological characteristics. The problem with morphological and growth characteristics is that they are highly influenced by environment. They reflect only part of the genetic variability since only functional genes are taken into consideration. With the advent of isozymes and other biochemical markers, the effectiveness in estimating genetic diversity increased tremendously. However isozymes are handicapped by low polymorphism and influence of isozyme variation by environment to certain extent. Now DNA based molecular techniques such as Random Amplified Polymorphic DNA (RAPD), Restriction Fragment Length Polymorphism (RFLP), Simple Sequence Repeat (SSR) are being increasingly used to detect genetic variation in populations. RAPD utilizes Polymerase Chain Reaction (PCR) to screen the DNA for detecting variability. Polymorphism i.e. difference between two genotypes with reference to particular location in the DNA results from changes that effect amplification of DNA located between two primers. These changes include most probably single base substitution as well as deletion or insertions that either changes the primer sequence or the size of the amplified DNA (Williams et al 1990).

CLONAL FORESTRY

Vegetative propagation is one of the important tools in tree improvement. It can capture both additive and non additive genetic variance thus maximizing gains in tree improvement programmes. It has been widely used in short rotation forestry (Singh et al 1999). Poplars, Willows, Eucalyptus etc. are mass produced through vegetative means. Besides its role in production forestry, it can be used to preserve genotypes in clonal banks (Thakur et al 2000), for establishing seed orchards and transferring traits which have low heritabilities. In species where vegetative propagation is feasible, best trees can be vegetatively propagated for substantial gains.Different types of vegetative propagation methods like grafting, rooting of cuttings, air layering, tissue culture, meristem and cell culture are available. Despite its advantages, vegetative propagation is dead end to breeding. Gains through clonal forestry decline over time unless backed by continued breeding and testing.

ACHIEVEMENTS IN TREE IMPROVEMENT

Tree improvement work in India was started by Prof. Champion who realized the importance of geographical variations and application of the knowledge of forest genetics during 1930. He established a provenance trial of chir pine at New Forest, Dehradun and found that spiral gains in chir pine are inherited. Laurie during the same period established an all-India Teak provenance trial, which remained the most important one in Asia. Forest Research Institute, Dehradun opened a Forest Genetics section during 1959-60 attached to Botany Branch under the then Directorate of Biological Research at FRI. In the year 1961 Prof. J.D.Mathews an expert from FAO suggested initially to work on some priority species viz. *Tectona grandis*, *Bombax ceiba*, *Pinus* spp. *Dalbergia sissoo*, *Santalum album*, *Morus alba* and *Pterocarpus santalinus*. During Vth plan period a central sector schemes were initiated in collaboration with Danish International Developmental Agency (DANIDA) as "Indo-Danish Project on Seed Procurement and Tree Improvement" under which tree improvement was carried out on *Pinus roxburghii*, *P.wallichiana*, *Cedrus deodara*, *Dalbergia sissoo*, *D.latifolia*, *Albizia lebbek*, *Tectona grandis*, *Bombax ceiba*, *Gmelina arborea* (Emmanuel et al 1992).

PROVENANCE TRIALS: Provenance trials have been carried out in number of species. A test consisting of 75 provenances *Tectona grandis* was established in 1971 by DANIDA on 50 sites in 16 countries. The trial was evaluated in 1983. A provenance trial consisting of 23 provenances from four states was established at Forest Research Institute, Dehradun in 1981. In 1995, trial of 30 neem provenances was established under International Neem Network where 21 countries are participating.

SEED STANDS AND SEED PRODUCTION AREA (SPA): Seed stands have almost selected in every state. Approximately 15,000 ha of seed stands have been selected [*Tectona grandis* (9358 ha), *Pinus roxburghii* (2934.2 ha), *Eucalyptus* spp. (99.44 ha) etc.]. The total area under SPA is 1225.6 ha [*Azadirachta indica* (87ha), *Dalbergia sissoo* (173 ha), *Tectona grandis* (433 ha), *Eucalyptus tereticornis* (42 ha) etc.].

PLUS TREE SELECTION: Selection of plus trees through comparison method have been reported in number of species such as *Pinus roxburghii*, *Populus ciliata*, *Pinus gerardiana* in Himachal Pradesh; *Dalbegia sissoo* in Punjab, Haryana, Maharashtra; *Tectona grandis* in Andhra Pradesh, Karnataka, Maharashtra, Kerala, Madhya Pradesh; *Casuarina equisetifolia* in Andhra Pradesh, Tamil Nadu; *Bombax ceiba* in Maharashtra and Kerala etc. A total of 1200 plus trees of different species have been selected (Emmanuel 2007). Prominent species are *Tectona grandis* (714), Bombax *ceiba* (62), *Dalbergia sissoo* (53), *Gmelina arborea* (60), *Pinus roxburghii* (186) etc.

SEED ORCHARDS: Seed orchards have been raised by different institutions. About 511 ha of seedling and clonal seed orchard of 17 priority species have been raised. In Andhra Pradesh, more than 250 ha of clonal orchard of *Tectona grandis* have been established. Bihar has established more than 134 ha of teak seed orchard. Similarly large scale orchards of teak have been raised in Karnataka, Maharashtra and Madhya Pradesh. Forest Department, Arunachal Pradesh has established clonal orchard of *Gmelina arborea*, *Bombax ceiba*, *Terminalia myriocarpa* and *Chukrasia tabularis*. In Himachal Pradesh, clonal seed orchards of *Grewia optiva*, *Bauhinia variegata* and *Dalbergia sissoo* have been raised by Department of Tree Improvement, Dr YS Parmar University of Horticulture and Forestry. Himalayan Bioresource Institute, Shimla has established seedling seed orchards of *Dalbergia sissoo*.

CLONAL FORESTRY: Wimco during 1980's introduced four clones of *Populus deltoides* from Australia, out of which two outstanding clones G-3 and G-48 laid the foundation of large scale poplar plantation in India. Since then large number of clones have been developed by FRI, Dehradun, State Agricultural Universities, Wimco etc. Wimco has got registered six new clones WSL-22, WSL-27, WSL-32, WSL-37, WSL A-26 and WSL A-49 with International Poplar Commission, Italy. Punjab Agricultural University, Ludhiana has released several clones of poplar (PL series) for cultivation in Punjab. Dr Y. S. Parmar University has released many clones of *Populus ciliata*, *P. deltoides* and hybrids of *P. ciliata* x *P. maximowiczii*.

The Department of Tree Improvement, Dr YS Parmar University, has been in the forefront of improvement and breeding of *Salix*. The clonal archive of Salix contains more than 250 clones of 40 species covering different countries. Seven promising clones namely J-7999, NZ-1140, 13/25, J-194, J-795, SI-63-007 and PN-731 have been recommended for cultivation. About 5000 hybrid plants have been raised through controlled breeding.

The genetic improvement for higher productivity of *Eucalyptus* through breeding approaches was initiated during 60's and has been continuing with strength till now. A series of F1 hybrids, popularly known as FRI 4, FRI 5 to FRI 15, were developed at FRI by using different species combinations. Some of the hybrids have shown pronounced and sustained degree of superiority in growth parameters. ITC, Bhadrachalam has pioneered clonal *Euclyptus* in India. Clones like 3, 6, 7, 10, and 27 developed at Bhadrachalam formed the basis of initial clonal plantation since 1992. Later on clones like 2045, 2070, 2155 were developed and are being planted on large scale. More than 40 million plants of fast growing *Eucalyptus* are being produced every year in India (Lal 2010).

GLOSSARY OF TERMS

Additive Genetic Variation: Average effect of alleles on all segregating loci. It results when there is lack of dominance. That part of the genetic variance of a quantitative character attributed to the average effects of substituting one allele for another at a given locus or at the multiple loci governing a polygenic trait.

Afforestation: To establish a forest by artificial means on an area from which forest vegetation has always or long been absent.

Allele: One of pair of genes located at the same locus in homologous chromosomes and controlling the same character.

Clonal Propagation: Asexual reproduction or multiplication of plants from a single original individual.

Clones: All plants (ramets) reproduced sexually from a common ancestor (ortet) and having identical genotypes.

General Combining Ability (GCA): The relative ability of an organism to transmit its genetic superiority to its offsprings when crossed with other organisms. High GCA implies presence of genes with additive effects.

Specific Combining Ability (SCA): The relative ability of an organism to transmit its genetic superiority to its offsprings when crossed with specific other organisms.

Combining Ability: The relative ability of an organism to transmit genetic superiority to its offsprings.

Dioecious: Dioecious species have male and female flowers disposed in separated plants or individuals **or** having the male and female elements on different individuals.

Dominance: Masking effect of one allele by another so that heterozygote and homozygous dominant are phenotypically indistinguishable.

Environmental, Phenotypic and Genetic Variances: The appearance of a tree, its phenotype determined by both genotype and the environment i.e. the interaction with environment produces the phenotype. The relative importance of genotype and the environment in determining the expression of the trait is measured in randomized and replicated experiments by analysis of variance. The population of phenotype has the variance resulting from addition of genetic and environmental variances.

Genotype (G) + Environment (E) $\longrightarrow$ Phenotype (P) *i.e.* $\sigma^2 P = \sigma^2 G + \sigma^2 E$

Epistasis: Dominance exerted by non allelic genes.

Even Aged: Applied to a stand consisting of trees of approximately the same age. Differences upto 25 per cent of the rotation age may be allowed in cases where a stand is not harvested for 100 or more years.

Exotic: An introduction from foreign land or not native to the area in question.

Full Sibs: Trees with both the parents in common.

Gene: The basic unit of inheritance. The gene consists of a segment of DNA chain with a definite nucleotide sequence that contains the instruction for a complete protein.

Genetic Advance: Progress in shifting the genotypic mean and gene frequencies of a trait in the population toward the desired direction as a result of selection.

GA = i × h^2 × Vp (i = selection differential; h^2 = heritability; Vp= phenotypic variance of base population)

Genetic Diversity: Variability between different genotypes of a species

Genetic Engineering: Genetic engineering, also called genetic modification, is the direct human manipulation of an organism's genome using modern DNA technology

Genetic Gain: Increase in productivity following change in gene frequency

Genetic Variation: Heritable portion of total or phenotypic variance. It is of three types, *viz.*, additive, dominance and epistatic variances

Genotype: The entire genetic constitution of an organism

Germplasm: The sum total of the genes and cytoplasmic factors governing inheritance.

Half Sibs: Trees with one parent in common.

Heritability: It is defined as degree to which a character is influenced by heredity as compared to environment or proportion of genetic variance in phenotypic variance or portion of total variance due to genetic factors. Heritability is of two types viz. broad sense heritability and narrow sense heritability. In a broad sense, the proportion of total variance due to genetic factors. In a narrow sense, the proportion of total variance due to genes with additive effects.

$$\text{Broad sense Heritability } (H^2) = \frac{\sigma^2 g}{\sigma^2 p} = \frac{\text{Genotypic Variance}}{\text{Phenotypic Variance}}$$

$$\text{Narrow sense Heritability } (h^2) = \frac{\sigma^2 a}{\sigma^2 p} = \frac{\text{Additive Variance}}{\text{Phenotypic Variance}}$$

Inbreeding: Mating between close relatives

Indigenous: Native to a specified area or region; not introduced. Anton. Exotic

Mass Selection: This method of selection depends mainly on selection of plants according to their phenotype and performance

Mating: All aspects of sex expression in plants that affect the relative genetic contribution to the next generation of individuals within a species.

Non Additive Genetic Variation: Variation in joint effects of dominance and additive × dominance and dominance × dominance × dominance epistasis

Outcrossing: Cross-pollination with other individuals that may or may not be related.

Over Dominance: The condition in which heterozygote is superior to best of homozygote.

Phenotype: The observable characteristics of an organism, often stated in reference to a single trait.

Progeny Test: An experiment usually replicated to compare the offsprings of different parents

Provenance: The ultimate natural origin of tree or group of trees

Qualitative Traits: Qualitative traits are the traits which are easily recognizable and divide individuals into distinct types

Quantitative Traits: The majority of economic traits such as height, diameter and volume production differ in degree rather than of kind.

Ramet: An individual member of a clone, descended through vegetative propagation.

Seedling Seed Orchard: A seed orchard composed of plantation of seedlings established primarily for production of seed of proven genetic quality.

Clonal Seed Orchard: A seed orchard composed of vegetative propagated trees established primarily for production of seeed of proven genetic quality.

Seed Production Area: A phenotypically superior stand or plantation rogued and managed to produce large quatities of seed.

Selection Intensity: The proportion of individuals in the population that are selected for breeding.

Selection: The choosing of individuals or populations with desirable characteristics to obtain genetic improvement.

Selective Breeding: Selection and crossing of individual trees.

Selfing: To place pollen from a male flower on female flower on same tree.

Sorrounds: An area maintained round an experimental or sample plot to ensure that the latter is not being affected by the treatment applied to the area outside them both.

Species: A group of similar individuals capable of interbreeding, more or less distinctly different from other species in the same genus.

Variance: A statistical measure of variability.

19 WOOD SCIENCE AND TECHNOLOGY

INTRODUCTION

Forests supply the raw materials for the lumber, plywood, paper, packaging and other wood-based materials that are staples of modern life. Wood and wood products are the main commercial products of forests. They include industrial roundwood, sawnwood, wood-based panels, wood pulp, paper and paperboard, semi-manufactures, wood articles, fuelwood and charcoal. Demand for wood products both industrial wood and fuelwood, has grown considerably in the last 25 years. Increasing demand for commercial timber and other wood and paper products has encouraged many countries to expand the area devoted to forest plantations, but pressure on rapidly diminishing natural forests have not been reduced. In the last 20 years, average global consumption of wood increased on average only 0.3 per cent per year.

Nearly 1.2 billion hectares of forests are primarily managed for the production of wood and non-wood forest products. The area designated primarily for productive purpose has decreased by more than 50 million hectares since 1990 as forests have been designated for other purposes. In 2005, global production of industrial roundwood was about 1.783 billion cubic meters or 53 per cent of the wood harvested globally. Global fuelwood production was about 1.576 which accounts 47 per cent of total wood removal. At the global level, reported wood removals amounted to 3.4 billion cubic meters annually, similar to the volume recorded for 1990 (FRA 2010). Logs and sawn wood for construction comprised the largest use of industrial roundwood followed by pulp for paper and paperboard and processed wood products such as veneers, chipboard and plywood. Wood removed from forests and other wooded land constitutes an important component of the productive function. The volume of annual wood removal indicates the economic and social utility of forest resources to national economies and dependent local communities.

Industrial roundwood production has grown largely because of rising incomes and it has reached 1.872 billion cubic meters in 2010. Most industrial roundwood production and consumption occurs in the high-income countries. Meanwhile, the rapidly growing economies of Asia and Latin America are where demand for construction timber, processed wood products and paper and paperboard is rising fastest. Of all industrial roundwood products, demand for paper and paperboard are growing the most rapidly. Globally, paper consumption increased by a factor of 20 in the 20th century and has more than tripled over the past 40 years.

Over the last 20 years, International trade of forest products increased from US$ 60 billion to US$ 257 billion with an average annual growth of 6.6 per cent and it is expected to reach US$ 450 billion by the year 2020. Commercial timber production is a major global industry. North America, Asia and Western Europe dominate industrial roundwood production but the timber industry is of relatively greater economic importance to some developing countries. The leading traditional countries in production and trade of wood products are Canada (13.8%), Finland (5%), Brazil (3.2%), Russia (3.0%), USA, Germany and Japan. China (7.2%) is leading in the emerging countries followed by Indonesia (3.2%), Malaysia (2.4%), Chile and Vietnam. The major factors influencing global forest product market are: global population, global income, bioenergy technology, climate change, technological change, prices of other goods and substitutes, consumer tastes and preferences, environmental concerns.

TREE GROWTH

PHOTOSYNTHESIS AND MERISTEMS

Like other plants, trees are *autotrophs* - meaning they produce their own food. Tree cells have rigid cell walls, a large central vacuole and chloroplasts. Trees grow by the action of photosynthesis in their leaves. Photosynthesis is where the energy from sunlight is combined with minerals dissolved in water drawn up from the roots and carbon dioxide from the atmosphere, to provide food. Areas within the plant that are capable of growth (cell division) are called meristems. Primary growth occurs only at the shoot and root tips in areas called apical meristems. Primary growth is responsible for elongating the plant. In areas that contain only primary growth, stem thickness increases by cell enlargement, not by the production of new cells.

Lateral meristems produce new cells that make the stems and roots thicker. This type of growth is called secondary growth. Secondary growth occurs only during the second and subsequent years and only in woody species. There are two kinds of lateral meristems, the vascular cambium and the cork cambium. These lateral meristems form as rings within the plant body as the stem increases in thickness. Secondary growth begins with the initiation of the vascular cambium, a cylinder of meristematic tissue that produces additional xylic and phloic tissues. The cells that eventually form the vascular cambium come from two sources, the procambium in the vascular bundles and the interfascicular parenchyma cells between vascular bundles. The vascular cambium divides to produce new xylem cells towards the inside of the vascular cambium and new phloem cells towards the outside.

VASCULAR CAMBIUM

The vascular cambium is a particularly significant tissue in both gymnosperms and dicotyledons where it plays the important role of providing for secondary growth resulting in an increase in the diameter (or girth) of stems and roots. Secondary growth results in the formation of both secondary xylem and secondary phloem. Cells of the vascular cambium are originally derived from procambium and develop into either fascicular (within vascular bundles) or interfascicular (developing in the ground tissues between vascular bundles) layers of cells that form a continuous ring. The vascular cambium is sometimes referred to as a lateral meristem, along with the cork cambium, or phellogen and it normally assumes a cylindrical form within stems and roots. Initially, vascular cambium is found between the xylem and phloem in the vascular bundles of dicots. After one year growth, it joins to form a continuous ring.

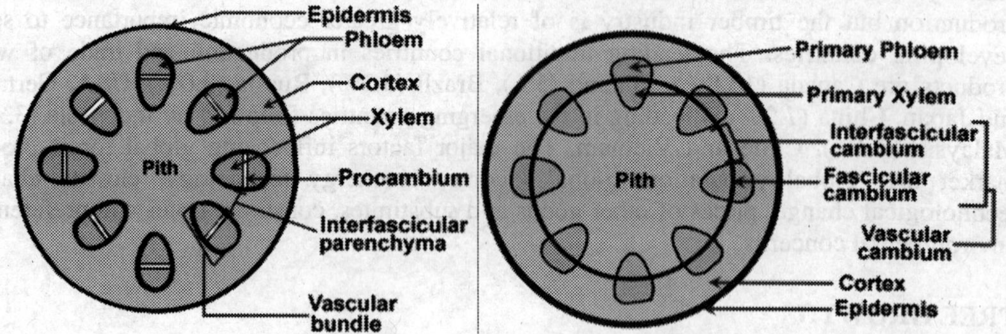

Fig. 19.1: Vascular Bundles and Vascular Cambium

In most cases, the vascular cambium is only one cell thick, but viewing with microscopy may make it appear to be several layers in thickness due to the fact that the immediate derivatives look similar. In fact, the thickness may be greater at times of the year when activity is greater. This range of apparent thickness has given rise to the use of the term cambial zone for the microscopic view of the cambium tissue and its immediate derivatives. Cells of the vascular cambium divide mostly periclinally and are said to be bifacial meristematic since they produce derivatives in two directions (the secondary phloem to the outside and the secondary xylem to the inside).

Vascular cambium cells are of two types: fusiform initials, which are elongated but broadly flattened on the tangential face and ray initials, which are shorter cells, somewhat elongated on their radial axis. Fusiform initials produce axial elements of secondary vascular tissues, such as fibres, tracheids, vessel members, companion cells, sieve-tube elements and axial parenchyma. Ray initials give rise to phloem and xylem rays--the radial (or horizontal) system of the secondary plant body. Vascular cambium initials have only primary cell walls that are never lignified. Their tangential walls are thinner than radial walls and lack primary pit fields. The radial walls, however, often have numerous orderly arranged primary pit fields that give the walls a "beaded" appearance. When the cells are dormant they have few and small vacuoles, but in the active state they become highly vacuolated and a central vacuole develops. Mitotic division and cytokinesis may take several days for completion from an initial.

Vascular cambium cells have no intercellular spaces and they grow intrusively. Sometimes, that intrusive growth goes through rays that have already been established. Cross divisions of fusiform initials may shorten the vertical axis of the cells to shorter cells that can change into ray initials. The growth of both secondary tissues and rays may be storied (stratified) in which the cells are arranged in uniform horizontal series or nonstoried (nonstratified) in which such a uniform arrangement does not exist. Growth in vascular cambium is stimulated by plant growth regulators (hormones), primarily auxin, but also including cytokinins and gibberellins.

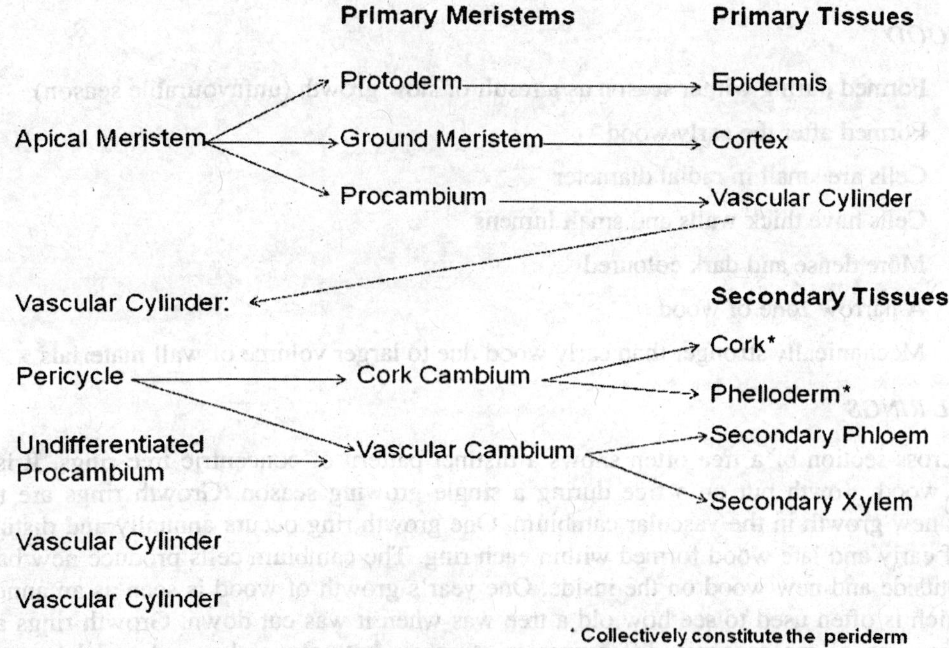

Fig. 19.2: Meristems and Vascular Cambium Growth

SECONDARY GROWTH

Trees and woody plants live for many years. Trees and shrubs grow radially as well as vertically. The trees increase in diameter or girth which results from activity of cambium. During favourable seasons, new growth is made from cells just beneath the bark. The cambium cells produce new bark on the outside and new wood on the inside. Wood is yielded by trees, which increases in diameter by the formation and this is technically known as secondary growth. It is as a result of cell division in the vascular cambium (lateral meristem) and subsequent expansion of new cells. Secondary growth is found in most woody dicots and gymnospersms and it is absent in monocots.

EARLY WOOD

- Formed during spring season as a result of rapid growth (favourable season)
- Formed early in a year
- Cells are large in radial diameter
- Cells have thin walls and large lumens
- Produced more in amount
- Less dense and light coloured

- A broad zone of wood
- Not as strong as late wood

LATE WOOD

- Formed during winter season as a result of slow growth (unfavourable season)
- Formed after the early wood
- Cells are small in radial diameter
- Cells have thick walls and small lumens
- More dense and dark coloured
- A narrow zone of wood
- Mechanically stronger than early wood due to larger volume of wall materials

ANNUAL RINGS

A cross-section of a tree often shows a distinct pattern of concentric tree rings. It is a layer of wood growth put on a tree during a single growing season. Growth rings are the result of new growth in the vascular cambium. One growth ring occurs annually and distinct layers of early and late wood formed within each ring. The cambium cells produce new bark on the outside and new wood on the inside. One year's growth of wood is seen as an annual ring, which is often used to see how old a tree was when it was cut down. Growth rings are visible because of macroscopic differences in structure between earlywood and latewood. Rings may be closely spaced or widely separated. Growth rings vary in width depending on species and site conditions. Many trees in temperate zone make one growth ring each year with the newest adjacent to bark. By comparison, trees in tropical regions may have more than one growth ring per year, or may appear to grow continuously and have no rings. For the entire period of a tree's life a year-by-year record or ring pattern is formed that reflects the climatic conditions in which the tree grows. The rings are counted and the number of rings is the age of the tree in years.

IRREGULARITIES IN ANNUAL RINGS

FALSE RINGS: In some trees, more than one growth ring is formed in a year which is termed as false rings. Duplicate rings are created by events such as drought, frost or defoliation, that cause terminal growth to slow or cease, reducing auxin production, within a season that are followed by favourable growth seasons. Slow growth may cause formation of latewood type cells.

DISCONTINUOUS RINGS: As a result of cambium remaining dormant in one or more regions around the stem, failing to form growth rings around the complete cross section is termed as discontinuous Rings. It occurs in trees with one-sided crowns, suppressed, heavily defoliated or overmature trees.

TREE GROWN IN TROPICAL CLIMATE: Almost continual growth can limit occurrence of rings. In some climates, stopping and restarting of growth can give more than one growth increment in a year.

WHAT IS WOOD?

Wood is a cellular material of biological origin. It is a hard, fibrous tissue found in many plants. It is an organic material, a naturai composite of cellulose, hemicelluloses and lignin. Wood is produced as secondary xylem in the stems of trees and woody plants. Wood is biodegradable and renewable natural resource. Wood is hygroscopic i.e. it has the ability to absorb the moisture from air. Wood is anisotropic which means that its structure and properties vary in different directions and wood is combustible. In living trees, it transfers water and nutrients to leaves and other growing tissues and wood provides support to the tree. Wood is considered as the most important raw material.

HEARTWOOD, SAPWOOD AND BARK

A stem cross-section shows a dark-coloured center portion surrounded by a lighter-coloured outer region. In mature trees, the xylem has both living and dead cells. Older wood in the middle of the trunk dies and becomes harder, darker and drier and is called heartwood. Heartwood is formed in the central part of the tree stem due to reduced water and oxygen availability which leads to death of parenchyma cells. It is followed by formation and deposition of extractives in the cells. This is the best wood in the tree. Heart wood is resistant to decay and insect and disease resistant. Heartwood is heavier, stronger and usually begins to form around age 14-18 years. It is difficult to penetrate with liquids and preservatives.

The wood formed immediately inside the bark of a tree is called sapwood. It is lighter in colour and a lot wetter than heartwood. Sapwood contains all of the live cells in the xylem. Rays provide water and nutrient transport from phloem. Nutrients are stored in specialized cells called parenchyma cells which also provide strength to the tree stem. Sapwood contains living cells that transport water from the roots to the branches and leaves at the top of the tree. Sapwood is permeable and readily accepts chemical treatments and preservatives. Bark is used for protecting the tree.

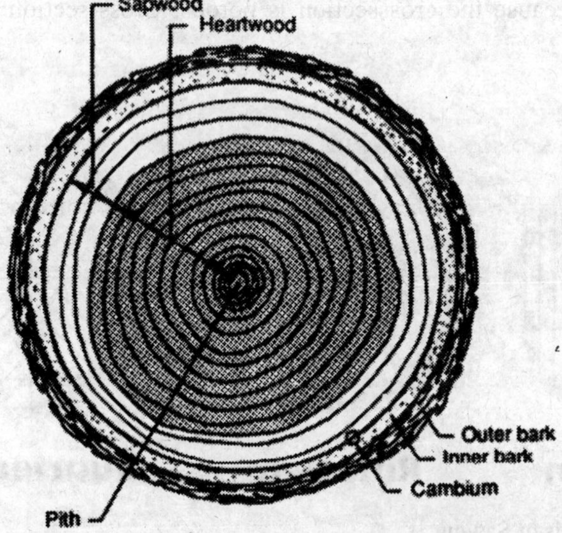

Fig. 19.3: Cross-section of Tree Stem

Bark is the outer layer and is composed of a dead outer phloem of dry corky material and a thin inner phloem of living cells. Its primary functions are protection and nutrient conduction. The thickness and appearance of bark vary substantially depending on the species and age of the tree.

SOFTWOODS AND HARDWOODS

The terms softwood and hardwood do not directly describe the hardness or softness of wood. Instead, they refer to the leaf form or mode of seed production on trees from which the wood is cut. The terms are especially confusing because some true hardwoods have softer or lighter wood than common softwoods. Softwoods are also very strong for their weight in comparison to the hardwoods. Coniferous trees in temperate regions provide soft wood. They have needle or scale-like leaves and maintain these leaves throughout the year (evergreens). They can grow quickly with straight trunks. Softwoods are most commonly used for construction lumbers, heavy timbers, poles and piles where strength is important. E.g. pines, spruces, firs, larch, cedars, hemlock, redwood, yew, cypress.

Hardwoods are trees such as oak, teak, sal and many others which have broad deciduous leaves. They grow slowly and sometimes have twisted trunks. They grow in tropical regions of the world. Hardwoods are often used for fine furniture and heavy timbers.

PLANES OR SURFACES OF WOOD

Visible characteristics, shrinkage and mechanical properties of wood are defined in terms of the three planes in which wood can be cut. Characteristics of these surfaces can also be useful in the identification of wood.

CROSS-SECTION OR TRANSVERSE: Cross-section is the surface exposed when wood is cut across the width of a log or board. Cross-section reveals the annual rings. It is the key surface for identification of wood. It absorbs preservatives more easily than radial or tangential surfaces because the crosssection is porous. Cross-section is also termed as end grain.

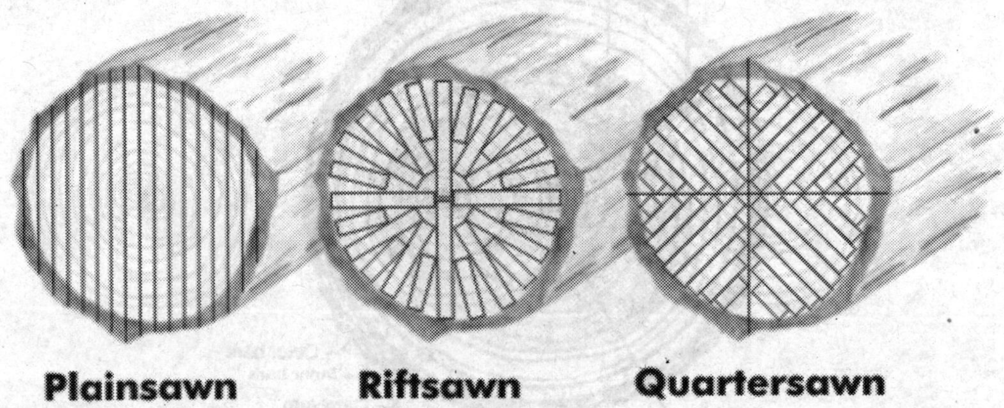

Fig. 19.4: Three Methods of Sawing

RADIAL: The radial surface is exposed when a log is cut longitudinally from its center to the bark (along the radius). In the hardwood industry, lumber cut this way is known as quarter sawn lumber. This surface has high strength and has high resistance to wear.

TANGENTIAL: The tangential surface is exposed when a log is cut parallel to the bark and tangent to the log diameter. The lumber cut this way is known as flat or plain sawn lumber. The method of this conversion is known as through & through sawing or slabbing. This is the cheapest, easiest and most common way of producing lumber today and it results in a characteristic U or V-shaped grain pattern in the softwoods with distinct earlywood and latewood brands and the coarse grained hardwoods.

GRAIN ORIENTATION

It is the direction of the long axis of longitudinal wood cells.

STRAIGHT GRAIN: If it is straight or parallel to the tree stem, it is called straight grain orientation.

SPIRAL GRAIN: When wood cells are spirally arranged about the stem axis, it is called spiral grain orientation. It is caused by abnormal cell division or anticlinal cell division (production of new initials by radial partitioning) in which new cambial cell formation occurs in one direction only. This is typically low in strength and stiffness and may tend to twist as it dries.

INTERLOCKED GRAIN: Woods with interlocked grain are produced when grain spirals in one direction for several years and then reverses direction to spiral oppositely. It is genetically controlled and wood is difficult to split, may shrink longitudinally upon drying and can warp unpredictably. It is sometimes desirable from an appearance standpoint as alternating grain direction cause light to reflect in varying patterns across radially cut wood, giving a 'ribbon stripe' figure.

FINE GRAIN: Wood is fine-grained if its annual rings are narrow. In wood industry, a fine-grained wood is capable of high polish.

COARSE GRAIN: Wood is coarse-grained if its annual rings are wide and not capable of high polish.

WOOD STRUCTURE

Wood is composed mostly of hollow, elongated, spindle-shaped cells that are arranged parallel to each other along the trunk of the tree. The characteristics of these fibrous cells and their arrangement affect strength properties, appearance, resistance to penetration by water and chemicals, resistance to decay and many other properties. The microscopic structure of wood resembles a bundle of straws glued together. Each straw represents a cell with a cellulose wall and a hollow centre (lumen) and lots of fine perforations through the wall. The cell walls are impregnated with lignin (phenolic compound), a natural polymer that glues the cells together and gives the wood its strength. Most cells in wood are oriented longitudinally (parallel to the tree trunk), with some cells radially aligned.

The primary structural building block of wood is the tracheid or fibre cell. Cells vary from 16 to 42 μm in diameter and from 870 to 4000 μm long. Thus, a cubic centimeter of wood could contain more than 1.5 million wood cells. When packed together they form a strong composite. Each individual wood cell is even more structurally advanced because it is actually multilayered, reinforced, close-end tube. Each individual cell has four cell wall layers (Primary, S_1, S_2 and S_3). Each layer is composed of a combination of three chemical polymers: cellulose, hemicelluloses and lignin. The cellulose and hemicelluloses are linear poly-saccharides (i.e., hydrophilic multiple-sugars) and the lignin is an amorphous phenolic (i.e., a three-dimensional hydrophobic adhesive).

Cellulose forms a long unbranched chains and hemicelluloses forms short branched chains. Cellulose is a polysaccharide of repeated glucose molecules which may reach 4 μm in length. These cellulose molecules are arranged in an orderly manner into structures about 10-25 nm wide called microfibrils. This microfibrils wind together like strands in a cable to form macrofibrils that measure about 0.5 μm in width and may reach 4 μm in length. This framework of cellulose macrofibrils is cross-linked with hemicellulose, pectin and lignin. Lignin encrusts and stiffens these polymers. These strands are very strong as like a steel cable. Lignin gives the cell wall rigidity and it is the substance that cements the cells together. Because carbohydrate and phenolic components of wood are assembled in a layered tubular or cellular manner with a large cell cavity, specific gravity of wood can vary immensely. Wood excels as a viable building material because the layered tubular structure provides a large volume of voids, is has an advantageous strength-to-weight ratio.

STRUCTURE OF SOFTWOODS

TRACHEIDS OR FIBRES: Tracheids are the principal longitudinal cell type in soft-woods. Tracheids or fibres comprise 90-95 per cent of the volume of the wood. They have the length diameter ratio of 100:1 and contribute greatly to the strength of the wood. Softwood fibres range from about 3 to 8 mm in length. The diameter of tracheids varies from one species to the next and useful for identification purpose.

PARENCHYMA: Parenchyma cells are generally short, thin-walled cells which are connected together in strands and serve primarily for storage and distribution of carbohydrates.

RESIN CANALS: Resin canals are tubular passageways lined with living parenchyma cells which exude resin or pitch. In pine, resin canals are easily seen with the naked eye on the endgrain as wells as the side grain.

PITS: Individual cells in softwoods are connected together by means of pits that allow movement of liquids between cells. Pits are important in softwoods since preservatives move from one cell to the next through them.

STRUCTURE OF HARDWOODS

The hardwoods have more cell types and the variation in the size and arrangement of these cells is greater than in the softwoods. As a result, the hardwoods are more varied in appearance and may have unique characteristics making certain species more desirable for selected end uses than others.

VESSELS (*PORES*): Hardwoods have specialized structures called vessels for conducting sap upward. Vessels are relatively large diameter, thin-walled, round cells with open ends that are connected end to end to form microscopic tubes that are ideal for sap conduction. In most hardwoods, the ends of the individual cells are entirely open; in others, they are separated by a grating. On the cross-section surface, vessels appear as holes and are termed as pores. The size, shape and arrangement of pores vary considerably between species. It is the vessels which constitute pores in hardwoods. The presence of large diameter unobstructed vessels makes some woods easy to treat.

FIBRES: Fibres are relatively smaller diameter, elongated cells with closed ends. They usually have small cavities and relatively thick walls. Fibres are primarily responsible for the strength characteristics of hardwood. Thin places or pits in the walls of the wood fibres and vessels allow sap to pass from one cavity to another.

PARENCHYMA: In the hardwoods, parenchyma cells vary widely in size and arrangement. In some trees, the rays (mostly parenchyma cells) are conspicuous to the naked eye and aid greatly in the identification of species. Wood rays are strips of short horizontal cells that extend in a radial direction. Their function is food storage and lateral conduction. Most of the rays in flat-grain surfaces are two to five cells wide, but their width and height vary in different species of hardwoods.

MISCELLANEOUS: Hardwoods contain other miscellaneous features which are often important in identification, as well as the end use of the product. For example, white oak (*Quercus alba*) contains tyloses. Tyloses are literally plugs in the vessels or pores which restrict sap movement. Thus white oak is used to make tight cooperages (alcohol barrels) and does not accept preservative treatment well.

PITS: Individual cells in hardwoods are connected together by means of pits that allow movement of liquids between cells.

GROUPS OF HARDWOODS

The distribution or large diameter cells (pores or vessels) within growth increments of hardwoods allows grouping of these hardwoods into three categories based on their cross sectional appearance viz. ring porous, semi-ring porous and diffuse porous.

RING POROUS WOOD: Group of hardwoods in which the pores (vessels) are formed comparatively large in size during favourable condition at the beginning of each growth increment. Pores decrease in size abruptly during unfavourable condition (outer portion of the ring). The pores can be easily seen with the naked eye. E.g. oak, ash, elm

SEMI-RING POROUS WOOD: Semi-ring porous woods have pores that are initially large and then gradually decrease in diameter throughout the growth increment. E.g. walnut, butternut

DIFFUSE POROUS WOOD: The pores are uniform in size across the entire growth increment and are generally too small to be seen without the use of a hand lens. E.g. maples, sweetgum, yellow poplar

CHEMICAL COMPOSITION OF WOOD

CELLULOSE: Cellulose comprises about 40 to 45 per cent of the oven-dry weight of wood. Long strands of cellulose molecules arranged more or less parallel to each other within the thick walls of structural cells (fibres and tracheids) contribute a high tensile strength to wood.

HEMICELLULOSE: Hemicellulose constitutes from 20 to 35 per cent of the oven-dry weight of the wood. The exact function of the hemicelluloses is not clear. Some possibility exists that they serve as a temporary matrix before lignifications.

LIGNIN: Lignin constitutes from 15 to 35 per cent of the oven-dry weight of the wood. Lignin is a complex polymer and it reinforces the cellulose portion of the cell wall, thereby contributing to the rigidity of wood.

ASH: The inorganic materials, or ash generally constitute less than one per cent of the oven-dry weight of the wood. The most common constituents are calcium, potassium, magnesium, carbonates, phosphates, silicates and sulfates.

EXTRACTIVES: Most woods usually contain some type of extractives which are located in the heartwood and are water soluble. They are often responsible for the general darkening of the heartwood, for the resistance of some heartwood to decay and insect attack, for specific odour and for good dimensional stability. Extractives make woods from some tree species more durable than others. If the extractives have a toxic or repellent effect, heartwood is more resistant than the sapwood. The more important organic extractives include the terpenes, resin acids, polyphenols, tannins and tropolenes.

WOOD PROPERTIES

COLOUR: Colour lends beauty to the wood and is of great value in the determination of its quality. It also aids in the identification of wood. The heartwood of tree species varies in colour viz. black colour of persimmon, dark brown of walnut, light brown of white oak, reddish brown of red oak, yellowish white of tulip and poplars, brownish red of redwood and cedars. They are all reliable marks of distinction and colour. Nearly formed wood in the outer few rings has little colour compared to old wood. The sapwood is generally light in colour. The different tints of colours are due to pigments or wood being denser.

ODOUR: Odour depends on chemical compounds or extractives and it is not due to the part of wood substance itself. When new surface is exposed in the wood during cutting, it gives characteristic odour and exposure to weather reduces and often changes the odour. Heartwood is more odouriferous than sapwood. Many kinds of wood are distinguished by strong and peculiar odours. This is especially true in the case of camphor tree, cedar, pine, oak, mahogany, teak etc. Decomposition of wood is usually accompanied by pronounced odours.

DENSITY AND SPECIFIC GRAVITY: Density is the single most important factor to indicate strength in defect-free wood and other characteristics such as ease of machining and hardness. Density is the weight of wood per unit volume. Moisture content will greatly affect the density. Generally as the density of wood increases, its strength also increases. Specific gravity is another important factor for comparing the relative weights of different woods. Specific gravity is the ratio of the oven-dry weight of a given volume of wood to that of weight of an equal volume of water at a standard temperature.

PERMEABILITY: The permeability of wood is the extent to which it allows fluid flow through a porous medium under the inflence of a pressure gradient. There is a good correlation beween wood permeability and treatability. The treatability describes the ease with which preservatives can be forced into wood under pressure and treatability varies with the different wood species.

MOISTURE CONTENT OF WOOD

The moisture content (MC) of green or fresh wood as found in the living tree or logs is highly variable which depends on species, location, season of the year, heartwood and sapwood content of wood. Amount of water in wood (moisture content) affects wood treatability, durability and stability. The moisture content of wood is defined as the weight or amount of water in wood given as s percentage of oven-dry weight. The following formula is used to calculate the moisture content in wood

$$\text{Moisture Content (\%)} = \frac{\text{Initial Weight - Ovendry Weight}}{\text{Ovendry Weight}} \times 100$$

Wood cut from a green log often contains as much or more than its oven-dry weight in the form of sap or water. Water is contained in wood as either bound water or free water. Bound water is held within cell walls by bonding forces between water and cellulose molecules. Free water is contained in the wood cell cavities.

FIBRE SATURATION POINT (FSP): When the wood has dried to about 30 per cent moisture content, it is at the fibre saturation point. In this state, the cell cavities are emptied of free water, but the cell walls are still saturated with bound water. Wood shrinks as it dries below the FSP. At moisture content above FSP, Wood can be attacked by decay fungi. The water remaining in the cell walls after wood has dried to the FSP is called bound water.

EQUILIBRIUM MOISTURE CONTENT (EMC): Wood is a hygroscopic material and the amount of water which the wood will lose depends on the relative humidity. Hence it responds to changes in atmospheric humidity. Wood loses or gains bound water until the amount it contains is in balance with that of surrounding atmospheric relative humidity. When this balance of moisture exchange is established, the amount of bound water eventually contained in wood is called equilibrium moisture content. EMC is always below 30 per cent.

SHRINKING AND SWELLING: Wood shrinks and swells due to the loss or gain of bound water from the cell walls. The amount of movement depends on the amount of water gained or lost, the orientation of the wood cells and species. When wood dries below fibre saturation point, it begins to shrink. Conversely, wood that is below FSP will swell as it takes on moisture and this will continue until FSP is reached. Changes in moisture content above FSP have no effect on shrinkage and swelling. Wood should be dried to its anticipated equilibrium moisture content to minimize problems due to changing moisture contents. Typical shrinkage values of wood on three surfaces are:

Maximum longitudinal shrinkage = 0.1 – 0.3%

Maximum radial shrinkage = 2.1 – 7.9%

Maximum tangential shrinkage = 4.7 – 12.7%

WOOD SEASONING: The process of drying of timber is called wood seasoning. The rate of drying of wood is dependent on temperature, relative humidity and air circulation. Wood may be seasoned in two ways: air drying and kiln drying. Air drying is simple method of drying the wood in open condition and it takes longer time upto 6-12 months. Kiln drying is the process of drying wood by keeping them in closed chamber in controlled temperature, relative humidity and air circulation. These large drying ovens are called kilns and it takes 4-5 weeks for drying wood in kilns. The kilns are categorized into progressive type, compartment type and solar kilns. Proper stacking of timber helps in uniform drying and reduces seasoning defects. Horizontal and vertical stacking methods are followed for drying different types of wood.

WOOD DESTROYERS AND PRESERVATIVES

WOOD DESTROYING AND STAINING FUNGI

BROWN ROT: Fungi are able to break down the cellulose of wood, leaving a brown residue of lignin and identified by dark brown colour of wood

DRY ROT: Brown rot when it is dry is called dry rot. Dry rot fungi decay relatively dry wood and they have water conducting strands that are able to carry water from damp soil to wood

WHITE ROT: It breaks down both lignin and cellulose, have a bleaching effect, which may make the damaged wood appear white

SOFT ROT: Soft rot fungi usually attack green wood (high MC) causing gradual softening from the surface inward that resembles brown rot

POCKET ROT: It describes the decay in wood that is characterized by small cavities of severe decay, scattered throughout the wood.

SAP STAINING FUNGI: Fungi penetrate and discolour sapwood and stain cannot be removed by brushing or planing

MOLD FUNGI: Fungi cause green, yellow, brown or black fuzzy or powdery surface growths on wood during warm & humid weather

INSECTS

TERMITES: Termites use wood for food and shelter and are the most destructive of all wood insects. Three major groups of termites are destroying the wood viz. subterranean or ground inhabiting termite, drywood termite, dampwood termite

ANT: Carpenter ant may be black or red and they use wood for shelter, not for food. They make galleries in wood and mounds of saw dust indicate the presence of carpenter ants.

BEETLES: The major wood destroying beetles include powder post or lyctus beetle, anobiid beetle, long horn beetle or roundhead borers, flathead borers or metallic wood-boring beetle

MARINE BORERS: The major marine borers are shipworm, pholad mollusks, crustacean borers

WOOD PRESERVATIVES

Wood preservatives fall into three categories: Creosote and creosote solutions (creosote and oily byproduct of bituminous coal), Oilborne Preservatives (Pentachlorophenol or Penta), Waterborne Preservatives (Various Metallic Salts, Inorganic arsenicals and compounds i.e. combinations of copper, chromium, arsenic and fluoride)

CREOSOTE AND CREOSOTE SOLUTIONS

- Coal-tar-creosote (Creosote) is a black or brownish oil made by distilling coal tar that is obtained after high temperature carbonization of coal. It has high toxicity to wood destroying organism. Relative insolubility in water and low volatility
- Coal tar or petroleum oil has been mixed with coal-tar-creosote in various proportions to prepare Creosote solutions
- Wood-tar-creosote, Oil-tar-creosote, Water-gas-tar-creosote are also used to some extent

OTHER OILBORNE PRESERVATIVES

Copper Naphthenate, Chlorothalonil (CTL), Chlorpyrifos (CPF), Oxine Copper (Copper-8-quinolinolate), Zinc Naphthenate, Bis (tri-n-butyltin) Oxide (TBTO), 3-Iodo-2-Propynyl Butyl Carbamate (IPBC), Alkyl Ammonium Compound (AAC) or Didecyldimethylammonium chloride (DDAC), Propiconazole, Tebuconazole (TEB)

OTHER WATERBORNE PRESERVATIVES

Acid Copper Chromate (ACC), Ammoniacal Copper Zinc Arsenate (ACZA), Chromated Copper Arsenate (CCA), Ammoniacal Copper Quat (ACQ), Copper bis (dimethyldithiocarbamate) or CDDC, Ammoniacal Copper Citrate (CC), Copper Azole-Type A (CBA-A), Inorganic Boron (Borax/Boric Acid)

APPLICATION OF PRESERVATIVES

PRESSURE PROCESS: It includes full-cell process and empty-cell process. The basic principle of pressure process involves the placement of wood material in an airtight, steel cylinder or retort and immersing it in a preservative under pressure to force the preservative into the wood.

NON-PRESSURE PROCESS: It includes brushing, spraying, pouring, dipping, cold soaking, steeping, hot and cold bath (thermal processing), double diffusion, vacuum process. In the vacuum process, wood products are enclosed in an airtight container from which air is removed with a vacuum pump and container is then filled with the preservative. The partial removal of air from the wood, by the vacuum, followed by addition of the preservative creates a slight pressure that drives the preservative into the wood.

ABNORMAL WOODS AND WOOD SEASONING DEFECTS

ABNORMAL WOODS

In both hardwoods and softwoods, depending on age and location in the tree, some woods are substandard and low quality. This wood is called abnormal wood. The strength and durability of this wood are significantly lower than that of normal wood. These abnormal woods are formed as a result of natural defects.

JUVENILE WOOD: Juvenile wood is that material formed near the center or pith of the tree and is prevalent in both softwoods and hardwoods. It is the wood produced during the first 5-15 years of growth. Juvenile wood is characterized by wide growth rings with shorter and thin-walled cells and fewer late wood cells, thus resulting in a lower density and reduced strength values. The juvenile wood will create warping defects during drying. The change from juvenile wood to normal wood is gradual, thus making identification of juvenile wood difficult.

KNOTS: A knot is the basal portion of a branch or limb which has been surrounded by subsequent growth of the tree. The grain of wood deviates around knot and considered as a weak point in the wood. Two types of knots are prevalent: tight knot and loose knot. Incorporation of living branches into the stem is termed as tight knot or intergrown knot. This knot is integral parts of the surrounding wood. Loose knots occur when stem growth encases the dead branch base. Branch stub is progressively covered by seasonal growth following pruning.

SPIRAL AND INTERLOCKED GRAIN: When longitudinal cells in wood are not arranged parallel to the main axis of the tree stem and slightly spiral about the stem, it is called spiral grain. Lumber sawn from these logs, will have a cross or diagonal grain pattern. These wood products are typically low in strength and stiffness and tend to twist upon drying. When spiral grain reverses direction to spiral oppositely following some years, it is called interlocked grain or reverse spiral grain. These wood products will show warping defects upon drying.

REACTION WOOD: Wood formed in the leaning tree or branches in order to correct growth irregularity in a stem is termed as reaction wood. It is an attempt by the tree to straighten itself out. Reaction wood comprises compression wood and reaction wood.

In softwoods, reaction wood is called compression wood and it is formed on the lower side of leaning trees. The part of the growth ring with reaction wood is usually wider than the rest of the ring and has high proportion of latewood. As a result, the tree develops an eccentrically shaped stem and the pith is not centered. It is often darker in colour and presents serious problems in wood manufacturing since it is much lower in strength than normal wood of the same density. Also, it tends to shrink excessively in the longitudinal direction causing cross grain checking.

In hardwoods, reaction wood is called tension wood and forms predominantly toward the upper side of the leaning tree. It may form irregularly around the entire stem and is often difficult to detect. When machined, tension wood may show a fuzzy, wooly or fibrous appearance in cut surface. Solid wood products have lower quality due to low strength. It has the tendency to collapse upon drying. It can produce good paper properties if pulping conditions are modified and also good for 'dissolving pulps' and cellulose source for making cellophane, rayon and nitrocellulose.

WOOD SEASONING DEFECTS

Wood is anisotropic as it dries below the fibre saturation point i.e. shrinkage is not equal in all directions. Longitudinal shrinkage is negligible, except in reaction wood. Tangential shrinkage is 1.5 to 3 times higher than radial shrinkage. This differential shrinkage sets up strains which cause ruptures or fractures in the wood tissues and warping defects, significantly devaluing the wood product. Rupture of wood tissues includes checks, cracks, splits, shake, collapse, honey comb etc. Warping defects include bow, crook, cup, twist, kink, diamond, etc.

SURFACE CHECKS: They are failures or very slight separations (shallow cracks) that usually occur in the wood rays on the flatsawn faces of boards and confined to the tangential surface. They result from the separation of the thinner-walled early wood cells. They occur because of drying stresses exceed the tensile strength of the wood perpendicular to the grain and they are caused by tension stresses that develop in the outer part of boards as they dry.

END CHECKS: They are failures or very slight separations usually occur in the wood rays but on the end-grain surfaces. End checks occur because moisture moves much faster in the longitudinal direction than in either transverse direction. Therefore, the ends of boards dry faster than the middle and stress develop at the ends. They occur in the early stages of drying and can be minimized by using high relative humidity or by end coating.

CRACK: Cracks are typically much deeper separations in a wood protruding deeply into the interior portion of wood.

SPLIT: Split is the deepest of separations in a wood, potentially going so far to completely separate a wood into two sections. It is the separation of the wood parallel to the fibre direction, due to the tearing apart of the wood cells.

BOX-HEART SPLIT: A split originating in the wood surrounding the pith during drying. It is caused by stresses set up because of the differences in tangential and radial shrinkage of the wood near pith.

RING SHAKE or RING FAILURE: It is the separation of the wood parallel to the grain along the growth rings. It can occur as a failure in the end grain in the initial stages of drying and extend in depth and length as drying progresses.

COLLAPSE: It is a distortion, flattening or crushing of wood cells. Collapse usually shows up as grooves or corrugations.

HONEYCOMB: Internal splitting in wood that develops in drying. It is an internal crack caused by a tensile failure across the grain of the wood and usually occurs in the wood rays. This defect occurs because of the internal tension stresses that develop in the core of the board during drying. It occurs when the core is still at relatively high moisture content and when drying temperatures are too high for long period.

CHECKED KNOTS: Checked knots are considered as defects and the checks appear on the end grain of knots in the wood rays. They are the result of differences in shrinkage parallel to and across the annual rings within knots.

LOOSE KNOTS: Due to differential drying, encased knots can become loose during drying since their wood is usually denser and shrinks more than the surrounding tissue. This dried dead knot is smaller than the knothole and frequently falls out during handling or machining.

CASEHARDENING: It results from too rapid drying, where the surface dries below the FSP first, but cannot shrink, putting tension on the surface. Nearly uniform moisture content but residual stresses, tension in the interior and compression in the outer layers of cells, causes casehardening.

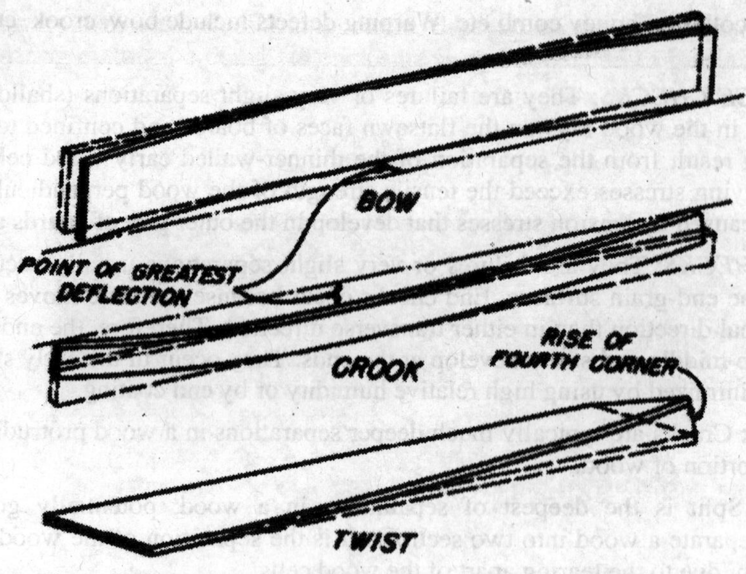

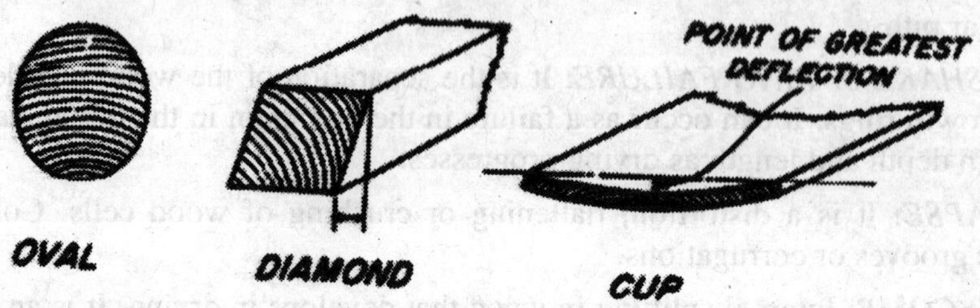

Fig. 19.5: Warping Defects

COMPRESSION FAILURE: It is the deformation of the wood fibres resulting from excessive compression along the grain either in direct end compression or in bending. In surfaced, lumber, compression failures appear as fine wrinkles across the face of the wood.

CROSS BREAK: A cross break is the failure of the wood cells across the grain and severely reduces the wood strength. Such breaks may be due to internal stress resulting from unequal longitudinal shrinkage.

BOW: It is the longitudinal curvature, flat-wise from a straight line. It is a deviation from edge-to-edge on the flat surface of board.

CROOK: It is the longitudinal curvature, edge-wise from a straight line. It is a deviation from edge-to-edge along the long surface of board.

CUP: It is the curving of the face of a plank so that it assumes a trough-like shape. It is a deviation from edge-to-edge on the end of board.

TWIST: It is the distortion of board sot that the two end surfaces do not lie on the same plane. One corner of a piece of wood twists out of the plane of the other three.

KINK: It is an abrupt deviation from the flatness or straightness due to localized grain distortion around knot or deformation caused by misplaced stickers in kiln.

DIAMOND: It is a form of warp found in squares or thick lumber end. The cross-section assumes a diamond shape during drying caused by difference between radial and tangential shrinkage in squares in which growth rings run diagonally from corner to corner.

WANE: It is also called as bark inclusion for bark incursion. Any area of a piece of lumber that includes the normally occurring edge of the tree it was cut from.

20 CLONAL FORESTRY

INTRODUCTION

Forest tree breeding aims to increase forest health, quality of trees and stand productivity using combination of different approaches. However, the application of science of genetics in improving yield from forests was quite late with large-scale tree breeding programs being initiated around the world only in the 1950s. Zobel and Talbert (1984) cite 23 papers from 14 countries published in the 1950s that describe tree improvement programs. Since then, substantial advances have been made in forest genetics and tree breeding research. Today, there are breeding programs for nearly all commercially important tree species in the world. Genetically improved plantations from such programs have made significant impacts on forest productivity, wood supplies and sustainability of forest resources.

Conventional tree breeding programs comprises several steps. The programme is initiated with provenance selection which aims to select well adapted genotypes to serve as base population. Within the base population, phenotypically superior (plus) trees in natural stands are selected. Trees are selected from these plus trees after the genetic tests for identification of individuals with good genotypes and high combining ability for economic and adaptive traits. The best parents may be selected, or the best progeny of the best parents may be selected. In order to use selected genotypes for reforestation, production populations are established. Based on genetic tests the seed orchard is rouged. Subsequent generations of tree breeding are initiated by treating the first generation seed orchard as the base population. These individuals are then inter-mated to generate new combinations and a new array of genetic variations to select in the second generation. Offspring's are then tested in second generation genetic tests and selections made to go into second generation orchards and a new breeding population. Thus efficient tree breeding requires long-term commitment, selection of appropriate methods and large scale application of results to provide practical benefits.

The incorporation of the improved genotypes that evolve out of the selection, testing and breeding cycle into operational forestry is an essential component of any successful tree improvement programme. The genetically superior plants that result from breeding programs are deployed into operational forestry through mass propagation. The objectives of breeding cycle and field deployment components of a tree breeding programme are quite distinct. The breeding cycle tries to improve the base population with every cycle, maintaining the broad genetic base, which is essential for any cross breeding crop and has a long vision that spans generations. The propagation component tries to utilize the best genotypes available in the operational forestry and has a focus on the current generation. Since, the success of any tree improvement programme is rated based on the realized benefits from the field; these two components will have to work in tandem.

VEGETATIVE PROPAGATION METHODS IN FOREST TREES

Vegetative Propagation was originally seen as a way to bulk up small amounts of rare and valuable material. Since, vegetative propagation is done from well-differentiated vegetative parts; the propagules maintain physiological condition of the propagated part and genetic consistence of the parent tree. It is possible to capture and transfer all genetic potential through use of vegetative propagation to the new tree. Vegetative propagation has been widely used in the establishment of clonal seed orchards and clone banks. Exceptional hybrids, sterile hybrids and selected plants can be propagated vegetatively (Quijada 1980). The original tree from which the parts to be propagated are taken is known as the ortet. Each plant propagated from an ortet is a ramet. A set of ramets from the same ortet is a clone. Differences observed between trees propagated by the same method are called clonal variation. Methods of vegetative propagation can be broadly classified as macro propagation and micro propagation techniques. There are three main methods of macro propagation used in trees: cuttings, grafts and layers. Latest in vegetative propagation of trees is the micro propagation through tissue culture techniques.

CUTTINGS

Cuttings form the earliest and easiest method of vegetative propagation. These are sections taken from the tree and put to root in an appropriate medium. Cutting as a method of propagation is cost-effective compared to other vegetative propagation methods and they result in good plant type. However, they need rooting. Multiplication rate of cuttings are low, need large areas for donor plant hedges and ageing problem complicates clone testing. Some tree species, such as the poplars and willows, are very easy to propagate by stem cuttings.

GRAFTS

Grafts are plants obtained by fusing a part from the tree to be propagated with another part which has its own root. This involves the removal of a vegetative part from the parent tree (the scion) and its attachment to a part with its own root (the stock) so that the tissues fuse. Heteroplastic grafts belong to different species, while homoplastic grafts belong to the same species. According to the position of the scion, there are top and side grafts, depending on whether the scion is inserted into the top or the side of the stock. Grafting is relatively simple technique and it avoids rooting problems. However, it is expensive and not applicable for mass propagation (Quijada 1980).

LAYERS

Layers are sections of the tree which are induced to root and then separated from the tree. The procedure consists of making a wound in a section of the tree and covering this with a medium which retains moisture (moss, earth, etc.). Healing causes the formation of a callus from which adventitious roots may grow. The most common type of layering is aerial layering in which the covering is done with a medium other than soil and the operation takes place rather high off the ground. When long, low, flexible branches are available, these can be inserted into the soil and it is known as ground layering. Usually the part propagated is a branch, but in some species interference with the roots can cause them to sprout, constituting a form of layering in that they do not separate from the stem. Layering is relatively simple technique and it avoids rooting problems. However, it is also expensive and not applicable for mass propagation.

MICRO PROPAGATION THROUGH TISSUE CULTURE

The latest among the technologies for vegetative propagation are micro propagation through tissue culture techniques. Tissue culture refers to the culture of cells with the potential mitotic activity, in an appropriate medium under aseptic conditions. Micro-propagation using tissue culture method provides unique opportunities to rapidly multiply the elite trees either through organogenesis or through somatic embryogenesis. In organogenesis, the explants under favorable media give rise to axillary or adventitious shoots which can be separated and rooted under another medium and hardened to plant out to the field. Tissue culture propagation by axillary shoots offers low risks of post-effects. However, the technique needs rooting and is expensive. *In vitro* propagation by adventitious shoots offers high multiplication rate but this technique also needs rooting. There are risks of post-effects and is an expensive propagation method. In somatic embryogenesis, embryo like structures arise from the explants or callus which can be germinated as a normal plant under specific media combinations. In somatic embryogenesis, embryos are induced either directly from the explant or from an intervening callus. Somatic embryogenesis is very promising because, the multiplication rate is very high and there is good prospect for automatic handling. However, the method is expensive, there are risks of post-effects, loss of genotypes and certain selection required during the process.

Vegetative propagation has been historically used for maintenance of genetic uniformity, since it offers the fastest way to get improved material into production by reproducing the genetic makeup of the selected individuals. However, consistency in phenotypic effect is not always ensured by vegetative propagation. Some of the vegetative propagation could have *somaclonal* variation, where the differences have genetic origin. Propagation methods can introduce effects that are non-genetic, but are confounded within the genetic unit. These non-genetic effects that can be confounded with genetic effects are called C-effects (Lerner 1958). In connection with vegetative propagation, the C effect can be caused either by differences in donor plant physiology or by differences in the environment during propagation (Libby 1976).

APPLICATION OF VEGETATIVE PROPAGATION IN FORESTRY

Vegetative propagation of trees was originally used for the species like *Poplar and Cryptomeria* which reproduced vegetatively under natural condition. By the development of other methods of vegetative propagation namely, cuttings and grafting it was made possible to multiply the selected superior trees for use in seed orchard and also to preserve germplasm in banks. Large-scale propagation from cuttings at costs similar to those of seedlings has been developed in many forest tree species.

Vegetative propagation has been used in forestry for production of quality planting stock. It has been used to develop fast and economical methods of raising superior planting stock in the intensive management of forests. These techniques become important in forestry owing to their potential in propagating superior trees to increase volume and quality of production. It has also been used for propagation of problem trees. Vegetative propagation could be effectively used in species, which are not able to produce seeds frequently like bamboos. Propagation by vegetative means may be easy, more rapid and economical than by seed in many cases. In some tree species, germination may be poor or slow or there may exist complex dormancy problems or the seed may lose its viability very quickly like in recalcitrant species. Moreover, seedlings of many species grow slowly and take a long time to reach marketable size. In all these cases, use of vegetative multiplication is a more convenient method of propagation.

Since, most of the tree species are cross-pollinated and highly heterozygous, asexual clonal propagation helps in maintaining the genetic characteristics of each species. Because of this property, vegetative propagation methods can be used to develop clonal repository. Genotypes are preserved in such clonal banks. In genetic testing's, vegetative propagation has been used to evaluate genotypes and their interaction with the environment through clonal testing. Clonal propagation will also help to conserve and multiply the valuable trees in a centralized area such as in a laboratory or green house for intensive study and breeding. Vegetative propagation has been used for maintenance of genetic gain. It helps to capture maximum genetic gains when used for regeneration in operational planting programme.

In addition to the above benefits, some forms of vegetative propagation can be used for production of disease free plants. In tree species that are often susceptible to some pests and diseases and while some may be partially or entirely resistant. Vegetative approaches like grafting will help to produce resistant clones. Vegetative propagation can also be used to speed up the reproductive cycle for accelerated breeding and testing. Vegetatively propagated plants are precocious in bearing than seed propagated plants. Early induction of flowering thus induced will help to reduce the rotation of the tree species and also to increase the productivity.

INCORPORATION OF VEGETATIVE PROPAGATION TECHNIQUES IN TREE BREEDING

Because of the various advantages of vegetative propagation in preserving the genetic characters of the propagated plants, it is used for the following uses in tree breeding programs.

ESTABLISHMENT OF CLONAL SEED ORCHARDS

One of the objectives of a tree improvement programme is to propagate the interim benefits of the programme at the fastest rate to meet the propagule requirement of operational forestry through appropriate technologies. In advanced breeding programmes seed orchards are established for production of superior quality seeds. A seed orchard is a collection of selected clones or families established in one physical location and then managed to produce genetically improved seed for operational forestation (White et al 2007). Seed orchard are established either through seeds (seedling seed orchards) from open pollinated seed or clones (clonal seed orchard) by the use of grafts, cuttings, air-layered plants, tissue culture plantlets or other methods of vegetative propagation. Clonal seed orchard is generally preferred when vegetative propagation is possible and the seed orchard only serves as a production area for seeds or vegetative propagules. Clonal seed orchard is raised in species in which flowering can be initiated at an earlier age through vegetative propagation. A clonal seed orchard typically contains many clones.

GENETIC TESTING IN A BREEDING PROGRAMME

In conventional tree breeding programmes, the usual approach for testing candidates for the next generation involves progeny tests involving progenies by controlled crossings. The candidates must reach the flowering stage, which can take several years in trees. Vegetative propagation allows the candidates to be directly tested as clones, instead of indirectly, as in progeny tests. Besides saving time, since it does not require trees to reach flowering competence, this strategy also increases the selection intensity, by allowing more candidates to be tested at the same cost. Studies on this subject, with varying breeding parameters, have consistently concluded that clonal candidate testing is efficient (Shaw and Hood 1985; Russell and Loo-Dinkins 1993; Rosvall et al 1998).

DEPLOYMENT OF GENETICALLY SUPERIOR FAMILIES THROUGH BULK PROPAGATION

Large scale deployment of genetically superior families selected based on genetic tests is one of the common steps in tree improvement programmes. After controlled pollination of selected parents in the breeding programme, families can be mass propagated without knowing the value of each progeny and without identifying the individual progeny. Quite often, it would be expensive to do controlled crosses to obtain large quantities of seed to meet requirement of operational forestry. In such situations, vegetative propagation using the limited seeds produced through such controlled crosses is often resorted to. Each seed produced through such crossing would represent a different genotype and if 100 seeds are obtained from one cross we will end up with 100 genotypes. If 30 such crosses are made, the lot would contain 3000 genotypes. This mode of using vegetative propagation is called bulk propagation and provides the same genetic gain as if the material was propagated via seeds.

CLONAL FORESTRY

Vegetative propagation can be used for large scale propagation of identified and tested clones with valuable properties. Such extended use of superior clones, is referred to as clonal forestry. Clonal forestry refers to the large-scale deployment of relatively few (10 to 50), known-superior clones that have proven their superiority in clonal tests (White et al 2007). Clonal forestry can be defined as 'the deployment of tested clones. The chief difference between clonal propagation and bulk propagation is in terms of the number of genotypes involved and the proven superiority of clone used in clonal forestry. The genotype effect consists of two components: additive and non-additive. Through sexual propagation, while it is easy to pyramid additive effects, due to recombination of the parental genotypes, the non-additive effects cannot be exploited. The potential to utilize non-additive effects is one of the attractive features of clonal forestry, since it offers possibilities of capturing additional genetic gain. Clonal forestry would then increase the genetic gain realized in the production forest compared with bulk propagation due to the use of tested genotypes. Clonal propagation offers a chance to propagate a superior genotype that would be lost through sexual reproduction.

APPLICATIONS OF CLONAL FORESTRY

Clonal forestry in tree breeding programme has been emerging as a strong attraction to the traditional seed orchard breeding system. It is now recognized that most efficient genetic improvement to trees can happen by combining the controlled breeding of highly selected parents with the clonal magnification of the very best offspring from each cross. Clonal forestry has been recognized as a method offering huge potential benefits in terms of tree improvement gains, uniformity and 'customization'. The key to clonal forestry is the ability to propagate selected, superior proven clones on a large scale. Through clonal forestry it is possible to deploy some unique reliable genotypes continually for many years, even several rotations and the possibility to adopt clone-specific management practices including matching clones to specific site types and optimizing silvicultural regimes for different clones.

Clonal forestry would be relevant under certain conditions namely, (1) when high performing individuals, proven as such through careful screening and evaluation in field tests are available; (2) when there is sufficient number of superior clones which can be mass propagated in equal numbers of each clone in order to maintain a "variation in space", new, possibly partly overlapping sets of clones, must be used each year; and (3) when more productive and more resistant clones are continuously injected into the planting programmes, through continuous genetic improvement programmes in order to maintain "variation in time" (Zabala 1993).

One of the fundamental requirements for initiation of a clonal forestry programme is sufficiently refined cloning technique for the species. Angiosperms have been mostly into clonal forestry, since, the cloning techniques were well established in many of them. Propagation of trees vegetatively was originally used for the species which reproduced vegetatively under natural condition such as Poplars (Zsuffa et al 1993), *Cryptomeria* (Ohba 1993) etc. Clonal plantations of *Cryptomeria* have been grown in Japan since the 1400s to provide an alternative to the use of natural forests. Later on, many researchers multiplied the selected superior trees, mostly through cuttings and grafting for use in seed orchard and also to preserve germplasm in banks. The interest in clonal forestry increased primarily due to improvement in vegetative propagation techniques in particular, the refinement of the procedures for rooting of cutting and other methods of somatic propagation (Park et al 1995). Recently, techniques have been developed in many forest tree species for large-scale propagation from cuttings at costs similar to those of seedlings. There has been growing interest in using the techniques of clonal forestry and thereby, exploiting the considerable amount of genetic variability already existing in the natural populations of forest tree species that have been amenable to vegetative propagation (Nanda 1970; Leaky 1987; Pal 1991) and partly because of tremendous yield improvement by adopting the clonal forestry approaches (Leaky 1987). The most widespread use of clonal plantings has been with the willow family (Salicaceae), involving willow (*Salix*) (Zsuffa et al 1993) and poplar (*Populus*) species and hybrids. Several species in the genus *Eucalyptus,* especially *Eucalyptus grandis* and several of its interspecific hybrids in many tropical and sub-tropical countries (Duncan et al 2000) have also been subjected to clonal forestry. *Cryptomeria japonica,* mainly in Japan (Ohba 1993) and *Picea abies* in Europe (Bentzer 1993) are further examples of successful clonal forestry programs. More applications of clonal forestry like in Christmas tree and landscaping in urban settings are being made (Kleinschmit et al 1993).

Clonal forestry is not a breeding strategy, but an operational option. Hence, an important requirement for a successful clonal forestry programme is the existence of an underlying breeding program of selection, breeding and testing to identify clones of ever-increasing genetic merit. Development of superior clonal lines which involves establishment of long term genetic testing of clonal lines developed through breeding programmes is a pre-requisite for initiating clonal forestry programme. This has to be constant process involving constant upgradation of clones in use, incorporating the results of the ongoing breeding programme. Techniques aimed at early selection will be an added asset in the development of a successful clonal forestry programme.

IMPLEMENTATION OF A CLONAL FORESTRY PROGRAMME

The important steps in implementation of a clonal forestry programme is as detailed below.

SELECTION OF TREES FOR CLONAL PROPAGATION: The selection criterion for clones varies with species, object of management and the end use. Selection of genetically true to type clone is done after correct identification and proper labeling. Generally, a dominant tree with good apical dominance, free from pests and diseases and free from genetical disorders with a history of superior phenotypic performance is selected. Clones can be specifically chosen to exploit different ecological niches on the merit of different mechanisms for resistance to a particular pest; and/or to increase genetic diversity of traits observed.

NUMBER OF CLONES TO BE SELECTED: The number of clones to be selected depends on rotation age, intensity of forest management, genetic variability of the species and clones involved, the likely risks and the acceptable loss levels (Libby 1987). The number of selected clones is based on the criteria that we should achieve the greatest improvement by selecting only the best clones within acceptable limits of the dangers of destructive losses in the forest. Hence, the safe and reasonable number of clones is recommended to be 7 - 30. In a stress affected area, more number of clones should be used. There are special conditions such as, severe insect or disease infested areas or very severe environment where in, very few clones may be justified (Zobel and Talbert 1984).

CLONAL MAINTENANCE: The selected clones have to be maintained and multiplied under strict sanitary conditions to prevent reinfection. Measures should also be taken for immediate detection of any reinfection or any change in the original stock.

CLONAL MULTIPLICATION: Identified superior clones of proven quality are assembled in an isolated area with proper labeling. These clones are vegetatively multiplied on a large scale after establishing under field condition. Large-scale plantations can be raised from clones obtained from clonal multiplication area.

CLONAL TESTING: To identify suitable clones for deployment, systematic screening of clones for productivity attributes and their adoptability to adverse environment is necessary.

Testing of clones should be done under multi locations with sufficient number of ramets from each clone. Testing of clones at multiple test sites provides information on adoptability of clones in varying environmental conditions.

White et al (2007) has given the following sequences for a typical clonal forestry programme. The program starts with the establishment of clonal trials on several sites in the planting zone to test the clones for suitability. Each test site can contain multiple ramets with large numbers of clones. These ramets are planted in randomized, replicated designs, usually in single tree plots to facilitate the precise ranking of clones. At appropriate selection age, measurements taken in the tests are used to predict clonal genetic values. The best ones are used for operational deployment. If there is substantial clone x site interaction, then specific clones are chosen for deployment to particular types of sites within the planting zone. The selected clones are then asexually propagated on a large scale, to produce plantlets for operational forestation. These plantlets have the same genotypes as the selected clones. Later on, with increased experience with selected clones, clone specific management prescriptions in all phases (nursery, plantation and processing facility) can be developed with concomitant cost savings and gains in efficiency of operation.

ADVANTAGES OF CLONAL FORESTRY

The greatest advantages in clonal forestry are derived from use of selected superior clones that are reproduced and deployed true-to-type in operational forestry. Clonal forestry offers the only option for complete capture of non-additive genetic variance and since, the deployed clones retain their entire genetic composition. Through clonal forestry, it is possible to exploit genotype x environment interaction through site-specific deployment of tested clones to edaphoclimatic conditions in which they are known to perform well. Hence, clonal forestry offers the greatest potential to deliver the benefits of the best genotypes from a controlled cross. Clonal forestry ensures maximum genetic gains from selecting and mass-producing best individuals. This is the best method for early deployment of improved material for commercial purpose and saves time.

Clonal forestry can ensure enhanced plantation productivity and economic benefits by mass propagating superior plants in the breeding programme. It can lead to increased productivity in the forests. There can also be substantial economic gain by ensuring better return on investment in both operational forestry and for breeding programme. This can also mean that less land need to be devoted to production forestry and more forest land available to meet environmental (e.g. increased biodiversity) and social (e.g. recreation) requirements.

Relative uniformity of clonal forests offers commercial advantages for the market, for the forest manager and for the plant producer. Overall performance of the clones can be very close matching with particular soil and cultural environments that optimize commercial utility of the clonal forests. Clonally forestry will ensure that all the trees reach harvestable stage at the same time and with the same morphological and chemical characteristics. There will be increased predictability and repeatability of the crop. Organizations desiring uniform plantations to reduce costs of harvest and produce more uniform products can even establish monoclonal plantations in which a single clone can be planted over the entire extent.

DISADVANTAGES OF CLONAL FORESTRY

Although there are many advantages for clonal forestry, there are a large number of disadvantages and concerns in adopting clonal forestry. The most important disadvantage is the problems created due to reduced genetic diversity. Reduction in genetic diversity can make the crop highly vulnerable stresses like to pests and diseases that the crop may not be exposed till then. The unpredictability in clone x environment interaction could be a serious disadvantage for a long term venture like forestry, where changes in ecosystem condition is possible and loss of adaptation to long term survival would be a major drawback. This disadvantage however, could be partly avoided by increasing the use of superior clones. Apart from this, genetic diversity has a better ecological advantage as different niches in a site is used up more efficiently. Moreover, such a diverse crop will offer more niches for other living beings too. Social and political risks, including public perceptions are some other disadvantages of clonal forestry. Worries about monoclonal forestry can limit the legally permitted and publically acceptable areas of vegetative propagation in forestry. Generally uniformity in forests is aesthetically poor and socially unacceptable.

21 WILDLIFE MANAGEMENT

WILDLIFE MANAGEMENT: DEFINITION

Wildlife includes any animal, aquatic or land vegetation which forms part of any habitat. Wildlife management can be defined as the manipulation of wildlife populations and habitat to achieve a goal (Sargent and Carter 1999). Aldo Leopold (1933) defined game management as the art of making land produce sustained annual crops of wild game for recreational use. Wildlife management is the term very often used by the environmentalists all around the globe. The best way to define wildlife management would be to say all about finding the balance between the needs of wildlife and the needs of people by using different scientific methods. Wildlife management is defined as the application of ecological knowledge to populations of vertebrate animals and their plant and animal associates in a manner that strikes a balance between the needs of those populations and the needs of people.

Wildlife management is the manipulation of wild plant and animal species behaviour or abundance for a specified goal. The reasons for such action include,

- *Conservation* of vulnerable or rare species and habitats
- *Exploitation* of wildlife resources such as harvesting game and fish stocks
- *Control* of species to address impacts such as preserving public health and safety and preventing serious damage to property including impacts on agriculture.

Wildlife management in these days is mostly focused on wildlife conservation and this requires help of many other scientific disciplines such as chemistry, biology, ecology, climatology and geography to get the best possible results. The ultimate goal of wildlife management is to stop extinction of many endangered species but in today's situation this is almost an impossible task due to many different ecological problems such as climate change and pollution. Wildlife management is multidimensional task that needs to view things from many different perspectives before choosing the perfect technique for given wildlife habitat. This makes wildlife management very complex science, which makes things very difficult for wildlife managers.

CONCEPT OF WILDLIFE MANAGEMENT

Can the mankind survive on its own? Obviously, No. Apart from air and water, the mankind is heavily dependent on animals and plants. In the beginning, their dependence on animals and plants were so high to an extent that they were inseparable. Ancient scriptures like Vedas, Puranas, Bible, etc., all have mentioned about the importance of animals, plants and their protection. For his survival, man started domesticating plants and animals. To protect himself, his plants (crops) and his animals (domestic animals), he started extirpating unwanted plants and animals. This lead to accelerated extinction rates of wild plants and animals. Thus there existed a love towards certain set of animals and plants and hatred towards another set of animals and plants. For our purpose, we will concentrate on only the animals.

Animals that were perceived as a threat to human life and crops were mercilessly hunted. Communal hunting was prevalent in many societies. Slow but steadily, the population of such animals started declining. With the advent of monarchy, hunting was either prohibited or regulated. Colonisation and world wars proved detrimental to most of the wildlife across the world. Decreasing wildlife also generated lot of interest about its conservation among the elite of many countries. Acts and treaties aiming to protect wildlife and regulate its utilization started coming up in the later part of the 20th century. Concept of wise use of the wildlife resources in the western countries paved the foundation for the science of "Game Management". Aldo Leopald considered as the father of modern wildlife management came up with his monumental work "*Game Management*" in 1933. Under Game management, wildlife was considered a resource that is to be managed in a way so that sustainable harvest was possible. From this anthropo-centric and consumptive view of the early 1900's, the approach towards wildlife management has evolved into an animal centric one with emphasis on non-consumptive use.

Wildlife conservation is very much part of the Indian society. As early as the 3rd century, King Asoka brought in laws prohibiting killing of animals and also created facilities like water holes and hospital for the animals. The ancient rulers, throughout the country in general, were always sympathetic over wildlife and had some kind of regulations. The Mughal period saw extensive hunting in their territory especially hunting tigers was very common. Cheetahs were kept under captivity for the use of them in their hunting expeditions. This in way combined with other reasons paved the way for the extinction of the Asiatic Cheetah.

Arrival of Britishers in the country saw establishment of forest department and scientific management of forest for the timber production. Exclusive acts like the Elephant Preservation Act (1873), World Birds Protection Act (1887), etc., also came into force. Also the first national park of the country Hailey National Park (present day Corbett National Park) was created in 1936. However, during the British rule large areas of forests were cleared and brought under agriculture. Both the herbivores and carnivores were allowed to be hunted for the protection of the human life and property. After the country became independent, the first major step towards wildlife conservation was the creation of Indian Board for Wildlife (IBWL) in 1950. The real contribution to the field of wildlife conservation in our country came in the shape of Wildlife Protection Act (1972).

WHAT TO CONSERVE?

India ranks sixth among the 12 mega biodiversity countries of the world. India with 2.4 per cent of the land area of the world accounts for 7-8 per cent of the total recorded species i.e. around 45,500 species of plants and 91,200 species of animals (MoEF 2009). Number of species documented in India as against the overall World figure is given in Table 21.1.

Table 21.1: Number of Documented Species in World and India (Source: Animal Discovery 2010; Zoological Survey of India)

Taxonomic Group	Number of Species in the World	Number of Species in India	% in India
Mammalia	4,629	397	8.58
Aves	9,026	1,232	13.66
Reptilia	5,817	460	7.91
Amphibia	5,162	289	5.60
Pisces	21,727	2,634	12.12
Prochordata	2,106	119	5.65
Animalia	11,95,567	89,220	7.46
Animalia + Protista	12,26,817	91,797	7.48

A brief look at the Table 21.1 demonstrates the biodiversity wealth of our country. The influences of three bio-geographic realms namely Ethiopean, Palearctic and the Oriental realms combined with a diversity of ecosystem have resulted in the richness of biodiversity in our country. Around 45 species of mammals, 50 species of birds, 197 species of reptiles, 135 species of amphibians and 341 species of fishes are endemic to our country. Of the charismatic species, half the population of the wild Asian Tiger and wild Asiatic Elephant is found in India. Around 80 per cent of the Great Indian Rhino is found in India. Such an important wealth needs protection for posterity.

HOW PROTECTED?

As mentioned earlier, protection of wildlife was part of our culture. Vedanthangal, one of the important birding areas in the country enjoys protection from the local people for the last 250 years. With the enactment of Wildlife Protection Act (1972), wildlife in the country is protected in the national parks, sanctuaries, conservation reserve and community reserves. These four areas comprise to form the protected area (PA) network of our country. Presently there are 102 national parks, 515 wildlife sanctuaries, 48 conservation reserves and 4 community reserves in India. Around 1,61,132 square km area (4.90 per cent) of the total geographical area of the country is under the protected area network. Protected area is defined as "an area of land and/or sea especially dedicated to the protection and maintenance of biological diversity and of natural and associated cultural resources, and managed through legal or other effective means" (IUCN 2004). Biological diversity means the variability among living organisms from all sources including, terrestrial, marine and other aquatic ecosystems and the ecological complexes of which they are part, this includes diversity within species, between species and of ecosystems (CBD 1992).

Table 21.2: Growth of Protected Areas in India (Source: Wildlife Institute of India)

Years	Upto 1970	1971-1975	1976-1980	1981-1985	1986-1990	1991-1995	1996-2000	2001-2005	2006-2010	2010 onward
No of PAs	69	104	194	285	495	539	577	602	617	669
Increase	-	35	90	91	210	44	38	25	15	52
% Increase	-	10.72	186.54	146.90	173.68	108.89	107.05	104.33	102.49	108.43

The period from 1971 to 90 saw a huge growth of protected areas in the country. This can be attributed to reasons like enactment of Wildlife Protection Act (1972), Forest Conservation Act (1980) and increased awareness among the policy makers and the people. Demand for land for various developmental works and increasing human wildlife conflict slowed down the pace of further expansion of the PAs after 1990. Of these four categories of protected area, conservation reserve and community reserve came into existence following the amendment of the Wildlife Protection Act (1972) in 2006. Unlike national park and wildlife sanctuaries both conservation reserve and community reserve do not put any restriction on the existing land use of the area concerned. This has helped in expansion of the PA network in the last 2 years. All the 52 PAs added from 2010 belong to either the conservation reserve or the community reserve.

NATIONAL PARK

National Parks (NP) are areas of land and / or sea designated to protect the ecological integrity of one or more ecosystems for posterity. Consumptive use of any resource from a national park is prohibited. However use of the area for educational, scientific, spiritual and recreational purposes is permitted but regulated. National parks get the highest degree of protection from the Wildlife Protection Act, 1972 e.g. Corbett National Park, Keoladeo National Park, etc.,

SANCTUARY

Wildlife Sanctuaries (WLS) are areas of land and / or sea designated to protect an area and managed by regular interventions aimed at a particular species. That is the reason why we find wildlife sanctuaries generally named with a species like Wild Ass Sanctuary of Gujarat, Musk Deer Wildlife Sanctuary of Uttarakhand, etc. Major differences between NP and WLS are:

- No rights of people is allowed in a national park.
- Removal of any resource from a national park is prohibited. However removal is permitted if helps in habitat management.
- Advisory committee can be formed for the management of a wildlife sanctuary only and not for a national park (see Section: 33 B of Wildlife Protection Act,1972).

CONSERVATION RESERVE

Conservation Reserve are designated by the State Government on areas owned by the Government, (after having consultations with the local communities) particularly the areas adjacent to national parks and sanctuaries and those areas which link one protected area with another, for protecting landscapes, seascapes, flora and fauna and their habitat.

COMMUNITY RESERVE

The State Government may, where the community or an individual has volunteered to conserve wildlife and its habitat, declare any private or community land not comprised within a national park, sanctuary or a conservation reserve, as a community reserve, for protecting fauna, flora and traditional or cultural conservation values and practices. Thus, these four categories namely national park, sanctuary, conservation reserve and community reserve are the four legal categories of protected areas in our country. Major differences between conservation reserve and community reserve are:

- Conservation area can only be declared on Government land whereas community reserve can be declared on a private or a community land.
- Protection of landscapes and seascapes is given emphasis in a conservation reserve, whereas protection of traditional and cultural values is given importance in a community reserve.

PROTECTED AREA

IUCN Categories of Protected Area (IUCN 2004) are:

Ia: Strict nature reserve

Ib: Wilderness area

II: National park

III: Natural monument or feature

IV: Habitat/species management area

V: Protected landscape/seascape

VI: Protected area with sustainable use of natural resources

PROTECTED AREA PLANNING

India is one of the few countries to use a scientific method in conservation planning both at the national and regional levels. The biogeographic classification developed by the Wildlife Institute of India (WII) in 1988 provides a scientific framework to have representative network of protected areas in the country. The entire country is divided into 10 biogeographic zones and 26 biotic provinces. The zone wise detail regarding the protected area is given in Table 21.3.

A look at the Table 21.3 reveals that the protected areas are not balanced across the different biogeographic zones. The biogeographic classification was developed in 1988 and by that time itself more than 400 protected areas were created in our country. Absence of any tool for planning would have resulted in this skewed distribution of protected areas across the zones. Further, the pace of protected area creation too slowed down after 1990s which also provided less opportunity for balancing this distribution. Gangetic plains, one of the most fertile parts of the country, has the lowest area i.e. 2.21 per cent under protected area when compared to its total area. The populations density in this zone is one of the highest in the country and creating new PAs in this zone would not only be difficulty but also would create lot of social unrest.

Table 21.3: Protected Area Details in various Biogeographic Zones (Source: Wildlife Institute of India)

Biogeographic Zone	Area (in Sq. Km)	% of Total Geographic Area	No. of Protected Areas (2009-10)	Area under Protected Area (in Sq. Km)	% of Zone Area
Trans Himalaya	1,84,823	5.62	7	16,247.56	8.79
Himalaya	2,10,673	6.41	77	23,230.39	11.03
Desert	2,14,014	6.51	6	16,075.38	7.57
Semi-arid	5,39,479	16.41	91	13,937.00	2.58
Western Ghats	1,32,179	4.02	63	1,361.43	10.30
Deccan Peninsula	13,80,339	41.99	151	53,536.53	3.88
Gangetic Plains	3,54,848	10.79	38	7,836.56	2.21
Coasts	91,319	2.78	25	4,690.63	5.14
North East	1,71,340	5.21	50	6,504.58	3.80
Islands	8,249	0.25	105	1,546.30	18.75

On the contrary, zones with lot of inaccessible mountains, islands, etc., are well protected e.g. 18.75 per cent islands and 19.82 per cent of the Himalayas. Creation of national parks in coming years would be very difficult for reasons like the rising demand for resources and rising population. However scope for establishing conservation reserve and community reserve is ample. If proper planning is done the country can have few more PAs under these two categories.

PROBLEMS FACED BY THE PAs

The protected areas across the country are not free of any problem. Problems are many and unique to each protected area. Broadly the major problems faced by these PAs are:

(1) Human pressure both from living inside and outside the PA

(2) Grazing pressure especially that by the unproductive cattle

(3) Absence of any clearly laid out policy on eco-sensitive zones around the PA

(4) Non-completion of the legal process in creation of PA which results in legal ambiguity about the PA status

Apart from these four categories of protected areas, there are few more areas protected/designated under various international treaties /conventions like the Ramsar Convention and World Heritage Convention.

RAMSAR SITES

Ramsar Convention on Wetlands of International Importance is an intergovernmental treaty that provides the framework for national action and international cooperation for the conservation and wise use of wetlands and their resources. This convention was adopted at Ramsar, an Iranian city in 2 February 1971 and came into force in 1975. At present there are 160 contracting parties with 2006 sites designated as Ramsar sites with a total surface area of 19,28,220 square km. Wise use of wetlands is the focal point of this convention and wise use is defined as "the maintenance of their (wetlands) ecological character, achieved through the implementation of ecosystem approaches, within the context of sustainable development" (Source: Ramsar Convention). Unlike the national park where resource use is prohibited, Ramsar convention advocates use of wetland resources in a sustainable way.

India joined the convention on 1 February 1982 and there are presently 25 sites designated as Ramsar sites from our country with a surface area 6,771.31 Square Km. There are 9 criteria given by Ramsar on which a site can be designated (for details visit www.ramsar.org). If for any reason, a site loses its ecological character or losing its ecological character for which it was initially recognized, then that site is put under a record called Montreaux Record. Once a site is put in the Montreaux Record, the contracting party will have to take such necessary steps so that the ecological character of the site is restored and the site is removed from the Montreaux Record. In India, two sites Keoladeo National Park (Rajasthan) and Loktak Lake (Manipur) are in the Montreaux Record.

Table 21.4: List of Ramsar Sites in India (Source: Ramsar Convention 2012)

S No	Name of the Site	State	Designated on
1	Chilka	Odisha	01.10.1981
2	Keoladeo National Park	Rajasthan	01.10.1981
3	Wular Lake	Jammu and Kashmir	23.04.1990
4	Harike Lake	Punjab	23.04.1990
5	Loktak Lake	Manipur	23.04.1990
6	Sambhar Lake	Rajasthan	23.04.1990
7	Kanjli	Punjab	22.01.2002
8	Ropar	Punjab	22.01.2002
9	Asthamudi Wetland	Kerala	19.08.2002
10	Bhitarkanika Mangrooves	Odisha	19.08.2002
11	Bhoj Wetland	Madhya Pradesh	19.08.2002
12	Deepaor Beel	Assam	19.08.2002
13	East Calcutta Wetlands	West Bengal	19.08.2002
14	Kolleru Lake	Andhra Pradesh	19.08.2002
15	Point Calimere Wildlife Sanctuary	Tamil Nadu	19.08.2002
16	Pong Dam Lake	Himachal Pradesh	19.08.2002
17	Sasthamkotta Lake	Kerala	19.08.2002
18	Tsomoriri Lake	Jammu and Kashmir	19.08.2002
19	Vembanad Kol	Kerala	19.08.2002
20	Chandertal Wetland	Himachal Pradesh	08.11.2005

21	Hokera Wetland	Jammu and Kashmir	08.11.2005
22	Renuka Wetland	Himachal Pradesh	08.11.2005
23	Rudrasagar Wetland	Tripura	08.11.2005
24	Surinsar – Mansar	Jammu and Kashmir	08.11.2005
25	Upper Ganga River (Brij Ghat to Narora Stretch)	Uttar Pradesh	08.11.2005

WORLD CULTURAL AND NATURAL HERITAGE SITES

UNESCO's Convention Concerning the Protection of the World Cultural and Natural Heritage was adopted in the General Conference of the UNESCO in Paris from 17 October to 21 November 1972. Under this convention a state party can nominate a site for its Outstanding Universal Value and get it designated either as a cultural or natural heritage sites. Outstanding Universal Value is defined as "cultural and/or natural significance which is so exceptional as to transcend national boundaries and to be of common importance for present and future generations of all humanity".

The protection of these sites for posterity i.e for both the present and future generations is the highest importance. For the purpose of identifying site, 10 criteria is given under the convention. Every State Party should prepare a tentative list of sites from which designation are made. At present there are 936 properties from 153 State Parties (countries). Of these 936 properties, 725 are cultural properties, 183 natural properties and 28 mixed properties. Presently, 189 countries have ratified this convention. In India, there are 23 cultural heritage properties and 5 natural heritage properties.

Table 21.5: List of Natural World Heritage Sites in India

S No	Property Name	State	Area	Year of Inscription and Criteria
1	Kaziranga National Park	Assam	42,996 ha	1985; ix and x
2	Keoladeo National Park	Rajasthan	2,873 ha	1985; x
3	Manas Wildlife Sanctuary	Assam	39,100 ha	1985; vii, ix and x
4	Sunderbans National Park	West Bengal	1,33,010 ha	1987; ix and x
5	Nandadevi and Valley of Flowers National Park	Uttarakhand	71,783 ha	1988; vii and x

Due to threats if any property is in danger of losing its outstanding universal value for which it was inscribed, then they are put in a danger list. It becomes the responsibility of the state party concerned to take such steps to reduce the damage and bring back the property to its original state. At present there are 35 properties in the danger list and none from our country. As many as 1559 properties from 166 state parties are in the tentative list. From India, 32 properties are in the tentative list, of which 7 properties are from the natural significance. They are Bhitarkanika Conservation Area, Desert National Park, Great Himalayan National Park, Kangchendzonga National Park, Western Ghats Cluster and Wild Ass Sanctuary (Little Rann of Kutch, Gujarat).

BIOSPHERE RESERVES

Man and Biosphere (MAB) programme is an intergovernmental scientific programme initiated under the UNESCO in 1970s. Man is recognized as an integral part of the ecosystem and it aims to set a scientific basis for the improvement of the relationships between people and their environment globally. This is a multidisciplinary programme which includes ecological sciences, social sciences and economics to develop systems that not only safeguard the ecosystem but also promotes development which is culturally and socially acceptable and environmentally sustainable.

Sites established by countries for the above intergovernmental scientific programme and recognized by the Man and Biosphere Programme is called as a Biosphere Reserve. Biosphere may include a terrestrial ecosystem or a coastal ecosystem or a marine ecosystem or a combination of these. Currently, there are 580 sites designated as biosphere reserves from 114 countries and India now has 17 biosphere reserves. Nilgri (2000), Gulf of Mannar (2001), Sunderban (2001), Nandadevi (2004), Nokrek (2009), Pachmarhi (2009) and Simlipal (2009) are the biosphere reserves identified by the MAB programme and is part of the World Network of Biosphere Reserves (WNBR).

TIGER RESERVES

Tiger occupies the centre stage of Wildlife conservation in our country. People are always attracted towards this majestic animal and at the same time also were hunted in large numbers. At the beginning of the last century, it was believed that there were around 40,000 (?) tigers in our country. E.P. Gee gave a figure of 4000 tigers in 1965, whereas Kailash Sankhala put the figure at 2500 in 1969. Some biologists believed that these magnificent animals might disappear and might not see the 21st century. Kailash Sankhala presented a report "Vanishing Tigers" in the parliament. During the IUCN General Meeting held in Delhi in 1969, decision on ban on killing of tigers was taken. The Indian Board for Wildlife advised all the State Governments to impose a complete ban on killing of tigers. Unfortunately, this period of 1969-79 witnessed large scale hunting of tigers in the country.

Guy Mountfort's (a trustee of WWF) continued effort on the issue of tiger conservation, lead to the constitution of an 11 member task force with Dr. Karan Singh as Chairman and Kailash Sankhala as Officer on Special Duty by then Prime Minister Mrs. Indira Gandhi. The task force gave the report in August 1972. First All India Tiger Census conducted during this period estimated the tiger population as 1827. This year also saw the enactment of Wildlife protection Act. Growing awareness and by the commitment of our leaders especially Mrs. Indira Gandhi, an ambitious project to protect tigers was launched on 1 April 1973 at Corbett National Park as a Central Sector Scheme (later made a Centrally Sponsored Scheme). Manas, Palamau, Simlipal, Corbett, Kanha, Melghat, Bandipur, Ranthambhore and Sunderbans were the first nine places declared as a tiger reserve. Now project tiger is renamed as National Tiger Conservation Authority.

The main emphasis under the Project Tiger was given to sound wildlife management practices, protection and site specific eco-development activities to reduce the dependence of the people on the reserve. Tiger Reserves are management units unlike National Park and Wildlife Sanctuaries which are legal categories of protected area. Tiger reserve may comprise a National Park/Wildlife Sanctuary or both or a forest area. Tiger reserves are divided into a core and a buffer area. In the core, disturbance is kept at a minimum so that the tigers are given a near natural environment to breed and survive. Extraction of resources from the core is totally prohibited whereas they are regulated in the buffer area. Eco-development activities are carried out in the buffer area so that the impact on the core is reduced. Further, these eco-development activities help the people to improve their living standards in way that their dependence on the resources of the tiger reserve is reduced. Management plans were drawn for every tiger reserve based on the following principles (Source: Project Tiger):

(1) Elimination of all forms of human exploitation and biotic disturbance from the core area and rationalization of activities in the buffer zone.

(2) Restricting the habitat management only to repair the damages done to the eco-system by human and other interferences, so as to facilitate recovery of the eco-system to its natural state.

(3) Monitoring the faunal and floral changes over time and carrying out research about wildlife.

Management of the individual tiger reserve is the responsibility of the State Government concerned. However, the Central Government through Project Tiger Directorate (Ministry of Environment and Forests) gives financial support to the tiger reserves in carrying out various activities including village relocation under a Centrally Sponsored Scheme (both the Central and State Government shares the expenditure). The number of tiger reserves increased rapidly and it is now 39 reserves in the 17 range states of our country.

Table 21.6: List of Tiger Reserves in India (Source: National Tiger Conservation Authority)

S No	Year of Creation	Name of Tiger Reserve	State	Total Area of the Core/Critical Tiger Habitat (In Sq. Km)
1	1973-74	Bandipur	Karnataka	872.24
2	1973-74	Corbett	Uttarakhand	821.99
3	1973-74	Kanha	Madhya Pradesh	917.43
4	1973-74	Manas	Assam	840.04
5	1973-74	Melghat	Maharashtra	1,500.49
6	1973-74	Palamau	Jharkhand	414.08
7	1973-74	Ranthambhore	Rajasthan	1,113.364
8	1973-74	Similipal	Odisha	1,194.74
9	1973-74	Sunderbans	West Bengal	1,699.62
10	1978-79	Periyar	Kerala	881.00
11	1978-79	Sariska	Rajasthan	681.1124
12	1982-83	Buxa	West Bengal	390.5813
13	1982-83	Indravati	Chhattisgarh	1,258.37
14	1982-83	Nagarjunsagar	Andhra Pradesh	2,527.00
15	1982-83	Namdapha	Arunachal Pradesh	1,807.82

16	1987-88	Dudhwa	Uttar Pradesh	1,093.79
17	1988-89	Kalakad-Mundanthurai	Tamil Nadu	895.00
18	1989-90	Valmiki	Bihar	840.00
19	1992-93	Pench	Madhya Pradesh	411.33
20	1993-94	Tadoba Andheri	Maharashtra	625.82
21	1993-94	Bandhavgarh	Madhya Pradesh	716.903
22	1994-95	Panna	Madhya Pradesh	576.13
23	1994-95	Dampa	Mizoram	500.00
24	1998-99	Bhadra	Karnataka	492.46
25	1998-99	Pench	Maharashtra	257.26
26	1999-2000	Pakke	Arunachal Pradesh	683.45
27	1999-2000	Nameri	Assam	200.00
28	1999-2000	Satpura	Madhya Pradesh	1,339.264
29	2008-09	Anamalai	Tamil Nadu	958.00
30	2008-09	Udanti-Sitanadi	Chhattisgarh	851.09
31	2008-09	Satkosia	Odisha	523.61
32	2008-09	Kaziranga	Assam	625.58
33	2008-09	Achanakmar	Chhattisgarh	626.195
34	2008-09	Dandeli-Anshi	Karnataka	814.884
35	2008-09	Sanjay-Dubri	Madhya Pradesh	831.25
36	2008-09	Mudumalai	Tamil Nadu	321.00
37	2008-09	Nagarhole	Karnataka	643.35
38	2008-09	Parambikulam	Kerala	390.89
39	2009-10	Sahyadri	Maharashtra	-NA-
		TOTAL		32,137.14

TIGER POPULATION ESTIMATION

Till few years back, Pug mark technique was followed to estimate the number of tigers in the country. This simple technique involves collection of the foot print of tigers by using plaster of paris and estimating the number of individuals after deducting the duplications of the foot print. This method was originally developed by Shri. S.R. Choudhry, a Forest Officer, in 1970's known as the "co-operation census" where in the pugmark evidences were combined with other evidences like sighting, etc., to arrive at a figure. Pug marks will not be identical between two individuals and this was used in calculating the number of animals in any particular area.

During a normal walk, the hind limb mark super imposes on the fore limb mark and the fore limb pug mark of both sexes almost look alike (whereas the hind limb pug mark of a male looks more squarish and of the female a rectangular one). For these reasons, it has been a practice to take impression of the left hind pug mark while doing a tiger census in the pug mark technique.

As mentioned above, no two individuals pug marks look alike. However, difference in the substrate like a wet surface and dry surface; sandy surface and hard surface; pug mark of an individual will look different leading into over estimation. Pugmark impression pads (PIP – surface prepared specifically for collection of pug mark impression) are used to reduce the error caused due to the difference in the surface. If careful selection of the site and proper training to the census workers is not done, the chances for error in arriving at a figure reasonably close to the actual population would be high. This was always an interesting point of debate in the wildlife circle.

The Sariska episode, in a way changed the way the census was carried out for estimating tiger population. The Sariska Tiger Reserve (Rajasthan) was claiming 16-18 tigers in the May, 2004. But by December it was found out that Sariska is devoid of any tigers (which was later confirmed by the Wildlife Institute of India). Following this crisis, a task force was constituted by the Prime Minister and the task force came up with a detailed report "Joining Dots" (available in the website of Project Tiger) dealing with almost every issue related to tiger conservation in India. After this episode, a new three phase method developed by the Wildlife Institute of India and the National Tiger Conservation Authority is used to estimate the tiger population.

In this new method, a ground survey is conducted at the field level to collect data related to presence or absence of tiger, sign density, etc., at the Phase I level. In the Phase II satellite data and statistical methods are used to stratify the areas into density categories. Finally, in the Phase III, camera traps are used to capture the images of tiger and the tiger numbers in a given area is estimated from this data (Jhala et al 2009; Jhala et al 2011).

PROJECT ELEPHANT

The Asian Elephant is found in 13 countries with an estimated population of 39,400 to 47,400 in wild and around 13,700-47,400 under captivity. In India, there are around 27,700 elephants in the wild (Source: Project Elephant) and 3,500 under captivity. This amounts to almost 50 per cent of the wild population and 20 per cent of captive population of the world. Elephants (like tiger) were always closely linked with humans. In India, conservation efforts in Elephant started as early as in 1873 with the enactment of the Madras Wild Elephants Preservation Act, followed by Elephant Preservation Act in 1879 and inclusion of Elephants in Wildlife Protection Act in 1972 (in Schedule II of Part I). In 1977, Elephant was included in the Schedule I of the Wildlife Protection Act. A ban on export and import of Indian ivory was brought in 1978 and a complete ban on domestic trade was brought in 1991. Project Elephant, an ambitious project for elephant conservation was launched in 1991.

Unlike other animals, elephants, the largest terrestrial animal differ in a variety of way. The requirement of food and water is so high that they themselves can degrade the habitat. They require a large area for movement and hence are difficult to manage in small protected areas. Their large home ranges also create a large interface with people, there by increasing the conflict situation. In spite of all these peculiarities/difficulties, the population of elephants in the country is steadily increasing (15,627 in 1960 to 27,700 in 2007). If population is on the rise, where is the problem? The problems are:

- Poaching of tuskers for their tusk over a continued period of time has resulted in a skewed male female ratio (e.g. 1 male bull to 29 adult cow in Mudumalai in 1999)
- The habitat degradation and change in land use pattern is pulling these magnificent animals into farms and villages and resulting in the damage to crops and life.
- Increasing hostility towards the animal is leading to poisoning and electrocution of these animals in a large scale.
- Development projects like railway networks, canal networks, roads, etc. have cut the once contiguous area into smaller fragments. This again has resulted in increased human elephant conflict.
- In few states there is a negative trend in the population.

In general, the following strategies would help in solving (mitigating) the problem mentioned above:

- Protection measures must be up scaled and enhanced by creating proper infrastructure like staying posts for the staff, strengthening the communication network and increasing the mobility of the staff involved in protection.
- Inter-state coordination committees must be made effective so as to enable easy flow of information about the movement of animals.
- Factors responsible for habitat degradation and fragmentation must be identified and appropriate action / steps to address them must be taken on a planned manner.
- Research work must be encouraged and must be carried out throughout the country to find out answers for the problems faced both by the forest department and people.
- Eco-development activities must be initiated in the elephant areas aiming at a larger involvement of people in the efforts of conservation and also in creating a platform for solving the conflict.
- Awareness programmes must be conducted extensively to educate people about the reason for the conflict and in the process enable them to find out an ecologically viable solution like change in crops, use of barriers, reduction in resource utilization from the elephant habitat, etc.
- Procedure to compensate for the damages done by elephants must be simplified and the delivery system must be made efficient and transparent.
- Special sensitization programmes must be conducted for the personnel from all line departments. This would help the personnel from other departments to understand the issue and design an ecologically sound development project.
- Use of a celebrity / public figure in advertisements regarding the importance of elephant conservation will generate lot of good will to the animal from the public.

HABITAT MANAGEMENT

Animals require food, water and cover for their subsistence in a place. The environment that provides these resources necessary for an organism to meet demands of existence or life requisites is called a habitat. Habitat is the specific set of environmental conditions under which species or communities exist. Thus, habitat management is an important component in wildlife management. Habitat is species specific and varies in the scale as macro-habitat and micro-habitat. The control over the factors related to the macro-habitat like wind movement, water current patterns, rainfall, etc., are not under control of a manager whereas the factors related to micro-habitat like availability of food, water, cover, etc., are under one's control and by planned intervention the requirements for species can be met out in an area. Habitat Ecology provides the required framework and scientific basis for management of such species specific habitat.

FOOD

The number of animals, an area can support and the health of the animals in an area basically depends on the above mentioned factors namely food, water and cover. The demand for food among species is different and even differs between seasons. There are general feeders and selective feeders like herbivores. Among predators too, preference towards a prey species varies e.g. tiger would prefer a sambhar than a small prey like a langur, because the effort a predator puts in hunting would be more or less same irrespective to the size of the prey. Site specific studies and experiences help in providing valuable inputs in managing this factor in a habitat. Generally the following activities are undertaken to improve the food availability in an area

- Grasslands are managed by activities like control burning, removal of weeds, seeding, regulation of grazing, etc.
- Woodlands are managed by removal of unwanted species, seeding, enrichment planting, regulation of grazing, helping establishment of the seedlings, etc.
- Attention is given to plant species which give edible fruits and fodder during the lean season.
- Removal of invasive alien species like *Lantana*, *Parthenium*, Water Hyacinth, *Prosopis*, etc. should be carried out on a priority basis.
- Providing salt licks at appropriate places like near a water source.
- The season of these operations varies with the area and the objective.

WATER

Water is equally important like food for any individual. Water management becomes a crucial limiting factor in arid and semi-arid areas. Providing water holes in adequate number and in areas of need thus becomes an important management intervention. For example, Sariska Tiger Reserve (Rajasthan) has developed a comprehensive water management regime by creation of a chain of water holes in the Kalighati Valley. Water for many of the water holes in the Kalighati valley is taken by underground pipes from the source through gravity (also pumps are used to take water to few higher reaches). Bereft of this arrangement it would be nearly impossible to sustain the wild animals there during the summers when temperature shoots above 45°C. The following activities are undertaken to ensure an appropriate water management in an area:

- Creation of water harvesting structures like reservoirs, ponds, etc.
- Creation of water recharging structures like percolation ponds, check dams, etc.
- Maintenance of natural water holes and older water harvesting structures, etc.
- Creation and maintenance of artificial water holes.
- Checking soil erosion by use of various techniques like contour bunds, contour dykes, etc.
- Creation of bore wells at places where other forms of making water available is difficult
- Water management in wetlands is an crucial aspect of management which involves a series of operation from filling the wetlands with water, maintenance of depth in a given area, draining excess water, de-silting, etc.

COVER

Cover in simple term means shelter. Shelter for an animal ranges from protection from natural factors like sunlight, wind, rain, etc., to protection from the predators. Cover could be of vegetation or could be natural feature like cave or a man made structure like a bridge, abandoned building etc. A thick vegetation may serve as a cover for a species when it is green and may not serve as a cover for the same species when it is dry. The cover is not a static one but it is dynamic. Knowledge on the ecological requirement of species concerned is vital in managing this habitat factor. Apart from activities like plantation and seeding done to improve the cover, riparian patches, forested area between adjoining areas must be protected because they help in animal movement from one area to another. Nesting Birds require adequate cover to protect their young ones. In a managed forests care should be taken to leave snag trees which are special habitats especially for the birds like parakeets, wood peckers, barbets, etc. Owing to the significance, habitat management interventions required in an area finds a detailed description in the management plans. The various covers of wild animals are,

Escape Cover: Cover used by most prey species to escape from the predators

Ambush Cover: Cover used by predators like tiger to stalk the prey

Thermal Cover: Cover used by animals to protect from high temperatures

Breeding Cover: Cover required to nurture the young ones

NICHE

Niche is defined as the functional status of an organism in an ecosystem. In E.P. Odum's term, habitat is the address of an organism and niche is the profession of an organism. A habitat may be occupied by many species at a point of time, with varying functions (niche). In other words, these related species have different roles to play in an ecosystem (ecological separation), thus avoiding any competition between them.

EDGE AND ECOTONE

When two or more communities meet, they give rise to edge. This is a tension belt or junction zone, linear in extent and generally narrower than the adjacent patches, e.g. zone in between a woodland and grassland; marshes in between lake and terrestrial ecosystem. Light demanders and other opportunistic plants dominate the edges.

ECOTONE

When two or more communities meet and intergrade they give rise to ecotones. Here the communities meet and exchange their components. Because of the diversity in the locality factors, diversity of plants and animals tend to be more in and about the edges and ecotones. This phenomenon is called "*Edge Effect*". Edges created due to geological changes, soil types, topographic changes, etc. are permanent and more stable whereas edges created due to fire and various biotic factors tend to be smooth and are less stable. Over a period of time, (in absence of the factor responsible for the change) the edge will slowly disappear.

SPECIAL AND UNIQUE HABITATS

Riparian zones and snags are examples of special habitats. Riparian zones are very sensitive and external influences can leave an irreparable damage on this habitat. Presence of water provides an environment (micro-climate) for a variety of species to inhabit the area and also use them in their transit. Any dead or partly dead tree of at least 10.2 cm in diameter (at breast height) and at least 1.8 meters tall is termed as a Snag. Snags play a vital role in the survival of wood peckers, owls, parakeets, barbets, etc. Certain species like wood peckers make their own hole in dead trees (snags) whereas owls and parakeets use a natural cavity in a snag or a cavity made by some other species.

Cliffs, talus and caves are examples of unique habitats (result of geo-morphic or topographic changes). Bats, raptors, swifts, reptiles, etc. are found in these unique habitats. Care should be taken to protect these habitats as they offer a great deal of diversity to an area.

WILDLIFE CENSUS

Resource inventory is a pre-requisite for any management process. This holds good for wildlife management too. Practice of making inventory is not new to the forest department. Scientific management of forests in our country started in the late 19th century with preparation of working plans for the forests. List of plants and animals found in an area was carefully documented in the working plans and were updated regularly during the plan revisions. During those days, timber production was given the top most priority of management and hence detailed account on population density of animals, their sex ratio, etc. were not made. However, certain areas like Corbett, Gir, etc. were managed with wildlife as the main focus.

For reasons mentioned earlier, wildlife received the much needed attention and the practice of managing areas exclusively for wildlife started growing across the world. This lead to the systematic inventory making of the wildlife resources at all levels (from a protected area to national to global level). Along with this, attempts were made to estimate the population of a species in an area. This process of estimating the population of a species/animal community is called as census.

Census started with simple counting of all individuals of a species in area to more complex methods involving statistical analysis. For a manager, is this number regarding a species enough? Not really, additional information about the population structure like age class and sex ratio helps the manager in proper decision making. Is information about just a species enough? Obviously No, because they live in communities along with various other species; all are inter-dependent and collectively they influence the ecosystem. This necessitates collecting information on all major species so that by appropriate management interventions, they can be perpetuated for posterity. Further, the number of animals occupying a place too varies by season. Hence, census of animals in an area is not conducted as an one time affair but done a periodical basis.

Census can be either conducted by direct counting or by indirect counting methods. In the direct count method the individuals are seen and counted whereas in indirect count method evidences left by animals are recorded to estimate the population.

DIRECT COUNT METHOD

Under the direct count method either all the animals present in an area can be counted (total count) or can be done by sample counts. Sampling could vary from a simple random sampling to complex stratified sampling. Counting all the animals in area particularly large tracts involves lot of resources and hence should be avoided. Among the direct count method Vehicle transect and King's census method is widely used in India.

VEHICLE TRANSECT OR ROAD COUNT METHOD: Vehicle transect involves counting of animals on roads by travelling in a vehicle at a fixed speed and during a fixed time. The distance for such transect is fixed and the counting is done on the transect regularly. Density of the animal is calculated using the formula given below,

$$\text{Density of Animal} = \frac{\text{Number of animals sighted}}{\text{Lenght of transect} \times \text{Width of transect}}$$

When the density figures of corresponding period are compared, they give a clear picture about the trend of the population. Say density of Peacock calculated in the month of April should be compared with the density figure of April the next year to arrive at a reasonable conclusion regarding the population trend. Transects should be placed in such a way that they cover all types of ecosystems found in an area. Adequate number of transects should be placed to capture the real picture of the area.

KING'S CENSUS: Under the King's census, data regarding sighting of animals is collected from transects by the observers walking on transects. Transects are laid to cover all the ecosystems present in an area. Any single transect laid should as far as possible cut across all the vegetation type/ecosystem. If there are grasslands, shrub land and forests, then the transect should run through all these three vegetation type and not just one. An observer along with one more person walks on the transect preferably in the early morning hours and record all the sightings on the transect with the angle of sighting and distance (angular distance). With the angular distance and angle of sighting the perpendicular distance of animals from the transect line is calculated. This perpendicular distance is used in calculating the area of the transect walked. The density of animals is calculated with the use of the formula given above.

Width of the transect could be varying like the one illustrated above or may be fixed. In the fixed width transect (generally done in areas with good ground visibility), the width of the transect is fixed and the observations beyond the fixed width is not recorded. Density calculated for a particular transect is then extrapolated for similar areas to arrive at density figures for the entire area.

INDIRECT COUNT METHOD

The pug mark technique explained previously is one of the indirect count method still widely used in our country for estimation of larger carnivores like tiger and leopard. Generally, Pellet count method is used to estimate the ungulate density in most of the protected areas in our country. Under pellet count method, pellet groups are counted in sample plots preferably laid out on the transects (laid for conducting King's Census) and the density of the animals (species wise is estimated) is calculated with the data collected on the number of pellet groups. Information on defecation rate for animal surveyed, knowledge on pellet size and pellet rate in a day is required.

OTHER METHODS

Apart from the methods discussed above there are certain methods which are not included in the direct or indirect count method viz. water hole census and capture-recapture method.

WATER HOLE CENSUS: In absence of trained and experienced staff, generally "Water Hole Census" is conducted to estimate the densities of animals. In this method, observations on sighting are collected at water holes (during summer) for 24 hours preferably from 6 PM to 6 PM on full moon days. The basic assumptions made in this method are that every animal has an equal opportunity to visit a water hole and every animal drinks water at least once in a day. Number of animals sighted are counted and recorded throughout the 24 hours and later on, the data is compiled species wise to arrive at the density figures. This method also gives an index or trend regarding the population and not the estimate of the population. If information on frequency of water consumption by animals for the area is collected, the figures compiled can be corrected for the repeat visitors and the trend can be refined a little. This is one of the easiest methods and with little training (along with volunteers from colleges, NGO's, etc) even the illiterate staff can collect some useful information. For these reasons, this is the most popular method in many parts of our country.

CAPTURE-RECAPTURE METHOD: This is neither a direct method nor indirect method but a combination of certain elements from both. This method is used in monitoring works of turtles, tortoises, crocodiles, birds, etc. First, certain numbers of animals are captured, marked and released. Second, again some animals are captured from the same area and checked for the marked animals. From the ratio of the marked and unmarked animals an estimate regarding the population can be made using the following formula,

$$\frac{\text{N}\,(\textit{Population Size})}{\text{S}\,(\textit{No. of animals captured \& marked})} = \frac{\text{n}\,(\textit{Number of animals captured})}{\text{s}\,(\textit{Number of animals maked earlier})}$$

REINTRODUCING RARE AND ENDANGERED SPECIES

The terms re-introduction, translocation, introduction, etc. are used loosely many a times. Knowing the difference and what exactly these term mean thus becomes important for a student of wildlife science. Intentional introduction or accidental dispersal of an organism outside its historically known native range is called Introduction. Re-introduction is defined as an attempt to establish a species in an area which was once part of its historical range, but from which it has been extirpated or become extinct (IUCN 1988). The major difference is about the organism's historical range (whether or not the species concerned is released outside or inside its historical range). As the definition says, re-introduction aims to establish a viable free ranging population of a species in its historical range. Such successful establishment of species in its historical range is called as re-establishment. Deliberate and mediated movement of wild individuals or populations from one part of their range to another is called as translocation (IUCN 1988). Some examples are,

- Re-introduction: Proposed Cheetah re-introduction; and movement of Gaur from Kanha to Bhandavgarh
- Introduction: Intentional introduction of an alien species *Lantana* as an ornamental plant in India.
- Translocation: Movement of Sambar from Sariska Tiger Reserve to Kumbalgarh Wildlife Sanctuary in Rajasthan.

The IUCN's (International Union for the Conservation of Nature) Re-introduction Specialist Group (RSG) prepares both generic and species specific guidelines regarding re-introduction. The RSG is a network of specialists whose aim is to combat the ongoing and massive loss of biodiversity by using re-introductions as a responsible tool for the management and restoration of biodiversity through actively developing and promoting sound inter-disciplinary scientific information, policy and practice to establish viable wild populations in their natural habitats (Source: IUCN). Details regarding various re-introduction programmes can be studied from the website of IUCN.

EX-SITU CONSERVATION

Practice of collecting animal species was common among the rulers across the world. Such collections were mostly for the private viewing of the royal families. However, modern day zoos came into existence as early as the 18th century with the establishment of Vienna zoo in 1752 followed by zoos at Paris (1793) and London (1826). That time they were referred to as menagerie. Zoos of those days tried and collected as much as species possible (generally exotics). The animals were kept under small cages and public were allowed to view them. The 19th century saw large number of zoos being established throughout the world. From this scenario of keeping animals in small cages, zoos have evolved towards organizations meant for wildlife conservation.

Keeping pace with the growth of zoos world wide, an organization called International Union of Directors of Zoological Gardens (IUDZG) was established in 1935 in Switzerland. This organization served as an international apex body regarding the zoo management until the end of World War II. In 1946, IUDZG was once again established in Rotterdam and in 2000 was renamed as World Association of Zoos and Aquariums (WAZA). The mission of this WAZA is to provide leadership support for zoos, aquariums and partner organization, of the world in animal care and welfare, conservation of biodiversity, environment education and global sustainability. At present 300 leading zoological institutions (from 50 countries), associations, affiliate organizations and corporate houses are members to this organization.

The Central Zoo Authority of India and Zoo Outreach Organization are affiliate members to WAZA.

Zoo is defined as "an establishment, whether stationary or mobile, where captive animals are kept for exhibition to the public and includes a circus and rescue centres but does not include an establishment of a licensed dealer in captive animals" (Wildlife Protection Act 1972). In India the Central Zoo Authority (CZA) created by an amendment to the Wildlife Protection Act in 1991 is the apex body regarding matters related to zoos. The aim of this autonomous statutory authority body is to complement the efforts of wildlife conservation in our country. The Recognition of Zoo Rules, 1992 helps in regulating the various activities related to the *ex situ* conservation activities. The National Zoo Policy, 1988 guides the zoo management in our country. It is mandatory for all the zoos in our country to obtain recognition from the authority. Presently there are 198 recognized zoos in our country. The zoos in our country are classified based on the area, number of collections, number of endangered species, visitation etc.

As mentioned before, zoos has evolved from just centres of recreation to more serious conservation related organizations. More emphasis is given on Conservation Breeding Programmes (CBP) in the zoos. Under conservation breeding programme, a zoo next to natural habitat of the species concerned is assigned the responsibility to act as a coordinating zoo. Two to four zoos in the range of the species selected is assigned the responsibility to as a participating zoo. Founders are identified and screened for their health either by the zoo veterinary hospital or by the Indian Veterinary Research Institute, Bareily. Also, the genetic health check up is done at the LaCONES (Laboratory for Conservation of Endangered Species), Hyderabad. After all these preliminary work, the zoos take up the breeding programme. To mention a few, conservation breeding programmes for Asiatic Lion, Lion Tailed Macaque and Red Panda are currently carried out by the zoos of Junagadh (Gujarat), Chennai (Tamil Nadu) and Guwahati (Assam) respectively.

22 FOREST NURSERY TECHNOLOGY

INTRODUCTION

During the past century, the world's land base has been subject to urban expansion, poor management practices and increasing pressure to provide resources for a growing population. As a result, tree seedling production has become a fundamental tool for addressing reforestation, restoration and conservation needs. Nursery practices for tree seedling production have evolved considerably over the past several decades in accordance with technological advances, increased understanding of seedling physiology and development, and changing customer demands (Dumroese et al 2005). Considerable research into plant physiology and nursery practices has led to a much greater understanding of seedling responses to environmental conditions, nursery treatments and growing regimes over the past several decades. For example, early nurseries had limited understanding about soil physical and chemical properties, amending soils to achieve maximum seedling growth, optimizing soil/media properties for seedling development. Now the scenario has changed and manufacture of chemical fertilizers and other products, as well as the ability to transport materials over great distances, have given plant growers more options for developing better planting stocks. Knowledge has also increased on the topic of seed management.

In addition to our better understanding of seedlings and the growing environment, technological advances in nursery equipment have greatly enabled forest nurseries to produce large volumes of high quality stocks. Now mechanical methods have replaced the manual labour works in transplanting seedlings, lifting of planting stocks, etc. Nowadays, there are custom machines for most nursery processes, from sowing to harvest, although smaller nurseries still rely on manual labour for many tasks. Additionally, most nurseries, small or large, carry out much of their weed control via hand weeding. The basic concepts regarding density, fertility, irrigation, pruning, grading, storage, handling and other seedling production practices have always been considered in nursery management. Whether it is 1912 or 2012, the underlying driving force is based on the Target Seedling Concept: "Targeting specific physiological and morphological characteristics that can be quantitatively linked to outplanting success" (Rose et al 1990).

Because nursery production is expanding more and more to encompass a larger variety of species, there is an accompanying need to develop and refine nursery techniques for non-traditional native species. Until the past few decades, nursery production was primarily focused on commercial/ industrial species for forest regeneration or horticultural cultivars for urban landscaping and gardens. The rise of the environmental movement in the late 20th century, the explosion of invasive non-native plants over the past century and increased land degradation has led to much attention being directed towards propagation and restoration of native or indigenous plants.

Nursery management represents the greatest concentration of technology and investment in the forest growth cycle. On an area basis, any effect on subsequent seedling or tree growth brought by nursery decisions and investment is multiplied many fold. This extraordinary leverage given to nursery manager's decisions not only makes those decisions the focus of land manager's attention but also provides an unusual opportunity for using research to improve productivity. Furthermore, because the nursery production cycle is short relative to many other forestry operations, the impacts of nursery manager's decisions usually can be quickly seen and the effects of new research information applied to nursery operations can be rapidly evaluated.

The National Action Plan on Climate Change (NAPCC) and Green India Mission aim at afforestation of 6 million hectares of degraded forest lands and expanding forest cover from 23 to 33 per cent of India's territory. The plantation forestry programmes viz. afforestation, reforestation and reclamation forestry, which focus on increasing productivity of forest, necessitates for production of quality planting stocks. These large scale plantation programmes require huge quantity of elite planting materials supplied through forest nurseries across the country. The forest nurseries thus play an important role in meeting the national objective of keeping 1/3rd of the country's geographical area under forest.

WHAT IS FOREST NURSERY?

Tree seeds germinate and trees grow readily under natural conditions. These delicate young seedlings and young trees are exposed to adverse dry season conditions and intense competition from other plants in natural forests or plantations. Forest nurseries can provide optimum care and attention to seedlings during their critical juvenile stage, resulting in the production of healthy, vigorous seedlings. In many cases, successful reforestation requires nursery grown seedlings, since degraded areas have unfavourable conditions making natural vegetation or direct seeding not feasible.

Forest nursery is a managed area, designated to produce tree seedlings grown under favourable conditions until they are ready for planting. It can be an informal, small-scale arrangement or a large commercial enterprise. Nurseries vary in size, facilities (supplies, tools, equipments, etc.), types of seedlings produced and nursery operations. They also differ significantly in quality and quantity of planting stocks produced. However, all nurseries primarily aim to produce sufficient quantities of high quality planting stocks to satisfy the needs of seedling users. Users include the nursery owners themselves, individuals, community organizations, farmer groups, government agencies, non-government organizations, corporate or private customers. Nurseries may often provide income generating opportunities for the operators and enhance social capital and technical capacity and leadership skills of communities. Smallholder tree nurseries also serve as important training and research ground for many smallholder farmers. Forest nurseries are defined in a numerous way:

- An area or place where forest seedlings are grown.
- An area where tree seedlings and plants are raised for eventual planting out.
- An area where trees, shrubs, or plants are grown for transplanting, for use as stocks for budding and grafting.
- A production unit that grows planting stock (seedlings and saplings) of forest trees and shrubs.
- A plot of land where young trees can be raised under more or less controlled conditions.

Forestry projects and large scale afforestation and reforestation programmes spawn the establishment of self-help and self-sustaining local nurseries, which remain functional even after project termination. These local nurseries may be linked in a network to meet project, community or market needs. Government and non-government agencies often provide modest support to enhance the development of local nurseries, in recognition of the spontaneous growth of local nurseries. These efforts strengthen local livelihood development and sustain public reforestation initiatives. In many areas, where nursery development has flourished, farmers also developed keen interests in collecting and propagating diverse species, including indigenous trees. Such forest nurseries have become rich repositories of tree biodiversity.

SIGNIFICANCE OF FOREST NURSERY

Good nursery operations incorporate the selection of the best quality seeds and seedlings, thus initiating tree improvement. Forest nurseries are advantageous in propagating recalcitrant seeds, dormant seeds and tree species with irregular flowering. Nurseries are also good venues for vegetative propagation operations. Some important species do not need seed every year. Plantation of these species can be raised annually only by collecting all available seed in years and sowing it in nursery to raise seedlings to be planted out in various years. Some species grow very slowly and if the seed of these species is sown directly in plantation, the seedlings are most likely to be swamped by weeds and killed. Therefore, slow growing species are generally raised in nursery and planted only when the seedlings are not liable to be damaged by weeds. Success of roadside avenue plantations depends largely on planting tall and sturdy plants which can only be obtained from nursery. Plantations of some species when raised by direct sowing are not so successful as when raised by planting. In such cases, nursery is an essential part of artificial regeneration of those species.

The best method of introduction of exotics, viz. tropical pines, poplars, eucalyptus, etc. is only by planting and therefore nursery is very essential for them. Planting of nursery-grown plants is surest method of artificially regenerating poor and barren sites. Causalities in plantations have to be replaced either in the year of formation or in the next year. Sowing done in the gaps are liable to be unsuccessful as a result of suppression from weeds and cannot catch up the plants from original sowing. Therefore, replacement of causalities is always done by planting nursery-grown plants or stumps and so nursery is very essential for causality replacement.

CLASSIFICATION OF FOREST NURSERIES

ON THE BASIS OF AREA SIZE

Based on size of area, forest nurseries are classified into:

- *Small Nursery*: Nurseries with a area of less than 3 ha
- *Medium Nursery*: Nurseries with a area of 3-10 ha
- *Large Nursery*: Nurseries with a area of more than 10 ha

ON THE BASIS OF IRRIGATION FACILITY

Based on the water availability, forest nurseries are classified into:

- *Dry Nursery*: It is a nursery that is maintained without any irrigation or other artificial watering.
- *Wet Nursery*: It is a nursery that is maintained by irrigation or other artificial watering during the dry periods.

ON THE BASIS OF PLANTING STOCK (Krasowski 2012)

Based on the planting stocks produced, forest nurseries are classified into:

- *Bareroot Nursery*: In bareroot nursery, plants grow directly in the nursery soil and the roots are separated from the soil at the time of lifting. The lifted planting stock is further handled and planted without soil surrounding the roots.
- *Container Nursery*: In container nursery, plants are grown with roots in a growing medium held in a container. Roots of lifted plants are covered by the adhering growing medium.

ON THE BASIS OF SIZE OF SEEDLINGS

Based on the seedling size, forest nurseries are classified into:

- *Seedling nursery*: A nursery which has only seedling beds i.e., in which seedlings only are raised, no transplanting being done is called seedling nursery.
- *Transplant nursery*: A nursery which has only transplant beds, in which seedlings are transplanted for preparation for forest planting, is called transplant nursery.

In general, separate seedling and transplant nurseries are seldom made.

ON THE BASIS OF OWNERSHIP (Roshetko et al *2010)*

Based on ownership, forest nurseries are classified into:

- *Institutional Nurseries*: They are commonly established, operated and maintained by research institutions, academic institutions, corporations and international organizations. They are permanent in nature but may be larger, medium or small, based on purpose.
- *Government Nurseries*: They are established and operated by national or local government to support reforestation and community tree planting programmes. They are large in size and are permanent in nature.

- *Industrial Nurseries*: They are commonly established and operated by large integrated wood based industries. They are large in size with high production capacities and permanent in nature. Seedling production complies with rigid quality standards or specifications.
- *Project Nurseries*: They are established and managed by projects and non-governmental organizations (NGOs) or development organizations to promote tree planting culture within target communities. The intended lifespan of project nurseries is usually 3-5 years, but may continue for longer periods.
- *Community Nurseries*: They are established by communities to support tree planting programmes for social benefits.
- *Group Nurseries*: They are also established to build technical and leadership capacities of group members and expand the number of species and quality of germplasm available to group members.
- *Individual Nurseries*: They are also known as household nurseries or backyard nurseries and are generally established and managed by individual farmer or household. These nurseries are usually initiated to produce seedlings for planting on operator's personal farms. The house compound or farmland provides the planting sites. They are normally small in size.

ON THE BASIS OF DURATION OF THE USE (Yadav 2012b)

Based on the duration of their use, forest nurseries are classified into:

- *Temporary Nursery*: It is a nursery that is maintained for supplying stock for a short period after which it is abandoned. Normally, it is constructed in the plantation area and usually small in size. It is suitable for hilly regions.
- *Permanent Nursery*: It is a nursery that is maintained for supplying nursery plants for a long time on a permanent basis. It is intended to meet the requirements of one or more ranges and it is relatively larger in extent.

CHARACTERISTICS OF TEMPORARY NURSERY

FEATURES: Temporary nurseries are constructed for a short period of time and smaller in size. Manuring is not necessary and mostly located near/inside the planting area which is appropriate chiefly for casualty replacement. Elaborate soil preparation is not necessary and cost of transportation of seedlings to the planting sites is low. Seedlings of different species for mixed crops are raised. Gap between lifting the stock from the nursery and actual planting is less.

ADVANTAGES: Temporary nurseries are usually constructed in newly cleared sites fairly rich in humus and so, manuring is not required. Eucalyptus nurseries are an exception to the general rules. Minimum trouble with the weeds, destructive insects and diseases are observed. It enables raising of species in their optimum altitudinal zone in hills. Cheap transport of planting stock without any serious damage or shock.

DISADVANTAGES: They are comparatively costly and difficult to supervise. Proper supervision is not possible as it is made out of way places. Due to lack of irrigation facility, the growth of seedlings is usually slow and heavy mortality is observed.

CHARACTERISTICS OF PERMANENT NURSERY

FEATURES: Permanent nurseries are fit for large and intensive work and intensively managed areas. It is established where all the facilities are available i.e., easy supervision, communication facilities, labours, etc. Intensive manuring and soil working are done in perpetuity. It is used for large scale afforestation works, or distribution to the villagers under community and private forestry programme. Large labour forces, tools and equipments are needed. Initial cost of formation is high but is cheaper in the long run. Regular skilled supervision is mandatory.

ADVANTAGES: It supplies varieties of planting stocks such as root-shoot cuttings, grafted plants, layering, budding, container seedlings, etc. Duration of service life is long and production cost is reasonable. It meets the requirement of larger areas/ranges/divisions. Supervision cost is low and can be easily supervised. Plants are raised year after year for a long time on same site.

DISADVANTAGES: Transportation of seedlings is difficult and costlier and establishment cost is high. Manuring of beds annually and intensive soil working are essential. It requires large labour forces throughout the year which is difficult during agricultural seasons.

SEEDLING QUALITY

The end user and the environmental conditions at the site where the seedlings will be planted determine what kind of seedling is needed. A researcher will require planning stock raised under uniform conditions to ensure that the research results are not confounded by the status of the seedlings. On the other hand, farmers will require seedlings that can survive under hard, dry and nutrient-deficient conditions. The good looking seedling at the nursery is worthless if it does not survive and grow after planting out. Targeting seedling production to the anticipated field site is important in producing strong healthy seedlings. Seedlings need to be hardened off and gradually prepared for real-life harsh conditions by withholding water from time to time after they have developed a strong and healthy root system. Insufficient care in the nursery leads to weak and retarded plants with small root systems that will not survive in a harsh environment.

Quality seedlings produced for different sites may look different from each other but they all have one thing in common: a well developed root system with many root tips from which new roots can quickly develop. In areas with adverse environments such as dry, flooded, saline or nutrient-deficient sites, only well developed plants have a good chance of survival. Seedlings with deeper root system are preferred for dry areas and larger plants are preferred for weedy sites so that they can outgrow weeds quickly. Many studies have proved that field survival and productivity are related to the quality of the seedling used. Seedling quality depends on the following parameters (Hannah Jaenicke 1999),

- The ability of plant to produce new roots quickly
- The speed with which seedlings get anchored in the ground and start assimilating and growing after planting out
- A well developed root system
- Sun-adapted foliage

- A large root collar diameter
- A high root : shoot ratio
- Good carbohydrate reserves
- An optimum mineral nutrition content
- The establishment of adequate mycorrhizal or *Rhizobium* infection

NURSERY PLANNING AND MANAGEMENT

Quality seedling production depends on a sound understanding and implementation of the key principles of nursery planning and management. As with the operation of any forest nursery, planning is essential for a successful and productive nursery. Factors to be taken into account are efficient use of inputs and resources, identifying problems and potential solutions, monitoring progress and working towards nursery objectives. Nursery plans vary in detail and in implementation. Nursery plan may be formal written documents or a list of informal management guidelines. Generally, project and group nurseries engage in participatory planning, working with stakeholders and members. Recommended elements of a nursery plan include: site selection, objectives, demand forecasting and target seedling production, inputs and scheduling, contingency plans to respond to unexpected problems, documentation, staff training, promotion and marketing and networking.

NURSERY SITE SELECTION

The selection of an appropriate nursery site is the most important decision affecting the efficient production of good quality plants. The following are some of the technical factors which need to be considered while setting a nursery.

WATER SUPPLY: In order to have the good irrigation facilities, it should be situated near or slightly below the source of adequate water supply. Reliable and adequate water supply is essential. Water source should be near and at the higher level than nursery site. Water should be available throughout the year.

AVAILABILITY OF SUITABLE SOIL: For raising seedlings in polythene bag (tube), soil of the actual nursery itself is not important. There should be suitable source of soil preferably forest topsoil and sand within easy reach of the nursery. Clayey soils should not be selected as their aeration and drainage is poor and they are liable to crack during summer. The symbiotic organism (*Mycorrhiza* and *Rhizobium*) should be present in soil because most of the trees grow best if their roots are associated with certain symbionts, which help in their nutrition. Mostly these organisms are found naturally but in some cases artificial inoculation are needed.

ACCESS: Nursery should be situated as centrally as possible with reference to the area to be planted. It should preferably be near residence area for close and regular supervision. Nursery should be near to the plantation site to facilitate easy transportation of seedlings and access road should be usable at all seasons of year.

ASPECT: Slope with South aspect is much warmer and chosen for hill nursery at high elevation while in low elevation north facing slope is chosen.

SLOPE: The ideal slope is about 5 degree which is steep enough to allow proper drainage. Complete flat land should be avoided; it is likely to become water logged during monsoon rains. Steeper slopes will have to be terraced and very steep slopes should be avoided as it may be difficult to make a nursery bed and path on each side to allow access to the beds.

EXPOSURE TO FROST, STRONG WINDS AND FLOODING: At high altitudes, sites which are particularly liable to frost damage should be avoided. Sites exposed to strong winds and with danger of flooding or landslides should be avoided.

LABOUR AVAILABILITY: Labour should be available without difficulty preferably near a village so that labourers need not walk much distance for work.

AVAILABILITY OF LAND: There should be enough land to raise the number of seedlings needed and if possible room for expansion. Legal enquiries should be made about the legal ownership of the land.

COLLECTION OF PLANTING MATERIALS (SEEDS, WILDINGS AND CUTTING)

SEED MANAGEMENT

SEED COLLECTION: There is a great variety of methods and equipment available for collection of seeds and the choice depends on the factors summarized as follows:

- Relative size and number of the fruiting units
- Characteristics of the fruit
- Characteristics of the tree
- Characteristics of the stand
- Characteristics of the site

The various methods of collection may be classified into the following: (1) collection of fallen fruits or seeds from the forest floor; (2) collection from the crowns of felled trees; (3) collection from standing trees with access from the ground; (4) collection from standing trees with access by climbing.

Collection of seeds from the forest floor includes natural seed fall, manual shaking, animal caches and use of rope, ladder, tree bicycle, rifles, bow & arrow, nets and sheets to collect falling seeds etc. Collection from the forest floor of fruits which have fallen after natural ripening and abscission is common practice with a number of large-fruited genera. It is cheap and does not require highly skilled labour. Important genera, from which seeds are commonly collected from the ground, include *Quercus, Fagus, Castanea, Tectona, Gmelina* and several genera among the Dipterocarps.

Trunks of small trees and low branches may be shaken directly by hand. Higher branches may be shaken by means of a long pole and hook or by a rope. This method has produced good results in several arid zone Acacias. One method of collecting large amounts of seed from the crowns of felled trees is to synchronize it with normal commercial fellings carried out during the seed ripening season. Collection from standing trees with access from the ground includes by hand; cutting, breaking and sawing;

SEED CLEANING: Once the seeds have been collected, they need to be cleaned thoroughly to remove all bits of fruit flesh, pod or husk, especially if the seed is to be stored. The flesh of soft fruit has to be brushed/rinsed off before the seed is planted or stored. Many fleshy fruits contain substances that inhibit germination. Seeds can be cleaned by hand or by winnowing in the wind for dry seeds e.g. Acacia species. Seeds may be sieved to remove dirt and low quality seeds. The best quality seeds are nearly always the biggest. Another way of separating seeds is to immerse them in water. The bad seeds and most dirt will float, while the good seeds sink. Good seeds are removed and dried carefully.

SEED STORAGE: If seeds cannot be planted immediately after collection, they must be stored. Some seeds (e.g. legumes) can be stored at normal temperatures for many years as long as they are kept dry. Most seeds, however, can only be kept for a limited period of time and recalcitrant seeds should be sown as soon as possible after extraction from the fruit. Well-stored seeds are kept dry at a constant cool temperature. Seeds are dried before storage to prevent infection by fungi and bacteria. However, extreme heat must be avoided as this destroys the seed's ability to germinate. Seeds should not be dried in direct sunlight, but in a shaded, well-ventilated place. Seeds, that have been properly dried, can be stored in containers such as pots, tins, boxes or bags. For insulation, seeds can be doubly sealed by first putting them in a thick plastic bag. Bag is closed tightly and put in a container with a lid. By burying the containers under a layer of dry soil in a shaded place, seeds can be kept at a constant cool temperature. It should be ensured that insects or rodents cannot get access to the containers. Each container should be labeled along with a note of treatment given and the date of storage. The contents should be regularly checked and at the same time the seeds need to be turned or container should be shaked.

PRE-GERMINATION TREATMENT OF SEED: Germination of seeds of some trees may take many months, the seed awaiting the low temperatures of winter or the heat of a forest fire to break seed dormancy. Germination may also be delayed by a hard or tough seed coat or by certain substances in the seed. Such seeds can be treated to speed up germination and even more important, to ensure more simultaneous germination. Wide range of treatments is used to accelerate germination: soaking in cold water, acid and hot water soaking; mechanical scarification; stratification; light burning with fire etc.

Soaking in cold water or acid is a very common method. Cold water soaking is a simple method involving soaking the seeds for 24 hrs in cold water prior to sowing. This makes the seed swell, breaking the coat and leaching the substances that delay germination. This method works for seeds from many tree species. In some cases, acid is added to make the liquid more abrasive. Soaking the seeds in boiled water for few minutes and subsequent cooling removes the hard outer cover of some seeds, leaving a softer covering that the sprout can break through more easily. This method is the right one for leguminous species such as *Acacia, Prosopis,* etc.

To facilitate germination, a shallow incision is made in the hard seed coat using a needle or knife, or the tip of the seed is cut off. A simple way of scarifying legume seeds physically is to rub them over a rough surface, e.g. sandpaper. This scratches the seed coat, but the scratches should not go right through the coat. This method works for *Leucaena* species and *Faidherbia albida*. Many seeds of trees from the temperate and subtropical zones require winter chilling to break seed dormancy and this is called stratification. It may be necessary to put the seeds in a pot with moist sand in the refrigerator for about 2 months.

WILDINGS

Wildings are planting material collected from natural vegetation, e.g. seedlings or cuttings. Wildings should have at least 2 to 4 fully formed leaves when you dig them up. Larger wildings have a stem about the diameter of a pencil. They can be collected when rain has moistened the soil, either with or without a root ball. Wildings with a root ball are dug up by making a cut on both sides of the plant at an angle of 45° using a spade or machete. The plant is then lifted on the spade, holding the stem with the other hand and removed with the root ball intact. It is difficult to dig up small wildings with a root ball, because the root system is too small to hold the soil. Bare-rooted wildings are uprooted by loosening the soil around the roots with a pointed stick. The plants are pulled up carefully and the soil is shaken off the roots.

CUTTINGS

Different parts of the plant - leaf, stem or root - may be cut for rooting. Propagation by leaf cuttings is virtually limited to some ornamental plants. Stem cuttings are most common and they are classified as softwood or hardwood cuttings. Softwood cuttings are taken from a growing shoot, not far below the shoot tip. Such tender, leafy cuttings require intensive nursing. Trees and shrubs of hardwood root well. Hardwood refers to the wood of resting shoots, after the termination of a flush. It includes older twigs and branches, formed in earlier flushes.

STEM CUTTINGS: The preferred time to take hardwood cuttings is generally during the dry or cool season, when shoot growth is minimal. In deciduous trees, the cuttings are best taken when the trees are leafless. Most woody plants are evergreen and the leaves have to be stripped from the part of the cutting that is inserted in the soil about two/third of the length of the cutting. Usually a few leaves - sometimes halved - are left at the tip of the cutting, the number depending on growing conditions (shade, humidity, etc.). Leaves generally stimulate root growth, but cuttings are likely to dry out if the leaf area is large.

The tip of the shoot or twig is usually discarded, but a vigorous shoot can still yield several cuttings of 15-50 cm, the recommended length. Commonly, the diameter of cuttings ranges from pencil-thick to about 3 cm. The upper cut is oblique so rainwater runs off. The lower cut is usually made just below a node, because rooting generally occurs mainly at the node. If a cut is not smooth and clean, rot may lead to failure of the cutting; it can also infect the wound on the mother tree. Upright branches and twigs are preferred for cuttings, because they grow upright after rooting, forming a tree with a proper trunk. Cuttings taken from horizontal or drooping branches often do not grow upright.

ROOT CUTTINGS: Tree roots can also be used to take cuttings, e.g. Casuarina species. Some trees even produce root suckers spontaneously. If these suckers form their own roots, they can start life on their own. To encourage the emergence of root suckers, roots are cut using a spade or small axe. The piece of root cut off may form a rooted sucker. These root cuttings are planted in a shaded spot in the nursery. If a high humidity is maintained rooted suckers will be ready after a couple of months.

LAYERING

Tree species, that cannot be propagated by cuttings because they do not readily form roots, may in some cases be induced to form roots on shoots before the shoots are detached from the mother tree. This propagation method is called layering. The simplest form is tips of drooping shoots of shrubs bend till they lay on the ground where they may root spontaneously.

BAREROOT NURSERY

DIRECT SOWING IN BEDS IN THE NURSERY

Growing bare-rooted seedlings is the simplest form of tree propagation in a nursery. This method is used if the species is easy to plant out, which means seedlings are not prone to heavy losses after field planting. Ample quantities of (cheap) seed and high germination percentage are also favourable for direct sowing. Seeds are sown in a well-prepared bed and the seedlings are grown till they are ready for field planting. Tree seedlings have to stay in the nursery bed for two months to one year.

SOWING IN SEEDBEDS AND FURTHER GROWTH IN PLANTBEDS

The above method is not used much for tree nursery work. It is only suited to seedlings which will be ready within a few months and will survive easily after field planting. A more common method is to sow the seed closely in seedbeds and the young seedlings are pricked out into plantbeds at the right spacing for their further growth in the nursery. Pricking out is the term used for transplanting young delicate seedlings by lifting them gently.

AIR-PRUNING OF ROOTS IN SEED TRAYS

A final improvement in the germination stage is the use of seed trays with an open-mesh bottom. Wooden trays with a wire mesh bottom (like a sieve) can be made or heavy netted plastic sheets can be cut and folded for using as windscreens. Cheap plastic trays with mesh sides and bottom are available in the market. Sand would fall through, so trays have to be filled with friable soil or - better still - with a good rooting medium such as shredded coconut fibre. The trays need to be placed on a bench with an open top, e.g. made of wire netting or slats, so that the underside of the trays is exposed to the air. The seeds are sown in the seed trays and often covered with loose material, such as coarse sand or very fine gravel, which are easier for the seedlings to break through.

PREPARATION OF SEEDBEDS AND SOWING IN NURSERY

The soil in seedbeds must be compacted to ensure good contact between the top layer in which the seeds are sown and the deeper layers. In loose soil, the topsoil may dry out quickly because moisture cannot move upwards between the soil particles. The soil is firmed, for instance using a board over which one can walk. The soil is sufficiently compacted in seedbed if only a slight imprint is left when pressed it with the knuckles of a fist. Light watering after sowing also helps to settle the soil around the seed. The soil should not be compacted in a bed when it is too wet as this can ruin the soil structure.

Broadcasting is scattering the seed by hand and it requires practice to achieve a fairly even distribution. If experience is lacking, it is better to sow thinly. The seedlings have to be thinned if there is overcrowding, which is a waste of planting material. Broadcasting is limited to small seeds, because large seeds cannot be placed at the right depth by broadcasting. Broadcast seed is either raked in or covered immediately with loose material, such as coarse sand or very fine gravel. Light watering of the covered seeds helps to settle the soil, ensuring good contact between seeds and soil.

Row-seeding makes it easier to ensure an even distribution, although dibbling also requires practice and some thinning out may be needed. Sowing in rows also facilitates weeding and pricking out. The rows can be made by pressing a slat across the bed to make small V-shaped furrows in which the seed is sown. If seeds are sown directly in their final position in the nursery (no pricking out) they are sown in pits at the proper spacing in a regular pattern. The spacing depends on the tree species and the required seedling size for planting out. A drilling board, fitted with pegs at the desired spacing is a very useful tool to make holes the correct distance apart and in the correct pattern. Both large and small seeds can be sown in this way.

Depending on the germination percentage, a single seed or several seeds (usually 3 to 6) are sown in each position. The spacing ranges from 7 x 7 cm (ending up with 200 seedlings per m^2) for slender conifers to 14 x 14 or 20 x 20 cm for species with large leaves such as teak (resulting in 50 or 25 seedlings per m^2). After emergence of the seedlings, they are thinned leaving only a single seedling in each grid position. If thinning is inadequate, the seedlings are crowded which grow spindly and may not survive after planting out.

Sowing depth is generally twice the diameter of the seed. For example, seed with a diameter of 3 cm is sown at a depth of 6 cm. If seed is sown too deep, the food reserves in the seed may be depleted before the seedling has formed green leaves for its nourishment. Seeds are covered immediately after sowing. If germination takes several weeks, it is worthwhile to mulch seedbeds. The mulch softens the impact of watering and keeps the topsoil moist. However, mulch should be removed as soon as the first signs of germination are observed, otherwise sprouts growing through the mulch will be blanched, spindly and prone to disease.

PRICKING OUT

Seedlings in seedbeds or seed trays have to be transferred to the plantbeds. Pricking out is usually done when the seedlings have one or two normal leaves. If seeds are sown at the right spacing, the seedlings fill the bed or tray by the time they have reached the stage for pricking out. For most species this is within a few weeks from emergence of the sprout. Simultaneous germination makes it possible to prick out all the seedlings at the same time. But even with proper pre-germination treatments, it is often necessary to prick out repeatedly at an interval of several days. In that case, the spacing of the seeds in the tray or bed needs to be wide enough to prick out seedlings without damaging the others. It should be noted that the seedlings are held by a leaf, not by the stem. This is to avoid damaging the tender stem or transmitting damping-off.

In most species, seedlings are large enough for pricking out about 10-20 days after germination, but conifers are often pricked out sooner, even 2-4 days after emergence. If the seedlings are too young they will be very delicate and likely to be damaged. However, waiting too long will result the seedlings become overcrowded and separating the roots becomes difficult. Only few seedlings should be pricked out at a time just to avoid drying out.

It should be made sure that the seedlings will have enough room when pricked out to attain the size desired for planting out in the plantbed. The planting distances may be the same as mentioned above for direct seeding, ranging from 7 x 7 cm to 20 x 20 cm. But to facilitate weeding and root pruning, rows are often spaced about 20 to 25 cm apart, and the seedlings are set closer together in the rows, for instance 5-10 cm apart. If wildings have to be nursed before field planting, they are set in plantbeds at the proper spacing. They can be graded according to size; paying particular attention to the amount of roots, to improve uniformity of the planting material.

CONTAINER NURSERY

POLYBAGS (POLYPOTS) AND POLYSLEEVES

Sowing or planting in container is an alternative to sowing or planting in beds. This requires extra work in the nursery and greatly increases the cost of transport if the seedlings are to be planted in a remote site. Making container from local materials is time-consuming and earthenware pots are expensive, but with the introduction of cheap plastic containers (polybag, poly sleeves, root trainers etc.) potting up has become a lot more popular. Polybags are supplied flat; filling with soil gives them their pot shape. Polybag and polysleeve are made of black or clear polyethylene and available in various volume sizes (polybag 0.3 to 45 L volume and polysleeve 0.3 to 1.5 L volume). Polybags have several drainage holes at bottom and the bottom of polysleeves is open.

A common problem with polybags is that plant roots tend to grow in spirals and this will lead to plants with restricted growth, poor resistance to stress and wind-throw. Seedlings growing in polybag for longer period may penetrate the bottom and roots go deep into the earth. So during shifting, the roots are damaged and result in low survival per cent in field planting. Poly sleeves allow for air pruning when placed on raised propagation beds. But they are less resilient in transport, because the potting substrate can easily fall out and damage the roots. Polybags and polysleeves are not very durable and can break prematurely. For plants with a long nursery cycle, this can be a problem. They are normally used only once. The discarded polybags and polysleeves are a problem for nursery waste management, as they do not decay and are often burned, producing serious air pollution.

ROOT TRAINERS

Root trainers are usually rigid containers with internal vertical ribs, which direct roots straight down to prevent spiral growth. The containers are set on frames or beds above the ground to allow air-pruning of roots as they emerge from the containers. The latest developments also encourage lateral air root pruning through vertical slits. Seedlings grown in root trainers have more vigorous and rapid root growth than seedlings grown in polybags. Outplanting survival and long-term survival are much better. Plants grown in root trainer systems are often ready for planting out when they are substantially smaller than those from conventional polybags. This helps to reduce space requirements in the nursery and transport costs to the field (Hannah Jaenicke 1999).

FILLING POLYBAGS

Soil for filling the polybags should be light and loose-textured for good aeration and drainage, so that root development is not impeded. But a ball of moist potting soil should not fall apart when picked up. The soil is sieved with wire mesh (with 5-10 mm square holes) to get rid off lumps and stones. Polybags should be well filled with sieved soil. The bag should be tapped on the ground to settle the soil, leaving no air pockets. Funnels should be used to fill narrow bags and this works well if the soil is sufficiently loose and dry to flow down the funnel. Filled bags should be watered and left to stand for a few days till the soil has settled. It is difficult to give a standard mixture for potting soil, because what is used depends on what is available locally. As a rule of thumb, when soil is to be part of the growing medium, the following mixtures are used (topsoil: fine gravel/sand: well decomposed organic matter such as manure or compost).

- Commonly used mixture – 3 : 1 : 1
- For heavy (clayey soils) – 1: 2 : 2
- For medium (loamy) soils – 1 : 1 : 1
- For light (sandy) soils – 1 : 0 : 1

BEDS FOR POLYBAGS

Polybags are placed close together in blocks of similar length and width as the plant beds in the nursery. It is very important that the ground has been properly levelled and firmed, so that the bags do not fall over and water does not erode the ground. A frame of wood, bricks or earth can be set up along the sides of the bed to keep the bags upright and to prevent overheating of the outer row of pots (black bags can get very hot in the sun). On a slope, the bags can be placed in trenches - again with carefully levelled bottom - to be inundated periodically by surface irrigation. The water cascades from the upper trench to the lower trenches. If water is scarce, the trenches can be lined with polythene.

The bags are placed together in straight lines of equal length and this makes counting and transplanting easier. The bags are always placed upright even though it is often easier to lean them against each other. If the bags lean to one side, even only slightly, water sprinkled over the bags tends to run off the lower edge instead of penetrating into the bags. It also results in a one-sided root system.

SOWING OR PLANTING IN POLYBAGS

Polybags can be seeded directly, or they can be planted with seedlings pricked out from seedbeds or seed trays. Wildings or cuttings can also be planted in bags. Following direct seeding, it is not possible to stimulate branching of the roots by (air) pruning. So this method is only suitable for species that naturally form a root system branching sufficiently to hold the potting soil together during field planting. Larger size bags must be chosen to enable the seedling to grow to the desired size. Also, the germination percentage of the seed lot should be known, because this determines how much seed can be sown in each bag. If nearly every seed produces a good seedling, as is the case for Acacia species, one seed per pot is enough and there will be no need to thin out.

For very small seeds, sand-seed mixture (2:1) should be used. Small (paint) brush is dipped in water, then in the sand-seed mixture and the mixture is spread carefully over the bags that are filled with soil. For Eucalyptus seeds, this will produce a maximum of 4 or 5 seedlings per bag. In case several seeds are sown per bag, only the strongest one is left after emergence. Thinned seedlings can be pricked out into bags in which none of the seeds germinated. For most tree species, pricking out seedlings into bags is much preferred to direct seeding in the bags for two reasons. Pricking out stimulates branching of the root system, so that the bags are better filled with roots to hold the soil during field planting. Because the pricked out seedlings are uniform, each batch of bag plants is uniform and problems with delayed germination are avoided.

GROWING MEDIUM OR SUBSTRATES

Good plant development depends to a large part on the growing medium used. Most nurseries use mixtures of topsoil with organic and inorganic additions. As an alternative to native soil, organic and inorganic soil-less substrates have been receiving increased attention in tropical countries. The soil-less materials have optimal physical and chemical properties and do not have weed seeds, fungal spores or insects or they can be heat sterilized without losing their properties. The two major groups and components of soil-less media are,

Inorganic Substrates: gravel, sand, vermiculite, perlite, tuff and pumice, polystyrene

Organic Substrates: peat, charcoal, softwood and hardwood barks, rice hulls, sawdust and other organic waste products

FERTILIZERS

Fertilizers provide plants with nutrients necessary for healthy growth. Apart from macronutrients N, P, K, Ca, Mg, and S, there is a known suite of micronutrients (Fe, Mn, B, Cu, Cl, Zn and Mo) that play important roles in the plant's metabolism. When soil or compost-based media is used, the substrate might contain enough nutrients for good plant growth. However, it is advisable to analyze the substrate for available plant nutrients regularly. The different types of fertilizers include,

- *Organic Fertilizers*: FYM, compost, vermicompost, animal waste (poultry litter, bone meal, fish meal, hoof and horn meal)
- *Inorganic Fertilizers*: Ammonium Sulphate, Urea, Ammonium Chloride, Ammonium Nitrate, Ammonium Sulphate Nitrate, Calcium Ammonium Nitrate (CAN), Sodium Nitrate, Calcium Nitrate, Potassium Nitrate, Rock Phosphate, Single Superphosphate, Double Superphosphate, Triple Superphosphate, Basic Slage, Dicalcium Phosphate, Muriate of Potash, Potassium sulphate, Ammonium Phosphate, Urea Ammonium Phosphate, NPK fertilizers
- *Biofertilizers*: Rhizobium (RHZ), Azotobacter (AZT), Azospirillum (AZS), Blue Green Algae (BGA), Azolla, Phosphate solubilising biofertilizer (PSB), Mycorrhiza such as VAM

ROOTED CUTTINGS

Cuttings rooted in a nursery are commonly set in raised, well-drained beds of friable soil. Before planting cuttings, the lower end may be cut again to ensure that the cut surface is clean. This cut at the base is made just below a node, because generally root formation is best at the nodes. If the bark is thick, it may be sliced or scraped off the base of the cutting to make it easier for new roots to break through. Push the cuttings into the soil until about 2/3 of the length of the cutting is covered. For cuttings of trees, two buds are allowed above-ground. If they both grow out, the weaker one is removed so that the most vigorous one will form the tree trunk. In the case of shrubs 3-5 buds may be left aboveground and all emerging shoots are allowed to grow. The cuttings may be set at an angle. A vertically placed cutting tends to produce one dominant shoot, whereas a more horizontally inclined cutting is likely to produce several more or less equivalent shoots. Hence a vertical position is generally preferred for trees and a slanted position for shrubs. The cuttings should not be planted upside down. The cuttings are planted in rows about 25 cm apart with a spacing of about 20 cm in the row. After planting, the bed is watered again.

Cuttings can also be set in pots, but this requires close attention to watering. The soil must be kept moist all the time, but if it becomes too wet the young roots quickly die because of lack of oxygen. In a bed, it is easier to keep the soil moisture between these limits. Shade helps to reduce moisture loss and is essential for cuttings that retain some leaves, also because shade reduces air temperature. Shelter is also important because it protects the cuttings against drying wind and tends to raise the humidity of the air. Whereas seedlings have a dominant taproot, cuttings form so-called adventitious roots which are basically all equal. In due course some adventitious roots may become much stronger than the others, but the root system does not become as strong as that of a seedling with a taproot.

STUMPS

They are root shoot cuttings of seedlings. Cutting root and shoot is known as stump. Plants with a collar diameter of 1.5-2.0 cm and taproot with a length of not less than 30 cm are selected. Shoots are cut off at 3 cm and all the lateral roots are pruned. The tap roots should be cut at a distance of 20-30 cm from the collage region. It is most common for propagating teak (*Tectona grandis*) and gamhar (*Gmelina arborea*). They are easily planted in the field and are very easy for handling and transportation.

PLANT PROPAGATION STRUCTURES (AgriLORE 2012)

The need for enclosed structures varies greatly with the climate. In humid locations, structures may be needed only to provide shade or to keep pests out, whereas under arid conditions enclosures are needed to maintain higher humidity for germinating seed, rooting cuttings and successful grafting. Although automated misting systems may be desirable, simple structures enclosed in polyethylene, such as polypropagator or polytunnel are low cost and sufficient in most cases where small number of plants are produced. Some of the important plant propagation structures are presented below,

GREENHOUSE/ GLASSHOUSE: A greenhouse (also called glasshouse) is a propagation structure in which plants are grown. These structures vary in size based on the purpose and size of plants grown. Greenhouses allow for greater control over the growing environment of plants. Depending upon the level of greenhouse, key factors which may be controlled include temperature, light, water, fertilizer and atmosphere. Greenhouses may be used to overcome the poor growing qualities of a piece of land, such as poor light levels, seasons, or soil infertility. Greenhouses allow certain plants to be grown throughout the year.

A greenhouse is a structure with different types of covering materials, such as a glass, fibreglass, double sheets of polyethylene (PE) film or plastic roof and frequently glass or plastic walls. Several new polycarbonate and acrylic materials are now made for greenhouse coverings. It heats up because incoming visible solar radiation (for which the glass is transparent) from the sun is absorbed by plants, soil and other things inside the building. Air warmed by the heat from hot interior surfaces is retained in the building by the roof and wall.

MIST CHAMBERS: Nursery plants propagated by cuttings are grown in mist chambers and sometimes for germinating seeds. In mist chamber, relative humidity is maintained artificially at high level (95 per cent) with the help of mist installations, which spray water under pressure. Artificial mircroclimate and high relative humidity facilitates better root initiation and development whereas cooling effect prevents the cutting from drying out. Higher success rate are found in propagation of hard wood cuttings under mist chamber conditions. Nowadays, mists chamber with fully automatic system and timers are installed which maintains temperature and humidity even in unfavourable environmental conditions.

NET HOUSES: Net houses are widely used as propagation structure in tropical areas, where artificial heating is not required and artificial cooling is expensive. The roof of net house may be covered with gunny cloth or even with live plant creeper to cut off the solar radiation and to keep the house cool. Different types of shade nets are available in the market for shade purpose. Net houses can be constructed as per need of propagation and therefore size of net houses varies.

POLY-HOUSES: Poly-house is widely used for propagating off season seedlings which fetch better price in the market. Modern poly-houses are automated for temperature and humidity control to assist propagation. In poly-houses, plants are propagated using pro-trays or poly-bags on ground or benches. In Poly-houses, thermostat, hygrometer, automatic ventilation system is maintained to facilitate microclimate. Irrigation and fertigation systems in poly-houses are automated to provide measured quantity of water and nutrients to plants.

PLASTIC TUNNELS: Plastic tunnel is a simple but effective method of protective cultivation, used by many nurserymen throughout the world. The three feet wide loops which support the polyethylene are made from 0.2 inch diameter wire and are erected at 30 inches intervals. A white translucent polyethylene sheet of 6 feet wide is then stretched over the loops and is fastened to the frame of the tunnel. Plastic tunnels prove effective for seed germination as well as vegetative propagation of nursery plants in winter season.

WALK-IN POLY TUNNELS: It is constructed for larger operations or larger plants. Walk-in structures like tunnel is made from semi-circular steel rods and covered with polyethylene sheeting.

PROPAGATION BOXES: For small operations and small plants, propagation boxes are sufficient. These boxes can be made to varying specifications according to the purpose. A recommended size is 2 m long, 1 m wide and 0.5-1.5 m high, with a slanting lid. Layers of stones, gravel and rooting substrate are arranged inside the box. The box is completely

enclosed in polyethylene, i.e. including the bottom, depends on the conditions and the main use. For propagating cuttings, a complete enclosure is recommended to guarantee very high air humidity.

PART 5:
MODERN FORESTRY APPROACHES

23 ECONOMIC VALUE OF FOREST ECOSYSTEM

INTRODUCTION

Sustained agricultural growth is a must, not an option, for most developing countries. Many developing countries have achieved impressive growth rates in agriculture in recent decades. India, which was threatened by hunger and mass starvation in the 1960s, is now self-sufficient in staple foods even though population had more than doubled. Inspite of this success in food front, serious issues remain unsolved for the future agricultural growth. The most important being the environmental problems associated with agriculture which could threaten the future levels of agricultural productivity. The aggressive agricultural production through use of modern technology and increasing food demand had driven the cultivation frontier into less favoured, often environmentally fragile areas, such as forests, hill sides and wetlands. Reduction of fallow periods has depressed the soil fertility. Therefore agricultural growth must be based on a sustainable basis so as not to jeopardize the underlying base of natural resources especially forest resources and it must ensure intergenerational equity so that future generations are not affected.

A sustainable agricultural system can only be achieved by not depleting the value of capital stock such as forest resources over time and replenishing any capital depreciation or losses incurred in the production process. For this purpose, capital is defined to include all natural resource, human and man-made capital assets. It is critical that natural resources like forest resources are accounted and valued properly to design sustainable policies.

Drawing the line between value and price is very fundamental to understand natural resources. Value of a commodity is a complex entity based on a theory, a philosophy and concepts of rationality. The philosophy can be existence (i.e. right to exist). The rationality can be equity (inter and intra generational) or pure market clearance. Economists capture the value through a concept of 'Willingness To Pay' by the consumer, user, non-user, or people at large. For various reasons, this value is not always revealed by people. Price for a commodity (or service) on the other hand, is based on a principle of scarcity. It reflects a balance between what a buyer or a consumer is 'Willing To Pay' (WTP) and a seller is 'Willing To Accept' (WTA). When it comes to natural resources, many resources do not have markets so they may not have a price, but they still have values.

REASONS FOR VALUING NATURAL RESOURCES

There are at least six reasons for valuing natural resources. Firstly, there is a situation of missing market. In the absence of a market, values of goods and services are not revealed. Secondly, even if there are markets they do not do their job well. For instance, market may be a regulated one and there may be restrictions on one's entry into it either to buy or sell. Thirdly, for most natural resources, it is essential to understand and appreciate its alternatives and alternative uses. Because of these, alternative value or opportunity costs are relevant while considering valuation of natural resources. Fourthly, there exists uncertainty of supply and demand of natural resources, specifically in the future. Fifthly governments may like to use the valuation against the restricted, administered or operating market prices for designing natural resource conservation programme. Finally, in order to arrive at natural resource accounting and for methods such as net present value methods, or for cost benefit analysis, valuation is a must (Kadekodi 2001).

VALUATION OF NATURAL RESOURCES

Present system of national accounts does not keep proper account of natural assets namely land, minerals, water and forest in the sense that they are included in the SNA boundary in so far as they are under the effective control of any institutional units. The cost of their use is not explicitly accounted for in production cost. It is well known that these natural resources are much more important than the man made resources since once they are used most of them are non-renewable. Day by day the need to have a set of accounts, which provide information about the utilization of these natural resources in a sustainable way, is being increasingly felt and a number of countries have hence initiated action to start with on an experimental basis for preparing estimates, which take into account the details of utilization of the natural resources.

The problem may acquire alarming proportions in the present era of liberalization wherein attempts are being made to release the growth potential of productive forces by relaxing several administrative and regulatory mechanisms. There are distinct signs of withdrawal of the state in many spheres of development activities that would have serious implications for the use of natural resources and the quality of physical environment. Unfortunately, there have not been systematic efforts to assess the impact of these developments quantitatively or to bring the costs of resource depletions.

In their anxiety to achieve a faster growth of national income, they have opened up the economy, removed some of the bureaucratic and legal hurdles and allowed the economy to respond to the forces of global market. Due to liberalization of economy, there has been political decentralization, giving powers and responsibilities of development planning to local governments and creating thereby a competitive environment for them. This will have serious impact on resource use. The encroachment in rural areas by establishing new industrial units is likely to have an impact on the rate of forest resource use in the entire region. The problem must be viewed as serious as many industries unable to cope with environmental legislation in European countries, are desperately looking for alternate locations in "soft states" like India. In order to tackle this problem, the first step would be to evaluate the impact of forest resource use in quantitative terms and sensitize the policy makers about the issue.

The marginal benefit (or cost) of a good refers to the incremental change in consumer welfare (or producer cost) resulting from an incremental change in the quantity of the good consumed (or supplied). In a competitive market the level of output is determined by prices, which will tend to equilibrate at the point where marginal benefit equals marginal cost and social welfare is maximized. This is the "efficient" level of output, for a given distribution of income. In principle, markets will allocate resources efficiently if prices reflect both the full marginal costs of production and the full marginal benefits of consumption, including all components of total economic value. Where prices do not reflect all costs and benefits, however, the so called "invisible hand" of the market does not work and resources may be used inefficiently, resulting in a loss of human welfare. Economists have identified various reasons why market prices fail to reflect environmental costs and benefits. Two of the most important reasons for market failure in forestry are the prevalence of "public goods" and "externalities".

Public goods are characterized by the fact that: (i) no one can be effectively excluded from consuming them and (ii) increased consumption of the good by one individual does not reduce its availability to others. For example, no one traveling on a public thoroughfare can be charged for enjoying a pretty view, even if the land in question is privately owned. Nor does one person's enjoyment of the view detract from that of another (provided there is no crowding). Such aesthetic value is among many public goods provided by forests, along with carbon storage and biodiversity conservation. Economic theory explains why the free market will systematically under-provide such goods, and why collective action, typically by the government, is usually required to ensure their adequate provision. Externalities are uncompensated costs or benefits arising from economic activity. A classic example in forestry is the decline in availability of other non-timber forest products due to logging. Unless the logging company (or landowner) pays compensation to hunters and gatherers for their loss of livelihood, the full economic cost of extracting timber will not have been paid. If similar conditions prevail elsewhere, market prices of timber products will tend to understate true economic costs and consumers will use timber inefficiently.

In addition to public goods and externalities, markets may fail to reflect non-timber forest benefits like ecosystem services due to lack of information about their contribution to economic welfare, distortions in prices arising from public policy and regulations, lack of clear or secure property rights over forest lands and other factors. In such cases, the question arises as to how decision makers can compensate for market failure and ensure that non-timber forest benefits are given sufficient weight in land use planning and management. The main advantage of cost-benefit analysis is that it places costs and benefits on a common conceptual basis, reducing them to comparable monetary quantities. This approach is not easily applied in the case of natural assets like forest resources, however, because many goods and services flowing from the environment are used by human beings in a way that cannot be likened to market consumption. No prices are established for these natural goods. Thus, if such forest resources are to be properly taken into account and allocated in optimal fashion, an economic valuation of these resources must be performed.

Two strands of research have been developed in valuation of forest ecosystem:

- The first endeavours to identify and formalise the various economic benefits expected from the forest resources. Aggregating these benefits gives the "total economic value" of a given natural asset namely the forest resources.

- The other strand of research focuses on techniques for assessing the monetary value of these benefits. Generally accepted practice is to distinguish between valuation methods based on observation of preferences as revealed in a real market, preferences revealed in a fictitious market and preferences expressed indirectly. These environmental valuation techniques have been used in Western countries since the 1970s, but in developing countries they have come into play only in the last few years.

FOREST ECOSYSTEM

There has been increasing realization that forests provide numerous benefits to mankind including improvement of the quality of environment. Forests provide goods and services and maintain life support systems like timber, fuel wood, fodder and a wide range of non-timber products. Further, forests are a source of natural habitat for biodiversity and repository of genetic wealth, provide means for recreation and opportunity for eco-tourism. In addition, forests help in watershed development, regulate water regime, conserve soil and control floods. They contribute to process of carbon sequestration and act as carbon sink, which is important for reduction of green house gases and global warming. In ecologically sensitive areas like mountains, as well as river catchments, forests play an important role for prevention of floods, etc. Degradation of forest resources has a detrimental effect on soil, water and climate, which in turn affects human and animal life. This has created global concern for protection and preservation of forests.

FUNCTIONS OF FORESTS FOR THE NATURAL SYSTEM

Protective Functions:

- soil protection by absorption and deflection, radiation, precipitation and wind.
- conservation of humidity and carbon dioxide by decreasing wind velocity.
- sheltering and providing required conditions for plants and animal species.

Regulative Functions:

- absorption, storage and release of CO_2, O_2 and mineral elements
- absorption of aerosols and sound.
- absorption, storage and release of water.
- absorption and transformation of radiant and thermal energy.

Productive Functions :

- efficient storage of energy in utilizable form in phyto and zoo mass.
- self regulating and regenerative processes of wood bark, fruit and leaf production.
- production of a wide array of chemical compounds, such as resins, alkaloids, essential oils, latex, pharmaceuticals etc.

FUNCTIONS OF FORESTS FOR THE SOCIAL SYSTEM

Protective Functions:

- sheltering agricultural crops against drought, wind, cold, radiation.
- conserving soil and water.
- shielding man against nuisances (noise, sights, smells, fumes).

Regulative Functions:

- improvement of atmospheric conditions in residential and recreational areas.
- improvement of temperature regimes in residential areas (roadside, trees, parks).
- improvement of biotype value and amenity of landscapes.

Productive Functions :

- supply of a wide array of raw materials to meet man's growing demands.
- source of employment.

VALUATION OF FOREST COVER

While the value of any resource can be assessed by its demand behaviour, in case of forests, markets may not exist for all types of products while forest 'services' providing public goods have no market place at all. Moreover, they have the peculiarity of inter-generational use. As a result of this, standard static economic analysis may not serve the purpose of decision making on issues regarding pricing and distribution of forest products. Also, such resources are subjected to a variety of property rights systems, different from individual or private property rights. In such a scenario alternative methods are required to value the forests.

There are externalities associated with forest resources and hence there is a gap between the value and notional price. The positive externalities are generally in terms of various ecological, biological and aesthetic benefits and very little price is paid. Most of the externalities are not accounted for and this results in gross under-estimation of environmental value of forests. It is mostly the value of timber that gets reflected in the contribution of forests in state domestic product. Such an under-valuation often leads to inadequate allocation of funds for maintaining the forests. It is due to these reasons that there is need to take complete stock of forest resources and assign economic value to all intangibles including goods and services, soil erosion and agricultural productivity, health, etc.

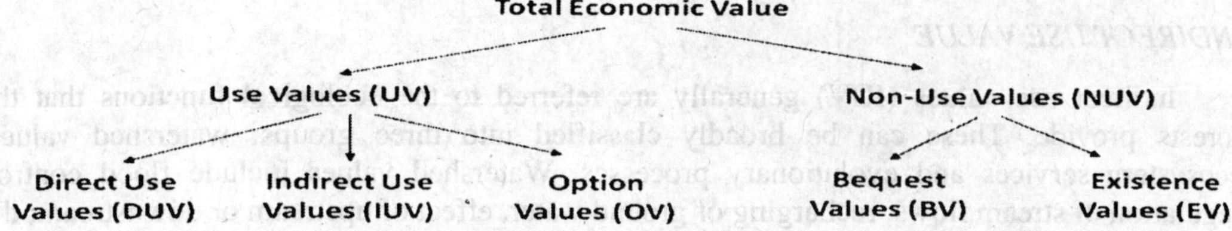

Fig. 23.1: Division of Total Economic Value

In the case of natural and environmental resources a concept of Total Economic Value (TEV) is perhaps the most complete measure to express the full range of value of benefits - both tangible and intangible. Natural resources provide a variety of goods and services to the users for their current or future benefits or welfare and are said to have use values. Examples include timber from forests, water from rivers or underground, coal from fossilized earth, etc. Use values (UV) can be further broadly classified into three groups; viz. direct, indirect and option values.

The total economic value of the resource may be classified into use and non-use values. Use value may further be segregated into direct, indirect and option values. The Total Economic Value is the sum of use values and non-use values. This can be expressed as:

$$TEV = UV+NUV = (DUV+IUV+OV) + (BV+EV)$$

DIRECT USE VALUE

Direct use values (DUV) refer to the current use (consumption) of the resources and services provided, directly by natural and environmental resources. Direct use value includes uses that are possible with extraction activity or direct interaction with the forest resources and have commercial or subsistence values. Direct use value can be either consumptive or non-consumptive. Forests provide timber, fuelwood, fodder, medicinal plants, fruits, etc., to the people and thereby generate direct consumptive use values. Recreation, education, research etc., are examples of direct non-consumptive use values. These values include the use of the forest resources for timber, non-timber forest products, educational purposes, agricultural cultivation and recreation, and tourism. The timber value of the forest has dominated extraction activities in the past but said use causes the drawdown or loss of other products and services as well. Further, there may be associated damages to society due to sedimentation of low-lying water bodies and farms. It is now recognized that the cost of consuming tree resources should also consider the value of the foregone products and resultant damages.

Non Timber Forest Products (NTFPs) include: fuel wood; extractives such as resins, latex, gums, bark and dyes; ornamental plants such as orchids; and food items such as seeds, spices, honey, fruits and bush meat. The local community consumes most of these products but certain commodities are used as raw materials by other sectors and hence are sold to local traders for further re-sale. The recreational value of the resource is now increasingly recognized both for domestic and international tourism. The development of this enterprise can lead to the growth of other local industries to cater to the demands of the tourists (e.g., food industry, handicrafts and the service sector) and hence has considerable economic potential.

INDIRECT USE VALUE

Indirect use values (IUV) generally are referred to the ecological functions that the forests provide. These can be broadly classified into three groups: watershed values, ecosystem services and evolutionary processes. Watershed values include flood control, regulation of stream flows, recharging of ground water, effect of upstream or downstream; the ecosystem services include fixing of nitrogen, assimilation of waste, carbon sequestration, gene pool; and evolutionary processes include life support, biodiversity preservation, etc.,

OPTION VALUE

The option value pertains to one's willingness to pay to keep the option open of being able to avail one self of the direct and indirect uses of the forest at some future time Actual value may or may not be realized. Option value (OV) is associated with the benefits received by retaining the option of using a resource in the future by protecting or preserving it today.

NON-USE VALUE

The non-use value of the resource includes existence and bequest values. Existence value is the desire to have the resource intact or preserved in its own right. It also includes wanting to have the resource available to mankind for some altruistic or humanitarian reasons. Non-use values (NUV) are generated without any direct link with the use of natural resource under question. These values are often revealed through peoples' perceptions and concerns towards conservation, culture and aesthetics and so on.

EXISTENCE VALUE

Existence vale (EV) is a concept associated with people's willingness to pay simply for the pleasure they derive from knowing that a natural area or particular species or characteristic exist, irrespective of any plans they may have to use these resources.

BEQUEST VALUE

Bequest value measures what one is willing to pay to protect a given area or resource for the benefits of his children and of the generations yet to come. The bequest value (BV) originates when people are Willing To Pay to conserve a resource for the use of future generations. People's Willingness To Pay for the preservation of endangered species is an example of existence value.

Economic value measures provide a common metric of value for the different services provided by the forests viz., timber, biodiversity, carbon sequestration, watershed values, etc. in monetary units, say Rs/hectare. All types of values mentioned above are converted into monetary terms in order to help in decision making on alternative scenarios for the use of forests.

COSTS ON FORESTRY

In most valuation studies, while the benefits from forestry have been well documented, the costs incurred on its cultivation, maintenance, restoration, depreciation and other miscellaneous costs to obtain the timber, non-timber forest products, eco-tourism and other benefits are often not accounted for. Here, opportunity costs are also ignored. Restoration costs are incurred for reproducing the original level of benefits. Certain expenditures are required to prevent degradation of the forests. Further, input costs are incurred on plantation, extraction cost, marketing expenses for forestry products, especially log wood and other NTFPs.

Direct expenditure on afforestation includes land preparation, pitting, planting, watering, fencing, trenching, apart from purchase of seedlings, fertilizers, after care cost, etc. Apart from the direct costs, there are certain indirect costs that are required to be incurred in order to reap the larger benefits associated with forests. Some of these costs would include infrastructure cost for promoting eco-tourism, markets for timber and NTFPs, etc. It is, however, important to note that most of these are in the nature of joint costs and attributing these costs exclusively to forestry activities could be misleading.

Most of the investment/expenditure in forests is undertaken by public agencies. Hence it is also important to look at the opportunity cost of land under forests or proposed for afforestation. This would require estimation of opportunity cost of alternative uses of the land such as grazing, agriculture, horticulture, habitation, industry, etc. The opportunity cost would reflect all economic outputs foregone or precluded by maintaining land under forest cover. Maintenance of existing natural forests would imply prospective benefits foregone from agriculture or other alternative uses of land as mentioned above. Thus, opportunity cost of sustainable forest management is the amount that could be earned from forest exploitation and various forms of land utilization. In case of converting wasteland under the afforestation programmes, the opportunity cost could be in terms of restricted availability of grazing and pasture land, etc.

Opportunity cost of benefits from forest would also indicate the shadow price of benefits arising from watershed catchment protection and soil conservation, nutrient cycling, maintenance of soil fertility amongst others. The other issue relates to social costs that are associated with existence of forests on the local people. Such social costs could take the form of damage to resilience of forest eco-system, natural resource depletion, pollution on account of forest fires, etc. Another widespread cost results from crop damage caused to local farmers by animals and birds dwelling in nearby forests.

VALUATION TECHNIQUES

The different techniques can be divided into five broad groups:

(a) Market price valuation, including methods to estimate the benefits of subsistence production and consumption

(b) Surrogate market approaches, including travel cost method, hedonic pricing and the substitute goods approach

(c) Production function approaches, which focus on biophysical relationships between forest functions and market activities

(d) Cost-based approaches, including replacement cost and defensive expenditure

(e) Stated preference approaches, mainly the contingent valuation method and variants

MARKET-BASED APPROACH

The simplest valuation methods are those, which rely on market prices. Many goods and services from tropical forest land uses are traded, either in local markets or internationally, including wood products (timber, pulp and fuel), non-wood forest products (food, medicine and raw materials). For the products that are commercially traded, market prices can be used to value. The market prices, however, do not always reflect the appropriate prices for the valuation of various forest products. Since many environmental commodities do not pass through the market system, the markets do not provide prices that can be used to value them. For those products that are commercially traded, market prices can be used to construct financial accounts to compare the costs and benefits of alternative forest land use options. In some cases, it may be necessary to adjust market prices to account for market or policy failures. The latter step, also sometimes called "shadow pricing".

In addition, there are many services provided by the forests for which there are no prices or it is difficult to have information about their prices. Consequently, a number of surrogate methods have been developed to value the forests.

SURROGATE MARKET-BASED APPROACHES

Where there are related goods in the market for non-marketed environmental commodity, then, the prices of the related goods may be used. The markets for property/land, labour, and recreational services are used as surrogate markets to infer the value of environmental characteristics of the area/labour market and natural parks, respectively. It is a second group of methods rely on the fact that certain non-market values may be reflected indirectly in consumer expenditure, in the prices of marketed goods and services, or in the level of productivity of certain market activities. These techniques statistically sophisticated methods, such as travel cost models and hedonic pricing, as well as simpler techniques such as the substitute goods method. The theoretical basis for all of these approaches is the household production function, which describes how households attempt to maximize their well-being by allocating time and resources to different activities.

HEDONIC PRICING APPROACH

More specific approaches in this group include the property value approach and the wage labour differential approach, which are used to value environmental amenities and dis-amenities. In both markets, the basic principle is the same: the property (worth) is characterized by several attributes that can be physical, location/access and environmental attributes, among other things. Hedonic pricing is based on the assumption that the value of an asset is related to attributes it possesses or the stream of net benefits derived from it. Hedonic pricing is based on the assumption that the market value of land or labour is related to the stream of net benefits derived from it. This stream of net benefits includes a range of factors, including environmental amenities. Therefore, the value of the environmental amenity can be imputed from the observed land or labour market. A particular buyer may be willing to pay for certain environmental attributes in some area that he may not be willing to pay in another area. The value of the property/labour is thus a result of the composite characteristics of the commodity. The specific contribution of the individual attributes can be obtained through a regression analysis to derive the implicit price of say, the environmental attribute (e.g., level of pollution in the area). Once implicit prices are obtained for all the observations, a demand function (an inverse) can be derived for the environmental characteristic and its value derived from an estimation of the area under the demand curve. The market value of a residential property, for example, is affected by many variables including its size, location, construction materials and also the quality of the surrounding environment. With sufficient data on property values and characteristics it may be possible to control for size, location, construction materials and other factors, such that any residual price differential may be imputed to differences in environmental quality. The hedonic pricing method requires large data sets, in order to account for and eliminate the influence of all other variables, which affect market prices. The approach also assumes that markets for land are competitive, and that both buyers and sellers are fully informed of the environmental amenity or hazard.

Hedonic pricing has been used in developed countries to estimate the negative impact of air and noise pollution, or the presence of waste disposal facilities, on the market prices of residential property and, conversely, the positive impact of proximity to water or public green space. It was found that no examples of hedonic pricing used to assess the environmental amenity value (or any other value) of forests in developing countries. One constraint on use of the technique in developing countries is that private property markets are often thin, uncompetitive and poorly documented. This is a particular problem at the frontier of forested areas, where formal title to land may be missing and where land is often essentially an open access resource.

This situation will probably change as incomes grow and land markets in developing countries become more efficient and discriminating (and as land transactions are better documented). It may already be possible to apply the hedonic pricing method to residential property markets in and around high-growth cities in developing countries, especially where new residential housing developments provide home buyers with the opportunity to reside in greener, forested areas, away from the smog and congestion of the city centre.

TRAVEL COST METHOD

This method is used to measure the recreational value of natural resource scenic areas that are normally unpriced or priced with very low entrance fee. The travel cost method is based on the assumption that consumers value a forest site in terms of the cost of getting there, including all direct transport costs as well as the opportunity cost of time spent travelling to the site (i.e. foregone earnings). The fee collected is usually based on cost recovery for the staff manning the area and may not even be enough to cover the man-made improvements made in some of the country's national parks. To obtain an idea of how much value people attach to the resource for recreation, the approach assumes that the amount that people are willing to spend to visit the site is a good indicator of the value of the resource to them. This survey-based method has been used extensively to estimate the value of recreational aspects of forests.

Three basic steps are involved in travel cost models. First, it is necessary to undertake a survey of a sample of individuals visiting the site to determine their costs incurred in visiting the site. These costs include travel time, any financial expenditure involved in getting to and from the site, along with entrance (or parking) fees. In addition, information on the place of origin for the journey and basic socio-economic factors such as income and education of the individual is required. The resulting data is manipulated to derive a demand equation for the site. This relates the number of visits to the site to the costs per visit. The third step is to derive the value of a change in environmental conditions. For this, it is necessary to determine how willingness to pay for what the site has to offer alters with changes in the features of the site. By comparing the willingness to pay for sites with different facilities it is possible to determine how the total benefits derived from the site change as the facilities of the site change. A demand function for recreation (measured in visit-days per year or visitation rate per 1,000 population) with price, income and prices of related goods as dependent variables, should be constructed. From this demand function, the value of the recreation benefits can be estimated.

SUBSTITUTE GOODS APPROACH

For those forest resources which are non-marketed or which are used directly by the harvester (e.g. fuelwood), value may be approximated by the market price of similar goods (e.g. fuelwood sold in other areas) or the value of the next best alternative or substitute good (e.g. charcoal). The extent to which the value of the alternative marketed good reflects the value of the nonmarket good in question depends, to a large extent, on the degree of similarity or substitution between them. That is, if the two goods are perfect substitutes then their economic values should be very close. As the level of substitution decreases so does the extent to which the value of the marketed good can be taken as an indication of the value of the non-marketed forest good.

Chopra and Kadekodi (1997) use the approach as an alternative to labour cost analysis for valuing fuelwood, in a study of deciduous forests in India. No examples were found of the substitute goods approach being used to estimate other forest values. In some circumstances, a substitute good may be non-marketed as well, but it may still be possible to infer a price for the good in question. In a cost-benefit analysis of a management programme for two forested watersheds in Nepal, fuelwood was valued in terms of the alternative uses of its closest substitute that is cattle dung which can be dried and burned when wood is unavailable. The opportunity cost of using cattle dung as fuel rather than as fertilizer is estimated in terms of losses in food grain resulting from lower dung inputs into agricultural production. Thus this approach involves two stages: first, estimating a conversion factor for equating the two substitutes (in this case, the energy value of dung and wood); and second, estimating the marginal effect on output and thus profits, of a change in the use of the substitute (in this case cow dung).

The main problem with the substitute goods approach is that it is difficult to find substitutes which are perfect and which do not provide additional side-benefits. Thus, such value estimates should be viewed with caution.

THE PRODUCTION FUNCTION APPROACH

Another valuation method involves an attempt to relate human well-being or the incremental output of a marketed good or service to a measurable change in the quality or quantity of a natural resource. Such method is variously called production function approach or dose-response or input-output method. The production function approach may be used to estimate the indirect use value of ecological functions of forests, through their contribution to market activities. The production function approach involves two steps. The first step requires establishing a dose-response function with the response reflecting measures of damages in physical measures (e.g., production losses, incidence of orbidity / mortality). The dose corresponds to the level of environmental stress (e.g., soil erosion, air pollution). The dose response function can be established from data obtained through experiments or survey work.

The second step involves translating the physical impacts to monetary measure. This stage is straightforward to the extent that market prices (adjusted for inefficiency) are available. In the case of loss production for example, the price of the crops may be used in the valuation. Where market prices are not available, cost-based techniques may be employed. To illustrate, morbidity impact may be valued using cost of illness approach. The cost of illness normally consists of hospitalization expenses and work loss days.

An essential requirement of the approach is the availability of reliable information on the physical relationship between the state of the environmental resource and the economic activity or asset it supports. In addition, market conditions and policy distortions affecting production decisions need to be taken into account. The production function approach has been used extensively in both developed and developing regions to estimate the impacts of changes in environmental quality (e.g. deforestation, soil erosion, air and water pollution) on productivity in agriculture, forestry and fisheries, on human health and on the useful life span or costs of maintaining economic infrastructure.

Surrogate market and production function approaches mentioned so far rely on price (Revealed Preference) to estimate the value of the forest goods and services. These methods try to estimate the value of an environmental good on the assumption that such good exists. An alternative is to ask consumers to state their preferences directly, in terms of hypothetical markets. Here, information on the value of an environmental benefit is obtained by posing direct question to consumers about their willingness to pay for it or, alternatively, their willingness to accept cash compensation for losing the benefit. The most widely used technique in this category is the contingent valuation method (CVM).

COST-BASED APPROACHES

The cost-based approaches measure the costs of ensuring the maintenance of the benefits provided by the environmental commodities. One must recognize that these cost measures are not really measures of benefits and should thus be viewed as proxy measures where benefit estimation is not feasible. The specific methodologies under this category include the opportunity cost method (i.e., the time to collect non-marketed products, usually valued using the wage rate, is used to value the benefits from the products collected). The other approaches are the replacement, restoration and relocation costs to restore the original condition of the environment. There are also the defensive or preventive expenditures incurred to prevent the change from the existing environmental situation.

In addition to the methods described above for estimating WTP or WTA for non-market forest benefits, some other cost-based approaches may be used to shed light on the costs of maintaining non-market forest benefits, or trade-offs with market values. Three alternative methods focus on the costs of providing, maintaining or restoring environmental goods and services. A thorough discussion of cost-based valuation can be found in Dixon et al (1986).

The most common methods are:

- Replacement cost methods, which measures environmental values by examining the costs of reproducing the original level of benefits;
- Preventive expenditure methods, which estimate the cost of preventing or defending against degradation of the environment; and
- Opportunity cost approaches, which use estimated production costs as a rough proxy for the value of non-market benefits.

Cost-based techniques are commonly used where there is limited time and resources for more rigorous estimation of environmental benefits. However, such techniques must be used with care, with particular attention to ensure that non-market benefits and costs are not confused. Because cost-based techniques do not directly measure WTP for environmental goods and services, the resulting estimates may over- or under-estimate forest benefits by a large margin. Problems arise when potential rather than actual expenditures are used, as it is not always clear that the environmental benefit in question justifies the costs of replacement, relocation, etc. On the other hand, while cost-based methods are inexact, they may be the only practical alternative in some cases, given resource and time constraints. Where such methods are used, key assumptions about the relationship between estimated costs and associated benefits should be stated clearly.

REPLACEMENT COST

The replacement cost technique generates a value for the benefits of an environmental good or service by estimating the cost of replacing the benefits with an alternative good or service. For example, where logging or road construction in upland forest areas leads to increased runoff and sedimentation, some studies use information on the costs of dredging or flood control as a rough estimate of the non-market benefit of watershed protection. The technique rests on the availability of such an alternative, which should - as nearly as possible - produce the same type and level of benefits as supplied by the resource or environmental function being valued. When developing a replacement cost scenario, it is normal practice to select the least cost option among all possible technologies, so as not to over-estimate the value of the environmental benefit.

Relatively few examples of the replacement cost method were found in the literature review. One application to valuing forest benefits is to estimate the value of soil nutrients lost due to increased erosion associated with logging or deforestation in terms of the cost of manufactured fertilizer needed to replace the eroded nutrients (Chopra and Kadekodi 1997).

PREVENTIVE EXPENDITURE

The preventive expenditure approach (also sometimes called "mitigation" or "defensive" expenditure) places a value on environmental goods and services by estimating the costs of preventing a reduction in the level of those benefits derived from a particular area. This approach may be most applicable for assessing the indirect use values of forests. No case studies using preventive expenditure were found in the literature review. However, a hypothetical example can illustrate how the technique could be used to estimate non-market forest values. For instance, projected expenditures on soil conservation measures aimed at halting or reversing soil degradation could be used as a rough proxy for the benefits generated by the natural nutrient cycling and watershed protection functions of forests. In the case of logging, the watershed protection benefits that might be diminished by building roads for the extraction of timber could be valued by examining the incremental cost of adopting less damaging extraction methods, such as non-mechanized extraction, helicopter logging, or alternative road layouts. As always, it is important to ensure that the benefits of the preventive expenditure match those originally provided by the environmental function, in order to obtain a realistic cost estimate.

The preventive expenditure method is sometimes confused with a variant of the production function approach known as "damage costs avoided." The difference is that the latter approach uses information on the costs of making good or repairing damages incurred as a result of some environmental change, whereas the preventive expenditure approach focuses on the costs of avoiding or mitigating damages before they occur. Where alternative valuation techniques yield different estimates of a particular forest benefit, it is generally preferable to use the lowest estimate so as not to over-estimate the non-market benefit in question.

OPPORTUNITY COST OF LABOUR

Another valuation approach focuses on the employment opportunities foregone in order to secure or protect a particular non-market benefit. As with other cost-based approaches, the focus is on the costs of providing a non-market benefit, rather than the magnitude of the benefit *per se*. The basic idea is that a non-market benefit is worth at least as much as the return that could be obtained by private producers if they were to devote the same effort (i.e. the labour used to secure the non-market benefit) in some alternative use.

The opportunity cost approach is most often used to value the subsistence benefits of NTFPs collection, where labour is the main input and prices are not available because all or most output is consumed directly by producers. In such cases, the implicit assumption is that a producer's decision to spend time collecting non-timber forest products is weighed against alternative uses of household labour. The opportunity cost of time spent harvesting NTFPs is thus taken as a proxy for the value of the product(s) in question. The only data required are the amount of time spent on the harvest, the resulting yield and the prevailing (rural) wage rate. Nevertheless, care must be exercised when using rural wage rates. First, it is important to ensure that the effort involved in harvesting a non-timber forest product is commensurate with the effort associated with the prevailing wage rate. For example, one day of a child's time spent collecting wild foods should not be valued at the same level as the wage received by an adult for physically demanding agricultural work. In such cases, it may be appropriate to adjust the wage rate used. Similarly, seasonal variations in rural wages should be considered, as the harvesting of NTFPs may be concentrated in periods of slack labour demand, when wages are lower. Third, gathering forest products is often combined with agricultural activities, for example when walking to or from the fields. This must also be taken into account when deciding how much labour time to include in NTFP harvesting costs.

STATED PREFERENCE APPROACH

MARKET CREATION TECHNIQUES /CONTINGENT VALUATION METHODS (CVM)

This method is particularly useful when data are unavailable to value the environmental effects of a particular project or when relevant market behaviour is not observable. These cases include species preservation; historical or cultural phenomena; scenic, ecological or other characteristics; genetic diversity and preservation of open spaces; unobstructed views or public access to amenity resources.

The CVM technique entails creating a hypothetical market for the good in question and asking the respondents how much they are willing to pay (WTP) for the good in question. The crucial elements of this approach are the definition of the product, the specification of the payment vehicle and the framing of the WTP questions. The hypothetical nature of the market makes the application of this technique quite open to a number of biases (e.g. starting point bias, strategic bias) and should thus be carefully considered. The CVM is applied in such areas as water quality, air quality, biodiversity and protection of endangered species and land/recreation facilities studies.

The contingent valuation method uses a direct approach, it basically asks respondents how much they are willing to pay for a benefit, or how much compensation they will be willing to accept to tolerate a cost. The individual's behaviour is inferred from the answers he provides in a survey framework. The aim of the contingent valuation method is to elicit valuations which are close to those that will be revealed if an actual market existed. This method is applicable for valuing a wide range of non priced environmental good or service. However, it is more effective when the respondents are familiar with the environmental good or service in question. The success in the use of this method also lies in the design, implementation and interpretation of questionnaires.

In forestry, CVM has been used to value wildlife and recreational benefits of protected areas. Several recent studies have demonstrated the feasibility of applying CVM to forest land use. Valuations produced by CVM are "contingent" because value estimates are derived from a hypothetical situation that is presented by the researcher to the respondent. The two main variants of CV are open-ended and dichotomous choice (DC) formats. The former involves letting respondents determine their "bids" freely, while the latter format presents respondents with two alternatives among which they are asked to choose.

Open-ended CVM formats typically generate lower estimates of WTP than DC designs. On the other hand, because no payment is made in most cases, some observers question the validity of stated preference techniques. Critics argue that CVM fails to measure preferences accurately and does not provide useful information for policy. Even practitioners accept that poorly designed or badly implemented CV surveys can influence and distort responses, leading to results that bear little resemblance to the relevant population's true WTP. Much recent attention has focused on overcoming potential sources of bias in CVM studies. Resolving these difficulties involves careful design and pre-testing of questionnaires, rigorous survey administration and sophisticated econometric analysis to detect and eliminate biased data.

It should be noted that in applications of CVM in developing countries WTP can be measured in non-monetary units if respondents are very poor. The use of CVM for valuing environmental resources originated and was largely developed in North America. In forestry, CVM has been used to value wildlife and recreational benefits of protected areas. Several recent studies have demonstrated the feasibility of applying CVM to forest land use in the developing world.

CONTINGENT RANKING

Contingent ranking (CR) is a variant of contingent valuation and it involves asking respondents to rank a series of alternative non-market goods. One advantage of contingent ranking is that monetary bids may or may not be used.

A variant of contingent valuation, this method involves asking respondents to rank a series of alternative non-market goods. One advantage of contingent ranking is that monetary bids may or may not be used. Some have suggested that the use of hypothetical cash payments in CVM may be inappropriate in remote rural communities in the developing world, where people may have relatively little exposure to the market economy. In such cases monetary values can be assigned indirectly, by including in the contingent ranking one or more "anchor" goods with known market values. The contingent ranking method is conceptually simple, easy to administer and able to generate rough estimates of value for a number of forest goods and services at once, without conducting separate WTP surveys for each use and non-use value. On the other hand, contingent ranking may not provide accurate estimates of WTP. A fundamental question is whether the scores generated by contingent ranking reflect an ordinal ranking of relative preference (i.e. first, second, third, etc.) as opposed to cardinal measurements of value (i.e. one, two, three...). In the former case, it may not be valid to use the cost/price of marketed "anchors" to impute monetary values to non-market goods and services, because they are on different (incommensurate) scales.

Table 23.1: Methods for Valuing Forests (Source: IIED 2003)

Valuation Method	Relevant Forest Benefits	Strengths &Weaknesses
Market Prices: Use data from surveys of producers and consumers, adjusted if necessary, to account for seasonal variation, value-added processing and/or public policy distortions.	Price-based valuation is commonly applied to non-timber forest products which are partly or informally traded, in order to estimate subsistence and / or unrecorded consumption.	Market prices clearly reflect consumer preferences, but often need adjustment to account for public policy distortions or market failures. Aggregation or extrapolation of values based on potential production is not valid effects (elasticity of demand).
Surrogate Markets: Travel cost – use survey data on direct cost (e.g. fares, accommodation) and, in some cases, opportunity costs of time spent traveling to and from a site, valuated at some fraction of the average wage rate.	Travel cost is often used to estimate demand for forest recreation at specific locations. Related methods used mainly in developing countries estimate the value of non-marketed, non-timber forest products in terms of the opportunity cost of time spent collecting and/or processing them.	Provided the relation between the benefit being valued and the surrogate market is correctly specified and prices in the surrogate market are not generally reliable. Travel cost estimates may need to account for various objectives (benefits) in a single trip.
Hedonic pricing – use statistical methods to correlate variation in the price of a marketed good to changes in the level of a related, non marketed environmental amenity.	Hedonic pricing is used to estimate the impact of proximity to forested land and/or logging on the prices of residential and commercial property.	Hedonic pricing requires large data sets, in order to isolate the influence of a non-market benefit on market price, relative to other factors.
Substitute goods – use market prices of substitutes for non-marketed benefits.	Substitute goods approaches may be used whenever close market substitutes for non-timber benefits exist.	
Production Function: Change in production method - use data on the physical relation between level (or quality) of non-market benefit and level (or quality) of output of a marketed good/service.	Change in production (or "input-output" or "dose-response") methods are used to estimate both on- and off-site impacts of land use change, e.g. the effect of logging on hunting, downstream water users, fisheries climate.	Change in production methods requires good data on biophysical relationships (dose-response).
Stated preference Contingent valuation method use consumer	Recreational values are often estimated using contingent	Contingent valuation estimates are generally considered reliable

surveys to elicit hypothetical individual willingness to pay for a benefit, or willingness to accept compensation for the loss of that benefit.	valuation. Stated preference methods such as CVM are the only generally accepted way to estimate non-use values, e.g. Landscape or biodiversity values, for which price data do not exist and/or links to marketed goods cannot easily be established.	if strict procedural rules are followed.
Contingent ranking /focus groups –use participatory techniques in group setting to elicit preferences for non-market benefits, either in relative terms (ranking) or in monetary terms.	Contingent ranking may be used where target groups are unfamiliar with cash valuation.	Participatory techniques are more experimental and not widely used to estimate non-market forest benefits. They are good at eliciting qualitative or "contextual" information, but there are doubts about their reliability for estimating willingness to pay.
Cost-based approaches: Uses data on the costs of measures taken to secure, maintain and/or replace forest goods and services.	Cost-based approaches include replacement/relocation cost, defensive expenditure and opportunity cost analysis; may be used (with caution) to value any type of forest benefit.	Cost based approaches are usually considered less reliable than other methods. One test of validity is evidence that people are prepared to incur costs to secure relevant benefits.

SELECTED ECONOMIC VALUATION METHODS FOR INTANGIBLE BENEFITS

For example, the direct non consumptive uses of forest ecosystem viz., ecotourism/ recreation education and research, human habitat, other non consumptive uses are valued by CVM (Vidhyavathi and Swaminathan 2007) and experiments method as direct methods and travel cost method can be used as indirect method. Indirect uses like watershed benefits, agriculture productivity, soil conservation, recharging of ground water, regulation of stream flows, other watershed benefits can be assessed CVM and Experiments method as direct methods and change in productivity approach; replacement cost approach can be used as indirect methods. Eco-system services like nitrogen fixing, waste assimilation, carbon store, microclimatic functions, other ecosystem services can be assessed CVM and Experiments method as direct methods. Replacement cost approach and indirect estimates derived from experiments are used as indirect methods. Evolutionary processes like global life support and biodiversity are assessed using CVM and indirect estimates derived from experimental data can be engaged. Non-use values will be assessed using CVM.

A study by Hadker et al (1997), entitled "Willingness-to-pay for Borivili National Park: Evidence from a Contingent Valuation" tried to assess the willingness to pay (WTP) of residents of Bombay (Mumbai) for conservation of the Borivilli National Protected Area. Average household WTP was estimated at US $ 0.23 per month or about US $ 31.6 million in aggregate present value terms, which far exceeded the current budget of US $ 520,000 to maintain the area. The main direct use value of the area is for recreation and the indirect use values may be associated with the area's role as a source of Bombay's drinking water and as home to many endangered animals. A number of studies on intangible benefits using various valuation techniques have been conducted for Indian forests. Chopra and Kadekodi (1997) examined the ecological functions of the Yamuna Basin using the contingent method and suggested an annual value of Rs.624 per hectare as the intangible benefit of forests. The same authors in another study of this region suggested watershed value of Rs.2 lakh per hectare metric soil using an indirect method of reduced cost of alternate technology.

Similarly, value of carbon store was examined by Kadekodi (2001) in a study at the All India level and suggested Rs.1.2 lakh per 17 hectare benefit using indirect estimates. Haripriya (2000) estimated value of carbon store at Rs.20,125 per hectare and aggregate of Rs.1292 billion from Indian forests using species miscellaneous forest inventory data.

A study by Madhu Verma (2000) examined the contribution of forests of Himachal Pradesh in sustaining livelihood of rural population, provisions for urban markets in local and external regions in terms of direct consumptive and non consumptive benefits. Apart from direct consumptive benefits from growing various species of forest stock, value of the forest growing stock, salvage, timber drawn by right holders, fuelwood requirement, grasses and grazing and minor forest produce. The study also accounted for direct non-consumptive and indirect benefits from Himachal Pradesh in terms of eco-tourism and recreation benefits, watershed benefits, micro-climatic factors, carbon sink, biodiversity, and employment generation through forestry works and suggested total economic value of forests of Himachal Pradesh at Rs.2.89 lakh per hectare. The economic value of forests of Himachal Pradesh is Rs. 2.89 lakhs per hectare of goods/services in terms of total geographic area of forests and Rs. 7.43 lakhs per hectare of goods/services in terms of area under tree cover and scrub forest. The study suggested that the annual indirect benefits far exceed the direct benefits of forests which are just Rs. 21,000 per hectare and Rs. 53,000 per hectare respectively. Hence, if the Gross State Domestic Product (GSDP) of the state is corrected for total economic value, as calculated in the current study, the contribution of forestry sector increases from 5.26 per cent of GSDP to 92.40 per cent of GSDP.

DISTRIBUTIONAL CONCERNS

The main objective of economic appraisal is to evaluate the costs and benefits of alternative activities in terms of aggregate economic efficiency, i.e. irrespective of the distribution of costs and benefits among different groups of people. Most policy-makers, of course, are concerned about distributional impacts, and we may expect alternative forest land use options to impose costs and confer benefits to varying degrees on different social and economic groups.

For example, the designation of forests as protected areas can be seen as a means by which certain interest groups (typically not the poor) secure recreational, amenity or non-use values. This may result in significant loss to another group, e.g. subsistence farmers who rely on forest land for extraction of non-timber forest products, or for shifting agriculture. Where the values of domestic and foreign consumers differ widely, the resulting conflict may be international in scale, as shown by recent heated debates about tropical forestry and timber trade policy in global forum. Similarly, efficiency and equity objectives in forest land use often conflict. Cost benefit analysis may lend weight to certain values and associated land use options that are unavailable to poorer groups, due to their limited access to capital and information. These groups must therefore confine themselves to "inferior" uses. For example, a study of a new national park in Madagascar estimated the value of additional benefits to international tourists at two to three times the loss, in terms of lost agricultural land, incurred by local villagers, despite measures allowing them access to buffer zones. While such a change in land use may be economically efficient, since the potential for compensation exists, it will aggravate poverty where compensation is absent or inadequate.

Distributional concerns may be incorporated in an economic appraisal of forest land use options in at least three ways:

- The distributional consequences of land use options can be made explicit by assigning costs and benefits to specific groups;
- Distributional weights can be used in the economic analysis to adjust the benefits and costs according to which groups enjoys or bears them; and
- The entitlements (use and access rights) of particular groups with respect to certain forest resources or benefits may be protected by defining minimum standards or guarantees.

To some extent the first approach is a prerequisite of the second and third. Unless the costs and benefits of a project or land use option can be linked to specific groups there is no way to know where distributional weights should be attached or what rights need to be protected. Therefore, the first step in any distributional analysis is to identify the different stakeholders in alternative forest land use options. Depending on the region and the particular land uses in question, this may include:

- Indigenous hunter/gatherer populations
- Subsistence farmers
- Commercial farmers
- Small-scale traders
- Industrial firms (owners and employees)
- Local, state and national government agencies
- Domestic and foreign consumers

The next step is to determine which groups are affected by the various impacts of alternative land use options. Some costs and benefits may be spread widely among a number of groups, while in other cases the impact on certain stakeholders will be more concentrated. For example, the benefits of timber harvesting will be spread among the owners of logging companies and their employees, as well as firms involved in providing equipment, wood processing, transport, distribution and sales. It may not be possible to single out every industry (let alone every firm) which benefits from a particular forest land use option. However, it is usually possible to distinguish impacts on broad sectors of the economy and on different categories of labour (e.g. skilled versus unskilled).

Finally, the link between costs and benefits and different groups needs to be quantified, if possible, to show the magnitude of the distributional impact. Ideally this will be in monetary necessary to describe their impact on different stakeholders in physical or qualitative terms. Costs and benefits may be expressed in financial terms, using market prices, or in economic terms, with adjustments made to account for market imperfections and/or policy distortions, although it is preferable to use the former when looking at distributional impacts. It may be more difficult to trace the distribution of non-marketed costs and benefits, although the techniques used to value these items can often be extended to distinguish different groups. For example, in the case of watershed protection benefits provided by an upland forest, it may be possible to identify those who stand to lose if these services are disrupted, e.g. land-owners and residents of the downstream flood plain, the regional water or irrigation management authority controlling a downstream reservoir subject to sedimentation, etc.

PRACTICAL DIFFICULTIES OF ECONOMIC VALUATION OF FORESTS

In theory, these valuation methods can be applied to biodiversity in order to calculate its total economic value, i.e. its contribution to human well-being. In practice, however, this exhaustive approach faces many obstacles, which explains why so few studies try to estimate the total economic value of an ecosystem or any other aspect of biodiversity. Thus, although the concept of total economic value seems valid theoretically, in the real world it provides only partial and often subjective information on the benefits that can be expected from use of biodiversity.

Four reasons are generally advanced to explain why this valuation exercise is so difficult:

- The total economic value of an ecosystem can be estimated directly only by using the contingent valuation method and this technique is difficult to apply in contexts having a low level of monetisation.
- Monetary quantification of natural assets is a tricky matter because we have only partial knowledge of how ecosystems work.
- It is very frequently observed that the assumptions used to estimate economic values are deliberately conservative: given the uncertainty of the valuation procedure, the analyst generally opts for a low estimate of the benefits derived from the environment. This choice indicates how much latitude the analyst has in producing the estimate.
- The literature shows that, in practice, estimates of total economic value result not from aggregating all the benefits drawn from the ecosystem concerned, but from aggregating only certain values that the analyst was able to quantify in monetary. The concept of total economic value thus corresponds to the sum of a few economic values selected subjectively by the analyst, rather than to the sum of all the values that actually constitute total value.

These practical difficulties do not, however, negate the legitimacy of economic valuation of tropical forests. Although estimating total economic value is indeed an ideal objective, monetary valuation of certain benefits of biodiversity often provides important information for users and/or managers of these resources.

Caution should be exercised in the use of shadow prices, due to the sometimes arbitrary and imperfect manner in which they are calculated. A recent review of experience at the World Bank, for example, found that shadow prices for conventional (as opposed to environmentally-related) policy failures were rarely used in project appraisals. In many cases the market price is taken as a rough proxy of the real economic value of a good or service, on the grounds that the effects of policy on prices are trivial, or simply due to lack of data or expertise to estimate accurate shadow prices. In practice, analysts often limit themselves to adjusting for the most visible policy distortions, such as those created by government intervention in foreign exchange markets, taxes and direct financial subsidies.

While the above classification would help to estimate the total economic value of natural and environmental resources, it is possible that some of the goods and services may fall in more than one category and hence it is in such cases that it is essential to avoid double counting. The aim of economic analysis or valuation is to ensure optimal use of the forest resources available to society, i.e. to allocate these resources where they will be best used and will maximize social well-being. To this end, cost-benefit analysis is currently the decision support tool most widely employed by policy makers, who use to measure economic efficiency in resource utilization. The central purpose of such analysis is to help decision-makers select the projects and strategies that use resources most efficiently.

24 APPLICATION OF REMOTE SENSING AND GIS IN FORESTRY

INTRODUCTION

The forest is a complex ecosystem consisting mainly of trees that have formed a buffer for the earth to protect life forms. The trees which make up the main area of the forest create a special environment which, in turn, affects the kinds of animals and plants that can exist in the forest. In the tropical and subtropical region, forests are further subdivided into plantations and natural forests. Natural forests are forests composed of indigenous trees, not deliberately planted. Plantations are forests established by planting or/and seeding in the process of afforestation or reforestation. There are about 16 major types of forests in India from the tropical type to the dry type. India has the total forest cover of 21.05 per cent (69.20 million ha) and tree cover of 2.76 per cent (9.08 million ha) respectively (FSI 2011). It has the recorded forest area of 76.95 million ha classified as reserved forest, protected forest and unclassed forest (FSI 2011).

Efficient and sustainable management of India's forests therefore, is critical, given the significance of biomass to rural communities. Materially, forests provide fuel and fodder, food, materials for construction, and agricultural implements, but crucially, forests also provide income-generating opportunities, supplying materials that can be employed in cottage industries such as bamboo, leaves, lac, honey, herbs and fruits etc. for direct sale. Using remote sensing, foresters can get more accurate and cost-effective information and can directly observe as large an area as necessary.

Due to the versatility and scale of remote sensing, it is invaluable in all stages of forest management. The forester's task begins with growing healthy forests. Remote sensing is a useful tool for assessment of environmental conditions, either in an existing forest or prior to planting a new one. Effective management of forest requires detailed local information on the status of forest. However, the traditional method of ground based surveys to obtain such detailed inventory data is currently expensive, time consuming and difficult to put into practice. Remote sensing and Geographic Information System (GIS) analyses have great potential for providing relatively cheap data to assist in the efficient management of forests. The Global Positioning System (GPS) technology is now being used extensively for field inventory. GPS in conjunction with remote sensing and GIS have been used by Forest survey of India, for implementation of suitable sampling design to carry out forest inventory in inaccessible areas.

REMOTE SENSING TECHNOLOGY

WHAT IS REMOTE SENSING?

Remote sensing in the science and art of obtaining information about an object, area or phenomenon through the analysis of data acquired by a device that is not in contact with the object, area, or phenomenon under investigation (Lillesand and Kiefer 2000). It is a technology for sampling electromagnetic radiation to acquire and interpret non-immediate geospatial data from which to extract information about features, objects and classes on the Earth's land surface, oceans and atmosphere (and, where applicable, on the exteriors of other bodies in the solar system, or, in the broadest framework, celestial bodies such as stars and galaxies). Without direct contact, some means of transferring information through space must be utilized. In other words, remote sensing refers to instrument-based techniques used in the acquisition and measurement of spatially organized (distributed) data/information on some property(ies) (spectral; spatial; physical) of an array of target points (pixels) within the sensed scene that correspond to features, objects and materials, doing this by applying one or more recording devices not in physical, intimate contact with the item(s) under surveillance.

The remote sensing techniques involve amassing knowledge pertinent to the sensed scene (target) by utilizing electromagnetic radiation, force fields, or acoustic energy by employing cameras, microwave radiometers and scanners, lasers, radio frequency receivers, radar systems, sonar, thermal devices, seismographs, magnetometers, gravimeters, scintillometers and other sensing instruments. All these advanced instruments gather different types of data that can be interpreted to derive accurate, large-scale information about the Earth's surface and atmosphere. Because these data and images are digital, they can easily be quantified and manipulated using computers. This makes remote sensing a uniquely versatile tool, since the same data can be analyzed in different ways for different applications. Some of the fields that use remote sensing are agriculture, forestry, geology, archaeology, oceanography, architecture, meteorology, etc.

CONCEPT OF REMOTE SENSING

When electromagnetic radiation falls upon a surface, some of its energy is absorbed while some is transmitted through the surface and the rest is reflected. Surfaces also naturally emit radiation in the form of heat. Photographic films or digital sensors in the satellite or aerial vehicle record the reflected and emitted radiation. Since the intensity and wavelengths of this radiation depend on the nature of surface, each surface is described as possessing a characteristic spectral signature. Specific instruments and softwares are used to identify and distinguish between different spectral signatures which will be ultimately useful in mapping the extent of surfaces. Satellite remote sensing is widely used as a tool in many parts of the world for the management of the natural resources and global activities. The remote sensing is divided into two major categories: satellite remote sensing and aerial photography.

The satellite may be geostationary (which permits continuous sensing of a portion of the Earth) or sun-synchronous with polar orbit (which covers entire Earth at the same equator crossing time. The LANDSAT series satellites have a repeat period ranging from 16-18 days, whereas in IRS satellite, it is 22 days. Sensor is a device used for making observations and uses satellite as platform and observes large areas of the Earth surface.

MAJOR COMPONENTS OF REMOTE SENSING TECHNOLOGY

The following are major components of remote sensing system,

1. Energy Source:

   Passive System: Sun, irradiance from earth's materials.

   Active System: Irradiance from artificially generated energy sources such as radar.
2. Platforms: These are vehicles to carry the sensor e.g. truck, aircraft, space shuttle, satellite, etc.
3. Sensors: Device to detect electro-magnetic radiation e.g. camera, scanner, etc.
4. Detectors: Handling signal data e.g. photographic, digital, etc.
5. Processing: Handling signal data e.g. photographic, digital etc.
6. Institutionalization: These are organizations for execution of all stages of remote sensing technology e.g. International and national organizations, research centres, universities, etc.

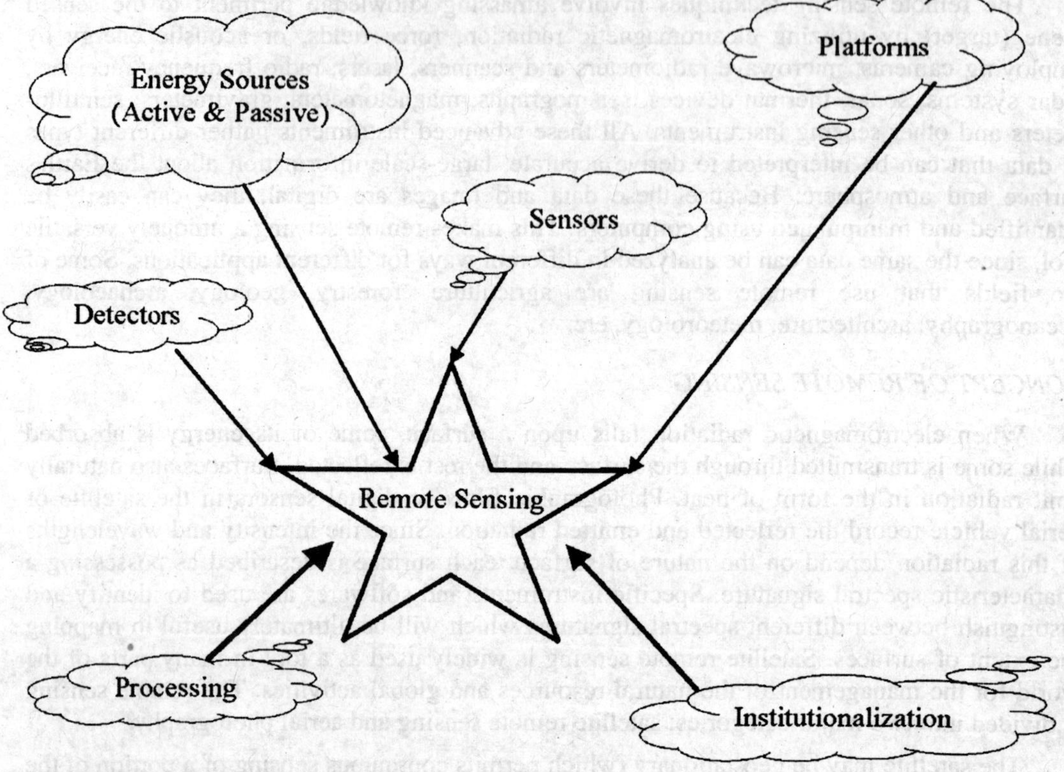

Fig. 24.1: Components of Remote Sensing

CLASSIFICATION OF REMOTE SENSING

With respect to the type of energy resources, remote sensing is classified into two categories: passive and active remote sensing.

Passive Remote Sensing: It makes use of sensors that detect the reflected or emitted electro-magnetic radiation from natural sources.

Active Remote Sensing: It makes use of sensors that detect reflected responses from objects that are irradiated from artificially-generated energy sources, such as radar.

With respect to wavelength regions, remote sensing is classified into three categories.

- Visible and Reflective Infrared Remote Sensing.
- Thermal Infrared Remote Sensing.
- Microwave Remote Sensing.

SATELLITE REMOTE SENSING

The Earth is constantly under observation by many satellites orbiting the planet and collecting data. They are engaged in an activity called "remote sensing", the act of obtaining information about something without being in direct contact with it. The satellite images, as well as the actual predictions, are obtained through remote sensing of the Earth. The satellites don't gather the information themselves; they simply orbit the Earth and provide platforms from which the sensors can observe large areas of the surface. Airplanes also provide platforms for remote sensing and some sensors operate from land. Remote sensing satellites are launched by government agencies of many countries including India, and are usually equipped with sensors that serve a particular purpose.

For the past four decades, Indian Space Research Organisation (ISRO) has launched more than 65 satellites for various scientific and technological applications like mobile communications, Direct-to-Home services, meteorological observations, telemedicine, tele-education, disaster warning, radio networking, search and rescue operations, remote sensing and scientific studies of the space. ISRO has established two major space systems, the Indian National Satellite System (INSAT) series for communication, television broadcasting and meteorological services which is geo-stationary satellites, and Indian Remote Sensing Satellites (IRS) system for resources monitoring and management which is Earth observation satellites. ISRO has launched many experimental satellites which are generally small comparing to INSAT or IRS.

IMPORTANT TERMINOLOGIES

ANALOG AND DIGITAL IMAGES: An image is a two-dimensional representation of objects in a real scene. Remote sensing images are representations of parts of the earth surface as seen from space. The images may be analog or digital. Aerial photographs are examples of analog images while satellite images acquired using electronic sensors are examples of digital images.

MULTILAYER IMAGE: Several types of measurement may be made from the ground area covered by a single pixel. Each type of measurement forms an image which carries some specific information about the area. By "stacking" these images from the same area together, a multilayer image is formed. Each component image is a layer in the multilayer image. Multilayer images can also be formed by combining images obtained from different sensors and other subsidiary data.

MULTISPECTRAL IMAGE: A multispectral image consists of a few image layers, each layer represents an image acquired at a particular wavelength band. For example, if a sensor operating in the multispectral mode detects radiations in three wavelength bands: the green (500-590 nm), red (610-680 nm) and near infrared (790-890 nm) bands, then a single multispectral scene consists of three intensity images in the three wavelength bands. In this case, each pixel of the scene has three intensity values corresponding to the three bands. While a Landsat TM multispectral image consists of seven bands: blue, green, red, near-IR bands, two SWIR bands and a thermal IR band.

SUPERSPECTRAL IMAGE: The more recent satellite sensors are capable of acquiring images at many more wavelength bands e.g. nearly 36 spectral bands, covering the wavelength regions ranging from the visible, near infrared, short-wave infrared to the thermal infrared. The bands have narrower bandwidths, enabling the finer spectral characteristics of the targets to be captured by the sensor.

HYPERSPECTRAL IMAGE: A hyperspectral image consists of about a hundred or more contiguous spectral bands. The characteristic spectrum of the target pixel is acquired in a hyperspectral image. The precise spectral information contained in a hyperspectral image enables better characterization and identification of targets.

SPATIAL RESOLUTION: It refers to the size of the smallest object that can be resolved on the ground. It is a measure of the smallest area identifiable on a digital image as a discrete separate unit (typically pixels) or measure of the smallest angular or linear separation between two objects that can be resolved by the sensor. A "High Resolution" image refers to one with a small resolution size. Fine details can be seen in a high resolution image. On the other hand, a "Low Resolution" image is one with a large resolution size, i.e. only coarse features can be observed in the image.

SPECTRAL RESOLUTION: Spectral resolution describes the specific wavelengths that the sensor can record within the electromagnetic spectrum. Narrow bandwidths in certain regions of the electromagnetic spectrum allow the discrimination of various features more easily.

TEMPORAL RESOLUTION: Temporal resolution is a description of how often a sensor can obtain imagery of a particular area of interest. It is based on the repeat period of a particular satellite. Ideally, the sensor obtains data repetitively to capture unique discriminating characteristics of the phenomena of interest.

RADIOMETRIC RESOLUTION: Radiometric resolution refers to the smallest change in intensity level that can be detected by the sensing system. It is the capability to differentiate the spectral reflectance/ emittance from various targets. This depends on the number of quantization levels within the spectral band. In other words, the number of bits of digital data in the spectral band will decide the sensitivity of sensor.

UTILIZATION OF SATELLITE REMOTE SENSING TECHNIQUES

Satellites have been providing multispectral images of the Earth continuously since the early 1970's. A unique 40 years data record of the Earth's land surface now exists. This unique retrospective portrait of the Earth's surface has been used across disciplines to achieve improved understanding of the Earth's land surface and the impacts of humans on the environment. Satellite data have been utilized in a variety of government, public, private and national security applications. Examples include land and water management, global change research, oil and mineral exploration, agricultural yield forecasting, pollution monitoring, land surface change detection, cartographic mapping, etc.

Satellite based remotely sensed digital data have been used to map forest resources since the inception of the LANDSAT satellite programme in 1972. Satellites provide synoptic images and homogeneous data, which can be geographically registered over time and which can be therefore an efficient tool for providing high quality forest management information. Since the 1980s, with the availability of scanners, improvements in computer software and development of image processing algorithms, there have been many studies of remote sensing applications in forestry.

AERIAL PHOTOGRAPHY

Aerial photography means photography from the air. Aerial photography is one of the most common, versatile and economic forms of remote sensing. It is a means of fixing time within the framework of space. Aerial photography was the first method of remote sensing and even used today in the era of satellite and electronic scanners. Aerial photographs will still remain the most widely used type of remote sensing data. Aerial photographs were taken from balloons and kites as early as in mid 1800s. In 1858, Gasper Felix Tournachon took the first aerial photograph from a captive balloon from an altitude of 1,200 feet over Paris. Aerial photographs can be made with any type of camera. Many successful applications have employed aerial photography made from light aircraft with handheld 35mm cameras. For the study of large areas, high geometric and radiometric accuracy are required and these can only be obtained by using cameras that are purpose built. An object looks smaller on a smaller scale photograph than on a larger scale photo. Larger scale aerial photograph will provide a detailed and high resolution view of a small area. Applications of aerial photography include land-use planning and mapping, geologic mapping, archaeology, species habitat mapping and integration of aerial photography into GIS.

CHARACTERISTICS OF AERIAL PHOTOGRAPHY

Synoptic Viewpoint: Aerial photographs give a bird's eye view of large areas enabling us to see surface features in their spatial context. They enable the detection of small scale features and spatial relationships that would not be found on the ground.

Time Freezing Ability: They are virtually permanent records of the existing conditions on the Earth's surface at one point of time, and used as an historical document.

Capability to Stop Action: They provide a stop action view of dynamic conditions and are useful in studying dynamic phenomena such as flooding, moving wildlife, traffic, oil spills and forest fires.

Three Dimensional Perspective: It provides a stereoscopic view of the Earth's surface and make it possible to take measurements horizontally and vertically - a characteristic that is lacking for the majority of remotely sensed data.

Spectral and Spatial Resolution: Aerial photographs are sensitive to radiation in wavelengths that are outside of the spectral sensitivity of the human eye. They are sensitive to objects outside the spatial resolving power of human eye.

Availability: Aerial photographs are readily available at a range of scales for much of the world

Economy: They are much cheaper than field surveys and are often cheaper and more accurate than maps.

AERIAL PHOTO INTERPRETATION

Aerial photographic interpretation is defined as the act of examining photographic images for the purpose of identifying objects and judging their significance. During the process of interpretation, the aerial photo interpreters usually make use of seven tasks: (1) detection, (2) recognition and identification, (3) analysis, (4) detection, (5) classification (6) idealization and (7) accuracy determination. One of the advantages of all aerial photographs is that when taken as overlapping pairs (called stereopairs) they can provide a 3D view or perspective view of the terrain. The most common instruments used are pocket stereoscope, mirror stereoscope, scanning stereoscope and interpreterscope. An interpreter uses the following basic characteristics of photograph,

Tone (also called Hue or Colour): Tone refers to the relative brightness or colour of elements on a photograph.

Size: The size of objects must be considered in the context of the scale of a photograph. The scale will help to determine the object.

Shape: It refers to the general outline of objects. Regular geometric shapes are usually indicators of human presence and use. Some objects can be identified almost solely on the basis of their shapes.

Texture: The impression of smoothness or roughness of image features is caused by the frequency of change of tone in photographs.

Pattern (Spatial Arrangement): The patterns formed by objects in a photo can be diagnostic e.g. random pattern formed by an unmanaged area of trees and the evenly spaced rows formed by an orchard.

Shadow: Shadows aid interpreters in determining the height of objects in aerial photographs. However, they also obscure objects lying within them.

Site: It refers to topographic or geographic location. This characteristic of photographs is especially important in identifying vegetation types and landforms.

Association: It refers to the occurrence of certain features in relation to others.

PHOTOGRAMMETRY

It is the science and technology of obtaining spatial measurements and other geometrically derived products from aerial photographs. Photogrammetric analysis procedures range from obtaining distances, area, elevations using hardcopy (analog) photographic products, equipments and simple geometric concepts to generating precise digital elevation models, use of digital images and analytical techniques. One of the most widespread uses of photogrammetry is in preparation of topographic maps.

GEOGRAPHIC INFORMATION SYSTEM

A Geographic Information System (GIS) is basically a computer based tool for mapping and analyzing geographic information. GIS is a system of hardware, software and procedures to facilitate the management, manipulation, analysis, modeling, representation and display of geo-referenced data to solve complex problems regarding planning and management of resources. GIS can be thought of as a map on a computer but it is more than just a map. It is possible to combine many maps or "layers" of information and then analyze and manipulate them to create a new map. These new maps are often used for making decisions about land use, resources, transportation, real estate, retailing, etc. Geographic Information System (GIS) is a computer based information system used to digitally represent and analyze the geographic features present on the Earth' surface and the events (non-spatial attributes linked to the geography under study) taking place on it.

This ability to incorporate spatial data, manage, analyze and answer spatial questions is the distinctive characteristic of a geographic information system. Map making and geographic analysis is not new, but GIS makes it possible to do this faster and more efficiently, because of the power and ease of using modern computers. It allows virtually anyone to create a map to explain historic events, plan for the future and predict outcomes. The Geographic Information System is explained in a following way:

GEOGRAPHIC: This is the part of GIS that explains "spatially" where things are such as the location of nations, states, counties, cities, schools, roads, rivers, lakes, forests and the list can go on and on. Spatially means where on the earth's surface an object or feature is located. This can be as simple as the latitude and longitude of a feature. The geographic feature or object can be anything of interest.

INFORMATION: GIS information is the "data" or "attribute" information about specific features that we are interested in. The name of the feature, what the feature is, the location of the feature and any other information that is important. An example could be the name of a city, where it is located, how big it is in square feet (area), its population, its population in the past and any other information that is important.

SYSTEM: The system in GIS is the computer and the software that is written to help people analyze the data, look at the data and combine it in various ways to show relationships or to create geographic models. A GIS can be made up of a variety of software and hardware tools, as long as they are integrated to provide a functional geographic data processing tool.

FORMS OF GIS DATA

GIS data is commonly of two forms: raster and vector.

Raster: A grid of rows and columns of cells. These might represent photographic or scanned images.

Vector: A geometric shape, including points, lines and polygons. These might represent buildings, roads and countries.

COMPONENTS OF GIS

GIS constitutes of five key components:

- Hardware: Computer system on which the GIS software will run.
- Software: Tools needed to store, analyze and display geographic information.
- Data: Tabular data either collected or purchased
- People: Technical specialists
- Method: Well designed models, plans, rules, etc.

GIS is an organized collection of computer knowledge and software with supporting data and personnel, which captures, stores, manipulates analyses and displays all forms of geographically referenced information. It involves the following steps:

- Compile source data
- Georeferencing by registration to a geographic coordinate system
- Digitization and deriving attributes
- Analysis of attributes
- Displaying results

Accurate and current information on land use and land cover is essential for many planning activities. GIS uses computer technology to merge the remote sensing images with other data sets and to produce spatially synthesized technological products which can handle and analyze spatially referenced data. It is reported that GIS has a tremendous potential of storing voluminous amount of spatial (maps) and non-spatial information (tabular data). Various options can be used for categorization of the forest areas as per the compartments (smallest management units) showing the growing stock which is very useful in preparation of the working plan. Nowadays, there are various softwares available for performing the GIS operations e.g. ArcGIS, ILWIS, SAGA GIS, GRASS, Manifold, GeoDa, Quantum GIS (QGIS), DIVA-GIS, MapWindow GIS, uDig, Kalypso, gvSIG, JUMP GIS, MapInfo, etc.

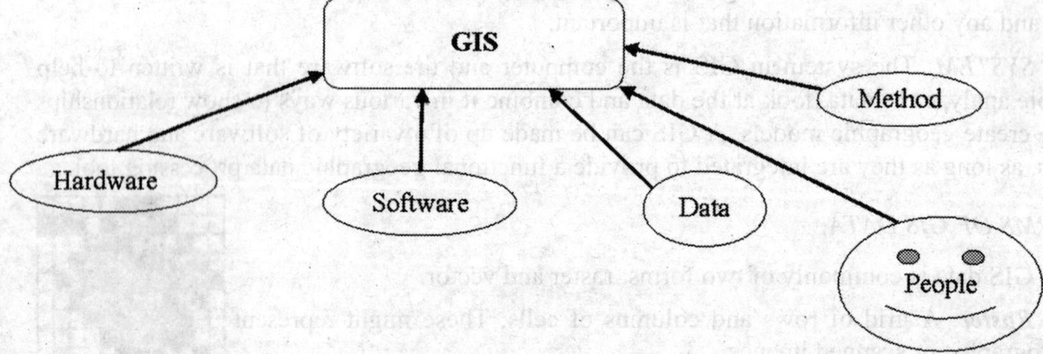

Fig. 24.2: Components of GIS

ROLE OF REMOTE SENSING AND GIS IN FORESTRY

Forest management is more complex as there are now multiple objectives to attain, as well as multiple criteria and constraints to be addressed. This makes remote sensing and GIS an important tool in decision making during policy formulation, planning and management. It can be established to provide necessary information about our forest and wildlife resources and can make planning and management of resources easier. In forestry, growing stock estimation is an important parameter required for forest management and planning purposes. General information about stock available per unit area is the key information desired for forest inventory. To evaluate, monitor and harvest the available natural resources efficiently and judiciously and for the scientific management of the forest for future planning forest inventory data and various information at national, regional and local levels are required. This gives insight for the availability of wood biomass, volume and forest resource as a whole.

The remote sensing data products include the aerial photographs on different scales; and the digital data in the form of computer compatible tapes. The satellite data is available either in individual bands or in the form of FCC (false colour composites). The satellite multispectral data characteristics and their possible application in forest / vegetation mapping have been highlighted by many scientists in the recent years. It has been found that the sensors capture the data, encompassing all degrees of forest dynamism. However it is essential to select the right kind of satellite data for specialized forestry studies. While selecting the data, therefore, type of satellite, sensors, resolutions, the phonological state of vegetation, phytogeographical region, land use practices, sun elevation & azimuth, cloud and snow coverage, etc. have to be considered. The technique is becoming increasingly important for mapping land use and land cover, mainly for the following reasons:

- Large areas can be imaged quickly and repetitively.
- Remote sensing agencies eliminate the problems of surface access.
- Image interpretation is faster and less expensive than conducting surveys.
- Good tool to monitor the changes in the forest areas.

The role of remote sensing and GIS in forest management can be broadly classified into four major groups.

(1) Forest Resource Monitoring

(2) Forest Resource Estimation and Harvesting

(3) Forest Resource Protection

(4) Wild Animals Management and Others

FOREST RESOURCE MONITORING

RECONNAISSANCE MAPPING AND FOREST COVER MONITORING

Objectives to be met by national forest/environment agencies include forest cover updating, depletion monitoring, measuring biophysical properties of forest stands, forest cover type discrimination, agroforestry mapping. With improvement in scale of interpretation and resolution of imagery, Forest Survey of India has made significant progress in forest cover mapping and assessment. In addition to forest cover mapping, these technologies have been extensively used of late to prepare classified tree outside forest (TOF) maps depicting block, linear and scattered patches of trees groups up to 0.1 ha.The forest cover at the national level is being biennially monitored using remote sensing data. This information is more relevant in the context of assessing rate of degradation grossly at the state/district level in terms of closed and open forests.

SPECIES IDENTIFICATION, FOREST TYPING AND BIODIVERSITY ASSESSMENT

Remote sensing and GIS provide a means of quickly identifying and delineating various forest types, a task that would be difficult and time consuming using traditional ground surveys. Data is available at various scales and resolutions to satisfy local or regional demands. Large scale species identification can be performed with multispectral, hyperspectral, or air photo data, while small scale cover type delineation can be performed by radar or multispectral data interpretation. Both imagery and the extracted information can be incorporated into a GIS to further analyze or present with ancillary data, such as slopes, ownership, boundaries, or roads. The extended application of forest inventory and enumeration including shrub layer and herb layer would enhance forest management system for preservation of ecologically rich zones and to account diversity of the areas. The initial probabilistic diversity zones should be narrowed down using satellite data and the consistency of patchiness as a function of landscape dynamics could be used as an element for biodiversity studies.

FOREST VEGETATION MAPPING, STOCK MAPPING AND PREPARATION OF WORKING PLAN INPUTS

In forestry, growing stock estimation is an important parameter required for forest management and planning purposes. Earlier, such information was collected by ground methods. At present, aerial photographs and satellite data is being increasingly used in forest inventory works. It is extremely useful in the application of sampling techniques as it helps in the correct delineation of strata depending on species, density, volume etc. and distribution of sampling units in them. The stratification of forests gives the estimate of volume at a very low cost and in the shortest possible time. In fact, the important work in the preparation of forest inventory is the estimation of growing stock. This is possible only when the forest is classified by species/type, so that the volume can be estimated precisely. The satellite data analysis is found adequate for differentiating three levels of stocking i.e. closed (40 per cent and above), open (10-40 per cent) and degraded (below 10 per cent) crown cover classes on 1:250,000 scale.

FOREST RESOURCE ESTIMATION AND HARVESTING

FOREST INVENTORY AND SAMPLING

The satellite data facilitate in generation of primary stratification units to stratify the area into homogenous forest strata. Based on the apriority knowledge, the minimum sample size can be selected and the field inventory can be accomplished by suitable proportional allocation of samples in all the strata. The stratified sampling procedures provide a reliable quantification of forest resources depending on the objective set during the inventory and to build volume from the enumerated data on a time and cost effective manner.

FOREST VOLUME ESTIMATION, GENERATION OF VOLUME TABLES AND YIELD TABLE AND HARVESTING

The satellite based multi-phase approach forest inventories with defined objectives provide ample data for further processing and computation of volume and yield tables. Spatial forest modelling using remote sensing and GIS can substantially enhance the planning, harvesting strategies, location of forest stand, timber inventory data, ecologically sensitive area, key attributes of the terrain and other important factors could be mapped and include for spatial analysis necessary in harvesting plan preparation. Spatial modelling tasks could help the forest manager see the economic, environmental and social impacts of the proposed harvest.

FOREST RESOURCE PROTECTION

DEFORESTATION

Remote sensing and GIS bring together a multitude of tools to better analyze the scope and scale of the deforestation problem. Multi temporal data provides for change detection analysis. Images of earlier years are compared to recent scenes, to tangibly measure the differences in the sizes and extents of the clear cuts or loss of forest. Data from a variety of sources are used to provide complementary information.With the rapid destruction of forests and encroachment, the use of multi-temporal satellite data using digit analysis procedure provide spatial change maps by image differencing methods and logical operations.

FOREST FIRE MANAGEMENT

Remote sensing and GIS can be used to detect and monitor forest fires, pest and diseases and human activities in the forests. In India, remote sensing and GIS is largely used for forest fire monitoring and the regrowth following a fire. As a surveillance tool, routine sensing facilitates observing remote and inaccessible areas, alerting monitoring agencies to the presence and extent of a fire. In India, the temporal data from IRS series have been effectively used to map and monitor incidence of fire in some of the wild life sanctuaries and national parks of the country. Thus remote sensing can help in three important ways in forest fire mapping, monitoring and management.

(1) Active fire mapping

(2) Burnt area mapping

(3) Fire prone area mapping i.e., Fire risk zonation

INVASIVE SPECIES MANAGEMENT

Biological invasions form a major threat to the sustainable provision of ecosystem products and services, both in man-made and natural ecosystems. Increasingly, efforts are made to avoid invasions or eradicate or control established invaders. It has long been recognized that remote sensing and geographical information system (GIS) could contribute to this e.g. through mapping actual invader distribution or areas at risk of invasion. Potentially GIS could also be used as a synthesising tool for management of interventions aiming at invasive species control. Remote sensing technology has many attributes that would be beneficial for detecting, mapping and monitoring invaders. Spatial heterogeneity complicates the study of seasonal and long-term trends of biological invasion.

ENVIRONMENTAL MONITORING

Conservation authorities are concerned with monitoring the quantity, health, diversity of the Earth's forests, deforestation (rainforest, mangrove colonies), species inventory, watershed protection (riparian strips), coastal protection (mangrove forests), forest health and vigour. Remote sensing and GIS are powerful tools which help greatly in environmental monitoring.

CONSERVATION OF WETLANDS, SOIL AND WATERSHEDS

Watershed spatial data base are prepared by using remote sensing and GIS application which can serve as the prime water resources database inside the forest and also used to do various soil and watershed protection measures.

WILD ANIMALS MANAGEMENT AND OTHERS

Remote sensing and GIS helps wildlife management professionals to examine and envision

- Habitat requirements and ranges
- Population patches and linkages
- Disease levels within populations
- Progress of management activities
- Historical and present wildlife densities

Researchers and managers undertaking wildlife habitat assessments commonly require spatially explicit environmental map layers such as those derived from forest inventory and remote sensing. Remote sensing and GIS are often used in habitat and vegetation mapping, monitoring, assessment and analysis of the progression of the conservation activities, ecological patterns in protected areas, monitoring of wildlife and marine populations and building management systems by comparing images from different data. Remote sensing not only provides spatial data but also allows to compare temporal variations in the spatial data, essential for wildlife management. Wildlife managers have been using topographic maps to generate management and other maps of their interest. Although technically complex, the remote sensing techniques have revolutionized the process of data gathering and map making. Remote sensing can be applied to wildlife habitat inventory, evaluation and wildlife census. Wildlife habitat mapping is similar to any type of land cover mapping. Both biotic and abiotic surface features including vegetation composition and/or density and landforms can be mapped and interspersion of habitat components, the extent of habitat types and the distance to other critical habitat components can be measured. The repetitive NOAA (National Oceanic and Atmospheric Administration), IKONOS, SPOT and IRS series of satellites add a temporal dimension to habitat mapping and change detection.

HUMAN ANIMAL CONFLICT ISSUES

Wildlife resources are very important components of a sustainable forest. All wildlife need food, water, forest cover and forest corridors for their survival. These basic biological needs are the most important components of the wildlife habitat in the forest. Carnivore is one of the groups, which need much wider forested areas and corridors for their survival and to continue to exist in a region. There is a need to see how human and wildlife can co-exist by identifying and assessing the condition of wildlife habitats both within and outside protected areas. Man-animal conflict problems have been recorded across India. Digital elevation model was generated for recording the information from low, medium and high conflict zones. Coded grids of 2 x 2 km having information about level of conflicts are used for calculating the geographical area and percent area of forest, non-forest and other habitat characteristics which will be useful to classify the area into low, medium and high conflict zone.

25 FOREST BASED INDUSTRIES

INTRODUCTION

With only two per cent of the world's total forest area and 15 per cent of its human and cattle population, India encounters a critical disequilibrium in its natural resource pool. Half of the country's legal forest is deplorably degraded and deforestation occurs at alarming rate of 1.5 million ha per year till the recent past. These have ushered in not only a total mismatch between supply and demand of both domestic and industrial wood requirements but also leads to degradation of the land surface. The forests have very low growing stock at 74 m^3 per ha compared to the world average of 110 m^3 per ha. Similarly, the mean annual increment is very low at less than 1 m^3 per ha per year compared to the world average of 2.1 m^3 per ha per year. The shortfall in forest area coupled with poor productivity resulted in dwindling supply of raw material requirements of various wood based industries. Most of the wood based industries like pulp, paper, match and veneers are largely dependent on forest department supply but due to government policies and the promulgation of Forest Conservation Act, 1980 restricted the supply of raw material to the wood based industries which necessitated massive import of sawn timber, pulp and even newsprints.

Substantial improvement in productivity of forest resources on sustainable basis and large scale expansion of industrial linked agroforestry plantations are critically important for meeting the industrial raw material requirements besides achieving the national goal of 33 per cent forest cover. Diversification of agriculture should receive high priority as water resources are depleting. However, tangible results can be achieved only if the farmers are offered practical, viable and economically attractive alternative land use options. Technology based farm forestry plantation with genetically improved, high yielding and fast growing clonal planting stock has tremendous potential for diversification and meeting growing shortages of industrial wood on sustainable basis. However, meticulous planning and integrated development of farm forestry and wood based industries are required to ensure regular demand and remunerative prices to the growers.

MAJOR FOREST BASED INDUSTRIES

Any industry which depends on forests for their raw material requirement is considered to be forest or wood based industry. India is one of the leading countries with mushrooming of wood based industries which include pulp and paper, match, saw wood, veneer and plywood, pencil and dendro biomass industries. The forest based industry is growing rapidly with the increasing demand for furniture, housing, construction material, packaging, agriculture good, sports goods, plywood, veneer, matches etc. Similarly the biomass based power generation industries are also on the raise across the country to generate electricity from forest biomass. This growing demand of wood and wood based industries will create a wood deficit of 20-70 million cubic meter by 2020. It is estimated that approximately 40 per cent of the forest products are supplied from outside forest areas and more than 95 per cent of fuel wood and major timber requirement are obtained from outside forest areas. The following are the major forest based industries which depend heavily on forest and agroforestry plantation to meet the raw material requirement.

PULP AND PAPER INDUSTRIES

The pulp and paper industry is one of the key industries in India and it is highly fragmented. Today, there are about 700 paper mills in India with 33 in the large scale sector. During 1990s, the *per capita* consumption of paper was 3.3 kg which has now escalated to 8 kg, but still lower compared to the global average of 47.7 kg. The current production of raw material for pulp and paper production is 2.76 million tons as against the demand of 5.04 million tonnes. The shortfall is as high as 45 per cent. The total installed capacity of pulp & paper mills in the country is estimated to be over 7.5 million tons which is likely to increase to 14 million tons by the year 2020. In India, paper is manufactured from wide range of raw material like wood, recovered paper, baggasse and other agro residues. Based on that the paper mills have been classified and furnished in Fig. 25.1.

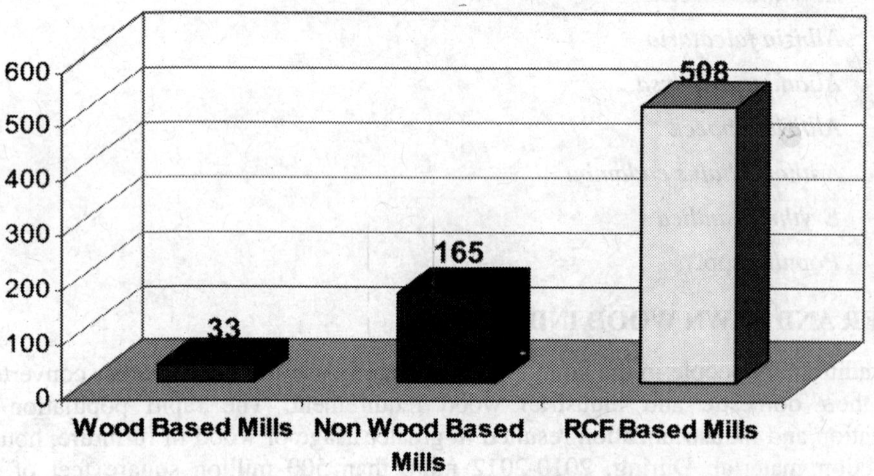

Fig. 25.1: Categorization of the Paper Mills Based on Cellulosic Raw Materials

Currently the paper industries in the country face serious problem in terms of raw material availability which became a major deterrent to competitiveness and growth of industry. Hence all the industries have started massive plantation programme with varying degree of success.

MAJOR PULPWOOD SPECIES

- *Bamboo*
- *Casuarina* spp.
- *Eucalyptus* spp.
- *Leucaena* spp.
- *Acacia* spp.
- *Populus* spp.

MATCH INDUSTRIES

Match wood industry is one of the oldest wood based industries in India. About 75 per cent of the total match wood industries in the country are located in the state of Tamil Nadu which comprises nearly 6,000 match industries with mechanized, semi mechanized and as cottage industry. The *per capita* consumption of matches in India increased steadily from 2.45 kg (1970) to 4.25 kg (1987). The current *per capita* consumption rose to 6.0 kg which is more staggering. The increasing demand for the matches coupled with declining wood resources is a major bottle neck faced by the entire match industries in India including the ancillary splint and veneer industries. In fact, there was a short fall of 9,00,000 m^3 in the year 2000. The veneer quality wood for match boxes, which accounts for 44 per cent of match wood used, is also in short supply.

MAJOR MATCHWOOD SPECIES

- *Ailanthus excelsa*
- *Albizia falcataria*
- *Alianthus triphysa*
- *Albizia lebbeck*
- *Anthocephalus cadamba*
- *Erythrina indica*
- *Populus* spp.

TIMBER AND SAWN WOOD INDUSTRIES

Traditionally people in the country predominantly use timber and other converted wood in all their domestic and industrial wood requirement. The rapid population growth, urbanization and industrialization resulted in greater usage of wood in furniture, housing and construction material. During, 2010-2012 more than 500 million square feet of space is estimated to be built in urban areas of the country and the wood products were valued around US Dollar 3 billion. With greater usage wood as a predominant material for housing and construction material in urban and semi urban areas there is going to be a great demand for timber and other sawn wood requirement. The Indian furniture market is estimated at 8 billion US Dollar and in most cases raw materials are imported from various countries (Durairasu et al 2011).

MAJOR TIMBER SPECIES

- *Tectona grandis*
- *Terminalia* spp.
- *Albizia* spp.
- *Gmelina arborea*
- *Azadirachta indica*
- *Pterocarpus* spp.
- *Mangifera indica*
- *Artocarpus* spp.
- *Dalbergia* spp.
- *Dipterocarpus* spp.
- *Eugenia* spp.

PLYWOOD INDUSTRIES

One of the fastest growing in India is the plywood industry. The industrialization and urbanization and the increased interest on interior decorations have made great usage of plywood in the country. Wide range of species have been found amenable for making face, core and inner veneers resulted in establishment of more than 2,000 small scale industries involved in plywood manufacture. The liberalization and privatization policy of government of India also significantly contributed towards establishment of new rural industries. These industries also depend heavily on various species which thereby attracted large scale promotion of plywood based industrial wood plantations.

MAJOR PLYWOOD SPECIES

- *Populus* spp.
- *Melia dubia*
- *Paulownia* spp.
- *Eucalyptus* spp.
- *Ailanthus* spp.

PARTICLE BOARD INDUSTRIES

Particle board is reconstituted constructional panel particularly developed as a substitute for natural constructional wood and is made from low grade waste woods or from ligneous agricultural residues. These particle boards are predominantly used for wall paneling and interior decorations in domestic and industrial wood sector. In India, the first particle board industry was set up in late 1950s at Sitapur in Uttar Pradesh and from then onwards large number of industries have been installed across the country.

MAJOR RAW MATERIALS

- All types of wood waste
- All types of pine needles
- All types of *Casuarina* needles
- *Ligneous* agriculture residues

FIBRE BOARD INDUSTRIES

Fibre board is constituted using sheet materials of widely varying diversities manufactured from refined or partially refined wood fibers or other vegetable fibers.

DENDRO BIOMASS POWER GENERATION INDUSTRIES

Biomass is an important fuel source in overall energy scenario. Biomass is produced through chemical storage of solar energy in plants and other organic matter as a result of photosynthesis. This biomass include plantation that produces energy crops, natural vegetable growth and other organic waste and residues. Among all these biomass, the role of dendro biomass is very significant due to their higher calorific value and increased fuel efficiency (Parthiban et al 2011). Hence, large number of dendro biomass based power plants have been established across the country to generate electricity.

MAJOR ENERGY CROP

- *Prosopis* spp.
- *Acacia* spp.
- *Albizia* spp.
- *Dalbergia sissoo*
- *Leucaena leucocephala*
- *Casuarina equisetifolia*
- *Eucalyptus* spp.
- *Gliricidia* spp.
- *Ceasalpinea* spp.
- *Chuckrassia* spp.
- Other hard woods

OIL AND BIODIESEL INDUSTRIES

The demand for edible and non edible oil is continuously on the rise due to industrial and economic development. But there is no concomitant effort to augment the production potential of oil seeds in the country. This resulted in a massive import of nearly 46 per cent of edible oil till the recent past thus claiming huge amount of country's exchequer. Similarly the crude oil requirement in the country is also increasing due to development in transport and industrial sector. Till the recent past the consumption of crude oil was about 184.68 million tones and the rate energy consumption is increasing at the rate of 6.5 per cent per annum. India's share of crude oil production is about one per cent of total world crude oil production while in consumption its share is about 3.1 per cent of total world consumption which necessitated massive imports of crude oil. The import of crude oil has increased from 63 per cent in 1971-1980 to nearly about 80 per cent in 2007-2008 which is an alarming issue for the country and warrants development of alternate renewable resources (Paramathma et al 2004). Under such circumstances, efforts have been taken by various departments of Government of India to promote non edible oil seeds in the country to augment the vegetable oil feed stock to generate biofuel. Simultaneously large number of private sector oil and biodiesel production industries have been established across the country but for want of sustainable raw material resource availability these industries are under great threat. This facilitated promotion of tree borne oil seeds across the country and their inclusion under farm and agroforestry system.

Table 25.1: Major Tree Borne Oil Seeds

Species	Oil Content (%)
Jatropha curcas	30-40
Pongamia pinnata	27-39
Madhuca latifolia	35-40
Callophyllum innophyllum	50-70
Shorea robusta	13-25
Hevea braziliensis	20-25
Aleurotis fordii	50-60
Azadirachta indica	35-40

VALUE ADDITION INDUSTRIES

The wood based industries have to store the harvested raw materials during rainy season in order to have sustainable raw material availability and to sustain the industrial process during lean season. The post harvest management of huge volume of industrial wood necessitates proper handling, storage and utilization which demand a scientific intervention in order to reduce post harvest losses due to biological agents particularly powder post beetles and pin hole borers. These biological agents are taking heavy toll of stored industrial raw materials which need to be addressed. Hence large number wood seasoning and preservative industries have been established to avoid post harvest losses. Similarly, the plantation and industrial processing activities accounts for 20-30 per cent of wood residues which are either unutilized or underutilized for want of suitable recycling technologies (Umesh kanna et al 2011). These plantation and industrial wood residues have been successfully value added into briquettes and as on today many industries have been established across the country and successful value addition using plantation residues have been evidenced. These value added briquettes acted as excellent feed stock for biomass power generation industry, boiler industries and other industries requiring biomass for meeting the energy demands. The entire value addition process of plantation residues is depicted in Fig.25.2.

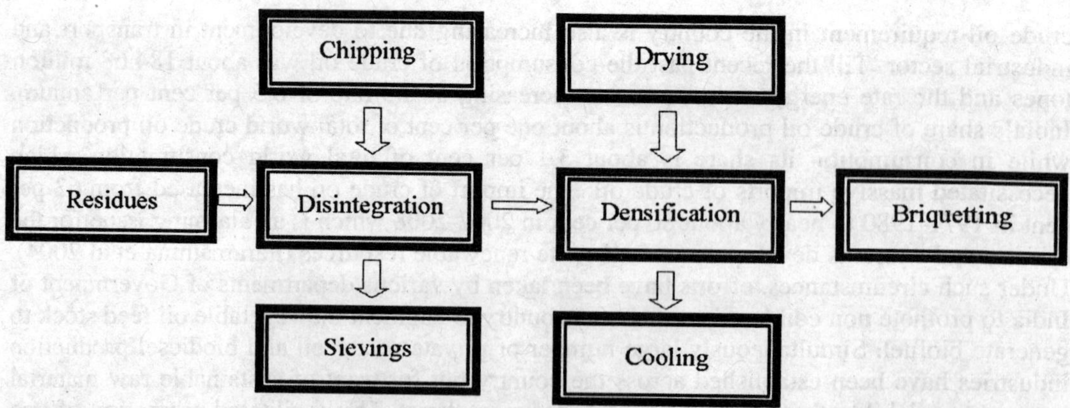

Fig. 25.2: Value Addition of Plantation Residues through Briquetting

PROJECTED WOOD DEMAND FOR VARIOUS INDUSTRIES

The wood requirement for the current situation and the projected demand for 2020 are furnished for various wood based industries (Parvez Ahmed 2005). It is estimated that for sawn wood based industries 52 million m^3 and 62 million m^3 of wood is required for the year 2015 and 2020 respectively (Table 25.2). This demand cannot be met from the natural forests and thus extend a wide scope for industrial wood plantations.

Table 25.2: Projected Demand of RWE for Sawn Wood Based Industries (million m^3)

Year	Housing	Packaging	Furniture	Automobile	Handicraft	Catamaran	Other Misc. Inds.	Total
2000	15.9	4.62	2.52	0.19	0.45	0.03	5.70	29.41
2005	19.4	5.54	3.36	0.28	0.54	0.05	6.70	35.87
2010	22.1	6.40	4.62	0.41	0.65	0.07	9.40	43.65
2015	26.3	7.55	5.90	0.60	0.78	0.11	11.20	52.44
2020	28.5	9.00	7.53	0.87	0.95	0.16	15.15	62.16

Similarly pulp and paper industries require huge volume of wood and it is estimated that 34.67 and 45.86 million m^3 of wood is required during 2015 and 2020 in order to meet paper and paper boards, news print and rayon grade pulp wood requirement (Table 25.3).

Table 25.3: Projected Demand of RWE for Pulpwood Based Industries (million m^3)

Year	Paper & Paper Board	News-Print	Rayon Grade Pulp	Total
2000	04.48	1.78	2.50	08.76
2005	08.96	2.56	2.80	14.32
2010	15.40	3.42	3.10	21.92
2015	26.64	4.63	3.40	34.67
2020	35.84	6.22	3.80	45.86

The other industries viz., agriculture implements, railway sleepers, sports goods, mining props and matchwood also require increased wood demand. A wood demand of 12.09 and 14.25 million m^3 is projected during the year 2015 and 2020 respectively (Table 25.4).

Table 25.4: Projected Demand of RWE for Sleeper, Round Wood and Mine-Prop etc. (million m^3)

Year	Agricultural Implements	Railway Sleepers	Sports Goods	Mining Props	Matchwood	Total
2000	2.12	0.03	0.35	3.20	2.30	08.00
2005	2.33	0.02	0.58	3.50	2.60	09.03
2010	2.50	0.02	0.98	4.00	3.00	10.50
2015	2.50	0.02	1.67	4.50	3.40	12.09
2020	2.50	0.02	2.73	5.00	4.00	14.25

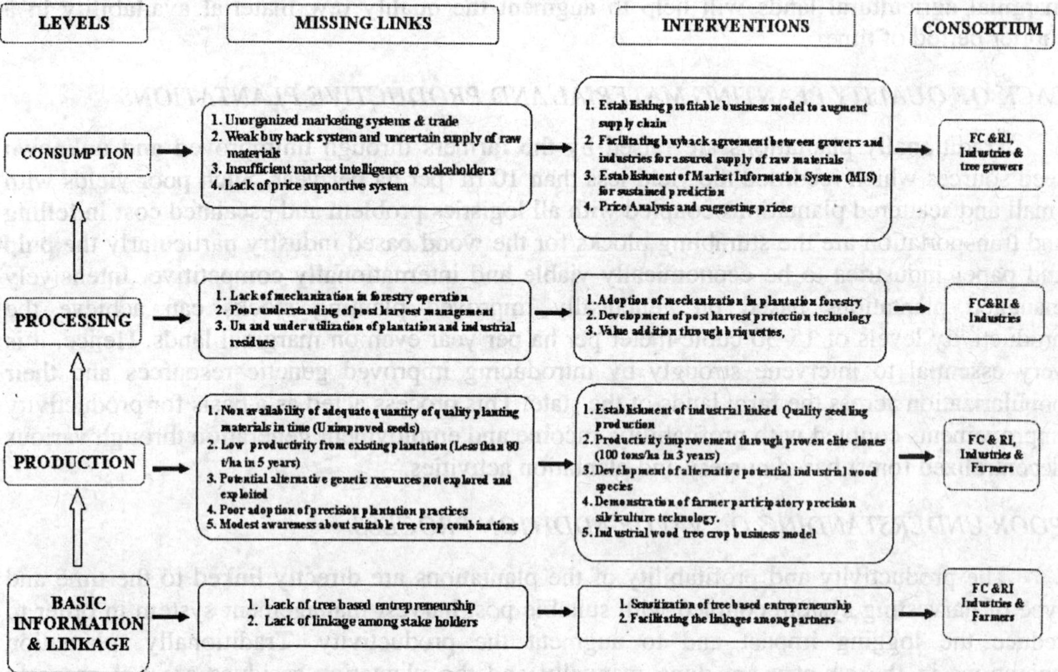

Fig. 25.3: A Value Chain Model for Industrial Forestry

The industries indicated are only a few but there are large number of primary and secondary wood based industries which demand enormous quantity of raw material and this cannot be met from the existing forest. This resulted in massive imports of raw material across the world to meet the raw material demands of various industries. Under such circumstances, most of the wood based industries in the country have started promoting industry specific plantation programmes and these activities are gaining momentum through contract farming system involving various stake holders.

CONSTRAINTS AND THE POSSIBLE INTERVENTIONS IN WOOD BASED INDUSTRIES

The industrial wood raw material requirement is traditionally met from the plantations of forest department till the recent past by the wood based industries. But due to expansion of wood based industries, diminishing supply from the unimproved plantation and uncertain availability are the major bottle neck in the current supply chain. The major constraints faced by the wood based industries and the possible interventions are depicted in Fig.25.3.

CHALLENGES OF FOREST BASED INDUSTRIES

NON AVAILABILITY OF ADEQUATE AND QUALITY RAW MATERIAL

The biggest challenge faced by the wood based industries is the spectare of raw material unavailability. The available raw materials are not only certain but also with varying degree of quality. Hence, promotion of farmer's linked clone based agroforestry activities in marginal agricultural lands will help to augment the quality raw material availability in a shorter period of time.

LACK OF QUALITY PLANTING MATERIAL AND PRODUCTIVE PLANTATIONS

Traditionally plantations are raised by the farmers through unimproved and unknown seed sources which recorded the yield less than 10 m^3 per ha per year. Such poor yields with small and scattered plantations coupled with all logistics problem and escalated cost in felling and transportation are the stumbling blocks for the wood based industry particularly the pulp and paper industries to be economically viable and internationally competitive. Intensively managed plantations based on genetically improved planting stocks can achieve the productivity levels of 15-30 cubic meter per ha per year even on marginal lands. Hence, it is very essential to intervene strongly by introducing improved genetic resources and their popularization across the farm lands of the state. This process acted as a basis for productivity improvements coupled with profitability, income and employment generation through various decentralized forest based nursery and plantation activities.

POOR UNDERSTANDING ON VALUE ADDITION PROCESS

The productivity and profitability of the plantations are directly linked to the time and type of harvesting system coupled with suitable post harvest management system in order to reduce the logging impact and to augment the productivity. Traditionally, plantation operations in the country are done manually and the plantation residues are not properly utilized. Rather the plantation residues are left in the field without any utilization. The intervention through introduction of mechanization in harvesting operation and established value addition process of plantation residues through briquetting technology proved successful in various plantations species particularly in Casuarina and Eucalyptus. The plantation and industrial wood processing activities accounts for 20-30 per cent of residues which are either unutilized or underutilized for want of suitable recycling technology and the associated supply chain system. The conversion of these resides into briquetting technology significantly increased the income of the growers and value addition industry.

UNORGANIZED SUPPLY CHAIN AND TRADE

The success of Industrial wood plantation schemes and the related plantation established is widely questioned. The reasons for the failures are numerous; but the key reasons are non involvement of local people, lack of assured buy back and minimal support price. Above all, the supply chain was multipartite as indicated in Fig. 25.4. To augment the supply chain system and to transform them as value chain system, the research group has designed a Quad-Partite Model Contract farming system through efficiently linking the growers, industries, research institutes and financial institutions. Currently the price information pattern of industrial wood species is not known to growers due to lack of price information system and its dissemination. This current research group resolved this issue by establishing a separate website to periodically disseminate the pricing pattern of industrial wood species through regular updating system.

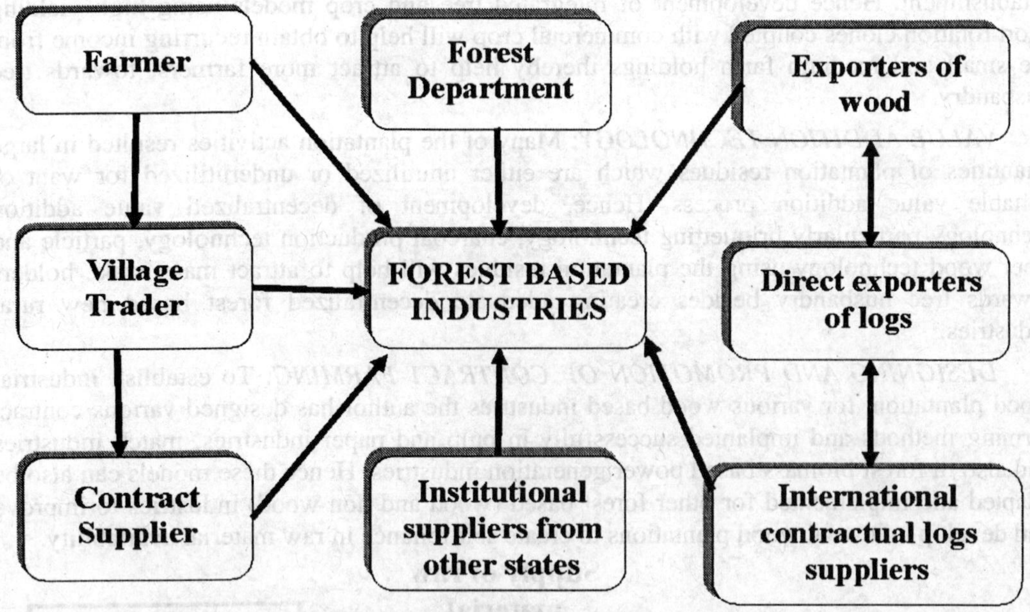

Fig. 25.4: Status of Industrial Wood Supply Chain

MEASURES FOR BETTER DEVELOPMENT OF FOREST BASED INDUSTRIES

The wood based industries have faced serious challenges in terms of raw material availability, unorganized supply chain and lack of synergy among various stake holders. The legal and policy changes also created major threat to the successful running of wood based industries. Under such circumstances, the following measures will help to augment the production potential of all wood based industries besides creating sustainability in their production process.

TECHNOLOGY DEVELOPMENT AND ADOPTION

DEVELOPMENT OF HIGH YIELDING VARIETIES AND CLONAL TECHNOLOGY: The success of wood based industrial operation depends directly on the availability of quality raw material in a short period of time. Hence continuous research and development support is essential to develop high yielding varieties (HYV) amenable for short rotation with multifarious utility.

PRECISION SILVICULTURE TECHNOLOGY: The knowledge on proper implementation of scientific and precision package of practices is most essential for exploiting the advantage of genetic superiority of developed variety. The technology development on optimization of nutritional and water requirement during different growth periods need to be developed for establishment of successful industrial wood plantations. Integrated development of pest and disease management for large scale industrial wood plantation needs to be developed for increased productivity.

COGENERATION OF WOOD AND FOOD: Cogeneration of industrial wood with other annuals will help to attract the small and medium farmers towards industrial wood plantation establishment. Hence development of integrated tree and crop models using high yielding short rotation clones coupled with commercial crop will help to obtain recurring income from the small and medium farm holdings thereby help to attract more farmers, towards tree husbandry.

VALUE ADDITION TECHNOLOGY: Many of the plantation activities resulted in large quantities of plantation residues which are either unutilized or underutilized for want of suitable value addition process. Hence, development of decentralized value addition technology particularly briquetting technology, charcoal production technology, particle and fiber wood technology using the plantation residues will help to attract many stake holders towards tree husbandry besides creating adequate decentralized forest based new rural industries.

DESIGNING AND PROMOTION OF CONTRACT FARMING: To establish industrial wood plantation, for various wood based industries the author has designed various contract farming methods and implanted successfully in pulp and paper industries, match industries and also in forest biomass based power generation industries. Hence these models can also be adapted and implemented for other forest based (wood and non wood) industries to improve and develop industrial wood plantations to create self reliance in raw material availability.

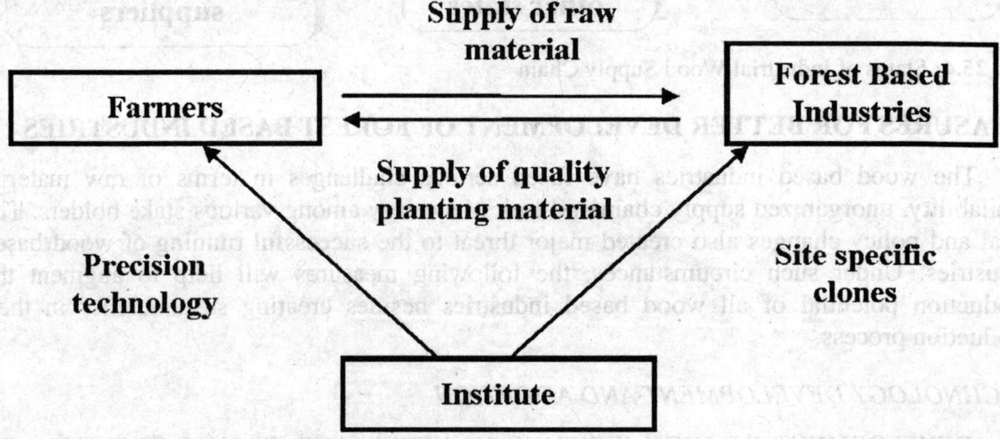

Fig. 25.5: Tri-partite Model contract tree farming

TRI-PARTITE MODEL

This model incorporates industry, growers and research institute. Under this system, the industry supplies quality planting material at subsidized rate and assures minimum support price or the prevailing market price whichever is higher. The farmer supplies the material to the contracting industry and the research institutes will advice on the site specific technology (Fig. 25.5).

QUAD-PARTITE MODEL

This system is similar to tri-partite model barring the involvement of financial institution (Fig. 25.6). The financial institutions viz., Indian Bank, State Bank of India and Syndicate Bank provide credit facilities to the growers at the rate of INR 15,000 to 20,000 per acre in three installments. For credit facilities, a simple interest rate at 8. 5 per cent is followed and the repayment starts after felling. The contract farming system extends no collateral security for loan amount up to INR 1,00,000 per farmer. The research institute particularly Forest College and Research Institute (TNAU) play a significant role for technological advancements through varietal development and also to advice site specific precision silvicultural technology to the growers. A pre and post-plantation scientific advice helps to develop human resources through on and off institute mode to farmers and plantation staff of the industries.

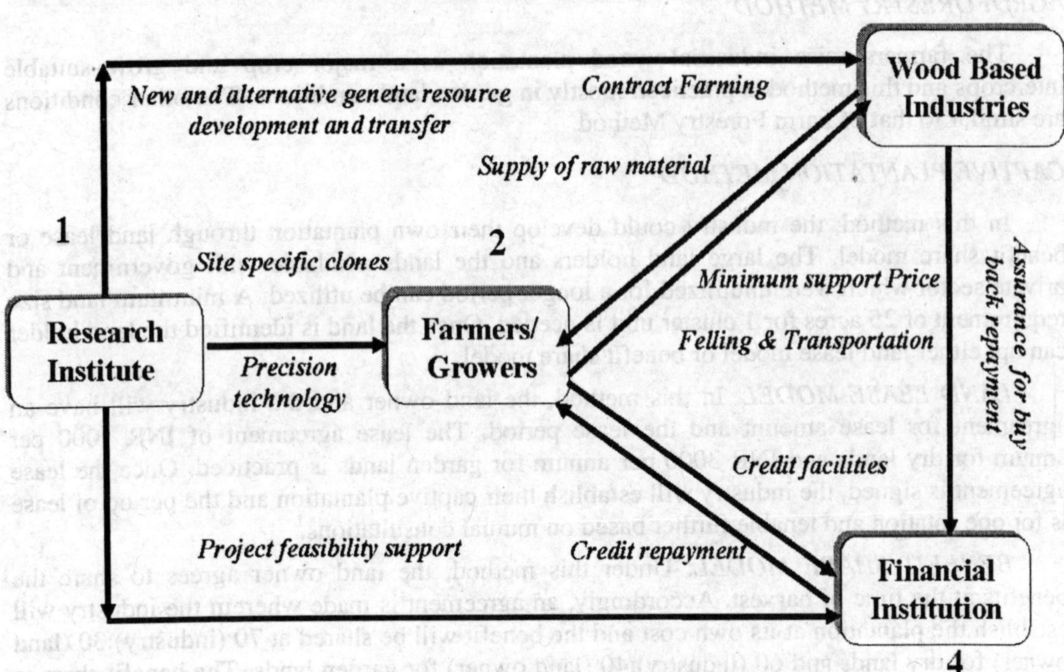

Fig. 25.6: Quad-Partite Model Contract Tree Farming

The industry mass multiply the potential genetic materials identified by the research institute in a decentralized manner and supply them at subsidized costs. The industry also facilitates felling and transport at their own costs, which resulted in strong linkage between industry and the farmers. The industry also help to repay the loan amount after felling of farm grown raw materials there by help the financial institutions for timely repayment, which resulted in strong institutional mechanism for sustainability of the contract tree farming system in the state.

CONTRACT FARMING METHODS

The following contract farming methods designed by Forest College and Research Institute of Tamil Nadu Agricultural University are successfully adopted and implemented by paper, match and biomass based power generation industries.

FARM FORESTRY METHOD

In this method, the industry supply the quality planting materials of site specific variety identified by the research institute to the interested farmers on a subsidized rate. The farmers in turn, develop his own plantation and obtain the needed technological support from the research institute. If needed, the farmers get credit facility from the financial institutions and final felling and transportation by the concerned industry. An agreement is made before the plantation establishment. In this method, the farmers grow only trees without any intercrop and mostly practiced in dry land condition.

AGROFORESTRY METHOD

The farmers raise industrial wood plantation as a major crop and grow suitable intercrops and this method is practiced mostly in garden land conditions. The other conditions are similar to that of Farm Forestry Method.

CAPTIVE PLANTATION METHOD

In this method, the industry could develop their own plantation through land lease or benefit share model. The large land holders and the lands available with government and private sector which were unutilized for a longer period can be utilized. A minimum land size requirement of 25 acres for 1 cluster unit is needed. Once the land is identified the land holder can opt either land lease model or benefit share model.

LAND LEASE MODEL: In this method, the land owner and the industry will have an agreement for lease amount and the lease period. The lease agreement of INR 1000 per annum for dry lands and INR 3000 per annum for garden lands is practiced. Once the lease agreement is signed, the industry will establish their captive plantation and the period of lease is for one rotation and tenable further based on mutual consultations.

BENEFIT SHARE MODEL: Under this method, the land owner agrees to share the benefits at the time of harvest. Accordingly, an agreement is made wherein the industry will establish the plantation at its own cost and the benefit will be shared at 70 (industry):30 (land owner) for dry lands and 60 (industry):40 (land owner) for garden lands. The benefit share is worked at the total yield of the produce.

PULPWOOD SPECIES

Casuarina

Eucalyptus

MATCHWOOD SPECIES

Ailanthus excelsa

PLYWOOD SPECIES

Melia dubia

Fig.25.7: Industrial Forestry Trees

BIOFUEL SPECIES

Jatropha curcas

Jatropha curcas

BIOENERGY SPECIES

Leucaena leucocephala

Prosopis juliflora

VALUE ADDITION TECHNOLOGY

Casuarina Residues

Briquette

Fig.25.8: Industrial Forestry

INDUSTRIAL POLICY AND FOREST BASED INDUSTRY

While establishing forest based industries, it is essential to consider precautions in order to avoid environmental disturbances. Establishment and development of forest based industries form an important part of National Forest Policy of 1988. The policy indicated the following guidelines towards establishment and sustainable functioning of forest based industries

- As far as possible a forest based industry should raise the raw materials needed for meeting its own requirement, preferably by establishment of a direct relationship between the factory and the individuals who can grow the raw material by supporting the individuals with inputs including credit, technological advice and finally harvest and transportation. The policy also indicated the following.
- No forest-based enterprise, except that the village or cottage level, should be permitted in the future unless it has been first cleared after careful scrutiny with regard to assured availability of raw material. In any case, the fuel, fodder and timber requirements of the local population should not be sacrificed for this purpose.
- Forest-based industries must not only provide employment to local people on priority but also involve them fully in raising trees and raw material.
- Natural forests serve as a gene pool resource and help to maintain ecological balance. Such forests will not therefore be made available to industries for undertaking plantation and for any other activities.
- Farmers, particularly small and marginal farmers, would be encouraged to grow, on marginal, degraded lands available with them, wood species required for industries. These may also be grown along with fuel and fodder species on community lands not required for pasture purposes and by Forest Department/Corporations on degraded forests, not earmarked for natural regeneration.
- The practice of supply of forest produce to industry at concessional price should cease. Industry should be encouraged to use alternative raw materials. Import of wood and wood products should be liberalized.
- The above considerations will, however, be subject to the current policy relating to land ceiling and laws.

CONCLUSION

Number of industries like pulp and paper, plywood, particle board, furniture, housing, sports goods, match, pencil, value addition industries and biomass based power plants are not only increasing its capacity but also expanding their production potential. All these industries are heavily demanding forest resources but due to policy and legal implication availability of these resources from the natural forests are not possible. Moreover, the existing plantations were raised through inferior seed sources with poor management practices resulted in poor yield. The unorganized supply chain system also played a major role in success or failure of many wood based industries. Hence, the forest policy of 1988 directed all the wood based industries in the country to look for their own raw material by establishing direct linkages

with farmers in order to create self reliance in raw material availability. However for want of suitable technology development and suitable contract farming model with organized supply chain system the achievement in industrial wood plantation development is so far dismally modest. This chapter clearly evidenced the issues and constraints and the possible intervention for successful development of industrial wood plantation to create self reliance in raw material requirement.

26 MODERN LOGGING METHODS

INTRODUCTION

Forest management concepts have changed over years from production forestry to conservation forestry and sustainable forest management. Yet logging is a very important aspect in forestry and forest management. There is one school of thought that the only way to prevent forest depletion is to discard all forms of timber extracting practices. Other feels that if the timber production stops, the economic value of forests diminishes and forest land would be diverted for more profitable and productive options. In between these two are some who believe that forests could be sustainably utilized. Logging is a very important economic activity in the temperate regions and tropical regions. Timber is an important economic product that contributes to the State exchequer. Hence it is very essential to understand logging operations and how best they could be carried out without resulting in depletion and degradation of our forests and forest resources.

Logging generally refers to the process of felling and extracting timber from the forests. In forestry, the term logging is sometimes used in a narrow sense where it refers to the log making only and consists of trimming off the branches, debarking and cutting into sections. In some instances, logging is used to refer to the whole logistics of moving wood from the site of felling to somewhere outside the forest, usually a saw mill or a depot from where it is further disposed. However, the term may be used to indicate a range of forestry or silviculture activities including the felling, on-site processing, extraction and transportation to saw mills or storage depots. Logging is distinct from timber harvesting in the sense that the latter includes all the stages of harvesting starting from pre-harvest planning to harvest, technical supervision and post harvest assessment and even the impact of the operations on the non-timber forest resources and the state of forests.

IMPACTS OF LOGGING ON FORESTS

No doubt logging and timber operations help to meet the ever increasing demand for timber and firewood. It adds to the revenue of the country and aids economic and industrial (forest based industries) development. It provides income and employment to people, especially those living in and near vicinity of the forests. But conventional logging is a cause of great concern as they cause significant damages to the forest ecosystems. Concerns about the depleting forest cover are the most dominant factors that evoke the feeling of absolute non interference with the forests and zero timber operations. The major impacts of logging on the forests and environment (Mehta 1981) are presented below:

LOSS OF BIODIVERSITY: The predominant practices of logging in our country are salvage felling and selective felling. Selective felling removes high priced and often huge trees from a natural forest which results in the loss of seed source and natural regeneration. Removal of tree cover and top canopy results in the loss of habitats of the wild fauna and several species of flora in the lower strata of the forest. Several examples of this can be cited across several countries. The consequences of forest cover destruction are grave in case of critical habitats and representative ecosystems. Protection of biodiversity and conservation of gene pools and genetic resources and commercial timber production are aspects of forest management that conflict each other.

SOIL EROSION AND SILTATION: The loss of tree cover exposes the soil and results in soil erosion. Soil erosion and depletion are considered to be a major factor that reduces forest productivity and agriculture productivity. In countries like India where agriculture is the major economic activity such changes could have far reaching effects on the economy and development of the country. The displaced soil causes sedimentation in the streams, rivers and reservoirs thereby reducing water availability for agriculture and other activities.

DAMAGES TO REGENERATION AND RESIDUAL TREES: As stated earlier, the removal of trees result in removal of seed sources. As the forest ecosystems are highly complex and diverse, unless adequate measures are not taken to restore the regeneration, the forest ecosystem itself may be altered. Apart from this, the use of logging equipment and machines for skidding and extracting of timber escalate soil erosion and cause serious physical damage to the young regeneration and residual vegetation. Large chunks of debris and slash produced due to logging increases the risk of fire incidence. Natural regeneration becomes inadequate where the forest department remains oblivious of such changes and where deliberate efforts are not made to reforest and it will result in the disappearance of the forest itself.

EFFECT ON WATERSHED: Watersheds are the major source of water for domestic and agricultural purposes. The removal of vegetation reduces the water retention capacity of the watersheds, increases runoff during peak precipitation and reduced water flows and water scarcity during the low precipitation periods. Soil erosion also affects the water quality.

OPENING UP OF FOREST LAND AND CONVERSION TO AGRICULTURE: Logging operations result in the opening up of the canopy and fragmentation of forest lands. One of the major factors responsible for the deterioration of forests is the popular belief that forests are of optimal land use and could be put to better use or the so called developmental activities for social welfare. The provision of open spaces in the forest will pave way for the conversion of forest lands to other purposes especially agriculture resulting in more encroachments and dwellings. The creation of roads for transport of timber within and outside the forest areas will make the forests more accessible for human migrants, encroachers, poachers and timber mafias which are all the more detrimental to the ecosystem.

EFFECTS ON NON-WOOD FOREST PRODUCTS: NWFPs are of equal importance and are given great weightage in the forest policies not only because of their economic potential but also due to the fact that there are considerable number of people in our country dependent on these resources for their livelihood. Though the potential of many of the non-wood forest products including medicinal plants are unknown, they can't be neglected since they form a major source of income for the indigenous tribal people and forest dwellers. Management of forests for commercial timber production is quite incompatible with the management of forests for NWFP and ecosystem services to the dependant population.

CLIMATE CHANGE AND CARBON SEQUESTRATION: Climate change is a global concern and has gained the world attention due to the visible and predicted consequences. Forest vegetation and tree cover are the most effective and easiest way of combating the climate change. Forest vegetation and the trees outside the forest capture the atmospheric CO_2 through photosynthesis and store it in the forest biomass and generate oxygen. This phenomenon which is referred to as the carbon sequestration reduces the atmospheric CO_2 and is increasingly covered under international conventions and agreements regarding the climate change. Natural forests have tremendous role in sequestering carbon and mitigating climate changes. The harvesting of timber results in the removal of forest biomass and reduction in the carbon sequestration.

EFFECTS ON SCENIC BEAUTY AND AESTHETICS: Pristine forests are the greatest feast for our senses. Nature can be easily interfered with but can hardly be replicated. The presence of trees and the residual population of fauna and flora are closely linked with the aesthetics of the site. The forest structures and natural topographic features of undisturbed forests are totally different from that of disturbed and over exploited areas. Logging removes the major components of the forest ecosystem and interferes with the natural, cultural and aesthetic aspects of the natural ecosystems.

METHODS OF LOGGING

Since logging is a very broad term, there is an array of methods which may be considered under the purview the term logging. The method adopted normally depends on the nature of forests, topography, accessibility, the local conditions, the technology available, etc. Logging is an age old practice. In the earlier part of the century, it was completely a hand process where trees were felled by axe and transported by sledges or carts or drawn by oxen, horses, mules or floated down river. Conventional logging operations were done manually with saws or chain saw where are the trees are felled followed by delimbing and topping. Bucking into log lengths, skidding to a landing, sorting the logs for specific products or purposes and finally loading onto a truck or trailer or removed with the help of animal power are the other important operations in timber harvesting. Several variations of this system are followed depending on site conditions and season. For example, the most popular method of transport of timber during the monsoon season in the mountainous regions is to float it on water where as in high altitudes and steep slopes a skyline may be much preferred.

The conventional logging systems have as much potential for low site impact as any alternative system if operators are trained and careful measures are taken. In fact, conventional logging systems using smaller skidding machines cause less damage to remaining trees and have lesser impact on the soil. The process from the 'tree to mill' can be broken down into 5 phases or five minor operations viz. the felling, processing, extraction, loading and transportation to saw mills or storage depots. Every phase has a set of operations that leads to the other and the whole series of operations starts from the standing tree (Mehta 1981).

LOGGING TERMINOLOGIES

- *Felling*: It is the process of cutting down a tree. It may be manual or mechanical.
- *Delimbing*: It is the removal of branches from the tree.
- *Topping*: It is a processing step to remove the upper part of a tree at some defined minimum stem diameter.
- *Bucking*: It is the division of a tree into defined log-length segments.
- *Chipping*: It is a processing function that reduces trees or parts of trees into small, regularly-sized pieces through a slicing action.
- *Sorting*: Arranging the desired logs segments for transport.
- *Loading*: Transferring the logs to the transport vehicle.
- *Trucking*: It is the secondary transportation function that moves logs from the landing to a processing facility
- *Cable Logging*: It involves a yarder which pulls one or several logs along the ground to platform where a truck is waiting.
- *Yarder*: It is a logging equipment which uses a system of cables to pull or fly logs from the stump to the landing.
- *Skyline Logging*: When the terrain is too uneven to pull logs on the ground, a skyline can lift logs off the ground vertically, similar to a ski lift.
- *Aerial Logging*: Heavy-lift helicopters may be used when cable logging is not allowed due to environmental concerns or when roads are lacking. Helicopters are the most expensive form of log transport.
- *Logging by Water stream*: Water stream or rivers are used to transport logs.
- *Logging by Animals*: Horse, Oxen or Donkeys are used to carry the logs.
- *Balloon Logging*: Gas filled balloon is used to transport logs.
- *Bunching*: Smaller and slower equipment, such as the crawler tractor, may be used to gather and bunch a load by a road or skid trail while a larger and faster piece of equipment, such as the wheeled skidder, may haul the load to the deck. This process is known as roading or bunching.
- *Primary Extraction*: It is the process of moving felled trees or logs from the stump to the landing.
- *Skidding*: It is the extraction by dragging logs or trees across the ground. Pulling logs with horses or mules (animal skidding) or mechanical skidding which substitutes either a tracked or wheeled machine to pull the load.

- *Forwarding*: Forwarding is a method of moving logs by carrying rather than dragging them. The forwarder is a heavy-duty trucks/ wheeled machine with bunks on the back for carrying short trees or logs.
- *Cable Skidding*: They are wheeled or tracked machines that use a winch and cable on the rear of the vehicle to gather and skid the logs to the landing.
- *Cable Yarding*: It is the process of transporting logs to the landing using a stationary winching machine (a yarder) and a system of heavy wire rope.

FELLING

Felling refers to cutting trees at the stump level or severing the tree from the stump and bringing it to the ground (Mehta 1981). Trees are felled to recover the log or timber volume and value. Historically this was carried out with an axe or saw or both. Throughout the major part of the Indian forests, felling is done by axe but the use of axe and saw combined is also practiced widely. The felling axe is usually made locally, the weight of the head varying from 0.7 to 1.8 kilograms. Felling is carried out by a team of two or more workers. Stumps are very often high up to 60 centimeters for ease of working. The most commonly used tools for felling and crosscutting are two-man manual crosscut saws with peg teeth. Later on came the chain saw which increased a person's efficiency to fell the trees. The process became much faster while using chain saw and the wastage produced while using an axe was considerably reduced.

The system of felling can also be categorized as monocyclic logging and polycyclic logging. In monocyclic logging, the entire growing stock is removed in a single operation (clearfelling or clearcutting) and the interval between two felling operations is typically equal to the maturation period of the main species of trees felled, the so-called rotation period. This may be as long as 60-80 years in natural forests. Monocyclic logging removes not only mature but also semi-mature trees and hence a relatively large proportion of the forest may be affected. The volume of timber removed during monocyclic operations is very high as 120 m^3 per ha in certain forests where as in polycyclic the harvested volumes are around 60 m^3 per ha. Such intense logging creates large gaps in the canopy and the potential for damage to both soil and remaining trees through monocyclic logging is relatively high (Bruijnzeel and Critchley 1994).

Polycyclic logging is the selective removal of only the largest individuals of desirable species. The objective is to wait for a sufficient number of trees to reach maturity and then to remove them. Compared with monocyclic logging, fewer trees and a lower volume of timber are harvested, but the intervals between harvests are shorter. Volumes of wood removed are typically 20-30 m^3 per ha per coupe. Disturbance to the forest flora and fauna occurs more frequently than under monocyclic cutting regimes and the amount of damage caused to the overall forest is, theoretically, considerably less for each operation due to the smaller amounts of timber being extracted. As smaller gaps are created, polycyclic logging alters the forests less and most forests in tropical regions are logged under some form of polycyclic system. This system is much similar to the selection felling where selected trees above a certain DBH are removed. In certain cases, only the dead and diseased trees are removed and are referred to as salvage felling where felling is restricted to the trees or the forest areas that have been damaged by wildfire, flood, severe wind, disease, insect infestation or other natural disturbance.

The three common timber harvesting methods are:

- *Tree-length Logging*: Trees are felled and then delimbed and topped at the stump. The log is then transported to the landing, where it is bucked and loaded on a truck. This leaves the slash (and the nutrients it contains) in the cut area.
- *Full-tree Logging*: Trees and plants are felled and transported to the roadside with top and limbs intact. The trees are then delimbed, topped and bucked at the landing.
- *Cut-to-length Logging*: It is the process of felling, delimbing, bucking and sorting (pulpwood, sawlog, etc.) at the stump area, leaving limbs and tops in the forest.

MANUAL FELLING

Manual felling is performed by a person with a chainsaw. Professional chainsaws range from 3 to 9 hp and can be equipped with various bar lengths to allow cutting a tree of any diameter. The important tools that may also be carried are axe, wedges, saw fuel and oil, first-aid kit, fire extinguisher and tools for minor repairs. Special tools such as jacks and felling levers have been designed to aid manual fellers in directing a tree's fall. Manual felling is the most dangerous task in timber harvesting and proper training and experiences are required to minimize the risk of injury. To cut a tree, the person must first evaluate a number of factors including lean, desired direction of fall, distribution of the crown weight, presence of any defects in the stem and anything else that may affect the dynamics of the falling tree. In manual felling of trees, four important points are considered: direction of fall, undercut, backcut and holding wood or hinge wood.

DIRECTION OF FALL: The feller has to decide which direction the tree has to fall. The tree's height should be the major determining factor when deciding where you want it to fall. It should be made sure that clear spot is present with no obstacles in the way. Any branches which can disturb the direction of fall should be cut down. Branches that are not too large should be cut down. A rope to the top of the tree may be tied to help it fall in the preferred direction.

UNDERCUT: After clearing around the tree and locating an escape path, the expert makes an undercut (face cut), and a back cut. The face cut/ undercut is a wedge-shaped cut that opens in the direction the tree is intended to fall. Undercut will be a V-shaped notch into the tree. After cutting, the notch should fall out. The undercut should be 1/3rd to 1/4th of the tree's diameter. The two cuts meet at the apex of the angle and neither cut bypasses the other. Along with controlling the direction of the fall, the undercut also allows the tree to slide rather than jump off the stump. Various forms of undercut have been developed for different types of felling applications.

BACK CUT: With the undercut wedge removed, the feller is ready for the back cut. Back cuts are designed to release the wood that holds the tree from falling over the undercut. This is a horizontal cut usually 1 to 3 inches above the face cut that is on the opposite side of the tree. Saw or chainsaw blade should be held parallel to the ground and horizontally sawn into the trunk. Wedges are used to prevent the tree from sitting back on the saw blade.

HINGE WOOD: The narrow strip of wood (holding wood) between the undercut and back cut is the hinge that controls the fall. While sawing, it should not be cut all the way through. Instead a hinge is left of about 1 inch (2.5 cm) or 10 per cent of the diameter of the tree. The hinge will stop the tree from immediately falling down, giving the feller a chance to get out of the way. The feller can "push" a tree to fall in direction other than way it is leaning using wedges or jacks.

In high-value timbers, it is important to minimize breakage during falling. Leaving low stumps, avoiding dropping trees across rocks or other obstacles and properly placing the cuts are necessary skills for timber fellers. Studies of manual felling productivity have found that the average time to fell a tree is affected by tree spacing, tree diameter and slope.

MECHANIZED HARVESTING SYSTEMS

Mechanized harvesting systems have taken over the conventional systems especially where the growing stock is uniform and even aged. The level of mechanization varies from low level to high level where felling and processing are done using highly specialized equipments and machines. High efficiency is the often quoted advantage of using machines in logging. There are three basic types of mechanical felling machines. Single-function machines are only capable of directionally felling a tree, while dual-function machines (called feller-bunchers) are able to fell a tree and move the cut stem to a pile. There are also multifunction machines (called harvesters) that can fell, delimb, buck and place the processed pieces. Mechanized felling is faster and safer than manual felling. For example, a shear feller-buncher can fell a tree in 3 to 6 seconds compared to 30 and 90 seconds for a person with a chainsaw. Feller-bunchers are capable of felling and bunching upto 200 trees per hour in favourable conditions. Manual felling cannot maintain steady levels of productivity comparable to mechanized felling. Mechanization has certain advantages and disadvantages.

ADVANTAGES OF MECHANIZED LOGGING

- High Efficiency in harvesting trees especially smaller trees
- Wide applicability in commercial thinning
- Reduced residual stand damage due to controlled felling and proper placement of felled stems.
- Higher wood utilization due to lower stump height
- Safely used in weather and lighting conditions that are unacceptable for manual cutters
- Improved safety
- Use of improved and modern techniques

DISADVANTAGES OF MECHANIZED LOGGING

- Very high initial and operational costs
- Soil disturbances
- Potential wood damage with shear-type cutters

- Stability limitations on steep slopes
- Equipment utilization needs technical knowhow and skill upgradation of the persons
- Tree stem size and terrain could be major limitations
- Mechanization may not be feasible in all types forests

Various mechanized harvesting systems and methods are practiced depending on the form in which the wood is delivered, the terrain, the size of the trees, etc. Harvesting systems refers to the whole set of tools, equipments and machines used to harvest an area where as the harvesting methods or logging methods refer to the form in which wood is delivered to the main access road and the amount of processing (e.g. delimbing, bucking, debarking, chipping) which occurs after felling. The different harvesting equipments are described under felling, extraction, processing and loading.

FELLING EQUIPMENTS

Where harvesting is done manually most of the operations are done by the axe, saws and billhooks. Modern logging equipments have now replaced them because of their high efficiency and are widely utilized in commercial forestry.

CHAIN SAW

Chain saws are the most commonly used equipment for felling and delimbing. It is powered by diesel, hydraulic or electric motor and has a saw with a loop of chain that is pulled around a bar which rotates at a high speed. The chain contains cutting teeth that are capable of sawing wood. A chainsaw can be used to fell and process trees. Chains come in a variety of pitches (the distance between the rivets in the chain), gauges and cutter teeth and a range of power sizes and bar lengths. Though the operations are faster while using chain saws, proper safety measures have to be ensured due to the inherent dangers associated with the working. The greatest advantage of chain saws is that felling is generally not limited by slope or soil conditions.

HARVESTER

"Harvester" is a commonly used term in forestry. It refers to a machine that fells, delimbs and bucks trees into logs. It is the first of a pair of machines, the second being a forwarder, that jointly makes the mechanized "cut-to-length" (CTL) system. It is a 4 or 6 wheeled articulated machine used to fell trees, remove branches and cut the logs to required lengths which help to maximize the value of the end product. It is a heavy machine powered by diesel engine and is usually fitted with a hydraulic loader onto which a harvester head is mounted. A harvester head consists of a hydraulically powered *chain saw,* two or more curved *de limbing knives* which reach around the trunk to remove branches, *feed rollers* to grasp the tree and to move the tree through the head, *diameter sensors* to calculate the volume of timber harvested and a *measuring wheel* to measure the length of the stem as it is fed through the harvesting head. The whole harvesting process is controlled by an on-board computer and an operator sitting in the cab of the vehicle. Harvesters are mostly employed for cut-to-length logging operations, for felling, delimbing and bucking trees.

The standard cut-to-length logging system consists of a harvester, forwarder and self-loading trucks. This system contains relatively few types or number of equipment, but the harvester and forwarder are very expensive to purchase and operate. A forest harvester is mostly used along with a forwarder that hauls the logs to a point from where it is finally transported outside the forests. The major limitation of harvesters is that it can be employed effectively only on level grounds and moderately steep terrains and not on very steep terrains and undulating surfaces. They are most effective for clear felling of even aged uniform stands and is seldom used for selective felling and in mixed stands. They operate best in forests consisting of single stem soft wood species such as conifers, aspen and birch. They are not suitable for hardwood species with multiple merchantable stems.

Harvesters may be tracked or wheeled. Tracked harvesters can operate on slopes up to 55 per cent and wheeled harvesters are generally limited to less than 40 per cent slope. For very steep hills and for felling individual trees chain saws are more preferred. Operating on steep slopes is risky and may cause roll over of the machine. For productivity reasons, harvesters are being widely used and prototype machines like walking harvesters or steep terrain harvesters are put to use in steep unfriendly terrains. The size of the tree that can be felled and processed is also limited by the capacity of the cutting head and the weight of the machine. Harvesters are not very effective to fell and process large trees with multiple merchantable stem.

FELLER BUNCHER

It is a motorized vehicle which runs on tracks or wheels, with an attachment that can rapidly cut and gather several trees before felling them. It is a type of harvester with a cutting head that can hold more than a single tree at a time. The cutting head is not only used for cutting but also for holding and placing the stems on the ground but feller bunchers cannot do any kind of processing. A feller buncher consists of a standard heavy equipment base with a tree-grabbing device furnished with a chain-saw or circular saw or shear (shear is a device designed to cut small trees off at the base). The machine keeps the cut trees on a stack facilitating the further removal of the loads by a skidder or forwarder or other means of transport. The feller buncher may be 'drive to tree buncher' or 'swing boom buncher.' A 'drive to tree feller buncher' is a machine with the cutting head mounted directly to the carrier rather than on a hydraulic boom. And as the name indicates the whole machine drives up to each tree to cut it. Where as a 'swing boom feller buncher' has the cutting head mounted on a boom and the machine does not have to drive to each tree.

Fig. 26.1: Types of Feller Buncher (Source: Forests and Rangelands 2012)

EXTRACTION EQUIPMENTS

Pulling or extracting the tree from the forest area to a landing or storage area is referred to as extraction. Any harvesting system that uses an extraction machine that drives into the forest or enters the forest on ground is referred to as ground-based harvesting. When the tree is dragged out of the forest it is called skidding, when they are carried out on a trailer it is called forwarding. Skidders, Forwarders, agricultural tractors and animal power are all ground-based extraction methods. Aerial logging and yarders (cable logging) are other methods of extraction practiced in specific locations where ground based extractions are nearly impossible.

SKIDDER

Specialized machines used for dragging the wood out of the forest area to the storage areas or road side are called Skidders. Wheeled skidders and tracked skidders can be used as per the terrain. Tracked skidders have slow ground speeds and should be limited to shorter extraction distances typically less than 500 feet. The wheeled skidders are capable of faster speeds and can extract material upto 1,000 feet. Very large grapple skidders, also called grapple forwarders, are also available that can extract material to greater distances. Cable skidder uses a main winch cable and choker set to assemble and hold a load. In Grapple skidder, a set of bottom-opening jaws assembles and holds the load. A cable skidder has skid line with chokers attached. The grapple skidder is capable of picking up more than one stem at a time and the main advantage over a cable-skidder is that the operator need not get out of the cab to accumulate the lot. It greatly increases productivity and safety, but is limited to sites where it can be drawn to the stump site.

A clam-bunk skidder is also a type of skidder used in logging. It has top-opening hydraulic jaws to hold a load of trunks for extraction. The advantage over a grapple skidder is typically a larger load and improved traction. But the disadvantage is that it needs to be loaded. Other commonly used machinery includes the shovel, a tractor and bulldozers. A shovel is a tracked excavator with a grapple on the arm for moving logs or trees from the woods to either directly to the landing or road, or it can also be used to shovel to a skid road, whereby it will pre-bunch the trees for more efficient extraction.

FORWARDER

They are extraction machines that remove material from the forests. They are basically tractors pulling a wagon load of wood. The cab may be fixed or capable of rotating on the chassis. Many forwarders have a boom mounted grapple for loading and unloading material. Compared to skidders, forwarders cost more to purchase. But the ability of forwarders to accumulate large loads makes the cost of forwarding per unit of extraction distance covered a lesser issue when compared to skidding.

SKYLINE AND CABLE LOGGING (YARDER)

Skyline and cable logging operations are preferred when logs or trees have to be extracted from steep terrains. In such systems, a skyline or wire rope or cable, which aids to pull the stems from the forest to a landing, is suspended across the slope with a yarder (basic component of the cable yarder system) at one end and a tail tree or tail holder at the other end. The yarders may be mobile yarder or a stationary yarder. The yarder helps to keep the cable off the ground and helps to lift the logs off the ground. Carriers or wagons facilitate the transport of materials. The yarder has mechanical or hydraulic winches or drums which facilitate the movement of the carriages and the main line along the skyline cable. The drum stores the main line cable. The logs are set and hooked by the chocker setter standing on the ground and the lots are pulled up and dropped at the landing.

An operator from a cab controls the yarder with the help of electronic switches and hydraulic levers. Where there are possibilities of using other machines to fell, bunch or process the trees, a cable yarder with a grapple is used. Loader booms and processing heads mounted on cable yarders are also in use.

Yarders may have either a swinging boom or a fixed boom. Most swinging booms have a limited height of 30 feet to 60 feet. Fixed boom yarders have towers as tall as 100 feet. Swinging booms permit a wider skyline corridor and thereby reduce the number of yarder moves. For increasing the productivity of cable operations, automated yarders are being used. Automated yarding systems make use of a computer to control the inhaul and outhaul of the carriage. Outhaul is either to the last location it stopped or to a pre-set distance and inhaul is automatically set to stop 10 or 20 m from the yarder for safety reasons. Automating the movement of the carriage frees up both the yarder operator and the choker-setters to carry out their tasks while the carriage is in motion.

AERIAL LOGGING WITH HELICOPTER

Helicopter can be used to extract trees in steep, remote areas and in areas that lack road infrastructure. Since the early 1950's, helicopter yarding is in practice due to its ability to avoid many of the obstacles that are encountered in the ground based and skyline systems, including site sensitivity, urgency to remove or deliver the product, lack of access and slope of the terrain. The use of helicopters in forestry still continues to expand in many countries. Helicopter operations have high operational costs as well as high initial costs. Though the cost involved is huge, it is highly productive. Advantages of aerial logging using helicopters are:

- Avoid road constructions and road transport.
- Applicable in any kind of terrain.
- It will reduce the impact on soil, vegetation and regeneration because the logs are lifted vertically.
- It is able to extract large volumes of logs within short period of time.

PROCESSING EQUIPMENTS

Once the tree is extracted to the landing, it is processed before loading onto the truck. Processing may be done on the stump site or at the landing. This process typically involves delimbing, topping the tree and bucking (cutting the stem into logs). These tasks can be carried out by chainsaw or delimbers. If the desired product is logs, then processing is done with delimbers. If the wood is to be used for paper and pulp, a wood chipper may be used. To remove the tree residues, mulchers and grinders are used.

DELIMBER

Delimbing and topping are the processes of removing the limbs (branches) and cutting the top of the tree from the stem. Delimbing can be done by chainsaw or mechanically using delimbers. There are different kinds of delimbers available. A stroke boom delimber consists of two grapples: a front grapple and a rear grapple and saw mounted on base machine. It is capable of delimbing and bucking the tree trunk. The tree trunk is normally grasped near the middle of the stem. The butt of the tree is placed in the rear grapple. The grapple arms functions as delimbing knives. The front grapple moves along the stem to remove limbs while the rear grapple holds the tree stem in position. The trunk is topped and bucked into log lengths by the saw on the boom. The rear saw is also used to trim the butt end of the tree and also to make bucking cuts. Delimbers are generally used with a whole tree harvesting system such as a feller buncher/skidder system. They are good at processing trees into sorted piles, but are not capable of loading trucks.

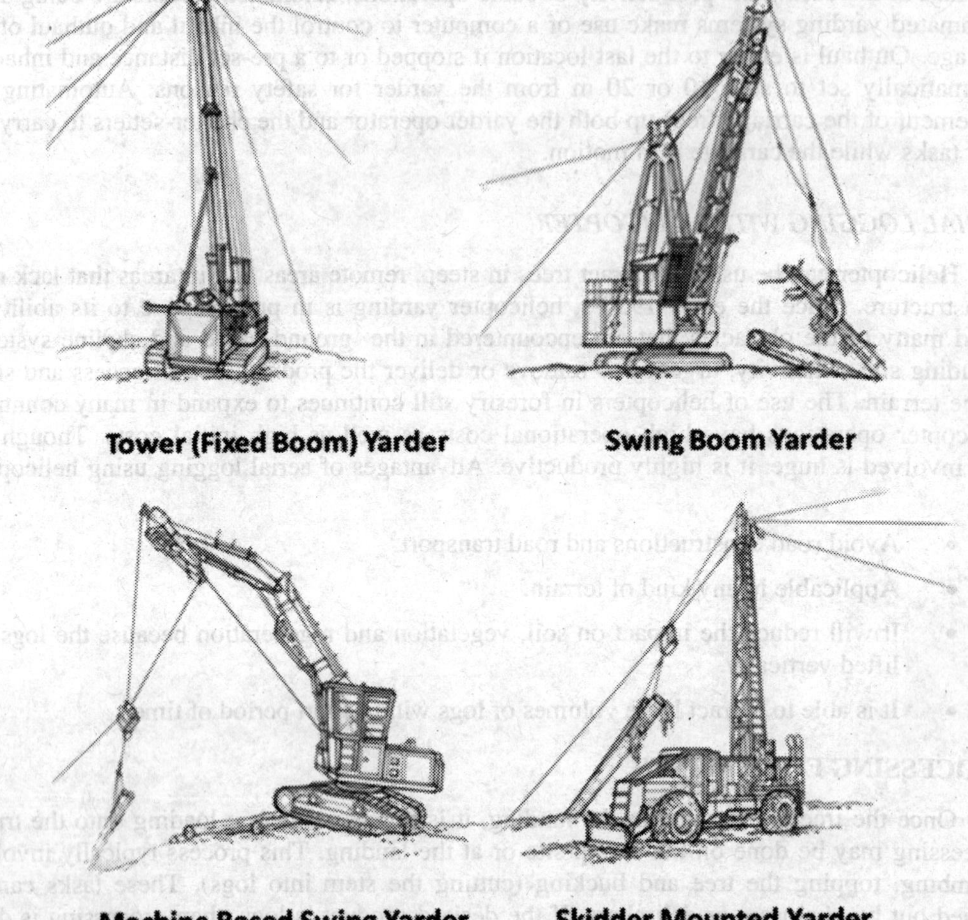

Fig. 26.2: Types of Yarder (Source: Forests and Rangelands 2012)

Gate, flail and pull-through delimbers are the other different types of delimbers available based on methods employed for removing limbs from the boles of trees. These are relatively low cost methods for delimbing. Gate delimbers are welded structures consisting of steel tubing welded in a grid pattern. Gate delimbers may have an integrated support structure, or consist of a simple flat grid that is tied to standing trees. A grapple skidder backs a load of trees through the gate, breaking the limbs off the bole in the grid. Pull-through delimbers lack feeding mechanism or measuring abilities. A pull through delimber can either be a stand alone unit, or it can be mounted onto a trailer as and when required. The tree has to be picked up and placed in the pull-through delimber. The delimber has three knives at the point where the loader pulls the stem through, removing the limbs from the tree. Most pull-through delimbers also have a topping saw in them which cuts off the top.

Flail delimber consists of chains mounted on rotating drums. The chains break the limbs off stems of the trunk fed through them. They may be self-feeding or may require a loader to pull the stems through the delimber. When a bunch of stems are pulled through the delimber, the chains not only 'beat and break off' the limbs, but also effectively removes most of the bark. Most efficient for relatively small trees from dense stands with small limbs and where the primary product is 'clean' chips (chips going to be used for making pulp and paper). Delimbers are used where full trees are being delivered to the delimber. The form of the trees, single bole trees, and the size of the tree that can fit through the throat of the delimber or grids in the gate generally limit their use to few specific species only. Flail delimbers are not limited by tree size, but also by the size of the branches.

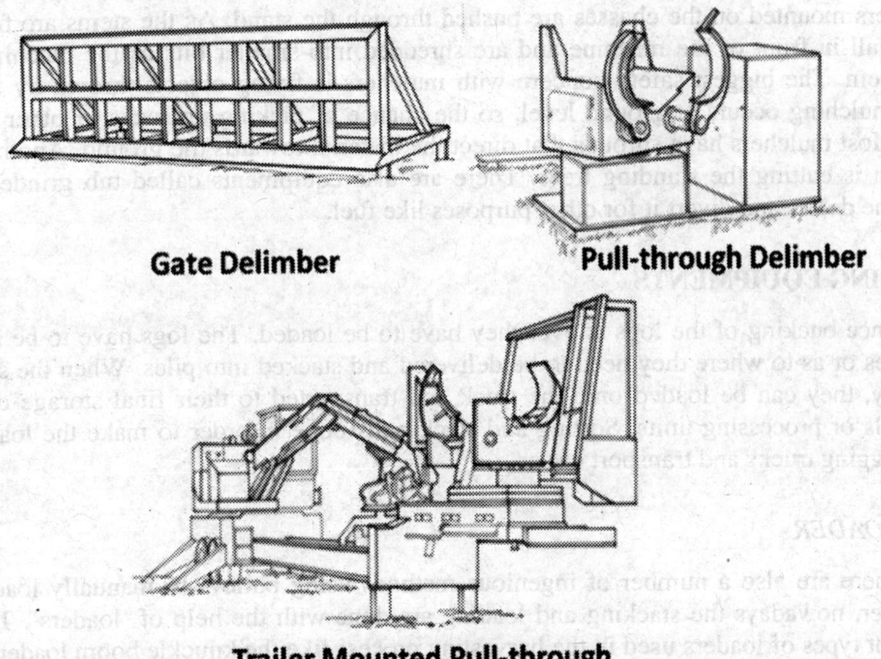

Fig. 26.3: Types of Delimber

CHIPPER

A chipper is a machine that is used to produce high quality chips for the pulp and paper industry. The machines are large and highly productive and are capable of reducing trees to wood chips. Chippers differ from grinders in that they produce uniform sized chips depending on the equipment settings. Common features include a feed mechanism, a set of rotating knives, which may be either drum or disc mounted, an anvil and a discharge chute. Chip size is determined by the distance from the anvil to the knives, the distance between the knives on the drum or disc, the feed speed and the speed the drum or disc rotates. The feed mechanism is usually a set of feed rollers at the throat of the chipper. Some chippers may have conveyor belts or trailers to assist in feeding the chipper. Chippers may be either trailer mounted or self propelled. Self propelled chippers are usually track mounted. Some chippers are also equipped with a loading arm and grapple for self loading. Some chippers have integrated delimbing/debarking functions. These are used to produce clean chips that may be used for fuel or pulpwood.

MULCHER (MASTICATOR OR BRUSHCUTTER)

Mulchers are equipments used to cut, chop or grind vegetation and left over residues after harvesting into smaller pieces that are usually left on-site as mulch or waste materials. It is as a part of alternative to chemical treatment of residues, for aesthetic treatments, right-of-way maintenance and site rehabilitation. Mulchers are mounted on a variety of chasses including skid steers, excavators, farm tractors, four wheel drive loaders and bulldozers. Mulchers mounted on the chasses are pushed through the stand. As the stems are felled, the stems fall in front of the machine and are shredded into smaller bits as the machine drives over them. The biggest safety concern with mulchers is flying objects thrown by the head. Most mulching occurs at ground level, so the chance of picking up rocks or other debris is high. Most mulchers have shrouds that direct the material towards the ground. Another safety concern is cutting the standing trees. There are also equipments called tub grinders which grind the debris and divert it for other purposes like fuel.

LOADING EQUIPMENTS

Once bucking of the logs is over they have to be loaded. The logs have to be sorted as per sizes or as to where they need to be delivered and stacked into piles. When the sorted lot is ready, they can be loaded onto the truck and transported to their final storage depots or sawmills or processing units. Sorting and stacking is done in order to make the loading out onto logging trucks and transport easier.

LOG LOADER

There are also a number of ingenious methods using pulleys to manually load trucks. However, nowadays the stacking and loading are done with the help of 'loaders'. There are different types of loaders used in the harvesting process like the knuckle boom loaders, trailer mounted knuckle boom loaders, tracked loaders, shovel type loaders, etc. Knuckle boom loaders are machines with specially designed booms for handling logs. They are hydraulically operated loading boom whose mechanical action imitates the human arm. They may be also track based, wheel based, or trailer mounted. They are versatile implements and a knuckle boom loader mounted on a semi trailer can be used for any or all of the following functions viz. delimbing, bucking, sorting, and/or loading. Knuckle boom loaders are often used where they are needed only at the landings and the tree size is not too large.

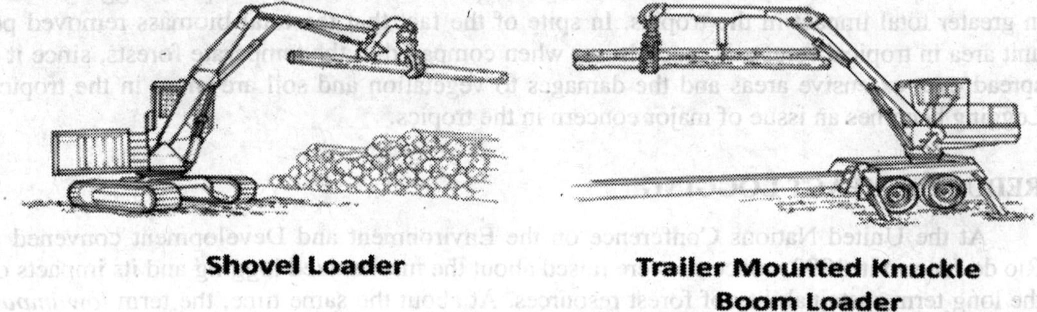

Fig. 26.4: Types of Loader

Truck mounted loaders are boom type loaders mounted directly on the log truck. Truck mounted loaders are only used to load small wood and have the disadvantage that it can handle logs only as much the truck can carry. There are also number of grapples like the bunching grapples, pulp wood grapples, log grapples, butt and top grapples, etc. that can be used for loading the logs depending on size and length of the lot. Bunching grapples have tongs for picking up and handling multiple stems. The alternative is a log grapple that has single tines. Shovel loaders are used where piece sizes and volumes are too large for knuckle boom loaders.

TRANSPORT

Trucks and trailers or tractors are the most common vehicles that are used to transport logs. The vehicle selection depends on the size of logs being transported, the condition of the roads and regional preferences. Trucking delivers the logs from the landing to the saw mill for further processing. The trucks used to transport are different from those used for transport of other goods and have special features that aid the movement of long logs. Full-length logs and poles are transported by log trailers attached to the trucks to their final destination points. The trailers may be for a fixed length log trailers or for flexible lengths. Straight trucks with a single log bunk or more than one log bunks are also used depending on the size of the lot.

LOGGING OPERATIONS IN TROPICS

Though timber is one of the major sources of revenue for the Government, unlike in temperate regions, timber operations are less mechanized and organized in tropics. This is partly due to the nature and type of vegetation and partly due to easy availability of human labour. Even aged uniform forests of single species is characteristic of the temperate regions, where as in the tropics the forests are mainly non-uniform and uneven aged with multi layers of vegetation and varied species. In spite of the high biodiversity and high biomass in the tropical forests, the volume of merchantable timber per unit area is much lesser than temperate forests. The volume of timber harvested per hectare of tropical forest world wide is less then $30m^3$ and involves usually 10 or less than 10 trees of different species where as a typical coniferous forests yields nearly 500 m^3 of timber per ha from 200-300 trees of the same species (Patrick et al 2001). Since the volume of merchantable timber per unit area is lesser, more area needs to be worked to meet the timber requirement resulting in construction of more number and additional length of roads. Extensive areas get disturbed and though the impact on unit area basis is much lesser than temperate forests in the tropics, logging results in greater total impact in the tropics. In spite of the fact that the total biomass removed per unit area in tropical forests is much lesser when compared to the temperate forests, since it is spread over extensive areas and the damages to vegetation and soil are more in the tropics. Logging becomes an issue of major concern in the tropics.

REDUCED IMPACT LOGGING

At the United Nations Conference on the Environment and Development convened at Rio de Janeiro in 1992, concerns were raised about the mechanized logging and its impacts on the long term sustainability of forest resources. At about the same time, the term *low impact logging* was introduced by the Tropical Forest Foundation. But the term was contradicted by many experts since they claimed that logging can never be of low impacts. In 1993, when the term Reduced Impact Logging (RIL) came into usage denoting the 'environmentally sound timber practices' (Putz and Pinard 1993), it was widely accepted and has gained great attention. The term has even become interchangeable in the vernacular with timber harvesting. RIL is not an entirely novel idea. It is nothing but a systematic approach to planning, implementation, monitoring and evaluation of harvesting techniques. Though most of the RIL technologies have been developed in the temperate regions due to the heavy mechanization there, it has substantial potential in the tropical regions also and code of practices for forest harvesting are prevalent in many of the counties in the temperate and Asia-Pacific regions.

Reduced impact logging can be defined as the intensively planned and carefully controlled implementation of timber harvesting operations to minimize the environmental impact on forest stands and soils (ITTO). It generally requires the following practical measures (Thomas Enters et al 2002):

- A pre-harvest inventory and the mapping of individual crop trees
- The pre-harvest planning of roads, skid trails and landings to minimize soil disturbance and to protect streams and waterways with appropriate crossings
- Pre-harvest vine-cutting in areas where heavy vines bridge tree crowns
- The construction of roads, landings and skid trails following environmentally friendly design guidelines
- The use of appropriate felling and bucking techniques including directional felling, cutting stumps low to the ground to avoid waste and the optimal crosscutting of tree stems into logs in a way that maximizes the recovery of useful wood
- The winching of logs to planned skid trails and ensuring that skidding machines remain on the trails at all times
- Where feasible, using yarding systems that protect soils and residual vegetation by suspending logs above the ground or by otherwise minimizing soil disturbance

- Conducting post-harvest assessment in order to provide feedback to the resource manager and logging crews and to evaluate the degree to which the RIL guidelines were successfully applied

APPROPRIATE LOGGING METHOD

Most of the equipments and operations described in this chapter are frequently used for the harvesting of soft woods in the temperate regions and less frequently used in the tropics and the developing countries. Natural forests dominate tropical regions unlike plantations in the temperate regions. It was in the natural forests that forest harvesting first began and in many countries, they still are the only or the main source of wood supply for the forest-based industries. Harvesting in these natural forests is always oriented towards a few large trees of high commercial value. Due to the unfriendly terrain in the natural forest, heavy mechanization may not be always cost effective where trees are huge, few in number and highly varied in size, age and species. But they are feasible options in the short rotation large scale plantations established for paper and pulpwood, fibre, etc. The logging equipments used for large trees in natural tropical forests, such as heavy-duty wheeled and crawler tractors, yarders, trucks and loading equipment, are usually costly and many of these technology alternatives are inappropriate. Many of the single or multipurpose modern logging machines currently used in the plantation forests of highly developed countries of temperate regions is inappropriate at present for use in our situations. With the various existing logging methods and techniques and the interaction of the climate, soil characteristics, the terrain and the socio-economic conditions prevailing in the countries choosing an appropriate logging technology is not easy.

27 ECOTOURISM

INTRODUCTION

According to the latest United Nations World Tourism Organization's (UNWTO) world tourism barometer, International tourist arrivals grew by 4.4 per cent in 2011 to a total 980 million, up from 939 million in 2010 and it is expected to cross 1 billion mark in 2012. The argument for the integration of tourism with conservation was first made widespread by Budowski (1976) in an article. However, the term "Ecotourism" has become prevalent concept since the mid 1980s and it has experienced the fastest growth of all sub-sectors in the tourism industry. The popularity shows a proof of change in tourist perceptions, increased environmental awareness and a desire to explore natural environments.

The tourism activities stretch from conventional tourism to ecotourism and there has been a lot of contention to the limit at which biodiversity preservation, local social-economic benefits and environmental impact should be considered in ecotourism. For this reason, environmentalists, special interest groups and governments define ecotourism differently. Environmental organizations have generally insisted that ecotourism is nature-based, sustainably managed, conservation supporting and environmentally educated. The tourist industry and governments, however, focus more on the product aspect, treating ecotourism as equivalent to any sort of tourism based in nature. As a further complication, many terms are used under the rubric of ecotourism. Nature tourism, nature travel, low impact tourism, green tourism, soft tourism, biotourism, ecologically responsible tourism (Laarman and Dust 1987; Wilson and Laarman 1988; Valentine 1992; Hall and Weiler 1992) and others have been used in literature and marketing, although they are not necessary synonymous with ecotourism. Presently the Ecotourism as industry is one of the fastest growing economic sector in the globe.

The concept of ecotourism is probably equally due to the widespread and growing interest in the natural environment and a corresponding recognition of the importance of conserving natural environmental quality. The idea of visiting and experiencing high-quality natural environments and also protecting them from harmful impacts is now an acceptable and marketable one. Ecotourism involves nature conservation by tourists who visit several attraction sites. This is aimed at securing threatened species of plants and animals that depend of nature to survive. This is done mainly by tourists with an aim of sustaining the beauty and attraction of the sites. It is the exercise that is worthy investing in since the outcome benefits the generations to come and creates a sustainable source of government revenue.

When the visiting tourists explore a certain site and encouraged to conserve the environment, it becomes a source of inspiration and fun to them. On the other hand, the tourists gain an experience of what makes the site attractive and they become eager to visit the same site in future in order to see the outcome of what their work. It is a partnership that enables all the parties to participate in sustaining the attractive sites. Ecotourism has numerous merits to the environment and nature in general. It has a long run benefits to the climate which in turn improves the favourable weather in the area. By adopting ecotourism as a sustainable model, the forest cover is increased, the canopy is created and in the long run, the climate improves. This reduces the effects of global warming and improves the water reservoir. It is a common knowledge that nature conservation is easier and cheaper compared to the painful outcome of the nature destruction. For this reason, the environment authorities across the world have adopted ecotourism as a sustainable model.

IMPORTANT TERMINOLOGIES

TOURISM: Tourism encompasses the whole world-wide industry of travel, hotels, transport and facilities that serves the needs and wants of travelers and the primary motive for travel is leisure. It is an extreme form of recreation and traveling in the expectation of pleasure which is distinguished by relatively long lengths of stay away from home and relatively large distances travelled.

ECOTOURISM: Ecotourism is tourism and recreation that is both nature-based and sustainable and it conserves environment and improves the socio-economic condition of local people. A walk through the forest is not ecotourism unless that particular walk somehow benefits that environment and the people who live there. A rafting trip is only ecotourism if it raises awareness and funds to help protect the watershed.

ALTERNATIVE TOURISM: Any type of travel that is not mass tourism (i.e. beach vacations or traditional sightseeing tours). This includes ecotourism, backpacking, volunteer tourism, adventure tourism, historical tourism, tornado chasing, couch surfing or any other form of travel that is atypical.

SUSTAINABLE TOURISM: Any form of tourism that does not reduce the availability of resources and does not inhibit future travelers from enjoying the same experience. If the presence of large numbers of tourists disturbs an animal's mating patterns so that there are fewer of that species in the future then that visit is not sustainable.

RESPONSIBLE TOURISM: Tourism which operates in such a way as to minimize negative impacts on the environment. A wilderness camping trip using "Leave No Trace" ethics would be considered responsible tourism.

NATURE-BASED TOURISM: A more generic term for any activity or travel experience with a focus on nature. Large jungle lodges fall into this category as do cruise ships to view penguins in Antarctica. These types of trips may or may not be environmentally sustainable or responsible.

GREEN TOURISM: It is a term often used interchangeably with ecotourism and sustainable tourism but more accurately described as any activity or facility operating in an environmentally friendly fashion. A lodge with composting toilets, gray water system and solar powered lighting is probably "green". There are varying degrees of "greenness"; an awareness of where resources are coming from and where wastes are going is at the heart of the idea.

ADVENTURE TRAVEL: While travel to another country is often adventurous but it is not necessarily Adventure Travel. Adventure is defined as an unusual experience including some level of risk and uncertainty. "Adventure Travel" includes the idea of risk and oftentimes some unconventional means of transport. There is sometimes a distinction made between "Soft" and "Hard" adventures. Soft adventures have a lower level of risk, greater comfort in accommodations and are less physically rigorous. Hard adventures often have very basic facilities, higher risk factor and greater physical challenge (e.g., mountain climbing, backpacking or river expeditions).

MULTI-SPORT ADVENTURES: These trips have a focus on physical outdoor activities. Rafting, mountain biking, climbing, surfing, diving, etc. all offered in the same package. Not necessarily sustainable or eco but might be since many companies want to protect the areas where these activities take place.

CULTURAL TOURISM: Interacting with and observing unique cultures are the focus of this style of trip. The concept of learning from other cultures to broaden ones perspective is usually a core value. An artisan showing you how to weave a tapestry and learning from them about their traditional dress would be a form of cultural tourism. Buying crafts in the market with no more interaction than the exchange of money does not provide the insight into another culture that is the central theme of cultural tourism.

GREEN WASHING: It is the commercialization of tourism schemes disguised as sustainable, nature based and environmentally friendly ecotourism. These schemes are environmentally destructive, economically exploitative, and culturally insensitive at its worst. This is actually a misuse of ecotourism by misleading the tourists and manipulating their concerns for the satisfaction of the needs of private agencies.

WHAT IS ECOTOURISM?

Defining ecotourism has proven to be a difficult task and there are many different definitions to what ecotourism really is. It is more feasible to treat ecotourism as a spectrum with a variety of products rather than attempting to define ecotourism from a specific stance or product (Wight 1993a, b). Ecotourism was first defined as travelling to relatively undisturbed or uncontaminated natural areas with the specific objective of studying, admiring and enjoying the scenery and its wild plants and animals, as well as any existing cultural manifestations found in these areas (Ceballos-Lascurain 1987). The International Ecotourism Society defines ecotourism as "responsible travel to natural areas that conserves the environment and improves the welfare of local people" (TIES 1990).

Honey (2008) defined ecotourism as travel to fragile, pristine and usually protected areas that strives to be low impact and often small scale and helps educate traveler; provides funds of conservation; directly benefits the economic development and political empowerment of local communities. It is also defined as nature-based tourism that involves education and interpretation of the natural environment and is managed to be ecologically sustainable. Ecotourism is responsible travel in areas containing natural resources that possess endemic characteristics and cultural or historical resources that are integrated into the area's ecological system. Its purpose is to create awareness among all concerned parties of the need for and the measures used to conserve ecosystems and as such is oriented towards community participation as well as the provision of a joint learning experience in sustainable tourism and environmental management. Another definition by The World Conservation Union's Commission on National Parks and Protected Areas, defines ecotourism as an environmentally responsible travel and visitation to relatively undisturbed natural areas, in order to enjoy and appreciate nature that promotes conservation, has low visitor impact and provides for beneficially active socio-economic involvement of local populations.

It is a purposeful travel to natural areas to understand the culture and the natural history of the environment; taking care not to alter the integrity of the ecosystem; producing economic opportunities that make the conservation of the natural resources beneficial to the local people. When compared to mass tourism, ecotourism not only stresses the appropriate use of all resources, but also emphasizes community development to meet the economic, social and cultural needs of the community. Mass tourism on the other hand creates initiatives in Third World countries that are directed towards satisfying the needs of the tourists. Ecotourism development is most likely to be at a smaller scale, locally owned, with low import leakage and a higher proportion of profits remaining in the local economy

Ecotourism is travel to fragile, pristine and usually protected areas that strives to be low impact and small scale. It helps to educate the traveler; provides funds for conservation; directly benefits the economic development and political empowerment of local communities; and fosters respect for different cultures and for human rights. Ecotourism appeals to ecologically and socially conscious individuals. Generally speaking, it focuses on volunteering, personal growth and learning new ways to live on the planet. It typically involves travel to destinations where flora, fauna and cultural heritage are the primary attractions. Ecotourism is a conceptual experience, enriching those who delve into researching and understanding the environment around them. It gives us insight into our impacts as human beings and also a greater appreciation of our own natural habitats.

WORLD TOURISM ORGANIZATION

The World Tourism Organization (UNWTO) was founded on 27 September 1970 and from 1980 onwards, this day is celebrated as "World Tourism Day". WTO is the United Nations agency responsible for the promotion of responsible, sustainable and universally accessible tourism. As the leading international organization in the field of tourism, UNWTO promotes tourism as a driver of economic growth, inclusive development and environmental sustainability and offers leadership and support to the sector in advancing knowledge and tourism policies worldwide.

THE INTERNATIONAL ECOTOURISM SOCIETY

Concept of TIES was launched by a team organized by Megan Epler Wood at a conference in Florida in 1989 as the world's first international non-profit dedicated to ecotourism as a tool for conservation and sustainable development. The International Ecotourism Society (TIES) was founded in 1990 and has been in the forefront of the development of ecotourism, providing guidelines and standards, training, technical assistance, research and publications. TIES global network of ecotourism professionals and travelers is leading the efforts to make tourism a viable tool for conservation, protection of bio-cultural diversity and sustainable community development.

Through membership services, industry outreach and educational programs, TIES is committed to helping organizations, communities and individuals promote and practice the principles of ecotourism. TIES currently has members in more than 120 countries, representing various professional fields and industry segments including: academics, consultants, conservation professionals and organizations, governments, architects, tour operators, lodge owners and managers, general development experts and ecotourists. TIES advocated for sound ecotourism practices at World Parks and Protected Areas Congress in Caracas, Venezuela (1992) and First World Congress on Tourism and the Environment in 1992 at Belize.

GLOBAL SIGNIFICANCE OF ECOTOURISM

The global importance of ecotourism is highlighted by several international agreements including: The UN Commission on Sustainable Development, 7th Session 1999; UN World Tourism Organization Code of Ethics (1999); The Guidelines on Biodiversity and Tourism Development issued by the CBD (2003); The Quebec Declaration on Ecotourism (2002); and the World Summit on Sustainable Development (2002). Guidelines and standards relating to sustainable/ responsible tourism (including specific reference to biodiversity) are developed by the Tour Operators' Initiative for Sustainable Development which has created environmental guidelines for hotels, resorts and tourist attractions.

Concrete UNEP projects include the Tour Operators Initiative for Sustainable Tourism development, with WTO and United Nations Educational, Scientific and Cultural Organization (UNESCO), promotion of the Convention on Biological Diversity's Guidelines for Sustainable Tourism in Vulnerable Ecosystems and the UNEP Principles for Sustainable Tourism; the further development and implementation of the tourism components of the International Coral Reef Action Network (ICRAN), the Great Apes Survival Project (GRASP) and the Mountain Commons project.

WORLD ECOTOURISM SUMMIT/ QUEBEC DECLARATION

In the framework of the UN International Year of Ecotourism, 2002, under the aegis of the United Nations Environment Programme (UNEP) and the World Tourism Organization (WTO), over three thousand representatives from 132 countries met at the World Ecotourism Summit, hosted in Quebec City, Canada by Tourism Quebec and Canadian Tourism Commission during 19-22 May 2002. The salient features of the event are:

- The first-ever Ecotourism Summit which has signaled that ecotourism, in practice, can contribute to poverty alleviation and environmental protection
- Ecotourism must be considered a privileged tool, leading the way and paving the road towards a tourism that is truly sustainable
- Sustainable tourism can contribute to the more global effort of protecting the sustainability of our planet's resources
- Major issues facing ecotourism have been debated: Concept of Ecotourism; Ecotourism Policy and Planning; Regulation of Ecotourism; Product Development,

Marketing and Promotion of Ecotourism; Monitoring Costs and Benefits of Ecotourism

- Ecotourism embraces the principles of sustainable tourism, concerning the economic, social and environmental impacts of tourism.
- Ecotourism also embraces the specific principles: ecotourism contributes actively to the conservation of natural and cultural heritage; includes local and indigenous committees in planning, development and operation and contributing to their well-being; interprets the natural and cultural heritage.

OBJECTIVES OF ECOTOURISM

The question of "why ecotourism" is one of the emerging questions in the present day context. There are two groups and hence two opinions regarding the subject. One group strongly suggests it as an effective tool for conservation while the other group denies it by terming it as the prominent threat to the existence of nature and natural resources. In this situation the objectives of ecotourism can be listed as follows:

- To cultivate environmental consciousness among the local population by educating them about the dangers of overexploiting resources and unrestricted number of tourists.
- To help preserve ecosystems and natural areas with a high tourism potential.
- To promote new economic incentives, alternate employment opportunities, etc.
- To ensure communal ownership and control, and that part of the profits flow into community development programs rather than into personal enrichment.
- To foster a feeling of pride and community through a revival or preservation of traditional practices and cultural techniques.

CHARACTERISTICS OF ECOTOURISM

Ecotourism aims to promote excellence in tourism by development and protection of natural areas, benefiting the local communities and encouraging commercially successful and environmentally sound tourism operations. Its focus is sustainability, brought about by building up the following characteristics.

- Tourism activity is carried out in a relatively undisturbed natural setting.
- All natural based forms of tourism in which the main motivation of the tourists is the observation and appreciation of nature as well as the traditional cultures prevailing in natural areas.
- Educational, appreciation and interpretation features.
- Generally, but not exclusively, organized by small groups by specialized and small, locally owned business.
- Minimum negative impacts upon the natural and socio-cultural environment.
- Conserves natural and cultural heritage.

- Contributes to sustainable development and is a profitable business.
- Actively involves local people in the process, providing benefits to them.
- Supports the protection of natural areas by generating economic benefits for host communities, organizations and authorities managing natural areas with conservation purposes.
- Provides alternative employment and income opportunities for local communities and increases awareness towards the conservation of natural and cultural assets, both among locals and tourists.

PRINCIPLES OF ECOTOURISM

Ecotourism is a tool to unite conservation, communities and sustainable travel. Hence, those who implement and participate in ecotourism activities should follow the following ecotourism principles:

- Minimize impact on environment.
- Build environmental and cultural awareness and respect.
- Provide positive experiences for both visitors and hosts.
- Provide direct financial benefits for conservation.
- Provide financial benefits and empowerment for local people.
- Provide healthy political, environmental and social climate for the host country.

POSITIVE IMPACTS OF ECOTOURISM

There are two important promises ecotourism put forward. One promise of ecotourism is that it increases the monetary value of standing biological systems and thereby hopefully encourages the preservation of biological diversity. Another promise is that it is supposed to bring people closer to nature and thus instill a greater desire to protect it. Ecotourism activities planned and established on the sound ecotourism principles will have the following advantages,

- It will provide maximum visitor satisfaction with minimal impact on the environment.
- It will build awareness and respect for the local culture and environment.
- Benefits the local people through employment which is a positive step for the socio-economic development of the local people.
- Educates visitors by an on-site visit about the local political, social and environmental issues.
- Money from the tourism activities go back into the conservation of the area.
- Visitors carry new ideas back to influence their own environment.

ECOTOURISM AND CONSERVATION

Is ecotourism a tool for conservation of nature and natural resources? This is debating question today. Both pros and cons are there, but it is, if planned and conducted properly. Ecotourism provides both direct and indirect benefit for conservation. Direct benefit is in terms of financial outturns. The revenue obtained through ecotourism programmes can be directly used for the conservation efforts. The indirect benefits are raised in terms of the awareness to the people regarding nature and natural resources, employment to the local people, etc. Lack of awareness regarding the importance of nature and natural resources is one of the important challenges faced by conservation efforts. Hence the chances of visiting natural and cultural heritage areas will enhance the ability of people to appreciate it and will make aware them regarding the importance of conservation. Hence ecotourism can be a simple way of communicating the idea of conservation to the local people. If planned properly, the extent of awareness they are explored can be made more.

ECOTOURISM AND LOCAL COMMUNITIES

By increasing local capacity building and employment opportunities, ecotourism is an effective vehicle for empowering local communities around the world to fight against poverty and to achieve sustainable development. In many areas, nowadays ecotourism has been selected as a powerful tool to compensate for the injustice done to the tribal people in the past. There is always a conflict between the existing rules and regulations and the traditional jobs of the forest dependent community. In this context, it is very important to have some tools for providing sufficient employment opportunities to meet their daily needs and also for improving their living standards. Another important advantage is that, a feeling among the people will arise, that the nature and natural resources are the means of their livelihood and hence they will put their maximum effort to conserve it. In many of the protected areas, the need of ecotourism is caused by the necessity of the local people for sufficient livelihood opportunities. The new ecotourism activities also provide an opportunity of getting new skills to them and the same can be imparted through proper training also. In short, the healthy relation between the nature and the local community can be achieved through properly planned ecotourism activities.

ECOTOURISM AND INTERPRETATION

Ecotourism can also be termed as a way of effecting nature interpretation to the visitors. If the packages are designed in such a way that they are getting chances of interpreting the nature, natural resources, natural phenomena etc., it will be an effective interpretation tool. The lack of awareness among the people is one of the emerging threats to the conservation of nature today. Hence effectively planned ecotourism activities will be an efficient method for making aware the people regarding the conservation of nature and natural resources.

THREATS TO ECOTOURISM

One definition of ecotourism is the practice of low-impact, educational, ecologically and culturally sensitive travel that benefits local communities and host countries. Many of the ecotourism projects are not meeting these standards. Even if some of the guidelines are being executed, the local communities are still facing other negative impacts. Some of the negative impacts are forcing people to leave their homes, gross violations of fundamental rights and environmental hazards - far outweigh the medium-term economic benefits etc. A tremendous amount of money is being spent and human resources continue to be used for ecotourism despite unsuccessful outcomes and even more money is put into public relation campaigns to dilute the effects of criticism. Ecotourism channels resources away from other projects that could contribute more sustainable and realistic solutions to pressing social and environmental problems. The money tourism can generate often ties parks and managements to ecotourism. But there is a tension in this relationship because ecotourism often causes conflict and changes in land-use rights, fails to deliver promises of community-level benefits, damages environments and has plenty of other social impacts. Indeed many argue repeatedly that ecotourism is neither ecologically nor socially beneficial, yet it persists as a strategy for conservation and development. While several studies are being done on ways to improve the ecotourism structure, some argue that these examples provide rationale for stopping it altogether.

IMPROPER PLANNING

Launching of ecotourism activities may cause harmful effects if not planned properly. All the possible demerits will cause direct environmental impacts and hence environmental degradation. Some of the demerits of the bad planned ecotourism activities are as follows,

- Overcrowding constructions
- Pollution of the habitat
- Unlimited numbers of tourists
- Traffic congestion
- Footpath erosion and soil loss
- Tourists don't understand or care what ecotourism really is
- Criticism as greenwashing where the environment is used as a bait to attract tourists

DIRECT ENVIRONMENTAL IMPACTS

Ecotourism operations occasionally fail to live up to conservation ideas. It is sometimes overlooked that ecotourism is a highly consumer-centered activity, and that environmental conservation is a means to further economic growth. Although ecotourism is intended for small groups, even a modest increase in population, however, temporary, puts extra pressure on the local environment and necessitates the development of additional infrastructure and amenities.

THREATS TO INDIGENOUS CULTURES IS ANOTHER NEGATIVE IMPACT

Ecotourism often claims that it preserves and enhances local cultures. However in many areas, evidence shows that with the establishment of protected areas, local people have illegally lost their homes, and most often with no compensation. Pushing people onto marginal lands with harsh climates, poor soils, lack of water, and infested with livestock and disease does little to enhance livelihoods even when a proportion of ecotourism profits are directed back into the community. The establishment of parks can create harsh survival realities and deprive the people of their traditional use of land and natural resources.

MISMANAGEMENT

While governments are typically entrusted with the administration and enforcement of environmental protection, they often lack the commitment or capability to manage ecotourism sites effectively. The regulations for environmental protection may be vaguely defined, costly to implement, hard to enforce and uncertain in effectiveness.

MANAGEMENT OF ECOTOURISM SITES BY PRIVATE ECOTOURISM COMPANIES

Private management of tourism sites offers an alternative to the cost of regulation and deficiency of government agencies. It is believed that these companies have a self interest in revenue generation and less concerned about environmental conservation. Tourists will pay more for pristine environments, which translates to higher profit. However, theory indicates that this practice is not economically feasible and will fail to manage the environment.

ECOTOURISM IN PROTECTED AREAS

Many of the protected areas in the world are also excellent ecotourism centres. Ecotourism activities have become prominent among the routine activities of the protected areas today. These activities have been planned and launched in a participatory manner which have benefitted the conservation efforts a lot. The need of ecotourism activities in protected areas should mainly aim conservation effort through public awareness, assistance for other activities, monetary benefit for assisting conservation activities and assisting local people through employment opportunities.

Many environmentalists are against the practice of ecotourism in protected areas. They argue that the ecotourism activities will not bring anything other than the depletion of the natural resources, especially in wildlife areas. As the protected areas are very much sensitive to human disturbances, the activities in these areas have to be planned with more care. Some of the facts we have to give special attention while designing an ecotourism package in protected areas are:

- Carrying capacity of the area has to be studied and the planning has to be done accordingly.
- Care has to be given for packages in which the resources are used sustainably.
- Protection oriented packages has to be planned.
- Wildlife monitoring oriented packages has to be planned.
- Local participation has to be ensured while planning the packages.
- Maximum employment opportunities to the local people have to be ensured.

ECOTOURISM – KERALA EXPERIENCES

Kerala popularly known as God's Own country is gifted with greatest biodiversity wealth in the entire stretch of Western Ghats. Ecotourism is very popular in Kerala with the successful implementation of ecotourism programmes in Parambikulam Tiger Reserve and Periyar Tiger Reserve.

ECOTOURISM EXPERIENCES FROM PARAMBIKULAM TIGER RESERVE

Parambikulam Tiger Reserve is the most protected ecological piece of Anamalai sub unit of Western Ghats, surrounded on all sides by protected areas and sanctuaries of Kerala and Tamil Nadu, the sanctuary is endowed with a peninsular flora and fauna which are excellently conserved due to total protection and minimal human interferences. It has a total area of 285 Square Km. Parambikulam Tiger Reserve is the 38th Tiger Reserve of India and second of Kerala. It is also well known for the sustainable ecotourism programmes.

Many innovative ecotourism programmes have been designed here with a view to give maximum visitor satisfaction and also to give maximum employment opportunities, but without affecting the ecological sustainability of the area. The emerging threat to this protected area was a huge number of unemployed tribal people and the pressure on forests for their livelihood activities. Tourism was there but was not in a controlled and sustainable manner. Efforts have been taken to design effective sustainable ecotourism packages without affecting the healthiness of the wildlife habitat existing here. Main objectives of the ecotourism project planned here are:

- To give maximum employment opportunities to the local people.
- To give effective nature interpretation programmes to the visitors.
- To provide effective ecotourism facilities which will help the visitors to appreciate the nature and natural resources.
- To use ecotourism as a tool of eco-development and hence to attain local support for protection activities.
- To make ecotourism as a tool of protection, wildlife monitoring, nature education, etc.

The packages designed here under the ecotourism projects are mainly protection and monitoring oriented. At the same time, attention has been given to have maximum income generating potential also. Some of the important characteristics of the ecotourism projects launched in Parambikulam area as follows.

STRICT CONTROL ON THE MOVEMENT OF VEHICLES INSIDE THE PARK: Vehicle safari programme have been launched to attain this goal. Day visitors have been allowed to enter the park only through safari vehicles. Private vehicles are directed to park at the entrance itself. This effort has helped a lot to have effective control on the vehicles moving inside the park.

ENGAGING LOCAL TRIBALS AS GUIDES/ NATURALISTS: Each visitor group has been provided with a guide or naturalist during their visit inside the park. This guide has been given direction to give proper guidelines to the visitors. At the same time, guide has the power to control on the visitors in violation of the rules and regulations of the park. Attention has also been given to give proper training to the guides, to provide reasonable wages, etc.

PROVIDING SUFFICIENT INFRASTRUCTURE TO THE VISITORS: Proper infrastructure and its maintenance have been given special care. Interpretation Centre, Information

Centre, Accommodation facilities, Canteen, Toilet blocks, etc. have been made available to the visitors through proper timely actions. New constructions have been done without affecting the natural beauty of the area and also without affecting the wildlife habitats. Special attention has been given to maintain the facilities by engaging local tribals for up keeping and maintenance.

EFFECTIVE ECOTOURISM PACKAGES: The ecotourism packages have been designed with a view to have minimum disturbance to the environment, but maximum revenue and also maximum employment to the local people. At the same time, visitor satisfaction has also been taken care. For e.g., in "*Tented Niche*" programme, seven tents have been provided to the visitors in an undisturbed area. They have also been provided food, safari, boating etc. under the same package and they will also be assisted by trained guides throughout the programme. For all these provisions under the package, they have to pay a fixed amount. The package shows an excellent example for conducting a variety of activities under a single programme, in a controlled manner, by gaining a reasonable fee and also by giving maximum employment to the local people.

PROTECTION ORIENTED ECOTOURISM PACKAGES: Some packages are designed to make people presence in offense sensitive areas such as sandal regeneration areas, illicit felling areas, poaching sensitive areas, etc. This helped a lot to have a check on the entry of offenders to these areas. Similarly trekking in these areas have a check on the illegal activities in park.

MONITORING ORIENTED ECOTOURISM PACKAGES: Some packages have given the chances to the visitors to put their effort in filling the wildlife monitoring datasheets during their visit. For e.g., "*Full Moon Census*" is an ecotourism package in which the visitors are allowed to stay in Machan near a vayal during full moon day. They are also directed to observe the animals in the vayal and to record the same in the data sheet provided. This data will be very valuable in wildlife monitoring.

NATURE CAMPS: Systematically arranged and properly conducting nature camps are another feature of the ecotourism here. Nature education is the main objective behind this activity. The classes are taken by experts according to the target groups.

ASSISTANCE THROUGH PROPER ECO-DEVELOPMENT ACTIVITIES: There are many eco-development units such as paper bag unit, plastic reduction unit, honey unit, Parambikulam dhara unit, etc. to aid the proper functioning of the ecotourism packages. For e.g., in Parambikulam dhara unit, a water filtering unit is functioning to produce pure bottled water for giving the visitors with the condition that they will not be allowed to use their plastic water bottles inside the park. This effort will help to check the menace of plastic bottles inside the park and help ecotourism. Similarly, paper bag unit will check the uncontrolled use of plastic bags inside the park.

ENSURING PEOPLE'S PARTICIPATION: All the activities are planned and controlled by a democratically constituted agency called Forest Development Agency (FDA). The Agency is the apex head of eco-development committees, which are the local committees formed in the area. All the decisions regarding ecotourism have been taken by FDA. People's participation has thus been ensured throughout the running of a package.

VISITOR FRIENDLY EFFORTS: Updating the facilities through new steps such as online booking, computerized billing, website services, etc. are also important characteristic of Parambikulam ecotourism.

ECOTOURISM EXPERIENCES FROM PERIYAR TIGER RESERVE

Periyar tiger Reserve, Thekkady, is an example of nature's bounty, with great scenic charm, rich biodiversity and providing veritable visitor satisfaction. Sprawled over an area of 925 Square km, Periyar is one of the 27 tiger reserves in India. Zealously guarded and efficiently managed reserve is a repository of rare, endemic and endangered flora and fauna and forms the major watershed of two important rivers of Kerala, the Periyar and Pamba. People oriented and park centered community based ecotourism is the hallmark of Periyar Tiger Reserve. These programmes are conducted by local people responsible for the surveillance of the vulnerable parts of the reserve.

PERIYAR TIGER TRAIL-ADVENTURE TREKKING AND CAMPING: There used to be a great threat to the Reserve from the illegal collection of Cinnamom bark from Periyar Tiger Reserve (PTR) from a group of smugglers. Lots of trees were getting killed because of these collections. In order to overcome this issue, PTR management had started an innovative ecotourism programme in order to rehabilitate these smugglers. They were asked to leave their illegal style of livelihood and PTR offered them employment in a programme called "*Periyar Tiger Trail*". PTR management has initiated this programme which is a trekking and camping programme inside the forest. One armed staff also accompanies a group of 5 tourists for this camping programme. The guiding will be done by these ex-smugglers since they have an intimate knowledge of the forest terrain in the reserve. Now because of this ecotourism programmme, a unique experience is being given to the visitors, the livelihood of the local people is improved who once used to be the smugglers and extra protection to the park.

BULLOCK CART DISCOVERIES: There were some poaching pressures from the people across the bordering areas of Tamil Nadu to Periyar Tiger Reserve. In order to bring them into the mainstream, an initiative was started in the form of an ecotourism programme called, "*Bullock Cart Discoveries*". These poachers were asked to stop the illegal activities in the PTR and management offered them a couple of bullock carts. Now the tourists coming to Periyar Tiger reserve are given the option of availing this interesting ecotourism programme by which one can have the feel of the grapevine farmyard, bird watching experiences on a bullock cart. Now because of this ecotourism programme, all the poaching pressures have been reduced drastically as the livelihood of poachers has been ensured through this programme. Moreover, the tourists are now getting one of the very interesting ecotourism experiences.

BAMBOO RAFTING: This is a dawn to dusk range hiking and rafting programme through some of the richest forest tracts of Periyar Tiger Reserve. A mosaic of habitats will be traversed before the party gets into rafts made of bamboos. The forests are rich in bird life and arboreal animals like giant squirrel and Nilgiri langur. The rafting is for about three hours and one gets a panoramic view of forest-clad hills reflected on the lake. Animals like elephant, gaur and sambar are sighted keeping close to the edges of the lake. An armed guard and guides will accompany the tourists.

NATURE WALK-THE GUIDED DAY TREK: Different nature trails traversing diverse habitats form the trekking routes, generally 4 to 5 km in length. This is an interpretive programme offering excellent opportunity to watch birds, butterflies and other wildlife. The trails often pass through evergreen and moist deciduous forests interspersed with marshy grasslands.

BORDER HIKING: This is a conservation oriented hard trek taking up a full day. The route passes through undulating terrains with altitudinal ranges of 900-1,300 meters and the trekkers could glimpse the lofty escarpments bordering the park watershed and the vast plains down below. Gaur, sloth bear, elephant, etc. are often sighted along this route apart from birds and butterflies. The trekkers will go with two guides and an armed forest guard.

GREEN MANSIONS/ JUNGLE INN: A picturesque land is one of the off-the-track destinations ideal for a wilderness retreat replete with trekking, birding, canoeing and facilities for boarding. Hosted and accompanied by trained local people, the visitors can go for trekking and bird watching in the many trails around, paddle in still waters to cool off or perhaps settle down for watching sunset in this oasis of peace and tranquility. The tourists have the pleasure of staying in jungle during night in green mansions.

BAMBOO GROVE/ ECO-LODGE: A hotel that is truly an "eco-lodge" is one that makes efforts to conserve resources and limit waste. The islet of elevated ground studded with bamboo thickets is the eco-lodge for experiential learning. Here a miniature habitat is being recreated for man and nature to co-exist in the already available milieu of grassy downs, sedges and bamboos and stream fringed with screw pine. Dwelling units made exclusively of natural materials like bamboo and grass and tents are slotted in the available spaces together with treetop huts to give the place an eco-friendly ambience. Experiential learning and nature sensitization camps are conducted for discerning aspirants. Lectures, discussions, audiovisual presentations, field visits etc. go with the programme.

JUNGLE PATROL: This programme is part of the regular night patrolling in the fringe eco-development zones of the tiger reserve. By participating in this, one is actually helping in the protection of the forests of Periyar. The trekking could be at any time between 7 pm and 4 am and could be taken by persons having proper physical and mental fitness to trek through the wilderness in the night. The maximum duration for a slot is 3 hours and protection watchers and armed forest guard accompany the tourists.

28 BIOFUEL FROM TREE BORNE OILSEEDS

INTRODUCTION

Oil consumption in India has been increasing at the rate of seven per cent every year. India is currently the fourth largest oil consumer in the Asia Pacific region, after Japan, China and South Korea. Efficient, reliable and competitively priced energy supplies are prerequisites for accelerating economic growth. For any developing country, the strategy for energy development is an integral part of the overall economic strategy. Efficient use of resources and long-term sustainability remains core objective of economic planning. Sustainability would take into account not only available natural resources and issues related to ecological balance but also established delivery mechanisms, the technological constraints that are prevalent in the system, immediate compulsion to meet the priority needs of the economy, economic equity and self-reliance.

Realisation of high economic growth aspirations by the country in the coming decades, calls for rapid development of the energy market. The fossil energy resources available indigenously are limited and may not be sufficient in the long run to sustain the process of economic development translating into increased energy import dependence. The base of the country's energy supply system is tilted towards fossil fuels, which are finite. This has serious long-term implications as the emerging patterns of energy consumption, which is heavily skewed towards oil and gas, bring to focus many ecological and environmental issues.

India produces nearly 30 per cent of its total oil energy requirements and the rest through imports. With the increase in share of hydrocarbons in the energy supply and use, the current share of imported energy is expected to increase. If the corrective measures are not taken in time, the India's maximum revenue will drain out in the import of petroleum products. The challenge is therefore, to secure adequate energy supplies at the least possible cost. With a view to reducing dependence on imported oil, it is imperative to encourage use of indigenous sources of energy, exploration of domestic production, equipping domestic refining industry, promoting use of ethanol-blended petrol and biodiesel throughout the country. Biofuel from tree borne oilseeds (TBOs) opens a potential avenue for meeting the energy need of the country.

BIOFUELS

Energy and fuels from biomass can provide tremendous economic, environmental and energy security benefits. As energy prices rise and become increasingly volatile and as evidence of global warming mounts, the case for clean, renewable, domestic sources of energy has never been clearer. Biomass or plant matter can be converted into heat, electricity and alternatives to gasoline and diesel. The bioenergy that results can be a clean, renewable and domestic alternative to fossil fuels. Bioenergy can improve our economy, our environment and our energy security. Biomass to make electricity (also known as biopower) can provide roughly the same economic and environmental benefits as making fuels (also known as biofuels). When energy security is considered, the balance of benefits tips strongly in favour of using biomass to help reduce our oil consumption by producing biofuels.

Biofuels are renewable source of energy derived from biological raw material as well as solid biomass, liquid fuels and various biogases. Although fossil fuels have their origin in ancient carbon fixation, they are not considered biofuels by the generally accepted definition because they contain carbon that has been "out" of the carbon cycle for a very long time. In order to be considered a biofuel the fuel must contain over 80 per cent renewable materials. It is originally derived from the photosynthesis process and can therefore often be referred to as a solar energy source.

Biofuels are gaining worldwide acceptance as one of the solutions for problems of environmental degradation, energy security, rural employment, agricultural economy, oil import, oil price hikes, savings in foreign exchange, concern over greenhouse gas emissions from fossil fuels and vehicular pollution. Two major biofuels for the transport sector, bio-ethanol and bio-diesel, have become popular in many countries across the world.

BIO-ETHANOL

Biologically produced alcohols, most commonly ethanol and less commonly propanol and butanol, are produced by the action of microorganisms and enzymes through the fermentation of sugars or starches (easiest), or cellulose (which is more difficult). Bio-butanol (also called bio-gasoline) is often claimed to provide a direct replacement for gasoline, because it can be used directly in a gasoline engine (in a similar way to biodiesel in diesel engines). The principle fuel used as a petrol substitute for road transport vehicles is bio-ethanol. Ethanol or ethyl alcohol (C_2H_5OH), commonly termed as alcohol, is close to be perfect fuel because of low molecular weight, high oxygen content, high combustion efficiency, non-toxic and non-polluting nature and being a renewable raw material from varied saccharine and starchy substrates. Ethanol is a clear colourless liquid which is biodegradable.

The raw material used for producing ethanol varies from sugar in Brazil, cereals in USA, sugarbeet in Europe and molasses in India. In India, bio-ethanol (ethanol) is produced from a vast array of raw material: sugarcane, sugarbeet, sweet potato, cassava, rice, millets, maize, wheat, cellulosic biomass, bagasse, cane trash and cane tops, crop residues etc. There is also ongoing research and development into the use of municipal solid wastes to produce ethanol fuel. Corn-to-ethanol and other food stocks have implications both in terms of world food prices and energy yield. Now technology has been developed for the production of cellulosic ethanol.

Ethanol is a high octane fuel and has replaced lead as an octane enhancer in petrol. By blending ethanol with gasoline we can also oxygenate the fuel mixture so it burns more completely and reduces polluting emissions. Ethanol fuel blends are widely common in world market. The most common blend is 10 per cent ethanol and 90 per cent petrol (E10). Vehicle engines require no modifications to run on E10 and vehicle warranties are unaffected also. Only flexible fuel vehicles can run on up to 85 per cent ethanol and 15 per cent petrol blends (E85). Brazil uses ethanol as 100 per cent fuel in about 20 per cent of vehicles and 25 per cent blend in gasoline in the rest of the vehicles. Australia and USA use 10 per cent ethanol gasoline blends, whereas 5 per cent blend is used in Sweden and Europe uses 15 per cent ethanol blended gasoline.

BIO-METHANOL

Methanol (CH_3OH) is currently produced from natural gas, a non-renewable fossil fuel. It is called as wood alcohol and it can also be produced from biomass through biomass gasification. The methanol economy is an alternative to the hydrogen economy, compared to today's hydrogen production from natural gas. Methanol can be produced from hydrogen-carbon oxide mixtures by means of the catalytic reaction of carbon monoxide and some carbon dioxide with hydrogen. Biosynthesis gas (bio-syngas) is a gas rich in CO and H_2 obtained by gasification of biomass.

Biomass sources are preferable for bio-methanol than for bio-ethanol because bio-ethanol is a high-cost and low-yield product. This is a promising alternative, with a diversity of fuel applications with proven environmental, economic and consumer benefits. Bio-methanol's biggest advantage is that as a fuel itself, it can be used in automotive engines very similar to those currently on the market, as well as being able to be stored and transported in much the same way that diesel and gasoline are today.

BUTANOL

Butanol (C_4H_9OH) is formed by ABE fermentation (acetone, butanol, ethanol) and experimental modifications of the process show potentially high net energy gains with butanol as the only liquid product. Butanol will produce more energy and can be burned straight in existing gasoline engines (without modification to the engine or car) and is less corrosive and less water soluble than ethanol and could be distributed via existing infrastructures. DuPont and Bharat Petroleum are working together to help developing Butanol. *Escherichia coli* have also been successfully engineered to produce butanol by hijacking their amino acid metabolism.

BIO-DIESEL

The term bio-diesel stands for neat vegetable oils used as diesel fuel (DF) as well as neat methyl esters prepared from vegetable oils or animal fats and blends of conventional diesel fuel with vegetable oils or methyl esters. With increasing emphasis on the use of esters as DF, however, the term "bio-diesel" increasingly refers to alkyl esters (especially methyl esters) of vegetable oils and animal fats. Bio-diesel is also defined as the mono alkyl esters of long chain fatty acids derived from renewable lipid feedstock, such as vegetable oils or animal fats, for use in compression ignition (diesel) engines. Bio-diesel commonly quoted as "alternative fuel" which refers to any "fuel, other than alcohol, that is derived from biological materials". It is produced from oils or fats using transesterification and is a liquid similar in composition to fossil/mineral diesel.

Bio-diesel is a clean burning alternative fuel, produced from domestically grown, renewable resources. Bio-diesel is safe to handle and transport because it is as biodegradable as sugar, 10 times less toxic than table salt and has a high flash point of about 148ºC compared to petroleum diesel fuel, which has a flash point of 52ºC. Bio-diesel contains no petroleum products, but can be blended at any concentration with diesel from fossil sources to create a biodiesel blend. It can be used in compression-ignition (diesel) engines with little or no modification. Bio-diesel is simple to use, biodegradable, non-toxic and basically free of sulphur compounds and aromatics.

When Rudolf Diesel designed his prototype diesel engine a century ago, he ran it on peanut oil. He planned that diesel engines would operate on a variety of vegetable oils. But when petroleum diesel fuel arrived on the marketplace, it was cheap, reasonably efficient and readily available and therefore quickly became the diesel fuel of choice. Blending of bio-diesel with diesel would result in the reduction of un-burnt hydrocarbons. The biomass-derived esters can be blended with conventional diesel fuel or used as a neat fuel (100 per cent bio-diesel).

Blending conventional DF with esters (usually methyl esters) of vegetable oils is presently the most common form of bio-diesel. The most common ratio is 80 per cent conventional diesel fuel and 20 per cent vegetable oil ester (also termed "B20" indicating the 20 per cent level of bio-diesel). There have been numerous reports that significant emission reductions are achieved with these blends. No engine problems were reported in larger-scale tests with urban bus fleets running on B20 and its fuel economy was comparable with DF. Ester blends have been reported to be stable, for example, a blend of 20 per cent level bio-diesel with 80 per cent DF did not separate at room temperature over a period of 3 months.

Pure bio-diesel (B100) has a solvent effect, which may well release deposits accumulated on tank walls and in pipes from operation. It will also attack paint and similar surfaces, given the chance. Using high blends of biodiesel, the release of deposits may clog filters initially and care should be taken to replace fuel filters until the build-up of deposits is eliminated. This issue is less of a problem with B20 blends and there is no evidence that lower-blend levels such as B20 have caused filters to become blocked.

Bio-diesel use has been showing remarkable increase since the last decade, particularly in Europe, Germany, France and Australia which have fairly large programs. However, now bio-diesel has also caught the serious attention of the USA, Canada, Malaysia, Indonesia, India, Thailand, Philippines, Ghana, South Africa, China, Zambia and many other countries. Bio-diesel can be stored just like the petroleum diesel fuel and hence does not require separate infrastructure. Feed stocks for bio-diesel include animal fats, vegetable oils, soya, rapeseed, mustard, flax, sunflower, palm oil, hemp, algae and Tree Borne Oilseeds (TBOs) like jatropha (*Jatropha curcas)*, karanja (*Pongamia pinnata)*, jojoba *(Simmondsia chinensis)*, neem *(Azadirachta indica)*, mahua (*Madhuca* species) etc.

Table 28.1: Comparison of Energy Species with Possible Potential

Species	Yield (tons /ha)	Fuel Production (In liters of fuel oil/ha)	Energetic Equivalent (kwh/ha)
Elaeis guineensis	18-20	3,600-4,000	33,900-37,700
Jatropha curcas	6-8	2,100-2,800	19,800-26,400
Aleurites fordii	4-6	1,800-2,700	17,000-25,000
Saccharum officinarum	35	2,450	16,000
Ricinus communis	3-5	1,200-2,000	11,300-18,900
Manihot eaculenta	6	1,020	6,600

CONSTRAINTS IN BIOFUEL PRODUCTION

Despite all bright aspects and potential of biofuel production, there are certain constraints in its production as mentioned below,

- Unorganized plantations/cultivation and collection/ processing of raw material
- Denatured ethanol is treated as a chemical and should not be controlled
- The future of biofuel programme depends on parity of ethanol pricing with that of petrol on ex-refinery basis
- The petrol prices vary with price of crude oil and ethanol prices vary with that of molasses and cane
- The collection of raw material for bio-diesel production coincides with monsoon and too with little duration
- Absence of a reasonable negotiated price of bio-fuel valid for 4 to 5 years
- Complicated procurement system of ethanol and bio-diesel, affecting regular supply to the oil depots
- Variable tax and duty structures on ethanol and bio-diesel from state to state
- Interference by the State Government in
    - putting procedural restrictions on the industrial alcohol/ethanol/methanol and bio-diesel production, sale and distribution mean for blending with petrol and diesel
    - levying imports-excise, entry tax, export or import fee or any other kind of levy
    - creating a transparent unambiguous and well defined mechanism against the possibility of evasion of excise duty and misuse or diversion of rectified spirit mean for ethanol blending in petrol

ADVANTAGE OF BIOFUELS

Apart from energy security, employment generation, economical gain, social security etc., the biofuels have enormous environmental benefits being superior in many aspects than conventional source of energy i.e. petrol and diesel. The worldwide study including India reveals the superiority of the biofuel over conventional fuels of many aspects enumerated as under:

BIO-ETHANOL

- Reduction in carbon monoxide and hydrocarbons by 35 to 50 per cent
- Blended fuels give same mileage as that of gasoline
- Smoother and cooler engine operation
- No change in the engine or the carburetor is required for use of ethanol blended petrol upto 20 per cent
- CO_2 emissions are 57 per cent lower than petrol
- No SO_2 emissions, no NOx emissions, no particulate emissions and hydrocarbon emissions are not carcinogenic
- Ethanol-gasoline (petrol) blend enhances octane numbers, reduces exhaust emissions, bio-degradable and non-toxic
- Lower energy content as compared to gasoline
- Ethanol blend to the extent of 20 per cent commercially viable
- Indigenously produced fuel
- It has a higher octane rating than ethanol-free gasoline available at roadside gas stations which allows an increase of an engine's compression ratio for increased thermal efficiency
- In high altitude (thin air) locations, some states mandate a mix of gasoline and ethanol as a winter oxidizer to reduce atmospheric pollution emissions
- Ethanol is also used to fuel bio-ethanol fireplaces. As they do not require a chimney and are "flueless", bio-ethanol fires are extremely useful for new build homes and apartments without a flue.

BIO-DIESEL

- Particulate matters emissions are 30 per cent lower than diesel
- Hydrocarbon exhaust emissions are 93 per cent lower than diesel
- Nitrogen oxide emissions can be effectively managed and efficiently eliminated from bio-diesel
- Bio-diesel reduces the health risk associated with diesel by release of decreased levels of polycycline aromatic hydrocabon (PAH) and nitrate PAH compounds by 75 to 80 per cent
- The overall ozone (smog) forming hydrocarbon emission is 50 per cent lower than diesel
- Carbon monoxide exhaust emissions are 50 per cent lower than diesel

- It reduces overall exhaust emissions
- Bio-degradable and non-toxic
- Fuel consumption comparable to diesel
- Higher flash point
- Clean burning and renewable fuel
- Indigenously produced fuel
- Biodiesel is less expensive than conventional diesel

TREE BORNE OILSEEDS (TBOs)

There are hundreds of species of plants/ trees/ shrubs found wild or cultivated inside as well as outside the forest areas containing sizeable amount of vegetable oil having domestic and industrial utility. Out of these species, the oilseeds of tree origin like Jatropha (*Jatropha curcas*), Karanja (*Pongamia pinnata*), Jojoba (*Simmondsia chinensis*), Neem (*Azadirachta indica*), Mahua (*Madhuca indica*) etc. have uses like illumination, lubricants and alternative source of biodiesel. They have the capacity to grow and establish in varied agro-climatic conditions and in the wastelands of the country. There is need to exploit the existing potential and also to augment the current potential to achieve the self reliance in biofuel requirement. The TBOs possess 20-60 per cent oil content in seeds/kernel, which are edible fat and non-edible for industrial uses. Some of them can be utilized as source of bio-diesel production.

Table 28.2: Botanical Names, Percentage of Oil/ Fat, Uses and Potential States of some of the TBOs as source of Bio-diesel production in India

S.N.	Common Name	Botanical Name	Oil/Fat (%)	Uses	Potential States in India
1	Neem	*Azadirachta indica*	20 (Dry Fruit)	Bio-pesticide, medicinal	All India
2	Karanja	*Pongamia pinnata*	27-39 (kernel)	Illuminant, lubricant, soap, bio-diesel	A.P., Chhattisgarh, Karnataka, M.P., Maharashtra, T.N., U.P.
3	Pilu	*Salvadora oleoides*	33 (Seed)	Medicinal, soap, candle making	Maharashtra, Rajasthan, Haryana
4	Ratanjyot	*Jatropha curcas*	30-40 (Seed)	Illuminant, lubricant, bio-diesel	All India
5	Jojoba	*Simmondsia chinensis*	50 (Seed)	Lubricant, illuminant, adhesive, pharmaceuticals	Rajasthan, Gujarat, Haryana
6	Wild Apricot	*Prunus armeniaca*	45-50 (Seed)	Medicinal, pharmaceuticals	J&K, H.P., Uttaranchal
7	Tumba	*Citrullus colocynthis*	21 (Seed)	Soap, candle making, industrial	Haryana, Rajasthan

8	Indian Zuzube or Ber	*Zizyphus mauritiana*	33 (Seed)	Bio-diesel	Punjab, Haryana, U.P., Rajasthan, Gujarat, Maharashtra, Madhya Pradesh
9	False Sandal Wood	*Ximenia americana*	49-60 (Kernel)	Lubricant, soap, candles	Dry forests of Deccan Peninsula and rocky coastal regions of Andamans
10	Paras Pipal	*Thespesia populnea*	20 (Seed)	Illuminant	Coastal regions of Konkan, Bengal and Andaman
11	Sainjna	*Moringa oleifera*	35 (Seed)	Illuminant, lubricant, cosmetics, medicinal	All India
12	Nahar	*Mesua ferrea*	45 (Seed)	Illuminant, lubricant, soap, medicinal	Forests of North-East India, Kerala, Karnataka, Andaman
13	Tung	*Aleurites fordii*	50-60 (Kernel)	Highly lubricating, industrial	Assam, Bengal, Nagaland, North-Eastern states
14	Mahua	*Madhuca indica*	35 (Seed)	Illuminant, soap, cocoa butter substitute	Central India Plain, North India
15	South Indian Mahua	*Madhuca longifolia*	35 (Seed)	Soap, illuminant, edible	Monsoon forests of Western Ghats, Deccan and other parts of South India
16	Amoora	*Aphanamixis polystachya*	35 (Kernel)	Soap, medicinal, illuminant	Eastern U.P., Western Ghats, Andaman, Bengal, Assam
17	Silk Cotton	*Bombax ceiba*	20 (Seed)	Soap, edible, illuminant	All India
18	Undi	*Calophyllum inophyllum*	50-63 (Kernel)	Soap, illuminant	Along Sea Coast
19	Siltimur	*Litsea cubeba*	22 (Seed)	Illuminant	East Himalayas, Assam, Manipur
20	Kusum	*Schleichera oleosa*	36 (Seed)	Soap, medicinal, illuminant	Eastern India, Forests of Sub Himalayan tracts in the North
21	False Nutmeg	*Myristica malabarica*	40 (Kernel)	Soap, illuminant	Western Ghat (Konkan Southward)
22	Kokum	*Garcinia indica*	25 (Kernal)	Soap, candle, kokum butter, medicinal	Maharashtra, Goa, Karnataka, Kerala, Tamil Nadu
23	Paradise Tree	*Simarouba glauca*	60-70 (Kernal)	Confectionary, lubricant, soap, candle	Odisha, Karnataka, Gujarat, Andhra Pradesh, Bihar, Chhattisgarh, Maharashtra, Tamil Nadu

There is need to exploit the existing potential of above TBOs. The oil extracted from some of these sources, not only form an essential part of human diet but also serve as important raw-material for manufacturing soap, paints, varnishes, cosmetics, medicines, hair oil, lubricants, textile, auxiliaries and also potential substitute of diesel and kerosene. Oil cakes are used as cattle feed and organic manures.

There are enormous wasteland areas available in India as well as in global level and developing these wastelands by undertaking plantations of tree borne oilseeds has great scope in future. Promotion of TBOs plantation can generate tremendous job opportunity among the rural masses, simultaneously augment the availability of edible and non-edible oils. The oils of *Jatropha curcas* and *Pongamia pinnata* are potential substitute of diesel and the indigenous production of bio-diesel will resist the outflow of valuable foreign exchange caused by the import of crude petroleum. It may lead to a revolutionary transformation of the gloomy economic and energy scenario with an era of economic bloom and prosperity of all sections of society.

CHARACTERISTIC FEATURES OF POTENTIALTREE BORNE OILSEEDS

NEEM *(AZADIRACHTA INDICA)*

Neem is a large evergreen tree, but deciduous in dry areas and occurs in the entire India. It thrives better in drier climate of North-West India. It grows well in any soil and climatic conditions below 3,500 meter altitudes. Generally it is found wild in deccan forests and in scattered form in Northern and Central parts of India. Andhra Pradesh, Gujarat, Haryana, Karnataka, Maharashtra, Madhaya Pradesh, Odisha, Punjab, Rajasthan, Tamil Nadu and West Bengal etc. are some of the potential states.

The neem can be propagated through seeds and vegetative methods like cutting and grafts. The flowering occurs in clusters during February-May and fruiting starts from May to August. The seed collection period in North India is June-July whereas in South India, it is May-June. The dried fruits contain about 20 per cent of oil and 30-35 per cent of proteins. About 25-28 per cent kernel is found in dried fruits and contains about 50 per cent oil. Neem oil is composed of various alkaloids such as azadirachtin, salanin, nimbin and nimbidin.

Every part of the plant like leaf, wood, bark, flower, fruit and seed have the medicinal, pesticidal, industrial and domestic uses. The neem wood is used in making of furniture, house and agricultural implements and it is also used as fuel. The leaf contains protein, fat, carbohydrates, fibre and minerals like calcium, phosphorus and iron. It is used as fodder for cattle like camel and goat as well as for medicinal purposes. The neem bark, which contains nimbin, nimbidin, tannin, amino acid, calcium, potassium and iron, is used in manufacturing of medicines. The flower is used in domestic purposes and it has also the ritual importance. The neem seeds are used for extraction of oil which is used in the manufacturing of bio-pesticides, soap and cosmetics. The cake, which contains nitrogen, phosphorus, potash, protein, fat, fibre and ash, is used as manure. The oil has many alternate uses such as bio-diesel and pesticide purpose.

MAHUA *(MADHUCA INDICA)*

Mahua has been categorized into two major species namely, *indica* and *longifolia*. *Madhuca indica* (syn. *Bassia latifolia*) is a deciduous tree which is generally found in dry-tropical deciduous forests in central plains of North India. *Madhuca longifolia* (syn. *Bassia longifolia*) is an evergreen or semi-evergreen tree which is mostly found in the monsoon forests of South India. Andhra Pradesh, Chhattisgarh, Gujarat, Jharkhand, Karnataka, M.P., Orissa, U.P. and Tamil Nadu are some of the potential states. It can be propagated through seeds in the month of July-August. Flowering starts from the end of February to April, fruits appear in May-June and the seed collection period starts from June-July.

Mahua wood has good timber value and the leaves can be used as green manure and animal fodder. The bark possesses tannin which is useful for dyeing and tanning and the bark has medicinal properties also. The flowers are used for distillation of liquor as well as essential oil. Being rich in sugar, vitamins and calcium, it is also used for edible purpose. The kernels which form about 70 per cent of the weight of seeds contain 45-50 per cent oil in *M. latifolia* and 50-55 per cent in *M. longifolia* which are smaller in size. The oil is used for edible purposes by tribals, in vanaspati and substitute for cocoa butter in confectionery. It has also the industrial (soap/candle making) and medicinal uses. The cake is used as cattlefeed and good manures with insecticidal properties. The cake contains saponius, a powerful fish poison, could be used to control predatory fish present in fish ponds and lakes. The cake is a cheap source of effective wormicide and could be used in maintenance of turf, lawn and golf grounds free from earth worms. The oil has potential use of edible as well as bio-diesel purposes.

KARANJA *(PONGAMIA PINNATA)*

Karanja, a medium sized glabrous tree with a short bole and spreading crown, is distributed throughout India along road side and railway tracks. It is drought resistant and tolerant to salinity and occurs on the water courses. It can be propagated by seeds or by root suckers during July-August. Flowering occurs during April-September depending upon locality and seed collection period starts from May to October.

The karanja produces hard wood for domestic/agricultural uses. The leaves are used as green fodder and as green manure. The seeds contain 27-39 per cent oil which is used for soap making, leather dressing, lubrication, illumination and for medicinal purposes. The seed cake is used as manures for coffee plantations and is also reported to reduce the incidence of root knot nematodes. The oil has many alternate uses and suitable for bio-diesel production.

JATROPHA *(JATROPHA CURCAS)*

Jatropha, a large soft wooded deciduous shrub, is found in India in semi arid conditions. Andhra Pradesh, Gujarat, Rajasthan, Karnataka, Maharashtra etc. are some of the potential states. It can be propagated by seeds and cutting during February-March. The flowering occurs twice in a year during March-April and September-October. The seed collection can be started during June-July and October-November, respectively.

Jatropha grows rapidly and is not browsed by cattle and therefore, it is generally used as hedge plant. All parts of the plant exude sticky, bitterly, pungent and astringent latex which can be used in making ink. The bark contains tannin, wax, resin and saponins which make it useful for industrial purposes. The kernel which forms 60-65 per cent of the weight of seed contains oil about 45-58 per cent of kernel weight and 30-40 per cent of seed weight. The oil is used for illumination without smoke, substitute of diesel and kerosene, lubricants, soaps and candle manufacturing. It can be used as hair oil and has also application to livestock against sores. The cake is used as good manure. Jatropha is grown worldwide for biodiesel. In India, more than 0.2 million hectare area is under Jatropha cultivation for bio-diesel purpose.

JOJOBA *(SIMMONDSIA CHINENSIS)*

Jojoba, pronounced as Ho-Ho-Ba, is an exotic hardy desert shrub which can withstand extreme arid conditions of temperature and very low annual rainfall. However, irrigation and manuring are needed for good yield and economic return. Further, shallow soils and temperature factor have major impact on flowering and fruit yields in A.P. and other Southern States. The possibilities of promoting jojoba in such areas may be examined more critically. Rajasthan, Gujarat, Haryana and A.P. are potential states wherein, it can be propagated through seeds and cutting during October and February. The jojoba plants start yielding consistently in the fourth year and it can yield upto 200 years.

The jojoba seeds contain about 50 per cent of oil. Jojoba oil and its derivatives have diversified uses viz. cosmetics, pharmaceuticals, lubricants, food, electrical insulators, foam control agents, plasticizers, fire retardants, transformer oil. It is also used for obtaining waxes which is used for polishing floors, furniture, automobiles and in candle and textile industries. The cake can be used as cattle feed.

WILD APRICOT *(PRUNUS ARMENIACA)*

Wild apricot, locally called chullu, is found in the dry temperate region of J&K, H.P. and Uttaranchal. It is distributed in the North Western Himalayas particularly in the valley of Kashmir, Chenab, Kullu, Shimla and Garhwal hills. It is propagated through seeds in the month of October and through cutting in February-March.

The fruit of wild apricot mixed with doze of cultivated types are utilized in production of apricot jam, apricot nectar and apricot papad. The fruit is also used by the tribals for the preparation of distilled alcoholic liquor. The kernel which constitutes about 20-40 per cent of fruit contains about 45-50 per cent oil. The oil is used as a substitute for almond oil. The oil is also used for food purposes, cosmetics and pharmaceutical preparations and in medicines for ear ache and other ailments. The cake is used as a feed for livestock.

TUNG (*ALEURITES* SPECIES)

Tung is an exotic deciduous rapid growing tree found in tropical and sub-tropical regions of Eastern Asia and Malaysia. Tung trees are grown for its valuable quick drying oil (A better drying oil than linseed oil). China and United States are the major tung oil producing countries. An elevation of 2,500-3,500 meter and rainfall of about 50-70 inches are best suited to its growth. Out of six species of tung, two species viz. *Aleurites montana* and *Aleurites fordii* have been well acclimatized in Mizoram. More than 5 million grown up tung trees are reported available in Mizoram State.

The plants can be raised from seeds, either by direct sowing or by transplanting. Propagation by budding and grafting has also been suggested. To increase the yield, it is advisable to select seeds from dominantly female flowering trees, but vegetative propagation of high yielding trees will give better yields.

The seedlings are transplanted after six months at a spacing of 3.8 m in rows 9 m apart. Ordinarily 200-250 trees could be maintained in a hectare. The tree is comparatively fast growing and can attain a height of 2.4 to 3 m in the third year, when some of them begin to flower and bear fruit. The full bearing capacity is attained at the 10th year and the tree continues to bear fruits for another 20 years. A ten year old tree generally yields 20-25 Kg of fruits per season. The fruits yield 50-60 per cent of nuts or seeds. In India, the seeds yield 55-60 per cent of kernels which in turn yield 49-65 per cent of oil. The wide range in oil content indicates adequate scope for selection and improvement in oil yield for biodiesel production.

INDIAN JUJUBE OR BER (*ZIZIPHUS MAURITIANA*)

The Indian jujube is native from the Province of Yunnan in southern China to Afghanistan, Malaysia and Queensland, Australia. It is cultivated to some extent throughout its natural range but mostly in India where it is grown commercially and has received much horticultural attention. *Z. mauritiana* is a medium sized tree that grows vigorously and has a rapidly developing taproot, a necessary adaptation to drought conditions. The species varies widely in height, from a bushy shrub of 1.5-2.0 m tall, to a tree of 10-12 m tall with a trunk diameter of about 30 cm.

It is hardy tree that copes with extreme temperatures and thrives under dry conditions with an annual rainfall of 15–225 cm. It also grows well on laterite and medium black soils with good drainage or sandy, gravelly, alluvial soil of dry river-beds where it is vigorously spontaneous. In India, there are 90 or more cultivars available depending on the habit of the tree, leaf shape, fruit form, size, colour, flavour, quality and fruiting season. Among the important cultivars, eleven are described in the encyclopedic Wealth of India: 'Banarasi' (or 'Banarsi'), 'Pewandi', 'Dandan', 'Kaithli' (or 'Patham'), 'Muria Mahrara', 'Narikelee', 'Nazuk', 'Sanauri-1', 'Sanauri-5', 'Thornless' and 'Umran' (or 'Umri').

It is widely grown from seeds which may remain viable for two and half years. The hard stone restricts germination and cracking the shell or extraction of seeds hastens germination. Without pretreatment, the seeds normally germinate within six weeks whereas extracted seeds only need one week to germinate. The germination can be improved by soaking seeds in sulfuric acid. Ber seedlings do not tolerate transplanting, therefore the best alternatives are to sow the seeds directly in the field or to use polythene tubes placed in the nursery bed. Seedlings are ready for budding in 3-4 months. In addition, seedlings from the wild cultivars can be converted into improved cultivars by top-working and grafting. Nurseries are used for large scale seedling multiplication and graft production. The seedlings should also be given full sunlight. The seedlings may need as long as 15 months in the nursery before planting in the field. Scientists in India have standardized propagation techniques for Ber establishment. Budding is the easiest method of vegetative propagation used for improved cultivars. Different types of budding techniques have been utilized with ring-budding and shield-budding being the most successful.

The major production regions for Indian jujube are the arid and semi arid regions of India. Trees in northern India yield 80-200 kg of fresh fruit per tree per year when the trees are in their prime bearing age of 10-20 years. The fruit is eaten raw or pickled or used in beverages. It is quite nutritious and rich in vitamin C and it is second only to guava and much higher than citrus or apples. The fruits are applied on cuts and ulcers and employed in pulmonary ailments and fevers. The fruits are mixed with salt, chilli and peppers and given for indigestion and biliousness. The dried ripe fruit is a mild laxative. The seeds are sedative and are taken, sometimes with buttermilk, to halt nausea, vomiting and abdominal pains in pregnancy. They check diarrhoea and used as poultice on wounds. Mixed with oil, they are rubbed on rheumatic areas. The fatty-acid methyl ester of *Z. mauritiana* seed oil meets all of the major biodiesel requirements. The average oil yield is 4.95 kg oil per tree or 13700 kg oil per hectare.

WILD PLUM *(XIMENIA AMERICANA)*

False sandalwood is synonymous for wild plum and it is also known as wild olive, sour plum and tallow nut in English and kadaranji and siruyilandai in Tamil. *Ximenia americana* is drought resistant tree and mostly available solitary. It is available in open country, savannah, gallery forest, along coastal areas, in the understorey of dry forests, in dry woodlands, or on riverbanks. It can be grown in the area of 0-2000 meter altitude, having mean annual temperature of 14-30ºC and mean annual rainfall of 300-1,250 mm. It can be grown in often poor and dry soils including clays, clay loam, loamy sands, sandy clay loam and sandy soils. It is a bush-forming shrub or small tree of 2-7 m high. Trunk diameter seldom greater than 10 cm and bark colour dark brown to pale grey and smooth to scaly. It is easily propagated through seed and cuttings. The tree regenerates naturally from seed and coppice. Fresh seeds should be sown for good germination.

The fruits being pleasant to eat raw and can be used to make juice, jams and jellies or an intoxicating drink. The pulp of seed and fruit contains hydrocyanic acid and it is advisable not to chew the seed. Seed oil contains oleic, linoleic, linolenic, arachidonic, eicosatrienoic, erucic and nervonic acids. Kernel oil is used as a vegetable butter and as a ghee substitute. Firewood and charcoal are the chief uses of the wood, because the trunk is usually too small to make it useful for timber. Volatile oil from leaf yields benzaldehyde, hydroxy benzyl cyanide and isophorone. Kernels are purgative and the oil is suitable for bio-diesel production.

INDIAN TULIP TREE (*THESPESIA POPULNEA*)

Thespesia populnea, commonly known as Portia Tree, is a flowering plant in the family Malvaceae. It is a small tree or shrub that has a pantropical distribution, found on coasts around the world. However, the Portia Tree is probably native only to the Old World and may have originated in India. It is possibly indigenous to the Hawaiian Islands and elsewhere in the Pacific, but may have been spread by early Polynesians for its useful wood and bast fibres. The Portia Tree reaches a height of 6-10 m tall and a trunk diameter of 20-30 cm. It grows at elevations from sea level to 275 meter and in areas that receive 500–1,600 mm of annual rainfall. The Portia Tree is able to grow in the wide range of soil types that may be present in coastal environments, including soils derived from quartz (sand), limestone and basalt. It favors neutral soils (pH of 6-7.4).

The heartwood of the Portia tree is used to make the *thavil,* a carnatic musical instrument of South India. *Milo* is popular in Hawaii for woodworking (commonly turned into bowls) because of the range of colors expressed (tan, through yellow, to red). The fruits, flowers and young leaves are edible. Other products extracted from the plant include tannin, oil and gums (a dark red resin exudes from the bark). Ground up bark is used to treat skin diseases (India), dysentery and hemorrhoids (Mauritius). Leaves are applied to inflamed and swollen joints (South India). When cut, the young fruit secretes a yellow sticky sap used to treat ringworm and other skin diseases (South India). There is some modern investigation of the plant's effects on high blood pressure and the oil is suitable for biodiesel production.

DRUMSTICK *(MORINGA OLEIFERA)*

Moringa oleifera is called as drumstick tree from the appearance of the long, slender, triangular seed pods. It is also known as horse radish tree from the taste of the roots which resembles horse radish or ben oil tree from the oil derived from the seeds. The moringa tree is grown mainly in semi-arid, tropical and subtropical areas. It is a fast-growing, drought-resistant tree that is native to the southern foothills of the Himalayas in northwestern India. The tree itself is rather slender, with drooping branches that grow to approximately 10 m in height. India is the largest producer of Moringa. Among the states, Andhra Pradesh leads in both area and production followed by Karnataka and Tamil Nadu. Tamil Nadu is the pioneering state as it has varied genotypes from diversified geographical areas and introductions from Sri Lanka.

It is propagated by planting 1-2 m long limb cuttings, preferably in June to August. The plant starts bearing pods 6-8 months after planting, but regular bearing commences after the second year, continuing for several years. It can also be propagated by seeds, which are planted an inch below the surface and can be germinated year-round in well-draining soil.

The Moringa pod is known as *murungai kai*, *saragwa* or *saragwe* in India and is often referred to as "drumstick" in English. In South India, it is used to prepare a variety of food items. It is considered as one of the world's most useful trees, as almost every part of the moringa tree can be used for food or has some other beneficial property. The leaves are highly nutritious being a significant source of beta-carotene, vitamin C, protein, iron and potassium. The leaves are cooked and used like spinach. Its leaves are commonly dried and crushed into a powder and used in soups and sauces.

Murungakai, as it is locally known in Tamil Nadu and Kerala, is used in Siddha medicine. The tree is a good source of calcium and phosphorus. In Siddha medicines, these drumstick seeds are used as a sexual virility drug for treating erectile dysfunction in men and also in women for prolonging sexual activity. The Moringa seeds yield 35-40 per cent edible oil (called ben oil from the high concentration of behenic acid contained in the oil). The refined oil is clear and odourless and resists rancidity. The seed cake remaining after oil extraction may be used as a fertilizer and the oil is a potential source of biodiesel.

INDIAN ROSE CHEST NUT (*MESUA FERREA*)

Mesua ferrea (Ceylon ironwood) is a species in the family Calophyllaceae. It is the national tree of Sri Lanka. The plant is named after the heaviness of its timber and cultivated in tropical climates for its form, foliage and fragrant flowers. It is native to tropical Sri Lanka but also cultivated in the state Assam in India. It is a tall tree reaching up to 30 m tall, often buttressed at the base with a trunk up to two meters in diameter. It has simple, narrow, oblong, dark green leaves of 7-15 cm long with a whitish underside. The emerging young leaves are red to yellowish pink and drooping.

The wood is very heavy, hard and strong. Weight is about 32.7 kgs per cubic foot and density is 1.12 metric ton per m^3. Colour is deep dark red and refractory in sawing and mechanics. It is used for railroad ties and heavy structural timber. Its resin is slightly poisonous, but many parts have medicinal properties. It enhances the complexion and leads to fragility transparency to the skin. The flowers are acrid and anodyne and used for digestive, constipating and stomach ache problems. They are useful in conditions like asthma, leprosy, cough, fever, vomiting and impotency. The seed oil is considered to be very useful in conditions like vata and skin diseases. Dried flowers are used for bleeding hemorrhoids and dysentery with mucus. Fresh flowers are useful remedy for itching, nausea, erysipelas, bleeding piles, metrorrhagea, menorrhagea, excessive thirst and sweating. Oil from the seeds is used for sores, scabies, wounds, rheumatism and for biodiesel production.

STATUS OF BIOFUEL PRODUCTION FROM TBOs IN INDIA

CONSTRAINTS IN UTILIZATION OF TBOs IN INDIA

The Government of India launched National Biofuel Mission in 2003, primarily with a view to explore the potential of biofuels as a cleaner source of energy and to partially offset the growing burden of crude oil import bills. There is need to explore the existing potential of more than 50 lakh metric tons of oilseeds of tree origin in the country. The main advantage of tapping the existing potential of tree borne oilseeds is that no additional land and inputs are needed as in case of agriculture crops, but only the collection of seed and fruits from trees occurring in nature is required. There are some constraints, which restrict the seed collection between 15-20 per cent of the total exploitable potential. Some of the important constraints are mentioned below,

- Most of the oilseeds of tree origin except for sal, occur scattered both in forest as well as in non-forest areas
- The collection period of most of the TBOs is short, about 4-6 weeks which precedes or even coincides the rainy season
- The collections are not confined to areas near inhabitation and easily accessible
- There are no storage facilities near to the collection points and no motorable roads to facilitate the transportation of seeds collected
- The primary seed collector usually does not get the remunerative price for the seeds collected, which discourage him/her to collect the seed from interior areas
- In some cases during the same period, when these oilseeds have to be collected, the tribals have alternative opportunities like collection of tendu leaves, tamarind etc. which seems to be more remunerative
- The task of collection of such oilseeds is quite complex and require all possible support of Government to enhance the collection. The Government's role in creating infrastructure would be important to increase the procurement of seed.

- There are several interlinked and complicated market channels available for disposal of seed collected by tribals or other weaker sections of the society in different States. Prevalence of monopolistic market is a discouraging factor in some cases.
- In some areas, absence of any type of market channel leads to the products rotting without sale

STRATEGY TO OVERCOME CONSTRAINTS IN INDIA

To overcome the constraints in seed collection, marketing, augmentation of potential etc., the following strategies may be adopted.

- Focus on increasing productivity
- Encourage exploitation of large untapped potential of non traditional vegetable oils from TBOs
- Establishment of model seed procurement centers and sub-centers in potential districts of the States
- Installation of pre-processing and processing facilities including oil expeller
- Establishment of elite nursery and compact model plantation
- Technology development, refinement and value addition
- Dissemination of technical know-how through awareness programmes namely trainings, field visits, study tour etc.
- Price support system to the primary seed collector and grower
- Financing for establishment of related commercial activities
- Encourage sale of blended oils

BIOFUEL PRODUCTION PROCESSES

METHODOLOGY FOR PRODUCTION OF BIO-METHANOL

Recently, a new method of gasification by partial oxidation and production of bio-methanol from carbohydrate resources has been developed. This process enables any source of biomass to be used as a raw material for bio-methanol production. In this process, the biomass feedstock must be dried and crushed to powder. When the crushed material are gasified at 900-1000ºC with gasifying agent (steam and oxygen), all carbohydrates are transformed to hydrogen (H_2), carbon monoxide (CO), carbon dioxide (CO_2) and vapour (H_2O). The mixture of gases is readily utilized for generating electricity. The mixture of gases is transformed by thermo-chemical reaction to bio-methanol under pressure (40-80 atm) with Cu or Zn-based catalalyst.

$$CO + 2H_2 \rightleftharpoons CH_3OH + Q \quad \text{(Radiation of heat)}$$

$$CO_2 + 3H_2 \rightleftharpoons CH_3OH + H_2O + Q \quad \text{(Radiation of heat)}$$

METHODOLOGY FOR PRODUCTION OF BIO-ETHANOL

Ethanol is produced by fermentation of any material that contains sugars either in a free form as in the case of molasses or in the form of starch as in the case of grains. Chemically starch is a polymer of glucose. The main processes in ethanol production are enzyme digestion (to release sugars from stored starches), fermentation of the sugars, distillation and drying. The distillation process requires significant energy input for heat (thermal energy).

Molasses contains sucrose, which is easily fermented by yeast to produce alcohol. However, for alcohol production from grains, the starch present in the grain has to completely broken down to glucose by a combination of two enzymes, viz. α-amylase and amyloglucosidase, before it is fermented by yeast to produce ethanol. This additional step of starch hydrolysis and low cost of molasses as compared to grains makes alcohol produced from molasses cheaper by 25-30 per cent. However, treatment of byproducts generated by the process of fermentation of molasses into alcohol is difficult leading to pollution of the environment. Typical yield of alcohol from molasses is 225 liters per ton. Considering molasses contains 45 per cent sugar, only 70-75 per cent of the sugars get fermented to alcohol and the remaining sugars go into spent wash.

On the other hand, in typical grain fermentation the alcohol yield is more than 360 liters per ton of grain, containing 60 per cent starch which corresponds to 85 per cent conversion of sugars to alcohol. Moreover, the remaining sugars/starch do not go into the effluent because the stillage is dried. The dried residue is called Distiller's Dried Grain and Solubles (DDGS), which is an excellent ingredient for animal feed.

Chemical equation for conversion of starch to glucose

$$(C_6H_{10}O_5)_n + nH_2O \rightarrow nC_6H_{12}O_6$$

Chemical equation for conversion of glucose to ethanol

$$C_6H_{12}O_6 \rightarrow 2C_2H_5OH + 2CO_2$$

METHODOLOGY FOR PRODUCTION OF BIO DIESEL

TRANSESTERIFICATION: Biodiesel is a methyl ester formed by a process called transesterification. The transesterification process is the reaction of a triglyceride (fat/oil) with an alcohol to form esters and glycerol. A triglyceride has a glycerine molecule as its base with three long chain fatty acids attached. The characteristics of the fat are determined by the nature of the fatty acids attached to the glycerine. The nature of the fatty acids can in turn affect the characteristics of the biodiesel.

During the esterification process, the triglyceride is reacted with alcohol in the presence of a catalyst, usually a strong alkaline like sodium hydroxide. The alcohol reacts with the fatty acids to form the mono-alkyl ester, or biodiesel and crude glycerol. In most production methanol or ethanol is the alcohol used (methanol produces methyl esters, ethanol produces ethyl esters) and is base catalyzed by either potassium or sodium hydroxide. Potassium hydroxide has been found to be more suitable for the ethyl ester biodiesel production, either base can be used for the methyl ester.

$$\begin{array}{l} CH_2O-\overset{O}{\overset{\|}{C}}-R \\ | \\ CH-O-\overset{O}{\overset{\|}{C}}-R \\ | \\ CH_2O-\overset{O}{\overset{\|}{C}}-R \\ \text{Glyceride} \end{array} + \underset{\text{Alcohol}}{CH_3OH} \underset{\text{Catalyst}}{\overset{OH^-}{\rightleftharpoons}} \underset{\text{Esters}}{3CH_3O-\overset{O}{\overset{\|}{C}}-R} + \begin{array}{l} CH_2OH \\ | \\ CH-OH \\ | \\ CH_2OH \\ \text{Glycerol} \end{array}$$

Fig.28.1: Bio-diesel Production Reaction

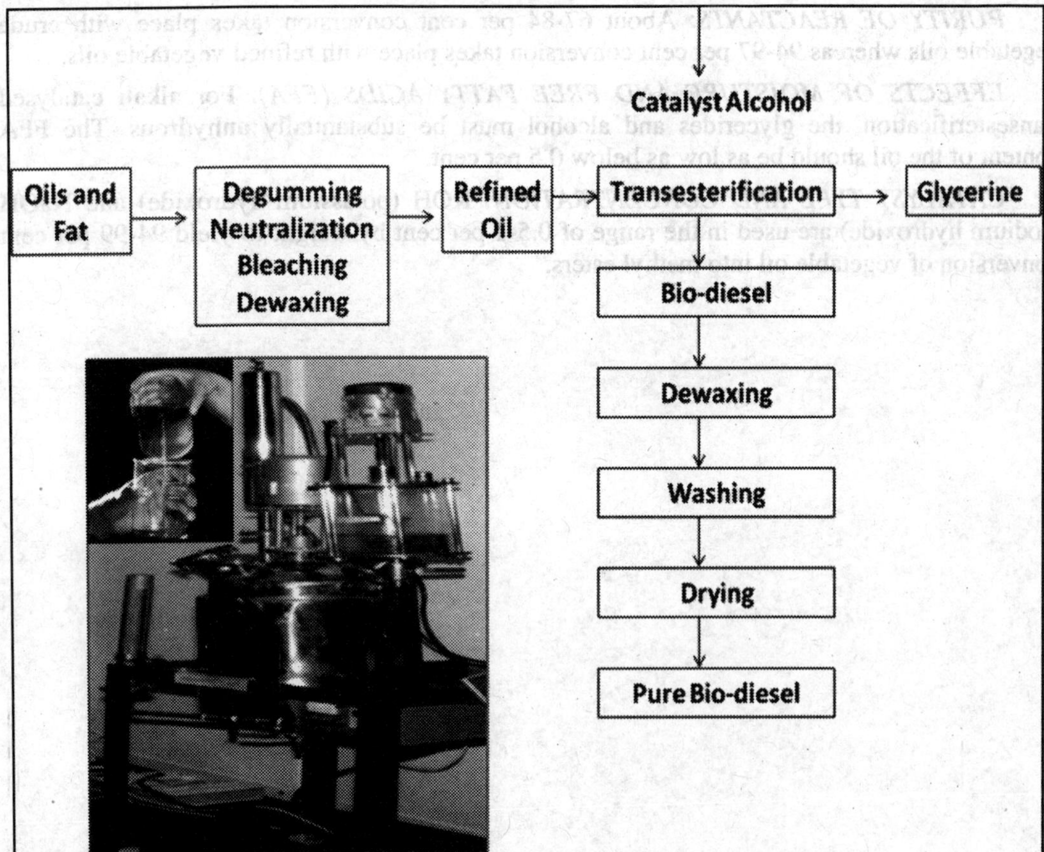

Fig.28.2: Flow Diagram for Production of Bio-diesel

Oil can be extracted from seeds of TBOs with a simple oil mill (as used for mustard seeds). The TBO oil is reacted with methanol in the presence of a catalyst to yield methyl esters and glycerol. Sodium hydroxide and potassium hydroxide are commonly used catalyst. The oil is heated at 65°C. Sodium hydroxide tablets are dissolved in methanol to make a solution. The quantity of NaOH and methanol should be 2.5 and 10 per cent respectively, of the total quantity of TBO oil. After mixing this solution into hot TBO oil, the solution should be stirred for 5 to 7 minutes. Then this solution is kept undisturbed at least for four hours. Glycerol being heavy will slowly settle down at the bottom and biodiesel can be easily separated from the top. To drain the impurities like sodium, 2-3 washings with water is given to oil. Water is added to the oil and after five minutes, the oil floating on surface is collected. This process is repeated and then finally oil is heated to evaporate the water and biodiesel is ready to use.

REACTION TEMPERATURE: Reaction is conducted close to the boiling point of methanol i.e. 60-70ºC under atmospheric pressure. This requires removal of FFA from the oil by refining or pre-esterification.

RATIO OF ALCOHOL TO OIL: A molar ratio of 6:1 (Alcohol:Oil) is normally used in industrial process to obtain Methyl Ester higher than 98 per cent yield.

PURITY OF REACTANTS: About 67-84 per cent conversion takes place with crude vegetable oils whereas 94-97 per cent conversion takes place with refined vegetable oils.

EFFECTS OF MOISTURE AND FREE FATTY ACIDS (FFA): For alkali catalysed transesterification, the glycerides and alcohol must be substantially anhydrous. The FFA content of the oil should be as low as below 0.5 per cent.

CATALYST TYPE AND CONCENTRATION: KOH (potassium hydroxide) and NaOH (sodium hydroxide) are used in the range of 0.5-1 per cent by weight to yield 94-99 per cent conversion of vegetable oil into methyl esters.

PART 6:

PRACTICES FOR INCREASING FOREST COVER

29 URBAN FORESTRY AND RECREATIONAL VALUES

INTRODUCTION

The world is fast becoming an urban place and almost two third of the world's population is expected to exist in urban areas by 2025 (Schell and Ulijaszek 1999). Cities occupy less than 3 per cent of the global terrestrial surface, but account for 78 per cent of carbon emissions, 60 per cent of residential water use and 76 per cent of wood used for industrial purposes. In 1800, there was only one city, Beijing, in the entire world that had more than a million people and 326 such cities exist 200 years later. Indeed, such rapid has been the pace of growth that in 1900 just 10 per cent of the global population was living in urban areas which now exceeds 50 per cent and is expected to further rise to 67 per cent in the next 50 years (Grimm et al 2008). In developed countries, the majority of the population is presently living in cities, e.g., more than 80 per cent of USA and 85 per cent of the Australian population are living in and around urban cities (Brack 2002). Urbanization is a global phenomenon, although the degree of urbanization and the rate of urban growth vary in different parts of the world. People move towards cities not only for employment but also for education of their children and due to increase in annual income. This uncontrolled urbanization resulted in deterioration of natural resources and environment in urban areas in many of the cities in India as well as in the world.

The rapid rate of urbanization causes more urban sprawl, noise, water and air pollution resulting urban areas becoming precarious to live. It is logical that scientists, planners and general public now urgently redesign urban systems that necessarily take into account the fact that 50 per cent of the people are now living in less than three per cent of the earth's urbanized surface. In addition, on the face of climate change, adaptation and mitigation actions for cities in India are critically required where the urban population is likely to grow by around 500 million in few decades. Dealing with multiple risks of climate change such as temperature and precipitation variability, drought, flooding and extreme rainfall, cyclone and storm surge, sea level rise and related environmental health menace is a stern public policy and adaptation management challenge for India (Revi 2008). Without vigilant production of knowledge and large investments to link that knowledge to action, cities will be overwhelmed with environmental challenges. Foremost among these challenges is maintaining human wellbeing by provisioning for clean air and healthy living through conservation and restoration of urban forests (Singh et al 2010).

Urban forestry is the scrupulous care and management of urban forests, i.e. tree populations in urban settings for the purpose of improving the urban environment. The urban forests involve entire cities and their environs and are an environment of trees and related organisms, structures and people. Urban forestry can be defined as the raising and management of trees in and around urban areas. Urban forestry includes tree planting, care and protection and the overall management of trees as a collective resource. Urban forestry is the art, science and technology of managing trees and forest resources in and around urban community ecosystems for sociological, economic and aesthetic benefits trees provide for society (Miller 1997). Urban forests are one of those green infrastructures comprising trees which are more known for their intangible benefits than tangible benefits in urban environment. In urban areas, trees are planted or maintained in back yards, parks, gardens and as avenue trees. These trees in urban areas contribute to an improved quality of urban life in many ways, even though these functions are often taken for granted by the public and urban authorities in most of the developing countries including India.

The urban area has many parts ranging from undistributed natural wood lands to open areas nearly void of trees. There is a great complexity of ownership, with responsibility for care of trees sometimes being legally transferred to others. Rights of ownership are not absolute, in deference to the needs of society. The urban forest is also in a constant state of flux as people continuously plant trees, push them aside, build within natural woodlands and fashion the forests to their needs. The practice of urban forestry is an attempt to manage these activities in such a way as to make trees and related organisms, structures and people as compatible as possible.

INCREASING IMPACT OF PROCESS OF URBANIZATION ON FORESTS

Urbanization refers to general increase in population and the amount of industrialization of a settlement. It includes increase in the number and extent of cities which symbolizes the migration of people from rural to urban areas. Urbanization occurs because people move from rural areas to urban areas in search of better life and facilities. Living in urban areas permits individuals and families to take advantage of the opportunities of proximity, diversity and market place competition. In most of the developing countries including India, urbanization is virtually pulling the people from rural areas with a notion of better standard of living in urban areas. Uncontrolled population growth and the rapid contemporary urbanization impact natural areas and forests both directly and indirectly. Flora and fauna are disrupted and sensitive ecological areas are destroyed. Urban areas are having an ever increasing impact on local, regional and global environments. These areas continue to extend outward, altering or displacing forests, agricultural fields and other valued open spaces. This process of urbanization is likely to be one of the greatest influences on forests in the 21st century.

IMPACT OF URBAN GROWTH ON INDIAN FOREST RESOURCES

Rapid urbanization in India is bringing complex changes to ecology, economy and society. During the last 50 years the population of India has grown two and a half times and the urban population has grown nearly five times (Taubenbock et al 2009). About 60 per cent of this urban population growth is attributable to natural growth and the remaining 40 per cent is due to migration and spatial expansion (Sivaramakrishnan et al 2005). Undoubtedly, urbanization will continue to have significant impact on forest resources and people at local, regional and global scales. Estimates of India's forest resources and rates of depletion are many and varied. The important cause for the forest resource depletion is growth in both urban and rural populations. Urban growth and forest degradation are happening since several decades. The requirements of forest products for essential development generate a pressure on forest resources like fuelwood, fodder, timber, lumber, paper, etc. which in turn triggers a deforestation process. Overexploitation of the forest resources as compared to its incremental and regenerative capacities, escalates the forest depletion and degradation process.

Population-driven pressure on croplands, pasturelands and forestlands coupled with abject rural poverty triggered rural outmigration to urban and industrial centres for wage employment. The pace of urbanization has always been faster than the pace of population growth in India. Furthermore, the process of metropolitization has always been more accelerated than that of urbanization and had further escalated since the 1980s. Agglomeration of the population in the metros and megapolises is primarily because of distress outmigration from rural areas and it has generated a growing demand for timber, lumber and paper which in turn has resulted into depletion and degradation of forest land in India. This process is expected to gather further momentum over the coming years (Gulati and Sharma 2000).

Along with urbanization, linear human land uses such as roads, power lines, water supply, drainage system, etc. are increasingly prevalent in the world. Most of these are built either in forest areas or passing through forests. The effects of linear land uses on forest should be considered in a landscape context. Linear corridors such as roads and power lines in forest areas create forest fragmentation which can exclude sensitive forests and wildlife from the adjoining habitat. Forest fragmentation has deleterious impact on reproductive success of rare and endangered flora. Clearance of forests for roads, railway lines, powerlines and waterlines facilitates the spread of invasive weeds and many exotic weeds to gain dominance in the forested areas. For example, spread of *Chromolaena odorata* along the rail track happened in 20th century in India. In India, urbanization is also associated with development of slums and more often people in slums depend on forest for firewood. Often the forests adjoining urban areas are subjected to deforestation if not degradation due to unscientific and unscrupulous collection of firewood.

Linear corridors act as barriers to the movement of some wildlife species and fragmenting populations of wildlife. Apart from this, road mortality has been well documented for many wildlife species, but the extent of the problem varies with a number of parameters including traffic speed and volume, road type, extent of cleared right-of-way, wild life species present and season. Road-related mortality is a serious problem for some endangered species, such as tiger, elephant especially in wildlife sensitive areas like Bandipur, Dandeli, etc. in Karnataka and Sathyamangalam, Nilgiris, Mudumalai, etc. in Tamil Nadu.

URBAN FORESTRY IN INDIA

In India since primeval period, flowers and plants have been admired and cultivated. There are many references to the urban forests or urban planting in old Sanskrit and Buddhist literature. However, the splendid garden traditions were introduced to India by various muslim conquerors from Persia, North and Central Asia. Some of the mughal gardens are still found in Delhi, Agra, Allahabad and Srinagar. After independence, in newly developed cities like Chandigarh, special care has been taken to include urban forestry in the city's planning.

From the global perspective, there are wide variations both in coverage as well as per capita availability of green spaces in cities. Cities renowned for their urban green spaces often have 20 per cent to 35 per cent coverage of total geographical area with per capita green space of 25 to 100 m^2. Most of the Indian cities, with the exceptions of Gandhinagar and Chandigarh, are far behind in per capita urban forest availability in comparison to European/Australian/US cities (Chaudhary and Tewari 2011). Per capita urban forests or green spaces in some cities and countries are presented in Table 29.1. Per capita urban green space is found to be highest in Netherlands followed by Gandhinagar in India and least is for Jaipur in India.

Table 29.1: Per Capita Urban Green Space or Urban Forests in Important Cities

Country/ City	Per Capita Urban Green Space (m^2)	Source
France/Paris	80.0	Konijnendijk (2003)
The Netherlands	228.0	Konijnendijk (2003)
Australia/Canberra	80.0	Brack (2002)
India/Delhi	21.52	Chaudhary et al (2011)
India/Chandigarh	54.45	Chaudhary et al (2011)
India/Gandhinagar	162.80	Chaudhary et al (2011)
India/Bangalore	17.32	Chaudhary et al (2011)
India/Jaipur	2.30	Chaudhary et al (2011)

URBAN FORESTS IN SOME CITIES OF INDIA

New Delhi, the capital city of India, has grown to be one of the greenest capitals in the world due to the consistent emphasis to grow more trees and strict monitoring of tree cutting permissions. According to Chaudhary and Tewari (2011), presently 20 per cent of Delhi's geographical area is under green cover. Municipal Corporation of Delhi (MCD) maintains nearly 14,000 parks, New Delhi Municipal Council (NDMC) has about 1,000 parks and gardens and Delhi Development Authority (DDA) has many parks, city forests, biodiversity parks and other green belts. Recently, the parks and garden society has been set up to coordinate the greening activities in Delhi. There are nine city forests and two biodiversity parks in Delhi.

Gandhinagar, the capital city of Gujarat has 57.13 per cent of the total geographic area under tree cover making per capita available green space of more than 160 m^2 (Chaudhary and Tewari 2011). Bangalore city of India is known as the Garden City of India due to the large number of parks and private gardens, roadside and avenue trees and the magnificent Lalbagh and Cubbon park. The city has 705 parks spread across the city in the form of small and medium sized parks as well as large parks. Besides these regular parks, there are around 200 open spaces and green areas, which are waiting to be developed as parks and are without any kind of infrastructure. These are basically community amenity sites earmarked for development of community infrastructure such as parks and gardens (Gowda et al 2007). Sudha and Ravindranath (2000) reported that majority of avenue species in Bangalore city are exotic, largely planted for their high growth rate and decorative appearance and there is need to plant trees that provide multiple benefits, particularly in house compounds for providing edible pods, flowers, fruits, leaves, etc., like *Mangifera indica, Murraya koenigii, Moringa oleifera, Tamarindus indica, Artocarpus integrifolia, Phyllanthus embelica* and *Syzygium cumini*. Estimated tree crown cover in the city is about 19.9 per cent of the geographical area (Sudha and Ravindranath 2000) which amounts to per capita green space availability to around 17 m^2(Chaudhary and Tewari 2011).

Chandigarh city, also known as 'city beautiful' has more than 35 per cent of its geographical area under forest and tree cover, making it one of the greenest cities of India and per capita availability of green space is around 55 m^2 (Chaudhary and Tewari 2011). The city has more than 2,000 big and small parks and gardens, besides two reserved forests and a wild life sanctuary.

BENEFITS OF URBAN FORESTS

Trees are an important part of the natural life support system and they have a vital role in the sustainability of towns and cities. There is a growing recognition that urban forests improve the quality of urban life in many ways by providing both tangible (food, energy, timber, fodder, etc.) and less tangible benefits to meet local necessities. Multipurpose urban forestry is especially important for the urban poor.

IMPORTANT TANGIBLE BENEFITS

FOOD: Food from trees in urban parks and gardens can contribute significantly to food security in developing countries (Kuchelmeister 1999). Low-care requiring wild edible plants are often excellent candidates for multipurpose use as ornamental roadside plantings.

FUELWOOD: Fuel wood is one of the important sources of energy in smaller urban centres in developing countries, especially in dry zones (Kuchelmeister 1998). Poor urban households spend a significant proportion of their cash income in obtaining wood energy. If the urban poor population continues to grow, an increase in the consumption of traded woodfuel is likely to be a consequence. Under favourable circumstances, fuel wood from urban forests can contribute significantly to fuel wood supply.

TIMBER: Availability of an adequate timber supply is a problem for a growing number of households in developing countries. Principle sources of timber in urban areas are plantations, street trees, shelterbelts or windbreaks and greenbelts, parks and gardens. In many cities, timber harvesting is combined with intensive outdoor recreation activities. Systematic planting of street trees for timber production is widely practised in China and Malaysia (Webb 1998).

IMPORTANT INTANGIBLE BENEFITS

Trees in urban systems provide a variety of environmental services including biodiversity conservation, removal of atmospheric pollutants, oxygen generation, noise reduction, mitigation of urban heat island effects, microclimate regulation, stabilization of soil, ground water recharge, prevention of soil erosion and carbon sequestration (Bolund and Hunhammar 1999).

BIODIVERSITY: Biodiversity in urban green spaces can be large. If decline in some species are to be arrested or reversed, conservation effort will need to focus much more strongly on understanding and managing urban populations, because these might buffer some species against regional population depletion. Green areas have a vital role in urban biodiversity. Suburban wetlands can be some of the most productive natural ecosystems and can provide important habitats for fauna. Incorporating green areas in networks will improve biological conservation and biodiversity. Greenbelts and greenways (linear parks) can serve as biological corridors. Trees present in urban areas can provide shelter to many important animals especially avifauna which are important from the point of enriching biodiversity in the urban area.

MICROCLIMATE AND AIR QUALITY IMPROVEMENT: Urban trees can help to improve the air quality by cooling and cleaning the air. Landscaping in urban areas involving strategic tree planting can conserve energy and maintain comfort without air-conditioning. Urban shade trees offer significant benefits in reducing air-conditioning demand and improving urban air quality by reducing smog. Since urban trees reduce the need for burning fossil energy, they are a cost-efficient investment for mitigation of greenhouse effects. Planting of vegetation is increasingly utilized as an effective approach for reducing air pollution. Trees mitigate pollution by reducing energy use, carbon dioxide emissions and ground-level ozone. The *Urban Heat Island* is a phenomenon whereby temperatures in urban areas are warmer than the surrounding rural countryside, often by several degrees. As urban green spaces and urban forests increase, evapotranspiration rate increases. Thus, a common measure to mitigate urban heat island effect is to increase urban green spaces. Studies on microclimate formation through built-up morphology and urban shade trees have clearly established the importance of urban trees in alleviating the heat island effect in a hot and humid summer (Shashua-Bar et al 2010).

WATER USE, REUSE AND CONSERVATION: Urban forests can help in the protection of urban water supply, wastewater treatment systems and storm water management. Most poor cities face significant wastewater treatment challenges and could integrate stabilization ponds into park systems and reuse wastewater for urban forestry. Reusing city wastewater not only recharges aquifers but also reduces the demand exerted on scarce water reserves. The greatest potential of wastewater reuse is in arid zones in developing countries (Kuchelmeister 1998).

SOIL CONSERVATION: Trees and forests are a means of soil conservation, preventing landslides in fragile ecosystems with steep terrain, little vegetation and harsh seasonal rains, and thus protecting people's lives and homes. Trees can help to reduce erosion in the different ways such as reduction of impact of falling rain drop on exposed soil, slowing wind and water flow, protection from wind and water erosion, increasing infiltration and soil aggregation.

SOLID WASTE AND LAND RECLAMATION: Recycling of waste from urban trees reduces waste disposal and secures new raw materials (Webb 1998). In poor cities, most wastes may be used as source of fuel and in recent days organic manures are produced from bio-wastes in many cities. Unused and degraded land and terminated landfill sites are increasingly being reclaimed through afforestation and converted to parks. Where land is contaminated, particularly with heavy metals, some trees are planted which are capable of absorbing the pollutants. Through repeated felling and removal of the timber, the level of contamination can gradually be reduced (Dickinson 1996).

SOCIAL BENEFITS

Urban forests play a crucial role in improvement of the society by benefiting the human beings. Some of the important recognizable social benefits are discussed hereunder.

HEALTH: Parks and green areas provide opportunities for healthy physical activity. In addition, the passive benefit to physical and mental health of an urban landscape with trees has been documented in industrialized countries (Ulrich 1984) and enjoyment of green areas may help people to relax or may give them fresh energy. Improving air quality through the planting of vegetation certainly has an impact on health, with such obvious benefits as decreased incidence of respiratory illnesses. Urban forests can also contribute to nutritional food security through edible fruits especially for urban poor in developing countries. Urban open green spaces play an important role in offering town-dwellers a more stress free environment, irrespective of sex, age or socio-economic background. According to Grahn and Stigsdotter (2003), people who spend more time outdoors in urban open green spaces, the less they are affected by stress and related complaints.

EMPLOYMENT: Tree planting and maintenance in urban forests can be labour intensive and provide work opportunities which may be especially important in poorer cities. In wealthier countries, the arboriculture industry is a significant business. Urban forests and green areas also provide opportunities for many kinds of formal and informal establishments related to recreation and related enterprises.

EDUCATION: Trees in urban forests have a pivotal role in environmental and forestry education in urban areas. Urban forests in the form of parks, botanical gardens, zoological gardens, avenue trees and other urban green spaces are the centres of education on flora and fauna. Easily accessible trees and woodlands provide a vital facility for both formal and informal learning. Trees in urban forests create awareness on role of trees in human life which is very much essential for conservation of tree resources.

PROPERTY VALUE IMPROVEMENT: Nowadays, the concept of green cities and eco-friendly attitude of the urban dwellers paved the way for improvement in property value around urban green spaces. Research from around the world indicates that property owners value the urban forest by the premium they pay to live in neighbourhood's urban green spaces and public parks. Many studies have shown increase in house prices where property is associated with urban trees, e.g. upto 5 per cent in Hong Kong (Webb 1998) and in the Finnish town of Salo (Tyrvainen 1999) and up to 18 per cent in the United States (Morales et al 1983). In Singapore and Kuala Lumpur, it has been recognized that a tree-rich urban landscape is an important attraction for new businesses and investors (Kuchelmeister 1998).

RECREATIONAL VALUE: Earlier people used to dwell in rural areas in close contact with natural environment. Rapid industrialization and urbanization coupled with population growth widened the distance between man and natural environment. Also, with the advancement in technology, the work became specialized and can be done quickly with less time. Compared to past, our transportation became reliable and faster. As a result, the general standard of living has gone up. With further advancement in technology, it is predicted that leisure and not the labour would be greatest problem in years to come. In recent past, the public attitude towards the relation of work and leisure has been greatly modified. These facts with their growth and changing composition of population indicate that the future would witness growing importance for recreation as a form of leisure. Forested landscapes or trees which have the ability to provide several goods and services to mankind can also provide recreation to human beings through their inherent nature and by supporting various fauna and other life forms.

Urban dwellers need areas close to nature so as to have a break from their busy, tiring, often repetitive and tedious routine jobs and works. Most of the time, the lower and middle class families cannot afford to travel distant places for recreation. Hence, it is essential to bring a part of nature closer to city through urban parks, gardens etc. which are part of urban forestry. Trees and shrubs found in urban parks or gardens not only provide shade but also enhance the beauty of the area and at the same time attract avifauna. Shade is an essential ingredient of the recreation because it provides coolness to users and contrast to the scenery. Apart from this, the aesthetic value coupled with elegance and fragrance of trees relieves the man from his tension. In other words, the man will be relaxed both physically and mentally from his routine work schedule.

Urban forests and trees in urban parks or gardens are used for morning and evening walks or physical exercises and recreation by all kinds of people. A study in nine cities of Sweden indicated that people of all categories, professions and age consider urban parks/gardens/urban forests as most effective means for stress-relieving and relaxation (Grahn and Stigsdotter 2003). A study undertaken in Guangzhou, China, indicated that more than 50 per cent of its residents use urban forests for recreational purpose and for stress-relieving (Jim and Chen 2006).

ESTABLISHMENT AND MAINTENANCE OF URBAN FORESTS

In urban areas, trees are planted or maintained for many purposes and different situations. Establishment of urban forest involves consideration on planting site, tree species, size of planting stock, maintenance etc.

PLANTING SITE

Several types of planting sites are unique to urban areas including street lawn, tree pit, avenue and cluster planting. These sites may require special considerations when selecting a species and choosing a proper planting technique. There are variety of places to plant seedlings or trees such as residential and business yards, parking lots, empty lots, highway medians, avenues, parks and courtyards between buildings. Depending on the requirement of the planting site and site conditions appropriate species need to be selected. For example, trees suitable for urban avenues may not be suitable for parking areas.

SELECTION OF SUITABLE TREE SPECIES

Trees may experience different types of stress and respond to stress in different ways. Some tree species can tolerate stress such as poor soils, adverse environmental conditions, etc. better than other species. Several important factors of tree species related to growth such as mature size and form, growth rate, branching pattern, foliage characters, bark texture, flowers, fruits, seeds, etc. should be considered when selecting a species.

The assemblage and display of floral trees can achieve very significant and telling effects on the landscape of the crowded cities, slum areas and dry roads. However, these trees should occupy only well thought positions in the landscape. Judicious selection of species for a particular landscape will have added ornamental value. The choice of species should camouflage with the type of habitation, building patterns, colour of the country side, nature of terrain (rocky, undulating, plains, etc.) and its texture (broken, gently slope, smoothly grading into each other or with abrupt transitions, etc.). Landform should be the guiding factors so that the trees fit in well and synchronize with the country side. Menninger (1962) has enumerated in detail, the species suited to different locations for landscape management in urban areas. He grouped the species into nine categories depending on the foliage, floral and other characters.

- Grandeur of the size
- Gracefulness of the branching patterns, tree form and stem character
- Harmony of line and symmetry of form
- Density or pattern of foliage in relation to shade
- Elegance of foliage
- Spectacular foliage colour
- Spectacular floral display colour
- Attractive fruits
- Luxury of fragrance

GRANDEUR OF THE SIZE: Grandeur of the size is a synthesis of several components and dimensions. Flora of India is full of trees of gigantic sizes. The grandeur may be chosen for robustness of stem, massive branching pattern, girth and height of trees and overall build of the trees. Features like buttressing (height and width of plant buttresses), ruggedness of proportion of clean stem, density of foliage, etc. also contribute to the total effect. The grandeur can be best judged when these trees are planted though in single numbers, occupy a dominant space in the landscape.

GRACEFULNESS OF THE BRANCHING PATTERNS, TREE FORM AND STEM CHARACTER: Trees can be chosen for their spectacular branching pattern, curiously distorted and variously blotched branches, twisted as unusual drooping branches, distinguished foliage or floral composition and shape of crown. The trees can be easily categorized into the following types:

Branching Form: Branching form may be round, broad, square and spreading.

Round form: *Ailanthus excelsa, Tectona grandis, Tamarindus indica, Azadirachta indica, Emblica officianalis*

Broad form: *Ficus infectoria, Mangifera indica, Saraca asoca*

Square form: *Alstonia scholaris, Mimusops elengi*

Spreading form: *Albizia procera, Casssia siamea, Plumeria alba*

Species with heavy foliage and thorny habits such as *Acacia nilotica* and *Prosopsis juliflora* should not be planted on road side as the heavy leaf shedding and thorns make them undesirable and cumbersome. Similarly, trees with large leaves and high foliage which permit light penetration and good circulation of air should be preferred for planting in parks. Round and broad tree forms can be planted on pedestrian paths.

Tree Form: Tree form may be conical, columnar, tapering and arrow.

Conical form: *Grevillea robusta, Polyalthia longifolia Bombax ceiba, Spathodia companulata*

Columnar form: *Mellingtona hortensis, Cedrus deodara, Pinus roxburghii*

Tapering: *Polyalthia longilolia, Araucaria cunninghamii*

Arrow forms: *Casuarina equisetifolia*

The broad conical forms are planted to provide colour and texture to the background and form a foreground natural screen. The evergreens keep on adding green when the deciduous have shed their leaves and they clearly define their areas and become dominant features of beauty in the gardens and avenues. Pyramidal forms in groups help to obstruct undesirable views very effectively sight upto the ground level. They also serve as good back ground for a flowering tree in the foreground, a character exploited in Japanese gardens.

Weeping form: *Callistemon* spp., *Salix babylonia, Putranjiva roxburghii, Hardwickia binata*

Weeping forms are famous for their beauty, gracefulness, faminity, sorrow and elegance. These forms should be used occasionally to produce the effects at special points such as small roads, relax spots, corners of park, etc. Generally, they are established in a single specimen. They can also be planted along the moving watersides, stagnant pool of water and in the middle of lawn to create loneliness and sculpturing value and sudden change of mood in the landscape.

Horizontal forms: *Araucaria* spp., *Terminalia arjuna, Chukrasia tabularis, Toona ciliata*

Contorted forms: *Cassia fistula, Jacaranda mimosifolia, Delonix regia, Butea monosperma*

Multi trunk forms: *Albizia lebbeck, Acacia nilotica, A.auriculiformis, A. planiformis, Adanasonia digitata*

Trees with irregular or contorted forms take shapes due to excessive biotic or climatic constraints. However, with the passage of time, they become source of individualistic beauty and convey meaning to different persons. They should be best planted near walking places, resting places and clubs.

Stem Characters: Many trees can be selected to suite urban forestry designs on the basis of various coloured and shaped stems, patterns of bark, natural ramifications of stems and branches.

Crocodile bark: *Terminalia alata, Cinnamomum camphora*

Ash coloured bark: *Eucalyptus* spp.

Pinkish gray bark: *Terminalia arjuna*

Yellowish bark: *Albizia procera*

HARMONY OF LINE AND SYMMETRY OF FORM: Trees can break monotony if they are depicted to show the symmetry of form and harmony when placed in line at variable distances.

Harmony of line: *Mimusops elengi, Barringtonia acutangula, Pinus oocarpa, P. wallichiana*

Symmetry of forms: *Alstonia scholaris, Acacia plantiformis, Michelia champaka, Agathis robusta, Populus nigra*

DENSITY AND PATTERN OF FOLIAGE IN RELATION TO SHADE: The following trees confirm to such characters and produce shade to the desired effects: *Alstonia scholaris, Barringtonia acutangula, Dillenia indica, Chukrasia tabularis, Ficus elastica, Ficus retusa, Mimusops elengi, Madhuca latifolia, Syzygium cumini, Terminalia bellerica, Tamarindus indica*

ELEGANCE OF FOLIAGE: In large open areas or tracts, large trees at greater distances can be planted. The following species are having elegance in their foliage: *Acrocarpus fraxinifolius, Jacaranda mimosifolia, Daubanga grandiflora, Araucaria cunninghamii, Melia composite*

SPECTACULAR FOLIAGE COLOUR: The foliage of some trees being spectacular in colour and changes varying with the seasons. These trees provide foliage colour which impart distinct characters at a particular time of the season. The foliage of some trees changes of tints to delicate pink, gray red, bronze or yellowish orange at the time of leaf fall as well as when the new leaves make their appearance. The variegated foliage also provides a graceful appearance at vantage points in a landscape. New leaves of many trees show spectacularly tinted foliage in the spring.

Colour Tints (Leaf Shedding)

Orange – *Bischofia javanica, Bridelia retusa*

Greyish red – *Lagerstroemia speciosa*

Red purple – *Careya arborea*

Colour Tints (Leaf Sprouts)

Reddish pink – *Saraca asoca*

Red – *Schleichara oleosa, Acer oblongum*

Species with Variegated Leaves

Ficus elastica – yellowish spots

Erythrina indica – yellowish spots

Psidium guajava – with blood red spots

SPECTACULAR FLORAL DISPLAY: A flowering tree provides a feast for eyes when they bloom and they change the colour of the entire landscapes. Brilliancy, colour harmony, fineness, sequence of blooming, bloom density, periodicity of bloom, the time and longevity of bloom are some of the characters which should be known for framing landscaping design.

Table 29.2: Important Urban Trees with Colourful Flowers

Tree Species	Flower Colour
Amherstia nobilis	Flower with golden spots
Bauhinia acuminata	White
Bauhinia purpurea	White
Cassia fistula	Yellow
Cassia javanica	Pink
Delonix regia	Scarlet red
Butea monosperma	Scarlet red or orange
Callistemon lanceolatus	Scarlet
Samanea saman	Pale red
Careya arborea	Pinkish white
Glirisidia maculata	Pale pink
Jacaranda mimosifolia	Violet blue
Millingtonia hortensis	Silvery white
Peltrophorum ferrugineum	Bright yellow
Pongamia glabra	Mauve
Saraca indica	Orange red
Spathodea companulata	Orange crimson
Thespesia papulnea	Yellow and crimson red
Wrightia tinctorea	Red

Well designed plans can be prepared, keeping in view the periodicity of flowering, deciduous habit and colour of flowers.

TREES WITH ATTRACTIVE FRUITS: If the objective of planting in urban forestry is providing shelter for avifauna attractive fruit trees should be planted.

Table 29.3: Urban Trees with Colourful Fruits or Seeds

Species	Seed or Fruit Colour
Adenanthera pavonina	Bright scarlet seed
A. microsperma	Bright red seed
Elaeocarpus robustus	Green fruits
Gelonium multiflorum	Deep orange fruits
Bixa orellana	Reddish seeds
Casimiroa edulis	Yellow fruits

LUXURY OF FRAGRANCE: The mild delicate fragrance brings the pleasure to urbanites. Following species produce highly aromatic flowers: *Aegle marmelos, Holarrhena antidysenterica, Michelia champaka, Millingtonia hortensis.*

Apart from these, important environmental conditions such as hardiness zone adaptation, wind and storm damage, light requirements, pollution tolerance, insect and disease resistance, fire resistance and maintenance requirement need to be considered for selection of species.

APPROPRIATE SIZE OF PLANTING STOCK

Size of planting stock to be planted in different planting site is of paramount importance in establishment of urban forests. In urban environments, forests are established with planting of seedlings or tall tree stocks. Factors to consider when selecting type of planting stock include site conditions and location, time of year, cost and protection against browsing etc. Generally, avenue planting is done with tall seedlings whereas normal seedlings are planted in urban parks. Under special cases, tall tree planting can also be followed to create immediate tree environs in the area.

URBAN FOREST MAINTENANCE

Maintenance of urban forest is an important aspect because the seedlings planted in urban condition should withstand various kinds of disturbances. Need based application of nutrients is essential for better growth and development of urban trees. Tree maintenance includes all those measures employed regularly to improve the health and safety of tree as well as that of its planting site. The important execution of maintenance work should be carried out by trained maintenance team. In contrast, the tree treatment includes all those measures, necessary to restore the long-term health and safety of the flowering tree and its planting site, such as reduction of the crown, the pruning of lopped crown, root treatment or the treatment of trees suffering from salinization and water-logging damage. Tree surgery mainly deals with the treatment of damage to the timber, including the necessary stabilizing measures required for the above long-term treatment. Examples of tree surgery include the opening up of the large areas of rot or decay, the insertion of root or cable braces near the top of the trunk. By the insertion of so called cable braces, wide spreading crown can be prevented from breaking. However, in such cases the correct pruning of endangered branches is often sufficient. In case of rotting due to fungus attack, affected areas and hollow can be treated with fungicides. Edges of the cuts should be treated with callus forming wood dressing. Some of the important treatments such as training and pruning are of paramount importance in urban forest management. Training trees should start at sapling stage for attaining desired form and size. Different kinds of pruning trees in urban tree management includes crown pruning, branch pruning for health, safety etc.

CROWN PRUNING: Pruning of trees and the subsequent treatment of wounds are probably the most important of all tree maintenance practices. Proper and systematic pruning helps trees better to withstand adverse climatic and environmental conditions. In addition, properly pruned trees require less fertilizing, branching and spray materials to keep them healthy. The properly pruned wounds reduce losses due to wood-decay fungi. Trees are pruned principally to preserve their health and appearance and to prevent damage to human life and property.

PRUNING FOR HEALTH: Broken, dead or diseased branches are pruned to prevent decay- producing fungi from penetrating into the part of the tree to which these branches are attached. Live branches are removed to permit penetration of sunlight and circulation of air through the canopy or to compensate for loss of roots.

AESTHETIC PRUNING: Only in formal plantings or to restore the characteristic form of a badly shaped tree, pruning for aesthetic purpose is recommended. Trees should be pruned back to preserve the tree's form.

PRUNING FOR SAFETY: Dead, split and broken branches are a constant hazard to human life and to the property. Low hanging live branches must be removed to a height of 3 to 4 m for avoiding interference with pedestrian and vehicular traffic. It is pertinent to mention that the intensity of pruning is species specific and it has to be carried out according to the species ability to withstand pruning intensity.

30 AGROFORESTRY PRACTICES

INTRODUCTION

India is endowed with a variety of soils, climate, biodiversity and ecological regions. About 228 million ha (69 per cent) of the country (329 million ha) fall within the dryland. The majority of cultivated area is rainfed with enormous unexploited agricultural production potential. An area of 46.70 million ha has been estimated under wastelands during 2008-09 which is 14.75 per cent of the total geographical area of the country (Wastelands Atlas of India 2011). At present, woodfuels account for 20-30 per cent of all energy used in India and more than 90 per cent of woodfuels are consumed in the domestic sector. These situations pose threat to crop yields and warrant for an alternate land husbandry and crop diversification for sustainable production. There is a great risk of growing food grains in degraded and culturable wastelands. The ever growing demands of the increasing population for food, fodder, fuelwood, fruit, fibers, timber, pulpwood, etc. requires emphasis on checking land degradation for which agroforestry practices are considered a most vital technology and a potential farming system for minimizing the land degradation.

The apparent high potential of agroforestry systems are its applicability to control soil erosion, soil improvement, creating congenial and conducive microclimate for tree and understory crops and reduction in the accumulation of greenhouse gases in the atmosphere. Agroforestry has generated rather high levels of enthusiasm in recent years concerning with rainfed land use and sustainable resource management system. It has the most apparent potential in rainfed areas and in resource limiting small holding farming systems where monoculture agriculture may not be the most feasible or desirable, based on the principles of self maintenance.

Keeping in view the increasing human and livestock population and decreasing per capita availability of land and grazing resources, there is an urgent need for the development of location specific agroforestry system at culturable wastelands and degraded lands, which will not only fulfill the futuristic requirement of wood, food, fodder, fuelwood, etc. but also conserve and provide solution to sustainably use the land resources. Therefore, the agroforestry should be considered as a dynamic, ecologically based natural resource management system that can sustain small holder production through the integration of trees in farm and/or in rangelands for increased social, economical and environmental benefits.

CONCEPT AND DEFINITIONS OF AGROFORESTRY

The gap between demand and supply and land degradation processes could be mitigated through diversification of agriculture by adopting alternate land use system by introduction of woody perennials in agricultural land as well as in degraded lands where woody species can only be grown successfully for their reclamation and further used for cultivation of arable crops. In this context, the concept of 'Agroforestry' comes in handy in fulfilling these deficiencies through integration of forestry with agriculture for increasing productivity per unit area per unit time. Agroforestry, as the term, is used for growing of woody perennials in close association with agricultural crops. It is a land use system that involves trees, arable crops and/or livestock seeking positive synergism on the same land unit in space or time. Agroforestry in arable lands can help substantially in meeting such demands and solve some of our pressing problems besides achieving national target of 33 per cent area under tree or forest cover (Tejwani 1994).

Agroforestry is not a new system or new concept, the practice is age old but the term is definitely new. People are rising together woody perennial species, crops and animals, traditionally on the same farm. The practice of mixed farming developed over centuries for meeting most of the requirement of a family. Modern Agroforestry establishes a symbiosis among agricultural crops, woody species and livestock. In other words, these are complementary and beneficial to each other.

AGROFORESTRY DEFINITIONS

- One of the earliest definition of agroforestry is that of Bene et al (1977) as well as King and Chandler (1978) which defined agroforestry as a sustainable land management system which increases the overall yield of the land, combines the production of crops (including tree crops) and forest plants and /or animals simultaneously or sequentially on the same unit of land and applies management practices that are compatible with the cultural practices of the local population.
- Nair (1979) defined agroforestry as a land use system that integrates trees, crops and animal in a way that is scientifically sound, ecologically desirable, practically feasible and socially acceptable to the farmers.
- One of the simple definition of agroforestry is that of Lundgren and Raintree (1982), who defined agroforestry as a collective name for land-use systems and technologies where woody perennials (trees, shrubs, bamboos, palms, etc.) are deliberately used on the same land management units as agricultural crops and/ or animals, in some form of spatial arrangement or temporal sequence. This definition implies that:
    - Agroforestry normally involves two or more species of plants (or plants and animals), at least one of which is a woody perennial
    - An agroforestry system always has two or more outputs
    - The cycle of an agroforestry system is always more than one year
    - Even the simplest agroforestry system is more complex, ecologically (structurally and functionally) and economically, than a monocropping system

- According to Lundgren (1987), agroforestry is a collective name for all land-use systems and practices where woody perennial plants are deliberately grown on the same land management unit as agricultural crops and /or animals, either in a spatial mixture or in temporal sequence. There must be significant ecological and economic interactions between the woody and non-woody components.'
- Young (1989) proposed and defined agroforestry as a collective name for land use system in which woody perennials (trees, shrubs etc.) are grown in association with herbaceous plants (crops, pastures) and /or livestock in spatial arrangement or rotation of both and in which there are both ecological and economic interaction between the trees and non tree components of the system.'
- Sanchez (1995) defined agroforestry as a collective name for land use system and practices in which woody perennials are deliberately integrated with crops and /or animals on the same land management unit.
- It has been argued that the above definitions are not sufficiently dynamic and the following alternative has been suggested by Leakey (1996): Agroforestry is a dynamic, ecologically based, natural resource management system that, through the integration of tree in farmland and rangeland, diversifies and sustains small holder production for increased social, economic and environmental benefits.

A scientific definition of agroforestry stresses on two characteristics common to all forms of agroforestry and separates them from the other forms of land use, namely:

- The deliberate growing of woody perennials in the same unit of land as agricultural crops and/or animals, either in some form of spatial mixture or sequence.
- There must be a significant interaction (positive and/or negative) between the woody and non-woody components of the system, either ecological and/or economical.

All agroforestry systems possess theoretically three attributes which are:

(1) *Productivity*: Agroforestry systems mostly aim to maintain or increase production as well as productivity of the land.

(2) *Sustainability*: Agroforestry can maintain and conserve fertility of soil and sustain the production.

(3) *Adoptability*: Traditional agroforestry practices already been accepted and adopted by the farming community but improved agroforestry system should be encouraged to adopt.

Agroforestry also has four traits, i.e. '4I'

- Intentional
- Intensive
- Interactions
- Integrated

These traits describe that agroforestry practices are *intentional* combinations of trees with crops and/or livestock that involve *intensive* management of the *interactions* between the components as an *integrated* agroecosystem. These key characteristics differentiate agroforestry from other farming or forestry practices.

IMPORTANCE AND IMPACT OF AGROFORESTRY

- Increased per unit production due to introduction of trees
- Growth of trees or woody perennials conserves soil and soil moisture
- Agroforestry improves soil fertility and can improve productivity by increased output of tree products, improved yields of associated crops, reduction of cropping system inputs and increased labour efficiency
- Provides more diverse products and services
- Provides higher revenue due to sale of tree products
- Reduced pressure on remaining natural forests for wood, fuelwood and other forest based products and pressure of grazing
- Enhanced microclimates by increase in atmospheric humidity (10-20 per cent)
- Reduces wind velocity by 40 per cent
- Protects field crops against the scorching and desiccating conditions
- Trees supply cheap fuelwood, fodder, small timber, poles
- Helps in reducing carbon loss from soils by sequestrating CO_2 from atmosphere
- Reduction of run-off, nutrient leaching and soil erosion by tree roots and stems
- Enhances recycling of nutrients and nutrient pump due to deep and tap rooted trees
- Addition of soil nutrients through litter fall and their decomposition
- Improves soil physical and chemical conditions through the accumulation and decomposition of organic matter
- Reduced chances of total crop failure which is common to monoculture system
- Increase in farm incomes due to improved production and sustained productivity
- Help in improvement of rural living standards from sustained employment and recurring incomes
- Improvement in nutrition and health due to increased quality and diversity of food outputs
- Better protection of ecological systems

SCOPE

Agroforestry has great potential to rehabilitate and reclaim degraded lands for sustainable production. It is one of the important alternatives farming system for improvement of degraded sites as it can play a very vital role in utilization of all the natural resources in a most effective manner for sustainable crop production by maintaining the long term production potential of land. It is an important tool for arresting land degradation and amelioration of lands. It is a land management system that optimizes the overall land productivity and reduces resources depletion through the utilization of positive interaction between crop-tree-animals in its temporal and spatial dimensions. Agroforestry can be seen as phases in the development of productive agro-ecosystem and an alternate land use system, wherein, farmers can manipulate and manage their land by growing trees for the service or products (Tewari 1995).

Demands of our rising population for food grains, fuelwood, timber, fodder, etc. are increasing day by day and the present availability and supply are in short. Moreover, over 200 million people, mostly poor, are highly dependent on forests for meeting their subsistence and livelihood needs (FSI 2009). The above needs cannot be fetched from the traditional farming practices. An alternative land use system has to be evolved to fulfillment of needs of escalating population and solve the land and climatic degradation. The agroforestry practice is the answer to release the pressure on land upto some extent. Agroforestry is a valuable and viable tool for increasing the tree cover as well as to meet the demand for wood and food. The several agroforestry practices already proved that it can meet out the demands of timber, fuel, fodder, non-timber minor forest produce and medicinal and aromatic plants. The sustainable development and management of these lands through agroforestry innovations not only meet the multiple needs for food, fodder, fuel wood, fibre, fertilizer, NWFPs, medicinal and aromatic plants etc. but also ensure economic and environmental security for human beings by synchronization of proper land uses. Agroforestry is the only realistic approach and tool to conserve natural resources and to meet the target of increasing forest cover to 33 per cent from the present level of 23.81 per cent (FSI 2011).

The increasing pressure on land as well as on forests, rising demand for fuel and fodder and finally the requirement of timber for multifarious purposes have strengthened the need of agroforestry as an alternate land use system especially at natural resource conservation front. Today, the main concern of agroforestry is to ensure optimum use of natural resources and develop viable multiple use alternatives which can improve the overall economic condition of the farmer. It is also important that the system what we are developing and promoting must be sustainable. The purpose of growing trees with crops and/ or animals in interacting combination has variety of objectives including biological control of soil erosion, where the emphasis is now forced upon. This system not only control physical erosion but also reduces losses of soil organic matter and nutrients. Today, the potential of agroforestry for soil conservation is generally accepted and widely propagated. The system also helps in reduction of atmospheric carbon by sequestering in tree and biomass development (Tewari 1995).

CLASSIFICATION OF AGROFORESTRY SYSTEMS

In the early stages of agroforestry development, several attempts were made to classify agroforestry systems (Nair 1993). However, the most obvious and easy-to-use criteria for classifying agroforestry systems are the spatial and temporal arrangement of components, importance and role of components, production aims or outputs from the system and the social and economic features. They correspond to the system's structure, function (output), socioeconomic nature, or ecological (environmental) spread. These characteristics also represent the main purpose of a classification scheme. Therefore, agroforestry systems can be categorized according to these sets of criteria (Tejwani 1994):

- *Structural basis*: It refers to the composition of the components, including spatial arrangement of the woody component, vertical stratification of all the components and temporal arrangement of the different components.
- *Functional basis*: It refers to the major function or role of the system, usually delivered by the woody components (these can be of a productive, service or protective nature, e.g., soil conservation, windbreak, shelterbelt, etc.).
- *Socioeconomic basis*: It refers to the level of inputs of management (low input, high input) or intensity of management and commercial goals (subsistence, commercial and intermediate).
- *Ecological basis*: It refers to the environmental condition and ecological suitability of systems based on the assumption that certain types of systems can be more appropriate for certain ecological conditions i.e., there can be separate sets of agroforestry systems for arid and semiarid lands, tropical highlands, lowland humid tropics, etc.

Several agroforestry systems have been evolved for different areas and one system differs from the other in respect of structure, function, arrangement of components and composition. Nair (1985) classified the agroforestry systems based on nature of components, structure, function, socio-economic and ecological factors, etc.

STRUCTURAL CLASSIFICATION OF AGROFORESTRY SYSTEMS

The structure of the system can be defined in terms of its components and the expected role or function. However, it is important to consider the arrangement of components in addition to their type. In this structural classification, the type of component and their arrangement both are important consideration. Hence, on the basis of structure, AF systems can be grouped into two categories: classification based on nature of components and arrangement of components.

CLASSIFICATION BASED ON THE NATURE OF COMPONENTS

Based on the nature of components, AF systems can be classified into the following categories:

(1) Agrisilvicultural systems

(2) Silvopastoral systems

(3) Agrosilvopastoral systems

(4) Other systems

AGRISILVICULTURAL SYSTEMS

It refers to combined production of agricultural crops and trees including shrubs/vines and forest crops. This system involves the conscious and deliberate use of land for the concurrent production of agricultural and forest crops both. Based on the nature of the components this system can be categorized into various forms:

- Improved fallow species in shifting cultivation

- The Taungya system
- Multispecies tree gardens
- Alley cropping (Hedgerow intercropping)
- Multipurpose trees and shrubs on farmlands
- Crop combinations with plantation crops
- Agroforestry fuelwood production
- Shelter-belts and Wind-breaks
- Soil conservation hedges etc.

SILVOPASTORAL SYSTEMS

It refers to combined production of trees and fodder crops/grasses and/or animals. This system deals with combined cultivation of timber yielding tree and fodder yielding tree plus grasses and /or animals. This system is further classified into three groups:

- Protein bank
- Living fence of fodder trees and hedges
- Trees and shrubs in pasture

AGROSILVOPASTORAL SYSTEMS

It refers to combined production of trees, agricultural crops and pasture and/or animals for production of food, fodder, timber and also grasses/grazing.

OTHER SYSTEMS

The following systems can be included in this category:

- ***Apiculture with Trees***: In this system, various honey (nectar) producing tree species frequently visited by honeybees are planted on the boundary, mixed with an agricultural crop. The main purpose of this system is the production of honey. Species preferred are *Eucalyptus* spp., Litchi, *Mangifera indica, Madhuca indica,* etc.
- ***Aquaforestry***: In this system, various trees and shrubs preferred by fish are planted on the boundary and around fish-ponds. Tree leaves are used as forage for fish. The main or primary role of this system is fish production and bund stabilization around fish-ponds. Species preferred are *Salix* spp., *Eucalyptus* spp., *Syzygium cuminii, Terminalia arjuna,* etc.
- ***Multipurpose Wood Lots***: In this system, special location-specific MPTs are grown mixed or separately planted for various purposes such as wood, fodder, soil protection, soil reclamation, etc.

Following are common agroforestry systems/practices prevailing in different agro-ecological regions of India classified on the basis of the nature of components (Dhyani 2009).

- Agri-silviculture (forest trees + crops)

- Boundary Plantation (trees on boundary + crops)
- Block Plantation (Block of trees + Block of crops)
- Energy Plantation (trees + crops during initial years)
- Alley cropping (hedges + crops)
- Agri-horticulture (fruit trees + crops)
- Agri-silvi-horticulture (forest trees + fruit trees + crops)
- Agri-silvi-pasture (forest trees + crops + pasture or animals)
- Silvi-olericulture (forest trees + vegetables)
- Horti-pasture (fruit trees + pasture and /or animals)
- Horti-olericulture (fruit trees + vegetables)
- Silvi-pasture (forest trees + pasture and / or animals)
- Forage forestry (forage trees + pasture)
- Shelter belts (trees for wind protection + crops)
- Wind breaks (rows of trees + crops)
- Live fence (shrub and trees on boundary)
- Silvi or Horti-Sericulture (forest trees or fruit trees + silkworm)
- Horti-apiculture (fruit trees + honeybee)
- Aqua-forestry (forest trees + fishes)
- Homestead (multiple combinations of forest trees, fruit trees, vegetables etc.)

CLASSIFICATION BASED ON ARRANGEMENT OF COMPONENTS

The arrangement of components i.e. trees, shrubs, crops and/or animals or multispecies combinations involve in the dimensions of space and time.

SPATIAL ARRANGEMENT: Spatial arrangements of plants in an agroforestry mixture may result in dense mixed stands (as in homegardens) or in sparse mix stands (as in most systems of trees in pastures). The species (or species mixtures) may be laid out in strips of varying widths either on boundary or inside the fields.

TEMPORAL ARRANGEMENT: Temporal arrangements of plants in agroforestry may also take various forms. An extreme example is the conventional shifting cultivation cycles involving 2-4 years of cropping and more than 15 years of fallow cycle. Similarly, some silvopastoral systems may involve grass leys in rotation with some species of grass remaining on the land for several years. These temporal arrangements of components in agroforestry are termed coincident, concomitant, overlapping (relay cropping), separate and interpolated.

FUNCTIONAL CLASSIFICATION OF AGROFORESTRY SYSTEM

The production and protection are theoretically two fundamental attributes of all agroforestry systems. This implies that agroforestry systems have a productive function yielding one or more products that usually meet basic needs, as well as a service role (i.e., protecting and maintaining the production). Productive functions of agroforestry systems are producing one or more products for food, fodder, manure, fuelwood, etc. Protective functions of agroforestry systems are protecting and maintaining production systems by the system functions e.g. wind-break, shelter-belt, soil conservation, moisture conservation, soil improvement, shade (for crop, animal and man). However, the emphasis on production of outputs should not diminish the importance of sustainability.

CLASSIFICATION BASED ON ECOLOGICAL GROUPING

Agroforestry system characterizations pertain to specific ecological conditions of different geographical regions. The most famous and intensive effort in describing several existing agroforestry systems were made by Nair (1985). Based on the major agroecological zones, agroforestry systems are grouped into the following categories:

HUMID/ SUB-HUMID LOWLANDS: This region is characterized by hot humid climate for all or most of the year and evergreen or semi-evergreen vegetation. The lowland humid and sub-humid tropics (commonly referred as the humid tropics) are by far the most important ecological regions. Because of climatic conditions that favour rapid growth of a large number of plant species, various types of agroforestry plant associations can be found in areas with a high human population, plantation of crops with combinations and multilayer tree gardens, trees on rangelands and pastures, etc.

SEMI-ARID/ARID LANDS: This region is characterized by less rainfall confined to 9-21 days in July-September, 2-4 wet months, solar radiation incidence (400-500 cal/cm^2/day), high wind velocity (20 km/hour), high potential evapotranspiration (6 mm/day) and high mean aridity Index (70-74.8 per cent).

HIGHLANDS: Variable rainfall, degraded and shallow lands at high altitude to deep rich soils in valleys and great climatic variations are the features of highlands. This area is a storehouse of great biological diversity. The Himalayan region is an excellent example of this type of area. Agroforestry has long encompassed many well-known land-use systems practices. A system based on the nature of the components and their major functional characteristics for specific purpose appears more logical, simple and practical purpose oriented approach to classification of Agroforestry systems. Again, the choice for a system may be based on many interacting considerations - social, ecological and economical. The right choice for the right situation is necessary.

CLASSIFICATION BASED ON SOCIO-ECONOMIC CONDITION

Socioeconomic criteria such as scale of production, level of technology, input and management have been used as a basis for classifying agroforestry systems. Based on such socioeconomic criteria as scale of production and level of technology input and management, agroforestry systems have been grouped into three categories:

Commercial System: The major aim of the system is production of the output (usually a single commodity) for sale.

Intermediate System: Intermediate agroforestry systems are those that are intermediate between commercial and subsistence scales of production and management, i.e. production of perennial cash crops and subsistence crops undertaken on medium-to-small-sized farms.

Subsistence System: Subsistence farmers those who produce most of what they consume, or consume most of what they produce.

Table 30.1: Classification of Agroforestry Systems and Practices (Nair 1985)

Categorization of Systems (Based on their structure and functions)			**Grouping of Systems (According to their spread and management)**	
Structure (nature and arrangement of components, especially woody ones)		**Function (role and/or output of components, especially woody ones)**	**Agro-ecological environmental adaptability**	**Socio-economic and Management level**
Nature of components	**Arrangement of components**			
Agrisilviculture (crops and trees incl. shrubs/trees and trees)	*In space (spatial)* *Mixed dense* (e.g., home garden)	*Productive function* Food Fodder Fuelwood	*Lowland humid tropics*	*Based on level of technology input* Low input (marginal)
Silvopastoral (pasture/animals and trees)	*Mixed sparse* (e.g. most systems of trees in pastures)	Other woods Other products	*Highland humid tropics* (above1,200 msl, Malaysia)	Medium input
Agrosilvopastoral (crops, pasture/ animal and trees)	*Strip* (width of strip to be more than one tree)	*Protective function* Windbreak Shelterbelt Soil conservation Moisture conservation	*Lowland subhumid* tropics (savanna zone of Africa, Cerrado of South America)	High input
Others (multipurpose tree lots, apiculture with trees, aquaculture with trees, etc.)	*Boundary* (trees on edges of plots/fields)	Soil improvement Shade (for crop, animal and man)	*Highland subhumid* tropics (tropical highlands) (e.g. in Kenya, Ethiopia)	*Based on cost/benefit relations* Commercial
	In time (temporal) Coincident Concomitant Overlapping Sequential (separate) Interpolated			Intermediate Subsistence

SELECTION CRITERIA FOR SUITABLE AGROFORESTRY TREES

Tree species that are chosen to be grown in conjunction with agricultural crops should have *inter alia* the following characteristics:

- Identifying the proper species suited to the soil and site conditions
- The selected species should be capable of providing the desirable products and services
- Species adoptability and compatibility with the companion crop
- Tree species should be fast to medium growing
- High value that a slower growth rate is acceptable
- Single stemmed with clear bole and tall
- Amenable to early wide spacing
- Deep-rooted with tap root system so the trees do not compete with companion crops for moisture and nutrient
- Drought resistance or capable of growing on a wet site
- Trees should have a low crown diameter to bole diameter ratio
- Light branching in habit
- Produce a light rather than a heavy shade
- Phyllotaxis should permit the penetration of light to the ground
- Tree phenology should be advantageous to the growth of the annual crops
- Litter fall and litter decomposition rate should have positive effect upon the soil
- Tree should have efficient nutrient pump system
- Preferably leguminous and nitrogen fixer and MPTs
- Possess self-pruning properties
- Tolerance to relatively high incidences of pruning
- Marketability of products (wood, fruit, fodder, fuel, etc.)

TREE CHARACTERISTICS FOR AGRISILVICULTURE SYSTEM

- Tree species medium to fast growing
- Straight, single stemmed with clear bole and tall
- Low crown and less branching habits
- Good survival and adaptability
- Tap and deep rooted species with efficient nutrient pump
- Preferably leguminous and nitrogen fixer and MPTs
- Less competitive with associated crops for water, light and nutrient
- Resistance to disease and pests
- Trees should not act as a common host for diseases and pests

TREE CHARACTERISTICS FOR SILVIPASTURE SYSTEM

- Tree species medium to fast growing
- Crown should be round with medium diameter
- Multiple stemmed and upright branch angle
- Ability to produce green fodder during lean period
- Leaf should have high palatability and good nutritive value
- Preferably leguminous and nitrogen fixer
- High coppicing power and liable to high lopping and pollarding habits
- Deep rooted with efficient nutrient pump
- Hardy and tolerance to adverse environments
- Resistance to disease and pests
- Trees should not act as a common host for diseases and pests.

In reality, no individual tree can be successful for all sites and can provide all type of services. However, the tree should be selected which has high adaptability to site, provide early and economic returns, multiple uses, maximize land use in time and space, complimentary to the system and non-competitive with associate crops or /and other component of agroforestry system. The suitable tree species to be used in agroforestry largely depend on soil and site conditions, nature of agriculture/horticultural crops, objectives of farming and agro-climatic conditions, market demand, industrial demand and need of the farmer.

MULTIPURPOSE TREE SPECIES IN AGROFORESTRY

The multipurpose trees are woody perennials that can provide multiple products and services in agroforestry systems. They may supply food in the form of fruit, fodder, leaves and at same time supplying wood, fuelwood, fiber, flosses, fertilizer (add nutrients to the soil) and supply some other products (silk, gum, lac etc.). In short it can be said that multipurpose trees are those trees which provide multiple products and perform various functions and the trees have a greater impact on a farmer's well being because they fulfill more than one basic need.

The MPTs can be integrated with agroforestry or any other form of tree farming system to improve total biomass and yields, to diversify products, to increase economic resiliency and to improve farm viability and sustainability in the long-term. The multiple products and functions of multipurpose trees can be summarized as follows:

- food (fruit/vegetables/root/shoot/oil)
- wood /timber
- small wood
- fuelwood
- fodder

- tannin/dyes/gums
- wood pulp
- Litter/green manure/ mulch
- pest control
- living fence
- medicinal uses
- honey and nectar production
- shade
- wind resistance (to use as a component in a windbreak)
- erosion control
- nitrogen fixation
- soil fertility improvement
- shade and shelter to wildlife habitat

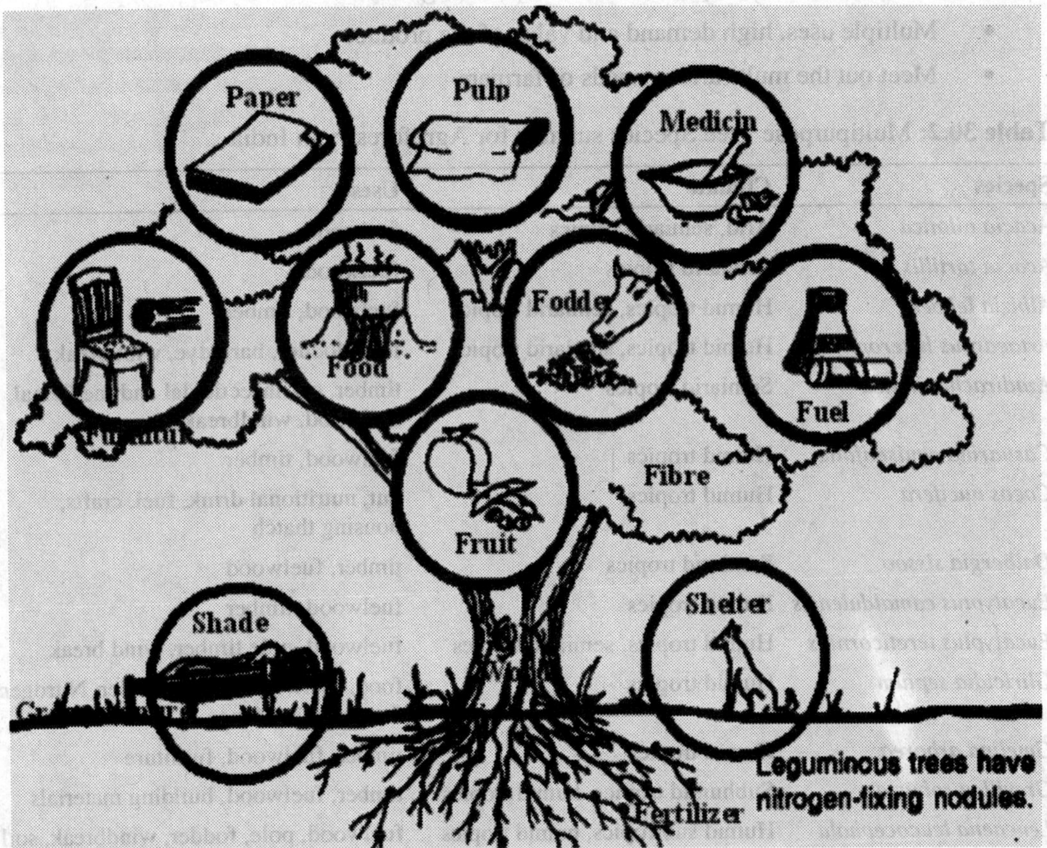

Fig. 30.1: Benefits of MPTs - '6Fs' - (food, fruit, fodder, fuel, fiber and fertilizer)

The most important benefit of multipurpose trees planting and integrating in other forms of tree farming system is that they can provide an assured amount of insurance in the event of a agricultural crop failure or market fluctuation. The desirable characteristics of MPTs to be considered while choosing the tree species are:

- Easy establishment and management
- Easy adaptability to local climatic condition
- Fast growing
- Short rotation
- Preferably nitrogen fixing ability or leguminous species which can contribute to soil nutrients
- Deep root system to draw nutrients and water from lower soil layers
- Easy availability of quality planting stocks and easy propagation
- Provide fodder and green manure
- Tolerance to drought and minor injuries and withstand against flood
- Resistant to insects, pests, diseases and strong winds
- Multiple uses, high demand and value of the produce
- Meet out the multifarious needs of farmers

Table 30.2: Multipurpose Tree Species suitable for Agroforestry in India

Species	Climate	Uses
Acacia nilotica	Arid, semiarid tropics	fuelwood
Acacia tortillis	Semiarid tropics	fuelwood
Albizia lebbek	Humid tropics, semiarid tropics	fuelwood, timber
Artocarpus heterophyllus	Humid tropics, semiarid tropics	fruit, fodder, bark dye, windbreak
Azadirachta indica	Semiarid tropics	timber, oil, insecticidal and medicinal, fuelwood, windbreaks
Casuarina equisetifolia	Humid tropics	fuelwood, timber
Cocos nucifera	Humid tropics	nut, nutritional drink, fuel, crafts, housing thatch
Dalbergia sissoo	Semiarid tropics	timber, fuelwood
Eucalyptus camaldulensis	Humid tropics	fuelwood, timber
Eucalyptus tereticornins	Humid tropics, semiarid tropics	fuelwood, pole timber, wind break
Gliricidia sepium	Humid tropics	food, fuelwood, poles, fodder, Nitrogen fixing, living fence
Gmelina arborea	Humid tropics	timber, fuelwood, furniture
Grevillea robusta	Subhumid tropics, humid tropics	timber, fuelwood, building materials
Leucaena leucocephala	Humid subtropics, humid tropics	fuelwood, pole, fodder, windbreak, soil improver

Mangifera indica	Semiarid tropics	fruit, medicine, leaves as mulch, timber, windbreak
Moringa oleifera	Semiarid tropics	fruit, medicine, fodder,
Prosopis cineraria	Arid and semiarid tropics	windbreak
Sesbania grandiflora	Humid tropics	fodder, fuelwood, food
Tamarindus indica	Semiarid tropics	shade, fruit, light timber, medicinal uses
Ziziphus mauritiana	Semiarid tropics, subhumid tropics	food, shade

SOIL FERTILITY IMPROVEMENT IN AGROFORESTRY

Tree legumes (nitrogen fixing trees-NFTs) are especially valuable trees integrated into an agroforestry system especially in subtropical and tropical agroforestry to restore nutrient cycling and fertility self-sufficiency. Nitrogen is an essential nutrient for plant growth and NFTs are a major source of nitrogen in tropical ecosystems. In legume, nitrogen fixation starts with the formation of a nodule. A common soil bacterium, Rhizobium, invades the root and multiplies within the cortex cells. Nitrogen fixing trees (NFTs) are trees and shrubs that have the ability, through a symbiotic association with certain soil bacteria, to take nitrogen from the air and use it for growth. NFTs can make substantial contributions to nitrogen inputs. The major group consists of leguminous tree species nodulating with *Rhizobium, Bradyrhizobium,* etc. these include many of the most widely used MPTs such as *Acacia, Erythrina, Gliricidia, Leucaena* and *Sesbania* species. In addition, a limited number of non-leguminous genera such as *Alnus* and *Casuarina* species nodulate with *Frankia.*

The NFTs can be integrated into a system in many different forms like plantings, alley cropping, contour hedgerows, shelter belts, live fencing, protein bank, etc. Apart from productive system, they can serve many functions: microclimate for shade-loving crops like coffee or citrus, support for vine crops like vanilla, pepper, and yam, mulch banks for home gardens and living fence and fodder sources from around fields. The important soil improvement trees and shrubs for agroforestry is presented in Table 30.3.

Table 30.3: List of Soil Improvement Tree Species for Agroforestry

SOIL IMPROVEMENT TREE SPECIES		
Acacia auriculiformis	*Casuarina equisetifolia*	*Paraserianthes falcataria*
Acacia cyanophylla	*Casuarina glauca*	*Parkia biglobosa*
Acacia mangium	*Centrosema pubescens*	*Paulownia elongata*
Acacia mearnsii	*Cordia alliodora*	*Peltophorum dasyrrachis*
Acacia nilotica	*Crotalaria* spp.	*Populus deltoides*
Acacia senegal	*Dalbergia sissoo*	*Prosopis chilensis*
Acacia seyal	*Dactyladenia barteri*	*Prosopis cineraria*
Acacia tortilis	*Dendrocalamus* spp.	*Prosopis glandulosa*
Albizia lebbeck	*Erythrina caffra*	*Prosopis juliflora*
Albizia saman	*Erythrina orientalis*	*Prosopis tamarugo*
Anacardium occidentale	*Erythrina poeppigiana*	*Schinus molle*
Alnus acuminata	*Faidherbia albida*	*Senna reticulata*
Alnus nepalensis		*Sesbania bispinosa*

Alnus spp.	*Flemingia macrophylla*	*Sesbania grandiflora*
Atriplex spp.	*Gliricidia sepium*	*Sesbania rostrata*
Azadirachta indica	*Grevillea robusta*	*Sesbania sesban*
Bactris gasipaes	*Inga edulis*	*Tamarix aphylla*
Bamboo genera	*Inga jinicuil*	*Tephrosia candida*
Cajanus cajan	*Leucaena diversifolia*	*Tephrosia vogelii*
Calliandra calothyrsus	*Leucaena leucocephala*	*Tithonia diversifolia*
Cassia siamea	*Melaleuca leucadendron*	*Ziziphus mauritiana*
Cassia spectabilis	*Melia azedarach*	*Ziziphus nummularia*
	Musanga cecropioides	*Zizyphus spina-christi*

The nutrient cycling consists of inputs into (gains), output from (losses) and internal turnover or transfer within the system. The paths of these gains, losses and transfers are also similar: inputs come through fertilizer, rain, dust, organic materials from outside the system and N_2 fixation (for N) as well as weathering of rocks (for other elements); the principal outputs are derived from erosion, percolation (leaching), crop harvest (for all nutrients), denitrification and volatilization (for N) and burning (for N and S). The general pattern of nutrient cycling in an agroforestry system, agricultural system and forestry system is given in Fig. 30.2.

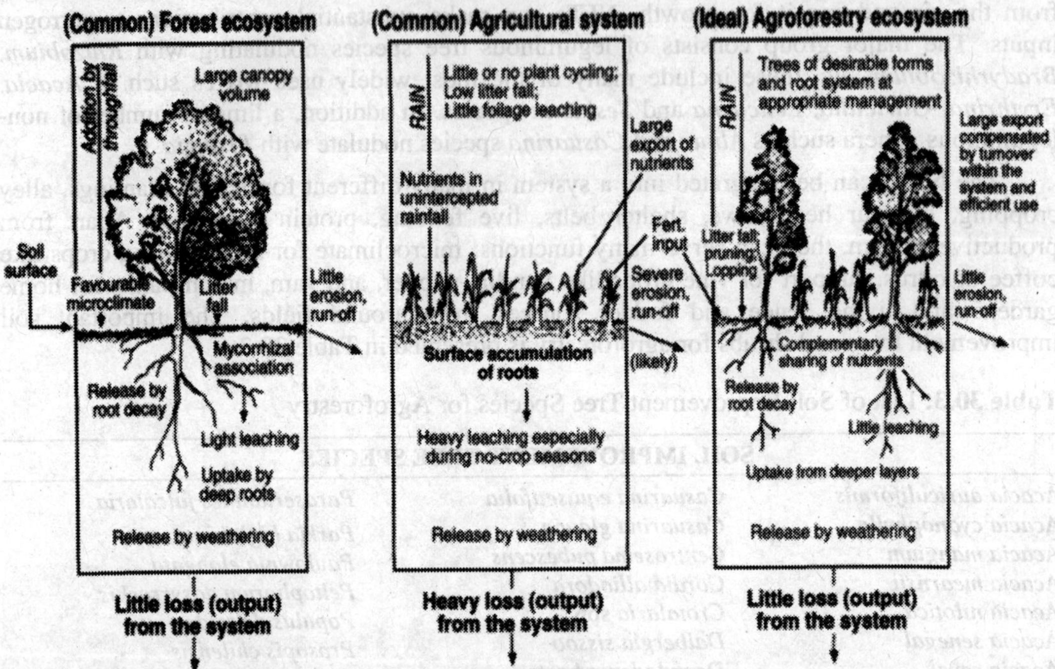

Fig. 30.2: Comparison of Pattern of Nutrient Cycling in Forest, Agriculture and Agroforestry (Source: Nair 1993)

Forest ecosystems represent closed and efficient nutrient cycling systems, meaning that they have high rates of turnover and low rates of out puts or losses from the system. In other words, they are self-sustaining. On the other hand, common agricultural systems are often open, meaning that the turnover within the system is relatively low and losses as well as inputs are comparatively high. The Sources of nutrient addition in agroforestry system that should be taken into account are:

- *Trees:* Litter (leaf fall/pods/fruits + small twig etc.)
- *Crop:* Manures and fertilizers, mineralization, N-fixation, weathering, crop litter, etc.
- *Rainfall* also accounts for nutrient additions via through fall, stem flow and atmospheric dust

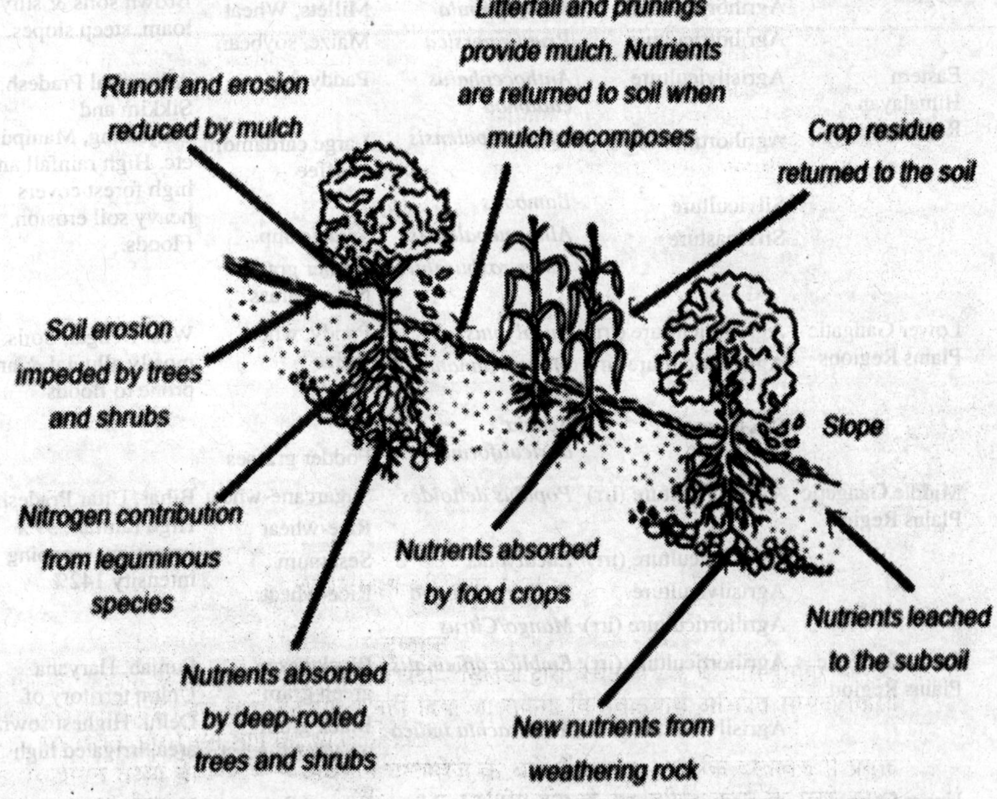

Fig. 30.3: Pattern of Nutrient Cycling under Agroforestry (Source: Nair 1993)

AGROFORESTRY MODELS FOR DIFERENT AGRO-CLIMATIC ZONES OF INDIA

Agro-climatic region is the land unit in terms of major climate, superimposed on length of growing period (moisture availability period) and an agro-ecological zone is the land unit carved out of the agro-climatic region, superimposed on the landform, which acts as a modifier to climate and length of growing period. In 1989, the Planning commission of India divided the country into 15 broad agro-climatic zones based on physiographic and climate. The emphasis was given on the development of resources and their optimum utilization in a suitable manner within the frame work of resource constraints and potentials of each region. Similarly, our country has been broadly grouped into 20 agro-ecological regions.

Table 30.4: Promising Agroforestry Models for Agro-Climatic Zones of India (Source: Pathak et al 2000)

S No	Agro-Climatic Zones	Agroforestry System	Tree component	Crop/grass	Description
1	Western Himalayan Region	Silvipasture (rf)	*Grewia optiva*	*Setaria* spp.	Ladakh, Kashmir, Punjab, Jammu etc. Brown soils & silty loam, steep slopes.
			Morus alba	*Setaria* spp.	
		Agrihorticulture	*Malus pumila*	Millets, Wheat	
		Agrihorticulture	*Prunus persica*	Maize, soybean	
2	Eastern Himalayan Region	Agrisilviculture	*Anthocephalus cadamba*	Paddy	Arunachal Pradesh, Sikkim and Darjeeling, Manipur etc. High rainfall and high forest covers heavy soil erosion, Floods.
		Agrihorticulture	*Alnus nepalensis*	Large cardamom / coffee	
		Silviculture	*Bamboos*		
		Silvipasture	*Alnus nepalensis*	*Setaria* spp.	
			Pinus roxburghii	Napier grass, Native grass	
3	Lower Gangatic Plains Regions	Agrisilviculture (irr)	*Eucalyptus*	Paddy, wheat maize	West Bengal, Soils mostly alluvial & are prone to floods.
		Agrihorticulture (irr)	*Mango/Banana, Litchi*		
		Silvipasture	*Acacia auriculiformis*	Fodder grasses	
4	Middle Gangetic Plains Region	Agrisilviculture (irr)	*Populus deltoides*	Sugarcane-wheat Rice-wheat	Bihar, Uttar Pradesh, High rainfall, 39% irrigation, cropping intensity 142%
		Agrisilviculture (irr)	*Eucalyptus*	Sesamum	
		Agrisilviculture	*Dalbergia sissoo*	Rice-wheat	
		Agrihorticulture (irr)	*Mango/Citrus*		
5	Trans Gangetic Plains Region	Agrihorticulture (irr)	*Emblica officinalis*	Black gram/ green gram	Punjab, Haryana Union territory of Delhi. Highest sown area, irrigated high
		Agrisilviculture	*Azadirachta indica*	Black gram-wheat/ Mustard	
6	Upper Gangetic Plains Region	Agrisilviculture (irr)	*Populus deltoides*	Wheat, Bajra fodder	North region of U.P. (32 Dists) irrigated by canal & tube wells good ground water
		Agrisilviculture (irr)	*Eucalyptus*	Rice-wheat	
		Silvipasture			
7	Eastern Plateaus & Hills Region	Agrisilviculture	*Gmelina arborea*	Paddy, linseed	Chota Nagpur, Garhjat hills, M.P, W. Bundhelkhand plateau, Odisha. Soils shallow to medium sloppy, undulating. Irrigation tank & tube wells.
		Agrisilviculture	*Acacia nilotica*	paddy	
		Silviculture	*Acacia mangium*		
		Silviculture	*Bamboos*		
8	Central Plateau & Hills Region	Agrihorticulture (irr)	*Psidium gujava*	Bengal gram	Madhya Pradesh, Rajasthan, UP,
		Agrihorticulture (irr)	*Emblica officinalis*	Black gram /	

				green gram	climate hot to sub-humid, rainfall varies from 500 mm-1000 mm with mixed red, yellow and black soils
		Agrisilviculture	*Acacia nilotica/ Azadirachta indica*	Soybean, black gram-mustard/ wheat	
		Silvipasture (RF-Degraded lands)	*Albizia amara, L.leucocephala D.cineraria S.scabra*	*Chrysopogon fulvus, Stylo santhes*	
		TBOs (RF)	*Jatropha curcas*	-	
9	Western Plateau & Hills Region	Agrihortisilviculture (irr)	*Tectona grandis, Achrus sapota*	Paddy, Maize	Maharashtra. M.P. Rainfall 904 mm Sown area 65%, forest 11%, irrigation 12.4%
		Agrihorticulture	*Areca catechu*	Black pepper, Cardamom	
		Silviculture	*P. juliflora*		
		Silviculture	*Acacia mangium*		
		Silvipasture	*Zyziphus mauritiana, D.sissoo, T.grandis*	Stylo, native grass	
10	Southern Plateau & Hills Region	Agrisilviculture (rf)	*Eucalyptus*	Cotton	T. Nadu, Andhra Pradesh, Karnataka, Typically semi arid zone, Dry land Farming 81% Cropping Intensity 11%
		Agrisilviculture (irr)	*Eucalyptus*	Chilli	
		Silviculture (rf)	*L.leucocephala*	-	
		Silviculture (rf)	*Eucalyptus*	-	
		Agrihorticulture (irr)	*Tamarindus indica*	Chilli	
		TBOs	*Pongamia pinnata*	-	
11	East Coast Plains & Hills Region	Agrisilviculture (rf)	*Ailanthus excelsa*	Cowpea	Tamil Nadu, Andhra Pradesh, Odisha. Soils- alluvial, coastal sand, Irrigation
		Silviculture	*Casuarina equisetifolia*		
		TBOs	*Pongamia pinnata*	-	
		Silvipasture	*A.leucocephala*	*C.ciliaris*	
12	West Coast Plains & Hills Region	Agrisilviculture (rf)	*Acacia auriculiformis*	Black pepper	Sourashtra, Maharashtra, Goa, Karnataka, T. Nadu, Variety of cropping Pattern, rainfall & soil types.
		Agrihorticulture (rf)	*Artocarpus heterophyllus*	Black pepper	
		Agrisilviculture (rf)	*Acacia auriculiformis*	Paddy	
		Agrihorticulture	*Cocos nucifera / Areca catechu*	Paddy	
		Agrisilviculture	*Casuarina equisetifolia*	Paddy	
		Silvipasture	*Casuarina equisetifolia*	Hybrid Napier	
13	Gujarat Plains & Hills Region	Agrisilviculture	*Azadirachta indica Prosopis juliflora*	Cowpea	Gujarat (19 dists) Low rainfall arid zone. Irrigation 32% well and tube wells.
		Silvipasture	*Ailanthus excelsa*	Green gram	
		Agrisilviculture			

14	Western Dry Region	Agrisilviculture TBOs Silvipasture	*Prosopis cineraria, A. indica* *Jatropha curcas* *Zizyphus numularia* *Capparis deciduas* *Panicum turgidum*	Pearl millet	Rajasthan (9 Dists) Hot sandy desert, rainfall erratic, high evaporation. Scanty vegetation, famine, droughts.
15	The Island Region	Agrihorticulture Silvipasture	*Cocos nucifera* *Areca catechu*	Paddy Hybrid Napier, Stylo	Eastern Andaman, Nicobar, Western Lakshdweep. Typical equatorial, rainfall 3000 mm (9 months) forest zone undulating.

rf: Rainfed; irr: Irrigated

Table 30.5: Agroforestry Practices in Different Agro-Ecological Regions of India (Planning Commission 2001)

S No	Systems/Practices	Agro-Ecological Adaptations
A	**Agrisilviculture Systems**	Woody Perennials + Crops
1	Shifting Cultivation	In Tropical forest areas
2	Taungya	In all regions
3	Homegardens	Mainly tropical regions
4	Plantation based cropping systems	Mainly humid tropical regions
5	Scattered trees on farm lands	In all regions specially semi-arid and arid regions
6	Shelter belts and wind breaks	In wind-prone areas, all regions viz. coastal, arid, alpine areas
7	Boundary plantation and live hedges	In all regions
8	Woodlots for soil conservation	In hilly areas, along sea coast and ravine lands
9	Trees on rangeland	In all regions
10	Plantation crops with pastures and animals	Mostly humid & sub-humid regions of South East Asia and South Asia with less grazing pressure on plantation lands (Trees + crops + animals/pasture)
11	Industrial Plantation with crops	Intensively cropped areas in northern India
B	**Agrosilvopastoral systems**	
12	Seasonal forest grazing	Semi-arid and mountain ecosystem
13	Aquaforestry	low lying land
14	Apiculture with trees	All through
C	**Silvopastoral**	
15	Silvipastures	Sub-tropics and tropics with bioedaphic subclimaxes.
16	Hortipastoral	In hilly orchards for soil conservation and forage.

TREE-CROP INTERACTIONS UNDER AGROFORESTRY SYSTEM

In agroforestry systems, very little information is available regarding tree-crop interactions. Agroforestry literatures repeatedly emphasize that the success of agroforestry relies heavily on the exploitation of component interactions. The success of agroforestry relies heavily on the utilization of component interactions. Interactions refer to the influence of one component of a system on the performance of the other components as well as the system as a whole. Ecologically the types of interactions in two species populations have often been described on the basis of net effect of interactions.

Table 30.6: Analysis of Interactions between Two Populations of Species A and B

Type of Interaction	Effect of the Interaction on the population (Species A & B)		Nature of the Interaction	Agroforestry Examples
	A	B		
Mutualism	+	+	Interaction favourable to the two species populations	Mycorrhizae, Rhizobium - legume
Facilitation	+	0	Interaction favourable for A but not obligatory; B not affected	Windbreaks, shade trees Alley cropping (well managed)
Commensalism	+	0	Interaction obligatory for A; B not affected	Support trees for vines; Improved fallows
Neutralism	0	0	None of the populations affects the other in crop land	Scattered trees
Parasitism/ Predation	+	-	Interaction obligatory for A; B is inhibited	Pest and disease
Amensalism	-	0	A inhibited; B not affected	Allelopathy
Competition and Interference	-	-	Each population is inhibited by the others use of (above or below ground) growth resources	Alley cropping (poorly managed)

Note: '0' = No significant interaction
'+' = Advantage for the population in question (growth, survival, reproduction, etc.)
'-' = Disadvantage for the population in question

Agronomists and agroforestry researchers have used the terms "below-ground" and "above ground" to describe interactions (mostly competitive) between components for growth factors observed through roots (nutrients and water), and intercepted through leaves (mainly radiant energy). These interactions were grouped into 'tree-crop interface' (TCI) and the 'tree-animal interface' (TAI) in all agroforestry systems. In simultaneous agroforestry systems, trees and food crops interact in many ways, leading to positive and negative impacts on the growth of both trees and crops. The main effects of tree crop interactions are:

- Negative Effect (or competition): due to shading, root competition for water and nutrient

- Positive Effect (or complementary): because of litter fall & decomposition and pruning biomass of trees increase C, N, P and other nutrients, deep rooted trees play a role as 'safety-net' for leached nutrients in the deeper layer or as 'nutrient-pump' for fertile soil.

FUTURE PROSPECTS OF AGROFORESTRY

Active agroforestry research has been going on for the last three decades wherein, the purposeful growing of trees and crops and/or animals in interacting combination for a variety of objectives has been studied. Concepts and potentials of different agroforestry systems and practices has already been practically proved that agroforestry can increase the productivity of land besides meeting the needs of food, fodder and fuel, especially in third world countries. The food, fodder and fuel production in developing countries will have to be increased by 60 per cent in the next 25 years to meet the needs of the growing population. The FAO study says that about 80 million hectares of culturable wastelands would have to be brought under cultivation/plantation in the next 25 years. The study also stresses that the trend of stripping forest cover to meet food/fodder and fuel need must be controlled and efforts be made to increase productivity of agricultural land by advocating agroforestry as one of the most practical way of meeting the needs for food, fodder and fuel.

Coming to recent times, the increasing pressure on land as well as on forests, rising demand for fuel and fodder and finally the requirement of timber for multifarious purpose have reinforced, as never before, the absolute and inseparable nexus between agriculture and forestry. In fact, the distinction sought to be made between agriculture as related to food, fibre and oilseed crops and forestry embracing tree culture in a general sense is, to a great extent, artificial and not well founded. For his survival, man has to ensure optimum use of agriculture, grasslands, forests and aquatic bodies, including oceans. Of these, the combination of first three systems i.e. agrisilviculture, silvopastoral and agrosilvopastoral are covered under agroforestry.

Although the term sustainable land use is at present widely used, the precise interpretation of this concept remains problematic. We consider ecological sustainability as one component of sustainable land use. Ecological sustainability refers to the need to maintain the long-term production potential of the land. Other components of sustainable land use refer to social, cultural and economical aspects, such as the ability of land managers to adopt proper management systems and the equitable distribution of inputs and outputs of production. Agroforestry systems are often claimed to be superior to agricultural systems without a woody component, because of their ability to contribute to sustained production of crops, livestock and wood especially on fragile land and in areas with prevalent low external input agriculture. The woody plants are assumed to contribute to both protective functions (enhancing ecological sustainability) and productive functions (taking the place of chemical fertilizers). It is also commonly expected that agroforestry can contribute to land rehabilitation and increased production on degraded lands (NRCAF 2007).

31 SOCIAL FORESTRY

INTRODUCTION

The growing requirement of rapidly increasing human population has put tremendous pressure on forest. Clearing of forest for agriculture, encroachment of people, shifting cultivation, unmanaged grazing, forest fires and fuel-wood removal are leading to reduction in forest cover and affecting the sustained supply of raw materials. Taking these factors into consideration, social forestry was implemented. Social forestry is a forestry system which aims to create a flow of protection and recreation to the community. The word "social forestry" was coined by Westoby at the 9th Commonwealth Forestry Congress during 1968 in Delhi. A large population depends on forest for their daily requirements, such as firewood, fuel wood or small timber. They also meet their energy demands from the traditional forest. This leads to severe depletion of forest causing other environmental problems as well. In order to solve these problems of rural areas, the concept of social forestry was evolved. As mentioned in National Forest Policy 1988, India should aim at maintaining one third of its total land area under forest. So promotion of the planting on non-forest land through social forestry programme is important to improve the green cover of the country. The proportion to be attained is 60 per cent in the hilly areas and 20 per cent in the plain areas. Planting of trees under social forestry programme generates employment to the rural poor and improves their livelihood. Planting may be done on government lands or community lands or on private lands. It is easy to implement and is a very good strategy towards reducing poverty and improving the food security (Tewari 1983).

Conceptually, the difference between traditional forestry and social forestry appear to be so thin, as both provide benefit in one form or the other. While, Westoby differentiated between the two by allotting production function to production forestry and protection as well as recreation function to social forestry. The National Commission on Agriculture brought the supply of fuel, fodder and small timber to the local populations also under the concept of social forestry. Social forestry has remained a rather broad subject, a statement of purpose rather than an operational model.

WHAT IS SOCIAL FORESTRY?

Social forestry is defined as "of the people, by the people and for the people." It means the management and protection of forests and afforestation on barren lands with the purpose of helping in the environmental, social and rural development, as against the traditional objective of securing revenue (Negi 1986).

SCOPE OF SOCIAL FORESTRY

- Creation of woodlots in the village commonlands, government wastelands and panchayat lands (Estimated at least 12 million ha).
- Planting of trees on the sides of roads, canals and railways. This, along with planting on wastelands is known as 'extension' forestry, increasing the boundaries of forests (Estimated at nearly 1 million ha).
- Afforestation of degraded government forests in close proximity to villages, which have experienced the unauthorised harvesting of biomass (Estimated at over 10 million ha).
- Planting of trees on and around agricultural boundaries and on marginal, private lands, constituting farm forestry, or agroforestry, in combination with agricultural crops (Agricultural land covers about 143 million ha) (Prabhakar 1998).

ECONOMIC AND ENVIRONMENTAL BENEFITS OF SOCIAL FORESTRY

1. *Employment Potential:* It has a great advantage in providing job to local people. The poorest section gets more benefit as the labour component is unskilled or semi-skilled which includes works such as nursery works, preparation of beds, sowing, planting, hoeing, watering, etc. Social forestry also provides benefits indirectly by providing raw materials to the cottage industries such as furniture, walking sticks, making sports goods etc.
2. *Capital Gain:* Besides providing benefit in the form of employment, plantations through social forestry also help to create economic capital to the rural poor.
3. *Increased Production of Food and Fodder*: When a forest is properly maintained, it increases food and fodder production.
4. *Cottage Industries*: Social forestry plantations open up a big storage and supply of raw materials for cottage industries which may be used in house-hold goods, house building, sports goods and industrial goods.
5. *Forest Industries*: It also provides forest based raw materials such as resin, pulpwood, match industry and minor forest products such as gums, resins, honey, lac, katha, etc.
6. *Regulation of Water Cycle*: It acts as an interception which prevents direct hitting of raindrops on the ground. Forests present near the catchment areas helps to maintain the water regime. Trees also traps rain water and prevent floods during rainy season.

7. *Carbon Sequestration*: It is an important way of sequestering atmospheric carbon, thereby mitigating greenhouse effect and climate change.

8. *Increased Life of Dams and Reservoirs*: Planting of trees through social forestry programme near the catchment areas protects the reservoir from being silted up. Soil erosion is prevented, reducing the flow of silt into the reservoirs.

9. *Flood Control*: Both human life and property are damaged by floods. Increased silt in the river causes river to overflow. So planting of trees along the river bed reduce surface runoff and control the rise of river bed.

10. *Energy Requirements*: Villagers collect fuel wood from forest for their daily use which results into depletion of forest. Apart from these villagers, even industries also need fuel wood. Thus planting more trees helps to reduce depletion and provides required amount of fuel wood both to villagers as well as industries. It will also reduce pressure on forest and solving the energy problems.

Social Forestry plantations are important to combat the environmental problems as well as meeting the daily needs of the people especially those residing in rural areas or near the forest which are totally dependent on forest. For example, Barodiya Panchayat in Jaipur district collaborated with the Forest Department and decided to plant trees such as *Acacia tortilis*, *Eucalyptus* spp and *Dalbergia sissoo*. This plantation provides the community with fuel wood, fodder and timber benefits. It also improves the soil moisture and groundwater situation. After the rotation period, trees are harvested and the Forest Department equally divides the benefits with local community. This creates a large benefit to the whole community (Negi 1986).

MISSION OF SOCIAL FORESTRY

1. To carry out a need based and time bound programme of afforestation with special emphasis on fuel wood and fodder development on all degraded and denuded lands/forests.
2. Afforestation of abandoned jhum lands and mined areas.
3. Linear strip plantation of fast growing species on sides of public roads, rivers, streams and irrigation canals.
4. Afforestation on unutilized lands under State, corporate, institutional or private ownership.
5. Green belts in urban and industrial areas.
6. Shelter belt (generally more extensive than the wind breaks) for the purpose of shelter from wind and sun covering areas larger than a single farm on a planned pattern.
7. Farm forestry in the form of raising rows of trees on bund or boundaries of fields and individual trees in private agricultural land as well as creation of wind breaks around a farm or orchard by raising one or two lines of trees.
8. Raising flowering trees and shrubs mainly to serve as recreation forests for the urban and rural population.

9. To elicit people's participation involving women and young people in conservation of forests, wildlife and environment.

10. Environmental awareness generation and celebration of vanamahotsava, environment day, wildlife week etc.

OBJECTIVES OF SOCIAL FORESTRY

Social forestry schemes that have been started all over the country have made a considerable difference in overall forest cover in a short time. Afforestations are made outside the conventional forest area for the benefit of rural and urban communities. The main objectives are:

- Improves the environment for protecting agriculture from adverse climatic factors
- Increases the supply of wood fuel for domestic use, small timber for rural housing, fodder for livestock and minor forest produce for local industries
- Increases the natural beauty of the landscape and creates recreational forests for the benefit of rural and urban population
- Provides jobs for unskilled workers
- Land rehabilitation and increase in forest cover
- Enhances the standard of living and quality of life of the rural and the urban people

PEOPLE'S PARTICIPATION

Social forestry also aims at raising plantations by the common man so as to meet the growing demand for timber, fuel wood, fodder, etc., thereby reducing the pressure on the traditional forest area. This concept of village forests to meet the needs of the rural people is not new. It has existed through the centuries all over the country but it was now given a new character. With the introduction of this scheme, the Government formally recognized the local communities' rights to forest resources, and is now encouraging rural participation in the management of natural resources. Through the social forestry scheme, the government has involved community participation, as part of a drive towards afforestation and rehabilitating the degraded forest and common lands.

CLASSIFICATION OF SOCIAL FORESTRY

National Commission on Agriculture (1952) has classified social forestry to include three broad classes, viz., farm forestry, rural forestry and urban forestry.

FARM FORESTRY

Farm forestry is defined as the practice of forestry in all its aspects on farm or village lands, generally integrated with other farm operations (BCFT 1953). Presently, the social forestry programmes have been taken up in the form of both commercial and non-commercial farm forestry. Farm Forestry is another name for agroforestry which is a part of social forestry.

Commercial farm forestry: Commercial farm forestry is defined as the process under which farmers grow trees on commercial basis on farm lands (Negi 1986). It is usually undertaken in areas where there is a ready market of wood or other forest based products. FAO describes this as turning peasants into entrepreneurs and producers.

Non-Commercial Farm Forestry: Non-commercial farm forestry is defined as increasing the number of trees raised by individual farmers for their own family uses (Negi 1986). Individual farmers are being encouraged to plant trees on their own farmland to meet the domestic needs of the family. In many areas, this tradition of growing trees on the farmland already exists. Non-commercial farm forestry is the main thrust of most of the social forestry projects in the country today. It is not always necessary that the farmer grows trees for fuel wood, but very often they are interested in growing trees without any economic motive. They may want it to provide shade for the agricultural crops as wind shelters and soil conservation or to effectively utilize wasteland.

RURAL FORESTRY (COMMUNITY FORESTRY)

Community forestry is defined as raising of trees on public or community land rather than on privately owned lands as in case of farm forestry (Negi 1986). The common feature of this programme is to provide benefits or services to the community as a whole. This form of forestry is one of the approaches for tackling the problem of deforestation of village common lands brought about by the local people through mutual cooperation. The government has the responsibility of providing seedlings, fertilizer but the community has to take responsibility of protecting the trees. Some communities manage the plantations sensibly and in a sustainable manner so that the village continues to benefit. Some others take advantage and sell the timber for a short-term individual profit. Common land being everyone's land is very easy to exploit. Over the last 20 years, large-scale planting of Eucalyptus, as a fast growing exotic, has occurred in India, making it a part of the drive to reforest the sub-continent and create an adequate supply of timber for rural communities under the augur of social forestry.

URBAN FORESTRY

It is the practice of growing trees on non-forest land in urban areas for recreational purposes and aesthetic value.

EXTENSION FORESTRY

Planting of trees on the sides of roads, canals and railways, along with planting on wastelands is known as extension forestry for increasing the boundaries of forests. Under this programme, there has been creation of wood lots in the village common lands, government wastelands and Panchayat lands. Schemes for afforesting the degraded government forests that are close to villages are being carried out all over the country.

SOCIAL FORESTRY PLANTATIONS IN SPECIFIC AREAS

Social forestry plantations are usually taken up in non-forest areas which are not suited for agriculture or where agriculture is not practiced e.g., denuded wastelands, erosion prone areas, road side strips, canal bank strips, bunds, etc (Negi 2005).

CRITERIA FOR SELECTION OF SUITABLE TREES FOR SOCIAL FORESTRY

- Greater photo synthetic efficiency: mainly for production of more biomass
- Improved capability of fixing biological nitrogen
- Inbuilt system for genetic improvement
- More resistance towards biological and environment stresses
- Better capability for nutrient and water uptake, reduced losses of nitrogen fertilizer from the process of nitrification and denitrification.
- Relatively fast growth
- Multiple uses

SEVERELY ERODED MOUNTAIN SLOPES

Locality Factors: Soil is poor, shallow and stony. Excessive erosion has washed away the topsoil. Very often, such areas are exposed to severe wind erosion. Excessive runoff results in low soil moisture content. Grazing incidence is very high, hence the seedlings need adequate protection.

Preparation: Contour trenches along the contours are usually made on slopes up to 20 per cent. Patches or pits are adopted in very rocky areas.

Choice of Species: Different species are adopted under varying conditions viz., *Pinus roxburghii, Acacia catechu* (Moist sub tropical); *Acacia catechu, A. modesta, Prosopis juliflora, Robinia psedoacacia* (Dry sub tropical); *Pinus wallichiana, Cedrus deodara* (Temperate) and *Eucalyptus, Prosopis* spp. (Peninsular India).

RAVINES

Extensive ravine lands occur in the Chambal and Yamuna river catchments, encompassing parts of U.P., M.P., Rajasthan and Haryana.

Locality Factors: Soil varies from sandy to stiff clayey with enough lime to lead to the formation of a kankar pan. This obstructs the rain water from infiltrating deeper into the soil. The average annual rainfall in the region is 700 mm with frequent drought years having precipitation of less than 200 mm. Soluble salts are concentrated just below the surface giving rise to chronic saline alkalinity.

Preparation: Bunds are constructed in the catchment areas for safe disposal of the excess water. Staggered trenches are dug at a spacing of 3.5 m x 4.5 m on the same contour line or lines which are 4.5 m apart. Continuous trenches are made in the valley areas with a safe passage for water provided at regular intervals.

Choice of Species: *Acacia catechu, Melia azedarach, Prosopis juliflora, Dalbergia sissoo, Pongamia pinnata, Albizia* spp. (sandy soil); *Acacia arabica, A. leucophloea, Terminalia ajuna, Prosopis spicigera* (Clayey soil).

GRASSLANDS

Locality Factors: Soil aeration, drainage and biological condition are largely poor. Dense roots may be formed by some grasses like *Imperata* and *Ischaemum*. These are difficult to eradicate. The fast growth rate of grasses makes it difficult for plantations to come up unassisted by frequent and heavy weeding. Grasses also increase the incidence of grazing, fire and frost.

Preparation: If possible, the entire area should be subjected to deep ploughing in strips.

Choice of Species: *Acacia catechu, A. arabica, Ailanthus excelsa, Eucalyptus hybrid, Dalbergia sissoo*, *Bombax ceiba* (U.P. terai); *Cupressus torulosa, Populus ciliata, Robina pseudoacacia, Grevillea oppositifolia, Pinus roxburghii, P. wallichiana* (Western Himalaya); *Morus laevigata, Terminalia myriocarpa* (N.E. India).

ABANDONED FIELDS

Locality factors: Soil may usually be poorly fertile because of repeated raising of agriculture crops without manuring and top soil being washed away. Drainage, aeration and moisture content of the soil are also poor. The condition becomes further worse, if the abandoned field was previously used for storing water and puddling for rice cultivation. Because of the fact that the land remained under cultivation for a long time, soil organisms beneficial for tree growth may have disappeared. Weeds may be heavy if the period of time between cultivation and tree raising is long.

Preparation: Soil digging may be carried out in strips. Care should be taken to ensure that the time lag between abandoning of cultivation and plantation is as short as possible. If conditions permit, ploughing of the entire area should be carried out by tractor.

Choice of species: Mixed plantations are preferred to monoculture. *Eucalyptus hybrid, Bombax ceiba, Dalbergia sissoo, Terminalia tomentosa, Acacia catechu, Prosopis juliflora* (plains of North India).

DRY AREAS WITH IRRIGATION

Dry areas with irrigation are available for social forestry plantation in parts of Punjab, Haryana, Western U.P., Rajasthan and Tamil Nadu.

Locality Factors: Soil conditions vary from sandy to clayey with high alkalinity. Kankar pan may also occur. Maximum temperature in the region is 45°C and hot, dusty winds blow during the summer. The temperature falls down considerably during the winter months, when conditions of frost may occur.

Preparation: The area to be planted will have to be leveled and stumps removed so as to bring about effective irrigation. In Punjab and Haryana, ploughing is done by caterpillar tractors, following which the area is divided into large, irrigation blocks. Trenches about 30 x 30 x 30 cm are formed at a spacing of 3 m. Irrigation is usually done by flooding and care is taken to ensure that there is no over flooding.

Choice of Species: *Tamarix* spp., *Acacia catechu* (poor kankar soil), *Prosopis juliflora, P. spicigera, Cassia* spp., *Eucalyptus* (soils free of salts); *Acacia catechu, A. arabica, Bombax ceiba, Morus alba, Syzygium cumini, Eucalyptus, Populus ciliata, Dalbergia sissoo* (Mixed plantation).

DRY AREAS WITHOUT IRRIGATION

Locality Factors: Dry areas are generally subject to serve wind and water erosion. Annual rainfall is below 900 mm, areas with minimum precipitation experience only a few occasional showers amounting to about 200 mm each year. During summers, the temperature may soar upto 45ºC, often accompanied by hot, dusty winds. Winter frosts are common in many areas. There is a great pressure of humans and animals on the land and this poses a severe threat to any plantation programme that may be taken up.

Preparation: Soil is dug as deep as possible. Contour trenches are formed in sloping areas. The dug up soil is heaped to form a ridge along the portion of the trench. Care is taken so as not to heap the soil on a hard surface, else on reaching hard ground, the roots will tend to curl sideways.

Choice of Species: Fast growing, drought and frost resistant species viz., *Acacia arabica, A. leucophloea, Dalbergia sissoo, Albizia lebbeck, Zizyphus jujuba, Cassia siamea, Casuarina equisetifolia, Pongamia* spp., *A. catechu, A. senegal, Eucalyptus, Ailanthus excelsa, Holoptelia integrifolia, Cassia fistula, Prosopis juliflora, Anacardium* spp. should be selected for such difficult areas.

COASTAL SANDY AREA

Locality Factors: Soil is in the form of pure sand which keeps on shifting under the pressure of wind. Moisture retentivity is zero and any precipitation seeps into the ground almost immediately. Water table occurs to within a few meters from the surface and wells can easily be dug up for irrigation. Underground water is usually brackish. Rainfall is fairly heavy and relative humidity high. Strong winds blow across these areas, often carrying salt with them. Further, there is considerable human and animal pressure on the land.

Preparation: Pits of 30 cm (diameter) x 30 cm (depth) with spacing varies from 1.5 m x1.5 m to 3.5 m x 3.5 m, depending upon the local conditions and the species being raised. If the soil is less sandy, ploughing by tractor may also be done.

Choice of Species: *Casuarina equisetifolia, Acacia auriculiformis, Eucalyptus hybrid, Pongamia pinnata.*

INLAND RIVER SANDS

Locality Factors: Soil comprises of almost pure sand which is of unstable nature and deficient in mineral nutrients. One redeeming feature is that the ground water table is not very low and hence there is a possibility of irrigation by underground water. Very high temperatures are experienced during the summer months, when the mercury may rise up to about 45ºC. Severe frost conditions occur during winters. Average annual rainfall is moderate, with bulk of the precipitation being received during the monsoons. Atmospheric humidity is low. There is a heavy human and animal pressure on the land. High grazing incidence leaves the land barren. Whatever vegetation exists is consumed either by man in the form of fuel or by animals as fodder. Riverine vegetation may occur along the fringes of the sandy wastes and these too are under heavy biotic pressure so much so that such forests have become highly depleted.

Preparation: Under certain conditions the sands are unstable and hence there is a need to stabilize it. This may be brought about by introducing certain grasses and shrubs (e.g. *Saccharum, Aristidis* etc) so as to stabilize any prospective planting site. Later on, trenches and pits are dug for planting.

Choice of Species: Species capable of growing under extremely harsh conditions viz., *Acacia catechu, Albizia procera, Melia azedarach, Ailanthus excelsa, Ricinus communis, A. arabica, Azadirachta indica, Pongamia pinnata, Prosopis juliflora, Haplophragma* spp. are selected.

ARID (DESERT) AREAS

Tropical desert comprises about 3.2 lakh sq. km area in the country. About three fifth of this area is in Rajasthan and the remaining in small pockets in Gujarat, Punjab, Haryana, Maharashtra, Andhra Pradesh and Karnataka.

Locality Factors: Extremely dry climatic conditions prevail. During summers, the temperature may rise above 45ºC. Very often, very strong, hot and dust laden winds blow across the countryside. On the other hand, during winters the temperature may fall to 3ºC and conditions of severe frost occur. There is usually a sudden diurnal variation of temperature. Precipitation is extremely low and erratic. Relative humidity remains very low for a major part of the year. Soil varies from pure sand to sandy. The land is subject to heavy human and animal pressure.

Preparation: Immediately after a particular area is available for planting, it is closed against entry by humans and animals. Mulching is usually done to stabilize the area.

Choice of species: Prosopis juliflora, P. spicigera, Acacia arabica, A. senegal, Ricinus communis, A. catechu, Zizyphus spp., *Albizia lebbeck*. However, before raising tree species, certain herbs and shrubs viz., *Zizyphus nummularia, Cassia auriculata, Lasiurus hirsutus, Tamarix articulata, Calligonum polygonoides, Panicum antidotale, Saccharum munja* may be introduced.

WIND BREAKS AND SHELTERBELTS

WIND BREAKS: Windbreak is a protective plantation in a certain area, against strong winds. It is usually comprised of a few rows of trees (or shrubs) spaced 0.5 to 2.5 m apart. Usually wind breaks are created around fields which are lashed by severe winds. In certain cases, 1 to 1.5 m high earthen ridges are formed and grasses like *Saccharum* spp are planted. Once these ridges have become stabilized, planting of trees is done and the spacing varies with the species.

SHELTERBELTS: Shelterbelt is a wide zone of trees, shrubs and grasses, planted in rows, usually at right angles to the direction of the prevailing winds. The main objectives are:

- To deflect the air currents
- To reduce the velocity of the prevailing winds
- To provide general protection to the leeward areas against the effects of wind erosion
- To protect the leeward areas from the desiccating effects of hot winds

The following are the main characteristics of shelterbelts:

Shape and Composition: Shelterbelts have a typical triangular shape. This can usually be brought about by raising tall trees in the centre.

Density and Width: A certain degree of penetration by winds is planned, as by raising a solid wall, the protection decreases very fast on the leeward side. Shelter belts of upto 50 meters width are considered ideal under Indian conditions.

Height and Spacing: The ratio of height and width should roughly be 1:10. However, this figure may vary with local conditions.

Orientation: Orientation of the shelterbelt depends upon the direction and velocity of the prevailing winds. Shelterbelts may be raised in quadrangles if wind directions tend to change very often.

Length: Length is an important consideration so far as shelter belts are concerned, because at the fringes of these zones eddies are formed and this results in an increase in the wind velocities. The minimum length of a shelterbelt should be about 25 times its height.

Choice of Species:

Grasses - *Saccharum spontaneum, Panicum antidotale, S. munja, Cenchrus* spp.

Shrubs - *Calotropis procera, Clerodendron phlomoides, Cassia auriculata, Dodonaea viscosa*

Trees - *Acacia Arabica, Capparis aphylla, A. leucophloea, Dalbergia sissoo, Lannea coromandelica, Tamarix articulata, Eucalyptus* spp., *Parkinsonia aculeata, P. spicigera*

ROADSIDE STRIPS

Large strips of land lies barren along the roads. Emperor Ashoka was the first Indian King who raised trees along roadsides. In modern India, Chaturvedi drew the attention of the government towards this aspect in 1938. Since then, millions of trees have been raised along national and state highways and other roads all over India.

Locality Factors: Soil conditions usually vary from sandy to stiff clay. Borrow pits upset the natural drainage and may result in conditions of water logging. Climatic conditions vary from area to area through which the roads pass. Usually very high summer temperatures are experienced. Bulk of the precipitation is received during the monsoons. Pressure of human population is very severe. Lopping for fodder and fuel wood is a common feature. Owners of cattle, goats, cows, sheep, buffaloes, camels, etc. let loose their animals to graze upon these areas.

Preparation: Pits of 60 cm x 60 cm x 45 cm are dug during the winter months when the soil is easy to dig. Ridges may have to be formed in area of water logging. Pits are dug in 6 to 10 rows on both sides of the road. In certain water logged areas of northern India, small mounds of earth are formed in which Eucalyptus is planted. The pits are so dug that they are exactly opposite to each other on either side of the road.

Choice of Species: Species to be raised should be hardy, evergreen or winter deciduous, wind firm and fast growing at least in the early stages. Some suitable species are *Acacia Arabica, Azadirachta indica, Albizia procera, Ficus bengalensis, Mangifera indica, Cassia fistula, Populus* spp., *Bauhinia variegata, Eucalyptus* spp, *Ficus glomerata, Tamarix articulata, Cassia javanica, Delonix regia, Jacaranda ovalifolia.*

RAILWAY LINE STRIPS

Locality Factors: Soil conditions vary from sandy to stiff clayey. There are also vast stretches of *usar* land, which have formed due to alkalinity. Biotic pressure on the neighbouring forests is severe. Lopping for fodder and fuel wood is very common.

Preparation: Trenches are dug up in areas where sowing is to be carried out whereas, pits are dug in areas to be planted. Care is taken to ensure that the first row is at a minimum distance of 6 m from the centre of the track or at the toe of the embankment (whichever is more). No preparation needs to be done on inner portions of curves or near level crossing as raising plantations in such areas will cause a traffic hazard and hamper visibility. Similarly, trees are not to be raised within a distance of about 200 m on either side of railway signals.

Choice of Species: Hardy species viz., *Acacia catechu, Azadirachta indica, Eucalyptus* spp., *Cassia fistula, Ficus bengalensis, Jacaranda ovalifolia, Madhuca indica, Prosopis juliflora, A. arabica, Albizia procera, Delonix regia, Cassia javanica, Ficus infectoria, Populus* spp., *Mangifera indica, Terminalia arjuna* are usually selected.

CANAL BANKS

Locality Factors: Soil conditions vary from area to area. Usually, these may vary from sandy to clayey. In certain localities, *usar* soil may have formed due to seepage from the canal. Seepage may also result in conditions of water logging in the canal bank strips. As the canals usually run through agricultural areas where there are practically no forests, biotic pressure on canal bank plantations is very severe.

Preparation:

Haryana technique: The first row is raised along the outer edge of the canal road, with plants about 6 m apart.

Uttar Pradesh technique: The strip on the left bank is planted whereas that on the right bank is sown. Pits for planting are of 60 x 60 x 60 cm size.

Rajasthan technique: Planting is to be done on either banks and spacing adopted is 3 x 4 m.

Choice of Species: Hardy species viz., *Acacia arabica, Delonix regia, Ficus bengalensis, Jacaranda ovalifolia, Robinia pseudoacacia, Swietenia macrophylla, Cassia javanica, Eucalyptus* spp., *Populus* spp., *Dalbergia sissoo, Parkinsonia aculeata, Vateria indica* are selected.

SALINE AND ALKALINE (KANKAR) SOILS

About 7 million hectares of land in the country are affected by varying degree of salinity and alkalinity, and it is estimated that the area is increasing at an alarming rate of 10,000 hectares per year (Khanna 1977). In a country, where the percentage of forest area is less than the desired level and where there are practically less forests in the predominantly agricultural zone, afforestation of saline and alkaline soils can go and long way to meet the requirement of grasses, fuel and small timber of the villagers.

Locality Factors: The value of pH is usually very high and usually there occurs a kankar pan in the soil. The tract is subjected to severe human and animal pressure in the form of overlapping, grazing, browsing, etc.

Preparation: Soil preparation is done by digging pits so as to break the kankar pan. Pits are usually 30 to 60 cm wide and up to 2 m deep. In areas where the soil is highly alkaline, pits need to be filled with non-alkaline soil brought in from elsewhere. Mounds have to be made in water logged areas. Soil ameliorants such as gypsum, farm yard manure, etc. may also be added.

Choice of Species: *Azadirachta indica, Eucalyptus hybrid, Terminalia arjuna, Acacia catechu, Albizia procera, Pongamia pinnata, Prosopis juliflora, Acacia arabica.*

LATERITIC SOILS

Locality Factors: Soil is shallow, highly ferruginous and lacking in essential nutrients.

Preparation: Soil preparation is done by tractor ploughing followed by ridging. Manual labour may also be used to dig trenches and form a ridge-ditch pattern.

Choice of species: *Acacia arabica, Alstonia scholaris, Anacardium occidentale, Eucalyptus* spp., *Casuarina equisetifolia, Dendrocalamus strictus.*

SOCIAL FORESTRY PROGRAMMES IN INDIA

India was one of the first countries to have launched social forestry. In 1950, the festival of tree planting was instituted by the Central Government as an annual feature. Since then, rapid developments have taken in the field of social forestry. The National Forest Policy (1952) strengthened the emphasis to be laid on these programmes. The implementation and progress of social forestry programme in different states across the country are presented below.

GUJARAT

Gujarat was the first state in India to set up a separate social forestry wing in its forest department in 1970. Since then, it continues to lead the field in social forestry, not only in India, but also in the entire S.E. Asian regions. In the beginning, efforts were concentrated towards raising trees along canal banks, railway line strips and roads. Fast growing species were raised by the forest department in a number of rows, having a rotation of 7 to 10 years. In 1980, this system was changed with a 50 : 50 partnership arrangement with the local Panchayats. By the end of 1982, about 29,500 ha equivalent strip plantation had been raised in the state along railway line strips, canal banks and roadsides.

Supervised wood lots scheme was introduced in the year 1974 in which the forest department, in collaboration with the local Panchayats, raised 4 hectare woodlots in different villages of fuel wood, fodder and fruit species on the community lands. In return, the villagers were allowed for collecting fuel, fodder and fruit in addition to a 50:50 share of the profit when the trees were harvested. By the end of 1982, such woodlots had been established in 4,000 of the total 20,000 villages of the State, with a total area of 28,000 ha. Another programme, known as the village self-help scheme was launched in 1980. The intention of this programme was that the forest department was to distribute free seedlings and provide technical advice to the people for raising trees on their own lands.

A farm forest farming project was launched in 1972, under which farmers were provided seedlings free of cost, for raising trees as commercial crops on their land. This project evoked immediate response, so much so that, people raising cotton on the fertile black cotton soil took to commercial farm forestry in a big way. They found that the return under this scheme was more and botheration lesser. Mr. Vithal bhai Patel of Vatwa farm near Ahmedabad a large farmer had switched over from cotton growing to tree farming. He maintains his own nursery and is earning huge profits since in 1972 (Negi 1986).

Table 31.1: Social Forestry Programmes Implemented in Gujarat (Source: Forest and Environment Department, Gujarat)

Duration	Schemes
1969-70 to 1979-80	Social Forestry plantations carried out through the State scheme
1980-81 to 1984-85	Word Bank Phase 1, first phase of Social Forestry called "Community Forestry Project" was implemented with assistance from the Word Bank.
1985-86 to 1992-93	Word Bank Phase 2, second phase was implemented with assistance from the Word Bank and USAID
1993-94 to 1995-96	Social Forestry plantations were undertaken under the State scheme
1996-97 to 2001-02	Integrated Forestry Development Project (IFDP) with assistance from the OECF (Japan)
2002-03 to 2003-04	Social Forestry plantations were undertaken under the State Scheme

Table 31.2: Cumulative Achievement of Social Forestry Works (1969-70 to 2003-04) in Gujarat (Source: Forest and Environment Department, Gujarat)

Programme	Area / Number
Strip plantations	0.96 lakh ha
Panchayat lands (Village forest)	1.22 lakh ha
Private degrade lands	1.65 lakh ha
Seedling distribution(VM)	443.84 crores

TAMIL NADU

The state forest department launched way back in 1956. Afforestation of wastelands was started in 1960 mainly with tank plantations. A new social forestry project combining all the previous farm and community forestry prgrammes was embarked upon in 1981, funded by the Swedish International Development Authority (SIDA) with the following main programmes:

- Farm forestry over a wide area
- Tree raising for family uses
- Tree raising on communal or common lands
- Strip plantations along roads, railway lines and canals

- Creation of a network of fuel-fodder plantations

The farm forestry scheme consists of two components: an incentive scheme aimed at poor farmers and an extension scheme for larger farms in which seedlings are issued at cost price. Up to the end of 1983, total of approximately 10 million seedlings had been distributed under these two schemes. Some reservations have however, been expressed about the practicability of the incentive scheme because of the administrative problems of checking survival rates and making small payments to a large number of farmers dispersed over a wide area.

WEST BENGAL

In 1981, a World Bank aided project was launched in the State. West Bengal has the lowest per capita forest area (0.02 ha) and average population density is 610 persons per km^2. The following targets have been set to be achieved by 1987:

- 20,000 ha of strip plantations along roads, canals and railway lines
- 1,500 ha of degraded forest land to be rehabilitated
- 6,000 ha of village woodlots to be created for supply of fuelwood and fodder.
- Raising of as many state owned wasteland plantations by allotting small plots to landless farmers

Large tracts of degraded land which was previously under forests, occur in the south west corner of the State. Soil conditions are very poor and the area is characterized by chronic drought conditions. Efforts are being made for the rehabilitation of such lands under social forestry programmes. In east Midnapore district, the State government is allotting plots of 0.4 ha of waste land to the landless people. Main species which are being raised in this area are *Acacia auriculiformis* and *Eucalyptus tereticornis.* Pit digging is done by the farmers during the dry season, so that it does not interfere with their main occupation as agricultural labourers. Free seedlings are supplied by the forest department along with a small quantity of fertilizer. A cash incentive of up to Rs. 220 is also given on basis of the survival per cent after the second and third years. The farmer gets bulk of the return on maturity of the trees.

UTTAR PRADESH

Social forestry was launched in a big way, in parts of eastern U.P. during the late sixties and early seventies. In 1979, a programme was launched with the aim of providing 45,000 ha of fuel and fodder forests in Land along roads, railway lines and canal banks; Common lands or Civil Soyam lands; and Degraded forest lands on which villagers enjoy certain rights such as fuelwood and fodder collection.

The main burden of making this programme a success was entrusted to the forest department. It was intended to make this project a model for demonstration. Commercial farm forestry was encouraged in Uttar Pradesh on the cost-price seedlings system being followed in Gujarat. This project met with astounding success particularly in western U.P. where the average size of holdings is larger and the people are economically more comfortable. By the end of 1982-83, a total of 150 million seedlings were distributed under this scheme which amounted to about 30 times the planned target. Other social forestry, projects, particularly those involving road, railway line and canal bank strip planting have also met with similar success in the state.

ODISHA

Though Odisha has 40 per cent of land area under forests, their distribution is unsatisfactory. The entire coastal tract and large pockets of inland have inadequate forests. Supply of forest materials to the populace in these areas is difficult and people here prefer to burn cow dung. Because of poor railway system, the transport of forest produce from distant forests to these scarcity pockets in uneconomical. Large scale afforestation of waste lands and denuded hill slopes will be very expensive. Farm forestry offers itself as a solution but cannot be practiced extensively in conventional form as practically no one owns a sizeable compact farm land.

People in general have no idea of farm forests and Government should set up at least one demonstration farm forest in each district and publicize its benefits. Planting for deep rooted trees with light foliage such as *babul* should be encouraged. In inland areas, bamboos and fast growing trees can be planted in backyards. It has been extensively adopted by people in coastal areas. Shelterbelts are not necessary in Odisha conditions except near the sea. *Casuarina* plantations in the sand dunes throughout the length of Orissa's coastline have already been planted and headway has been made to establish government plantation. In Ganjam, people themselves have planted sand dunes extensively with *Casuarina* and coconut. Collective farming will assist establishment of farm forests and land legislation is therefore necessary.

PUNJAB AND HARYANA

Punjab and Haryana occupy one of the most fertile areas of India. They are considered to be the food basket of the country. Social forestry has made much progress in these two states with the following main thrusts:

- Raising plantation along roads, railway lines and canal banks. This work was carried out in an extensive scale during the seventies and today almost all the strips along roads and canals have been brought under social forestry plantations. The work of afforestation along railway line strips is also considerable progress.
- Tree plantation along boundaries of fields and wastelands with a view to meet the local demand for fuelwood and fodder. Ample land is available along field boundaries for raising trees, so that the local demand can be met.
- Raising trees on agricultural land for commercial purposes. The idea of changing over from agricultural crops to forestry crops seems to have caught on in both these states in a big way..

Thousand of marginal and big farmers have stopped raising agricultural crops and are now growing forestry crops with a 5 to 7 years rotation. They sell wood to the paper-pulp mills or match factories. Eucalyptus and Poplars are favourite species for commercial farm forestry in these states. In Siwalik foothill belt in the north (mainly Hoshiarpur & Ambala districts) along the Himachal Pradesh border, social forestry schemes have somewhat different objects;

- Afforesting bare hills and improving the existing stock.
- Preventing excessive runoff and increasing percolation of water into the soil, thereby raising the ground water table.
- Training of hitherto uncontrolled torrential beds in order to prevent them from damaging cultivated fields and endangering the existence of settlements, roads, canal, etc.
- Increasing the economic value of these areas by producing for local use (and also for commercial use wherever possible) the following forest product: timber, fuelwood, fodder, resin, katha, etc.
- Improving the aesthetic and recreational value of these areas and also helping to ameliorate the environmental conditions.

RAJASTHAN

Rajasthan comprises mainly of a semiarid and arid tract and a part of which is occupied by Rajasthan desert. The climate is characterized by extremes and large variations of annual and diurnal temperature, low and erratic rainfall and extreme aridity. Largely due to unwise exploitation of natural resources, a peculiar pattern of soil erosion and desert formation have been set up and it is believed that the desert is advancing at a very fast rate, engulfing hundreds of hectares of land in the process. Thus, one of the main measures for arresting the march of the desert is to increase the forest area.

Social forestry has a prominent role to play under peculiar conditions of advancing of the desert, chronic shortage of fuel and fodder, dearth of forest resources and conditions of severe environmental degradation. For this, the following programmes are being taken up:

- Shelterbelts and wind breaks, rows of trees along field boundaries are being raised to check the spreading desert.
- Afforestation of strips along roads, railway lines and canals (mainly the Rajasthan canal).
- Fuel and fodder plantations are being raised on village waste lands and in fact, wherever suitable site is available.

MADHYA PRADESH

In 1955 and 1956, the fuel cum fodder reserve at Pacheda over an area of 288 acres was established. Following the example of Pacheda, it has been proposed to raise similar fuel cum fodder reserves in other part of the Chhattisgarh plains.

The existing position about this subject in the other regions of the state may be considered as practically of no value except that in the Madhya Bharat region, a fairly considerable and creditable work of afforesting the ravines of Chambal and its tributaries has been achieved since 1924. The total area thus afforestated is about more than 9,000 acres and the work is still progressing at a rapid pace. This work is important not only as an anti-erosion measure, but it has aided in raising the standard of rural economy of the neighboring villages.

HIMACHAL PRADESH

Himachal Pradesh is a mountainous state, occurring in the western Himalaya. There is a high level of biotic pressure on the forests mainly in the form of removal of fuel wood, fodder, timber for house construction and wood for making packing cases. Social forestry projects have been going on in the state for quite some time. The Indo-German Dhauladhar Project was launched in 1980 in Palampur sub division of Kangra district. This project is perhaps, one of the most successful integrated social forestry projects in north India and the project has a total outlay of Rs. 10.5 crores of which Rs. 65 crore is provided by the Federal Republic of Germany. The following are the main thrust areas of this project:

- Creation of village woodlots to serve as fuel and fodder reserves
- Providing seedlings for farm forestry purposes to the villagers
- Introducing energy conservation measures such as a smokeless Dhauladhar Chula which has 50 per cent more efficiency.

ASSAM

Most of the unutilized Govt. waste land, VGRS & PGRS have been identified and brought under social forestry plantation with the objective to provide fuel wood, fodder, fruits, small timber for construction of huts and agricultural implements to the rural people. Social forestry wing of Assam has also given full importance to support the traditional cottage industry. For example, to support the "Eari & Muga" silk industry in the rural areas, some plantations have been taken up extensively. The plantations raised under social forestry schemes are approximately one lakh hectares. Under Extension Forestry, road side parks and recreation centres have been created to improve the quality of life of the people. Until now, 32 such recreation parks have been created.

Under social forestry wing during 2007-08, an area of 67.45 ha with 2,29,33,000 seedlings were raised with State Govt. fund for Rs. 183.04 lakh under SFG, TSP and SCCP schemes. Maintenance was done for covering an area of 2,133 ha and 20 lakh Agar and 10 lakh Maha Neem seedlings are being raised in the nurseries in Assam. With the Central Govt. fund under NAP scheme for Rs 2,366.35 lakh, an area of 19,666 ha plantation was done through 550 of JFMCs under 28 FDAs in Assam which includes artificial regeneration, natural regeneration, bamboo plantation and medicinal & aromatic plantation (Source: Forest and Environment Department, Assam).

SUPPORT FOR SOCIAL FORESTRY

National Commission of Agriculture (NCA), World Bank, Swedish International Development Agency (SIDA), United States Agency for International Development (USAID) and Canadian International Development Agency (CIDA) provide support for social forestry development. In 1956, social forestry started in Tamil Nadu and Uttar Pradesh in the riverbank. They adopted *Eucalyptus* species and *Acacia nilotica.* Gujarat was the first state where forest department also collaborated in social forestry. In the 6th five year plan, 78 per cent of the total budget allocated for state forest department was spent on social forestry programme. By 7th five year plan (1985-90), Wasteland Development Project through social forestry was implemented in India.

SOCIAL FORESTRY FOR RURAL DEVELOPMENT

Due to increase in population pressure, demands of firewood in rural areas have led to degradation of forest. Many foresters have been blamed that they are not able to protect and manage forest properly. For example, in North India, canal system was made because of scarcity of water caused by depletion of forest. The presence of small village forest prevents the area from water logging, thus people can grow agricultural crops without facing any problem of soil erosion. Raising multiple lines of trees along the canals also prevents seepage. Social forestry programme in rural areas play a very vital role in promoting their livelihood apart from other direct and indirect benefits, such as stabilizing the environment which is very badly affected by many pollutions. Social forestry programme therefore not only improves the livelihood of people but is also important from ecological point of view. It also helps in stabilizing the hill slopes of the Himalayas.

WATERSHED MANAGEMENT AND SOCIAL FORESTRY

The watershed resources are developed and managed in such a manner to achieve optimum production without disturbing the ecological balance and this is known as watershed management. A good watershed has negligible soil erosion and a good vegetation and sediment-free stream flow except in severe flood peaks. But with rapid increase in population, demands of fuel wood, fodder, timber is increasing thus leading to a severe destruction of forest resources from the watershed. In arable land, there is severe soil erosion, thus turning the watersheds into wastelands.

Due to anthropogenic activities, soil erosion and sediment discharge leads to scarcity of fuel, grass, water, etc. which ultimately affects the socio-economic conditions of people. For example, Kandi tract of Haryana and Punjab and the rivers of the Chambal basin are badly managed watersheds (Khosla and Kohli 1987). Trees, bushes and grasses are used for soil conservation .With increasing demand for fuel wood, fodder and small timber, the vast area of wasteland have to be fertilized through the social forestry programme. By implementing these programmes, the areas can be adequately protected from soil erosion and damaging effects of floods.

ROLE OF BANKING SECTOR IN SOCIAL FORESTRY

India is one country which is known for its rich natural resources. But the excessive anthropogenic interference has led to a tremendous pressure on forest which leads to depletion of forest. Deforestation is an activity which is most adverse as it leads to reduction of many productive lands through floods and drought. This makes the rural people migrate to urban areas which cause the breakdown of rural economy. So it is important to develop the natural resources to stop migration and to create employment in the area. Social forestry is an effective way to offer job opportunities in the rural areas. But the investment in this sector is very low.

NABARD takes up initiative to formulate bankable projects to develop social forestry so that the production may be increased. NABARD provides loan to the local people and they can repay the loans in easy installments. People can start setting up of nurseries and plantation fields to improve their income and livelihood. But collaboration from both sides is required. When this development bank extends its financial support to rural people, it is essential for them to regenerate and maintain a continuous supply of raw material so as to increase production and to improve the living conditions of the rural people which ultimately leads to improvement in the rural economy as a whole.

ROLE OF NGOs AND PANCHAYATS IN SOCIAL FORESTRY

Some lands are left barren for a long time. These lands, known as marginal lands, may be managed by Forest Department but there are times where management is poor. So NGO's takes up initiatives in utilizing these lands so as to increase the forest cover as well as maintaining the ecological balance. Apart from this, rural livelihood is also enhanced. NGOs encourage the community to take part in managing the land and forests. Thus NGOs play a very effective role in extending its lands and improving social forestry programme which helps in development of rural areas and in reducing poverty.

Similarly, a Panchayat or a village council also plays an important role in social forestry. As we know, Panchayat takes the responsibility in upliftment of the village or a particular community. The Panchayat can involve itself in promoting education in the village. Similarly it can also create awareness among the people about the importance of forest, how to manage the forest and what are the species to be planted in order to meet the demands of the rural poor, such as fuel wood, fodder or timber. So Panchayat has a role to lead the people to make sincere efforts in social forestry development. It is through these Panchayat that people become aware of the social forestry programme which brings them benefits directly and indirectly. Through these programmes, people set up their small industries in the village provided there is a continuous supply of raw materials. It is through the efforts of Panchayat that social forestry is developed in many rural areas thus reducing poverty, illiteracy and disunity from the villages, there by improving the livelihood condition of the rural poor.

DIFFERENT SOCIAL FORESTRY SCHEMES IMPLEMENTED IN INDIA

1. **Strip Plantation**: This activity involves planting of trees along roads and railways for aesthetic value as well as shelterbelts and windbreaks. It also helps in controlling dust pollution and noise pollution.
2. **Village Woodlot**: This takes place in barren lands. Forest department and Panchayat make a resolution in which Forest department plants trees in the village whereas irrigation is done by the Panchayat. On maturity, 75 per cent of profit is given to concerned Gram Panchayat and 25 per cent is used again for afforestation works in the village.
3. **Rehabilitation of Degraded Lands**: Trees are planted on the boundary of the farm. The ownership of trees and benefits lies with the owner.
4. **Environmental Plantation**: These are done in schools, government offices, college campuses, etc. Protection work is done by the organizations where plantations have been carried out.

5. **Kisan Nursery**: Many marginal farmers and families belonging to below poverty line involve in nursery activity.
6. **Van Mohotsava**: In this scheme, people are encouraged to plant trees to increase the forest cover and are motivated to have their own nurseries so that the income of people of lower economic strata is improved.
7. **Biogas/ Solar Cooker Distribution**: In order to save firewood, beneficiaries are given solar cookers and they are also taught to use biogas. Cattle excreta is kept in a cemented tank and allowed to rot. The gas generated is transported through pipes and is used for cooking. The residue is used as manure in agricultural field.
8. **Improved Crematoria**: This saves about 40 per cent firewood in the village.
9. **Special Component Plan**: This scheme is mainly for the benefit of Scheduled Caste population in the State. It includes rehabilitation, creation of nursery and distribution of seedlings.
10. **Afforestation of Barren Hillocks**: Seedlings are distributed and planted to increase forest cover as well as to conserve the soil.
11. **Planting of Fruit Trees on the sides of Public Roads with participation of Farmers and Land Labourers**: Forest department distributes fruit trees. A beneficiary gets 100 per cent benefits from fruit trees and 50 per cent from timber.
12. **Different Prize Schemes**: These are awards given by the Central Government.

Table 31.3: List of Completed Social Forestry Projects (Source: MoEF)

S No	Name of the Project	Aid Agency	Project Period	Project Cost (Rs. in Cr)	Actual Expense (Rs.in Cr)	Area (in ha.)
1	Social forestry Project Uttar Pradesh	World bank	1979-80 to 1983-84	40.00	50.00	76,000
2	Social forestry Project Maharashtra	USAID	1982-83 to 1990	56.40	72.80	75,726
3	Social forestry Project Andhra Pradesh	CIDA	1983-84 to 1990-91	38.38	42.76	45,217
4	Social forestry Project Bihar	SIDA	1985-86 to 1991-92	53.85	48.60	53,375
5	Social forestry J & K and Haryana	World Bank	1982-83 to 1990-91	57.07	106.19	1,86,281
6	Social Forestry West Bengal	World Bank	1981-82 to 1990-91	34.75	64.00	2,42,578
7	Social Forestry Karnataka	World Bank/ ODA	1983-84 to 1991-92	124.55	85.21	53,351
8	Social Forestry Project Kerala	World Bank	1984-85 to 1992-93	59.51	89.68	1,31,000
9	National Social forestry Project (UP,HP,RAJ,GUJ)	World Bank/USAID	1985-86 to 1992-93	387.29	698.18	11,98,742

10	**Gujarat Community Forestry Project**	**World Bank**	**1980-81 to 1984-85**	**66.65**	**67.64**	**1,08,355**
11	**Orissa Social Forestry Project, (Phase I)**	**SIDA**	**1983-84 to 1987-88**	**28.17**	**27.06**	**33,592**
12	**Social Forestry Project Tamil Nadu (Phase I)**	**SIDA**	**1981-82 to 1988-89**	**65.68**	**56.96**	**1,40,363**
13	**Social Forestry Project Orissa, (Phase II)**	**SIDA**	**1988-89 to 1995-96**	**78.34**	**136.80**	**1,19,450**
14	**Social Forestry Project Tamil Nadu, (Phase II)**	**SIDA**	**1988-89 to 1995-96**	**85.40**	**154.86**	**1,08,176**
15	**West Bengal Forestry Project**	**World Bank**	**1992-93 to 1997-98**	**114.00**	**136.79**	**2,05,711**
16	**Rehabilitation of Common lands in Aravallis, Haryana**	**EEC**	**1990-91 to 1999-00**	**48.15**	**NA**	**NA**
	TOTAL			**1,338.19**	**1,837.53**	**26,37,554**

32 JOINT FOREST MANAGEMENT (JFM)

INTRODUCTION

Forest is an area having woody as well as non-woody vegetation, dominated largely by trees. It provides land for growing food, plays important role in economic development by providing employment and provides wood as well as non-wood products. In addition, forest plays important role in stabilizing the environment and checking soil erosion. Forest is also important as a carbon sink, thereby reducing the greenhouse effect. But to meet the needs of the increasing population, deforestation is on rise. Many forest lands are being cleared for agriculture and other developmental works such as construction of roads, bridges, dams, etc. resulting into loss of forest cover. In developing countries like India, where a large population depends on forest for their daily requirements, importance of forest is being increasingly recognized and more so particularly in the context of ongoing changes in global climate (Gupta 2006).

Although afforestation programmes are being carried out through various governmental and non-governmental bodies like World Bank, SIDA, CIDA, USAID and DANIDA, the total area under forest cover is still declining. This is due to not only to deforestation, but also to the failure of afforestation programmes (Venkatesh et al 2003). The success of afforestation depends on identifying suitable multipurpose trees for the specific location (Venkatesh et al 2001), using quality planting stock (Venkatesh et al 2000) and after care or management of the plantations on participatory mode. In order to reduce the deforestation and forest depletion, a new paradigm has recently emerged, called "Joint Forest Management" (JFM). This is the activity which involves participation of people in managing forest. Village committees take part in safe guarding the forest. Forest department collaborate with the village people and share ownership and benefits. Thus JFM is a forestry practice or a management which sustains forest and produce social, environmental and economic benefits.

ORIGIN AND EVOLUTION OF JFM

To understand the concept and process of Joint Forest Management (JFM) in India, one needs to delve into the evolution of the forest policy and legislations in the country. Though the initial set of policies and laws on forestry date back to the colonial period and the immediate post-independence period, one notices a paradigm shift in India's forest policy and legislations in the 1980s, with the passage of the Forest (Conservation) Act 1980. This Act highlighted the primacy of conservation of forests over the previous emphasis on utilizing 'forests' for meeting the requirements of agriculture and industry.

The first forest policy of India was enacted during the British period in 1894. This policy was centrally influenced by the Volcker Commission Report, which posited forests to be a biomass provider for the agricultural sector. After independence, the Government of India enacted a new forest policy in the year 1952 that, while largely subscribing to the philosophy of the 1894 policy, nevertheless highlighted the functional classification of forests. The 1952 policy classified forests into protection forests, national forests, village forests and tree lands. The policy recommended weaning of tribes from shifting cultivation practices and controlling of grazing and other activities in forest areas. Functional classification presupposed that forests had to be typified in terms of their relative ability to subserve agricultural and industrial systems in India. Accordingly, forests in hilly regions had to be preserved and protected on account of their possible role in preventing soil erosion and water runoff from agricultural catchments. Forests that had the potential for timber and related raw materials required for the industry were to be exploited on the basis of 'scientific working plans' to yield raw materials. Meanwhile, in 1973, the National Commission on Agriculture came up with the idea of production forestry based on 'high productivity' man-made plantations. A string of forest development corporations was accordingly set up in the late 1970s to 'corporatize' the process of production forestry in India (Damodaran 2002). This development was in consonance with the prescriptions of the National Forest Policy 1952. In the late 1970s, the Government of India and the State Governments initiated the social forestry movement with a view to carrying out tree planting in and around village areas. This was ostensibly designed to meet the growing demand for firewood and small wood required by the local communities. It was also in line with the thinking implicit in the National Forest Policy of 1952.

In the middle and late 1970s, the spread of the 'Chipko' movement in the U.P. hills led to a situation where the accepted tenets of the 1952 National Forest Policy were questioned by the environmental movements in the country. The adverse consequences of large-scale diversion of forestlands to non-forestry purposes (which hit an astounding rate of 1,50,000 hectares per year prior to the 1980s), were too glaring to be ignored by the policy makers (Damodaran 1992). Accordingly, Forest Conservation Act of 1980 was enacted by the Government of India to check diversion of forestlands for non-forestry purposes. This act made it obligatory for State Governments to seek prior approval of the Central Government for undertaking diversion of forestlands for non-forestry purposes. Meanwhile, in pursuance of the 1972 Wildlife (Protection) Act, the Government of India set up an extensive network of protected areas in the country. By the end of the 1980s, protected areas accounted for 13.6 million hectares with the constitution of nearly 70 national parks and 411 wildlife sanctuaries in different parts of the country.

The other major development was the growing protest against forest plantations in different parts of the country. The movement against Eucalyptus plantations in the early 1980s raised serious questions about the ecological desirability of raising industry-oriented monoculture tree plantations in Indian forest areas. The key issue was how to resolve the growing biomass shortage for agriculture. The social forestry programme initiated in mid-

1970s had aimed to raise fuel wood and biomass generating plantations in non-forest lands. But track record of the program was dismal. Further the program suffered for want of the participatory element. These concerns caused the Government of India and the States to shift their policy towards a massive afforestation programme in the wastelands of the country, which were estimated to be of the order of 175 million hectares. The National Wasteland Development Board (NWDB) was set up in the year 1985 to promote the afforestation process in community and private lands, with the involvement of stakeholders. The NWDB schemes included establishment of rural fuel wood plantations, treatment of micro-watersheds in Himalayan States, promotion of tree growers cooperatives, establishment of people's nurseries and farm forestry activities. In due course, afforestation programs of the NWDB were restructured to cover degraded forestlands as well.

Despite these initiatives, the trend of depletion of forest cover in India continued unabated. Working plans were not successful in conserving forests. The Forest Survey of India came up with the finding that only in 15 per cent of the forest area covered by working plans, adequate regeneration was noticed. Fire and grazing were held to be the culprits in this regard. Further, it was noted that in over 60 per cent of the area covered by the Working Plans, the annual cut exceeded the increment on account of unauthorized felling (Government of India 1991). Between 1983 and 1987, the country lost forest cover at the rate of 47,500 hectares per annum. However, these developments were not without their exceptions. The success of two community-driven "greening" movements in India opened the eyes of the policy makers to the immense potential afforded by people's participation in the management of forests. The first one was a community-based forest conservation movement initiated in Araberi in Midnapur District of West Bengal State during 1971-72, while the second one was a grassroots movement in the Sukhomajri village in Haryana in the 1980s, to rejuvenate forests and agricultural systems in the village. In Araberi, the movement was triggered by a Silviculturist of the State Forest Department, who by offering incentives to local communities induced them to protect and regenerate degraded Sal forests. The local communities, which were organized into 'forest protection committees' (FPCs), successfully protected the degraded forests from illegal felling, overgrazing, fire and encroachment. In Sukhomajri of Haryana State, the movement started from amongst the people. Construction of earthen dams stabilized agricultural output in the village. Forests in and around Sukhomajri village regenerated as a result. The regenerated forests in turn provided valuable biomass, including bhabar grass to local communities.

These instances awakened the policy makers at the Central Government to the need to go beyond the legalistic "Forest Conservation Act 1980". Institutional measures for arresting the alarming trend of forest depletion in the country were actively considered. The National Forest Policy 1988 was accordingly enacted by the Government of India with a strong focus on conservation, environmental stability and ecological balance through association of tribals and local communities in protection, regeneration and development of forests. In pursuance of the National Forest Policy of 1988, the Government of India issued a 'circular' in June 1990 for involvement of village communities and village associations (VAs) in the regeneration of degraded forest lands. This marked the birth of the Joint Forest Management (JFM) movement in India. Events generated by the 1990 circular forced the pace for the formation of the National Afforestation and Eco-Development Board (NAEB) in the year 1993, which was given the mandate of focusing its activities on degraded forest lands. The NWDB focused on its original mandate i.e., afforestation of community and non-forest wastelands. Subsequently while the NAEB functioned under the Ministry of Environment and Forests, the NWDB was shifted to the Ministry of Rural Development.

JFM – DEFINITION

JFM is defined as a concept of developing partnerships between fringe forest user groups and the Forest Department (FD) on the basis of mutual trust and jointly defined roles and responsibilities for forest protection and development (Anonymous 2001). Another definition of JFM states 'JFM is a forest management strategy under which the Government (represented by the Forest Department) and the village community enter into an agreement to jointly protect and manage forestlands adjoining villages and to share responsibilities and benefits' (Government of India 2002). JFM has several objectives:

- To analyze the history of forest use and management
- To analyze the social and economic characteristics of forest users and interrelationships of these factors
- To analyze the relevance of appropriate cultural factors
- To appraise the perspective of forest users and forestry staff

PRINCIPLE OF JFM

The main principle of JFM is to motivate the forest communities to protect the forest from over exploitation so that it meets the subsistence needs of the rural poor. JFM represents a significant shift in India's forest policy that balances community and government interest while remaining sensitive to local conditions, institutions and forest dependence.

SCOPE OF JFM

The scope of joint forest management has been largely confined to degraded forest areas, though in certain cases such as Haryana, Madhya Pradesh and West Bengal, non-degraded forestlands have also been covered under the JFM.

STRUCTURES OF JFM

Until recently, Forest Protection Committees (FPC) were by and large not legal bodies as they were not recognized or registered under the Societies Registration Act or through related, enabling legislations. However, States such as Gujarat, Karnataka and Rajasthan had provided for registration of FPCs under the Cooperative Societies Act in the early stages itself (Murali et al 2000). Structures and nomenclatures of JFM have varied from State to State. In general, one can delineate the following structure for JFM. The base of the JFM structure, which comprises village level institutions, is referred to as forest protection committees. These bodies include a "General Body" and a "Core" or "Executive" or "Management Committee (MC)" elected by the General Body for discharging the assigned functions. The General Body comprises eligible members of the village. The eligibility criteria for membership vary from State to State. In most States, adults are eligible for joining the General Body.

The General Body elects the local community representatives in the Executive Committee. The Executive Committee has elected members ranging in numbers from 5 to 15. It also includes ex-officio, non-elected members drawn from the Forest Departments, local NGOs, village schools, village administrative and development officers and in some cases representatives of the Gram or Mandal Panchayats. The States of Bihar, Himachal Pradesh, Karnataka, Jammu and Kashmir, Kerala, Madhya Pradesh and Punjab give representation for underprivileged communities and castes and women in the Executive Committee. Andhra Pradesh has moved towards compulsory 30 per cent representation for women in the Executive Committee. The States of Gujarat, Jammu and Kashmir, Karnataka and Madhya Pradesh require inclusion of two women representatives in the Executive Committee and provide for both 'husband' and 'wife' from a given household to be members of a General Body. The term of the Executive Committee is generally for two years in most of the States. However, in Arunachal Pradesh, Jammu and Kashmir, Madhya Pradesh, Maharashtra, Tripura and West Bengal, the tenure is one year. In Karnataka, Nagaland and Tamil Nadu, the tenure of the Executive Committee is five years. The Executive Committee undertakes decision-making, planning and implementation of management plans.

POWERS OF FOREST PROTECTION COMMITTEES

The functioning of FPCs in the post-1990 phase indicates that there are widespread inter-State differences, when it comes to grant of execution of powers by Forest Protection Committees. Different States give different flexibilities to FPCs in the matter of administrative, executive and financial powers. Nevertheless, the standard parameters of power granted to forest protection committees can be broadly categorized as under:

- Framing of rules governing management of forests
- Administrative powers for convening meetings of General Body and Management Committee meetings
- Voting rights in Management / Executive Committees
- Punitive powers – i.e., powers for punishing and imposing fines on offenders
- Rights to collect specific resources and sell them
- Distribution of benefits arising from conservation and regeneration
- Cancellation of membership of recalcitrant members
- Financial powers – to incur expenses, maintain accounts

SALIENT FEATURES OF JFM

It encourages the development of partnerships between local people and forest department to manage forest lands jointly. It provides a communication structure for consultation and negotiation between forest department and forest users and provides legalized access for the local communities to forest lands. JFM encourages local people to protect forest areas in order to prevent free grazing of livestock and to assist in preventing illegal activities by outsiders. JFM assures local people that a certain proportion of the intermediate and final harvests from forest lands will accrue to them (Gupta 2006).

TRADITIONS AND EXPERIENCE OF JFM

JFM is an old concept and a traditional way of managing forest lands. But with rapid increase in population, there is high pressure on forest. By the middle of the 19th century, Indian forest used to be managed scientifically. The royal commission on agriculture examined the role of forest. It favoured establishing plantation of fuel wood along the river banks, streams etc. and recommended that forest lands be given to village people. After that, taungya system evolved in Bihar, Orissa, Madhya Pradesh, West Bengal, Uttar Pradesh and Northeast India.

TAUNGYA SYSTEM

Three types of Taungya system are being practiced in India

- *Departmental Taungya*: Forest department engages local people for planting trees and raising agricultural crops and management is done by forest department.
- *Leased Taungya*: In this system, the local communities are given land on lease and management is done by the local communities who have got lease.
- *Village Taungya*: A land is given to people for planting trees and management is also done by them (Negi 2001).

Advantages: It is a cheap means of obtaining artificial regeneration of forest lands. It involves local communities in managing plantation and helps in protection of forest. It also generates employment to local people.

Disadvantages: It reduces soil fertility and causes soil erosion. It is more prone to epidemics or insect attack. Allotting forest land to local communities creates problems of ownership and tenurial rights. It also causes exploitation of human labour.

Traditional forestry practices such as home gardens and tree farms have been in practice since time immemorial to meet the demands of forest resources by the local communities. Some trees are planted and managed for commercial purposes which are known as cash crops.

JFM AND FOREST POLICIES

With reference to JFM, forest policies had an influence on people's participation in forestry. Some forest policies evolved and provisions were laid out in different forest policies. People in olden days lived in harmony and did not have any problems in accessing the forest resources. National Forest Policy 1894 was enacted which stressed on conserving forest and maintaining environment stability and to the basic needs of those people residing near the forest. The policy was then revised in 1952 which give importance to social forestry and agroforestry. The policy proposed that the forest cover should be 33 per cent of the total geographical area. The national forest policy was once again revised in 1988 which encourages the community to participate in protection and regeneration of forests and to manage the forest resource sustainably.

PARTICIPATORY ASSESSMENT AND PLANNING

It is a process for determining the availability of natural resources, identification of problems, gathering information and identifying the activities which can be implemented.

Participatory assessment has multiple objectives such as finding the status of natural resources in the village and how sustainable management is done, and level of dependence of the community on these natural resources. This helps in building cooperation and strengthening relationship between the community and the staff of forest department. The local community could understand better about the JFM activity through participatory assessment and planning and the information generated by participatory assessment can be used in base line survey and evaluation of the activity.

PARTICIPATORY PLANNING

It is a developmental plan generated by the local communities. It consists of the following aspects (after FARM programme, Negi 2000):

1. Natural resource management: This plan provides improvement, regeneration and sustainable management
2. Enhancing production: This is a plan for sustainable enhancement of forest produce
3. Enterprise and market support: Communities make a plan for developing forest enterprise as well as forest based products
4. Management of capital and funds: This is very important before implementation of the plan
5. Social/ Institutional support: This is also important for implementation of the plan
6. Phased plan: Activities should be planned according to seasons
7. Flexibility: This is important so that any change can be made in future if necessary
8. Monitoring and evaluation: Provisions are made for this

PARTICIPATORY RURAL APPRAISAL (PRA)

It is a process of learning about the conditions of the rural areas in a very intensive way. It is a tool which involves local community in order to acquire information and to enhance understanding of rural conditions and combines with scientific expertise. It is based on the assumptions that the communities have more knowledge about the problems faced by them and motivates the local community to participate in this exercise so that their problems are mitigated. Following are the tools used in PRA (Negi 2000):

(1) Stakeholder analysis- with the help of Venn diagram (also known as chapatti diagrams).

(2) Gender analysis- shows the role of women and needs for involving in the exercise.

(3) Community needs assessment- depending on the local conditions needs, problems and solutions are assessed.

(4) Time/history- History of the village, year of development of schools, hospital are recorded.

(5) Participatory mapping- includes resource maps and social maps.

(6) Wealth and well-being ranking

(7) Transect work

(8) Cropping pattern

(9) Seasonal calendar

(10) Matrix scoring /ranking- tree species, crop variety and uses of land.

(11) Livelihood analysis

(12) Mobility mapping

RAPID RURAL APPRAISAL

It is an important tool for creation, collection, analysis and management of information. Following are the salient features. It is seen as a learning process and less time consuming. It is site specific and information can be analyzed during the process of information generation. Results are reliable. The constraints are:

- New technique that has to be widely adopted under a diverse range
- Methods varies by different experts
- Personal bias if professional is inexperienced

The tools include semi-structured interviews, group level interviews and observation. The methods consist of Triangulation, pre-existing data, interviews and multi-disciplinary team interactions (Negi 2000).

JFM TRAINING

Training is an activity to develop and upgrade the skills and knowledge of an individual, person or group of persons or a community. There are different types of training viz., Induction training, professional training, field training etc. the duration depends on the local conditions. Extension tools for JFM consist of processes in which the information is transferred to the people which gain knowledge of skills and utilization of the information given to them. Different extension tools and methods (Negi 2000) are:

- Farm and home visits
- General meetings
- Group contacts
- Method demonstration
- Results demonstration
- Campaign
- Voluntary local leadership
- Conducted tours
- Exhibitions
- Publications
- Enquiries
- Mass media

- Audio and visual aids

JFM AND MICROPLANS

Micro-planning is a process of linkage of local people with local programmes. This is a plan for linkage of resource available or to be generated of identified area with specific individuals/ groups through consultation with the planners. The main aim of JFM micro-planning is to prepare a forest management plan. Micro-planning should take care of the followings:

- Group facilitation
- Options regarding choice of species
- Views of weaker sections (passive group women)
- Using terms as local practice
- Avoiding lead question
- Avoiding those answers already known
- Avoiding sensitive questions directly, example encroachment, politics etc.
- Format should be very simple to understand.

NEEDS OF JFM MICRO-PLANNING

JFM micro-planning needs to ensure people participation in management of forest. It should help to access the needs of the participating community and scope for development of the region or resource. It needs to plan, project and regulate the benefit flow to the participant in consultation with them. It should identify and highlight the local resources and set priorities for resource development based on site conditions and available funds. The process of micro-plan formulation includes:

- To explain the purpose to Forest Protection Committee in preliminary meeting.
- To identify resources on map with the help of villagers (men, women and young people).
- Beat officer prepares a draft micro-plan in consultation with the villagers on the basis of need assessment and resources.
- Plan is then sent to Ranger who compiles sector wise programmes and financial requirements, and sends it to DFO.
- DFO, having small committee including ACF and Rangers, finalizes all plan allocation and draft plans are revised.
- They discuss about revised draft plan and finalize it.
- Each copy is sent to DFO, Beat Officer and forest protection committees.

BENEFITS OF MICRO-PLANNING IN JFM

- Two way flow of information during planning

- People's participation in resource management
- Problem solving at local level
- Site specific and need based plan preparation
- Fixing realistic objectives
- Linkage of users and planners
- Better identification of local resources
- Keeping aspirations and expectation within achievable limit by consent
- Reduction of conflict
- Better chance of success
- Ensure grassroots monitoring and evaluation

ROLE OF NWFPs IN JFM

Non-wood forest products have significant importance and attraction in JFM system. NWFPs are integral to the lifestyle of forest-dependent communities. They fulfill basic requirements, provide gainful employment during lean periods and supplement incomes from agriculture and wage labour. Medicinal plants have an important role in rural health (Prasad and Bhatnagar 1991). In parts of West Bengal, communities derive as much as 17 per cent of their annual household income from NWFP collection and sale (Malhotra et al 1991). Small-scale forest-based enterprises, many of which rely on NWFPs, provide up to 50 per cent of the income for about 25 per cent of India's rural labour force.

Second, NWFPs have a decided advantage over timber in terms of the time needed to achieve significant volumes of commercially valuable production. Timber production is a long-term endeavour and in many areas, timber harvesting may not be ecologically desirable. Moreover, many NWFPs become available even in the earliest stages of rehabilitation of degraded forest areas. Third, at the national level over 50 per cent of forest revenue and about 70 per cent of forest export revenue comes from NWFPs, mostly from unprocessed and raw forms (Prasad et al 1996).

Thus NWFP management has clear ecological, social and economic benefits. Managing forests for multiple products including NWFPs and adding value to them at the local level are two of the most pressing challenges facing the JFM programme. In attempts to optimize the production of multiple products to meet the objectives of the various stakeholders, due attention should be paid to the potential for sustainable production of NWFPs in forest management efforts, including JFM arrangements. The true spirit of JFM gets translated only when forests are also managed to meet the people's needs.

CONTROL OF NWFPs BY STATE

Traditionally, the collection of NWFPs has been of low intensity and generally sustainable. However, as the economic potential of NWFPs has become apparent, the intensity of collection has increased and more significant infrastructures for trade and processing have developed. This has raised concerns for the sustainability of the resources and the distribution of the benefits derived from them. In reaction to these concerns, a number of State Governments have taken over the control of a number of NWFPs. The state regulations bringing certain NWFPs under monopoly trade are summarized in Table 32.1. Some of the explicit objectives for state monopoly of NWFP trade are:

- to prevent unscrupulous intermediaries and their agents from exploiting NWFP collectors
- to ensure fair wages to collectors
- to enhance revenue for the state
- to ensure quality
- to maximize the collection of produce (Prasad et al 1996)

Table 32.1: State Trading Regulations Promulgated by State Governments (Source: MoEF 1998)

State	Regulations	Implications
Andhra Pradesh	Andhra Pradesh Minor Forest Produce (Regulation of Trade) Act, 1971 Andhra Pradesh scheduled areas	Trade in NWFPs is declared state monopoly whether ownership is with government or not
Bihar	Bihar Kendu Leaves (Control of Trade) Act, 1973 Bihar Forest Produce (Regulation of Trade) Act, 1984	Bihar State Forest Development Corporation operates as state government agent for collection and marketing of kendu leaves, sal seed, mahua (*Madhuca latifolia*) and harra
Gujarat	Gujarat Minor Forest Produce (Regulation of Trade) Act, 1979	Minor forest products identified include timru leaves (tendu leaves), mahua flowers, fruits, seeds and gum
Himachal Pradesh	Himachal Pradesh Resin and Resin Produce (Regulation of Trade) Act, 1981	Resin, bamboo and *Acacia catechu* (khair) collection through Himachal Pradesh Forest Development Corporation Ltd
Madhya Pradesh	Madhya Pradesh Vanopaj (Vyapar Viniyam) Adhiniyam, 1969	Items under monopoly include tendu leaves, sal seed, harra and gums; Madhya Pradesh Minor Forest Produce (Trade and Development) Federation acts as agent of state government
Rajasthan	Rajasthan Tendu Leaves Act, 1974	Rajasthan Tribal Area Development Federation collects and markets NWFPs
Odisha	Orissa Forest Produce (Control of Trade) Act, 1981 Orissa Kendu Leaves (Control of Trade) Act, 1981	Collection and sale of NWFPs are monopolized by Forest Department and leased to Tribal Development Cooperative Society which in turn delegates to an individual; collection and trade of leaves are handled by Forest Corporation

In most cases, trading is controlled through State-owned institutions such as State forest development corporations, federations, cooperatives and tribal societies. In Odisha, however, where the Forest Produce (Control and Trade) Act of 1981 provides the scope for a state monopoly on certain selected forest products. The State also has the option to give monopoly leases for collection and trade of forest products. In fact, the State has granted monopoly rights for 29 NWFP items to a private company, Utkal Forest Products Ltd. (Prasad and Saxena 1996; Agragamee 1997). Under this agreement, the local people who collect NWFPs are required to sell their collected materials to the company's agents at preset prices that are lower than those they could have obtained by selling directly to processors. It is noteworthy that some of the 29 items yield very insignificant amounts of revenue yet have nevertheless been taken under the state monopoly.

EXTENT OF JFM IN INDIA

On 1st June 1990, Government of India issued guidelines to the State Governments for involving local communities in the protection and development of the degraded forest. Based on this, JFM programme has been widely adopted in several States. It is a powerful tool and aims at empowering people to participate actively in the management of forest resources to derive benefits and to improve their livelihoods. The forest cover and the area under JFM are shown in Table 32.2. It shows that West Bengal has more than half of its forest area under JFM followed by Madhya Pradesh with more than 60 per cent of its forest area under JFM. There are 84,632 JFM committees covering the states in India. About 83,84,788 families are involved in JFM process (Table 32.3). Odisha has the highest number of JFMCs (15,985) and Madhya Pradesh, Uttaranchal, Chhattisgarh, Andhra Pradesh and West Bengal also have large number of JFM committees.

Table 32.2: JFM as Percentage of Forest Cover in different States of India (2004)
(Source: Center for monitoring Indian Economy (CMIE) 2006)

State	Forest Cover in 1000 ha	Area under JFM in 1000ha	JFM % in Forest Cover
Andhra Pradesh	6,199	1,887	33.44
Arunachal Pradesh	5,154	80	1.55
Assam	1,933	79	4.08
Bihar	622	267	42.92
Chhattisgarh	6,300	2,846	45.17
Goa	125	13	10.40
Gujarat	1,706	160	9.38
Haryana	45	56	12.44
Himachal Pradesh	1,099	291	26.44
Jammu & Kashmir	2,023	49	2.42
Jharkhand	2,333	848	36.34
Karnataka	3,070	233	7.59
Kerala	1,082	171	15.80
Madhya Pradesh	8,681	5,500	63.35
Maharashtra	5,214	1,411	27.06
Manipur	1,693	94	5.56
Mizoram	1,542	10	0.64
Nagaland	863	23	2.67

Odisha	5,813	82	14.14
Punjab	277	56	4.08
Rajasthan	2,651	377	14.22
Sikkim	265	60	0.22
Tamil Nadu	2,132	446	20.91
Tripura	606	34	5.61
Uttaranchal	3,466	859	27.78
Uttar Pradesh	1,689	113	6.69
West Bengal	1,194	604	50.59
India	67,777	17,329.6	25.57

Table 32.3: Participants involved in JFM in different States of India (as on 31-03-2004)
(Source: Forests & Wildlife statistics of India 2004)

State	No. of JFM Committee	No. of Families involved			Total No. of Families
		SC	ST	General	
Andhra Pradesh	7,245	1,36,789	1,88,621	2,85,685	6,11,095
Arunachal Pradesh	308	-	20,474	-	20,474
Assam	503	4,145	31583	21,613	57,341
Bihar	493	53,377	31,618	1,57,644	2,42,639
Chhattisgarh	6,881	2,51,012	7,60,892	3,48,347	13,60,251
Goa	26	36	-	300	336
Gujarat	1,424	5,748	1,23,347	20,096	1,49,191
Haryana	875	31,000	-	1,34,500	1,65,500
Himachal Pradesh	835	62,915	7,024	1,89,130	2,59,069
Jammu & Kashmir	935	4,789	5,324	38,347	48,460
Jharkhand	3,358	53,617	76,615	1,65,903	2,96,135
Karnataka	3,470	50,465	37,303	1,55,589	2,43,357
Kerala	323	3,828	11,371	25,425	40,624
Madhya Pradesh	13,698	2,91,000	7,09,000	6,41,000	16,41,000
Maharashtra	5,322	1,07,409	3,08,934	5,05,291	9,21,634
Manipur	205	57	13,883	5,511	19,451
Mizoram	249	1	32,240	-	32,241
Nagaland	306	-	57,492	-	57,492
Odisha	15,985	-	-	-	-
Punjab	287	20,029	9	17,984	38,022
Rajasthan	3,667	33,229	1,52,017	1,15,049	3,00,295
Sikkim	158	669	835	1,764	3,268
Tamil Nadu	1,816	72,290	11,484	1,87,160	2,70,934
Tripura	234	2,896	11,291	6,052	20,239
Uttaranchal	2,030	1,96,973	7,240	4,16,152	6,20,365

Uttar Pradesh	10,107	87,500	15,000	3,97,480	4,99,980
West Bengal	3,892	1,13,304	1,15,836	2,36,255	4,65,395
Total	84,632	15,83,078	27,29,433	40,72,277	83,84,788

CAUSES FOR POOR PERFORMANCE OF JFM

Different studies regarding the implementation of JFM in different parts of the country attribute the following negative features that have hampered the functioning of FPCs/VFCs:

- Lack of legal status and financial and executive powers for FPCs.
- Forest Department in certain States vested with arbitrary powers to dissolve FPCs.
- Faulty design of micro-plans and management plans.
- Absence of participation by women in spite of their formal representation in management committees.
- Unreasonable controls over the duration of exploitation of admitted resources leading to low level of exploitation of admitted NTFPs.
- Excessive rule and regulations.
- Inadequate remuneration for local communities from JFM activities.
- Inter and intra-community conflicts that hamper FPC functioning.
- Denial of rights on disposal over valuable NTFPs to local communities.

EXTERNALLY AIDED PROJECTS FOR JFM

According to Ravindranath et al. (2000), JFM has been particularly successful in States, which have received external assistance for forestry projects. The States of Karnataka, Maharashtra, Madhya Pradesh, Haryana, Himachal Pradesh, Andhra Pradesh and West Bengal have received external assistance, which the authors attribute as the reason for the States bringing in a higher percentage of open forests under JFM. The details of the different donor agencies and the projects they have funded in these States are given in Table 32.4.

Table 32.4: Financial Assistance Received from External Donor Agencies for JFM
(Source: Saxena 1997)

Name of the Donor Agency	Name of the Projects and Period	Total Financial Assistance (In Million Rs.)	Allocation for Participatory Component
World Bank	(1) West Bengal Forestry Project (1992-97)	1,140	464 (41%)
	(2) Maharashtra Forestry Project (1992-98)	4,310	208 (5%)
	(3) Andhra Pradesh Forestry Project (1994-2000)	3,540	101 (3%)
	4) Madhya Pradesh Forestry Projects (1995-99)	2,460	1,468 (60%)
	Total	1,1450	2,241 (20%)
OECF-Japan	(1) Rajasthan Afforestation Project for Aravalli Hills (1992-97)	1,670	971 (58%)

	(2) Rajasthan Forestry Development Project (1995-2000)	1,390	503 (36%)
	(3) Rajasthan Afforestation and Pasture development along Indira Gandhi Nahar Project (1990-2000)	1,070	NA
	Total	4,130	1,474 (36%)
DFID, UK	(1) Karnataka Western Ghats Forestry Project (1993-99)	1,050	534 (51%)
	(2) Himachal Pradesh Kullu Mandi Forestry Project (1994-97)	130	54 (42%)
	Total	1,180	588 (50%)
EU	Haryana rehabilitation of common lands in the Aravalli Hills (1990-97)	481	481 (100%)
Grand Total		17,241	4,784 (28%)

However there are problems in assuming that JFM has worked well in all States that have benefited from externally aided projects. The States of Kerala and Maharashtra, which have benefited from externally aided projects, have not really performed well in terms of the multiple criteria for evaluating JFM success. While Maharashtra has done well in terms of coverage but the structuring of FPCs in this State has not been in accordance with the objective of empowering communities. Similarly the State of Kerala, despite having the benefit of World Bank assisted Social Forestry Project, has not extended the JFM concept beyond "pilot zones" (Damodaran 1997).

EVALUATION CRITERIA AND FACTORS CONDUCIVE TO SUCCESS OF JFM

JFM, in its essence, is an institutional incentive for biodiversity conservation. UNEP (1996) categorizes three broad diagnostic tools for adjudging the effectiveness of different incentives (including institutions) in the matter of biodiversity conservation. These tools center on "constraints" which could be "formal" (i.e. arising from laws, legislation policies), "social" (involvement or non-involvement of social groups and individuals having a stake in the conservation of the resource) and/or "compliance" (arising from actual 'non-compliance' to the scheme of conservation for which the incentives have been instituted). Smith (1997) views incentives from the point of view of their efficiency as reflected in administration costs. In the context of JFM, various studies have viewed JFM performance in terms of multiple criteria. These include frameworks developed by Sarin (1993), Raju et al (1993) and SPWD (1992). Following Sarin (1993), the framework for analysis of community level institutions can be considered as including the following parameters:

- Viable, social unit of organization
- Organizational norms and procedures
- Accountability mechanism (transparancy, equity and democratization)
- Conflict resolution mechanisms
- Autonomous status

Raju et al (1993) consider six basic traits of community level institutions, namely:

- Transparency
- Clarity of rules
- Level of awareness
- Initiative and independence
- Tenure security
- Satisfaction of basic needs

The Society for Promotion of Wastelands Development (1992) considers JFM to be successful where:

- Existence of strong community forest management group
- High degree of homogeneity within the community and this may be highest where the community is a homogenous tribal group
- Effective leadership
- Easy access to rules and regulations
- High degree of environmental concern among the members
- Community attaches a high level of importance to social / religious roles of forests
- Shared perception of acute resource scarcity

All these approaches serve as useful diagnostic parameters. However, what is common to them is their emphasis on the 'efficiency', 'effectiveness' and 'ramification' criteria for assessing JFM. The policy-intent or the philosophical approach is not scrutinized.

RECENT POLICY CHANGES ON JFM

THE FEBRUARY 2000 GUIDELINES

Ten years after the issue of the 1990 circular, the Government of India issued a new set of guidelines on 21 February 2000 for strengthening the joint forest management programme in the country. The guidelines were framed in the light of the experience with the JFM process in India since 1990. The new guidelines provide for the following:

Legal Backup to the JFM Committees: Realizing that JFM committees did not have any legal backup, the guidelines called upon all State Governments to register JFMs or village committees under Societies Registration Act by 31 March 2000. It was also laid down that nomenclatures for JFM committees might be made uniform in all States by naming them as Joint Forest Management Committees (JFMC). The State Governments were further required to sign Memorandum of Understanding (MoU) with these committees which defined their roles and responsibilities in different areas. Further, all adults of the village were required by these guidelines to become members of the JFMC.

Participation of Women in the JFM Programme: The guidelines called for 50 per cent of the JFMC general body members to be women and also require 50 per cent of the women members to be present for a general body meeting. Further, the guidelines also laid down that the management committee or the executive committee should have 33 per cent of their membership from women general body members. The quorum for holding meetings of the committees should be one third of women executive members or minimum of one, which ever is more. Further, one of the office bearers of the MC have to be a woman.

Extension of JFM to Good Forest Areas: The circular laid down that JFM programme should cover both degraded and non-degraded areas. Micro-plan and the MoUs have to be different for degraded and good forest areas (crown density greater than 40 per cent). In good forest areas, JFM would concentrate on NTFP management and would not deviate from working plan prescriptions. The benefit sharing mechanisms would be different for the good forest areas and timber would be distributed only if the committees had satisfactorily protected the forests for at least ten years and the sharing percentage was stipulated as not to exceed 20 per cent of the revenue. The extent of good forest areas to be included under JFM would depend upon number of village households, but have to observe the twin limits of 100 hectares maximum and 2 km radius from the village boundary. For degraded forest areas, the radius was fixed at 5 km from the village boundary. The implementation of JFM in good forest areas has to be done on pilot basis, which has to be monitored carefully.

Preparation of Micro-plan in JFM Areas: In case of new working plans, the guidelines called for JFM overlapping working circles to incorporate broad provisions for micro-plans. The micro-plans have to be prepared carefully, after detailed Participatory Rural Appraisal (PRA) exercises by the forest officers and JFMCs, so that it reflects the consumption and livelihood needs of the local communities and also utilizes locally available knowledge and strengthened local institutions. The micro-plan is also supposed to account for market linkages for better returns on NTFPs. Where working plans are already in force, the guidelines call for incorporation of micro plans into these plans and permitted deviations that did not change the basic principles of silviculture.

Conflict Resolution: State Governments were requested to constitute divisional and state-level representative forums or working groups comprising of all stakeholders including NGOs for resolving disputes amongst and within communities.

Recognition of Self-initiated Groups: The guideline called upon the State Governments of Odisha, Bihar, Gujarat, Andhra Pradesh and Karnataka to identify, recognize and register self interest groups and give proper benefits to them after evaluating their performance.

Contribution for Regeneration of Resources: The Guidelines called upon State Governments to ensure that not less than 25 per cent of the share of village communities arising from sale of resources from regenerated areas should be deposited in the village development fund for meeting conservation and development needs of forests. It also called for a matching contribution to be made by the Forest Department from its share of such sales. Similarly, it was laid down that mechanisms for computing income for sharing of benefits should be transparent.

Monitoring and Evaluation: The guidelines called for concurrent monitoring and progress of the programme at an interval of 3 to 5 years at the Division and State levels.

THE DECEMBER 2002 GUIDELINES

These guidelines have the following elements:

- JFM Committees should be treated as the 'basic forest management units' and a MoU should be signed between them and the Forest Department.
- JFM Committees may be given authority to act and provide adequate monetary and other incentives to participate as genuine stakeholders.
- The plan of action of JFM Committees should be holistic, covering forest lands, village commons and private lands.
- The unique and separate non-political identity of the JFM Committees as 'guardian of forests' should be maintained and ensured, though the Committees should take advantage of the administrative and financial position and organizational capacity of the Panchayats.
- NTFP management is mentioned as crucial for the success of JFM.
- Capacity building and institutional reforms of stakeholders should be accorded priority.

CHANGES AT THE MACRO LEVEL

As the JFM process gained momentum at the micro-level, the need for instituting macro structures for overseeing and monitoring the progress became essential. Three steps were taken in this regard. The steps are as follows:

Setting up of a JFM Monitoring Cell: Responding to the long-standing demand for creation of a nodal agency for JFM at the national level, the Ministry of Environment and Forests (MoEF) created a JFM Monitoring Cell within its Forest Protection Division. The cell would collect information on various aspects of the JFM programme, analyze it and generate appropriate policy responses (Government of India 2002).

JFM Network and Stakeholder's Forum: A multi-stakeholder JFM Network was created at the national level, which acts as a regular mechanism for consultation between various stakeholders, and to assist policy makers in obtaining constant feedback for proper policy formulation. The Director General of Forests is the Chairman and the Deputy Inspector General in charge of JFM is the Member-Secretary of the JFM Network. A neutral stakeholders forum called Resource Unit for Participatory Forestry (RUPFOR) was also created to support the functioning of the network and to promote interaction among various stakeholders. RUPFOR is currently housed within the New Delhi office of Winrock International India (Anon 2001; Government of India 2002).

National Afforestation Programme: Steps have already been initiated to set up Forest Development Agencies (FDAs) at the divisional level in different States. As on March 2003, 285 FDAs have been established at the division level in different States. In the remaining phase of the Tenth Five-Year Plan it is expected that 775 FDAs would be established to cover the remaining divisions.

33 LAND DEGRADATION AND WASTELAND MANAGEMENT

INTRODUCTION

Land degradation threatens the livelihoods of Earth's many inhabitants, the so called environmental refugees, including many of the world's poorest and most marginalized populations. Degradation caused by overcultivation, overgrazing, deforestation and inefficient irrigation affects most of the world's drylands. Desertification occurs in drylands, which span a third of the Earth's land surface in over 110 countries. All continents are under threat including 37 per cent of arid areas in Africa, 33 per cent in Asia, 14 per cent in Australia and some areas in America and the southern fringes in Europe. According to United Nations Convention to Combat Desertification (UNCCD), over 250 million people are directly affected by land degradation. In addition, some one billion people around world are at risk. Each year 12 million hectares are lost to deserts. That is enough land to grow 20 million metric tons of grain.

The land degradation issue for world food security and the quality of the environment assumes a major significance when one considers that only about 11 per cent of the global land surface can be considered as prime or Class-1 land, and this must feed the 6.5 billion people today and the 8.2 billion expected by the year 2020. Hence land degradation will remain high on the international agenda in the 21st century. Desertification is often the result of human activity and can therefore be prevented or controlled by human effort. Because of its magnitude, desertification is a global issue, and such can only be addressed through a global partnership.

WHAT IS LAND?

It is an area of the Earth's surface including all reasonably stable or predictably cyclic attributes vertically above and below it. Land includes not only the soil resource but also the water, vegetation, landscape and microclimatic components of an ecosystem.

LAND RESOURCES

Land, a non-renewable resource, is central to all primary production system. Land is considered as the most important basic natural resource. It is a dynamic and complex combination of geology, topography, hydrology, soil, flora and fauna and influences every sphere of human activity. Different sectors including agriculture, industries, infrastructure and power projects put forth competing demand for land. Subsistence farming practices, accelerated soil and water erosion, erratic rainfall, increasing population and high density of livestock population have all contributed to unsustainable land use that has led to degradation of this valuable resource.

Land is classified into two categories viz. arable and non-arable. Non-arable land comprises of area under forests, permanent pasture land, current fallow, cultivable wasteland and land put to non-agricultural use. Arable land includes area sown with crops (net sown area), area sown more than once and gross sown area.

DEFINITIONS

DEGRADED LAND: According to International Fund for Agricultural Development (IFAD), it is defined as the land which due to natural processes or human activity is no longer able to sustain properly an economic function and/or the original natural ecological function; or, the loss of the productive capacity of the land to sustain life.

LAND DEGRADATION: The definition of land degradation, under Article 1 of the UNCCD, is the reduction or loss, in arid, semi-arid and dry sub-arid areas, of the biological or economic productivity and complexity of rain fed cropland, irrigated cropland, or range, pasture, forest and woodlands resulting from land use or from a process or combination of processes, including processes arising from human activities and habitation patterns, such as:

(i) Soil erosion caused by wind and/or water

(ii) Deterioration of the physical, chemical, and biological or economic properties of soil and

(iii) Long-term loss of natural vegetation

The Global Environment Facility (GEF) defines land degradation as any form of deterioration of the natural potential of land that affects ecosystem integrity either in terms of reducing its sustainable ecological productivity or in terms of its native biological richness and maintenance of resilience.

DROUGHT: The UNCCD definition of drought is the naturally occurring phenomenon that exists when precipitation has been significantly below normal recorded levels, causing serious hydrological imbalances that adversely affect land resource production systems.

DESERTIFICATION: The UNCCD definition of desertification is the land degradation in the arid, semi-arid and dry sub-humid areas resulting from various factors, including climatic variations and human activities.

LAND-USE POLICY: An expression or perception of the direction to be taken on major issues related to land use and the proposed allocation of the national land resources over a fixed period of time. It has a production and conservation components (FAO/UNEP).

SUSTAINABLE LAND MANAGEMENT: It combines technologies, policies and activities that are aimed at integrating socio-economic principles with environmental concerns so as to simultaneously maintain or enhance production, reduce the level of production risk, protect the potential of natural resources and prevent (buffer against) soil and water degradation, be economically viable and be socially acceptable.

ARIDITY INDEX: Aridity, the characteristic nature of an arid climate, is measured using an aridity index (AI). AI is a numerical indicator of the degree of dryness of the climate at a given location. A number of aridity indices have been proposed to identify, locate or delimit regions that suffer from a deficit of available water, a condition that can severely affect the effective use of the land for such activities as agriculture or stock-farming.

In 1948, C.W.Thornthwaite proposed and defined AI as,

$$AI_T = 100 \times \frac{d}{n}$$

(d = water deficiency which is calculated as the sum of monthly differences between precipitation and potential evapotranspiration for those months where normal precipitation is less than the normal evapotranspiration; n = sum of monthly values of potential evapotranspiration for the deficient months)

In the preparations leading to the UN Conference on Desertification (UNCOD), the UNEP issued a dryness map based on a different aridity index, proposed originally by M.I. Budyko in 1958 and defined as follows:

$$AI_B = 100 \times \frac{R}{LP}$$

(where R is the mean annual net radiation (also known as the net radiation balance); P is the mean annual precipitation; L is the latent heat of vaporization for water)

More recently, the UNEP has adopted yet another index of aridity, defined as:

$$AI_U = \frac{P}{PET}$$

(where PET is the potential evapotranspiration; P is the average annual precipitation)

Table 33.1: Classification of Land based on Aridity Index

Classification	Aridity Index	Global Land Area
Hyper-arid	$AI < 0.05$	7.5%
Arid	$0.05 < AI < 0.20$	12.1%
Semi-arid	$0.20 < AI < 0.50$	17.7%
Dry sub-humid	$0.50 < AI < 0.65$	9.9%

LAND DEGRADATION

Land degradation is recognized as the main outcome of desertification. Land degradation refers to a temporary or permanent decline in the productive capacity of the land or its potential for environmental management. Net Productivity change reflects both natural and human induced processes of degradation and improvement. Some types of land degradation are, for all practical purposes, irreversible. Examples are severe gullying and advanced salinization. In these cases, the long-term biological and environment potential of the land has been compromised. Displacement of soil material (erosion) is also irreversible, although its long-term effects on productive capacity depend on the depth and quality of soil remaining.

Most types of land degradation, however, can be prevented or reversed viz. adding nutrients to nutrient-depleted soil, rebuilding topsoil through soil amendments, re-establishing vegetation (reforestation), buffering soil acidity. The practicality of rehabilitating degraded landscapes depends on the costs relative to the value of output or environmental benefits expected. Where farmers wish to intensify agricultural production on a sustainable basis, it may be necessary to undertake such land-enhancing or land-protecting measures even in non-degraded landscapes.

Degradation is a process of change over time. From policy perspective, it is critical to distinguish those lands that are currently undergoing degradation, in order to assess the need for action to stabilize or reverse the process. Other lands may have reached a degraded state relative to their natural condition many decades or centuries ago, but are currently in a stable or improving condition. For practical purposes, the land degradation problem refers to the former areas.

GLOBAL SCENARIO OF LAND DEGRADATION

The international community has long recognized that desertification and land degradation are major economic, social and environmental problems of concern to many countries in all regions of the world. In 1991, the United Nations Environment Programme (UNEP) concluded that the problem of land degradation in arid, semi-arid and dry sub-humid areas had intensified. As a result, the United Nations Conference on Environment and Development (UNCED) supported a new integrated approach to the problem emphasizing action to promote sustainable development at the community level. As a result, the United Nations Convention to Combat Desertification (UNCCD) was launched in 1996 to tackle desertification and drought.

Land degradation imposes an important constraint in achieving sustainable development. The World Summit on Sustainable Development (WSSD) in 2002 reaffirmed land degradation as one of the major global environment and sustainable development challenges of the 21st century. It called for action to address causes of desertification and land degradation in order to restore land and to address poverty resulting from land degradation. Addressing land degradation also contributes significantly to the Millennium Development Goal (MDG) of reducing by half the proportion of people in poverty by 2015 and ensuring environmental sustainability.

By taking into account and realizing the importance of preventing desertification through land rehabilitation measures and sustainable land management practices and noting that these concerns were critical, the UNCCD expanded its programme to encompass activities to combat land degradation by making land degradation a Global Environment Facility (GEF) Focal Area. The second assembly of the GEF in October 2002, designated land degradation, primarily desertification and deforestation, as a new focal area for the purposes of supporting the implementation of UNCCD.

RIO CONVENTIONS

The three Rio conventions, which include UNCCD, Convention on Biological Diversity (CBD) and the United Nations Framework Convention on Climate Change (UNFCCC) are all relevant to the prevention and control of land degradation. The CBD recognizes the importance of addressing land degradation for the conservation and sustainable use of biodiversity. Furthermore, the relationship between climate change and land degradation is also recognized. Climate change contributes to desertification and deforestation through changes in temperatures which in turn adversely affects the respiration rate of vegetation. Human created deforestation releases carbon dioxide to the atmosphere and the loss of sequestered carbon in biomass and soils, further contributing to greenhouse gases in the atmosphere. The work programme of the UNFCCC, therefore, emphasizes the role of conservation and sustainable management of forests and woodland in carbon sequestration and reducing carbon dioxide emissions. The UNFCCC also recognizes the need to develop strategies to adapt the impacts of climate change. Many of these adaptation strategies are linked to combating land degradation and drought.

Joint work programmes for the three Rio conventions are now starting to achieve multiple global benefits, such as poverty alleviation, sustainable ecosystems and minimising greenhouse gas emissions. Since all of the Rio conventions are closely linked to each other, consolidating the policy and planning work required to carry out the responsibilities of the conventions allows for maximum efficiency and synergy in planning.

UNCCD

The United Nations Convention to Combat Desertification in countries experiencing serious drought and/or desertification, particularly in Africa (UNCCD), was adopted on 17 June 1994. The CCD aims to combat desertification and mitigate the effects of drought, particularly in Africa. The convention recognizes that the people of the drylands themselves hold the keys to combating desertification, and that the eradication of poverty is a precondition for success. The CCD entered into force in 26 December 1996; today 179 countries acknowledge it as a legally binding framework to tackle land degradation and promote sustainable development in fragile ecosystems. The global mechanism was established under CCD to promote actions leading to the mobilization and channelling of substantial financial resources, including for the transfer of technology, on a grant basis, and /or on concessional or other terms, to affected developing country parties.

TYPES OF LAND DEGRADATION

The land degradation is classified according to specific phenomenon prevalent in the land with respect to physical, chemical and biological condition.

(i) Physical: compaction, crusting, sealing, loss of structure, accelerated erosion

(ii) Chemical: salinization, toxicities, nutrient depletion and imbalances, acidification

(iii) Biological: reduction in biodiversity, loss of ecological component

Various types of soil and land degradation have been explained by many authors including FAO/UNDP/UNEP. As only to be expected, authors differ in their approach to describing and classifying land degradation. It is generally recognized that land degradation has five main components. (i) soil degradation, (ii) vegetation degradation, (iii) water degradation, (iv) climate deterioration (v) loss due to urban/industrial development. Each of these major components could be subdivided into more specific types of degradation.

FRAGILE ECOSYSTEMS

Fragile ecosystems are important ecosystems, with unique features and resources. Fragile ecosystems include deserts, semi-arid lands, mountains, wetlands, small islands and certain coastal areas. Most of these ecosystems are regional in scope, as they transcend national boundaries. For Small Island States, these fragile ecosystems may encompass the entire nation. Comprehensive plan of action should be taken globally, nationally and locally by international communities of the UN Systems, Governments and major groups in every area in which human's impact on their environment. It also addresses managing of fragile ecosystems: combating desertification, land degradation and drought.

CAUSES OF LAND DEGRADATION

The causes of land degradation has been grouped into three categories viz. natural degradation hazards, direct causes of land degradation, underlying causes of land degradation

(i) Natural degradation hazards: cyclones, drought, volcanic activities etc.

(ii) Direct causes of land degradation: deforestation, overcutting of vegetation, shifting cultivation, overgrazing, non-adoption of soil conservation practices, extension of cultivation in the lands of lower potential and /or high natural hazards, improper crop rotation, unbalanced fertilizer and pesticide use

(iii) Underlying causes of land degradation: population increase, attitude, economic pressure, land tenure, land shortage, poverty

MAJOR CAUSES OF LAND DEGRADATION

LACK OF LAND USE PLANNING: Road and dam construction projects have a number of significant environmental impacts. It results in destruction of vegetation, hence increasing vulnerable lands to soil erosion and degradation. The establishment of land use and management planning will reduce the land degradation

SEA LEVEL RISE: Sea level rise is a significant threat to many countries. In recent times, land has become subject to inundation and saltwater intrusion during tide events. Sea level rise degrades coastal areas including nearby vegetation. Inland vegetation is also affected as the sea water percolates up through the ground and forms large pools of saltwater on the land. Sea level rise not only affects vegetation, but also affects ground water. The biotic component of the watershed depends on ground water. Intrusion of salt water into ground water is a problem that has been identified in many countries.

DROUGHT: Drought causes damage to water supply, food security systems and the degradation of terrestrial ecosystems.

UNSUSTAINABLE AGRICULTURAL PRACTICES: The infertile nature of soils has led to an increase in the use of agricultural chemicals. Fertilizers and pesticides have been used in excess to enhance agricultural productivity. While higher crop yields have been gained from regular and intensive use of these chemicals, they have also caused problems. Overuse of chemicals causes land to become unsuitable for agriculture through changes in the physical and bio-chemical composition of the soil. Consequently farmers abandon their land.

UNCONTROLLED WASTE DISPOSAL: Overpopulation and overcrowding contribute to the problem of solid and liquid waste disposal. When population exceeds the natural carrying capacity of the area, it puts severe pressure on the limited available resources. Uncontrolled waste disposal systems result in land degradation.

ASSSESSMENT OF LAND DEGRADATION

GLASOD

The land degradation concept aims at covering a range of climatic and man induced processes that lead to a decline of the soil potential to sustain plant productivity. The first attempt to produce a global assessment occurred two decades before (1990), resulting in a 'Global Assessment of Soil Degradation' (GLASOD) by the International Soil Reference and Information Centre (ISRIC), currently World Soil Information. GLASOD was a qualitative assessment, largely based on expert judgement that distinguished the main processes and drivers leading to soil degradation, such as water and wind erosion-sedimentation, soil and water salinization, loss of soil organic carbon and soil nutrients, loss of soil structure etc. GLASOD data were used in the preparation of World Atlas of Desertification, published by UNEP in collaboration with some national and international institutions.

ASSOD

The Assessment of the Status of Human-induced Soil Degradation in South and South East Asia (ASSOD), which is also made by ISRIC, is a follow-up activity of GLASOD for regional studies. The same methodology, slightly refined, was used on a more detailed scale (1:5M) for South and South East Asia. The study provides data for 17 countries and includes data on several degradation types including water and wind erosion and their subtypes (e.g. loss of topsoil and terrain deformation) and the dominant subtypes of chemical deterioration (including salinization).

LADA

GLASOD approach was recently upgraded by a new worldwide project 'Land Degradation Assessment in Drylands' (LADA) in 2002 sponsored by UNEP, GEF and FAO. While retaining the original GLASOD categories of soil degradation, LADA takes a step forward by (i) attempting at delivering quantitative results, (ii) formally including socio-economic drivers and (iii) enlarging its scope to carbon balances and biodiversity as components of the functional land system and its degradation process. LADA aims at developing and testing an effective methodology to assess causes, status and impact of land degradation in drylands.

MA

A third global initiative with implications in land degradation assessment is the Millennium Ecosystem Assessment (MA). It evaluates the condition of desertification in drylands, by asking pointed questions and providing answers based exclusively on the reports generated by the MA. It provides a consistent picture of the links between land degradation, climate change and biodiversity loss. It also supplies guidelines for improving assessment and monitoring approaches taking into account the role of human actions and climate variability.

The three afore-mentioned projects reveal an historical trend of increasing complexity in land degradation assessment approaches. They go from soil to land, from considering effects to explicitly include drivers (climate variability and human activity) and to be more concerned with global interactions of desertification, climate change and biodiversity.

GLADA

Another global project in land degradation assessment is the GEF-UNEP-FAO program Land Degradation in Drylands, the Global Assessment of Land Degradation and Improvement (GLADA). GLADA will identify the state and trends of land degradation, areas suffering severe degradation or at severe risk and places where degradation has been arrested or reversed. Biomass is an integrated measure of productivity and its deviance from the norm may indicate land degradation or improvement. Biomass can be assessed by remote sensing of the normalised difference vegetation index (NDVI). Norms may be established by stratifying the land area according to climate, soils and terrain and land use/vegetation. The deviance may then be calculated regionally and then combined globally to allow universal comparisons.

DeSurvey

The DeSurvey is the last newcomer of the efforts in this line. Its objective is to develop and implement a cost-efficient procedure for assessing and monitoring of land degradation status and trend over large areas in an appropriate format to provide a link to the 'land suitability' attribute that is used classically by economists to allocate land use changes in space.

EFFECTS OF LAND DEGRADATION

The most important on-farm effects of land degradation are declining potential yields. The threat of degradation may also be reflected in the need to use a higher level of inputs in order to maintain yields. Serious degradation sometimes lead to temporary or permanent abandonment of land. In other cases, degradation induces the farmers to convert land to lower-value uses. FAO/UNDP/UNEP has grouped the effects of land degradation as effects upon production and consequences for the people.

EFFECTS ON LAND DEGRADATION UPON PRODUCTION

- Reduced vegetation cover to the soil and biodiversity loss
- Top soil loss
- Watershed destruction with subsequent water shortages
- Reduced return of organic matter and less biological activity in the soil
- Increased pollution from increased use of agrochemicals

- Land is abandoned (where degradation is severe)
- Crop yields are reduced
- Inputs and costs of prodution are increased (where farmers attempt to combat reduced yields by increased inputs)
- Responses to inputs are decreased
- Flexibility of land management is decreased
- Risk is increased
- Labour, technical and financial resources are diverted to reclamation

CONSEQUENCES OF LAND DEGRADATION FOR THE PEOPLE

- Landlessness is increased
- Food supplies are reduced or less reliable
- Labour requirements are increased
- Incomes are decreased
- Migration of young people in search for employement opportunities

WASTELANDS

WASTELAND DEFINITION

Wastelands are the degraded and unutilized lands except current fallows due to different constraints. Poor land practices have led to malnutrition and decline in production capacity of the soil. It is estimated that in wastelands the biomass production is less than 20 per cent of its overall potential. It includes areas affected by water logging, sheet and gully erosion, ravine, riverine lands, shifting cultivation, salinity and alkalinity, wind erosion, shifting sand dunes, extreme moisture deficiency, coastal sand dunes etc. These degraded lands are ecologically unstable with almost complete loss of top soil and are unsuitable for cultivation due to decline in their quality and productivity.

The term wasteland designates various types of unutilized land. It also indicates small pieces of land, usually on less fertile tract, which fail to yield a return to the farmers. It is also defined as the land which is lying uninhabited, uncultivated, land left after some use or land which is no longer serving any purpose. The most widely accepted definition of wasteland is the degraded lands which can be brought under vegetative cover with reasonable efforts and which are currently underutilized and the land which is deteriorating for lack of appropriate water and soil management or on account of natural causes. In general, wastelands may be put under two categories: cultivable wasteland and uncultivable wasteland.

CULTIVABLE WASTELAND: The land which is capable or has potential for the development of vegetative cover and not being used due to different constraints of varying degrees such as erosion, water-logging, salt effects etc.

UNCULTIVABLE WASTELAND: The land which cannot be developed for vegetative cover e.g. rocky areas, deserts and snow-covered glacial areas.

WASTELAND ASSESSMENT

As early as mid sixties scientists working on wasteland management realized the need to categorize the wasteland in accordance with their intrinsic characteristics and causative factors. Among various characteristics, data on soil texture, soil depth, pH values, slope, erosion status and other inhibiting factors like salts, water-logging, flooding, rockiness, stoniness etc. are essential. Complete inventory of reasons, responsible for lands lying waste, were assessed by collecting information from various sources. They serve as points for assessment of any wasteland.

- Poor fertility of the soil due to rocky, gravely, sandy, saline, alkali, water-logging, shallowness and eroded nature of the soil
- Steep and undulated slopes
- Shifting cultivation
- Frequent droughts
- Lack of irrigation facilities
- Frequent submergence and flooding hazards
- Lack of resources
- Poor economic conditions
- Lack of labour during peak periods
- Uneconomical return under cultivation,
- Certain domestic and legal difficulties
- Absentee landlords holdings

WASTELAND MAPPING TECHNIQUES

Through mapping, the findings of these questions are recorded in a pictorial form under a specified scale. These maps will serve anyone who wants to study about a particular area of wasteland or to utilize it for some purpose or other. The following important mapping systems are available in India: conventional, aerial photo-interpretation and remote sensing. The first system can be stated as a real time system where as the latter two are the tools to improve the time and cost efficiency.

CONVENTIONAL SYSTEM: It is based primarily on ground transect using suitable base maps for plotting boundaries. For detailed investigations, the surveyor transverses from field to field with a base map to collect the soil and land-use data after closely spaced field observations to identify and map wasteland.

This approach is very effective but very time consuming. One field surveyor can collect information from only 16,000 ha per annum. For mapping the wastelands on regional basis, the reconnaissance type of survey could be adopted with Survey of India topographical maps at 1:50,000 scale as base material. Detailed data on texture, depth, profile, morphology, physical and chemical characteristics are gathered. The ownership details and other socio-economic factors have, however, to be collected by actual visits to the village and local enquiries.

AERIAL PHOTO-INTERPRETATION (API): Black and white panchromatic aerial photographs of 1:25,000 scale and smaller are very suitable for wastelands mapping. False colour APs may improve the level of mapping as far as wastelands are concerned since the cropped lands, orchards and forested areas can be segregated more efficiently with their varied colour manifestation. The derived map could be at a semi-detailed level. However, the data on socio-economic factors, ownership details and other relevant information have to be collected by field visits personal enquiries and verification of village records only.

REMOTE SENSING: The satellite era ushers new scopes in land use mapping. The periodicity and larger perspective view have their own advantages in mapping wastelands and monitoring the changes using satellite data in a big way involving several organizations. Mapping with satellite data is quicker and cost-effective as compared to any other methods. Ability to effectively monitor the changes is the special advantage in this system because of the periodicity of the data. It is expected to gain the accuracy of 74.1 to 95.3 per cent. Ownership details cannot be included in the remote sensing data unless resorted to other methods such as grid survey, revenue records etc.

LAND DEGRADATION IN INDIA

India has a total geographical area of 328.2 million hectare with a cultivated area of 141 million hectare. The country is situated between 8°04' to 37°06' N latitudes and 68°07' to 97°25' E longitudes. Over the years, the country's landmass has suffered from different types of land degradations. Land degradation is caused by biotic and abiotic pressures. An ever increasing population places enormous demands on the land resources. This is particularly true in India, which has only 2.4 per cent of the world's geographical area but supports over 16 per cent of the world's population. It has 0.5 per cent of the world's grazing area but has over 18 per cent of world's cattle population. These pressures have led to drastic changes in the proportion of land utilized for agricultural activities, urbanization and industrial development.

Table 33.2: Wasteland Category/ Classes in India 2008-2009 (DoLR and NRSC 2011)

S No	Wasteland Class	Area (Sq. Km)	Percentage to TGA
1	Gullied and/ or ravinous land- medium	6,145.96	0.19
2	Gullied and/ or ravinous land- deep/very deep	1,266.06	0.04
3	Land with dense scrub	86,979.91	2.75
4	Land with open scrub	93,033.00	2.94
5	Waterlogged and marshy land- permanent	1,757.07	0.06
6	Waterlogged and marshy land- seasonal	6,946.31	0.22
7	Land affected by salinity/ alkalinity- moderate	5,414.53	0.17
8	Land affected by salinity/ alkalinity- strong	1,391.09	0.04
9	Shifting cultivation area- current jhum	4,814.68	0.15
10	Shifting cultivation area- abandoned jhum	4,210.46	0.13
11	Under utilised/degraded forest- scrub dominated	83,699.71	2.64
12	Agricultural land inside notified forest land	15,680.26	0.50
13	Degraded pastures/ grazing land	6,832.17	0.22

14	Degraded land under plantation crops	278.53	0.01
15	Sands- riverine	2,111.96	0.07
16	Sands- coastal sand	654.47	0.02
17	Sands- desert sand	3,934.80	0.12
18	Sands- semi-stabilized to stabilized (>40m) dune	9,279.75	0.29
19	Sands- semi-stabilized to stabilized moderately high (15-40m) dune	14,273.03	0.45
20	Mining wastelands	593.65	0.02
21	Industrial wastelands	58.00	0.00
22	Barren rocky area	59,482.29	1.88
23	Snow covered and/or glacial area	58,183.44	1.84
	Total	4,67,021.16	14.75

Intensive agricultural practices that rely heavily on water, chemical fertilisers and pesticides have caused waterlogging and salinity in many parts of the country. The expansion of the irrigation system without adequate steps for treatment of the catchment areas has exacerbated this. The quest for increased agricultural productivity has led to intensive cultivation of marginal lands causing their degradation. These pressures on land are compounded by the fact that over 69 per cent of our geological area falls within dry zone as per the Thornwaite classification. Land degradation has a direct bearing on the productivity of soil, its vulnerability to rainfall variations, scarcity of drinking water, fodder and fuel wood. Given the inter-linkages of crop production, livestock economy and environment, land degradation has a major impact on the livelihoods of the people, especially in rural areas.

Some of the most degraded lands in the country are the common property resources and these resouces include community pastures, community forests, wastelands and common dumping and threshing grounds. In spite of concerted efforts to check deforestation and the various afforestation schemes taken up during successive Plan periods, large tracks of forest continue to be classified as degraded. Of the total forest area, 31 million hectare suffers from some form of degradation and 14.06 million hectare suffers from extreme degradation and are part of 63.85 million hectare of wastelands reported by the NRSA. Desertification and Land Degradation Atlas of India has reported that total area under land degradation in India is 105.48 million ha (32.07 per cent) and total area under desertication is 81.45 million ha (24.78 per cent) (Space Application Centre 2007)

In addition to the wastelands identified by the NRSA, other areas such as deserts, drought-prone, flood-prone and tribal areas have been subjected to severe forms of degradation. The capacity of these lands is limited due to environmental factors. Pressures of human and livestock population have further compromised them.

National Commission on Agriculture made the first step for assessment of area affected by the land degradation and reported 148 million ha land area are affected in India. Other organizations have also taken up the assessment of land degradation in India viz. Ministry of Agriculture, ICAR's National Bureau of Soil Survey and Land Use Planning (NBSS & LUP), National Wasteland Board, National Remote Sensing Agency-Hyderabad (NRSA), Space Application Centre of ISRO. NBSS & LUP followed process based degradation mapping methodology in agriculture and non-agriculture areas, derived from 1:250,000 soil map prepared from the satellite data supported with soil profile studies. The NRSA followed remote sensing techniques, identifying land use and physical condition of the surface features for mapping non-agriculture areas (non-arable) on 1:50,000 scale using satellite image with adequate field checks.

Table 33.3: Land Degradation Assessment by Different Organizations
(Source: *Gautam and Narayan 1988; Space Application Centre 2007; DoLR and NRSC 2011*)

Agency	Estimated Extent (M ha)	Criteria for Delineation
National Commission on Agriculture(NCA 1976)	148.09	Based on secondary data
Ministry of Agriculture (1978-Soil & Water Conservation Division)	175.00	Based on NCA's estimates. No systematic survey undertaken
Society for the Promotion of Wasteland Development (SPWD) – (Bhumbla and Khare 1984)	129.58	Based on secondary data
NRSA (1985)	53.28	Mapping on 1:1 million scale based upon remote sensing techniques
Ministry of Agriculture (MOA 1985)	173.64	Land degradation statistics for the states
Ministry of Agriculture (MOA 1994)	107.43	Elimination of duplication of area. Area reclaimed counted
NBSS and LUP (1994)	187.70	Mapping based upon Global Assessment of Soil Degradation (GLASOD) guidelines
NBSS and LUP (2004) revised	146.82	1:1 million scale soil map
Dept of Environment (Vohra 1980)	95.00	-
National Wasteland Development Board (1985)	123.00	-
NRSA (2000)	63.85	Remote sensing techniques
Space Application Centre (2007)	105.48	Remote sensing techniques
DoLR and NRSC (2010)	47.23	Remote sensing techniques
DoLR and NRSC (2011)	46.70	Remote sensing techniques

PROCESSES OF LAND DEGRADATION IN INDIA *(Space Application Centre 2007)*

VEGETAL DEGRADATION: It is observed as deforestation, forest blanks, shifting cultivation area and also degradation in grazing area / grassland as well as in scrubland. At places, agriculture is observed within forest land, this has been classified under vegetal degradation within forest area. Vegetal degradation processes are found in both cold and hot regions.

WATER EROSION: It is observed in both hot and cold areas, in various land covers and with varying severity levels. The sheet erosion, rill erosion and shallow gullies mostly within agricultural lands are categorized as low category of severity. While deep & wide gullies and ravines are categorized as high category.

WIND EROSION: It pertains to the Aeolian activities. It denotes the spread of sand by virtue of lift and drift effect of wind, even upto lofty altitudes of Himalayas. Various categories of sand cover and their severity are classified based on the depth and spread of sand sheet, sand dunes and barchans.

SALINIZATION AND ALKALINIZATION: It is fundamentally the chemical property of soils. It occurs mostly in cultivated lands, especially in the irrigated areas. At places salinity is clearly observed on satellite images, while the alkalinization is not seen and mostly inferred based on ground truth data and published maps.

WATERLOGGING: The undrained land areas tend to accumulate standing water for longer duration of time on the surface, this condition is called waterlogging. The severity of waterlogging is determined based on the period for which the water remains stagnant.

MASS MOVEMENT OR MASS WASTING: It is predominantly observed in mountainous areas, especially in cold dryland regions and it is defined as the process of desertification which leads to the down slope movement of rock, regolith and debris through the action of gravity viz. landslides, scree slopes

FROST HEAVING: It is defined as the process of intense frost and freezing of water operating in glacial and periglacial environment and evolves peculiar forms of rock, regolith and soil, for example, a typical irregular pattern ground are seen.

FROST SHATTERING: It is defined as the freeze and thaw action operating mostly in periglacial environment. When water, passing through the crevices and pores within the rock freezes, it expands by almost ten times. This puts enormous pressure on the surrounding rocks which shatters them.

MAN MADE: All those land degradation processes which is included directly or indirectly by human intervention and are not natural, are categorized as man made desertification processes. It includes mining/quarrying, brick kiln, industrial effluents, city waste, excess use of fertilizers and pesticides. It occurs across various land use/ land cover classes.

EXAMPLES OF LAND DEGRADATION IN INDIA

- Mining sites are abandoned after excavation work is complete leaving deep scars and traces of over-burdening.
- In states like Jharkhand, Chhattisgarh, Madhya Pradesh and Orissa, deforestation due to mining have caused severe land degradation.
- In states like Gujarat, Rajasthan, Madhya Pradesh and Maharashtra, overgrazing is one of the main reasons for land degradation.
- In the states of Punjab, Haryana, Western Uttar Pradesh, over irrigation is responsible for land degradation due to water logging leading to increase in salinity and alkalinity in the soil.
- The mineral processing like grinding of limestone for cement industry and calcite and soapstone for ceramic industry generate huge quantity of dust in the

atmosphere. It retards the process of infiltration of water into the soil after it settles down on the land

- In recent years, industrial effluents as waste have become a major source of land and water pollution in many parts of the country.

HISTORICAL BACKGROUND OF WASTELAND DEVELOPMENT IN INDIA

(Source: DoLR)

CAZRI

The first step towards a systematic effort to tackle the problem of drought and desertification was the establishment of Research Centre at Jodhpur in 1952 to carry out research on certain core needs of desert areas such as sand-dune stabilization, shelter-belt plantation, afforestation, etc. In 1959, the entire responsibility of research on arid areas was entrusted to the Centre which was then designated as Central Arid Zone Research Institute (CAZRI).

RWP

Rural works programme (RWP) was launched in 1970-71 with the object of creating assets designed to reduce the severity of drought in affected areas. In all, 54 districts in the country together with parts of another 18 districts contiguous to them were identified as drought-prone. The programme covered 12 per cent of country's population and nearly one-fifth of the area in the country.

CRIDA

For the purpose of dryland farming technology, All India Coordinated Research Project for Dryland Agriculture, later renamed as the Central Research Institute for Dryland Agriculture (CRIDA), was set up. It focuses on soil management, water harvesting, improved agronomic practices, drought-resistant crops etc.

DPAP

The Drought Prone Areas Programme (DPAP) is one of the areas development programmes launched by the Government in 1973-74 to tackle the special problems faced by those fragile areas. The programme was implemented on Watershed basis and covered 13 States viz. AP, Bihar, Gujarat, HP, J&K, Karnataka, MP, Maharashtra, Orissa, TN, Rajasthan, UP and WB.

DDP

On the recommendations of the National Commission on Agriculture in its Interim Report (1974) and the Final Report (1976), the Desert Development Programme (DDP) was started in the year 1977-78. The programme was started both in the hot desert areas of Rajasthan, Gujarat and Haryana, and the cold desert areas of Jammu & Kashmir and Himachal Pradesh. From 1995-96, the coverage has been extended to few more districts in Andhra Pradesh and Karnataka. It focuses on land development, water resource development and afforestation/pasture development.

NWDB

National Wasteland Development Board (NWDB) was established in 1985 under the Ministry of Environment and Forests mainly to tackle the problem of degradation of lands, restoration of ecology and to meet the growing demands of fuel wood and fodder at the national level. During the Seventh Five Year Plan, the strategy adopted by the National Wasteland Development Board emphasized more on tree planting activities rather than community participation for wasteland development.

IWDP

Integrated Wasteland Development Programme (IWDP) scheme has been under implementation since 1989-90. From 1st April, 1995, the scheme is being implemented on watershed basis under the common guidelines for Watershed Development. The Scheme also helps in generation of employment in rural areas besides enhancing people's participation in the wastelands development programmes at all stages. The major activities taken up under the scheme are soil and moisture conservation, planting and sowing of multi-purpose trees, shrubs, grasses, legumes, pasture land development, encouraging natural regeneration, promotion of agro-forestry and horticulture, wood substitution and fuel-wood conservation measures, dissemination of technology, training and extension, encouraging people's participation etc. 247 IWDP projects in 25 states in a total outlay of Rs. 778.12 crore to treat total project area of 15.98 lakh hectare are at various stages of implementation.

DoLR

Govt. of India had created the Department of Wasteland Development during July, 1992 under the Ministry of Rural Development and the National Wasteland Development Board was placed under it. It has been subsequently reorganized and renamed as Department of Land Resources (DoLR), with a broader mandate. DoLR has been responsible for mainly development of wastelands in non forest areas in totality by involving local people at every stage of development. It aims at creating a scenario where the Government acts as a facilitator and the people at the grass root level become the real executioner of the programme.

TDET

Technology support is extremely vital for the success of a land based programme, especially in the development of wastelands. Realising this, a central sector scheme namely Technology Development, Extension and Training Scheme (TDET) was launched during 1993-94 to develop suitable technologies for the reclamation of wastelands for sustained production of food, fuel-wood, fodder etc. The Scheme is being implemented through ICAR, State Agricultural Universities, DRDAs and Government institutions having adequate institutional framework and organizational back-up.

IPS

Investment Promotional Scheme (IPS) was launched in 1994-95 in order to stimulate involvement of corporate sectors and financial institutions etc. for the flow of funds for development of non-forest wastelands. The projects promoted by Nationalized Banks, Regional Rural Banks, Land Development Banks and Cooperative Banks are eligible for promotional subsidy of Department of Land Resources under the Scheme. The scheme, which has now been restructured and the guidelines revised, has a major thrust for development of degraded lands belonging to small and marginal farmers including SCs/STs.

IWMP

Integrated Watershed Management Programme (IWMP) is a modified programme of erstwhile Drought Prone Areas Programme (DPAP), Desert Development Programme (DDP) and Integrated Wastelands Development Programme (IWDP) of the Department of Land Resources. This consolidation is for optimum use of resources, sustainable outcomes and integrated planning. The scheme was launched during 2009-10. The programme is being implemented as per Common Guidelines for Watershed Development Projects 2008. The main objectives of the IWMP are to restore the ecological balance by harnessing, conserving and developing degraded natural resources such as soil, vegetative cover and water. The outcomes are prevention of soil erosion, regeneration of natural vegetation, rain water harvesting and recharging of the ground water table. This enables multi-cropping and the introduction of diverse agro-based activities, which help to provide sustainable livelihoods to the people residing in the watershed area.

SLEM

Sustainable Land and Ecosystem Management (SLEM) Programme is a joint initiative of the Government of India and the Global Environmental Facility (GEF) under the latter's Country partnership Programme (CPP). The objective of the SLEM programme is to promote sustainable land management and use of biodiversity as well as maintain the capacity of ecosystems to deliver goods and services while taking into account climate change.

NATIONAL ACTION PROGRAMME TO COMBAT DESERTIFICATION

India became a signatory to the UNCCD on 14 October 1994 and it came into effect on 17 March 1997. One of the obligations of all developing country parties to the convention, including India, is to prepare the National Action Programme to Combat Desertification and to mitigate the effects of drought. The Ministry of Environment and Forests, as the National Focal Point for the implementation of the Convention, has prepared the National Action Programme to Combat Desertification (NAPCD) through the setting up of a High-level Inter-Sectoral National Steering Committee (NSC) in July 1999. The NSC constituted four Working Groups (WG) on various relevant to desertification viz. Desertification Monitoring and Assessment, Sustainable Land Use Practices for Combating Desertification, Local Area Development Programme, Policy and Institutional Issues.

Similarly Desertification Cell has been established under Ministry of Environment and Forests comprising multi-disciplinary team including agriculture, forestry, environment science, geo-hydrology, rural development and social sciences. It would service on all matters like preparation of NAP, monitoring its implementation, organizing reviews, feasibility studies and evaluation and impact studies.

FEATURES OF NAPCD

- Water is a diacritical element for building people's confidence and satisfaction level, reclamation of degraded lands for sustainable biomass production ultimately leading to a better quality of life and enabling conditions through empowerment of the local communities.

- Bottom up approach and project planning, evaluation and monitoring by *Panchayat Raj* Institutions (Grass root level elected local self Governments in which women have at least 30 per cent representation) on all aspects of land development including rehabilitation of degraded areas
- Convergence of resources and services. All resources available under different schemes will be channelled through *Panchayats* through single window service to the communities.
- Gaps in all the on-going schemes of different departments will be identified and resources will be provided to fill them up for generating good impact at the ground level.
- The first five-year period will be experimental and include pilot projects and activities.
- Concurrent and continuous monitoring will be done and mid-course correction will be effected as the experiences are gained
- The problem faced by farmers will be reported to R&D institutions for finding solutions. Thus both lab-to-land and land-to-lab flow of information will be ensured

WASTELAND MANAGEMENT

Wasteland management includes measures for land conservation, land rehabilitation and land improvement.

(i) Land Conservation: It is the measures taken to maintain the functions of the land. The land is allowed for productive use, but within limits. Land is maintained to stay within system tolerance and to allow system to reach an equilibrium. If it is necessary, land is removed from use, or even positive measures are taken to rehabilitate.

(ii) Land Rehabilitation: It is the restoration of the previous land functions of the degraded land

(iii) Land Improvement: It is an increase in the environmental function or productive potential of land.

WHAT IS ECOLOGICAL RESTORATION?

Ecological restoration can be defined as the process of assisting the recovery and management of ecological integrity, including a critical range of variability in biodiversity, ecological processes and structures, regional and historical context and sustainable cultural practices. Restoration can be passive, in which the degrading agent(s) is identified and removed or active, in which management techniques such as planting, weeding, burning and thinning are undertaken with a particular image of desired structure, composition, or pattern in mind.

The need for restoration assumes some level of impairment in an ecosystem because of low ecological integrity. Although natural disturbances can cause impaired conditions, often human activities are responsible and thus a change in the management can theoretically bring about improved conditions. There are number of land restoration options that, if applied wisely, could boost agricultural production and slow down land and environmental degradation.

MULTI-FACETED ACTIONS TO COMBAT LAND DEGRADATION

The integrated approach is necessary to combat land degradation. The following approaches are considered necessary for the land improvement.

- Integrated management of natural resources
- Soil conservation and restoration
- Improving agricultural production and stock-farming systems
- Agricultural water management
- Sustainable forest management
- Agroforestry
- Watershed management
- Environmental monitoring, drought early warning and forecasting/response system
- Promotion of sustainable land management techniques geared to climate change adaptation
- Biodiversity preservation
- Adding value to agricultural, forest, farmed products
- Diversifying activities so as to relieve pressure on resources
- Increasing land and labour productivity with available technology
- Developing technology to overcome agricultural production constraints

Table 33.4: Types of Land Degradation and Improvement (Sara J. Scherr and Satya Yadav 1996)

Component	Degradation	Improvement
Physical soil management	Crusting, compaction, sealing, wind erosion, water erosion, devegetation, overtillage	Soil conservation barriers (live, inert), terracing, revegetation of denuded lands, tree protection, soil decompaction, breaking up of pans, covercrops, windbreaks, soil deposition, improved tillage methods
Soil water management	Impeded drainage, waterlogging, reduced waterholding capacity, reduced infiltration, salinization	Irrigation, water harvesting, field drainage, draining of waterlogged areas, filter strips
Soil nutrient and organic matter management	Alkalinization, acidification, nutrient leaching, removal of organic matter, burning of vegetative residues, nutrient depletion	Fertilization, composting, green manuring, animal manuring, Flushing of saline and alkaline soils, liming of acid soils

Soil biology management	Overapplication of agrichemicals, industrial contamination	Introduction of biotic organisms, nitrogen fixing microorganisms
Vegetation Management	Decline in vegetative cover, decline in biodiversity, decline in species composition, decline in availability of valuable species	Increased vegetative cover, increased species diversity, increased species composition, Improved availability of valuable species

TREES, SHRUBS AND GRASSES FOR WASTELAND DEVELOPMENT

It is well recognized that a large area in the country is degraded due to soil erosion by water. Soil erosion is caused mainly by lack of vegetative cover, slope, nature of soil, rainfall and its intensity, cropping system and land management. It is also caused by deforestation, shifting cultivation, overgrazing and improper cropping of undulated lands. If the intensity and degree of soil erosion are high, it may lead to the formation of deep gullies and ravines. Appropriate measures are required to check runoff and soil loss which include planting vegetation (trees, grasses, shrubs), agroforestry, mulches, bunding, terracing, contour cultivation, graded trenching, crop rotation practice with different crops, no tillage strategy etc.

Similarly drylands need to be reclaimed using drought tolerant species. Sand areas and sand dunes should be stabilized using appropriate tree and grass species. Area affected by wind erosion need to be protected by establishing wind breaks and shelter belt. Salt tolerant grasses and tree species are planted in lands affected by salinity and alkalinity.

Species selection is the important criteria for the success of any wasteland development programme which is based on compatibility to local site conditions, survival and productivity. The genetic quality of native species plays a key role for local adaptability and to withstand adverse environmental conditions. It is also important for suitability to soils with different depth and water retention capacities. The plant establishment mainly depends on the development of deep and good root system. The inherent characteristics of a species, to propagate or regenerate itself vegetatively viz. root suckers, when damaged, are also important for survival.

Tree Species: *Acacia albida, Acacia tortilis, Acacia senegal, Acacia catechu, Acacia auriculiformis, Acacia ampliceps, Acacia bivenosa, Acacia nilotica, Prosopis cineraria, Prosopis juliflora, Parkinsonia articulata, Tamarix articulata, Ziziphus mauritiana, Zizypus nummularia, Calligonum polygonoides, Cassia siamea, Albizia lebbeck, Leucaena leucocephala, Gliricidium sepium, Sesbania sesban, Erythrina* spp., *Casuarina equisetifolia, Inga dulcis, Grewia tenax, Anogeisus rotundifolia, Euphorbia caducifolia, Tamarix aphylla, Cercidium floridum, Eucalyptus tereticornis, Eucalyptus camaldulensis, Hardwickia binata, Pongamia pinnata, Tecomella undulata, Azadirachta indica, Dendrocalamus strictus, Bambusa bambos, Agave sisalana, Agave americana, Calliandra* spp., *Gmelina arborea, Grevillea robusta, Dalbergia sissoo, Paulownia* spp., *Terminalia* spp., *Anthocephalus* spp., *Madhuca* spp., *Pterocarpus* spp., *Emblica officinalis, Aegle marmelos*

Grasses and Forage Crops: *Vetiveria zizanioides, Cynadon dactylon, Dichanthium annulatum, Panicum repens, Panicum antidotale, Panicum maximum, Cenchrus setigerus, Cenchrus ciliaris, Lasiurus sindicus, Sehima nervosum, Bracharia brizantha, Bracharia mutica, Sporobolus marginatus, S. airoides, S. helvolus, Fagonia cretica, Eragrostis* spp.,

Taverniera cuneifolia, Barleria acanthoides, Eremopogon foveolatus, Heteropogon contortus, Lepidagathis spp., *Aristida* spp., *Oblong folia, Tribulus terrestris, Brachiaria ramosa, Cymbopogon jwarancusa*

Medicinal Plants: *Aloe vera, Balanites aegyptiaca, Commiphora wightii, Euphorbia antisyphilitica, Haloxylon* spp., *Cassia angustifolia*

WASTELAND DEVELOPMENT THROUGH AGROFORESTRY

Agroforestry refers to land use systems in which trees or shrubs are grown in association with agricultural crops, pastures or livestock and in which there are both ecological and economic interactions between trees and other components. An agroforestry practice is a distinctive arrangement of components in space and time, characterized by environment, plant species and arrangement, management and social and economic functioning. Agroforestry ensures maximum productivity per unit of land on sustainable basis. It improves the soil fertility and productivity, reduces the soil run off and provides more income to land owners than monocropping system. Adoption of agroforestry practices such as agrisilviculture, silvopastoral and agrisilvipastoral systems with an appropriate combination of trees, grasses and agricultural crops develop the wastelands very fast. It ensures moisture improvement, soil improvement and biomass improvement in the wastelands.

PART 7:

SOCIAL VALUES AND BENEFITS

34 FOREST PRODUCTS UTILIZATION

INTRODUCTION

The pattern of utilization of forest resources has been influenced by growing rural population, with limited income opportunities and the related widespread poverty and migration of landless people in the forest areas. This pattern has intensified agricultural cultivation and inappropriate exploitation of forest resources, resulting in depletion and degradation of natural asset. Governments of many countries have responded to these emerging conflicts through a series of initiatives that have addressed the need of the rural population without compromising environmental protection and conservation policy. Forest provides a range of wood and non-wood products as well as social and environmental services. Wood and wood products as the main commercial products of forests include fuelwood and charcoal, industrial round wood, sawn wood, wood-based panels, wood pulp, paper and paperboard and wooden articles. The abundance of raw material resources in certain regions of world makes the wood industry important to the region's economic development. Besides providing employment in the sawmilling and furniture manufacturing industries, wood also provides building materials to the local construction industry and contributes to the growth of non-traditional exports through increased exports of wood products.

Non-wood forest products (NWFP) have an important role to play in the livelihoods of many rural communities, particularly in developing countries, where they provide a broad range of subsistence and commercial livelihood opportunities. While much of the trade is domestic, for some NWFP species and products, the international trade is significant and generates income for the resource harvesters and collectors as well as many other actors in the commodity chain. The dearth of information on the trade of NWFP makes it difficult to estimate the total and relative levels of use for both domestic and commercial purposes, and this is complicated by the difficulty in distinguishing between subsistence use and trade for commercial purposes. The past thirty years have witnessed a significant growth in domestic and international trade in practically all products, including NWFP. This growth has had accompanying implications, both positive and negative, for the species concerned as well as the numerous individuals operating within the commodity chains of the NWFP. Accordingly various instruments have been developed and implemented in attempts to regulate trade, normally with the intention of ensuring sustainable trade and equitable distribution of the benefits of such trade. These instruments range from multilateral environmental agreements such as CITES and CBD to regional trade and other agreements in national controls and voluntary instruments such as eco-labeling and certification schemes.

A forest product is any material derived from a forest for commercial use e.g. wood material such as timber, fuelwood, paper and pulp logs, plywood logs or non-wood material such as fodder, gums, tannin, dyes, resins, fibre and flosses, fragrance and flavours, medicine, etc. Wood is the dominant commercial forest product but non-wood products derived from forest resources are also having more subsistence and commercial value being diversified in nature. Wood has been an important construction material since humans began building shelters, houses and boats. Wood products are categorized in different types but major wood products include lumber or timber, plywood, laminated forest products, particle based products, fibre based products, fuelwood, pulp and paper etc. They all cater the need of constructional and structural timber, household items, furniture, cabinets as well as industrial need of paper & pulp, matchwood, pencil, sports goods and panel products industries. Wood is also commonly used as shuttering material, pit props in mining and poles for different purposes. The churches of *Kizhi* in Russia are among a handful of World Heritage Sites built entirely of wood, without metal joints.

LUMBER/ TIMBER

Lumber is a general term which includes board, dimension lumber and timber. Thinnest sizes are referred as Boards (thickness ranges from 1" to 1.5"). Board is used for range of rough or finished products such as packaging, crating, paneling, siding, flooring, moulding. Similar but thicker sizes are referred as Dimension Lumber (thickness ranges from 2" to 4"). It is used for framing and structural applications such as joists, planks, rafters, studs, small posts or beams. Larger sizes are called Timbers (thickness 5" or more). Lumber is the wood in any of its stages from felling to readiness for use as structural material for construction or wood pulp for paper production. It is supplied either rough or finished. Besides pulpwood, rough lumber is the raw material for furniture-making and other items requiring additional cutting and shaping. Finished lumber is supplied in standard sizes mostly for the construction industry primarily from softwoods or coniferous species but also from some hardwoods for high-grade flooring. In the United Kingdom and other Commonwealth countries such as India, Australia and New Zealand, timber is a term also used for sawn wood products e.g. timber floor boards, whereas generally in the United States and Canada, the product of timber cut into boards is referred to as lumber.

In the United States and Canada, timber often refers to the wood contents of standing, live trees that can be used for lumber or fibre production. It can also be used to describe sawn lumber whose smallest dimension is not less than 5 inches (127 mm) such as the large dimension and often partially finished lumber used in timber-frame construction. The word lumberjack is used in the UK and Australia to refer to North Americans who fell standing trees and so the word lumber conjures images of what North Americans call timber, and *vice versa*. Wood to be used for construction work is commonly known as lumber in North America. Elsewhere, lumber usually refers to felled trees and the word for sawn planks ready for use is timber. Dimensional lumber is a term used for lumber that is finished/planed and cut to standardized width and depth specified in inches. Examples of common sizes are 2×4, 4×2, 2×6 and 4×4. The length of a board is usually specified separately from the width and depth. It is thus possible to find 2×4s that are four, eight or 12 feet in length. The standard lengths of lumber are 6, 8, 10, 12, 14, 16, 18, 20, 22 and 24 feet (Smith and Wood 1964).

In India, main timber species are Deodar (*Cedrus deodara*), Sal (*Shorea robusta*), Teak (*Tectona grandis*), Babool (*Acacia nilotica*), Chir pine (*Pinus roxbughii*), Toon (*Toona ciliata*), Mahua (*Madhuca indica*), Neem (*Azadirachta indica*), Haldu (*Adina cordifolia*), Silver fir (*Abies pindrow*), Spruce (*Picea smithiana*), Irul (*Xylia xylocarpa*), Chickrassy (*Chukrassia tabularis*), *Calophyllum* spp., Gurjan (*Dipterocarpus indicus*), *Artocarpus* spp, *Terminalia* spp., etc. Teak, also called king of timber, is still placed high value in furniture, cabinet and flooring. Among fast growing industrial woods, *Eucalyptus, Casuarina, Poplar, Robinia, Ailanthus, Melia, Salix* and *Morus* are common in different ecologies of India.

COMPOSITE WOOD/ ENGINEERED WOOD

Composite wood or engineered wood is a general term for built up bonded products, consisting either wholly natural wood or in combination of wood fibres, strips, metal, plastic and veneer sheets. Smaller diameter trees of the same hard and softwood used to manufacture lumber are also used. In engineered wood products, the materials are bonded together using an adhesive resin which makes the finished product resist warping. Some different products made from engineered wood include flooring, several types of lumber and particle board. Engineered flooring is commonly used in both residential and commercial properties because it is typically less expensive and often stronger than traditional hardwood. An engineered hardwood floor is constructed using a thin layer of sealed and sanded hardwood glued to a thicker bottom layer of plywood. In recent years, various methods and processes have been developed for building larger piece from relatively smaller pieces or treating and modifying wood by means of pressure, heat and chemicals. These processes help in utilizing waste wood and wood of inferior species.

Engineered lumber is broad term used to describe several types of engineered wood products. Engineered wood products are becoming a bigger part of the construction industry. They may be used in both residential and commercial buildings as structural and aesthetic materials. In buildings made of other materials, wood will still be found as a supporting material, especially in roof construction, in interior doors and their frames and as exterior cladding. Engineered wood products, glued building products "engineered" for application-specific performance requirements, are often used in construction and industrial applications. Glued engineered wood products are manufactured by bonding together wood strands, veneers, lumber or other forms of wood fibre with glue to form a larger, more efficient composite structural unit. These products include glued laminated timber (glulam), wood structural panels (including plywood, oriented strand board and composite panels), laminated veneer lumber (LVL) and other structural composite lumber (SCL) products, parallel strand lumber and I-joists. The trends suggest that particle board and fibre board will overtake plywood.

Engineered wood products display highly predictable and reliable performance characteristics and provide enhanced design flexibility on one hand, these products allow the use of smaller pieces and on the other hand, they allow for bigger spans. They may also be selected for specific projects such as public swimming pools or ice rinks where the wood will not deteriorate in the presence of certain chemicals and are less susceptible to the humidity changes commonly found in these environments. Engineered wood products prove to be more environmentally friendly and, if used appropriately, are often less expensive than building materials such as steel or concrete. These products are extremely resource-efficient because they use more of the available resource with minimal waste. In most cases, engineered wood products are produced using faster growing and often underutilized wood species from managed forests and plantations.

Wood unsuitable for construction in its native form may be broken down mechanically (into fibres or chips) or chemically (into cellulose) and used as a raw material for other building materials, such as engineered wood, as well as chipboard, hardboard and medium-density fibreboard (MDF). Such wood derivatives are widely used in many industries viz. wood fibres are an important component of most paper and cellulose is used as a component of some synthetic materials. Wood derivatives can also be used for kinds of flooring e.g. laminate flooring. Major forms of composite wood are plywood, laminated wood, core boards, sandwich board, fibre boards, particle boards, etc.

TYPES OF ENGINEERED WOOD

1. Glued Laminated Timber
2. Multilaminar Veneer
3. Veneer-based Wood
   - Plywood
   - Glued laminated timber (Glulam)
   - Laminated veneer lumber (LVL)
   - Stamina wood
4. Particle-based
   - Waferboard
   - Particleboard (Chipboard)
5. Fibre based
   - Fibreboard
   - Parallel strand lumber (PSL) or Oriented strand borad (OSB)
   - Laminated strand lumber (LSL)
   - Insulation board
   - Medium-density Fibreboard (MDF)
   - Hardboard
6. Wood-plastic composite

TYPES OF ADHESIVES USED IN ENGINEERED WOOD

- Urea-formaldehyde resins (UF) : Most common, most cheap and not waterproof
- Phenol-formaldehyde resins (PF) : Yellow/brown, and commonly used for exterior exposure products

- Melamine-formaldehyde resin (MF) : White, heat and water resistant and often used in exposed surfaces in more costly designs
- Methylene diphenyl diisocyanate (MDI) or Polyurethane (PU) resins : Expensive, generally waterproof and do not contain formaldehyde

VENEERS

In woodworking, veneer refers to thin slices or sheet of wood, usually thinner than 3 mm (1/8 inch) to as low as 0.05 mm, that typically are glued onto core panels (wood, particle board or medium-density fibreboard) to produce flat panels such as doors, tops and panels for cabinets, parquet floors and parts of furniture. They are also used in marquetry. Plywood consists of three or more layers of veneer, each glued with its grain at right angles to adjacent layers for strength. Veneer beading is a thin layer of decorative edging placed around objects, such as jewelry boxes. Veneer is also a type of manufactured board. Veneer is obtained either by rotary peeling log or by slicing large rectangular blocks of wood known as flitches. In peeling process, veneers come out as unrolled paper roll and length of log is about 2.5- 3.0 m and thickness of veneers is 1.25 mm, 1.50 mm or 2.5 mm. In slicing, even thinner veneer with thickness 0.6 mm to 1.25 mm are obtained and mainly done for decorative veneers. The appearance of the grain and figure in wood comes from slicing through the growth rings of a tree and depends upon the angle at which the wood is sliced.

TYPES OF VENEERS

There are a few types of veneers available and each serves a purpose.

- *Raw veneer* has no backing on it and can be used with either side facing up. It is important to note that the two sides will appear different when a finish has been applied, due to the cell structure of the wood.
- *Paper backed veneer* is that are backed with paper. The advantage is that it is available in large sizes, or sheets, as smaller pieces are joined together prior to adding the backing. This is helpful for users that do not wish to join smaller pieces of raw veneers together. This is also helpful when veneering curves and columns as the veneer is less likely to crack.
- *Phenolic backed veneer* is less common and is used for composite, or manmade wood veneers. Due to concern for the natural resource, this is becoming more popular. It too has the advantage of being available in sheets and is also less likely to crack when being used on curves.
- *Laid up veneer* is raw veneer that has been joined together to make larger pieces. The process is time-consuming and requires great care, but is not difficult and requires no expensive tools or machinery.

PATTERNS OF VENEERS

There are a number of patterns (the way the veneers are laid up) common to veneered work (Benson Jonathan 2012).

- *Book matched*: where the veneers are opened from the flitch much like the pages of a book
- *Slip matched*: where the pieces are joined together in the order they come from the flitch and have the same face kept up
- *Radial matched*: where the veneer is cut into wedge shaped pieces and joined together
- *Diamond matched*: where the pattern formed is diamond shaped

PLYWOOD

Plywood is a manufactured wood panel made from thin sheets of wood veneer. It is one of the most widely used wood products. It is flexible, inexpensive, workable, re-usable and can usually be locally manufactured. Plywood is used instead of plain wood because of its resistance to cracking, shrinkage, splitting and twisting/warping and its general high degree of strength. Plywood layers (called veneers) are glued together with adjacent plies having their grain at right angles to each other. Cross-graining has several important benefits: it reduces the tendency of wood to split when nailed at the edges, it reduces expansion and shrinkage equating to improved dimensional stability and makes the strength of the panel consistent across both directions. There are usually an odd number of plies so that the sheet is balanced-this reduces warping. Because of the way plywood is bonded (with grains running against one another and with an odd number of composite parts), it is very hard to bend it perpendicular to the grain direction. Plywood was invented around 3500 B.C. by the Egyptians, who attached several thinner layers of wood together to make one thick layer. They originally did this during a shortage of quality wood, gluing very thin layers of quality wood over lesser-quality wood. Modern plywood was invented by Immanuel Nobel, father of Alfred Nobel. Nobel realized that several thinner layers of wood bonded together would be stronger than one single thick layer of wood.

STRUCTURAL CHARACTERISTICS

A typical plywood panel has higher grade face veneers than core veneers. The principal function of the core layers is to increase the separation between the outer layers where the bending stresses are highest, thus increasing the panel's resistance to bending. As a result, thicker panels can span greater distances under the same loads. In bending, the maximum stress occurs in the outer layers, one in tension and the other in compression. Bending stress decreases from the maximum at the face layers to nearly zero at the central layer. Shear stress is consistent throughout the depth of the panel.

Plywood production requires a good log, called a peeler, which is generally straighter and larger in diameter than one required for processing into dimensioned lumber by a sawmill. The log is laid horizontally and rotated about its long axis while a long blade is pressed into it, causing a thin layer of wood to peel off. In this way the log is peeled into sheets of veneer which are then cut to the desired dimensions, dried, patched, glued together and then baked in a press at 140°C (284°F) and 1.9 MPa (280 psi) to form the plywood panel. The panel can then be patched, re-sized, sanded or otherwise refinished, depending on the market for which it is intended. Plywood for indoor use generally uses the less expensive urea-formaldehyde glue which has limited water resistance, while outdoor and marine-grade plywood are designed to withstand rot and use a water resistant phenol-formaldehyde glue to prevent delamination and to retain strength in high humidity. The adhesives used in plywood have become a point of concern. Both urea formaldehyde and phenol formaldehyde are carcinogenic in very high concentrations. As a result, many manufacturers are turning to low formaldehyde-emitting glue systems denoted by an "E" rating. Plywood produced to "E0" has effectively zero formaldehyde emissions. In addition to the glues being brought to the forefront, the wood resources themselves are becoming the focus of manufacturers, due in part to energy conservation, as well as concern for natural resources. Forest Stewardship Council (FSC), Leadership in Energy and Environmental Design (LEED), Sustainable Forestry Initiative (SFI) and Greenguard are all certification programs that ensure that production and construction practices are sustainable.

Plywood is used in many applications that need high-quality, high-strength sheet material. Quality in this context means resistance to cracking, breaking, shrinkage, twisting and warping. Exterior glued plywood is suitable for outdoor use, but because moisture affects the strength of wood, optimal performance is achieved in end uses where the wood's moisture content remains relatively low. On the other hand, subzero conditions don't affect plywood's dimensional or strength properties, which makes some special applications possible. Plywood is also used as an engineering material for stressed-skin applications. Plywood-based furniture and designing a wide range of boats built primarily of plywood are famous in major wood consuming nations. Plywood is often used to create curved surfaces because it can easily bend with the grain. Skateboard ramps often utilize plywood as the top smooth surface over bent curves to create transition that can simulate the shapes of ocean waves.

OTHER COMPOSITE WOODS

LAMINATED WOOD

It is built up product made of wood layers (laminae), all laid with their grain parallel and glued or fastened together. Laminae may be the thin veneers or boards cut in uniform width and length arranged in proper order and fed into glue spreader and pressure is applied. Laminated are wood used in furniture parts, cores of veneered panels, sports goods, areoplane hangers, auditorium, exhibition halls, gymnasia, planetariums, theaters, warehouses, etc. They are also used in flooring, picker sticks, boat and ship timber, aircraft parts, diving boards, etc.

GLULAM

It is a glue laminated timber. They are manufactured from solid sawn lumber. The lumber is glue jointed end to end and then these "long laminations" are then glued face to face resulting in beams upto 90' in length. They are used in commercial constructions and residential constructions.

CHIPBOARD

It is made by compressing and gluing small chips of waste wood. When in use, it is normally covered by a plastic coating for hardwood veneer.

CORE BOARD

It is a composite board built up of a core composed of strips of wood of various dimensions glued together or jointed together to form slab which in turn glued between two or more outer veneers with the direction of the grain of the core strips running at right angles to that of the adjacent veneers. It is batten board when the strips of the wood are not more than 7.5 cm wide and Block board when size is not more than 2.5 cm. When each strip of wood has a thickness of not more than 7 mm, it is called lamin board. Core boards also covered with outer layers of fibreboard, synthetic resin fibres, metal sheets, sunmica, etc. Due to their low weight, better stability, good acoustic and heat insulation properties, they are used for doors and partition in residential and office construction.

SANDWICH BOARD

They are built up board having a core of light material, faced on both side with relatively thin layer of material having high strength properties. Sandwich constructions are composites of different materials bonded together in a unit to achieve a combination of desirable properties. Two thin facing (skins) are usually of a strong dense material for load bearing. They are used in aircrafts, motor boats, table tops, flush doors and containers.

FIBRE BOARD

It is a sheet of material made up from fibres of wood. The wood is first defibrated or pulped and fibres are then interfelted into a mat and consolidated by pressure and heat. Bonding agent and supplementary material may be added at the felting stage to improve the mechanical properties. Fibre boards are used as core material in core board and sandwich board. Such boards are designated as building board, insulating board, wall board etc. As per densities, they can be grouped hard boards, semi hard boards, medium hard boards and soft boards. Fibre boards are made up from agricultural or forest waste material (lops & tops), waste veneers, etc. They are used in panels, insulating and cover material in buildings and office cabinets. Medium density fibreboard (MDF) is made from fine wood fibres, compressed and glued together. When in use, it is normally convered by a plastic coating or hardwood veneer.

PARTICLE BOARD

It is board or sheet constituted from fragments of wood and other lignocellulosic material, bonded with organic binders with the one or more agents like glue, heat, pressure, humidity, catalyst, etc. It uses chips, flanks, splinters, sawdust and as per material used these are called chip board, flake board, shaving board, bamboo board, etc. They are used only for interior housing or furniture and not suited for exterior use.

HARDBOARD

It is an engineered wood product manufactured from wood chips that are mechanically reduced to wood fibres and then bonded together into panels through heat and pressure. These panels are grainless, dense, uniformly textured, strong and bone dry. They are humidified and trimmed to size.

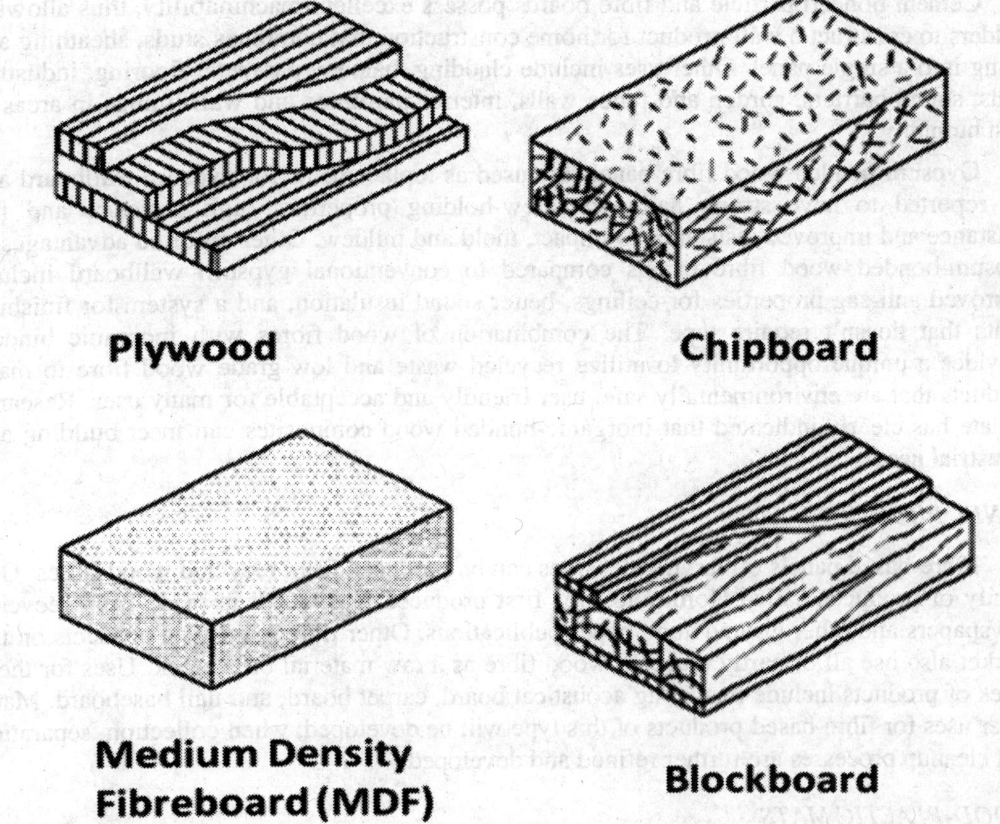

Fig. 34.1: Composite Woods

WOOD FIBRE COMPOSITES

Recycled wood fibre can be used to make composites from particles or fibres of wood held together with an inorganic matrix, such as Portland cement or gypsum. Such composites can be used in a variety of structural and industrial applications and have unique advantages over conventional building materials. Some composites are water resistant and can withstand the rigours of outdoor applications. Almost all are either fireproof or highly fire resistant and are very resistant to decay by fungi. Inorganic-bonded wood composites are molded products or boards that contain between 10 and 70 per cent (by weight) wood particles or fibres and conversely 30 to 90 per cent inorganic binder. Many different types of material can be incorporated with the inorganic binder matrix, including recycled paper fibre, fiberized demolition waste or scrap pellets, industrial waste wood, wood residues, non-commercial wood species and very low grade, non-merchantable wood. The wood particles must be fully encased with the binder to make a coherent material. Thus, the amount of inorganic binder required per unit volume of composite material is much higher than that in resin-bonded wood composites. The properties of inorganic-bonded wood composites are significantly influenced by their density as well as the amount and nature of both the inorganic binder and the woody material.

Cement-bonded particle and fibre boards possess excellent machinability, thus allowing builders to construct a wall product for home construction that combines studs, sheathing and siding into a single panel. Other uses include cladding, balcony parapets, flooring, industrial walls, sound barriers, garden and fence walls, interior partitions and wall linings in areas of high humidity.

Gypsum-bonded wood fibre panels are used as replacements for gypsum wallboard and are reported to have strong nail and screw-holding properties, high moisture and fire resistance and improved resistance to impact, mold and mildew. Other reported advantages of gypsum-bonded wood fibre panels compared to conventional gypsum wallboard include improved anti-sag properties for ceilings, better sound insulation, and a system for finishing joints that doesn't require tape. The combination of wood fibres with inorganic binders provides a unique opportunity to utilize recycled waste and low-grade wood fibre to make products that are environmentally safe, user friendly and acceptable for many uses. Research to date has clearly indicated that inorganic-bonded wood composites can meet building and industrial needs.

PANEL PRODUCTS

Fibre-based panels of varying densities can be produced from recycled wood fibres. One family of products, called Homosote, was first produced in 1916. It is made from recycled newspapers and other groundwood paper publications. Other fibreboard-type products on the market also use all or partly recycled wood fibre as a raw material base stock. Uses for these types of products include insulating acoustical board, carpet board, and nail baseboard. Many other uses for fibre-based products of this type will be developed, when collection, separation and cleanup processes are further refined and developed.

WOOD-PLASTIC MATS

Another potential use for recycled wood fibre is in low-density mats. Wood and plastic fibres are introduced into a turbulent air stream, transferred via the air stream to a moving support bed and subsequently formed into a continuous, low-density mat. The fibres in the mat are further intertwined and strengthened through needling. The ratio of wood to plastic in this matrix can be in the range to 95 to 5 per cent by weight. The plastic can also be replaced with a long lignocellulosic fibre such as jute or kenaf. An interesting application for fibre mats is for mulch around newly planted seedlings. The mats provide the benefits of natural mulch. Application for low-density wood fibre mats is for air filters or other types of filters.

OTHER SYNTHETIC PRODUCTS

Wood-based clear packaging film: It is a wood-based material with properties like plastic, suitable to be used as packaging material. These include food packaging, where the material - nano fibrillated cellulose film (NFC) - can help increase the shelf life of a product.

Wood fibre bioplastic packaging: Biodegradable plastic food packaging is made using up to 25 per cent wood fibres.

IMPROVED WOOD/ MODIFIED WOOD

The wood is treated in various ways to improve its strength, dimensional stability (hygroscopicity) and service life. It includes impregnation with synthetic resins or other materials and /or compression and / or heating and may be applied to solid wood or the veneers that are subsequently bonded together. Depending upon treatment, improved woods are categorized as follows (Mehta 1981; ICFRE 1990):

- *Impregnated Wood*: wood impregnated by wax, paraffin, resins, oils, etc. with the objective of improving dimensional stability and service life.
- *Heat Stabilized Wood*: Stabilization of dimension and decay resistance by simple heat treatment (260-315ºC for few minutes under non-oxidizing condition) to reduce hygroscopicity.
- *Compressed Wood*: Wood is compressed to increase the specific gravity of wood at pressure of 90-140 kg/sq. cm without any impregnation.
- *Compregnated Wood*: Wood is both compressed and impregnated with synthetic resins.
- *Heat Stabilized Compressed Wood*: wood compressed with heat but without impregnation.
- *Chemically Modified Wood*: Use of chemicals for impregnation for different resistance e.g. fire, termite, water, etc.

FUELWOOD

According to IEA report, Global Energy Demand is nearly 467 exajoules (EJ) and 88 per cent of which is met by fossil fuels. With continuing population growth, global energy demand is expected to grow an additional 50 per cent by 2030 mostly in rapidly industrializing countries. Many countries are developing renewable energy resources (bioenergy) to offset fossil fuel use. Bioenergy can be obtained from:

- *Biofuel*: It is defined as any fuel produced directly or indirectly from biomass e.g. liquid fuels such as ethanol and biodiesel.
- *Woodfuel*: It encompasses all types of biofuel derived directly or indirectly from trees and shrubs (woody biomass) grown on forest and non-forest land.
- *Fuelwood*: It is used to describe woodfuel where the original composition of the wood is preserved.

Fuelwood is a type of fuel in the form of wood derived from trees and shrubs grown on forest and non-forest land. Firewood remains the dominant source of both domestic and commercial energy requirements for Africa's rural population. Substantial increase in human and bovine population as well as decreasing forest areas have led to fuelwood shortages in many countries. In most African countries and some Asian countries, indigenous woodlands still provide both urban and rural populations with by far the greatest proportion for their fuelwood requirements.

Fuel wood is a wood used as fuel and the burning of wood is currently the largest use of energy derived from a solid fuel biomass. Wood fuel can be used for cooking and heating and occasionally for fueling steam engines and steam turbines that generate electricity. Wood fuel may be available as firewood, charcoal, chips, sheets, pellets and sawdust. The particular form used depends upon factors such as source, quantity, quality and application. Sawmill waste and construction industry by-products also include various forms of lumber tailings. Some consider wood fuel bad for the environment, however this is not the case if proper techniques are used. One might increase carbon emissions using gas powered saws and splitters in the production of firewood, but when wood heat replaces carbon-producing fuels such as propane, heating oil or electricity from a coal-burning plant, then wood burning has a positive impact on the carbon footprint. Wood may be used indoors in a furnace, stove or fireplace. Wood also may be burned outdoors in a campfire or bonfire.

In India and other developing countries fuelwood is major wood product utilized by the masses to meet out their routine energy needs. The fuelwood demand is about 275 million m^3 while supply is only 50 million m^3 and forest is still a major source besides agroforestry plantations. Characteristics of ideal fuelwood tree are: fast growing, multipurpose, good coppicing and pollarding, nitrogen fixing, high calorific value of wood, good combustibility with less smoke, easy to split (fissility). But most of the woods are used irrespective of these attributes. The major woods used as fuelwood include *Acacia* spp, *Eucalyptus* spp, *Tamrindus indica,* Oak (*Quercus* spp.) Chirpine (*Pinus roxbughii*), Neem (*Azadirchta indica*), Mango (*Mangifera indica*), Axlewood (*Anogeissus pendula*), Subabul (*Leucaena leucocephala*), Anjan (*Hardwikia binnata*), Palas (*Butea monosperma*), *Albizia* spp, Mahua (*Madhuca indica*), Jamun (*Syzygium cumini*), *Gliricidia* spp., *Casurina equisetifoila, Prosopis juliflora,* Khejri (*Prosopis cineraria*), Pilu (*Salvadora persica*), Maharukh (*Ailanthus excelsa*), Persian liliac (*Melia azedarach*), *Celtis australis, Alnus nepalensis,* Ash (*Fraxinus micrantha*), Black Locust (*Robinia pseudocasia*), *Salix* spp., *Lyonia* spp., etc.

PULP AND PAPER WOOD

Origin of paper making can be traced back to China using vegetable wastes but using wood to make paper is a fairly recent innovation. In India, first paper industry was established in 1830 in West Bengal. In the 1800s, fibre crops such as linen fibres were the primary material source and paper was a relatively expensive commodity. The use of wood to make pulp for paper began with the development of mechanical pulping in Germany by F.G. Keller in the 1840s and by the Canadian inventor Charles Fenerty in Nova Scotia. Chemical processes quickly followed, first with J. Roth's use of sulfurous acid to treat wood, followed by B. Tilghman's U.S. patent on the use of calcium bisulfite, $Ca(HSO_3)_2$, to pulp wood in 1867. Almost a decade later the first commercial sulphite pulp mill was built in Sweden as a process developed by Carl Daniel Ekman. By 1900, sulphite pulping had become the dominant means of producing wood pulp, surpassing mechanical pulping methods. The competing chemical pulping process, the sulfate or kraft process was developed by Carl F. Dahl in 1879 and the first kraft mill started (in Sweden) in 1890. The invention of the recovery boiler by G.H. Tomlinson in the early 1930s allowed kraft mills to recycle almost all of their pulping chemicals. This along with the ability of the kraft process to accept a wider variety of types of wood and produce stronger fibres made the kraft process the dominant pulping process starting in the 1940s.

MANUFACTURE OF WOOD PULP

The timber resources used to make wood pulp are referred to as pulpwood. Wood pulp comes from softwood trees such as spruce, pine, fir, larch and hemlock and hardwoods such as eucalyptus, aspen and birch. A pulp mill is a manufacturing facility that converts wood chips or other plant fibre source into a thick fibre board which can be shipped to a paper mill for further processing. Pulp can be manufactured using mechanical, semi-chemical or fully chemical methods (kraft and sulphite processes). The finished product may be either bleached or non-bleached, depending on the customer requirements. Wood and other plant materials used to make pulp contain three main components (apart from water) cellulose fibres (desired for papermaking), lignin (a three-dimensional polymer that binds the cellulose fibres together) and hemicelluloses, (shorter branched carbohydrate polymers). The aim of pulping is to break down the bulk structure of the fibre source, be it chips, stems or other plant parts, into the constituent fibres. Different pulping processes and types of pulps are presented below (Christopher Biermann 1993; Sojostrom 1993).

MECHANICAL PULPING

Manufactured grindstones with embedded silicon carbide or aluminum oxide can be used to grind small wood logs called "bolts" to make stone groundwood pulp (SGW). If the wood is steamed prior to grinding, it is known as pressure groundwood pulp (PGW). Most modern mills use chips rather than logs and ridged metal discs called refiner plates instead of grindstones. If the chips are just ground up with the plates, the pulp is called refiner mechanical pulp (RMP) and if the chips are steamed while being refined the pulp is called thermomechanical pulp (TMP). Steam treatment significantly reduces the total energy needed to make the pulp and decreases the damage (cutting) to fibres. Mechanical pulps are used for products that require less strength such as newsprint and paperboards.

HYBRID PULPING

The various mechanical pulping methods, such as groundwood (GW) and refiner mechanical (RMP) pulping physically tear the cellulose fibres one from another. Much of the lignin remains adhering to the fibres. Strength is impaired because the fibres may be cut. There are a number of related hybrid pulping methods that use a combination of chemical and thermal treatment to begin an abbreviated chemical pulping process, followed immediately by a mechanical treatment to separate the fibres. These hybrid methods include thermomechanical pulping and chemi-thermo-mechanical pulping, also known as CTMP. The chemical and thermal treatments reduce the amount of energy subsequently required by the mechanical treatment and also reduce the amount of strength loss suffered by the fibres.

THERMOMECHANICAL PULPING

It is a pulp produced by processing wood chips using heat (thermo) and a mechanical refining movement (mechanical). It is a two stage process where the logs are first stripped of their bark and converted into small chips. These chips have a moisture content of around 25-30 per cent and a mechanical force is applied to the wood chips in a crushing or grinding action which generates heat and water vapour and softens the lignin thus separating the individual fibres. The pulp is then screened and cleaned, any clumps of fibre are reprocessed.

This process gives a high yield of fibre from the timber (around 95 per cent) and as the lignin has not been removed, the fibres are hard and rigid.

CHEMICAL PULPING

Chemical pulp degrades the lignin and hemicellulose into small, water-soluble molecules which can be washed away from the cellulose fibres without depolymerizing the cellulose fibres (chemically depolymerizing the cellulose weakens the fibres). Chemical pulp is produced by combining wood chips and chemicals in large vessels known as digesters where heat and the chemicals break down the lignin, which binds the cellulose fibres together, without seriously degrading the cellulose fibres. Chemical pulp is used for materials that need to be stronger or combined with mechanical pulps to give a product different characteristic. The kraft process is the dominant chemical pulping method, with sulfite process being second. Historically soda pulping was the first successful chemical pulping method.

CHEMI-THERMO-MECHANICAL PULPING

Wood chips can be pretreated with sodium carbonate, sodium hydroxide, sodium sulfite and other chemicals prior to refining with equipment similar to a mechanical mill. The conditions of the chemical treatment are much less vigorous (lower temperature, shorter time, less extreme pH) than in a chemical pulping process since the goal is to make the fibres easier to refine, not to remove lignin as in a fully chemical process. Pulps made using these hybrid processes are known as chemithermomechanical pulps (CTMP).

RECYCLED PULPING

It is also called deinked pulp (DIP). DIP is recycled paper which has been processed by chemicals, thus removing printing inks and other unwanted elements and frees the paper fibres. The process is called deinking. DIP is used as raw material in papermaking. Many newsprint, toilet paper and facial tissue grades commonly contain 100 per cent deinked pulp and in many other grades, such as lightweight coated for offset and printing and writing papers for office and home use, DIP makes up a substantial proportion of the furnish.

OTHER PULPS AND PULPING METHODS

Organosolv Pulping: It uses organic solvents at temperatures above 140^0 C to break down lignin and hemicellulose into soluble fragments. The pulping liquor is easily recovered by distillation.

Alternative Pulping Methods: Research is under way to develop biological pulping, similar to chemical pulping but using certain species of fungi that are able to break down the unwanted lignin, but not the cellulose fibres. This could have major environmental benefits in reducing the pollution associated with chemical pulping. The pulp is bleached using chlorine dioxide stage followed by neutralization and calcium hypochlorite. The oxidizing agent in either case oxidizes and destroys the dyes formed from the tannins of the wood and accentuated (reinforced) by sulfides present in it. Steam exploded fibre is a pulping and extraction technique that has been applied to wood and other fibrous organic material.

Market Pulp: It is any variety of pulp that is produced in one location, dried and shipped to another location for further processing. Important quality parameters for pulp not directly related to the fibres are brightness, dirt levels, viscosity and ash content. Pulp has high market potential in International market.

Air Dry Pulp: It is the most common form to sell pulp. This is pulp dried to about 10 per cent moisture content. It is normally delivered as sheeted bales of 250 kg. The reason to leave 10 per cent moisture in the pulp is that this minimizes the fibre to fibre bonding and makes it easier to disperse the pulp in water for further processing to paper.

Roll Pulp or Reel Pulp: It is the most common delivery form of pulp to non-tradtitional pulp markets. Fluff pulp is normally shipped on rolls (reels). This pulp is dried to 5-6 per cent moisture content. This will be used in further processing.

Flash Dried Pulp: It is prepared by pressing the pulp to about 50 per cent moisture content and then let it fall trough soils that are 15-17 m high. Gas fired hot air is the normal heat source. The temperature is well above the char point of cellulose, but large amount of moisture in the fibre wall and lumen prevents the fibres from being incinerated. It is often not dried down to 10 per cent moisture (air dry). The bales are not as densely packed as air dry pulp.

GLOBAL PAPER INDUSTRY

All kinds of paper are made out of 100 per cent wood with nothing else mixed into them (with some exceptions, like fancy resume paper which may include cotton). This includes newspaper, magazines and even toilet paper. Most pulp mills use good forest management practices in harvesting trees to ensure that they have a sustainable source of raw materials. The FSC (Forest Stewardship Council), SFI (Sustainable Forestry Initiative), PEFC (Programme for the Endorsement of Forest Certification) and other bodies certify paper made from trees harvested according to guidelines meant to ensure good forestry practices. The number of trees consumed depends on whether mechanical processes or chemical processes are used. It has been estimated that based on a mixture of softwoods and hardwoods, 12 meters (40 ft) tall and 15-20 cm (6-8 inch) in diameter, it would take an average of 24 trees to produce 1 metric ton of printing and writing paper, using the kraft process (chemical pulping). Mechanical pulping is about twice as efficient in using trees since almost all of the wood is used to make fibre therefore it takes about 12 trees to make 1 metric ton of mechanical pulp. The pulp and paper industry comprises companies that use wood as raw material and produce pulp, paper, board and other cellulose-based products. The industry is dominated by North American (United States and Canada), Northern European (Finland, Sweden and North-West Russia) and East Asian countries (China, Japan and South Korea). Australasia and Brazil also have significant pulp and paper enterprises.

Table 34.1: Production Level of Top Ten Paper Producing Countries in 2010 (Source: JPA 2011)

Rank	Country	Production (1,000 ton)	Rank	Country	Production (1,000 ton)
1	China	92,599	6	Finland	11,789
2	USA	75,849	7	Sweden	11,410

3	Japan	27,288	8	South Korea	11,120
4	Germany	23,122	9	Indonesia	9,951
5	Canada	12,787	10	Brazil	9,796

Table 34.2: The world's Top 20 Paper and Paperboard Company Groups in 2010 (Source: RISI 2011)

Rank	Company Group	Country	Production (1,000 ton)	Rank	Company Group	Country	Production (1,000 ton)
1	International Paper	USA	11,922	11	Abitibi Bowater	Canada	5,318
2	Stora Enso	Finland	10,812	12	New Page	USA	4,400
3	UPM	Finland	9,914	13	Norske Skog	Norway	3,998
4	Svenska Cellulosa Aktiebolaget	Swedan	8,948	14	Mondi	UK	3,697
5	Smurfit Kappa Group	USA	7,650	15	Temple Inland	USA	3,660
6	Nippon Paper	Japan	7,292	16	Lee and Man Paper	China	3,350
7	Nine Dragons Paper	China	7,280	17	Domtar	Canada	3,482
8	Sappi	South Africa	6,900	18	Shandong Chenming	China	3,350
9	Oji Paper	Japan	6,861	19	Cascades	Canada	3,330
10	Surfit-stone	USA	5,896	20	Siam Cement	Thailand	3,191

SUITABLE INDIAN TREE SPECIES FOR PAPER AND PULP

- Bamboos: *Bambusa bambos, B. nutans, B. balcooa, B. tulda, B.vulgaris, Dendrocalamus strictus, D. hamiltonii, Melocanna bambusoides, Oxytenanthera monostigma, Ochlandra rhedii* and *O. travancoria*
- Softwoods: Silver fir (*Abies pindrow*), Spruce (*Picea smithiana*), Chir pine (*Pinus roxburghii*), Blue pine (*P. wallichiana*), *Cryptomeria japonica*
- Hardwoods: Salai (*Boswellia serrata*), paper mulberry (*Broussonetia papyrifera*), *Kydia calycina, Eucaylptus* spp, *Poplar* spp., subabul (*Leucaena leucocephala*), Beef wood (*Casuarina equisetifolia*)

CHARCOAL

Charcoal is a formless mass of carbon and can be made from most carbonaceous materials. It is one of the oldest man-made fuels. Charcoal is prepared by firing wood in the absence of air in pyrolysis drum and the product is quenched, collected and packed in gunny bag. Charcoal is inexpensive compared to peroleum fuels (kerosene, LPG, natural gas) and electricity. Charcoal making is termed wasteful because only 30-40 per cent of the wood is actually converted to charcoal. The rest is released into the atmosphere as gases.

Commercially grown trees can produce 18 tons of charcoal from one hectare. Charcoal is a traditional fuel as like fuelwood and it can be purchased in preferred quantity. It burns without smoke. It does not decompose even after extended storage and does not create dangerous flames around cooking vessels. It requires a simple stove whose heat output is relatively easy to control

CHARCOAL MANUFACTURE TECHNIQUES

There are many techniques for making charcoal. In the simplest method, wood is burned in an open fire and the charred remains recovered as charcoal. A more controlled process is by restricting the supply of air during carbonization. Three main methods are traditionally used.

- Wood is stacked on the ground and covered with soil, forming what is often referred to as an earth mound kiln.
- Second method involves digging a hole in the ground into which the wood is placed and then covered with soil. This one is commonly referred to as a pit-kiln.
- Charcoal is also produced by heating wood in airtight ovens or retorts or pyrolysis drums or chambers or in kilns supplied with limited and controlled amounts of air. High-temperature heating by all methods breaks down the wood into gases, a watery tar mixture and the familiar solid carbon material commonly known as charcoal.

USES OF CHARCOAL

Charcoal is still a major source of energy throughout the world. Charcoal plays a considerable role in African economies and countries in Asia. Nearly 80 per cent of the African population and majority of Asian rural population continue to depend on traditional biomass fuels (fuelwood and charcoal) for their energy need. Charcoal industry provides a considerable amount of employment in rural areas. The major uses of charcoal are:

- Charcoal is used as fuel for cooking purposes.
- It is used in certain metallurgical purifying treatments and as filter to remove organic compounds such as chlorine, gasoline, pesticides and other toxic chemicals from water and air.
- Activated charcoal is used in purifying and refining metals.
- Activated Charcoal is used in Gas masks that are used during war.
- Activated charcoal is used as an antidote for many types of poisons and it is an effective anti-flatulent.

TREE SPECIES FOR CHARCOAL PRODUCTION

- *Prosopis juliflora*
- *Acacia spp.*
- *Eucalyptus spp.*

- *Leucaena leucocephala*
- *Casuarina equisetifolia*
- *Sesbania sesban*
- *Albizia spp.*
- *Pterocarpus spp.*
- *Terminalia spp.*
- *Tamarindus indica*

OTHER MULTIPLE USES OF WOOD

Wood has long been used as an artistic medium. It has been used to make sculptures and carvings for millennia. Certain types of musical instruments, such as those of the violin family, guitar, clarinet and recorder, xylophone and marimba are made mostly or entirely of wood. The tree species widely used for manufacture of musical instruments, sports goods and agricultural implements are presented below (Mehta 1981; ICFRE 1990).

MUSICAL INSTRUMENTS

The choice of wood may make a significant difference to the tone and resonant qualities of the instrument, and tonewoods have widely differing properties, ranging from the hard and dense african blackwood (used for the bodies of clarinets) to the light but resonant European spruce (*Picea abies*) (traditionally used for the soundboards of violins). The most valuable tonewoods, such as the ripple sycamore (*Acer pseudoplatanus*), which is used for the backs of violins, combine acoustic properties with decorative color and grain which enhance the appearance of the finished instrument. Indian woods used in musical instruments are:

- Violins: Maple (*Acer* spp.), Teak (*Tectona grandis*) Rosewood (*Dalbergia latifolia*), Eboney (*Diospyrus* spp) Sundari (*Heritieria minor*)
- Sitar: Toon (*Toona ciliata*), Deodar (*Cedrus deodara*), Teak (*Tectona grandis*), Sissoo (*Dalbergia sissoo*)
- Vina and Tambora: Jackfruit (*Atrocarpus heterophyllus*), Gamari (*Gmelina arborea*), Bijasal (*Pterocarpus marsupium*)
- Guitar: White dhup (*Canarium euphyllum*),
- Harmonium: Teak (*Tectona grandis*)
- Pianao Cases: Mahogany (*Swietenia* spp.), Walnut (*Juglans regia*), Satin wood (*Cholroxylon swietienia*), Paduk (*Pterocarpus dalbergioides*)
- Drums: Ash (*Fraxinus* spp.), Siris (*Albizia* spp.) , Sissoo (*Dalbergia sissoo*)

SPORTS GOODS

Many types of sports equipment are made of wood, or were constructed of wood in the past. For example, cricket bats are typically made of white willow. The baseball bats which are legal for use in major league baseball are frequently made of ash wood or hickory, and in recent years have been constructed from maple even though that wood is somewhat more fragile. Many other types of sports and recreation equipment, such as skis, ice hockey sticks, lacrosse sticks and archery bows, were commonly made of wood in the past, but have since been replaced with more modern materials such as aluminum, fibreglass, carbon fibre, titanium, and composite materials. Indian wood suitable for sports goods are:

- Cricket bats, stumps and bails: Cricket bat Willow (*Salix alba var. caerulea*), Mulberry (*Morus alba*), Persian lilac (*Melia azedarach*) Gutel (*Trewia nudiflora*), Chinese tallow tree (*Sapium sebiferum*), Sandan (*Ougeinia oojenensis*)
- Hockey sticks: Mulberry (*Morus alba*), Celtis (*Celtis australis*)
- Bows and Arrows: ***For Bow*** - Yew (*Taxus baccata*), Parrotia (*Parrotia jacmontiana*), Dhaman (*Grewia tilaefolia*), Bijasal (*Ptercarpus marsupium*), Khair (*Acacia catechu*). ***For Arrow*** - Sissoo (*Dalbergia sissoo*), Poon (*Callophyllum* spp), Bijasal (*Pterocarpus marsupium*), Reeds and bamboos
- Skis: Sissoo (*Dalbergia sissoo*), Mulberry (*Morus alba*), Axelwood (*Anogeissus latifolia*)
- Fishing rods: Sagopalm (*Caryota urens*), *Haplophragma adenophyllum*, Balck chuglum (*Terminalia manii*), Chooi (*Sageraea elliptica*), Eboney (*Diospyrus* spp)

AGRICULTURE IMPLEMENTS AND HANDLES

Most of the strong quality small woods are used in different agricultural tools and implements still in rural India and important species used for tool handles are: Bijasal (*Ptercarpus marsupium*), Khair (*Acacia catechu*), Babul (*Acacia nilotica*), Amaltas (*Cassia fistula*), Hopea (*Hopea odorata*), Mesua (*Mesua ferrea*), Oak (*Quercus* spp), Jamun (*Syzygium cumini*), Irul (*Xylia xylocarpa*), Ber (*Zizyphus maurtiana*), Ash (*Fraxinus* spp), Siris (*Albizia* spp.), Sissoo (*Dalbergia sissoo*), Sandan (*Ougeinia oojenensis*)

NON-WOOD FOREST PRODUCTS

'Non-wood forest products' is a relatively new term used generally to mean forest products other than wood. NWFP may be gathered from the wild, or produced in forest plantations, agroforestry schemes and from trees outside forests. The other synonymous terms used for NWFPs are byproducts of forests, minor forest products, non-timber forest products, non-wood goods and benefits, non-wood goods and services, other forest products, secondary forest products, special forest products, etc. by different groups and countries. NWFP include products used as food and food additives (edible nuts, mushrooms, fruits, herbs, spices and condiments, aromatic plants, game), fibres (used in construction, furniture, clothing or utensils), resins, gums, and plant and animal products used for medicinal, cosmetic or cultural purposes. The use of NWFPs is as old as human existence. Several million households world-wide depend heavily on NWFP for subsistence and/or income. Some 80 per cent of the population of the developing world use NWFP for health and nutritional needs. At a local level, NWFP also provide raw materials for small and cottage scale industrial processing. Some NWFP are also important export commodities. At present, at least 150 NWFPs are significant in terms of international trade, including honey, gum arabic, rattan, bamboo, cork, nuts, mushrooms, resins, essential oils and plant and animal parts for pharmaceutical products. NWFP have also attracted considerable global interest in recent years due to the increasing recognition of their contribution to environmental objectives, including the conservation of biological diversity.

Work on NWFP assessment is an important activity in the FAO Forestry Department, involving expertise from various technical units, in particular the Forest Resources Division (and its flagship programme for the Global Forest Resources Assessment - FRA) and the Forest Products Division (through its Non-Wood Forest Products Programme). FAO's Forestry Department established NWFP Programme in 1991 for the promotion and development of non-wood forest products with a mission to improve the sustainable utilization of NWFP in order to contribute to the wise management of the world's forests, to conserve their biodiversity and to improve income generation and food security. Over the past two decades an increasing number of governmental and non-governmental organizations (NGOs) and institutions, as well as the private sector, have become involved with the promotion and utilization of non-wood forest products (NWFPs). A lot of new information has been collected on the socio-economic importance and potential of NWFP utilization and its effects on the environment.

NWFP DEFINITION

In 1995, FAO made a first step towards a harmonized definition of NWFPs by organizing the International Expert Consultation on Non-Wood Forest Products in Yogyakarta, Indonesia, as "NWFPs consist of goods of biological origin other than wood, as well as services, derived from forests and allied land uses." In 1999, a new FAO working definition of NWFPs has been adopted: "Non-wood forest products consist of goods of biological origin other than wood, derived from forests, other wooded land and trees outside forests." According to this definition, the three components of the term "non-wood forest products" are interpreted as follows:

- *Non-wood*: The term NWFP excludes all woody raw materials. Consequently, timber, chips, charcoal and fuelwood as well as small woods such as tools, household equipment and carvings are excluded. Non-timber forest products (NTFPs), in contrast, generally include fuelwood and small woods and this is the main difference between NWFPs and NTFPs.
- *Forest*: NWFPs should be derived from forests and similar land uses (plantations). The final definition of "trees outside forests" includes trees growing outside the forest and not belonging to the category of forests, forest lands, or other wooded land.
- *Products*: In the proposed definition, the term "product" corresponds to goods that are tangible and physical objects of biological origin such as plants, animals and their products. Forest services (e.g. ecotourism, grazing, bioprospecting) and forest benefits (e.g. soil conservation, soil fertility, watershed protection) are excluded.

SOCIO-ECONOMIC BENEFITS OF NWFPs

A better understanding of NWFPs' overall socio-economic contributions to the subsistence and income of local communities and to national economies facilitates better programme response. NWFPs play an important role in food security, nutrition and community health. Their benefits are relatively more important for poorer households, women and disadvantaged groups and this has important consequences for planning NWFP activities. Interventions need to involve people's participation in the development process and the equitable distribution of benefits to different groups.

CHALLENGES FOR NWFPs DEVELOPMENT

Most NWFPs are often associated with traditional uses that are not widely known. Transactions related to NWFPs largely take place in households and small-scale units, mostly outside the established marketing systems/channels, thus forming part of unorganized, informal sector. NWFPs are often seasonal and for these reasons, they are often overlooked by planners and their local uses go unrecorded. Timber-orientation of forestry profession, and the bias of planners in favour of large-scale enterprises, often leaves NWFPs at a disadvantage. Sustainable management of NWFPs, especially of those occurring among the biological richness and ecological diversity of natural forests is extremely complex. This has resulted in its being left out from management prescriptions and preference being given to comparatively easier timber management. There is also overlapping of uses and sources. Same product can be produced from different non-wood raw materials; and same non-wood raw materials can provide different products. This adds to the complexity of managing and utilizing the resource. Lack of knowledge about the potential of NWFPs to support sustainable and remunerative enterprises is a major constraint (FAO 2012; IUFRO 2012).

MANAGEMENT OF NWFPs

Management of NWFP resources is critical to the future of remaining natural forests. A first step is better assessment of the resource and what represents a sustainable harvest level of different forest produces. NWFPs can play a role in more sustainable forestry through multiple use management for wood and non-wood products. Sustainable NWFP harvesting can also take place in forest reserves and buffer areas. Resource management needs to include mechanisms for resolving conflicts in land-use objectives. Domestication of species (e.g. through agroforestry) is one alternative for minimizing the impact of increasing NWFP demand. Research is needed to assess and learn from local knowledge and to improve management systems and to improve harvesting and post-harvest technologies.

MARKETING AND TRADE OF NWFPs

Information on NWFP resources, harvests, processing and trade is scarce and dispersed. It is generally maintained that NWFPs are important locally, nationally and internationally. Many of the products contribute significantly to the creation of economic benefits and cash income at the community level (e.g. forest foods marketed at village markets). Some of the products contribute to the national economy (e.g. tendu leaves in India) while many enter international trade and are vital elements in many industries worldwide (e.g. essential oils sold to the fragrance and flavour industries). Most failures of NWFP programmes result from inattention to markets. With increasing pressure on forest resources, well-informed NWFP marketing strategies could be crucial for maintaining the resource. Producers need better information on the nature and volume of existing NWFP trade, markets and product standards. With the increase in green consumerism, knowledge of international market concerns and quality standards of products is needed. More rational and transparent market transactions throughout the production/market chain are needed for producers to receive a more equitable share of the product value. A greater appreciation of marketing and market information is needed by producers, NWFP-programme planners and NGOs (FAO 2012; IUFRO 2012).

NWFPs IN INDIA

India is a country of vast diversity lying at the juncture of the big-geographic provinces of Afro-Eurasia and the Orient. Because of the country's diversified climatic and physiographic factors, India is blessed with all types of vegetation: tropical, subtropical, temperate and alpine. Due to its wide-ranging environmental regimes and diverse biological communities, the country is one of the world's top 12 "megadiversity" nations. Non-wood forest products (NWFPs) are derived from over 3,000 species. These products are classified as leaves, bamboos, gums, resins and oleoresins, oil seeds, essential oils, fibres and flosses, grasses and fodder trees, tans and dyes, medicinal plants (drugs) and spices, animal products, edible products etc. It is estimated that 60 percent of all NWFPs are consumed locally and are not accounted in the calculation of revenues. There also are many products which are not extracted fully or which go to waste because of insufficient knowledge of their use or because they occur in inaccessible locations.

The basic objectives of the National Forest Policy of 1988 include conserving the national flora and fauna, meeting the needs of rural and tribal populations and encouraging efficient utilization of all forest produce. The policy states that NWFPs which provide sustenance to local communities should be protected and improved. It provides for research into the conservation and management of forest resources and for increasing productivity through the application of modern scientific and technological methods. The survey of forest resources is to be given high priority. The Indian Forest Policy of 1988 and the subsequent government resolution on participatory forest management (MoEF 1990) emphasize the need for people's participation in natural forest management. The policy document asserts that local communities should be motivated to identify themselves with the development and protection of the forests from which they derive benefits. Thus, the policy envisages a process of joint management of forests by the state governments (which have nominal responsibility) and the local people, which would share both the responsibility for managing the resource and the benefits that accrue from this management. Non-wood forest products (NWFPs) have a key role in JFM efforts. With the increasing awareness of their economic potential and growing concerns for the sustainability of the resources and the distribution of the benefits derived from them, various state governments have taken over control of a number of NWFPs.

Traditionally, the collection of NWFPs have been of low intensity and generally sustainable. However, as the economic potential of NWFPs has become apparent, the intensity of collection has increased and more significant infrastructures for trade and processing have developed. This has raised concerns for the sustainability of the resources and the distribution of the benefits derived from them. In reaction to these concerns, a number of state governments have taken over the control of a number of NWFPs. The state regulations bringing certain NWFPs under monopoly trade with the objective of preventing unscrupulous intermediaries and their agents from exploiting NWFP collectors, to ensure fair wages to collectors, to enhance revenue for the state, to ensure quality and to maximize the collection of produce. In most cases, trading is controlled through state-owned institutions such as state forest development corporations, federations, cooperatives and tribal societies (Gupta 1994; Prasad 1999; IUFRO 2012).

Numerous literatures are available regarding different tree species yielding NTFPs in India and world. Tree species yielding flavours and fragrances (essential oils), gums, resins, latex, fibres and flosses, tannins and dyes, edible plant products and important bamboos, rattans and palms are presented below (Mehta 1981; ICFRE 1990; GFU 2008; MPSMFPCF 2012; NTFP 2012; MSG 2012; IUFRO 2012; FAO 2012).

FLAVOURS AND FRAGRANCES OF PLANT ORIGIN

Taste and smell are the two most sensitive human senses. The natural world - especially that of higher plants - provides a multitude of flavours and fragrances, either directly or indirectly, which impinge upon these senses. Essential oils are used in everyday life in foods and drinks and confectionery items; products for personal use such as perfumes, deodorants, shampoos, bath lotions, toilet soaps, toothpastes and mouth washes; pharmaceutical preparations where flavours are added to make the product more appealing or to mask the taste of less agreeable ones; items used about the house or office or in industry such as air fresheners, laundry soaps, detergents, cleaning agents and tobacco products.

Grass oils: Lemon grass oil (*Cymbopogon flexuosus*), West Indian lemon grass oil (*Cymbopogon citratus*), Palmarosa oil (*Cymbopogon martinii* var. *motia*), Ginger grass oil (*Cymbopogon martinii* var. *sofia*), Citronella oil (*Cymbopogon nardus* var. *winterianus*), Vetiver oil (*Vetiveria zizanioides*)

Wood oils: Sandalwood oil (*Santalum album*), Agar oil (*Aquillaria agalocha*), Deodar oil (*Cedrus deodara*), Pine oil/ Turpentine oil (*Pinus roxburghii*), Linaloe oil (*Bursera delpechiana*), Cedar oil (*Juniperus macropoda*)

Leaf oils: Eucalyptus oil (*Eucalyptus globulus*), Lemon scented gum oil (*Eucaylyptus citriodora)*, Camphor oil (*Cinnamomum camphora*), Pine needle oil (*Pinus roxburghii*), Mint oil (*Mentha* spp.), Patchouli oil (*Pogostemon patchouli*), Geranium oil (*Pelargonium graveolens*), Wintergreen oil (*Gualtheria fragrantissima*).

Root/Rhizome oils: Costus oil (*Saussurea lappa*), Indian valerian oil (*Valeriana wallichii*), Curcuma oil (*Curcuma aromatica*), Cyperus oil (*Cyperus scariosus*), Sweet flag oil (*Acorus calamus*)

Flower oils: Jasmine oil (*Jasminum officinale*), Rose oil (*Rosa damascena*), Keora oil (*Pandanus odoratissimus*), Tagtes oil (*Tagetes indica*), Champa oil (*Michelia çhampaca*), Lavander oil (*Lavandula officinalis*), Cassie perfume (*Acacia fernesiana*), Clove oil (*Syzygium aromaticum*),

Other Essential oils: Citrus oil/Lime oil (*Citrus aurantifolia*), Petgrain oil (*Citrus aurantium* and *Citrus limettioides*), Orange oil (*Citrus sinensis*), Mandarin oil (*Citrus reticulata*), Cinnamoum oil (*Cinnamomum zeylanicum*), Nutmeg oil (*Myristica fragrans*), Salai gum oil (*Boswellia serrata*)

GUMS, RESINS AND LATEX

Gums, resins and latexes are perhaps the most widely used and traded category of non-wood forest products. A recent estimate put the value of the world market for gums used as food additives at about US$ 10 billion. Of which the two largest forest gums viz. gum arabic and locust bean accounted for just over 12 per cent. Indonesia, India and the People's Republic of China are among the world's biggest producers of gums and resins. Sudan and Indonesia are the world's two biggest exporters of natural gums and resins.

Gums are translucent, amorphous substances which are degradation products of the cell wall of woody species. They exude spontaneously from trees and are soluble in water. The gums are useful in food, pharmaceutical and miscellaneous technical applications. In the food industry, advantage is taken of their thickening, stabilizing, emulsifying and suspending properties and they are employed in a very wide range of products, both foods and drinks. In the pharmaceutical industry, they are used as binding agents in tablets and as suspending and emulsifying agents in creams and lotions. It also has some specific applications in the dental and medical fields. Miscellaneous end users include the printing and textile industries.

Resins, including oleoresins and balsams, have an equally diverse range of applications. They are used in paints, varnishes and lacquers. Some resins, especially the soft resins and balsams, are used as sources of fragrances and pharmaceuticals, usually after preparation of a suitable solvent extract or distillation of a volatile oil.

Acacia Gums: Babul (*Acacia nilotica*), True gum arabic (*A. senegal*), Khair (*A. catechu*), *A. modesta, A. fernesiana, A. luecophloea.*

Other Gums: Karaya gum (*Sterculia urens*), *Cochlospermum religiosum*, Semal (*Bombax ceiba*), *Astragalus heratensis,* Kapok (*Ceiba pentandra*), Hog tragacanth (*Prunus amygdalus*), Cherry gum (*Prunus armeniaca*), Bengal kino (*Butea monosperma*), Gum kino (*Pterocarpus marsupium*), Jhingan gum (*Lannea coromandelica*), Salai gum (*Boswellia serrata*), Moringa gum (*Moringa pterygosperma*), Dhaura/ Ghatti gum (*Anogeissus latifolia*), Semla gum (*Bauhinia retusa*), Madiata gum (*Adenanthera pavonina*), Gutta percha gum (*Dichopsis polyantha*), Eucalyptus kino (*Eucalyptus globulus*), Lemon scented gum (*Eucalyptus citriodora*).

Minor Gums: *Albizia chinensis, A. odoratissima, A. procera, Anacardium occidentale, Azadirachta indica, Bauhinia purpurea, B. racemosa, B. variegata, Chloroxylon swietenia, Feronia limonia, Mangifera indica, Terminalia alata, T. bellerica, Agele marmelos, Ailanthus excelsa, Pithecellobium dulce, Spondias pinnata, Borassus flabellifer, Buchanania latifolia, Calophyllum apetalum, C. elatum, C. inophyllum, Chukrasia tabularis, Delonix regia, Elaeodendron glaucum, Gardenia turgida, Gaurga pinnata, Macaranga peltata, Prosopis juliflora, P. cineraria, Zizyphus sativa.*

True Resins: Balck dammar (*Canarium strictum*), Rock dammer (*Hopea odorata*), Green dammar (*Shorea tumbuggaia*), White dammar (*Vateria indica*), Amber (*Pinus succinfera*), Lacquer (*Rhus vernicifera*), Shell lac (*Laccifer lacca*), Mastic (*Pistacia lentiscus*), Sand arac (*Callitris quadrivalvis*), Storax (*Atingia excelsa*) .

Other True Resins: *Abies excela, A. webbiana, Ailanthus malabarica, Cannabis sativa, Carica papaya, Daemonorops kurzianus, Juniperus communis, Liquidambar orientalis, Styrax benzoin*

Oleo Resins: Salai gum (*Boswellia serrata*), Gurjan oil (*Dipterocarpus turbinatus*), *Kingiodendron pinnatum, Cedrus deodara, Erythroxylon monogynum, Melanorrhoea usitata, Pinus insularis, Pinus roxburghii, Pinus wallichiana, Shorea robusta*

Gum Resins: Gamboge (*Garcinia morella*), Myrrah/Gugal (*Commiphora mukul*), Galbanum (*Ferula galbaniflua*), Asafoetida (*Ferula asafoetida*), *Diospyros peregrina*, Dikamali/ Cumbi gum (*Gardenia gummifera*)

Latex: Chilte (*Cnidoscolus* spp.), Chicle (Manilkara zapota), Rubber latex (*Hevea brasiliensis*)

FIBRES AND FLOSSES

Fibres: *Agave sisalana, Streculia villosa, Abroma augusta, Abutilon* spp, *Ananas cosmosus, Antiaris toxicaria, Boehmeria nivea, Borassus flabellifer, Cannabis sativa, Cordia dichotoma, C. rothii, Giradinia heterophylla, Grewia glabra, G. elastica, G. optiva, Hibiscus* spp, *Malachra capitata, Marsdenia tenacissima, M. volubilis, Phormium tenax, Sansevieria roxburghiana, Sesbania bispinosa, Sida rhombifolia, Sterculia foetida, S. urens, Themeda arundinacea, Trema orientalis, Typha elephantina, Urena lobata, Oreocnide integrifolia*

Flosses: Indian Kapok (*Bombax ceiba*), Kapok or Silkcotton (*Ceiba pentandra*)

TANNINS AND DYES

TANNINS

Bark Tans: *Acacia mearnsii, A. dealbata, A. nilotica, Acacia auriculiformis, Cassia fistula, Cassia auriculata, Rhizophora mucronata, Ceriops roxburghiana, Terminalia arjuna, Shorea robusta*

Fruit Tans: *Terminalia chebula, T. bellerica, Acacia nilotica , Emblica officinalis, Zizyphus xylocarpa, Caesalpinia coriaria*

Leaf Tans: *Anogeissus latifolia, Emblica officinalis, Carissa spinarum*

Gall Tans: *Tamarix* spp.

DYES

Wood Dyes: Kutch dye (*Acacia catechu*) and other dyes from *Artocarpus heterophyllus, A. lakoocha, Pterocarpus santalinus, Caesalpinia sappan*

Bark Dyes: *Terminalia tomentosa, Acacia concinna, A. farnesiana, A. leucophloea, Alnus* spp, *Casuarina equisetifolia, Manilkara littoralis, Myrica esculenta, Ventilago madraspatana*

Flower and Fruit Dyes: *Mallotus philippensis, Woodfordia floribunda, Bixa orellana, Butea monosperma, Toona ciliata, Nyctanthes arbortristis, Mammea longifolia, Wrightia tinctoria, Crocus sativus*

Root Dyes: *Berberis aristata, Datisca cannabina, Morinda tinctoria, Punica granatum, Rubia cordifolia*

Leaf Dyes: *Indigofera tinctoria, Lawsonia inermis*

EDIBLE PLANT PRODUCTS

Edible Fruits: Chironji (*Buchanania lanzan*), Aonla (*Emblica officinalis*), Tamarind (*Tamarindus indica*), Bael (*Aegle marmelos*), Wood apple (*Feronia elephantum*), Barhal (*Artocarpus lakoocha*), Jamun (*Syzygium cuminii*), Custard apple (*Annona squamosa*), *Carissa opaca*, Karaunda (*Carisa carandas*), Ber (*Zizyphus jujuba*),

Edible Nuts: Cashew nut (*Anacardium occidentale*), Chilgoza (*Pinus gerardiana*), Walnut (*Juglans regia*), Almond (*Prunus dulcis*)

Edible Flowers: Mahua (*Madhuca indica*), *M. longifolia*, *Musa* spp, *Bombax ceiba*, *Bauhinia purpurea*, *Alangium salvifolium*, *Ficus glomerata*, *Sesbania grandiflora*, *Rhododenron arboreum*

Roots and Tubers: *Amorphophallus campanulatus*, *Dioscorea belophylla*, *D. oppositifolia*, *Ipomoea aquatica*, *Bamboos*

Leaves: *Agave americana*, *Aloe vera*, *Moringa oleifera*, *Antidesma diandrum*, *Fern species*

Fungi: *Agaricus campestris*, *Morchella esculenta*, *Volvaria terastius*, *Collybia albuminosa*

BAMBOOS, CANES AND PALMS

BAMBOO

Bamboo is a major non-wood forest product and wood substitute. There are about 1,250 species of bamboo in some 75 genera in the world. It is found in all regions of the world and plays an important economic and cultural role. India is the second richest country in bamboo genetic resources after China. These two countries together have more than half the total bamboo resources globally. It is reported that 136 species of bamboos are occurring in India. Used for housing, crafts, pulp, paper, panels, boards, veneer, flooring, roofing, fabrics, oil, gas and charcoal (for fuel and as an excellent natural absorbent), it is also a healthy vegetable (the bamboo shoot). Bamboo industries are now thriving in Asia and are quickly spreading across the continents to Africa and America. Because of its versatile uses, it is referred as "Green Gold" and "Cradle to Coffin Timber." Over 100 species of bamboos occur naturally in India. The major bamboos in India are *Bambusa bambos*, *B. tulda*, *B. nutans*, *B. vulgaris*, *B. pallida*, *B. polymorpha*, *Bambusa balcooa*, *Dendrocalamus strictus*, *D. hamiltonii*, *D. brandisii*, *D. giganteus*, *Melocanna baccifera*, *Ochlandra* spp., *Gigantochloa* spp., *Melocalamus* spp., *Phyllostachys* spp., *Sinarundinaria* spp.

CANES/ RATTAN

They are stems of climbing palms of genus *Calamus* consisting about 30species. Climbing often attain a length of even 100 m. They are usually cylindrical and have uniform thickness, solid and pale white to yellow in colour. Outer surface is hard, smooth and shinning. Mature culms are cut at the base and dried in sun or fire. Silica layer is removed by rubbing the freshly harvested cane over a knife or sharp piece of bamboo or rubbing with soapstone after steeping the cane in water. After desilication, cane is bleached by fumigation with burning sulphur. After bleaching, natural colour lustre is lost and it is again polished with woolen rag or soapstone. To get reddish colour, canes are treated by smoking over fire and then polished with coconut oil. Cane is used in variety of ways due to its pliability, strength and long length. It is used in walking sticks, polo sticks, umbrella handles, baskets and cane furniture. The important species of cane are *Calamus acanthospathus, C. guruba, C. tenuis, C. viminalis, C. rotang, C. andamanicus, C. brandisii, C. flagellum, C. gambeli, C. gracilis, C. latifolius, C. pulstris, C. rheedei, Daemonorops kurzanius, Plectocomia himalyana.* Among these, *Calamus tenuis* is the common cane of north India.

PALMS

Palms are monocots, included in the section of Angiosperms characterized by bearing a single seed leaf. Scientifically, palms are classified under the family Palmae (or Arecaceae) and are perennial, distinguished by having woody stems. Palms are important NWFP with numerous uses such as food, beverages, oil, fuel, feed, fertilizer, building material, furniture, games and toys, ornamental uses, medicinal uses etc. The important palm genus are *Areca, Arenga, Bentinckia, Corypha, Hyphaene, Licuala, Livistona, Loxococcus, Oncosperma, Phoenix, Pinanga, Wallichia, Caryota, Nypa* and *Borassus.*

35 FOREST POLICY AND LEGISLATIONS

WHAT IS POLICY?

Policy means "certain guidelines" for the achievement of "certain objectives." According to Webster's Dictionary, policy means "a settled course adopted and followed by a Government, institution, body or individuals." Kenneth E. Boulding, in his book *Principle of Economic Policy* defines the term policy which refers generally to the principles that govern the action directed towards given ends. The widely accepted definition for policy is, "Policy is a statement of intent" by the concerned organization.

DEFINITION

- A course of action adopted and pursued.
- Rational system based on deliberate aims and plans or as a consequence of a political activity arising from a series of decisions.
- Policy has to provide guidance and a sense of direction over a certain period of time in order to be useful.
- A negotiated agreement between Government and stakeholders.

ELEMENTS OF POLICY

- A set of aspirations, goals or objectives.
- An outline of a course of action to achieve them.
- It is a policy for the society not only for forest administration.
- Needs to be consistent with policies issued by other Govt. agencies and has to be in line with the country's forest-related commitments.

DYNAMICS OF POLICY

- Policy evolves over time as the actions of different bodies change in response to changing circumstances.

- Structure and contents is determined by needs of decision makers, others who develop, agree on and aim to use it.
- It resonates the history, culture, resources and aspirations.

LAW VS POLICY

- Policy can be adopted and amended through different procedures and bodies according to specific situation.
- Law can be adopted and amended only by parliament or head of state through procedures determined by legislation or constitution.
- Policy specifies visions, goals and how to reach them.
- Law specifies rights and duties.
- Policy can be general to adapt changing circumstances.
- Law must be specific to enable judicial dispute settlement.
- In Policy, soft mechanisms to deal with non-compliance.
- In Law, judicial powers to punish non-compliance.

FOREST POLICIES OF INDIA

THE 1894 FOREST POLICY

This is the Imperial Forest Policy and it proclaims that it wants to direct the policy of Forest Department such that it shall serve agricultural interests more directly than at present. It states that the sole object of administering forests is for public benefit. Public benefit is the prime importance and benefit is to the whole body of tax payers. Regulation of rights and restriction of privileges of users have been advocated. This is justified only when the advantage gained by public is great. Forest policy classified the forests of India into four major classes viz. preservation forests, commercial forests, minor forests and pasture lands.

The first class of forests was generally situated on hill slopes and was deemed essential for the protection of cultivated plains from damage caused by landslides and hill torrents. They served a conservation role for the benefit of agriculture in plains. The second class of forests included the vast reserves of valuable timber trees including *Cedrus deodara*, *Shorea robusta* and *Tectona grandis*. Driven by commercial interests, forest management measures were developed to promote natural regeneration of these first two species and artificial regeneration of the third. The third class of forest was to meet the people's requirements for timber, fuelwood, fodder, etc. in the form of minor forests. Fourth class of forests i.e. 'pastures and grazing grounds' to which certain restrictions were applied.

In case of protected forests, interests to be protected are great. It states that the protection from devastating action of hill torrents is necessary and production forests have to be managed on commercial lines. People should have full and easy satisfaction from these forests. The produce from these forests are not for free but at low and not at competitive prices. Considerations of forest income are to be subordinated to that satisfaction. It advocates minor forests should be protected against people's own improvidence. They should supply to the greatest advantage and convenience of the people. Distinction between income generation and convenience of local populace has to be made.

Formation of fuel and fodder reserves was recommended and periodical closures and strict preservation were advised in the policy. Pastures are forests only in name and it is a question of agency based on economy and expediency. When the tenure is ryotwari, trespass is forbidden. This is the only area from which requirements of grazing of local population can be met. They lead to rent seeking and extortion. Best is to lease or otherwise manage through the agency of community. Value to be fixed for lease through a moderate estimate since herds and flocks cannot exist without grazing.

In case of fuel and fodder reserves and grazing areas, if it is in midst of cultivated areas and if it is decided to retain them as forests it is better to classify them as protected forest rather than reserved forest. They provide for local requirements consistent with imperial interests and working plans should embody these principles. Concessions are made by way of privilege and grace and granted only for a limited period and they will be revised from time to time. The policy stipulates that forests which are the reservoirs of valuable timbers should be managed on commercial lines as a source of revenue to the States and that wherever an effective demand for culturable land exists that can only be supplied by a forest area, the land should ordinarily be relinquished without hesitation. People's interests were made subservient to the State's commercial interests with regard to forest during colonial rule.

THE 1952 FOREST POLICY

This is independent India's first Forest Policy. It talks about physical conditions amelioration, world wars and dependence on forests. Vital needs of the nation have been identified as balanced and complementary land use which produces most and deteriorate the least. It includes need to check denudation in mountains, erosion of banks of rivers which lead to ravine formation and checking invasion of sea sands. Establishing tree lands, increasing grazing supplies and small wood, more firewood supply, sustained supply of timber and other forest produce and maximum annual revenue in perpetuity are prime importance. This policy classified the forests of India into protection forests, national forests, village forests and tree lands. The focus and attention were on the kind and object of management since the country as a whole had a vast stake in conservation of forests irrespective of functions and ownership. They should be administered from national well being point of view.

Protection forests have to be preserved for climatic and physical considerations. National forests have to meet the needs of defense, communications and industry. Village forests have to provide for small timber, firewood and grazing. Tree lands outside the scope of ordinary forest management are essential for amelioration of physical conditions. It recommends that one third of the geographical area has to be under forests. It also recommends 60 per cent area in hills and 20 per cent in plains have to be under forests. Emphasis is on the need for special laws and creation of protected area network. Creation of central board of wildlife is another feature. This policy mentions elimination of natural enemies as the reason for increase in number of wild pigs, game and porcupine. It states that uncontrolled grazing is incompatible with scientific forestry. Grazing does take place and it has to be regulated as regards to time and place and also the number of cattle. Cheap grazing increases unhealthy cattle. Grazing is not a source of revenue to government, but to regulate grazing, a reasonable fee has to be levied. No grazing in regeneration areas and young plantations. Other salient points are as follows:

- Restriction on sheep and total exclusion of goats
- Damage to forests by shifting cultivation to be guarded against
- Regulating shifting cultivation through Taungya method
- Sustained yield principles to be followed in the management of all class of forests
- Only annual increment of forests needs to be cut
- Planned schemes for replacing inferior trees by species of commercial importance
- Permanent organization to deal with working plans compilation, revisions, etc. and also for research and statistics
- Efficiency of forest administration to be improved
- A common forest education system to develop common outlook and ensuring integrated policies throughout the country
- Decentralization of forest education to improve economy and efficiency and to ensure planned development of forests resources
- Contiguous states should cooperate to establish well equipped training schools
- Each State to maintain its own research organization
- Goodwill of people is necessary
- Intermediaries to be eliminated and to supplemented by forest cooperative societies
- Forestry to be free from vagaries of annual budget
- Creation of a sinking fund by investing a portion of forest revenue
- Rotations which yield large sized timber to be adopted rather than rotations which yield maximum rate of interest in forest investments

THE 1988 FOREST POLICY

The issue of revision of the Forest Policy formulated in 1952 had been under the active consideration of the Government. The serious depletion of forest resources due to biotic and industrial pressure and other reasons made the policy makers review the situation in the late eighties and evolve a new strategy for conservation of forests. In view of the social, economic and ecological importance of forests, considerable thought was given to the multifarious implication of the forest policy. Although the 1952 forest policy aimed at forest coverage of one-third of the total land area of the country, due to various constraints, this could not be attained. Rather extensive diversion of forest land had taken place for non-forest use. The genetic diversity had also been considerably affected by the destruction of flora and fauna. The Government had increasingly realized the great importance of forests in contributing to the ecological stability of the country. This had necessitated a re-examination of the forest policy and giving emphasis to the conservation and ecological aspects. The National Forest Policy was once again revised in 1988, which envisaged community involvement in the protection and regeneration of forests. It accorded highest priority to sustainable management of the forest resource.

OBJECTIVES

- Maintenance of environmental stability through preservation
- Restoration of Ecological balance
- Preserving remaining flora and fauna and its biodiversity and genetic resources
- Checking soil erosion and denudation in the catchment areas
- Checking the extension of sand dunes in desert areas
- Massive afforestation and social forestry programmes in denuded, degraded and unproductive lands
- Meeting the requirements of rural and tribal populations
- Increasing productivity of forests to meet essential national needs
- Efficient utilization of forest produce and maximizing substitution of wood
- Creating a massive people's movement with the involvement of women

ESSENTIALS OF FOREST MANAGEMENT

- Existing forests to be fully protected and productivity improved
- Diversion of good and productive agricultural lands to forestry to be discouraged
- Network of protected areas to be strengthened and extended
- Providing sufficient fodder, fuel, pasture to prevent depletion of forests beyond sustainable limits
- MFP to be protected, improved and production enhanced to generate employment and income

STRATEGY

- One-third of the geographical area to be under forest or tree cover
- Two-third of the area in hills under forest cover to ensure stability of fragile ecosystems
- Massive need-based and time bound afforestation with particular emphasis on fuelwood and fodder development in degraded and denuded lands
- Trees to be planted alongside of roads, railway lines, rivers and streams, canals and on other unutilised lands.
- Green belts to be raised in urban/industrial areas
- Village and community lands, including those on foreshores and environs of tanks, not required for other productive uses, should be taken up for the development of tree crops and fodder resources

- Laws to be modified to undertake tree-farming and grow fodder plants, grasses and legumes
- Degraded lands to be made available for tree farming on lease or on tree-patta scheme

MANAGEMENT OF STATE FORESTS

- All projects on steep slopes, catchments or geologically unstable terrain to be severely restricted
- Tropical rain/moist forests to be totally safeguarded
- No working in the forests without Government approved management plan
- Forest cover to be enhanced and productivity to be improved
- No such effort should entail clear felling or introduction of exotics

RIGHTS AND CONCESSIONS

- The rights and concessions, grazing, should always remain related to the carrying capacity of forests.
- Stall feeding of cattle to be encouraged
- The rights and concessions are primarily for the bonafide use of communities dependent on forests
- Tribals are the first charge of forest produce. The rights and concessions enjoyed by the tribals and other poor living within and near forests should be fully protected
- To change habits, substitute materials should be made available through convenient depots
- Wood substitution to be given importance

DIVERSION OF FOREST LANDS

- Most careful examination by the specialists from the standpoint of special and environmental costs and benefits
- Projects involving diversion of forest lands should at least provide funds for regeneration or compensatory afforestation
- No mining lease to be granted without proper mine management plan appraised from environmental angle
- Provision of corridors linking protected areas to maintain genetic continuity of wildlife

TRIBES AND FORESTS

- Contractors to be replaced by tribal cooperatives, labour cooperatives, government corporations, etc.

- Protection, regeneration and optimum collection of minor forest produce along with institutional arrangements for the marketing
- Development of forest villages on par with revenue villages
- Integrated area development programme to meet the needs of the tribal economy
- Alternative avenues of income with the right land use practices to contain shifting cultivation
- Introduction of improved agricultural practices
- Damaged area to be rehabilitated by social forestry and energy plantations

ENCROACHMENT AND GRAZING

- Effective action to prevent encroachment and no regularization of existing encroachments
- Improved and modem management practices to be adopted to deal with forest fires
- Grazing to be regulated by involving community
- Adequate grazing fee to be levied to discourage people maintaining non-essential livestock

FOREST BASED INDUSTRIES

- Industries should raise their own raw material
- No new forest-based industries unless assured availability of raw materials
- Natural forests not to be made available to industries for plantations
- Farmers to be encouraged to grow wood species required for industries
- Supply of forest produce to industry at concessional prices should cease
- Industries to be encouraged to use alternative raw materials and importing of wood and wood products

FORESTRY EXTENSION AND EDUCATION

- Involvement of educational institutions and mass media to inculcate the people a direct interest in forests, their development and conservation
- Farmers and interested people to be provided opportunities through institutions like Krishi Vigyan Kendras, Trainer's Training Centres to learn agri-silvicultural and silvicultural techniques
- Forestry to be recognized as a scientific discipline as well as a profession
- Curricula & PG courses in forestry to meet the manpower needs of the country
- Academic and professional qualifications in forestry to be given importance in IFS & SFS

FORESTRY RESEARCH AND FINANCIAL SUPPORT

- Increasing the productivity of wood and other forest produce by the application of modem scientific and technological methods
- Revegetation of barren/ marginal/waste/mined lands and watershed areas
- Effective conservation and management of existing forest resources
- Social foresry and rural development research
- Development of substitutes for wood and wood products
- Wildlife and management of PAs
- Forests not to be looked upon as a source of revenue
- A renewable natural resource
- Forests are national asset to be protected and enhanced for well-being of the people and nation

INDIAN FOREST ACT 1927

The Indian Forest Act was drafted first in 1865, placing most forests under state ownership. It was further revised in 1878 and consolidated in 1927. The National Forest Policy of 1894 evolved from the objectives of forest management as outlined in the Forest Act of 1865 and 1878. The Indian Forest Act, 1927 was "an Act to consolidate the law relating to forests, the transit of forest produce and the duty leviable on timber and other forest produce." The Act was divided into 13 chapters and 86 sections with a plethora of rules and regulations, penalties and procedures aimed at extending the Government's control over forests as well as diminishing the status of people's rights to forest use. This act applies to whole of areas in erstwhile states of Bihar, Bombay, Coorg, Delhi, Orissa, MP, UP Punjab and WB. The salient features of the act are presented below:

RESERVED FORESTS (Chapter II)

- *Power to reserve forests*: The State Government may constitute any forest-land or waste-land, which is the property of Government, or over which the Government has proprietary rights, a reserved forest (section 3)
- *Notification by State Government*: State Government shall issue a notification in the Official Gazette declaring to constitute a reserved forest, specifying the situation and limits of such land and appointing the Forest Settlement Officer (FSO) (section 4)
- After the notification under section 4, no right shall be acquired in or over the land except by succession or under a grant or contract (section 5)
- FSO to publish in the local vernacular about the situation and extent of proposed reserved forests, consequences of reservation, fixing a period not less than 3 months requiring every person claiming right to present in before FSO (section 6)
- FSO to inquire into claims at a convenient place (section 7)

- FSO has powers to enter upon any land, survey, demarcate and map the same and powers of a civil court in the trial of suits (section 8)
- Any right not claimed before final notification gets extinguished (section 9)
- Shifting cultivation if prevalent, claimants should produce any local rule or order under which practice is allowed. FSO to send his opinion to government and Government may permit it or prohibit in whole or part (section 10)
- In the case of a claim to a right in or over any land, right of way or right of pasture, or a right to forest produce or a water-course, the Forest Settlement-officer shall pass an order admitting or rejecting the same in whole or in part (section 11 and 12)
- FSO to pass such orders as will ensure the continued exercise of the rights so admitted (section 15)
- The State Government shall publish a notification in the Official Gazette declaring forest reserved (section 20)

ACTS PROHIBITED IN A RESERVED FOREST (section 26)

- Clearing of areas notified under sec 5
- Sets fire to a reserved forest, kindles any fire, or leaves any fire burning, in such manner as to endanger such a forest; or who, in a reserved forest
- Kindles, keeps or carries any fire except at such seasons as the Forest-officer may notify
- Trespasses or pastures cattle, or permits cattle to trespass
- Causes any damage by negligence in felling any tree or cutting or dragging any timber;
- Fells, girdles, lops, burns any tree or debarks
- Quarries stone, bums lime or charcoal, collects or removes any forest-produce
- Clears or breaks up any land for cultivation
- Hunts, shoots, fishes, poisons water or sets traps or snares
- Kills or catches elephants
- Punishment for above acts imprisonment up to 6 months fine up to Rs.500 or both
- In case of fire caused willfully or by negligence in addition to punishment, all rights of pasture or forest produce shall be suspended for such period by State Govt.
- State Government empowered to declare forest no longer reserved, but rights which have been extinguished shall not revive as a consequence (section 27)

VILLAGE FORESTS (Chapter III)

Formation of Village Forests: The State Government may assign to any village-community the rights of Government to or over any land which has been constituted a reserved forest and all forests so assigned shall be called village-forests. The State Government may make rules for regulating the management of village forests, for providing timber or other forest-produce or pasture and their duties for the protection and improvement of such forest (section 28).

PROTECTED FORESTS (Chapter IV)

The State Government may, by notification in the Official Gazette, declare any forest-land or waste-land which is not included in a reserved forest, a "protected forest"(section 29). The State Government may, by notification in the Official Gazette, declare any trees or class of trees in a protected forest to be reserved (section 30). The State Government may make rules to regulate the following matters (section 32),

- The cutting, sawing, conversion and removal of trees and timber and the collection, manufacture and removal of forest-produce, from protected forests
- Granting of licences to the inhabitants
- Examination of forest-produce passing out of such forests
- Clearing and breaking up of land for cultivation or other purposes
- Protection from fire of timber lying in such forests and of trees reserved under section 30
- Cutting of grass and pasturing of cattle
- Hunting, shooting, fishing, poisoning water and setting traps or snares in such forests and the killing or catching of elephants

Punishment for prohibited acts in protected forests includes imprisonment up to 6 months fine up to Rs.500 or both (section 33).

PROTECTION OF FOREST FOR SPECIAL PURPOSES (section 35)

The State Government may, by notification in the Official Gazette, regulate or prohibit in any forest or wasteland: breaking up or clearing of land for cultivation; pasturing of cattle; firing or clearing of the vegetation; when such regulation or prohibition appears necessary for any of the following purposes,

- Protection against storms, winds, rolling stones, floods and avalanches
- Preservation of the soil on the ridges and slopes and in the valleys of hilly tracts, the prevention of land slips or of the formation of ravines and torrents, or the protection of land against erosion, or the deposit thereon of sand, stones or gravel
- Maintenance of a water-supply in springs, rivers and tanks
- Protection of roads, bridges, railways and other lines of communication
- Preservation of the public health

PENALTIES AND PROCEDURE (Chapter IX; section 52 to 69)

- Seizure of forest produce together with all tools, boats, carts or cattle in case of forest-offence by any forest-officer or police-officer

- A seizure mark to be made on it and a report of seizure to be made to magistrate
- Property if it belongs to Government and offender unknown, it is sufficient to inform official superior (section 52)
- Ranger and above can release such seized property on execution of a bond for production of the property before the magistrate (section 53)
- Upon receipt of report, magistrate to expedite arrest and trial of offender and disposal of property (section 54)
- All timber or forest produce which is not the property of Govt. in respect of which a forest offence is committed, and all tools, boats, carts and cattle used for the same shall be liable for confiscation. Such confiscation may be in addition to any other punishment prescribed for such offence (section 55)
- On completion of trial, forest produce if it belongs to Govt. or if it is confiscated, be taken charge by forest officer or disposed as directed by court (section 56)
- When offender not known or cannot be found and if magistrate finds that an offence has been committed, order property to be confiscated and given to forest officer or a person whom he deems entitled. No such order shall be made until the expiration of one month from the date of seizing such property, or without hearing the person and evidences if any (section 57)
- Perishable property can be disposed by magistrate and proceeds dealt as property (section 58)
- Section 55, 56 and 57 orders can be appealed within 1 month to courts which otherwise entertain appeals from that court (section 59)
- When confiscation orders are issued and no appeals preferred or appellate court confirms order, such property to vest with government (section 60)
- Government can empower any officer to direct immediate release of any property seized under section 52 notwithstanding anything told before (section 61)
- For wrongful seizure of forest or police officer under pretence of seizing property liable for confiscation vexatiously and unnecessarily shall be punishable with imprisonment upto 6 months or Rs.500 or both (section 62)
- For counterfeiting government marks or alters defaces or obliterates any such mark and alters removes, destroys or defaces any boundary mark of any forest or wasteland to which this act applies is punishable with imprisonment upto two years or fine or both (section 63)
- Forest or police officer can arrest any person without warrant against whom a reasonable suspicion exists of his having been concerned with any forest offence punishable with imprisonment of one month and upwards. After arrest either to release on bond or produce before magistrate or to the officer in charge of the nearest police station without unnecessary delay (section 64)

- Any Forest-officer of a rank not inferior to that of a Ranger, may release a person arrested under the provisions of section 64, on his executing a bond before the Magistrate or the officer in charge of the nearest police station (section 65)

POWERS OF FOREST OFFICERS (Chapter XI; section 72 to 75)

The State Government may invest any Forest-officer with all or of the following powers.

- Enter any land, survey, demarcate and make map of the same
- Powers of a civil court to compel the attendance of witnesses and production of documents and material objects
- Power to issue a search-warrant under the Code of Criminal Procedure, 1898
- Power to hold inquiry into forest offences, to receive and record evidence
- Any such evidence taken in the course of proceedings shall be admissible
- In subsequent trial if it has been taken in the presence of accused person
- All Forest-officers shall be deemed be public servants within the meaning of the Indian Penal Code, 1860
- No suit shall lie against any public servant for actions done in good faith
- Forest officers shall not trade in timber or forest produce as principal/agent or be a party in lease of forest land or in contract for working any forest either inside or outside territories to which this act extends

WILDLIFE (PROTECTION) ACT 1972

Wildlife (Protection) Act 1972 was "an Act to provide for the protection of wild animals, birds and plants with a view to ensuring the ecological and environmental security of India." The Act was divided into seven chapters and 66 sections with a plethora of rules and regulations, penalties and procedures for protection of wildlife in India. It also has six schedules comprising various animals, birds and plants which give varying degrees of protection. The act has been amended from time to time e.g. wildlife protection (amendment) act 2002, wildlife protection (amendment) act 2006, wildlife protection (amendment) act 2010.The Salient features are as follows:

DEFINITIONS

- Animal includes pisces, amphibians, reptiles, birds and mammals and their young or eggs (in case of birds)
- Animal article means an article made from any captive animal or wild animal other than vermin
- Captive animal means any animal specified in Schedule I to IV which is captured or kept or bred in captivity
- Wild animal means any animal in Schedule I to IV found wild in nature
- Vermin means any wild animal in Schedule V

- Specified plants mean any plant in Schedule VI
- Trophy means whole or part of any captive or wild animal other than vermin kept or preserved naturally or artificially and includes rugs, skins, specimens, antlers, shell, horn, hair, feather, nail, tooth, tusk, musk, eggs, nests, honey comb etc.
- Uncured trophy is trophy that has not undergone taxidermy
- Wildlife includes any animal, aquatic or land vegetation which forms part of any habitat
- Hunting means killing, poisoning, capturing, coursing, snaring, trapping, driving, baiting any wild or captive animal and every attempt to do so. It also includes injuring, destroying or taking any part of body of any such animal and in case of wild birds or reptiles, damaging, injuring, destroying, disturbing of eggs or nests and also other chordates and invertebrates and their young ones and eggs.
- Zoo means an establishment, whether stationary or mobile, where captive animals are kept for exhibition to the public but does not include circus and an establishment of a licensed dealer in captive animals.
- Recognized zoo means a zoo recognized under section 38H
- Circus means an establishment, whether stationary or mobile, where animals are kept or used wholly or mainly for the purpose of performing tricks or manocuyres
- Protected area means a national park, a sanctuary, a conservation reserve or a community reserve notified under section 18, 35, 36A and 36C of the Act
- National park means an area declared, whether under section 35 or section 38, or deemed under sub-section (3) of section 66, to be declared as a National park
- Sanctuary means an area declared, whether under section 26A or section 38, deemed under sub-section (3) of section 66, to be declared as a wild life sanctuary

HUNTING OF WILD ANIMALS (Chapter III)

- Hunting of any wild animal specified in Schedule I to IV is prohibited except as provided under section 11 and 12 (Section 9)
- Hunting permitted when CWLW is satisfied that schedule I wild animal has become dangerous to life and property or diseased or disabled beyond recovery and may order in writing stating reasons and permit any person to hunt such animal. No wild animal shall be ordered to be killed unless the CWLW is satisfied that such animal cannot be captured, tranquilized, translocated or rehabilited in the wild (section 11; sub-sec 1a)
- Animals in schedule II to IV if it becomes dangerous to life or property, CWLW may order in writing stating reasons to permit any person to hunt such animal in a specified area (section 11; sub-sec 1b)
- Killing or wounding in good faith of any wild animal in defense of self or any other person is not an offence and such killed or wounded animal to be Government property (section 11; sub-sec 2&3)

- For the purpose of education, scientific research, scientific management including collection of specimens, collection of snake venom for preparation of life saving drugs, CWLW to grant a permit in writing stating reasons to any person on payment of requisite fees. For Schedule I animal Govt. of India previous permission and for other animals previous permission of State Govt. is needed

SANCTUARY (Section 18 to 34)

- The State Government may, by notification, declare its intention to constitute any area other than an area comprised within any reserve forest or the territorial waters as a sanctuary if it considers that such area is of adequate ecological, faunal, floral, geomorphological, natural or zoological significance, for the purpose of protecting, propagating or developing wild life or its environment. The notification will specify the situation and limits of such area (section 18)
- The State Government shall make alternate arrangements required for making available fuel, fodder and other forest produce to the persons affected, in terms of rights as per Government record (section 18A)
- The State Government to appoint an officer to act as Collector under the act within 90 days of coming into force of the Wild life (Preservation) Amendment Act 2002, or within 30 days of the issue of notification under section 18 (section 18B)
- Collector shall inquire and determine the existence nature and extent of rights of any person in and over the land (section 19)
- After the notification under section 18, no right shall be acquired in or over the land except by succession, testamentary or intestate (section 20)
- Collector to publish in the regional language about the situation and extent of sanctuary, fixing a period not less than 2 months requiring every person claiming right to present in before Collector (section 21)
- Collector to inquire into claims and existence of right and ascertain from the records of State Government (section 22)
- Collector has powers to enter upon any land, survey, demarcate and map the same and powers of a civil court for the trial of suits (section 23)
- Collector to pass an order admitting or rejecting the claim to a right in whole or in part (section 24)
- Collector to complete the land acquisition proceedings within a period of two years from the date of notification of declaration of sanctuary (section 25A)
- State Government to issue a notification specifying the limits of the area which shall be comprised within the sanctuary and declare that the said area shall be Sanctuary on and from such date as may be specified in the notification (section 26A)

NATIONAL PARK

- The State Government may, by notification, declare its intention to constitute any area other than an area comprised within any reserve forest or the territorial waters as a national park if it considers that such area is of adequate ecological, faunal, floral, geomorphological, natural or zoological significance, for the purpose of protecting, propagating or developing wild life or its environment (section 35; sub-sec 1)

- State Government to issue a notification specifying the limits of the area which shall be comprised within the national park and declare that the said area shall be National Park on and from such date as may be specified in the notification (section 35; sub-sec 4)

CONSERVATION RESERVE

- State Government may declare, after consultations with the local communities, any area owned by the Government, particularly the areas adjacent to the national parks and sanctuaries and those areas which link one protected area with another, as a Conservation Reserve for protecting landscapes, seascapes, flora and fauna and their habitat. Where the conservation reserve includes any land owned by the Central Government, its prior concurrence shall be obtained before marking such declaration (section 36A)

COMMUNITY RESERVE

- The State Government may, where the community or an individual has volunteered to conserve wild life and its habitat, declare any private or community land not comprised within a national park, sanctuary or a conservation reserve, as a Community Reserve for protecting fauna, flora and traditional or cultural conservation values and practices (section 36C)

SANCTUARY AND NATIONAL PARK DECLARED BY CENTRAL GOVERNMENT

- Where the State Government leases or otherwise transfers any area under its control, not being an area within a sanctuary or national park, to the Central Government, The Central Government may, if it is satisfied under relevant sections of the sanctuary or national park related to conditions fulfilled in relation to the area so transferred to it, declare such area by notification to be a sanctuary or national park (section 38)

TRADE OR COMMERCE IN WILD ANIMALS, ANIMAL ARTICLES AND TROPHIES

- Every wild animal other than vermin which is hunted or dead or killed, animal article, trophy, uncured trophy, meat from any wild animal in respect of which an offence has been committed, vehicle, vessel, weapon, trap or tool used for committing offence and seized shall be the property of Govt. Such property has to be handed over to authorized officer or nearest police station.
- Any person who obtains the possession, by any means, shall within 48 hours from obtaining such possession, make a report to the nearest police station or authorized officer and shall hand over such property to the officer-in-charge

- No person without previous permission of CWLW or authorized officers, can acquire, keep, transfer, destroy or damage such property (section 39)

PREVENTION AND DETECTION OF OFFENCES

- Any forest officer or police officer not below the rank of sub-inspector believes that a person has committed an offence under this act and require any such person to produce for inspection any captive or wild animal, meat or animal article, trophy in his control, custody and possession (section 50)
- May stop any vehicle or vessel to conduct search or enquiry or enter upon and search any premises, land, baggage or other things in his possession (section 50)
- May seize any captive animal, wild animal, meat, trophy or uncured trophy together with any trap, tool, vehicle, vessel or weapon used for committing such offence
- May arrest him without warrant and detain him (section 50)
- Power to any forest officer not below the rank of ACF to entrust such seized property on bond before Magistrate
- Seizures to be produced before magistrate and failure to produce is an offence
- Investigating officers vested with powers of civil court and any evidence recorded if taken in the presence of accused is admissible (section 50)
- Offences punishable with imprisonment upto 3 years or fine upto Rs. 25,000 or both (section 51; sub-sec1)
- In case of Schedule I or part II of Schedule II animal or hunting in a PA or altering boundaries of a PA, imprisonment not less than three years and may extend upto seven years and fine not less than Rs. 10,000 (section 51; sub-sec1)
- For second or subsequent offence, imprisonment not less than two years and extend upto six years and fine not less than Rs. 10,000 (section 51; sub-sec1)
- For contravention of Chapter V-A provisions, imprisonment not less than three years and may extend upto seven years and fine not less than Rs. 10,000 (section 51; sub-sec 1A)
- For contravention of Section 38J, imprisonment upto 6 months and fine upto Rs.2,000 or both (section 51; sub-sec 1B)
- Section 360 of Code of Criminal Procedure or Probation of Offenders Act not to apply for hunting in a PA under provisions of Chapter V-A unless the person is less than 18 years
- Hunting, altering boundary of national park or sanctuary or offence related to Schedule I or part II of Schedule II involve no bail not withstanding anything in Code of Criminal Procedure (section 51A)
- Attempts to contravene and abets the contravention of any provisions of this act are an offence (section 52)

- Vexatious and unnecessary seizure punishable with imprisonment upto six months or with fine upto Rs. 500 or with both (section 53)
- Any officer not below Asst. Director of Wild Life Preservation (Central Govt.) or Deputy Conservator of Forests (State Govt.) can accept payment of money as compensation for offence
- Suspect to be discharged from custody and no further proceedings in respect of offence
- Sum of money accepted in no case to be more than Rs. 25,000 (section 54)
- No offence for which a minimum imprisonment is prescribed in section 51 shall be compounded

FOREST (CONSERVATION) ACT 1980

Forest conservation Act is a regulatory act and not prohibitory. The Act is an interface between conservation and development. In 1980, the Forest (Conservation) Act was enacted for providing a higher level of protection to the forests and to regulate diversion of forest lands for non forestry purposes. It permits judicious and regulated use of forest land for non-forestry purposes. It extends to the whole of India except the State of Jammu and Kashmir and came into force on 25 October 1980. The salient features are:

RESTRICTION ON THE DERESERVATION OF FORESTS (section 2)

No State Government or other authority shall make, except with the prior approval of the Central Government, any order directing-

(i) Dereserve any reserved forests

(ii) Divert any forest land or any portion for any non-forest purpose

(iii) Lease forest land or any portion to any private person or to any authority, corporation, agency or any other organization not owned, managed or controlled by Government

(iv) Clear trees which have grown naturally in forest land or any for the purpose of reafforestation

Non-forest purpose means the breaking up or clearing of any forest land or portion for the cultivation of tea, coffee, spices, rubber, palms, oil-bearing plants, horticultural crops or medicinal plants and any purpose other than reafforestation. It does not include any work relating or ancillary to conservation, development and management of forests and wildlife, namely, the establishment of check-posts, fire lines, wireless communications and construction of fencing, bridges and culverts, dams, waterholes, trench marks, boundary marks, pipelines or other like purposes.

CONSTITUTION OF ADVISORY COMMITTEE (section 3)

The Central Government may constitute a Committee consisting of number of persons to advise the Government with regard to the grant of approval under section 2 and any other matter connected with the conservation of forests which may be referred by the Central Government.

PENALTY FOR CONTRAVENTIONS AND OFFENCES

Whoever contravenes or abets the contravention of any of the provisions of section 2, shall be punishable with simple imprisonment upto 15 days (section 3A). Where any offence under this Act has been committed by any department of Government, the head of the department or by any authority, every person who was responsible to the authority for the conduct of the business of the authority as well as the authority shall be deemed to be guilty of the offence and shall be liable to be proceeded against and punished accordingly (section 3B).

POWER TO MAKE RULES

The Central Government may, by notification in the Official Gazette, makes rules for carrying out the provisions of this Act. Every rule made under this Act to be laid before each House of Parliament (section 4).

SCHEDULED TRIBES AND OTHER TRADITIONAL FOREST DWELLERS (RECOGNITION OF FOREST RIGHTS) ACT or FOREST RIGHTS ACT 2006

This act has finally paved the way to undo the injustice to tribals and other forest dwellers. The Act recognizes and vests the forest rights and occupation in forest land in forest dwelling Scheduled Tribes and other traditional forest dwellers who have been residing in such forests for generations but whose rights could not be recorded. After the enactment of the Act, now tribals and other forest dwellers will have the right to cultivate forest land to the extent under occupation, the right to own, collect, use and dispose of minor forest produce, rights inside forests which are traditional and customary like grazing.

FOREST RIGHTS

Forest rights include

- Right to hold and live in forest land
- Community rights
- Right of ownership access to collect, use and dispose off NTFP
- Right over disputed land
- Right to convert forest villages into revenue villages
- Right to protect and manage community forest reserves
- Right to access biodiversity and intellectual property rights
- Right to customarily enjoyed traditions excluding hunting or trapping or extracting part of body of any species of wild animal
- Right to *in situ* rehabilitation

SALIENT FEATURES OF THE ACT (Source: Ministry of Tribal Affairs)

- Tribals who have been living in and depending on forests, for their livelihood prior to 13 December 2005, and other traditional forest dwellers who have been living in and depending on forests for their livelihood, for three generations prior to 13 December 2005 will have these rights. A generation for this purpose would mean a period comprising of 25 years.
- The cut off date for recognition and vesting of forest rights under the Act will be 13 December 2005.
- The Act provides for the ceiling of occupation of forest land for purposes of recognition of forest rights to the area under actual occupation and in no case exceeding an area of four hectares.
- The Act provides for conferring rights in the National Parks and Sanctuaries also, renamed as critical wildlife habitat on regular basis.
- The Act provides for the right to hold and live in the forestland under the individual or common occupation for habitation or for self-cultivation for livelihood by a member or members of a forest dwelling Scheduled Tribe or other traditional forest dwellers.
- The Act recognizes the right of ownership access to collect, use and dispose of minor forest produce which has been traditionally collected within or outside village boundaries.
- The Act has defined the term "minor forest produce" to include all non-timber forest produce of plant origin, including bamboo, brush wood, stumps, cane, tussar, cocoons, honey, wax, lac, tendu or kendu leaves, medicinal plants and herbs, roots, tubers and the like.
- The Act recognizes the right to in situ rehabilitation including alternative land in cases where the Scheduled Tribes and other traditional forest dwellers have been illegally evicted or displaced from forest land of any description without receiving their legal entitlement to rehabilitation prior to 13 December 2005.
- The Act provides for the forest right relating to Government providing for diversion of forest land for the purpose of schools, hospitals, anganwadis, drinking water supply and water pipelines, roads, electric and telecommunication lines, etc.
- The rights conferred under the Act shall be heritable but not alienable or transferable and shall be registered jointly in the name of both the spouses in the case of married persons and in the name of the single head, in the case of a household headed by a single person and in the absence of a direct heir, the heritable right shall pass on to the next of kin.
- The Act provides that no member of a forest dwelling Scheduled Tribe or other traditional forest dwellers shall be evicted or removed from forest land under his occupation till the recognition and verification procedure is completed.
- As per the Act, the Gram Sabha has been designated as the competent authority for initiating the process of determining the nature and extent of individual or

community forest rights or both that may be given to the forest dwelling Scheduled Tribes and other traditional forest dwellers.

INTERNATIONAL TROPICAL TIMBER AGREEMENT

The International Tropical Timber Agreement 1983 (ITTA 1983) is an agreement to provide an effective framework for cooperation between tropical timber producers and consumers and to encourage the development of national policies aimed at sustainable utilization and conservation of tropical forests and their genetic resources. It was signed in Geneva on 18 November 1983 and entered into force on 1 April 1985. Fifty eight parties signed up to the agreement.

International Tropical Timber Agreement 1994 (ITTA 1994) is a multilateral agreement or treaty signed on 26 January 1994 in Geneva and came into force on 01 January 1997. It replaced the International Tropical Timber Agreement, 1983. With an initial duration covering up to 01 January 2001, the Agreement had twice been extended by periods of 3 years: from 01 January 2001 to 31 December 2003 and from 01 January 2004 to 31 December 2006. Its 59 members represented about 80 per cent of the world's tropical forests and 90 per cent of the global tropical timber trade. International Tropical Timber Organization (ITTO) was established by the ITTA 1994. ITTO is an intergovernmental organization promoting the conservation and sustainable management, use and trade of tropical forest resources.

The International Tropical Timber Agreement 2006 (ITTA 2006) is the multilateral agreement or treaty that governs the International Tropical Timber Organization. ITTA 2006 was signed on 27 January 2006 in Geneva and came into force on 11 December 2011. The ITTA 1994 has been superseded by the International Tropical Timber Agreement 2006. The contracting parties are European Community, Australia, Belgium, Cameroon, Central African Republic, China, Colombia, Congo, Czech Republic, Ecuador, Finland, Ghana, Greece, Guatemala, Honduras, India, Indonesia, Italy, Japan, Lithuania, Madagascar, Malaysia, Mexico, Netherlands, New Zealand, Norway, Panama, Peru, Philippines, Portugal, Romania, Spain, Sweden, Switzerland, Togo, United Kingdom, United States of America. The main objectives of ITTA 2006 are:

- To promote the expansion and diversification of international trade in tropical timber from sustainable managed and legally harvested forests
- To promote the sustainable management of tropical timber producing forests
- To provide an effective framework for consultation, international cooperation and the development of timber economy policies
- To contribute to the process of sustainable development
- To promote the expansion and diversification of international trade in tropical timber from sustainable sources
- To improve forest management and the efficiency of wood utilization
- To provide members with new financial resources and to encourage information-sharing on the international timber market, through the International Tropical Timber Organization (ITTO)

36 FOREST TRIBES IN INDIA

INTRODUCTION

"Indigenous People" is a collective term referring to the native people who have close ties with the land they inhabit but each person (or tribe) has their uniqueness. They are groups or tribes with different historical and social backgrounds, language families, racial stocks and religious moulds. Indigenous peoples constitute about five per cent of the world's population. The United Nations Permanent Forum on Indigenous Issues (UNPFII) reported that about 300 to 370 million people belong to the world's indigenous groups. There are around 4,000 languages used by indigenous people. The International Fund for Agricultural Development (IFAD) reported that there are more than 5,000 different groups of indigenous people living in more than 70 countries. Indigenous people inhabit in every region of the world, but about 70 percent of them are concentrated in Asia. Two centuries ago, indigenous people could be found in most parts of the world. However, nowadays, they have legal rights to use only about six per cent of the planet's land and in many cases their rights are partial or disqualified.

The term "Adivasis" (original inhabitants) refers to the Indigenous People of India who possess distinct identities and cultures often linked to certain territories. The term is derived from the Hindi word "adi" which means "of earliest times" or "from the beginning" and "vasi" means inhabitant or resident, and it was coined in the 1930s. Officially they are termed as "Scheduled Tribes" (STs) which is a legal and constitutional term specifying the tribal groups with distinctive cultures, geographical isolation, shyness of contact with the community at large, traditional beliefs and practices such as indigenous arts of dance and music, unique way of life, nature worshipping and living in unreachable areas. STs also refer to the groups living in unreachable areas with social and economical backwardness and highly depending on forest resources.

Forest tribes in India have their own tradition, cultural heritage and folklore. They are living in this country from the time immemorial and also living in harmony with nature. In India, the life and economy of the tribal people are intimately connected with the forests. Majority of the tribal population in India actually lives inside the forests and make a living out of the forest produce collected by them, mainly edible roots and tubers and by hunting small animals. These primitive hunting and gathering tribal communities who are numerically very small live mostly in Odisha, Bihar, Jharkhand, Madhya Pradesh, Chhattisgarh, Andhra Pradesh, Karnataka and northeastern States. According to W.S Perry, tribe is a common group speaking a common language or dialect and inhabiting common territory. Tribes are social group with territorial affiliations, endogamus, with no specialization of functions, ruled by tribal officers, heredity or otherwise, united in language or dialect, recognizing social distance from other tribes, following tribal traditions, beliefs and costumes, illiberal of naturalization of ideas from alien sources. Normally, forest tribes in India are peace loving people. Forests have played a vital role in the socioeconomic and cultural life of the tribal people of India.The tribal community and their habitats constitute very significant parts of our country.

Tribal people constitute 8.2 per cent of the nation's total population and it is over 84 million people according to the 2001 census (Ministry of Tribal Affairs 2012). In the northeastern states of Arunachal Pradesh, Meghalaya, Mizoram and Nagaland, more than 90 per cent of the population of State is tribal. However, in the remaining northeastern states of Assam, Manipur, Sikkim and Tripura, tribal people form between 20 and 30 per cent of the population. Tribal communities have variety and complexity in the belief systems as well as religious practices. The survival of forest tribes is dependent on minor forest produce like mahua flowers, sal seeds, sal and tendu leaves, edible roots, tubers, bamboo and wild fruits, etc. Tribals in Odisha, Madhya Pradesh, Chhattisgarh, Bihar and Jharkhand indicate that over 80 per cent of the forest-dwellers collect 25-50 per cent of their food from the forests. Thus for the hunting and gathering tribal communities like Birhor, Korwa, Pahariya, Asur, Birjia, Chenchu, Kadar, Paliyan etc., minor forest produce is the major support for their survival.

Many Mongoloid tribes living in the northeastern States of India e.g. the Naga-Kuki tribes make their living by practising shifting cultivation (jhum cultivation) within the forest area. The Maria of Baster and the Gonds of Odisha calls the shifting cultivation penda. Shifting cultivation degrades the forest to some extent but whether this is wasteful and totally avoidable in the present context is debatable, because the very existence of these tribals depends on shifting cultivation. Before the advent of the British rule in India, the forest-dwellers and other indigenous communities enjoyed freedom to use forest or exploit forest resources for their livelihood. The British colonial administration realized the commercial value of forests and began to use them to augment revenue and in the process tried to regulate the rights of the forest dwellers and other indigenous people over forests.The recent forest policies made a greater impact on the forest in India. The modern culture and changing economic situation made a diifent kind of scenario in the tribal life (Sociology Guide 2012).

CONCEPT AND DEFINITION OF FOREST TRIBES

The word tribe has several meanings in deferent languages, in Greek it means fraternities and in English it is referred as communities having common surname. The word tribal or adivasi brings to our mind a picture of men and women in half-cloth, with arrows and spears in their hands, feathers in their heads, and speaking an unintelligible language, their lives often combined with myths of savagery and cannibalism. However, any person having visited a tribal village will be surprised and thrilled to see a community living close to nature, peace-loving, equitable and with advanced cultural/social forms. Regarding origin of tribes, various theories have been found from Vedic literature to the modern works. Different scholars have attempted to provide definitions of a tribe. Some of the definitions provided by eminent scholars are as follows:

- *The Imperial Gazetteer of India* (1911): A tribe is a collection of families bearing a common name, speaking a common dialect, occupying or professing to occupy a common territory and is not usually endogamous though originally it might have been so.
- *D.N. Majumdar*: A tribe is a collection of families or group of families bearing a common name, members of which occupy the same territory, speak the same language and observe certain taboos regarding marriage, profession or occupation and have developed a well-assessed system of reciprocity and mutuality of obligations.
- *Gillin and Gillin*: Any pre-literate local group may be termed as tribe, whose members reside in a common area, speak a common language and have common culture.
- *Dr.W.H.R. Rivers*: Tribe is a simple type of social group whose members speak a common dialect and work together at the time of war.
- *R.N. Mukherjee*: A tribe is that human group, whose members have common interest, territory, language, social law and economic occupation.

Our knowledge about the tribals is very limited, leading us to believe many myths at the cost of their dignity. Even when majority of the communities in the world keep changing their life-styles, compete with each other and develope materialistic instincts to keep pace with the progress of the world, there are communities still living in line with their traditional values, customs and beliefs.The exploitative mindset of the mainstream society have made these communities recede often into forests and high-altitude mountains, where they could continue to live in peace with nature and their unpolluted surroundings. As the so-called civilized communities of the mainstream society neither could comprehend the values and ideals of these communities nor have the patience to understand their lifestyles and the mainstream world have branded them variously as natives, uncivilized people, aboriginals, adivasis, tribals, indigenous people etc. In India, we mostly refer them as adivasis. In spite of the merciless treatment by the civilized men and the socio-economic perils faced by these communities all over the world, the tribals continue to live in the continents of Africa, Asia, North and South America and Australia.

RACES OF INDIAN TRIBES

Tribes in India have got both Pre-Aryan and Aryan racial affinities and origin and these people have been living since time immemorial and they are distributed throughout India. According to A.W. Green, a race is defined as a large, biological human grouping with a number of distinctive and inherited characteristics which vary within a certain range. In general, a race is a group of intermarrying people who are born of a common ancestor and possess similar physical traits and similar feelings. Most of them lay stress on certain physical characteristics or traits as determinants of a race. Some of these physical traits can be listed as follows:

- The form, colour or distribution of hairs on head, face or body.

- The body concentration, size of chest and shoulders.
- Form of skull, its height and width.
- Facial characteristics such as nasal form, lip form, eyelid, cheek bones, ear, jaws, etc.
- Complexion of skin and eye.
- Length of arms and legs.
- Blood type.

The most common races are:

(1) The Negritor or Dravidian strains

(2) The proto-Australoid or Austric strain

(3) The Mangoloid strain (further divided into two-sub types)

(i) Palaeo-Mangoloids

(a) Long-headed

(b) Broad-headed

(ii) Tibeto-Mangoloids

The Negrito strain is regarded as the earliest strain found in India and elements of this strain are found among the tribes of southern India like the todas, kodra, irula, soligas, jenukurbas, paniyana and bettakurubas.

HISTORY OF FOREST TRIBES IN INDIA

The Indian tribes may be said to be the earliest of the present inhabitants of the country. In the ancient literatures there have been references of 'Janas' or communities like Savaras, Kallutas, Bhils, Kinnouris, Soligas, Todas, etc. Many tribal communities of today still bear similar names and the communities cited above may be the ancestors of such tribes. An analysis of the long list of 'Janas' in the religious and secular literatures of the early times makes it evident that there was hardly any distinction made, until very late in history, between the then tribal communities cited above the other communities like Gandhara, Kambojas, Kasis, Kosalas, Kurus, Panchals etc. The sources suggest that these communities belonged to a different ethnic category with their own social-religious and economic organizations different from the traditional Hindu one. They fell outside the four-fold 'Jati' system of the society. The tribals may thus be said to be the indigenous people in the sense that they had long settled in different parts of the Indian subcontinents before the Aryans invaded India from the northwest. The Aryans with their superior socio-economic structures occupied the fertile plains of Indus and Ganga central India. Some of the nomadic pastoral tribes adapted to the Aryans ways of life and gradually evolved a superior organization and a superior techno-economy matching with the Aryans.

Aryans were not the only people to invade India. Dravidians came from west, Mongoloids from northeast, of course, at different times. This process continued inexorably for centuries and millennia. Thus the tribals in India are found mostly in the sub-montane regions of the north and northeast India, on the continuous hilly tracts of central India with lateral extensions in Aravalli, Kaimur and Chhattisgarh hills and Eastern Ghats. Down South isolated concentrations are met within the Nilgiris, Annamalon and Cardomon hills. Thus topography can be said to have played a determining role in the history and settlements of tribes in India. Their movements only speak of their history which was full of retreats. The movements of Santhalas, Mundas and Oraons of Chhotanagpur, Saoras of central and eastern India, Bhils of central and western India, Kurumbas of central Himalayas may be cited as some examples. The Indian tribes, thus for centuries, persisted in isolation with little or no contact with the population of open plains and centers of civilization. According to archeological department of Karnataka state, megalathical burial sites are excavated at BRhills and it has been revealed that in and around Biligiriranagan hills sanctuary, even during 800 BC (Pre-historical times) Soliga tribes were living. These burial sites are locally called *Morea Kalli* (Veena et al 2006). According to one such oral tradition, Garos tribes of Meghalaya, first immigrated to Garo hills from Tibet (referred to as Tibotgre) around 400 BC under the leadership of Jappa Jalimpa, crossing the Brahmaputra river and tentatively settling in the river valley. It is said that they were later driven up into the hills by other ethnic groups in and around the Brahmaputra river till they finally settled the erstwhile uninhabited Garo Hills. The Garo had the reputation of being fierce headhunters, the social status of a man being decided by the number of heads he owned. In December 1872, the British sent out battalions to Garo Hills to establish their control in the region. Most of the earlier literatures indicate that adivasis known as Janas, Savaras, Kollas, Kallutas, Bhils, Soligas, Kinnories, Kurumbas and Mundas are the tribal community from South East Asia. There are evidences that they also were in the Pre-British times. Munda tribe mainly inhabits the region of Jharkhand and the various corners of the states like West Bengal, Chhatisgarh, Odisha and Bihar.

CHARACTERISTICS OF THE TRIBE

The geneal charactersic features of forest tribes of India are as follows:

- Tribe is a group of families with specific name and members speak common language or dialect
- Members of the tribe reside in a common territory and observe taboos related to marriage
- Members of a tribe have a common occupation and have well developed system of reciprocal exchange.
- A tribe has a common culture and members work together at the time of war.
- Tribes are usually endogamous groups, though certain tribes have become exogamous after coming into contact with external world. Within the tribes, there are sibs or clans which may be exogamous. The reason for endogamy could be to keep the territory intact.
- Tribes are generally homogenous, cultural, social and political organizations which have their own laws and justice, customs, rituals, rites etc.
- Each tribe is headed by a tribal chief or headman who commands the tribe and whose decisions are final. The selection to this office is either hereditary or tribal of

strength. This kind of organization is required for purposes of protection from external forces and also to inculcate discipline among members.

- Most of the tribal languages do not have scripts.
- Each tribe claims descent from a common ancestor which may be real or mythical. This common ancestry serves to unite them in strong bonds.
- Tribes have territorial identity, i.e. each tribe occupies a common and well defined territory which serves as their homeland.
- Because of the nature of their life, tribes do not adopt easily to changing situations. It is generally difficult to bring changes in their life style, cultural and social customs, etc.
- Tribes believe in taboos which are certain forbidden things or restrictions, generally communicated through verbal do's and don'ts.
- Tribes believe in totems which are certain objects having mystic relationship with members of tribe.
- Tribes believe in certain myths, a rudimentary type of religion and magic.

The Constitution of India, Article 366 (25) defines Scheduled Tribes as "such tribes or tribal communities or part of or groups within such tribes or tribal communities as are deemed under Article 342 to the scheduled Tribes (STs) for the purposes of this Constitution." According to Article 342 of the Constitution, the Scheduled Tribes are the tribes or tribal communities or part of or groups within these tribes and tribal communities which have been declared as such by the President through a public notification. In Article 342, the procedure to be followed for specification of a scheduled tribe is prescribed. However, it does not contain the criterion for the specification of any community as scheduled tribe. An often used criterion is based on attributes such as:

- *Geographical isolation*: They live in cloistered, exclusive, remote and inhospitable areas such as hills and forests.
- *Backwardness*: Their livelihood is based on primitive agriculture, a low-value closed economy with a low level of technology that leads to their poverty. They have low levels of literacy and health.
- *Distinctive culture, language and religion*: Communities have developed their own distinctive culture, language and religion.
- *Shyness of contact*: They have a marginal degree of contact with other cultures and people.
- *Indications ofPrimitive Traits*

CLASSIFICATION OF INDIAN FOREST TRIBES

There are 663 tribal communities living in India as per the notified Schedule under Article 342 of the Constitution of India.The Indian forest tribes are classified into three categories: primitive tribes, nomadic tribes and semi-nomadic tribes.

PRIMITIVE TRIBAL GROUPS (PTGs) or PARTICULARLY VULNERABLE TRIBAL GROUPS (PVTGs)

Among the Scheduled Tribes, there are some tribes who are more backward than others. They have been classified as Primitive Tribes or PVTGs. They are characterized by low level of literacy, declining or stagnant population, pre-agricultural level of technology and economic backwardness. PVTGs are spread over 17 States / UTs and are 75 communities in number. Their total population is 24,12,664 as per 1991 census. In order to develop them to the level of other scheduled tribe communities, a central sector scheme was introduced in 1998-99. The scheme is very flexible and provides for integrated development of these communities through convergence of education, health and income generating schemes. PVTG women are extremely fond of ornaments and decorate themselves with tattoo marks on their body. The women wear metal collars, necklaces made of coins, todas etc., made of German silver. Simplicity, beauty and earthliness are the dominant factors of the arts. To acquaint this famous art to the public, Government of India, Ministry of Tribal Affairs for the first time made efforts to publish the Calendars highlighting the rich cultural heritage of PVTGs of India. Major primitive tribes are extremely poor and their population is also stagnant. Some of the major primitive forest tribes (75) living in various States (Chatrubuj Shau 1998) are presented below:

Table 36.1: Major Primitive Forest Tribes of India

S No	States	Name of the Tribes
1	Andra Pradesh	1.Bodo Gadaba, 2.Bondo Poraja, 3.Chenchu, 4.Dongaria Khonds, 5.Gutob Gadaba, 6.Khond Poraja, 7.Kolam, 8.Kondareddis, 9.Konda Savaras, 10.KuttiyaKondhs, 11.Parangiperj, 12.Thoti
2	Bihar and Jharkhand	13.Asur, 14.Birhor, 15.Birjia, 16.Hill Kharia, 17.Korwa, 18.Mal Pharia, 19.Paharias, 20.Sauria Paraharia
3	Gujarat	22.Kathodi, 23.Kotwalia, 24.Padhar, 25.Siddi, 26.Kolgha
4	Karnataka	27.Jenu Kuruba, 28.Koraga
5	Kerala	29.Cholanaikan (a section of Kattunaikan), 30.Kadar, 31.Kattunaikan, 32.Kurumbas, 33.Koraga
6	Madhya Pradesh	34.Abujh Maria, 35.Baiga, 36.Bharia, 37.Hill Korwa, 38.Kamar
7	Chhattisgarh	39.Sahariya, 40.Birhor
8	Maharashtra	41.Katkaria (Kathodi), 42.Kolam, 43.Maria Gond, 44.Maram Naga
9	Manipur	45.Birhor
10	Odisha	46.Didayi, 47.Bondo, 48.Dongria-Khond, 49.Juang, 50.Kharia, 51.Kutia Khond, 52.Lanjia Saura, 53.Lodha, 54.Mankirdia, 55.Paudi Bhuyan, 56.Saura, 57.Chuktia Bhunjia
11	Rajasthan	58.Seharias
12	Tamil Nadu	59.Kattunaikans, 60.Kotas, 61.Kurumbas, 62.Irulas, 63.Paniyans, 64.Todas
13	Tripura	65.Reangs
14	Uttar Pradesh	66.Buxas, 67.Rajis

15	West Bengal	68.Birhor, 69.Lodha, 70.Totos
17	Andaman & Nicobar Islands	71.Reat Andamanese, 72.Jarawas, 73.Onges, 74.Sentenelese, 75.Shompens

NOMADIC TRIBES

The word nomad has been derived from the Greek word 'nemo' which means to pasture. The dictionary of anthropology defines nomad as a person who lives completely of his flock and does not domicile. Nomadism is said to be seasonal or cyclic movements of groups of people for sustenance. They are the persons who rely largely on the live stock as a source of livelihood. They do not settle at any location for prolonged periods. This type is most frequently among the forest tribes of India. The major nomadic forest tribes are Gujjars, Kanjars, Birhors, Bhoi and Lambanii.

SEMI-NOMADIC TRIBES

Common semi-nomadick tribes are Akkipikki, Soligas and Bettakurubas and several other tribes of western and northern parts of India.

DEMOGRAPHY AND DISTRIBUTION OF FOREST TRIBES IN INDIA

India is one of the few nations in the world with a thriving tribal population in different parts of the country. There are 663 different tribal communities spread all over India. As per official data, (only) 258 tribal communities speaking about 106 different languages are notified as Scheduled Tribes. Out of total tribal population of 8.43 crore in the country, 91.7 per cent lives in rural areas and 8.3 per cent in urban areas. The population of tribes had grown at the growth rate of 24.45 per cent during 1991-2001. About 82 per cent of the total tribal population is concentrated in central and western parts of the country, whereas only 11 percent is dispersed in small pockets in the southern states.

Table 36.2: State-Wise Demographic Status of Forest Tribes in India (Ministry of Tribal Affairs 2010)

S No	States	Tribal Population (2001 Census)	% of Tribes in the State to Total State Population	% of Tribes in the State to Total ST Population of India in 2001
1	Andhra Pradesh	50,24,104	6.6	5.96
2	Arunachal Pradesh	7,05,158	64.2	0.84
3	Assam	33,08,570	12.4	3.92
4	Bihar	7,58,351	0.9	0.9
5	Chhattisgarh	66,16,596	31.8	7.85
6	Goa	566	-	0.001
7	Gujarat	74,81,160	14.8	8.87
8	Haryana	-	-	-
9	Himachal Pradesh	2,44,587	4.0	0.29

10	J&K	11,05,979	10.9	1.31
11	Jharkhand	70,87,068	26.3	8.4
12	Karnataka	34,63,986	6.6	4.11
13	Kerala	3,64,189	1.1	0.43
14	Madhya Pradesh	1,22,33,474	20.3	14.51
15	Maharashtra	85,77,276	8.9	10.17
16	Manipur	7,41,141	34.2	0.88
17	Meghalaya	19,92,862	85.9	2.36
18	Mizoram	8,39,310	94.5	1.0
19	Nagaland	17,74,026	89.1	2.1
20	Odisha	81,45,081	22.1	9.66
21	Punjab	-	-	-
22	Rajasthan	70,97,706	12.6	8.42
23	Sikkim	1,11,405	20.6	0.13
24	Tamil Nadu	6,51,321	1.0	0.77
25	Tripura	9,93,426	31.1	1.18
26	Uttarakhand	2,56,129	3.0	0.3
27	Uttar Pradesh	1,07,963	0.1	0.13
28	West Bengal	44,06,794	5.5	5.23
29	Andaman & Nicobar Islands	29,469	8.3	0.03
30	Chandigarh		-	-
31	Dadra & Nagar Haveli	1,37,225	62.2	0.16
32	Daman & Diu.	13,997	8.8	0.017
33	Delhi	-	-	-
34	Lakshadweep	57,321	94.5	0.07
35	Puducherry	-	-	-
	India	8,43,26,240	8.2	-

More than half of the tribal population is concentrated in the States of Madhya Pradesh (14.51%), Maharashtra (10.17%), Odisha (9.66%), Gujarat (8.87%), Rajasthan (8.42%) and Jharkhand (8.4%). The tribal population is increasing along with the Indian population but at a rate lower than the general population. The concentration of tribal people is in all States and Union Territories in India except the State of Haryana, Punjab, Delhi and Chandigarh. The predominant tribal populated States of the country (tribal population more than 50 per cent of the total State population) are Arunachal Pradesh, Meghalaya, Mizoram, Nagaland and Union Territories of Dadra & Nagar Haveli and Lakshadweep (Ministry of Tribal Affairs 2012).

Table 36.3: Trends in Growth and Proportion of Forest Tribes in India (Planning Commission 2005)

Census Year	Total Population (in millions)	Scheduled Tribes Population (in millions)	Proportion of STs Population
1961	439.2	30.1	6.9
1971	547.9	38.0	6.9
1981[#]	665.3	51.6	7.8
1991[@]	838.6	67.8	8.1
2001[$]	1028.6	84.3	8.2

[#] Excludes Assam in 1981 [@] Excludes J&K in 1991

[$] Excludes Mao-Maram, Paomata and Parul sub-divisons of Senapati District of Manipur in 2001

The areas inhabited by the tribals constitute a significant part of the under-developed areas of the country. The tribals live mostly in isolated villages or hamlets. A smaller portion of their population has now settled in permanent villages as well as in towns and cities. Tribal communities live in about 15 per cent of the country's areas, in various ecological and geo-climatic conditions ranging from plains and forests to hills and inaccessible areas. Tribal groups are at different stages of social, economic and educational development. For the tribal population in India, the literacy rate increased from 8.53 per cent in 1961 to 47.10 per cent in 2001 while the corresponding increase for the total Indian population was from 28.30 per cent in 1961 to 64.84 per cent in 2001.

For the purposes of the study of the tribal population, the whole country can be divided into seven principal tribal regions namely:

- The oceanic tribal groups of India
- The forest tribes of southern India
- The forest tribes of central India
- The western tribal region
- The forest tribes of eastern India
- The forest tribes of northeastern India
- The forest tribes of north India

THE OCEANIC TRIBAL GROUPS OF INDIA

The main areas under this region include the Andaman Island, the Nicobar Island and the Lakshadweep Island. The Andaman Island is the homeland of six tribal groups, inhabiting for thousands of years in the verdant tropical rain forests. Those living in the Andaman Islands are short in stature, dark skin colour with peppercorn hair and identify themselves as negrito (a sub-group of negriod people). They are Great Andamanese, Onges, Jarawas and Sentinelese. They are however, quite distinct from the African tribes. Each of this tribe has a different dialect and conventional territory division. In spite of their close proximity to certain foreign countries and lying close to the main commercial routes in the Bay of Bengal, their continuing primitive ways of life remains a riddle.

On that basis, the Constitution of India provided for their statutory recognition. Like in much other case in India, the Census was one of the earliest to document them from as early as 1901. In a pioneering attempt, Sir Richard C. Temple, first Census Superintendent of this territory gave a detailed account of the ethnographic, anthropological, demographic, social and cultural aspects of the lives of the various primitive tribes found in these islands in the census report of 1901. The census report of 1931 by Bonnington also gave an account of the tribals of these islands. Dr. Hutton, the Census Commissioner of India in 1931 listed these primitive tribes in the 1931 Census Report. The count of tribal groups of A & N Islands is however, debatable since 100 per cent enumeration in the true sense of them was not possible but their numbers were estimated. In few cases, only headcount was possible for many years. However, all these tribes were variedly estimated between 5,000-8,000 around 1780s when the British first attempted to colonize these islands. As a result of contact with the advanced and civilized people, these tribes started dwindling very fast due to killing by the colonizers coupled with diseases and dissipation which resulted from such contacts. Till the 1961 census, enumeration of the Andamanese of the Andaman group and the Nicobareses of Nicobar group of islands was made only. At the 1961 census, the Shom Pens of Great Nicobar and the Onges of Little Andaman were also enumerated for the first time. At the 1981 census, Jarawas were enumerated for the first time during the contact tour, though their population was estimated to be 200 persons.

The major tribal communities that reside in Lakshadweep are Aminidivi, Koyas, Malmis and Malacheris. Aminidivis are believed to be the first group of people to have started inhabiting these islands. They started living in an island named Amini and today it is teeming with Aminidivi people. They are most well known for their handmade crafts. They are also acknowledged for their different rituals and unique way of celebrating festivals making their festivals a grand affair. Malacheris are considered to be the main working community on the island. Their main profession is to gather nuts from coconut trees. Koyas are the land owners of Lakshadweep and Malmis are the one who work for them. Malayalam is the most commonly used language in Lakshadweep.

THE FOREST TRIBES OF SOUTHERN INDIA

This particular region consists of Karnataka, Andrapradesh and Tamil Nadu and Kerala. Most of the forest tribes are concentrated in Western Ghats and hilly forested areas some of them are considesd cave dwellers and mostly live in forest. The state wise forest tribes are described here under.

TAMIL NADU

The majority of the forest tribes in Tamil Nadu belong to the following tribes: Todas, Kadars, Irulas, Kotas, Kattunaikas, Kurumans, Muthuvans, Paniyans, Pulayans, Malayalees, Kurumbas, Eravallans, Kanikkars, Mannans and Paniyans. According to the Census Report 2001, the total tribal population in Tamil Nadu is 6,51,321 (1.04 per cent of the total population in Tamil Nadu). There are six particularly vulnerable tribal groups (PVTGs) in Tamil Nadu, such as Irular, Kattunaikan, Kotas, Kurumbar, Paniyan and Todas. Majority of tribals are living under the poverty line and have lost their indigenous identities. Government's schemes have not reached them and still many villages do not have basic amenities like drinking water, schools, electricity, road and transport. Exploitation, ignorance and discrimination are some of the stumbling blocks for the tribal development (Thurston 1909).

KARNATAKA

Toal tribal population in the State is 3.46 million which constitutes 6.6 per cent of the total State population. 49 tribal communities are currently identified in this state. Raichur and Bellary contain large number of tribal communities. Toda, Beda, Soliga, Hakki-Pikki, Konda Kapu, Koraga, Bhils, Chenchu, Gonds, Maleru, Badaga, Hasala, Meda, Iruliga, Jenu Kuruba, Erava and Siddis are some important tribes of Karnataka. All of them profess some religious beliefs and practice many rituals. A series of monographs published by Karnataka Sahitya Academy under thè stewardship of Baraguru Ramachandrappa provide ample information about these tribes. Tribal religions and rituals are under constant pressure from the major organized religions. In general, the tribes that remain geographically isolated are able to retain their traditional cultures and religions longer. On the other hand, communities that are either nomadic or live in the periphery of civilized life are prone for drastic changes. Most of the tribal beliefs and rituals are highly localized and they are not influenced by the major religions. Most of the tribal communities in Karnataka worship their own deities in addition to the Gods imposed from the main stream religions.The major forest tribes in karntaka are Kadu Kurumbas, Koragas, Soligas, Betta Kurumbas, Pani Yeravaa, Male Kudiyas, Kudiyas, Jenu Kurubas, Hallakki Okkaligaru, Siddi, Akkipikki Soligaru, Kadugoll and Mukuria (Thurston 1909).

Soligas - The bamboo children: Soligas are the semi nomadic tribes living in and around the BR Hills and MM Hills areas. They live in hornmony with nature and speak proto Dravidian type of ancient Kannada (Morab 1977). The soliga forest tribes worship Doddasampige, the gigantic Champak tree (*Michelia champaca*) believed to be over thousand years old, revered by the Soligas as a manifestation of their God Madeshwara. Soligas believes to have originated from a forest god, Karappa. There is a myth that the lord Karappa will protect the forests, man and animals. The Soligas speak directly to their deities or celebrate them in the most lyrical of songs and poetry (Somasundaram and Kibe 1990; Sudarshan 1998). Since time immemorial, Soligas were practicing shifting agriculture. Pastoral, shifting and settled agriculture are the traditional agricultural systems followed by the Soligas. Soligas follow two types of agroforestry systems they are mixed and multistoried cropping systems (Jadeyegowda 2000). Soligas depend on forest for all their needs without greed. When they have to cut the small trees for the construction of huts, they seek the blessings and permission of the god. After cutting, they confess their mistake and pray for forgiveness. After cutting, they place a stone on the stump of the tree because they believe that the soul of dead tree will transmigrate from the leaves to the stone. And this stone on weathering form soil which will provide nutrients to plants and other forms of life (Veena et al 2006).

ANDHRA PRADESH

There are 33 Adivasi communities were identified in this state. The majority of them belong to the following tribes: Chenchus, Koyas, Gadabas, Konda, Doras, Konda Kapur,

Konda Reddis, Sugalis/Lambadis, Yenadis, Yerukulas, Bhils, Gonds, Kolams, Pradhans and Valmikis. According to the Census Report 2001, the state share of tribals in the state is 6.6 per cent which has 5 million tribals. Andhra Pradesh state has both SAs and non-SAs. The Northern Andhra Pradesh is entitled as SA under the Fifth Schedule, where most of Adivasis depends on non-timber forest produce for their livelihoods and basic needs. However, in Southern Andhra Pradesh, tribals have lost their forest rights and identities.

KERALA

There are 36 tribal communities listed in this state. The majority of them belong to the following tribes: Paniyas, Malakuravans, Malayarayans, Malavetans, Malayans, Mannans, Ullatans, Uralis, Vishawans, Arandans, Kattunaykans, Koragas, Kadar, Kurichians, Kurumans, Pulayans, Malsars and Kurumbas. The state has the tribal population of 3,64,189 which is 1.1 per cent of total State population. The primitive vulnerable tribal groups in Kerala are Chola Naikker, Kattunaikker, Kurumbar, Kadar and Korugar. Wayanad is the most backward district of Kerala. More than one third of the tribals of Kerala State reside in Wayanad district. Major communities found in the district are Paniyan, Mullu Kuruman, Kurichian, Kattunaickan, Adiyan and Urali Kuruman. Adiya, Paniya and Kattunaikan are the three tribal communities which are still seriously underdeveloped comparing with the others. Tribal labourers can be generally categorized into three groups, namely agricultural labourers, marginal farmers and forest dependants.

THE FOREST TRIBES OF CENTRAL INDIA

MADHYA PRADESH

Major population of Indian tribals lives in Madhya Pradesh (14.51 per cent). The population of tribals in Madhya Pradesh is 12.2 million constituting 20.3 per cent of the total population of Madhya Pradesh. There are 46 recognized scheduled tribes and three of them have been identified as "Special Primitive Tribal Groups" in the State. Major tribal communities in MP are Gond, Bhil, Baiga, Korku, Bhariya, Halba, Kaul, Mariya and Sahariya. Dhar, Jhabua and Mandla districts have more than 50 per cent tribal population. In Khargone, Chhindwara, Seoni, Sidhi and Shahdol districts 30 to 50 per cent population is of tribes. Maximum population is that of Gond tribes (Sadguru Foundation 2012).

CHHATTISGARH

Chhattisgarh state is a repository of minerals and diverse cultural legacy. Chhattisgarh and tribal culture are two tautological terms since one third of the state's populace is dominated by tribals. The state has the tribal population of 6.61 million which is 31.8 per cent of total State population. The tribes of Chhattisgarh are unique race who mainly inhabit the dense forests of Bastar. In fact, more than 70 per cent of Bastar's population is composed of tribals. The lifestyle of the tribal people is unique and imbibed with traditional rituals and superstitions. They are a friendly and jovial lot who are industrious and diligent. Although shrouded in poverty, they live life to the hilt and love to celebrate every joyous occasion. Food, drink, music, dance, mirth and merriment add colour to their simple lives. The tribal women also love to adorn themselves in ethnic jewelry. One of the Chhattisgarh's eminent tribes is the Gonds or the Kotoriya tribe. The etymological connotation of their name comes from the Telegu term "Kond" meaning hills. The Gonds dominate most of Chhattisgarh's tribal population and primarily depend upon agriculture, forestry, cottage industries, hunting and fishing for their subsistence. Another tribe, the Abuj Maria lives in isolation in the dense and secluded enclaves of the forests of Narayanpur Tehsil in Basir. They are a ferocious and barbaric tribe who believe in primitive customs and are hardly tempted by the material pleasures. The other major tribes are Bhil, Katkari, Koli, Oraon, Warli, Bisonhorn Maria, Muria, Halbaa, Bhatra and Dhurvaa. The Bhils are considered as the third largest and most widely distributed tribal groups in (Indianetzone 2012).

THE WESTERN TRIBAL REGION

The tribal region including states like Rajasthan, Gujarat, central and western Maharashtra and northwestern Madhya Pradesh, have the highest tribal population in and around the hill regions of Udaipur, Doonangarpur and Banswara in Rajasthan, Sabarkantha, Panchmahals and Ahwa Dangs in Gujarat and Khandwa in Madhya Pradesh. In western Indian states like Gujarat, Rajasthan and Madhya Pradesh, there is heavy concentration of tribals. There are 34 predominantly tribal districts in these three states with the tribal population 26.81 million as per the census 2001.

MAHARASHTRA

The nomadic tribes constitute about 8.57 million in Maharashtra which is nearly 8.9 per cent of total State population. They are spread over in large areas, mainly on hilltops. Bhil, Gond-Madia, Katkari, Koli, Oraon, Warli are the major tribes of Maharashtra. A greater part of the tribes still follows the primitive characters. They live in groups moving from one place to another in caravans in search of livelihood. Their dress, dialect, folklore, customs and practices exhibit that they have a rich cultural heritage that is distinctly different from other social groups. Old traditions like worship of nature in different forms, animal sacrifices during religious ceremonies, adorning head with horn, etc. are still in practice. As they keep on wandering from one place to other, these tribes are in lack of education, which makes their survival difficult. So, they are forced to continue with this tradition of moving from place to place in search of work, which hurdles them to fit into the settled society.

Lambani Tribes: Lambanis lead a gypsy life and mainly inhabit the western Indian states including Maharashtra and Andhra Pradesh. A part of this tribal community is also found in the northern region of Karnataka. The Lambani tribe of India speaks a language which is believed to have been originated in the state of Rajasthan. Earlier, Lambanis used to supply grains to armies. In the olden days, the Lambani people carried grain, salt, bamboos and firewood. Women of Lambani tribes wear decorated clothes or fabrics. Jewellery is also an essential part of these tribal women's lifestyle. Women wear bone-made finger rings, bangles and bracelets. In addition, they adorn their hair with flowers and balls. People of Lambani tribes keep moving from place to place. Noticed usually in groups, people of this tribal community earn their daily bread by selling clothes at local markets. When the Lambani tribe inhabited the forest region of the southern India, they were also referred to as the bee-keeping caste.

GOA

Goa too has been invaded by Aryans just like the other parts of India. Still some tribes that exist in the state were the original settlers of the Konkan region much before the Dravidians invaded and occupied it. After sometime the Aryans came into the Konkan region area and settled down. Major tribes of Goa are the Gowdas, but there are other tribes too, who are the Kunbis, the Velips and the Dhangars. The tribal people live in segregated villages. The lifestyles, customs, practices, traditions, ceremonies, religious beliefs and superstitions have remained an integral part of the tribal people.

GUJARAT

7.48 million tribals live in Gujarat which constitute 14.8 per cent of the total state population. The Gujarat tribes are Bhils, Kuknas, Warlis, Naikas, Dublas, Gamits, Dhankas and Dhodias and the largest being Bhil. Forest tribes of this region have more unique lifestyles and are largely dependent on the forest. Habits and beliefs have an irresistible charm because they have maintained time honoured and fascinating traditions. Tribals have an inborn resistance to modernization. While it retards progress, it keeps them nearer to nature. Tribal population is concentrated in the Panchmahals, Baroda, Broach, Surat, Dangs, Bulsar, Sabarkantha and Banaskantha.

RAJASTHAN

The tribes constitute about 7.09 million in Rajasthan which is nearly 12.6 per cent of total State population. Southern part of Rajasthan is surrounded by hills and mountains called aravalis, which is forest region. This part is populated by a tribal community called bhils, who are most backward and poor. So, this southern part of the tribal region, which includes five districts i.e. Udaipur, Dungarpur, Banswara, Sirohi and Chittor. 23 blocks of these 5 districts are declared as tribal sub plan (TSP) area by State Govt. for speedy development of tribes of the region. Festivals have been a part of life for tribals since ancient time. Sarhul and karma are the two main important festivals of these tribals. They are also very fond of music and dance (Indianetzone 2012).

THE FOREST TRIBES OF EASTERN INDIA

ODISHA

In the state of Odisha, there are around 62 different tribes totaling up to 8.1 million. The antiquity of Odisha is endorsed by her ancient people who continue to inhabit their traditional dwelling places in remote areas in the deep forests and hilly interiors. Many tourists meet some of these tribal groups and see them in their natural habitat, though access to some tribal areas is still strictly restricted.The tribal economy is by and large based on activities around the jungles. Food gathering, hunting and fishing continue to be the main source of livelihood though some of the larger tribes such as the Santhals, Mundas and Gonds have become agriculturists. The Juang, Bhuyan, Bondo, Saora and Dharua tribes follow the shifting cultivation practice. The Koya tribals are cattle breeders while the Mohali and Lohara are simple artisans involved in basket weaving and tool making. The Santhal, Munda and Ho tribals have also become involved in the mining and industrial belts of Odisha. Though their economy is a bit shaky, the Odisha tribals enjoy a rich and varied cultural heritage, the most powerful instance of this being in their music and dance which are as colourful as they are rhythmical. The Gonds are the warrior caste, who have travelled the vast tracts of central and south India. The Oraon tribals are economically better placed because of their more progressive ways and interaction with the modern world, in the field of agriculture. An Oraon marriage partner can contest for divorce on the grounds of not only adultery but also for bad temper and laziness.

JHARKHAND

The state has the tribal population of 7.08 million which is 26.3 per cent of total State population. The Asur is the primitive iron smelting tribe of netarnat plateau (Jharkhand). The Asur is divided into three sub tribes viz. Bir (Kol) Asur, Birjia Asur and Agaria Asur. The Bir is also known by different names that is Solka, Thuppu, Kol, Jat etc. In official records the Bir section is known by tribal name. In Bihar, Birjia has been classified as a district scheduled tribe while Agaria is a scheduled tribe in Madhya Pradesh. Asur lives primarily in the Indian State of Jharkhand, mostly in the Gumla, Lohardaga, Palamu and Latehar districts. Asurs have traditionally been iron-smelters. The Asur religion is a mixture of animism, animatism, naturalism and ancestral worships. They also believe in black magic like bhut-pret (spirits) and witchcraft. Their chief deity is Singbonga. They celebrate festivals like Sarhul, Karma, Dhanbuni, Kadelta, Rajj karma, Dasahara Karam. Another important tribe in Jharkhand is Birhor. Bir means 'Jung' and Birhor means jungle people. The Birhors are of short stature, long head, wavy hair and broad nose. They belong to the Proto-Australoid racial stock. They claim they have descended from the Sun and believe that the Kharwars, who also trace their descent from the Sun, are their brothers. Ethnologically, they are akin to the Santals, Mundas and Hos.

THE FOREST TRIBES OF NORTHEASTERN INDIA

Whenever we talk of tribes and tribal culture in India, the north-eastern region of India at once comes to mind. And there are valid reasons for that. North-East India covers the states of Meghalaya, Assam, Manipur Arunachal Pradesh, Sikkim, Tripura, Nagaland and Mizoram. The region in the northeast is still untouched by modernisation. Apart from sheltering a magnificent tribal life, northeast is also rich in terms of landscape, flora and fauna. Visit to these states will introduce to a rich tribal culture which also includes a wonderful legacy of art and craft. A wide range of unique handicrafts are made by the tribal groups living in the northeastern part of India. The tribes mainly inhabit the hilly and forested areas of the states mentioned above. Each of the tribes in North-East India depends mainly on collecting, hunting and fishing for their livelihood. These tribes have their own cultural features, customs, languages and dress.

MEGHALAYA

Meghalaya tribes can mainly be classified into three groups - Garos, Khasis and Pnars or Jaintias. However the major tribes of this region are Garos, Khasis, Jaintia, Bhoi, Khynriams, War, Hynniewtrep Achiks Labangs, Nangphylluts and Nangtungs. Garos are believed to be the descendants of Tibeto-Burmar race who came down all the way from Tibet to the north eastern states, while the Khasis and Pnars or Jaintias are the descendants of Proto-Austroloid Monkhmer race. While Garos can be located in the Garo hills of Meghalaya, the Khasis are scattered all across the State and are known by different names. Mainly the pure Khasis can be sited at the Khasi hills, the Pnars or Jaintias can be located on Jaintia hills.

ASSAM

The population of Assam largely comprised of numerous Assam Tribes with their varied customs and beliefs. Starting from the plains to the hilly areas, the land of Assam is inhabited by different tribes. Some of the prominent tribes of Assam are Bodo, Singpho, Santhal, Dimasa people, Karbi, Khamti, Khamyang, Mishing, Nishi, Phake and Rabha. The Rabhas are one of the most important tribes of Assam as well as other parts of northeast India. The Karbis are one ethnic group who live mainly on the hilly slopes of Assam. They are also called as the Mikir, according to the Indian Constitutional Order. In accordance to some experts, the Phake tribe is believed to be migrated from Thailand and they still speak in the Thai language. One of the early settlers of the State, the Bodos are among the largest tribes of the State. They are the integral part of the larger ethnic group of the Bodo. Kachari. The Bodos reside mostly on the plains and have a great influence on the overall Assamese culture. Inhabiting the Darrang and Lakhimpur Districts of Assam, the Nishi Tribes are mainly farmers. Belonging to the Tibetan-Burmese clan, the Mishing tribes of Assam are another important ethnic group of the State. The Khamtis have their own language, which is very close to Thai and they practice Buddhism. Assam is the cradle ground for many civilizations. Numerous ethnic groups travelled to Assam through different routes in different ages, as the land was connected to several States and countries. Negritos, Dravidians, Alpines, Tibeto Burmese and Aryans had been the major races that came and dwelled in ancient Assam. They were considered as the aborigines of Assam and even today they are an integral part of the Assamese diaspora (Indianetzone 2012).

MANIPUR

The Meiteis (Meeteis), who live primarily in the state's valley region, form the primary ethnic group (60 per cent of the total population) but occupy only 10 per cent of the total land area. Their language, Meiteilon (Meeteilon) (also known as *Manipuri*), is also the lingua franca in the State, and was recognized as one of the national languages of India in 1992. The Muslims (Meitei-Pangal) who live in the valley, Nagas, Mizos-kuki and other smaller groups form about 40 per cent of the population but occupy the remaining 90 per cent of the total land area of Manipur State. The communities of Manipur in alphabetical order are: Aimol, Anal, Chiru, Chothe, Gangte, Hmar, Koirao, Koireng, Kom, Lamgang, Mao, Maram, Maring, Meitei, Monsang, Moyon, Paite, Tangkhul, Tarao, Thadou, Vaiphei, Zeliangrong (Zemei, Liangmei and Rongmei) and Zou. All tribes have their own distinct languages but in conversing with others they speak Meiteilon. Rice is the staple food for all ethnic groups of Manipur and meat, fish and seasonal vegetables are favourites.

ARUNACHAL PRADESH

Arunachal Pradesh is the home of 26 different tribal communities. The way that these tribes lead their lives are more or less the same and have the similar occupation and the societies are casteless. The tribes form their institutes and also select leaders according to their age for distinct social functions and for better administration. Experienced leaders of the tribal groups are revered for their position in the community. All the tribes follow endogamy and strictly observe the rule of clan exogamy. Festivals are an integral part of all the tribes of Arunachal Pradesh thereby reflecting the cultural exuberances and artistry. The artic acumen of the tribes of Arunachal Pradesh are furthered by their musical and dance bonanza as well as the vistas of exquisite crafts. The tribal people of this region are adept in making baskets, smithy work, weaving, pottery, wood carving, painting etc. The various tribal communities are Monpa, Hills Miris, Adis, Apatani, Wancho, Adi (Bogum and Bomis), Aka, Singpho, Khamba, Memba, Tagin, Khamti, Bugun or Khowa, Khamyang, Mishmi, Nishing, Monpa, Sulung or Puroik, Nocte.

SIKKIM

The tribes in Sikkim have really contributed a lot in the transformation of Sikkim into one of those States in India that are recurrently visited by the tourists. The tribal folks of Sikkim are basically of the most demure and simple nature and they take a lot of pride in their glorious culture and beliefs. One main similitude that lies between the tribes across Sikkim and other hilly people is the fact that all of them are quite soft-spoken and simple. The major tribal communities are Lepchas, Bhutias and Nepalese.

NAGALAND

Nagaland is entirely tribal populated state. Nagas majorly belongs to the Indo-Mongoloid family. Nagaland is the home of 14 major tribes. They are the Angami, Ao, Chakhesang, Chang, Khemungan, Konyak, Lotha, Phom, Pochury, Rengma, Sangtam, Sema, Yimchunger and Zeliang. The Chakhesangs were earlier known as Eastern Angamis and are a combination of the Chakri, Khezha and Sangtam sub-tribes. The Naga's have different stories about their origin. The Angamis, Semas, Rengams and Lothas subscribe to the Kheza-Kenoma legend.

A striking characteristic of the Naga tribes is their hospitality and cheerfulness. To be greeted with a smiling face while traveling on the roads is a common experience. A visitor to Naga village is heartily received and entertained with a surfeit of rice-beer, which is generally served by the lady of the house or her young daughter with warmth which is unforgettable. The word 'Naga' was given to these hill tribes by the plains people. This proved to be a great unifying force to the tribes now classified as Naga.

TRIPURA

Tribes of Tripura represent a range of human communities that are living in the State from quite a long time. The Tripuri is one of the oldest tribes living in the state of Tripura. They constitute a large percentage (above 50) of the total tribal population of Tripura. The main economic activity of the Tripuri tribe in Tripura is agriculture. Jhum cultivation is the widely practiced agricultural activity of the Tripuri groups of Tripura. The people of Tripuri community generally make their homes on the slopes of the hilly terrains of Tripura state. The Lebang Boomani and Garia are the two prominent folk dance forms of Tripuri tribe. Reang tribe of Tripura, although has a large population, yet it lacks sufficient educational and economic independence. Chakma (Changma) tribal group are the fourth largest tribal group of Tripura and they came to Tripura in the 5th century. People of Mog community follow the ritual customs of Buddhism. Halam tribe is yet another ancient clan of Tripura which has many sub-sections. There are a total of 19 tribes in Tripura and all these tribal groups of Tripura have their own festivals, rituals and customs. The 19 tribes of Tripura are Bhil, Bhutia, Chaimal, Chakma, Garo, Halam, Jamatia, Khasia, Kuki, Lepcha, Lushai, Mog, Munda, Noatia, Orang, Reang, Santhal, Tripuri and Uch (Wikipedia 2012)

MIZORAM

The origin of the Mizo people, like those of many other tribes in the northeastern India, is shrouded in mystery. Mizo history in the 18th and 19th century is marked by many instances of tribal raids and head hunting led by the village chief. The Mizo people are an ethnic group native to north-eastern India, western Burma (Myanmar) and eastern Bangladesh who speak Mizo. The Mizo are recognized as Scheduled Tribe under the 6th Schedule of the Constitution of India. The present Indian State of Mizoram (literally Mizoland) was called the Lushai Hills and was a district of Assam, before it became a Union Territory and afterwards a full-fledged state. The Lushai were the first Mizo tribe to have an external exposure and hence the ethnicity was initially known as the Lushai people. The various ethnic tribal communities are Chakma, Dimasa (kachari), Garo, Hajong, Hmar, Khasi and Jaintia, Any Juki tribes, Man (Tai-speaking), Any Mizo (Lushai) tribes, Mikir, Any Naga tribes, Pawi, Synteng and Paite.

WEST BENGAL

The tribes of West Bengal have occupied a large section of the total population of West Begal. Numerous tribal people reside in rural areas of the the West Bengal. The tribal population of Wes Bengal is 4.4 million which constitutes 5.5 per cent of the total population of the State. Santals constitute more than half (51.8 per cent) of the total tribal population of the state followed by Oraons (14 per cent), Mundas (7.8 per cent), Bhumij (7.6 per cent) and Kora (3.2 per cent) with sizeable population. The Lodhas, Mahalis, Bhutias, Bedias and Sabars are the remaining tribals having population of one per cent or more. More than half of the total tribal population of the state is concentrated in Medinipur, Jalpaiguri, Purulia and Bardhaman districts. Of the remaining districts, Bankura, Malda, Uttar Dinajpur and Dakshin Dinajpur have sizable ST population.

THE FOREST TRIBES OF NORTH INDIA

JAMMU AND KASHMIR

Bakarwal tribal community of Jammu and Kashmir is found in every hilly nook of the region. Also known as Bakharwal, they are special nomadic tribes, mainly found in the Pir Panjal range of mountains located between the two states of Jammu and Kashmir and Himachal Pradesh. Bakarwal tribes are also found in every corner of Northern provinces of the Himalayan Range, namely the states of Uttarakhand, Himachal Pradesh and Punjab. This tribe is also known as Dhangar in several parts of India. Bakarwal mainly are goatherds and shepherds by profession.

HIMACHAL PRADESH

The Kinners or Kinnaure inhabit the border district of Kinnaur. Physically they are closer to the Aryan races in their tall well-built bodies, their high foreheads, large eyes and fair complexions. Temperamentally they are a gentle and soft-spoken people, quite content to live in poverty. Their main occupation is rearing sheep and goats and raising wool. Some are engaged in agriculture and horticulture. They live in joint families and men and women have more than one mate.Their marriage customs are very interesting. All the brothers in a family share a wife. They call it the Pandava marriage. Due to this, lots of girls remain unmarried. But these systems are being abandoned in the changing socio-economic conditions. Their mongoloid features are evidence of the intermixing of ideas on the borders. Kinnaur women are beautiful, modest and homely and spend most of their time in the fields.

UTTARAKHAND

The tribal population of Uttarakhand is 2,56,129 which constitutes three per cent of the total population of the State. In Uttarakhand, there are five major tribal communities, namely Bhotia, Jaunsari, Raji, Tharu and Buksa. The habitation of a particular tribe is confined to a particular tribal location in the Uttarakhand. Despite their habitation in different zones, there prevail few common features in the economic and social life of all the tribal communities with minor variations necessitated to maintain a harmonious coordination between the resource availability and needs of the society. Agriculture forms the basic livelihood of Tharu, Buksa and Jaunsari tribes. Trade is the main source of livelihood of the Bhotia tribe. Raji (Vanrawat) are mainly dependent on animal husbandry and daily wages in various developmental activities.

TRIBAL ECONOMY

The primitive societies have passed through several stages of economic development everywhere in the world. Thus one finds the stages of food gathering, hunting and fishing, farming, etc. Among the Indian tribes, the tribes living in the forests and hills usually earn their livelihood by means of food gathering, hunting and fishing. This type of the life is noticed in case of Kadar of Kerala, Birhor and Kharia of Bihar and other tribes. The tribes living in dense forests, full of wild beasts, live by hunting. This is observed in the tribes of Naga, Kuki, Bhil, Santhal and Gond. The hunters leave the females to carry out household activities in the morning and return in the evening after hunting. In some tribes, there is a custom of hunting collectively. The Nagas use spears, arrows and bows. The Bhils are very much specialized in shooting by arrows. The tribes living near rivers and seas usually earn their livelihood by catching fish. The hilly tribes rear the cattle e.g. Goojars and tribes of Chamba. The Todas of Nilgiri rear buffaloes. Some tribes also carry out cultivation, but they are generally shifting cultivation. Important cultivating tribes are the Santhals and Gonds (Rajeeva 1989). Cottage industries, such as weaving cloths, preparing ropes, skins and utensils of different metals are prevalent in many tribes. The Kharia people are very much specialized to cottage industries.

On record, India has nearly 5,30,000 cooperatives with almost 230 million members. These cooperatives are spread across the country and at least one fourth of these are rural agricultural cooperatives, which include those in pockets with a high tribal population.

Almost every village is covered by at least one cooperative, usually an agricultural credit cooperative. Large-scale Adivasi Multi-Purpose Cooperative Societies (LAMPS) in tribal areas are based on the co-operative values of self-help, self-responsibility, democracy, equality, equity and solidarity. These socities were set up with the following seven cooperative principles.

- Voluntary and open membership
- Democratic member control
- Economic participation by members
- Autonomy and independence
- Education, training and information
- Cooperation among cooperatives
- Concern for community

POLICY RELATED TRIBAL DEVELOPMENT

In adopting the Nehruvian Panchasheel, Government of India formed a Ministry of Tribal Affairs for the first time in October 1999 to accelerate tribal development. The Ministry of Tribal Affairs has come out with National Policy on Tribals. The National Policy recognizes that a majority of Scheduled Tribes continue to live below the poverty line, have poor literacy rates, suffer from malnutrition and disease and are vulnerable to displacement. Five principles spelt out in 1952, known as Nehruvian Panchasheel, have been guiding the administration of tribal affairs. They are:

1. Tribals should be allowed to develop according to their own genius
2. Tribals' rights in land and forest should be respected
3. Tribal teams should be trained to undertake administration and development without too many outsiders being inducted.
4. Tribal development should be undertaken without disturbing tribal social and cultural institutions
5. The index of tribal development should be the quality of their life and not the money spent

NATIONAL POLICY ON TRIBALS (2006)

The National tribal Policy aims at addressing each of these problems in a concrete way. It also lists out measures to be taken to preserve and promote tribals' cultural heritage. For the first time after the country became Independent, the Government of India formulated National Policy on Scheduled Tribes. The policy seeks to bring Scheduled Tribes into the mainstream of society through a multi-pronged approach for their all-round development without disturbing their distinct culture. The policy has taken into consideration of the formal education, traditional wisdom, health, displacement and resettlement, forest villages, shifting cultivation, land alienation, intellectual property rights, tribal languages, the languages spoken by Primitive Tribal Groups (PTGs), Scheduled Tribes and Scheduled Areas, administration and research (Chittaranjan Kumar Paty 2007). The National Policy recognizes the importance of participatory approach to development of tribals. Non- Governmental Organizations (NGOs) and Voluntary Agencies (VAs) act as catalysts in reaching benefits of Government programmes and policies to the grass-root level and thus optimize the desired accomplishment.To bring the tribals into country's mainstream, the National Policy envisages the following:

- Identification of tribal groups with 'primitive traits' shall be done away with on a priority basis.
- The 'distinct culture' of the tribes reflected in their folk art, folk literature, traditional crafts and ethos shall be preserved. Their oral traditions shall be documented and art to be promoted.
- Opportunities shall be provided for tribals to interact with outside cultures.
- Their geographical isolation shall be minimized through development of roads, transport and means of communication and provision of concessional travel facility.
- National tribal policy supports the protective safe guards, political safeguards, developmental safeguards

ACTS AND REGULATIONS

Moreover, there are several laws and amendments, enacted in the tribal areas, which are working at cross purposes or have no linkage to the Fifth Scheduled of the Constitution. But whatever rules or legislations made in these areas have to strictly fall within the ambit of the Fifth Scheduled and not allowed to dilute it. For example, the Panchayatraj Extension to Scheduled Areas Act of 1996 (PESA) clearly supports the Fifth Scheduled and the rights of the Gram Sabhas (Village assembly) in the SAs. Also, the Land Acquisition Act, the Mines and Minerals Development Act, the Forest Management Act, the Environment Protection Act and others are to be superseded by the Land Transfer Regulation Act or the Fifth Schedule. As per the PESA Act 1996, the Gram Sabha has the supreme authority to make decision over the natural resources, including non-timber forest produces (NTFP). Those monopoly restrictions over NTFP should be immediately removed while ensuring social protection through providing support price to tribal. Tribals are allowed to collect, process, transportation and market NTFP freely (Sarkar 2006).

FOREST RIGHT ACT (FRA) 2006

The Scheduled Tribes and Other Traditional Forest Dwellers (Recognition of Forest Rights) Act, 2006 is a key piece of forest legislation passed in India on 18 December 2006. It has also been called in various names as The Forest Rights Act or The Tribal Rights Act or The Tribal Bill or The Tribal Land Act. The law concerns the rights of forest-dwelling communities to land and other resources, denied to them over decades as a result of the continuance of colonial forest laws in India. Supporters of the Act claim that it will redress the "historical injustice" committed against forest dwellers, while including provisions for making conservation more effective and more transparent. A little over one year after it was passed, the Act was notified into force on 31 December 2007. On 1 January 2008, this was followed by the notification of the Rules framed by the Ministry of Tribal Affairs to supplement the procedural aspects of the Act. This act protects the rights of the forest tribes on forest land and forest produce. After the enactment of the Act, tribals and other forest dwellers will hold the right to cultivate forest land, the right to own, collect, use and dispose of minor forest produce, rights inside forests which are traditional and customary like grazing. Tribals who have been living in and depending on forests, for their livelihood prior to 13 December 2005will have these rights.

37 SOIL AND WATER CONSERVATION

SOIL EROSION

Soil erosion has been our nation's most devastating environmental disaster. Soil in quantities of incomprehensible magnitude has been lost from the land where its forfeited productivity could have contributed to our national wealth. Soil erosion is the detachment and transportation of soil material from one place to another through the action of wind, water in motion or by the beating action of raindrops. Soil erosion is caused by detachment and removal of soil particles from land surface. With the advent of modem civilization, the pressure on land increased which leads to it's over exploitation and subsequently, its degradation.

TYPES OF SOIL EROSION

The process of soil erosion goes through two main stages of activities, namely the detachment of soil particles and their transportation. The principal agents responsible for these two stages of activities are wind, water and gravity. Wind erosion occurs in arid and semi-arid areas where rainfall is scanty and the day temperatures are very high. Water erosion occurs in areas where rainfall is high and gravity erosion in areas that are near to pits, rivers, roads, etc. The types of soil erosion caused by three natural agents-winds, water and gravity are described below.

CLASSIFICATION OF EROSION

Erosion may be broadly classified into two groups:

GEOLOGICAL EROSION: Geological erosion, sometimes referred to as natural or normal erosion, represent erosion under the cover of vegetation. It takes place as a result of the action of rainfall, runoff, topography, wind velocity, atmospheric temperature and gravitational forces. It is continuous and slow but constructive process. Soil removal is fairly balanced with soil formation. Erosion caused by biological activities is called as normal erosion.

ACCELERATED EROSION: Accelerated erosion of soil is caused mainly by managerial errors such as raising of crops without adopting any soil conservation practice, deforestation, etc. It leads to erosion in excess of the threshold value of the new soil formation causing severe deterioration of the top surface of the land. It occurs in faster rate and considered destructive process. It is the erosion in which rate of soil loss exceeds the rate of soil formation.

CAUSES OF SOIL EROSION

The causes of soil erosion are as follows:

Destruction of Natural Vegetation

(A) Deforestation

- Shifting cultivation
- Forest fire

(B) Over grazing

- Excess grazing
- Non-rotational

(C) Burning of grass land

Mismanagement of Land

(A) Improper land management

- Up and down hill cultivation on slopy land
- Un-successive growing of grass, crops
- Removal of organic matter and plant nutrients
- Faulty irrigation systems i.e. flood irrigation on slopy land
- Growing of wider row crops on slopy land

(B) Pressure on agriculture

- Animal
- Human being

WATER EROSION

Erosion of soil by water is caused by its two forms: liquid as the flowing water and solid as the glaciers. The impact of rainfall causes splash erosion. Runoff water causes scarping and transport of soil particles leading to sheet, rill and gully erosion. Water waves cause erosion of bank sides, reservoirs, lakes and ocean. The sub-surface runoff causes soil erosion in the form of pipe erosion, which is termed as tunnel erosion. The glacial erosion causes heavy landslides. In India, glacial erosions are mainly confined to Himalayan regions. Movement of soil, by rain water running rapidly over the exposed land surface, causes severe erosion. This is most commonly found in humid and sub-humid areas where rainfall is excess and no vegetation over the land. In water erosion, greater soil fertility loss occurs.

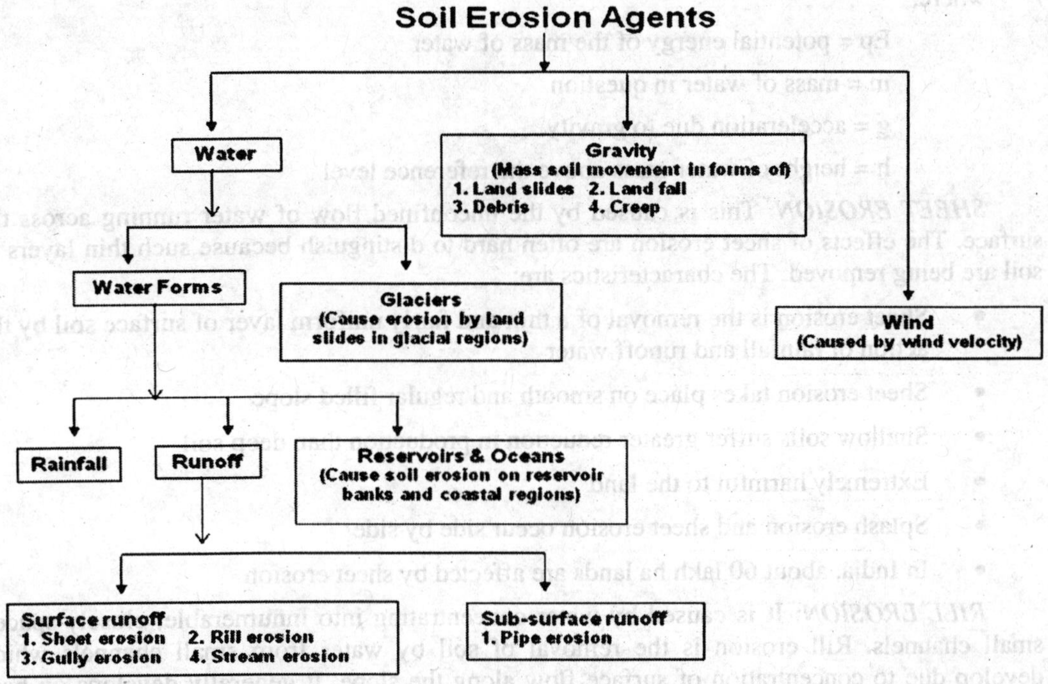

Fig. 37.1: Soil Erosion Agents

FORMS OF WATER EROSION

RAINDROP EROSION: The raindrop characterized by rainfall has a significant effect on the magnitude of splash erosion. The other factors which affect splash erosion are type of soil, land topography and vegetative cover. Splash erosion is caused by the impact of falling raindrops on the soil surface. Splash erosion is influenced by rainfall intensity and rainfall size. Splash erosion takes place, when rainfall intensity exceeds infiltration rate. On steeper slopes splash erosion is higher as soil particles are pulled downward to very large distance by the gravitational forces. The mechanical energy is manifested in two forms: kinetic and potential. Kinetic energy is the energy possessed by a substance by virtue of its motion and is proportional to the product of the moving mass and half of the square of the velocity of the mass, i.e.

$$E = 1/2\, mv^2$$

where,

E = kinetic energy

m = mass of water or drop in question

v = velocity of the mass of water

Potential energy is the energy possessed by a substance by virtue of its position and is expressed as,

$$Ep = mgh$$

where,

Ep = potential energy of the mass of water

m = mass of water in question

g = acceleration due to gravity

h = height of water mass above the reference level

SHEET EROSION: This is caused by the unconfined flow of water running across the surface. The effects of sheet erosion are often hard to distinguish because such thin layers of soil are being removed. The characteristics are:

- Sheet erosion is the removal of a thin and fairly uniform layer of surface soil by the action of rainfall and runoff water
- Sheet erosion takes place on smooth and regular filled slope
- Shallow soils suffer greater reduction in production than deep soil
- Extremely harmful to the land
- Splash erosion and sheet erosion occur side by side
- In India, about 60 lakh ha lands are affected by sheet erosion

RILL EROSION: It is caused by water concentrating into innumerable, closely-spaced small channels. Rill erosion is the removal of soil by water from small channels which develop due to concentration of surface flow along the slope. It generally develops on bare and unprotected land where soil conservation measures are not adopted. In advanced stage, it may take the shape of gullies. The rill erosion is an intermediary stage between sheet erosion and gully erosion. Removal of surface soil by running water with the formation of narrow shallow channels can be leveled by normal cultivation. Rill erosion is more serious in soils having a loose shallow top soil.

GULLY EROSION: Gullies are steep-sided trenches formed by the coalescence of many rills. Once they start, it is difficult to stop. It is the advanced stage of rill erosion which cannot be obliterated by normal tillage operations. Rills with more than 30 cm depth are generally called gullies. Gullies obstruct the operation of farm machinery and tend to deepen and widen with every heavy rainfall. They cut up large fields into small fragments and in course of time, make them unfit for cultivation. The rate of gully erosion depends on several factors, viz. runoff producing characteristics of the watershed, drainage area, soil characteristics, channel alignment, slope, vegetative cover on the ground, etc. The gullies are classified based on size, shape and state of gully. The development of gully occurs due to following four stages:

(1) *Formation Stage*: Channel erosion takes place by downward scour of the topsoil. If the topsoil can provide resistance, this stage proceeds slowly.

(2) *Development Stage*: Upstream movement of the gully head and simultaneous enlargement of width and depth take place. The weak parent material is rapidly removed by cutting C horizon.

(3) *Healing Stage*: Vegetation begins to grow in the channel and further erosion ceases.

(4) *Stabilization Stage*: The gully bed and sides reach a stable slope and sufficient vegetation grows to anchor the soil and to trap soil flowing from upstream. New topsoil develops and vegetative cover grows.

Table 37.1: Classification of Gullies based on the Size

	Very Small Gullies (g_1)	Small Gullies (g_2)	Medium Gullies (g_3)	Deep and Narrow Gullies (g_4)
Deep	upto 3 m	upto 3 m	3-9 m	>9 m
Bed Width	< 18 m	> 18	> 18 m	Varies
Side Slope	vary	8-15 %	8-15%	Mostly steep

Table 37.2: Classification of Gullies based on the Shape

Particulars	U-shaped Gullies	V-shaped Gullies
Location	Alluvial plains where surface and sub-surface soils are easily erodible	Hilly regions where land surface is tough to resist the cutting
Cross section	U shaped	V shaped
Slope	Slopy field	High slopy field
Velocity of flow	Less	High
Catchment's area	Large	Small
Runoff volume/ unit time	More	Less
Lateral spacing of gullies	Large	Small
Control measures	Permanent drop structure	Check dams

Table 37.3: Classification of Gullies based on the State

Active Gullies	Inactive Gullies
Active gullies are those whose dimensions are enlarged with time	Inactive gullies are those whose dimensions are constant with time
These gullies are found in plain	These gullies are found in rocky areas

OTHER FORMS OF EROSION

Besides the four main water erosions described above, there are some special forms of erosion also occur.

LANDSLIDES OR SLIP EROSION : During monsoon season or heavy down-pour, the hills get saturated with water resulting in instability of the land mass. The complete surface portion of soil may slip down from its actual position, causing landslides. Landslides are very common in hills and huge quantities of soil are washed away in streams and rivers. Hills devoid of vegetation are more prone to slip erosion, causing damage to agricultural lands, orchards, communication systems and human lives. Minor landslips occur during the drying phase in the months of May and June, especially in the sandy hills of Shiwaliks.

STREAM BANK EROSION: It is caused by streams and torrents, also called *chos,* cutting their banks and adjoining fields. It is mainly due to runoff flowing over the side of the stream banks or by scouring and undercutting below the water surface, the latter process being more serious. It is influenced by deforestation and overgrazing. Scour erosion is influenced by the velocity and direction of water flow, depth and width of channel and soil texture.

RIVER BANK EROSION: This is a common phenomenon in the rivers of the eastern and western Himalayan regions. During floods, the rivers undermine their banks and big chunks of land mass are engulfed by flood water. This leads to broadening of river beds and changes in their courses.

COASTAL EROSION: The strong waves of sea strike against the seashore causing large scale soil erosion. In this case, erosion is a combined effect of wind and water.

HARMFUL EFFECT OF WATER EROSION

- Loss of top fertile soils
- Accumulation of sand or other unproductive coarse soil materials on other productive lands
- Silting of lakes and reservoirs
- Silting of drainage and irrigation channels
- Decrease in water table
- Fragmentation of land especially in gully erosion
- Increase in flood (heavy floods in the river)
- Adverse effect on public health
- Loss of nutrients
- Destruction of soil structure
- Adverse effect on crop yield
- Hindrance of farming operations
- Affecting transportation
- Destruction of vegetation

WIND EROSION

Wind erosion is the detachment and transportation of soil particles by the forces of moving wind. In regions, where the precipitation is low and the atmospheric temperature during the day is high, the velocity of wind is invariably very high. Such climatic conditions are generally prevalent in arid and semi-arid regions where the wind velocity is also very high. Wind erosion is caused by mismanagement of land resources such as intensive farming, over grazing, deforestation, etc. It is characterized by movement of soil by wind rapidly over the exposed land surface. It is most commonly found in arid and semi-arid areas where rainfall is insufficient and no vegetation over the land is available.

MECHANISM OF WIND EROSION

Wind erosion is characterized by transport of soil particles through wind by three processes depending on the size of soil particles that are moved by wind: surface creep, saltation and suspension.

SURFACE CREEP: The movement of larger size particles (> 0.5 mm) occurs by the process of surface creep, where the movement is by rolling and sliding of soil particles on the land surface caused by direct push of wind and collisions between the saltating particles, i.e. the particles which are already moving by the process of saltation. Surface creep process destroys soil structure.

SALTATION: It is the next step of movement taken by soil particles which are in process of surface creep, after these particles have gained sufficient speed. On gaining speed, these particles start to hop, leap and bounce off the ground and the movement by this process is called saltation. Soil particles of medium size (0.1-0.5 mm) are carried by wind in series of short jump/bounces/hop/leap/lift caused by direct pressure of wind on soil particles. Saltation process is less in rough surface. Soil particles can jump upto the height of 1.0 m (average 30-60 cm) and jump particle moves 200-1000 round/minute.

SUSPENSION: Very small-sized particles of soil (0.1 mm) are carried by suspension in the air. It is the floating of small sized soil particles in air. The soil particles stay in suspension, because the downward movements of soil particles due to their weight are balanced by the lifting forces of eddies in the fluid. When soil particles are very small (<0.1 mm), they are carried over long distance upto 100 km. The particles come down only with rains or when wind velocity is decreased and it makes smooth surface.

FORMS OF WIND EROSION

There are five different forms of wind erosion:

(1) *Abration/ Corrasion*: Abration form is found in clody soils. With the effect of strong winds, the clods are contacted and broken up into small pieces.

(2) *Efflation*: This form is found in silty and clay particles dominant in soil.

(3) *Efflution*: This form of wind erosion is found in calcareous soil where clay and sand particles are dominant. With the effect of strong wind clay particles carry away a long distance by suspension process and big particles or sliding i.e. sand carry away a short distance by rolling or sliding.

(4) *Detrusion*: This form is found where bunds are constructed in cultivated soil with the effect of strong wind, soil particles of bund slip down and deposited in lower place.

(5) *Extrusion*: This form of wind erosion is found in sandy soils where sand particles are dominant. With the effect of strong wind, sand particles carry away from one place to another by rolling/ sliding.

HARMFUL EFFECTS OF WIND EROSION

- Loss of soil and plant nutrients
- Loss of production

- Change of soil textural class
- Reduction in water holding capacity of the soil
- Exposure of hard rocks
- Formation of sand dunes
- Destruction of soil structure

WIND EROSION FACTORS

The following factors affecting soil erosion by wind are assessed by the Wind Erosion Equation (WEE). A Wind erosion equation is an erosion model designed to predict long-term average annual soil losses from a field having specific characteristics. The equation is:

$$E = f(IKCLV)$$

where,

E = Estimated average annual soil loss expressed in tons per acre per year

I = Soil erodibility index

K = Soil ridge roughness factor

C = Climatic factor

L = Equivalent unsheltered distance across the field along the prevailing wind erosion direction

V = Equivalent vegetative cover

OTHER FACTORS

Tillage	: Soils are loosened by the excess tillage	= More erosion
Structure	: Single grain structured soil	= More erosion
Organic matter	: Low organic matter in soil	= More erosion
Vegetation	: Soil covered with vegetation	= Reduce erosion
Weather	: Continuous dry weather	= Increase erosion
	Wet and moist soils	= Decrease erosion
Wind velocity	: High wind velocity	= Increase erosion
Soil	: Rough surface	= Less wind erosion
	Smooth surface	= More wind erosion
	Clody soil	= Less wind erosion
	Without clods	= More wind erosion

FACTORS AFFECTING SOIL EROSION

Numerous factors affect soil erosion depending upon the local conditions with regard to physical, chemical and biological nature of the soil. The major factors affecting soil erosion are:

- Climate, especially rainfall and wind velocity and their intensity in time and space
- Topography, slope of land, its nature, degree and length
- Physical, chemical and biological characteristics of the soil
- Nature of ground cover and land use pattern
- Biotic interference due to human beings and animals
- Conservation practices
- Mass erosion, such as earthquakes, landslides, torrents, floods, etc.

Mathematically, soil erosion may be depicted by functional equation:

$$\text{Erosion} = f(cl, t, v, s, h, \ldots\ldots)$$

CLIMATE: The major climatic factors influencing runoff and erosion are rainfall, temperature and wind. Rainfall is by far the most important factor causing soil erosion by water. The soil erosion depends on the intensity, kinetic energy, amount, duration and frequency of rainfall. The intensity, duration and frequency of rainfall govern the rate and the volume of runoff. A light rain can be easily absorbed in the soil and causes no runoff and soil loss. When the intensity of rain is more than infiltration rate, it causes runoff and soil loss. The following are the factors in the soil erosion process:

- The detaching capacity of the erosive agent
- The detachability of the soil
- The transporting capacity of the erosive agent
- The transportability of the soil

TOPOGRAPHY: On flat lands, erosion is usually not a problem. It is the sloping lands that experience increasingly greater problems of erosion. The degree of slope and the length of slope are the two main features of topography affecting erosion. Velocity of the runoff water is influenced mainly by the degree of slope. If the land slope is increased four times, the velocity of water flowing over it is approximately doubled. When the velocity is doubled, the erosive capacity as represented by the kinetic energy of the flowing water, is increased about four times. The land slope mainly its degree and length, are important in determining the extent of soil erosion. Higher the degree of slope and longer is its length, more is the soil erosion due to increased velocity of water running down the slope.

VEGETATION: Vegetative cover retards the surface velocity of runoff water and transportation capacity reduces. Also the water remains on the surface for longer period and more amount infiltrates into the ground. The root system of the plants acts as binding agent for the soils and the detachability reduces. The decaying of roots, leaves, etc. increases the porosity of soil and the capacity to absorb water increases. Addition of humus increases the biological activity inside the soil. Accelerated biological activity is favourable for growth of bacteria, earthworms, etc. This also creates a favourable condition for absorption and holding of more moisture inside the soil. Thus vegetative cover helps to conserve both soil and moisture in several ways. The nature of vegetative cover on the soil surface is a major factor in influencing soil erosion. The following are the major effects of vegetation on soil erosion.

(1) *Interception of Rainfall*: A part of the rainfall intercepted by the canopy of vegetation never reaches the soil, but is evaporated directly from the leaves and stems. This part of the rainfall does not contribute to the runoff. In addition, the vegetative canopy absorbs the impact of the raindrops and thereby minimizes the dispersion of soil by the raindrops.

(2) *Runoff Velocity*: Vegetative cover is a hindrance to runoff. A well-distributed and close-growing vegetation not only slowdown the rate at which water travels down the slope but also tends to prevent a concentration of water. These two effects greatly reduce the erosive capacity of the runoff water.

(3) *Root Effects*: The knitting and binding effect of root systems in the surface layer of soils, aggregates the soil into granules and increases its resistance to erosion.

(4) *Biological Influence*: The soil fauna are most active in soils having ample vegetative cover. The soil under a thick forest covers increase aeration and good environment for the activities of beneficial bacteria and earthworms.

SOIL CHARACTERISTICS: Erodibility of soil is influenced by its physical and chemical properties, including soil texture, structure and organic matter, nature of clay, amounts and kinds of salts present. Soil structure, texture, organic matter infiltration and permeability influence the runoff and soil loss. Fine soils are more susceptible to erosion than coarse soils. The organic matter helps in binding the soil particles and improves the water holding capacity of soil. The three most significant soil characteristics which influence soil erosion are: (1) infiltration capacity (ii) structural stability (iii) antecedent soil moisture.

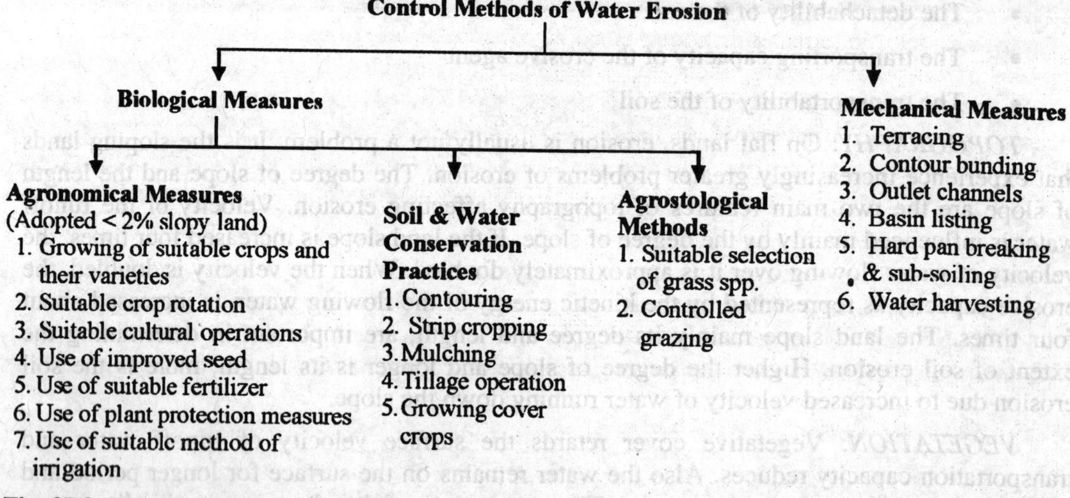

Fig. 37.2: Water Erosion Control Measures

BIOLOGICAL FACTORS: Biological factors that influence the soil erosion are the activities of man and animal. Felling of trees and faulty cultivation practices such as up and down cultivation in hilly areas, overgrazing by cattle, etc. contribute much to the soil erosion.

Without such activities, the soil, climate, vegetation, etc. will remain in a balanced condition and no erosion will take place.

SOIL AND WATER CONSERVATION METHODS

Planning soil conservation is to develop large array of techniques and practices to reduce soil erosion. The object of planning soil conservation is to make up a system by selecting a set of individual items which are each relevant to the conditions and which can be combined into a workable system. Looking at the large choice of mechanical works, the main factor in deciding which to select must be to define the objective. The main objective may be to modify the soil slope, to influence the surface run-off or to allow the agricultural use of steep slopes. Various methods adopted for the soil and water conservation (Norman 1987) in many countries are presented below.

CONSERVATION TILLAGE

This umbrella term can include reduced tillage, minimum tillage, no-till, direct drill, mulch tillage, stubble-mulch farming, trash farming, strip tillage, plough-plant (Mannering and Fenster 1983). In countries with advanced soil conservation programmes, the concept of conservation tillage is the main theme of the recommendations for cropland. The application is mainly in mechanized high production farming with good rainfall, or for the control of wind erosion where there is large-scale mechanized production. It is less applicable to low input level crop production or subsistence agriculture. The principles are equally effective in any conditions to maximize cover by returning crop residues and not inverting the top soil and by using a high crop density of vigorous crops. Conservation tillage also has the advantage of reducing the need for terraces or other permanent structures (Norman 1987).

DEEP TILLAGE

One of the reasons for low yields in semi-arid areas is the limited amount of moisture available to crop roots. The available moisture will be increased if the rooting depth is increased with deep tillage. Deep tillage is beneficial for some crops and on some soils. Also deep tillage requires greater draught power which is usually in short supply in semi-arid areas. Ripping or subsoiling can be beneficial, either to increase the porosity of the soil, or to break a pan which is reducing permeability. The application of this technique to subsistence farming will be difficult.

CONSERVATION FARMING

Like conservation tillage, this title covers many different farming techniques. It includes any farming practice which improves yield, or reliability, or decreases the inputs of labour or fertilizer, or anything else leading towards improved land husbandry, which gives foundation of good soil conservation. Farming on a rade is well established in India (Swaminathan 1982). Cultivations and planting are done on a gentle gradient, sometimes together with graded channel terraces. This encourages infiltration but permits surplus run-off at low velocities. Sometimes this may be combined with simple practices to encourage infiltration such as returning crop residues. Strip cropping is most useful on gentle slopes, where it may reduce erosion to acceptable levels without any banks or drains. Rotations are another well established and simple practice. Fallowing is well established and successful in some circumstances but not others.

Mixed cropping and interplanting are widely applied traditional techniques. A combination of crops with different planting times and different length of growth periods spreads the labour requirement of planting and of harvesting and also allows mid-season change of plan according to the rain in the early part of the season (Swaminathan 1982). Another possible advantage may arise from the use of legumes to improve the nitrogen status for the cereal crop. Theme of mixed cropping, intercropping and relay cropping are better farming systems. Surface mulching has the advantage of providing protective cover at a time when crop cover is not practical. It improves infiltration and may also beneficially reduce soil temperature.

TERRACING

Terraces have widespread application in semi-arid areas. Level terraces may be appropriate where irrigation is available and intermittent level terraces are used for run-off farming. Bench terracing has also been used in many countries for soil and water conservation. There may also be circumstances where a combination of shallow soils with limited storage capacity and heavy rain, results in frequent surface run-off which calls for a system of graded channel terraces, either without storage or with some storage and a designed overflow. The problem is that any such system is likely to be expensive in relation to the productivity of the land and it is difficult to maintain grassed waterways as disposal channels when rainfall is limited and unreliable. Level terraces for dry land farming have been extensively used in the past in many countries and nowadays are increasingly not maintained or abandoned as the maintenance becomes uneconomic or impossible because of labour shortages.

CONSERVATION BENCH TERRACES (CBT)

Conservation bench terraces is another type of rainfall multiplier, using part of the land surface as a catchment to provide additional run-off onto level terraces on which crops are grown. The method is particularly appropriate for large- scale mechanized farming. There is a soil erosion hazard during high-intensity summer storms, but CBTs are as effective at controlling erosion and more effective at reducing the overall run-off. The gentle slopes of 0.5-1.5 per cent are most suitable although the system has been used up to 6 percent in some countries. As with all terracing, a steeper slope requires more earth moving.

A deep soil is required, both to provide sufficient soil moisture storage and also to lessen the effect of cutting during the construction of the terraces. Good permeability is also required so that the contained flood water can be absorbed quickly. Smooth slopes are an advantage where large mechanized farming can be made more convenient by constructing all the terraces parallel and of equal width. Precise levelling of the bench terrace is important to ensure uniform build up of soil moisture. The main design factors are the width of the levelled terrace and the ratio of the area of the terrace to the catchment area. Typical widths are from 10 meters on land of 5-6 per cent, to 30 meters at 2 per cent and 50 meters or more at one per cent. The most usual practice is for the bench terrace to be level along its length, but there have been a few trials with the terraces on a gentle gradient of 1:400 to encourage safe disposal of excessive run-off (Norman 1987).

WATER DISPOSAL

In semi-arid conditions, it is seldom appropriate to divert surface run-off from arable lands and the same arguments largely apply to cut-offs or diversion drains put in at the top edge of arable land in order to protect it from surface run-off from uncultivated higher land. There could be special circumstances, such as a shallow saturated soil which would be less damaged if water coming down from above could be diverted. The difficulty is that the drain may also divert run-off during gentle storms which might have been usefully absorbed by the arable land. The use of diversions will therefore be limited to cases where there is uncontrolled flood run-off in a channel or gully which will be wasted unless it is diverted to some useful purpose. When there is a risk that any structure intended to gather run-off may be overtopped in heavy storms, it is important to deliberately provide planned overspills which can act as safety valves and make sure that the run-off is discharged in places where it will do least damage.

BROAD BED AND FURROW SYSTEM (BBF)

The Broad Bed and Furrow system has been mainly developed at the International Crops Research Institute for the Semi-arid Tropics (ICRISAT) in India (Krantz 1981; Pathak et al 1985). It is a modern version of the very old concept of encouraging controlled surface drainage by forming the soil surface into beds. The recommended ICRISAT system consists of broad beds about 100 cm wide separated by sunken furrows about 50 cm wide. The preferred slope along the furrow is between 0.4 and 0.8 percent on vertisols. Two, three, or four rows of crop can be grown on the broad bed, and the bed width and crop geometry can be varied to suit the cultivation and planting equipment. In India, the system has been used mainly on deep vertisols (heavy black clay soils sometimes called cotton soils). Wide beds are used on a gentle grade and they are formed by ox-drawn wheeled tool carriers.

RIDGING AND TIED RIDGING

This method is also known as furrow blocking, furrow damming, furrow diking and basin listing. The principle is to increase surface storage by first making ridges and furrows, then damming the furrows with small mounds, or ties. Graded ridges alone will usually lead to an increase of surface run-off compared with flat planting, while tied ridges will decrease the run-off and increase the storage. In different seasons either of these two effects may be preferable. An interesting possibility is to use tied ridging in connection with sprinkler irrigation to allow higher application rates at low pressure (Stewart et al 1981). Tied ridging is usually associated with mechanized farming. There have been some attempts at achieving it with ox-drawn implements, but the system really needs high draught for speed and precision, which is required if the ridges are to be re-ridged or split in subsequent years. Either ridging alone or tied ridging has occasionally been practised using hand labour, but the high labour requirement usually makes this unpopular with subsistence farmers. Also hand-made ridges are usually less efficient (Norman 1987).

CONTOUR FURROWS/ CONTOUR BUNDS

They are also known as contour bunds and strip farming. These are variations on the theme of surface manipulation which require less soil movement than conservation bench terracing and are more likely to be used by small farmers, or in lower rainfall areas. The cropping is usually intermittent on strips or in rows, with the catchment area left fallow. The principle is the same as conservation bench terraces, that is to collect run-off from the catchment to improve soil moisture on the cropped area. On heavier soils, contour bunds may be less effective because of the lower infiltration and prevalent of waterlogging (ICRISAT 1976).

Contour bunds are also used for a combination of soil conservation and water conservation. If the contour furrows are not laid out precisely on the contour, or are built with some irregularities, there may be a danger of uneven depths of ponding behind the bank. This can be reduced by smaller bunds at right angles, but as with tied ridging, these bunds should be lower in height than the main ridges so that if there should be any overtopping it will be laterally along the contour and not over the bund and down the slope. Sometimes the emphasis is on the excavated furrow which collects water, so that in exceptional storms the run-off can overflow without damage.

LOW-COST MEASURES

The discussion of terracing and conventional conservation works clearly points to the use of simple and easily applied measures. The first of these should always be farming on the contour. This alone can reduce soil loss to approximately half of what it would be with cultivation up and down the slope. Although rainfall in semi-arid areas will be less in total, it can still include very damaging storms and so it will usually be beneficial to have some form of structure which will slow down the surface run-off, encourage the deposition of suspended material and reduce the concentration of surface run-off in minor depressions.

Structures on the contour are simpler and cheaper than graded channel terraces. A general term for simple structures on the contour is 'stop-wash lines'' which correctly defines their purpose. The form of such lines will depend on what materials are available. On stony ground, using the stones to build rock lines serves the dual purpose of clearing them from the field as well as building the stop-wash lines. Where stones are not available, lines can be formed by piling up crop residues, perhaps with a few shovels of soil and progressively built up later by adding weeds from hand hoeing. No design is necessary, but the general principle is that there is not much point in building large or high structures, particularly if built from stone, since they will be very permeable and in general a larger number of small barriers will be more effective than a small number of large structures.

Grass strips can also be used as stop-wash lines and live hedges with drought-resistant species are sometimes planted for the same purpose. In areas with higher rainfall, grass may be densely planted to cut for fodder and cause a terracing effect. When stop-wash lines are intended to divert water out of small channels, it is desirable to reduce the permeability at this point. This is done using the principle of the reverse filter. The main structure is composed of large stones, then on the upstream side smaller stones are packed, but large enough so that they cannot be washed through the gaps in the large stones. Upstream of the small stones a layer of gravel is added. Water will still flow through the structure, but slowly and it will build up in the depression and flow out on either side eventually finding a way through the rock barrier and continuing its path down the slope. This same principle can be used on a larger scale for gully control structures. Some applications of stone lines have the primary objective of water harvesting rather than soil conservation (Norman 1987).

CONTROL OF WIND EROSION

Any practice or measure that reduces the wind velocity or improves the soil characteristics is helpful to control wind erosion. Improved soil characteristics should have better structure, improved cohesive property and good moisture holding capacity. Some of the measures may provide both the requirements. Vegetation improves the soil structure and at the same time retards the surface wind velocity. In general, the following practices may be followed to control the wind erosion.

- The soil should be covered with vegetation or crop residues as far as possible
- Limited cultivation should be done
- Dry soils should not be tilled
- Permanent vegetation may be established on unproductive soils
- After the rains, the soil may be tilled so that clod formation takes place
- Tillage implements should be selected in such a manner that rough surface is formed and crop residue is not buried
- Overgrazing should be avoided

Principal methods of reducing surface wind velocity are vegetative control, tillage practices and mechanical method. Vegetative control consists of cultivated crops, field and strip cropping, stubble mulching, shrubs and trees to act as mechanical barrier to wind. Windbreaks may also be used as a mechanical barrier. Some important practices are discussed below:

STRIP CROPPING

Field and contour strip cropping consists of alternate strips of row (erosion-susceptible) crops and close growing (erosion-resistant) crops in the same field. The strip cropping is laid out generally parallel to the field boundary or perpendicular to the erosive wind direction. The main benefits of strip cropping are:

- Vegetation provides physical protection against blowing of soil
- Soil erosion is limited to a distance equal to the width of the erosion susceptible crop
- Better conservation of moisture takes place
- Particles carried in saltation are trapped

In a mechanized farm, movement of machinery becomes difficult due to narrow strips. In case of attack by insects, there is more number of edges for protection. The width of the strips should be selected in a way such that the farming operation is not hampered and at the same time much erosion does not take place. For example, in a sandy soil the width of the erosion susceptible crop should be limited to 6 m. But for movement of machinery, the width has to be increased. In a sandy loam soil the width can be increased up to 30 m.

WINDBREAK

A windbreak is defined as type of barrier for protection from winds and refers to any mechanical or vegetative barriers of buildings, gardens, orchards and feed lots. Windbreaks made up of just mechanical barriers are not very useful for field crops. However, they are frequently used for the protection of farm sheds and small areas. The mechanical barriers include brush fences, board walls, vertical burlap or paper strips. Brush matting, rock or gravel barriers are also used as windbreaks. Some of these barriers are impermeable and others are semi-impermeable. Generally the semi-impermeable barriers are more useful as they provide better diffusion and eddying effects on the leeward side of the barrier. When vegetable crops in organic soils are required to be protected, vertical burlap or paper strips are often used. Brush matting, debris, rock, gravel etc. are more useful for stabilizing sand dune areas.

SHELTERBELTS (SHRUBS AND TREES)

A shelterbelt usually consisting of shrubs and trees is a longer barrier than windbreak. It is primarily used for protection of field crops, soils and conservation of moisture. The shelterbelt provides larger protection and wind erosion control. It also saves fuel, increases livestock production, reduces evaporation, prevents firing of crops from hot winds. In addition, it may provide better fruiting in orchards, make spraying of trees for insect control more effective manner.

TILLAGE PRACTICES

Tillage practices if properly adopted can reduce the soil blowing to a great extent. Similarly, faulty tillage operations increase the soil erosion by wind. If the soil is pulverized and the crop residues are buried due tillage operations, erosion problem increases. The effective way of prevention of wind erosion is by producing a rough, cloddy surface and exposing the crop residues on the surface. If the land is ploughed at optimum moisture content after the rains, big clod and large aggregate formation takes place, which ultimately decreases the process of erosion.

PREDICTING SOIL EROSION

UNIVERSAL SOIL LOSS EQUATION (USLE)

By taking all factors into account, a prediction equation was developed for calculating the soil loss, called as Universal Soil Loss Equation (USLE). It accounts all parameters affecting the soils loss and predicts the annual soil loss. It computes sheet erosion. It is not used when slope are steeper than 20 per cent.

$$A = RKLSCP$$ (Wischmeier and Smith 1965)

Where,

A = Estimated gross erosion (t/ha/year)

R = Rainfall erosivity factor

K = Soil erodibility factor

L = Slope length factor

S = Slope gradient factor

C = Crop cover or vegetation management factor

P = Support conservation practice factor

Demerits of USLE: This equation does not compute the sediment yield from the watershed directly. Annual sediment yield can be determined which is less important for design of water storage structures. This equation is used in small watershed to prediction of soil loss.

MODIFIED UNIVERSAL SOIL LOSS EQUATION (MUSLE)

This equation computes sediment yield either monthly or seasonal. This equation is used in large watershed for prediction of soil loss.

$$Y = 95\ (Qq)^{0.56}\ KLSCP \qquad \text{(Williams 1975)}$$

Where,

Y = Sediment yield for an individual storm (t/ha)

Q = Volume of runoff (ha-m)

q = Peak flow rate (m^3)

$$E = f\ (IKCLV) \qquad \text{(Chepil and Woodruffridowy 1965)}$$

Where,

E = Average annual soil loss (t/ha/year)

I = Soil erodibility

K = Roughness

C = Control of soil conservation measure

L = Slope length

V = Vegetation

[Note: Soil erosion = f (Erosivity of rain) x (Erodibilty of soil)]

PART 8:

FORESTS AND CLIMATE CHANGE

38 CLIMATE CHANGE AND FORESTS

INTRODUCTION

Climate actually means the average weather experienced in a region over a long period of time. Climate in a narrow sense is usually defined as the average weather, or more rigorously, as the statistical description in terms of the mean and variability of relevant quantities over a period of time ranging from months to thousands of years. The classical period is 3 decades, as defined by the World Meteorological Organization (WMO). These quantities are most often surface meteorological variables such as temperature, precipitation and wind. Climate in a wider sense is the State, including a statistical description of the climate system. The climate on earth has undergone many changes in the past and this is entirely natural. However, the rate at which the climate has been changing over the past 50 years has led to a consensus amongst scientists that this recent change is likely to be as a result of man's activities.

Earth's average temperature has risen drastically over the past century and is projected to rise further over the next hundred years. Global average temperature has increased by 0.6 ± 0.2ºC in the past 100 years and global average precipitation has increased slightly (Barnett 2001). Small changes in the average temperature of the planet can translate to large and potentially dangerous shifts in climate and weather. Rising global temperatures have been accompanied by changes in weather and climate. Many places have seen changes in rainfall, resulting in more floods, droughts or intense rain, as well as more frequent and severe heat waves. The planet's oceans and glaciers have also experienced some big changes - oceans are warming and becoming more acidic, ice caps are melting and sea levels are rising. As these and other changes become more pronounced in the coming decades, they will likely present challenges to our society and our environment.

Over the past century, human activities have released large amounts of carbon dioxide and other greenhouse gases into the atmosphere. The majority of greenhouse gases come from burning fossil fuels to produce energy, although deforestation, industrial processes and some agricultural practices also emit gases into the atmosphere. The most important of the greenhouse gases is carbon dioxide (CO_2), which has risen from a concentration of 275 parts per million (ppm) prior to 1850 to the current value of 390 ppm. Levels of CO_2 are expected to continue rising and current estimates predict a concentration of 525-950 ppm by the end of this century. The enhancement of the 'greenhouse effect' by rising concentrations of these greenhouse gases is largely responsible for the issue of global warming. However, rising temperatures are only one consequence of the enhanced greenhouse effect, with changes to weather patterns and many aspects of the climate such as rainfall and humidity predicted. Collectively, these are referred to as 'climate change'.

Human societies have adapted to the relatively stable climate we have enjoyed since the last ice age which ended several thousand years ago. A warming climate will bring changes that can affect our water supplies, agriculture, power and transportation systems, the natural environment and even our own health and safety. Carbon dioxide can stay in the atmosphere for nearly a century, so Earth will continue to warm in the coming decades. The warmer it gets, the greater the risk for more severe changes to the climate and Earth's system. Although it's difficult to predict the exact impacts of climate change, what's clear is that the climate we are accustomed to is no longer a reliable guide for what to expect in the future.

WHAT IS CLIMATE CHANGE?

According to Intergovernmental Panel on Climate Change (IPCC), climate change in usage refers to a change in the state of the climate that can be identified (e.g. using statistical tests) by changes in the mean and/or the variability of its properties and that persists for an extended period, typically decades or longer. It refers to any change in climate over time, whether due to natural variability or as a result of human activity. United Nations Framework Convention on Climate Change (UNFCCC) defines climate change as a change of climate that is attributed directly or indirectly to human activity that alters the composition of the global atmosphere and that is in addition to natural climate variability observed over comparable time periods.

Climate change refers to any significant change in the measures of climate lasting for an extended period of time. In other words, climate change includes major changes in temperature, precipitation, or wind patterns, among other effects, that occur over several decades or longer. IPCC prediction indicates that the increase in global temperature in the year 2100 will be minimum of 1.4ºC and maximum of 5.8ºC from that of existed during the year 1900. This will owe mainly to the fact that CO_2 concentration in the atmosphere will increase to the predicted level of 970 ppm during 2100. The phenomena by which the temperature increases due to increase in CO_2 concentration is called greenhouse effect which was first described by Joseph Fourier in 1824.

CLIMATE CHANGE TERMINOLOGIES

Greenhouse Effect: Greenhouse gases act like a blanket around Earth, trapping energy in the atmosphere and causing it to warm. This phenomenon of trapping and build-up of heat in the atmosphere (troposphere) near the Earth's surface is called the greenhouse effect and is natural and necessary to support life on Earth. Some of the heat flowing back towards space from the Earth's surface is absorbed by water vapour, carbon dioxide, ozone and several other gases in the atmosphere and then reradiated back towards the Earth's surface. If the atmospheric concentrations of these greenhouse gases rise, the average temperature of the lower atmosphere will gradually increase.

Greenhouse Gas (GHG): Any gas that absorbs infrared radiation in the atmosphere. Greenhouse gases include carbon dioxide (CO_2), methane (CH_4), nitrous oxide (N_2O), ozone (O_3), chlorofluorocarbons (CFCs), hydrofluorocarbons (HFCs), hydrochlorofluorocarbons (HCFCs), perfluorocarbons (PFCs), sulfur hexafluoride.

Global Warming: It refers to the recent and ongoing rise in global average temperature near Earth's surface. It is caused mostly by increasing concentrations of greenhouse gases in the atmosphere. Global warming is causing climate patterns to change. However, global warming itself represents only one aspect of climate change.

Climate Change Mitigation: Climate change mitigation may be defined as any attempt to reduce the rate at which greenhouse gasses are accumulating in the atmosphere. The two options that can be considered are: reducing emissions and carbon sequestration.

Carbon Capture and Sequestration: Carbon capture and sequestration (CCS) is a set of technologies that can greatly reduce carbon dioxide emissions from new and existing coal- and gas-fired power plants, industrial processes and other stationary sources of carbon dioxide. It is a three-step process that includes capture of carbon dioxide from power plants or industrial sources; transport of the captured and compressed carbon dioxide (usually in pipelines); and underground injection and geologic sequestration, or permanent storage, of that carbon dioxide in rock formations that contain tiny openings or pores that trap and hold the carbon dioxide.

Climate Change Adaptation: UNFCCC defines climate change adaptation as the actions taken to help communities and ecosystems cope with changing climate condition. The IPCC describes it as adjustment in natural or human systems in response to actual or expected climatic stimuli or their effects, which moderates harm or exploits beneficial opportunities. The UNDP calls it a process by which strategies to moderate, cope with and take advantage of the consequences of climatic events are enhanced, developed, and implemented.

Carbon Dioxide Fertilization: The enhancement of the growth of plants as a result of increased atmospheric CO_2 concentration. Depending on their mechanism of photosynthesis, certain types of plants are more sensitive to changes in atmospheric CO_2 concentration.

Alternative Energy: Energy derived from non-traditional sources (e.g., compressed natural gas, solar, hydroelectric, wind).

Climate Model: A quantitative way of representing the interactions of the atmosphere, oceans, land surface and ice. Models can range from relatively simple to quite comprehensive.

El Nino - Southern Oscillation (ENSO): El Nino, in its original sense, is a warm water current that periodically flows along the coast of Ecuador and Peru, disrupting the local fishery. This oceanic event is associated with a fluctuation of the intertropical surface pressure pattern and circulation in the Indian and Pacific Oceans, called the Southern Oscillation. This coupled atmosphere-ocean phenomenon is collectively known as El Nino-Southern Oscillation. During an El Nino event, the prevailing trade winds weaken and the equatorial countercurrent strengthens, causing warm surface waters in the Indonesian area to flow eastward to overlie the cold waters of the Peru current. This event has great impact on the wind, sea surface temperature and precipitation patterns in the tropical Pacific. It has climatic effects throughout the Pacific region and in many other parts of the world.

La Nina: La Nina is a coupled ocean-atmosphere phenomenon that is the counterpart of El Nino. During a period of La Nina, the sea surface temperature across the equatorial Eastern Central Pacific Ocean will be lower than normal by 3-5ºC. La Nina, sometimes informally called "anti-El Nino", is the opposite of El Nino and its effects are often the reverse of those of El Nino.

Heat Island: An urban area characterized by temperatures higher than those of the surrounding non-urban area. As urban areas develop, buildings, roads and other infrastructure replace open land and vegetation. These surfaces absorb more solar energy, which can create higher temperatures in urban areas.

Heat Waves: A prolonged period of excessive heat, often combined with excessive humidity.

STERN REVIEW

Though the greenhouse effect was described in 1824, only recently this impact is being talked and discussed both by developed and developing countries in the world. This may mainly be due to a report prepared by Sir Nicholas Stern on Economical Implications of Climate Change. The report is commonly known as 'Stern Review.' The Stern Review on the Economics of Climate Change is a 700-page report released for the British Government on 30 October 2006 by economist Nicholas Stern, chair of the Grantham Research Institute on Climate Change and the Environment at the London School of Economics and also chair of the Centre for Climate Change Economics and Policy (CCCEP) at Leeds University. The report discusses the effect of global warming on the world economy. Although not the first economic report on climate change, it is significant as the largest and most widely known and discussed report of its kind (Francis Cairncross 2006).

According to the Review, without action, the overall costs of climate change will be equivalent to losing at least 5 per cent of global gross domestic product (GDP) each year, now and forever. A wider range of risks and impacts could increase this to 20 per cent of GDP or more (Fig. 38.1). The Review proposes that one percent of global GDP per annum is required to be invested in order to avoid the worst effects of climate change. In June 2008, Stern increased the estimate for the annual cost of achieving stabilization between 500 and 550 ppm CO_2e to 2 per cent of global GDP to account for faster than expected climate change (Stern 2006).

EXECUTIVE SUMMARY OF THE STERN REVIEW

- The benefits of strong and early action on climate change outweigh the costs.
- The scientific evidence points to increasing risks of serious, irreversible impacts from climate change associated with business-as-usual (BAU) paths for emissions.
- Climate change threatens the basic elements of life for people around the world: access to water, food production, health and use of land and the environment.
- The impacts of climate change are not evenly distributed. The poorest countries and people will suffer earliest and most. And if and when the damages appear, it will be too late to reverse the process. Thus, we are forced to look a long way ahead.

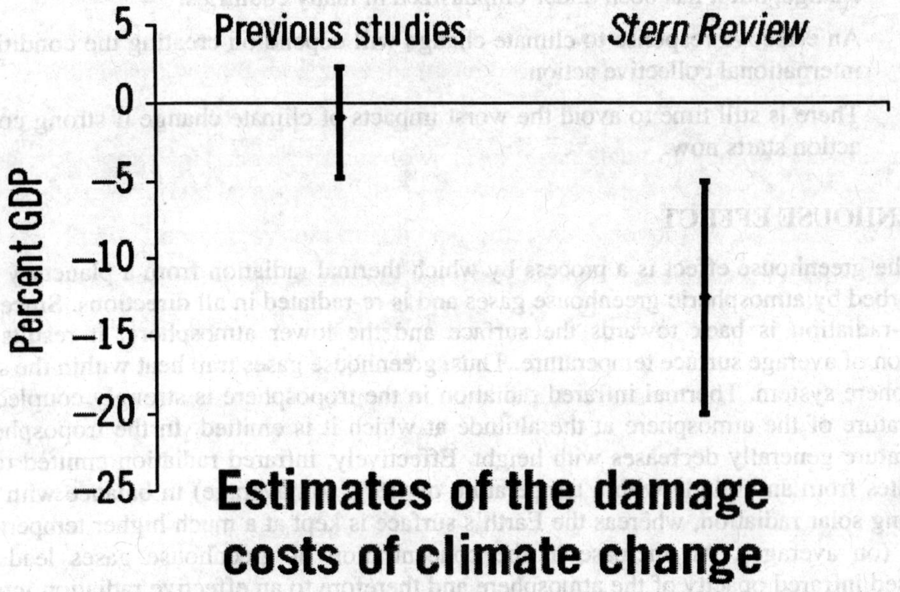

Fig. 38.1: Costs of Climate Change as percentage of GDP

- Climate change may initially have small positive effects for a few developed countries, but it is likely to be very damaging for the much higher temperature increases expected by mid-to-late century.
- Integrated assessment modelling provides a tool for estimating the total impact on the economy and estimates suggest that this is likely to be higher than previously suggested.
- Emissions have been, and continue to be, driven by economic growth and yet stabilization of greenhouse gas concentration in the atmosphere is feasible and consistent with continued growth.
- Central estimates of the annual costs of achieving stabilization between 500 and 550 ppm CO_2e are around 2 per cent of global GDP, if we start to take strong action now. It would already be very difficult and costly to aim to stabilise at 450 ppm CO_2e. If delayed, the opportunity to stabilize at 500-550 ppm CO_2e may slip away.
- The transition to a low-carbon economy will bring challenges for competitiveness but also opportunities for growth. Policies to support the development of a range of low-carbon and high-efficiency technologies are required urgently.
- Establishing a carbon price, through tax, trading or regulation, is an essential foundation for climate change policy. Creating a broadly similar carbon price signal around the world and using carbon finance to accelerate action in developing countries are urgent priorities for international cooperation.

- Adaptation policy is crucial for dealing with the unavoidable impacts of climate change, but it has been under-emphasized in many countries.
- An effective response to climate change will depend on creating the conditions for international collective action.
- There is still time to avoid the worst impacts of climate change if strong collective action starts now.

GREENHOUSE EFFECT

The greenhouse effect is a process by which thermal radiation from a planetary surface is absorbed by atmospheric greenhouse gases and is re-radiated in all directions. Since part of this re-radiation is back towards the surface and the lower atmosphere, it results in the elevation of average surface temperature. Thus, greenhouse gases trap heat within the surface-troposphere system. Thermal infrared radiation in the troposphere is strongly coupled to the temperature of the atmosphere at the altitude at which it is emitted. In the troposphere, the temperature generally decreases with height. Effectively, infrared radiation emitted to space originates from an altitude with a temperature of -19ºC (on average) in balance with the net incoming solar radiation, whereas the Earth's surface is kept at a much higher temperature of +14ºC (on average). An increase in the concentration of greenhouse gases leads to an increased infrared opacity of the atmosphere and therefore to an effective radiation into space from a higher altitude at a lower temperature. This causes a radiative forcing that leads to an enhancement of the greenhouse effect, the so called enhanced greenhouse effect (IPCC 2007). This greenhouse effect only keep the earth warmer at 33ºC (from -18ºC to +15ºC) thus making it habitable and allowed the present form of life to evolve. Without this greenhouse effect, the world would be freezing at about -17ºC.

The greenhouse effect is caused not exclusively by CO_2 and there are other gaseous elements which also contribute to this effect such as water vapour, methane, nitrous oxide and ozone. Moreover, there are a number of entirely human-made greenhouse gases in the atmosphere, such as the halocarbons and other chlorine and bromine containing substances. These gases are collectively called green house gases (GHGs). However, 72 per cent of the totally emitted greenhouse gases are carbon dioxide (CO_2), 18 per cent methane and 9 per cent nitrous oxide (N_2O). The potential of each GHG to cause warming is highly varying.

The concept of a global warming potential (GWP) was developed to compare the ability of each greenhouse gas to trap heat in the atmosphere relative to another gas. The definition of a GWP for a particular greenhouse gas is the ratio of heat trapped by one unit mass of the greenhouse gas to that of one unit mass of CO_2 over a specified time period. The Table 38.1 presents the potential of various GHGs to cause global warming, if we consider the potential of CO_2 as one.

Table 38.1: Tropospheric Concentration, GWP, Lifetime, Increased Radiative Forcing of Greenhouse Gases (Blasing 2012)

GAS	Pre-1750 Tropospheric Concentration	Recent Tropospheric Concentration	GWP (100 yr time horizon)	Atmospheric Lifetime (years)	Increased Radiative Forcing (W/m^2)
Concentrations in parts per million (ppm)					
Carbon Dioxide (CO_2)	280	390.5	1	~ 100	1.79
Concentrations in parts per billion (ppb)					
Methane (CH_4)	700	1,871	25	12	0.50
Nitrous Oxide (N_2O)	270	323	298	114	0.18
Tropospheric Ozone (O_3)	25	34	-NA-	hours-days	0.35
Concentrations in parts per trillion (ppt)					
CFC-11 (Trichlorofluoromethane) (CCl_3F)	zero	241	4,750	45	0.060
CFC-12 (CCl_2F_2)	zero	534	10,900	100	0.17
CF-113(CCl_2FFClF_2)	zero	75	6,130	85	0.024
HCFC-22($CHClF_2$)	zero	220	1,810	12	0.041
HCFC-141b(CH_3CCl_2F)	zero	22	725	9.3	0.0025
HCFC-142b(CH_3CClF_2)	zero	22	2,310	17.9	0.0031
Halon 1211 ($CBrClF_2$)	zero	4.3	1,890	16	0.001
Halon 1301 ($CBrClF_3$)	zero	3.3	7,140	65	0.001
HFC-134a(CH_2FCF_3)	zero	64	1,430	14	0.0055
Carbon tetrachloride (CCl_4)	zero	87	1,400	26	0.012
Sulfur hexafluoride (SF_6)	zero	7.41	22,800	3200	0.0029

Though the naturally occurring greenhouse effect is beneficial, in the recent past this effect is magnified to manifold mainly due to ever increasing concentration of GHGs more particularly CO_2. Atmospheric CO_2 is accelerating upward from decade to decade. For the past ten years, the average annual rate of increase is 2.07 parts per million (ppm). This rate of increase is more than double the increase in the 1960s (Table 38.2).

Table 38.2: Total Increase and Annual Rate of Increase in CO_2 during different Decades (Source: co2now.org)

Decade	Total Increase (ppm)	Annual Rate of Increase (ppm per year)
2002 – 2011	20.72	2.07
1992 – 2001	16.00	1.60
1982 – 1991	15.10	1.51
1972 – 1981	13.95	1.40
1962 – 1971	8.88	0.89

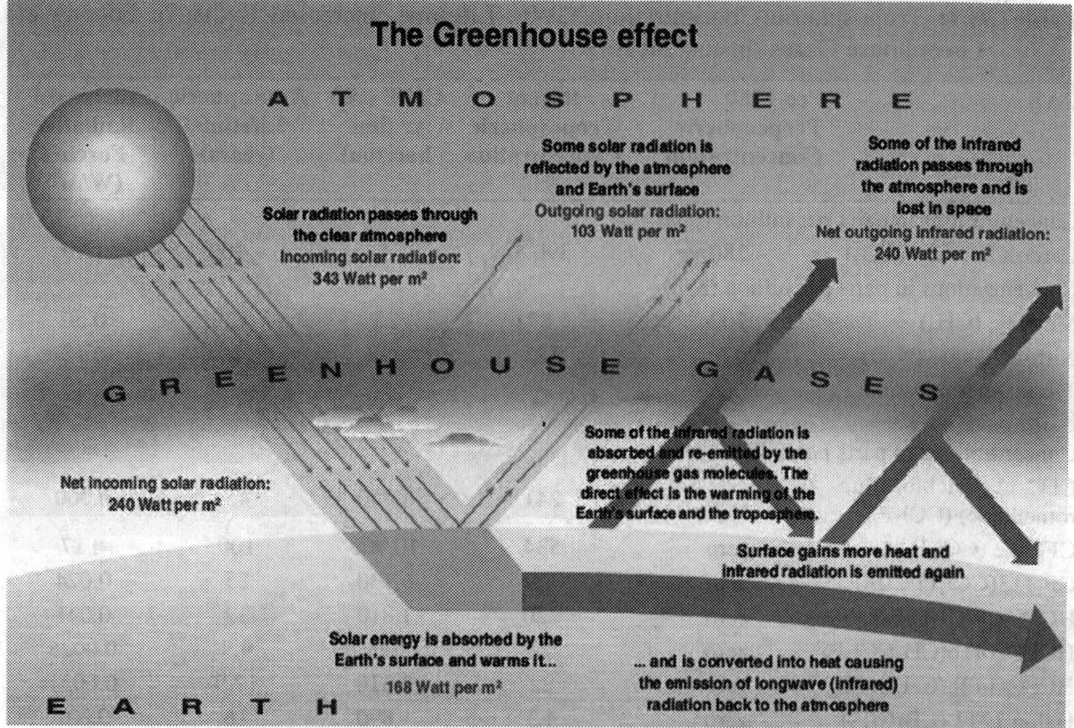

Fig. 38.2: The Greenhouse Effect (Source: UNEP, Oknagan University College, University of Oxford, EPA, IPCC)

DENDROCLIMATOLOGY AND CLIMATE CHANGE

Dendroclimatology is the science of determining past climates from trees (primarily based on properties of the annual tree rings). Tree rings are wider when conditions favour growth, narrower when times are difficult. Other properties of the annual rings, such as maximum latewood density (MXD) have been shown to be better proxies than simple ring width. Using tree rings, scientists have estimated many local climates for hundreds to thousands of years previous. By combining multiple tree-ring studies (sometimes with other climate proxy records), scientists have estimated past regional and global climates. The climate variables which can be analyzed from the tree-ring record are: (1) precipitation, (2) temperature, and (3) Palmer drought severity index (PDSI). Indian Institute of Tropical Meteorology (IITM), Pune has been actively involved in dendroclimatological studies and major highlights of the studies conducted so far are:

- 41 ring width and 6 ring density chronologies from Western Himalaya have been made available
- Longest chronology of 725 years has been studied from wood of *Cedrus deodara* from Gongotri
- It is observed that pre-monsoon summer (March-May) climate is the most responsive parameter for tree ring variations (Borgaonkar et al 1999)

- **Higher growth in recent years observed in high altitude near glacier tree-ring chronologies of western Himalaya has been associated with increasing temperature trend**
- **38 ring width chronologies have been established from wood of Teak (*Tectona grandis*) from central and peninsula India**
- **Teak tree-ring chronologies from central and south India indicate positive relationship with monsoon and annual rainfall and PDSI proposes the important role of moisture in tree growth and climate relationship**
- **Tree-ring drought records of Indian monsoon rainfall have been created since past five centuries.**

CLIMATE CHANGE AND ITS INFLUENCES ON FORESTS

CLIMATE CHANGE EFFECTS ON FOREST GROWTH

Climate influences the structure and function of forest ecosystems and plays an essential role in forest health. A changing climate may worsen many of the threats to forests such as pest outbreaks, fires, human development and drought. Climate changes directly and indirectly affect the growth and productivity of forests: directly due to changes in atmospheric carbon dioxide and climate and indirectly through complex interactions in forest ecosystems. Climate also affects the frequency and severity of many forest disturbances. In conjunction with the projected impacts of climate change, forests face impacts from land development, suppression of natural periodic forest fires and air pollution. Although it is difficult to separate the effects of these different factors, the combined impact is leading to changes in forests. As these changes are likely to continue in the decades ahead, some of the valuable goods and services provided by forests may be compromised. Many aspects of projected climate change will likely affect forest growth and productivity. Three major factors are considered to influence the forests more viz. increase in carbon dioxide (CO_2), increase in temperature, and change in precipitation (USEPA 2012).

CARBON DIOXIDE: It is required for photosynthesis, the process by which green plants use sunlight to grow. Given sufficient water and nutrients, increases in atmospheric CO_2 may enable trees to be more productive (CCSP 2008). Higher future CO_2 levels could benefit forests with fertile soils. However, increased CO_2 may not be as effective in promoting growth in some forest areas, where water is limited (USGCRP 2009).

TEMPERATURE: Warming temperatures could increase the length of the growing season. However, warming could also shift the geographic ranges of some tree species. Habitats of some types of trees are likely to move northward or to higher altitudes. Other species may be at risk locally or regionally if conditions in their current geographic range are no longer suitable (CCSP 2008). For example, species that currently exist only on mountaintops in some regions may die out as the climate warms since they cannot shift to a higher altitude.

RAINFALL: Climate change will likely increase the risk of drought in some areas and the risk of extreme precipitation and flooding in others. Increased temperatures would alter the timing of snowmelt, affecting the seasonal availability of water. Although many trees are resilient to some degree of drought, increase in temperature could make future droughts more damaging than those experienced in the past. In addition, drought increases forest fire risk, since dry trees and shrubs provide fuel to fires. Drought also reduces tree's ability to produce sap, which protects them from destructive insects (USGCRP 2009).

CLIMATE CHANGE EFFECTS ON FOREST DISTURBANCES

Climate change could alter the frequency and intensity of forest disturbances such as insect outbreaks, invasive species, forest fires and storms. These disturbances can reduce forest productivity and change the distribution of tree species. In some cases, forests can recover from a disturbance. In other cases, existing species may shift their range or die out. In these cases, the new species of vegetation that colonize the area create a new type of forest. Disturbances can interact with one another, or with changes in temperature and precipitation, to increase risks to forests. For example, drought can weaken trees and make a forest more susceptible to forest fire or insect outbreaks. Similarly, forest fire can make a forest more vulnerable to pests.

INSECT OUTBREAKS: Insect outbreaks often defoliate, weaken and kill trees. Insects sensitive to weather conditions may migrate to conducive areas and cause huge destruction to forest areas. A lack of natural controls, such as predators, or pathogens, or inadequate defences in trees, can allow insects to spread. Climate change could contribute to an increase in the severity of future insect outbreaks. Rising temperatures may enable some insect species to develop faster and expand their ranges.

INVASIVE PLANT SPECIES: Invasive plant species can displace important native vegetation because the invasive species often lack natural predators. Climate change could benefit invasive plants, since they are generally more tolerant to a wider range of environmental conditions than are native plants (USGCRP 2009).

FOREST FIRE: In recent years, forest fires cause huge destruction to more forest areas in dry areas. Warm temperatures and drought conditions during the summer contribute to the occurrence of forest fires. Climate change is projected to increase the extent, intensity and frequency of forest areas in fire prone areas. Warmer temperature coupled with decrease in water availability dry out woody materials in forests and increases the risk of forest fire. Fires can also contribute to climate change, since they can cause rapid, large releases of carbon dioxide to the atmosphere (CCSP 2008).

STORMS: Hurricanes, ice storms and wind storms can cause damage to forests. They cause release of huge amount of carbon by these trees.as they decay in a single year.

CLIMATE CHANGE EFFECTS ON FOREST PRODUCTIVITY

Forests respond to both short-term and longer-term variations in the environment (Innes and Peterson 2001). Although all biological activity in plants is ultimately dependent on absorbed solar radiation, it is obvious that solar radiation alone does not determine primary productivity. All plants require sunlight, carbon dioxide and water for photosynthesis. Beyond these basic requirements, the amount of foliage, the light-use efficiency of the foliage, water availability, ambient temperature, availability of soil nutrients, the adaptations of species to extreme temperatures and efficient use of water and nutrients are finer controls of forest productivity. The main abiotic controls of primary production (temperature, radiation and water) interact to impose complex and varying limitations on vegetation activity in different parts of the world (Nemani et al 2003). Physiological responses to changes in climate are highly dependent on the limiting factors of a particular site to forest growth. For example, increasing temperature may also increase vapour pressure deficit (VPD) of the air and thereby increase transpiration rates, resulting in adverse effects on dryer sites, unless stomata closes in response to other changes such as an increase in CO_2, or if increase in night-time temperature exceeds increase during the day.

Temperature (heat) controls the rate of plant metabolism, which in turn determines the amount of photosynthesis that can take place. Most biological metabolic activity takes place within the range of 0-50°C (Hopkins and Huner 2004). There is little activity above or below this range. The optimal temperatures for productivity coincide with 15-25°C which is the optimal range for photosynthesis (Hopkins and Huner 2004) and lethal levels are in between 44-52°C (Schulze et al 2002). In dry regions, there is a linear increase in NPP with increased water availability. With reference to aboveground biomass, low water availability in dry areas decreases productivity.

A large part of the recent increase in temperature records reflects a rise in minimum temperatures, whereas maximum temperatures may remain stable or actually decrease, especially during the summer season. Hence, systems limited by maximum temperature may not have shown any corresponding changes in productivity. Tropical forest regions show temperature increases averaging 0.26 ± 0.05°C since the mid-1970s. Overall precipitation appears to have declined in tropical rainforest regions at a rate of 1.0 ± 0.8 per cent per decade since 1960 (Malhi and Wright 2004). This pan-tropical decreasing trend in land surface measurements of rainfall, however, is primarily driven by a strong and significant decline in rainfall in the northern African tropics (3-4 per cent per decade). Rainfall only declined marginally in tropical Asia and showed no significant trend in Amazonia (Malhi and Wright 2004).

A review by Boisvenue and Running (2006) has reported positive, negative or no trend in tree growth or forest productivity for different time periods and geographic areas after analyzing many research papers reporting forest production levels with reference to climate change (Table 38.2). The Intergovernmental Panel on Climate Change have concluded that forests are highly sensitive to modern climate change. Although the details of expected change in forests on a regional scale are unclear, the scenarios of general circulation models (GCMs) predict major shifts in the area occupied by forest biomes. For example, globally, the area occupied by tropical and temperate forests is projected to expand by upto 20 per cent, whereas boreal forests may decline by 50 per cent, if other causes of change are ignored. Moreover, the rate of climate change over the next century may be faster than most historic changes, suggesting that adjustments forests have made to changes in the past may be more difficult today.

Noss (2001) suggests the following land-use and management practices likely to maintain forest biodiversity and ecological functions during climate change: (1) representing forest types across environmental gradients in reserves, (2) protecting climatic refugia at multiple scales, (3) protecting primary forests, (4) avoiding fragmentation and providing connectivity, especially parallel to climatic gradients, (5) providing buffer zones for adjustment of reserve boundaries, (6) practicing low-intensity forestry and preventing conversion of natural forests to plantations, (7) maintaining natural fire regimes,

(8) maintaining diverse gene pools, and (9) identifying and protecting functional groups and keystone species.

Table 38.3: Summary of Published Change in Forest Productivity under recent Climate Change (Source: Boisvenue and Running 2006)

Variable	Change	Time Scale	Spatial Scale	Data Type	Reference
ANPP	+	1982–1999	Globe	Satellite	Nemani et al (2003)
ANPP	-	ENSO years	Tropics	Satellite	Nemani et al (2003)
ANPP	-	ENSO years	Tropics	Satellite	Braswell et al (1997)
ANPP	-	ENSO years	Tropics	Satellite	Asner et al (2000)
ANPP	-	ENSO years	Tropics	Satellite	Los et al (2001)
NDVI	+	1982–1999	Northern hemisphere	Satellite	Slayback et al (2003)
ANBP	+	Growing-season	Boreal aspen	Land-flux tower	Chen et al (1999)
ANPP	+	Growing-season	North-western Europe	Satellite	Lucht et al (2002)
ANPP	+	1982–1999	China	Satellite	Fang et al (2003)
ANBP	-	1949–1980	China	Land biomass	Fang et al (2001a)
ANBP	+	1970s–1998	China planted forests	Land biomass	Fang et al (2001a)
ANBP	+	1982–1998	North America	Satellite	Hicke et al (2002)
ANBP	+	1945–1990	USA	Combined types	Houghton et al (1999)
NBP	+	1980–1989	USA	Atmospheric and land-flux tower	Pacala et al (2001)
NBP	+	1990s	European forests and grasslands	Combined types	Janssens et al (2004)
ANPP	+	Recent	Canadian & Alaska boreal forests	Combined types	Innes and Peterson (2001)
ANPP	-	1951–2000	Canadian prairies boreal aspen	Land biomass	Hogg et al (2005)
ANPP	+	Since 1970	Boreal/tundra forest	Land biomass	Gamache and Payette (2004)
NBP	+	1950–1999	European forests	Land biomass	Nabuurs et al (2003)
ANPP	+	20th century	Canadian Cordillera	Land biomass	Luckman et al (2004)
ANBP	+	During PDO events	Pacific Northwest	Land biomass	Peterson and Peterson (2001)
ANBP	+	During PDO events	Pacific Northwest	Land biomass	Peterson et al (2002)
ANPP	-	1926–2001	Northwest Russia	Land-soil fluxes	Lawrence et al (2005)
ANPP	+	20th century	Scots Pine in Lapland	Land biomass	Mielikainen and Sennov (1996)
ANPP	0	20th century	Scots Pine southern Finland	Land biomass	Mielikainen and Sennov (1996)
ANPP	+	20th century	Sweden	Land biomass	Elfving et al (1996)

ANPP	+	1953–1992	Swedish forest	Land biomass	Ericksson and Karlsson (1996)
ANPP	+	Since 1960s	Sweden	Land biomass	Ericksson and Karlsson (1996)
ANPP	+	1920s–1990	Danish beech forest	Land biomass	Skovsgaard & Henriksen (1996)
ANPP	+	1975–2000	Eastern Germany	Land biomass	Untheim (1996)
ANPP	+	Last 150 years	France	Land biomass	Badeau et al (1996)
ANPP	+	Since 1905	Toppwald Switzerland	Land biomass	Braker (1996)
ANPP	+	20th century	Norway spruce in the Swiss Jura	Land biomass	Schneider and Hartmann (1996)
ANBP	+	1947–1995	Japan	Land biomass	Fang et al (2005)
ANPP	+	100 years	Various parts of Switzerland	Land biomass	Zingg (1996)
ANPP	+	Since 1961	Austria	Land biomass	Schadauer (1996)
ANPP	+	Since 1961	Austria	Land biomass	Hasenauer et al (1999)
ANPP	+	1947–1990	Slovanian beech forests	Land biomass	Kotar (1996)

[Variables: net primary production (NPP), net ecosystem exchange (NEE), net biome production (NBP), above-ground NPP (ANPP), above-ground NEP (ANEP) and above-ground NBP (ANBP)]

CLIMATE CHANGE AND BIODIVERSITY

Biodiversity is the term given to the variety of life on Earth and it provides goods and services that sustain our lives. Human pressures on ecosystems are causing changes and losses at rates not seen historically. People are changing ecosystems more rapidly and more extensively than over any other period in human history. Climate change adds yet another pressure on natural ecosystems. According to the Millennium Ecosystem Assessment, climate change is likely to become the dominant direct driver of biodiversity loss by the end of the century. Projected changes in climate, combined with land use change and the spread of exotic or alien species are likely to limit the capability of some species to migrate and therefore will accelerate species loss. The Millennium Ecosystem Assessment (MA 2005) assessed the consequences of ecosystem change for human well-being. From 2001 to 2005, the MA involved the work of more than 1,360 experts worldwide. Their findings provide a state-of-the-art scientific appraisal of the condition and trends in the world's ecosystems and the services they provide, as well as the scientific basis for action to conserve and use them sustainably.

As per the MA report, over the past few hundred years, humans have increased species extinction rates by as much as 1,000 times background rates that were typical over Earth's history. There are approximately 100 well documented extinctions of birds, mammals and amphibians over the last 100 years, a rate 100 times higher than background rates. The most important direct drivers of biodiversity loss and change in ecosystem services are:

- Habitat change such as land use change, physical modification of rivers or water withdrawal from rivers, loss of coral reefs and damage to sea floors due to trawling, etc.

- Climate change
- Invasive alien species
- Overexploitation of species
- Pollution

With reference to climate change and biodiversity, MA reports that observed recent changes in climate, especially warmer regional temperatures, have already had significant impacts on biodiversity and ecosystems. It includes causing changes in species distributions, population sizes, the timing of reproduction or migration events and an increase in the frequency of pest and disease outbreaks. Many coral reefs have undergone major, although often partially reversible, bleaching episodes when local sea surface temperatures have increased during one month by 0.5-1ºC above the average of the hottest months. By the end of the twenty first century, climate change and its impacts may be the dominant direct driver of biodiversity loss and changes in ecosystem services globally.

The Intergovernmental Panel on Climate Change projects an increase in global mean surface temperature, increased incidence of floods and droughts and a rise in sea level of an additional 8-88 cm between 1990 and 2100. The impact on biodiversity will grow worldwide with both increasing rates of change in climate and increasing absolute change in climate. Some ecosystem services in some regions may initially be enhanced by projected changes in climate (such as increases in temperature or precipitation). Thus these regions may experience net benefits at low levels of climate change, as climate change becomes more severe, the harmful impacts on ecosystem services are likely to outweigh the benefits in most regions of the world. The balance of scientific evidence suggests that there will be a significant net harmful impact on ecosystem services worldwide if global mean surface temperature increases more than 2ºC above preindustrial levels or at rates greater than 0.2ºC per decade.

Climate change is projected to further adversely affect key development challenges, including providing clean water, energy services and food; maintaining a healthy environment; and conserving ecological systems and their biodiversity and associated ecological goods and services. The major projections reported in the MA report are:

- Climate change is projected to exacerbate the loss of biodiversity and increase the risk of extinction for many species, especially those already at risk due to factors such as low population numbers, restricted or patchy habitats and limited climatic ranges (with medium to high certainty).
- Water availability and quality are projected to decrease in many arid and semiarid regions (with high certainty).
- The risk of floods and droughts is projected to increase (with high certainty).
- The reliability of hydropower and biomass production is projected to decrease in some regions (with high certainty).
- The incidence of vector-borne diseases such as malaria and dengue and of waterborne diseases such as cholera is projected to increase in many regions (with medium to high certainty).

- Heat stress mortality and threats of decreased nutrition in many regions, along with severe weather traumatic injury and death (with high certainty).

KEY MESSAGES IN MA REPORT

- Biodiversity benefits people through more than just its contribution to material welfare and livelihoods.
- Biodiversity contributes to security, resiliency, social relations, health and freedom of choices and actions.
- Changes in biodiversity due to human activities were more rapid in the past 50 years than at any time in human history and the drivers of change that cause biodiversity loss and lead to changes in ecosystem services are both steady and show no evidence of declining over time, or increasing in intensity. Under the four plausible future scenarios developed by the MA, these rates of change in biodiversity are projected to continue or accelerate.
- Many people have benefited over the last century from the conversion of natural ecosystems to human-dominated ecosystems and from the exploitation of biodiversity. At the same time, however, these gains have been achieved at growing costs in the form of losses in biodiversity, degradation of many ecosystem services and the exacerbation of poverty for other groups of people.
- Improved valuation techniques and information on ecosystem services demonstrate that although many individuals benefit from biodiversity loss and ecosystem change, the costs borne by society of such changes are often higher.
- For achieving greater progress towards biodiversity conservation to improve human well-being and reduce poverty, it will be necessary to strengthen response options that are designed with the conservation and sustainable use of biodiversity and ecosystem services as the primary goal. These responses will not be sufficient, however, unless the indirect and direct drivers of change are addressed and the enabling conditions for implementation of the full suite of responses are established.
- Short-term goals and targets are not sufficient for the conservation and sustainable use of biodiversity and ecosystems. Given the characteristic response times for political, socioeconomic and ecological systems, longer-term goals and targets (such as for 2050) are needed to guide policy and actions.
- Improved capability to predict the consequences of changes in drivers for biodiversity, ecosystem functioning and ecosystem services, together with improved measures of biodiversity, would aid decision-making at all levels.
- Science can help ensure that decisions are made with the best available information, but ultimately the future of biodiversity will be determined by society.

GLOBAL INITIATIVES ON CLIMATE CHANGE

The Intergovernment Panel on Climate Change (IPCC) was established in 1988 for the purpose of controlling GHGs in the atmosphere and to provide the world with a clear scientific view on the current state of climate change. IPCC is the leading body for the assessment of climate change, established by the United Nations Environment Programme (UNEP) and the World Meteorological Organization (WMO). It is a scientific body which reviews and assesses the most recent scientific, technical and socio-economic information produced worldwide relevant to the understanding of climate change. Since 1988, the IPCC has produced four comprehensive Assessment Reports. The first assessment report was published during 1990, the Second Assessment Report during 1995 and the Third Assessment Report during 2000. The last Assessment Report (AR4) was released in November 2007.

United Nations Organization also created a global forum called "United Nations Framework Convention on Climate Change (UNFCCC)" during Rio Earth Summit in 1992 to tackle the problems of climate change. The ultimate objective of the Convention was to stabilize GHGs concentrations in the atmosphere at a level that will prevent dangerous human interference with the climate system. Recognizing that developed countries are principally responsible for the current high levels of GHG emissions in the atmosphere as a result of more than 150 years of industrial activity, the Protocol placed a heavier burden on developed nations under the principle of "common but differentiated responsibilities."

As a part of solution to this problem of increasing levels of CO_2 and other GHGs in the atmosphere, Kyoto Protocol was evolved in the COP 3 meeting held in Japan during 1997 by UNFCCC. The detailed rules for the implementation of the Protocol were adopted at COP 7 in Marrakesh in 2001 and are called the "Marrakesh Accords." The most important aspect of the Kyoto Protocol is its legally binding commitments for 39 developed countries to reduce their greenhouse gas (GHG) emissions by an average of 5.2 per cent relative to 1990 levels. These emission reductions must be achieved by 2008-2012: the so called 'first commitment period'. The developed countries with emission reduction targets are called the Annex 1 countries, whereas those without targets are the non-Annex 1 countries.

The Kyoto Protocol allows developed countries to achieve their targets in different ways through 'Flexibility Mechanisms'. Three mechanisms were suggested in this protocol viz. (1) Emissions Trading (trading of emission allowances between developed nations); (2) Joint Implementation (transferring emission allowances between developed nations, linked to specific emission reduction projects) and (3) Clean Development Mechanism (CDM). CDM is the only Flexibility Mechanism that involves developing countries. It allows developed nations to achieve part of their reduction obligations through projects in developing countries that reduce emissions or fix/sequester CO_2 from the atmosphere. Under the CDM, mitigation and adaptation are the two major solutions for control of GHGs in atmosphere. At present, major activities in CDM are biogas plants, solar photovoltaics, solar water heating, wind power, small hydro power and biomass power. With reference to forestry sector in CDM, a decision was made in the COP-7 of UNFCC in Bonn (Germany) in July 2001 to include "Afforestation and Reforestation (A/R)" as an effective way to reduce atmospheric carbon by building up terrestrial carbon stocks and to produce Certified Emission Reductions (CERs) in the first commitment period of the Kyoto Protocol (2008–2012).

A UN climate change conference in Durban concluded during December 2011 after negotiators from more than 190 countries agreed to consider a new document that would carry "legal force" and apply to both the industrialized countries and large emitters among emerging economies, like India and China. This Durban conference agreed to extend the Kyoto Protocol for a second period from January 2013 to December 2017 and decided that the new pact would be implemented from 2020 onwards. The conference delegates agreed to raise "the level of ambition" to work towards emissions management that would keep the average rise in global temperatures below 2^0 C in a bid to reduce the risk of catastrophic impacts of global warming and climate change caused by the greenhouse gas emission.

REDUCTION OF EMISSION FROM DEFORESTATION AND FOREST DEGRADATION IN DEVELOPING COUNTRIES (REDD)

The Intergovernmental Panel on Climate Change (IPCC) estimates that the cutting down of forests is now contributing close to 20 per cent of the overall greenhouse gases entering the atmosphere. In this context, the UNFCCC agenda item on "Reducing emissions from deforestation in developing countries and approaches to stimulate action" was first introduced at the Conference of the Parties (COP11) in December 2005. In response to the COP13 decision in Bali, requests from countries and encouragement from donors, FAO, UNDP and UNEP have developed a collaborative REDD programme. "REDD+" goes beyond deforestation and forest degradation and includes the role of conservation, sustainable management of forests and enhancement of forest carbon stocks. It is predicted that financial flows for greenhouse gas emission reductions from REDD+ could reach up to US$30 billion a year. This significant North-South flow of funds could reward a meaningful reduction of carbon emissions and could also support new, pro-poor development, help conserve biodiversity and secure vital ecosystem services.

OBSERVED CLIMATE CHANGES IN INDIA

In Indian scenario, the total GHGs production is 1,228 metric tons per year and out of which CO_2 contribution is 64 per cent and of methane and N_2O is 32 and 4 per cent respectively (Sharma et al 2006). India's Initial National Communication-2004 (NATCOM 1) to UNFCC has consolidated some observed changes in Climate in India.

- At national level, increase of 0.4^0 C has been observed in surface temperature over the past century. A warming trend has been observed along the west coast, in central India, the interior peninsula and the north-eastern India. However, cooling trends have been observed in north-west India and parts of south India.
- A trend of increasing monsoon seasonal rainfall has been found along the west coast, northern Andhra Pradesh and north-western India (+10 to +12 per cent of the normal over the last 100 years), while a trend of decreasing monsoon seasonal rainfall has been observed over eastern Madhya Pradesh, north-eastern India and some parts of Gujarat and Kerala (-6 to -8 per cent of the normal over the last 100 years).
- The records of coastal tide gauges in the Indian Ocean for more than 40 years showed a sea level rise between 1.06 to 1.75 mm per year.

INDIA'S INITIATIVES ON ADAPTATION AND MITIGATION STRATEGIES FOR CLIMATE CHANGE

A "National Action Plan on Climate Change" was released by Prime Minister of India on 30 June 2008, which was prepared by Prime Minister's Council on Climate Change. It proposed eight National Missions including National Mission for a Green India which is directly relevant to forestry sector. In brief, Green India Mission is a major initiative of Government of India to green 6 million ha of degraded forests. It is one of the largest global efforts and part of initiative of Indian Government on response to climate change which is capable of capturing 1,435 million tonnes of CO_2.

Green India Mission (GIM) puts the "greening" in the context of climate change adaptation and mitigation mainly meant to enhance ecosystem services like carbon sequestration and storage in forests and other ecosystems. Among three fold objectives of the GIM, one objective is to increase the greenhouse gases (GHGs) removals by India's forests to 6.35 per cent of India's annual total GHG emissions by the year 2020 (an increase of 1.5 per cent over what it would be in the absence of the Mission). This would require an increase in above and below ground biomass in 10 million ha of forests or ecosystems, resulting in increased carbon sequestration of 43 million tons CO_2e annually. Out of nine sub-missions, which will be implemented under GIM, the following three sub-missions needs special attention owing to their greater contributions.

Sub-Mission 1 Enhancing climatic resilience in moderately dense forests in 2 m ha: The key interventions mentioned are better protection, fire management (both prevention, detection and control), regulated grazing, invasive species eradication, management of insects and other pathogens, soil and moisture conservation (on ridge to valley basis).

Sub-Mission 2 Eco-Restoration of open forests in 4m ha: Plantation of indigenous species and/or regeneration of root stock and indigenous grasses and shrubs will be priority agenda with focus on in-situ moisture conservation/rain water harvesting/run off reduction activities on ridge to valley principle.

Sub-Mission 7 Agro-forestry and social forestry (increasing biomass and creating carbon sink) in 1.5 m ha: Under agroforestry, 0.80 million ha would involve improved agroforestry practices on the existing lands under agroforestry and that 0.70 million ha would involve additional lands under agroforestry. The Mission will support a programme of nurseries for raising quality seedlings to meet the demands of farmers. GIM would also support putting in place a system for certification of seed and genetically improved clonal planting stock or registration of clones and nurseries in India.

CORPORATE RESPONSES TO CLIMATE CHANGE

Between 1990 and 2004, energy consumption rose 37 per cent in India and 53 per cent in China. Besides International and National initiatives to curtail vagaries of climate change, the following are the corporate responses which every individual of this planet can reflect in their day-to-day activities by change in life style.

- Use of solar devices to heat water.
- Use of Compact Flourescent Lamps (CFL).
- Construction of green buildings which will have insulation to retain heat and opening of windows in the direction of wind to replace air-conditioning.
- Computers should be turned off when not in use.
- All electrical gadgets should be turned off when not in use.
- Use of cars for short distances should be avoided.

- **Air travel should be reduced if the work could be done by the use of telephone or e-mail.**
- **Switch to wind power is necessary. In India, the company "Suzlon" earned $1.5 billion in wind turbine revenue in 2006.**
- **Plastic bags should be avoided.**
- **Use of paper should be cut down. The Indian habit of putting everything down on paper should be avoided. Use of e-mail and phone can save thousands of trees.**
- **Public transport system should be used as far as possible.**
- **Power-generating plants based on fossil fuel will have to modify their ways.**
- **Switching over to biofuels: ethanol made from corn or sugarcane.**
- **All industries have to demand clean technology. Developed countries should sell clean technology to developing countries at a cheaper rate.**

39 FOREST CARBON SEQUESTRATION AND CARBON TRADE

OVERVIEW ON GLOBAL CARBON CYCLE

CARBON

Carbon is derived from the Latin word *carbo*, meaning coal or charcoal. Carbon is a chemical element with the symbol C and atomic number six. Carbon occurs in nature as the sixth most abundant element in the universe and it is an element found in all forms of life. Carbon makes up 18 per cent of our bodies and is a major component of trees and plants. It also exists in the environment in non-living things like rocks, oil, natural gas, coal and air. In short, it the basic building block of all life and the environment we live in. Both diamond and coal are nearly pure carbon. Carbon is a key element for life, composing almost half of the dry mass of the earth's plants.

Carbon occurs in organic as well as inorganic forms and this is a simple way of grouping different forms of carbon into biologically derived compounds (complex substances produced only by the growth of living organisms) and mineral compounds that can be formed in the absence of biological activity (but can sometimes be formed with the assistance of living things, as in the case of sea shells). Organic compounds includes such things as sugars, fats, proteins and starches and are contained in both living organisms and the material that remains after their death and partial decomposition. Mineral forms of carbon include carbonates contained in rock and seawater as well as carbon dioxide. Complete decomposition of organic matter results in a return to mineral forms, often as CO_2.

CARBON CYCLE

The carbon cycle is a bio-geochemical cycle which exchanges carbon between various reservoirs or storage places within the earth system viz. the land, the oceans and the atmosphere. The atmosphere has the smallest pool of actively cycling carbon and the terrestrial biosphere is the next largest reservoir of carbon which comprises soil and plants. The oceans are the earth's largest active carbon reservoir.

The carbon cylce is vitally important to life on earth. Through photosynthesis and respiration, Earth produces food and other renewable resources. Through decomposition, it serves as the Earth's waste disposal system. Globally carbon cycle plays a key role in regulating earth's climate by controlling the concentration of carbon dioxide in the atmosphere. Carbon dioxide contributes to the greenhouse effect and increased CO_2 in the atmosphere has been responsible for more than half of the climate warming observerd in recent decades.

The processes by which carbon moves through Earth's reservoirs take place on very different time scales. The short-term carbon cycle includes processes that transfer carbon from one reservoir to another in a matter of years, including photosynthesis, plant and animal respiration and the movement of CO_2 across the air-sea interface. On longer time scales, carbon from dead plant material can be incorporated into soils, where it might reside for years, decades or centuries before being broken down by soil microbes and released back to the atmosphere. Other processes, such as the transformation of carbon into limestone and its subsequent release as the rock is weathered, occur very slowly (thousands to millions of years). Similarly organic matter that became buried in deep sediments (and protected from decay) was slowly transformed into deposits of coal, oil and natural gas, the fossil fuels we use today. These slower transfers of carbon among reservoirs are called long-term carbon cycle.

CARBON SOURCES AND SINKS

The Earth's carbon reservoirs naturally act as both sources, adding carbon to the atmosphere and sinks, removing carbon from the atmosphere. A reservoir or component of the carbon cycle that releases more carbon than it absorbs is called carbon source. Anthropogenic emissions are a source of carbon. A carbon reservoir that takes in and stores more carbon than it releases is termed as carbon sink. Carbon sinks can serve to partially offset greenhouse gas emissions. Forests and oceans are both large carbon sinks.

Maintaining a steady amount of CO_2 in the atmosphere helps maintain stable average temperature at the global scale. However, because fossil fuel combustion and deforestation have increased CO_2 inputs to the atmosphere without matching increases in the natural sinks that draw CO_2 out of the atmosphere (oceans, forests, etc.), these activities have caused the size of the atmospheric carbon pool to increase. This has been responsible for the present build-up of CO_2 and is believed to cause the observed trend of increasing global temperatures. How far will CO_2 levels rise in the future? The answer depends both on how much CO_2 humans continue to release and on the future amount of carbon uptake and storage by the Earth's natural sinks and reservoirs. In short, it depends on the carbon cycle.

CARBON POOLS

Carbon is present in the earth's atmosphere, soils, plants, fossil fuels, oceans and crust. When viewing the Earth as a system, these components can be referred to as *carbon pools* or *stocks* or *reservoirs* because they act as storage houses for large amounts of carbon. In order to understand how carbon is cycled and how atmospheric CO_2 will change in the future, one must carefully study the places in which carbon is stored (pools), how long it resides there and processes that transfer it from one pool to another (fluxes). The amount of carbon in major reservoirs is so large that it is expressed in Gigaton (Gt), also known as Pentagram

(Pg), is equal to 10^9 metric tons or a billion metric tons (1 metric ton equals 1,000 kilograms). Collectively, all of the major pools and fluxes of carbon on Earth comprise what we refer as the global carbon cycle. Earth's carbon pools can be grouped into four categories that have the greatest relevance to the overall carbon cycle.

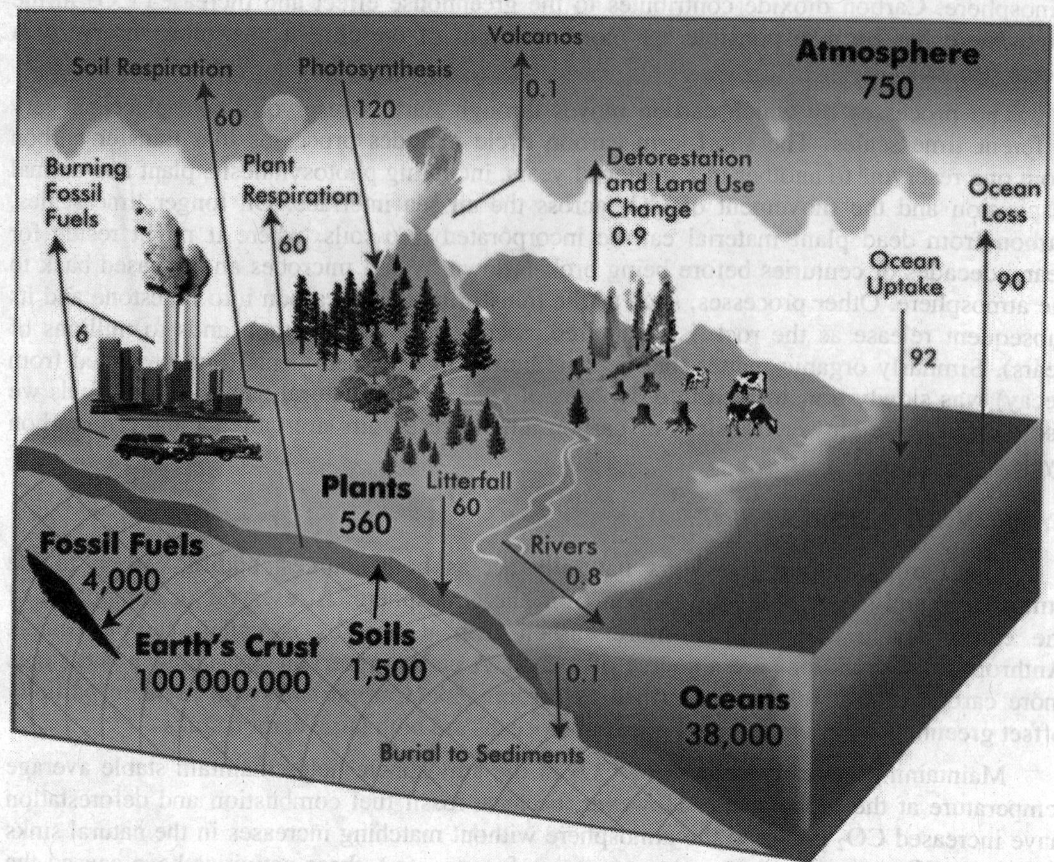

Fig. 39.1: Global Carbon Cycle – Units in Gigaton Carbon (Source: The Globe Program)

THE EARTH'S CRUST: The largest amount of carbon on Earth is stored in sedimentary rocks within the planet's crust. These are rocks produced either by the hardening of mud (containing organic matter) into shale over geological time or by the collection of calcium carbonate particles, from the shells and skeletons of marine organisms, into limestone and other carbon-containing sedimentary rocks. Together all sedimentary rocks store 100,000,000 GtC (Gigatons of carbon). Another 4,000 GtC is stored in the Earth's crust as hydrocarbons formed over millions of years from ancient living organisms under intense temperature and pressure. These hydrocarbons are commonly known as fossil fuels.

OCEANS: The Earth's oceans contain 38,000 GtC, most of which is in the form of dissolved inorganic carbon, stored at great depths where it resides for long periods of time. A much smaller amount of carbon, approximately 1,020 Gt, is located near the ocean surface. This carbon is exchanged rapidly with the atmosphere through both physical processes, such as CO_2 gas dissolving into the water and biological processes such as the growth, death and decay of plankton. Although most of this surface carbon cycles rapidly, some of it can also be transferred by sinking to the deep ocean pool where it can be stored for a much longer time.

ATMOSPHERE: The atmosphere contains approximately 750 GtC, most of which in the form of CO_2, with much smaller amounts of methane (CH_4), carbon monoxide (CO), and chlorofluorocarbons (CFCs). Although this is considerably less carbon than in the oceans or crust, carbon in the atmosphere is of vital importance because of its influence on the greenhouse effect and climate. The carbon increase in atmosphere has occurred in the past several centuries due to the result of C fluxes to the atmosphere from the crust (fossil fuels) and terrestrial ecosystems (via deforestation and other forms of land clearings).

TERRESTRIAL ECOSYSTEMS: Terrestrial ecosystems contain carbon in the form of plants, animals, soils and microorganisms. Unlike the Earth crust and oceans, most of the carbon in terrestrial ecosystems exists in organic form. The term 'organic' refers to compounds produced by living things including leaves, wood, roots, dead plant material and the brown organic matter in soils (which is the decomposed remains of formerly living tissues). Collectively, the Earth's plants store approximately 560 to 650 GtC, with the wood in trees being the largest fraction and the total amount of carbon in the word's soils is estimated to be 1,500 Gt.

CARBON FLUXES

The movement of any material from one place to another is called a flux and carbon flux is the transfer of carbon from one pool to another. Carbon fluxes are usually expressed as a rate in which an amount of carbon being transferred over a certain period of time. In the integrated system, fluxes connect reservoirs together to create cycles and feedbacks. A single carbon pool can often have several fluxes both adding and removing carbon simultaneously. For example, the atmosphere has inflows from decomposition (CO_2 released by the breakdown of organic matter), forest fires, fossil fuel combustion and outflows from plant growth and uptake by the oceans. The size of various fluxes can vary widely and some of the more important carbon fluxes are presented below.

PHOTOSYNTHESIS: During photosynthesis, plants use energy form sunlight to combine CO_2 from atmosphere with water from soil to create carbohydrates (*carbo* and *hydrate* signify carbon and water). In this way, CO_2 is removed from the atmosphere and stored in the structure of plants. Virtually all of the organic matter on Earth was initially formed through this process. Current estimates suggest photosynthesis removes 120 GtC per year from the atmosphere.

PLANT RESPIRATION: Plants also release CO_2 back to the atmosphere through the process of respiration. Respiration occurs as plant cells use carbohydrates, made during photosynthesis, for energy. Plant respiration represents approximately half (60 GtC per year) of the CO_2 that is returned to the atmosphere in the terrestrial portion of the carbon cycle.

LITTERFALL: In addition to the death of the whole plants, living plants also shed some portion of their leaves, roots and branches each year. Because all parts of the plant are made up of carbon, the loss of these parts to the ground is a transfer of carbon (a flux) from the plant to the soil. Dead plant material is often referred to as litter and once on the ground, all forms of litter will begin the process of decomposition.

SOIL RESPIRATION: When dead organic matter is broken down or decomposed (consumed by bacteria and fungi), CO_2 is released into the atmosphere at an average rate of about 60 GtC per year globally. Because it can take years for a plant to decompose (or decades in the case of large trees), carbon is temporarily stored in the organic matter of soil.

OCEAN-ATMOSPHERE EXCHANGE: Inorganic carbon is absorbed and released at the interface of the ocean's surface and surrounding air, through the process of diffusion. Carbon is readily exchanged between the atmosphere and ocean. In regions of oceanic upwelling, carbon is released to the atmosphere. Conversely, regions of downwelling transfer carbon (CO_2) from the atmosphere to the ocean. When CO_2 enters the ocean, carbonic acid is formed:

$$CO_2 + H_2O \rightleftharpoons H_2CO_3$$

This reaction has a forward and reverse rate that it achieves a chemical equilibrium. Another reaction important in controlling oceanic pH levels is the release of hydrogen ions and bicarbonate. This reaction controls large changes in pH:

$$H_2CO_3 \rightleftharpoons H^+ + HCO_3^-$$

The formation of carbonate in seawater allows oceans to take up and store a much larger amount of carbon. Carbonate is also important to a vast number of marine organisms that use this mineral form of carbon to build shells. Carbon is also cycled through the ocean by the biological processes of photosynthesis, respiration and decomposition of aquatic plants. Because ocean plants don't have large woody trunks that take years to breakdown, the process happens much more quickly in oceans than on land - often in a matter of days. For this reason, very little carbon is stored in the ocean through biological processes. The total amount of carbon uptake (92 GtC) and carbon loss (90 GtC) from the ocean is dependent on the balance of organic and inorganic processes.

FOSSIL FUEL COMBUSTION AND LAND COVER CHANGE: The modern-day carbon cycle includes several important fluxes that stem from human activities. The most important of these is combustion of fossil fuels *viz.* coal, oil and natural gas which releases CO_2 into the atmosphere. At present, fossil fuel combustion represents a flux to the atmosphere of approximately six to eight GtC per year. Another important activity that has caused a flux of carbon to the atmosphere is land cover change largely in the form of deforestation. Native forests in many areas have been cleared for timber or burnt for conversion to farms and grasslands and these changes have resulted in a net flux to the atmosphere of about 1.5 GtC per year.

GEOLOGICAL PROCESSES: It represents an important control on the Earth's carbon cycle over time scales of hundreds of millions of years. Collectively, these processes slowly convert carbon that was initially contained in living organisms into sedimentary rocks within the Earth's crust. Once there, these materials continue to be moved and transformed through the process of plate tectonics, uplift of rocks contained in the lighter plates and melting of rocks in the heavier plates as they are pushed deep under the surface. These melted materials can eventually result in emission of gaseous carbon back to the atmosphere through volcanic eruptions, thereby completing the cycle. Volcanic eruptions add the net flux to the atmosphere of about 0.1 GtC per year.

CARBON BUDGET

The Earth's carbon cycle is in a constant state of motion. Through processes that take place over seconds, days, years and millennia, carbon is constantly being transferred between all the various pools discussed above. If all sources are equal to all sinks, the carbon cycle can be said to be in equilibrium or in a state of balance and there is no change in the size of pools over time. The global carbon budget is the balance of the exchanges (incomes and losses) of carbon between the carbon reservoirs.

Just as in the constructing of a financial budget, a carbon budget is simply a list of all pools along with estimates of their size and of all the fluxes that constitute inputs and outputs. Presently, a budget of the Earth's carbon cycle shows that it is far from being in a state of balance. While random variation in natural processes such as climate and forest fires often results in some degree of imbalance on an annual basis, the very large imbalance in today's carbon cycle is due to the processes of fossil fuel combustion and land cover change.

CARBON DIOXIDE EMISSIONS

GLOBAL SCENARIO OF CO_2 EMISSIONS

Human activities, especially the burning of fossil fuels such as coal, oil and gas, have caused a substantial increase in the concentration of carbon dioxide in the atmosphere. This increase in atmospheric CO_2, from about 280 to more than 390 ppm over the last 250 years, is causing measurable global warming. There is a consensus that under existing emission scenarios, global mean temperature will increase by one degree Celsius by the year 2025 and three degree Celsius by the year 2100. It is expected that atmospheric CO_2 levels will continue to rise and may exceed 500 ppm by 2050. Not all gases are equal in their green house potential. CO_2 is most common and most important per unit of emissions than other GHGs. We need to distinguish between atmospheric carbon (C), atmospheric CO_2 and atmospheric carbon dioxide equivalent (CO_2e) warming potential.

The concentration of CO_2 in the atmosphere is now about 390 ppm and the concentration of CO_2e is about 450 ppm. Each ppm of atmospheric CO_2 corresponds to about 2.12 Gt carbon and 7.78 Gt of CO_2. Over the past 50 years, about 55 per cent of each emitted Gt of carbon has stayed in the atmosphere and the rest has been absorbed by land or oceans. Thus, it takes about 3.8 Gt C per one ppm of atmospheric CO_2. The rate at which atmospheric CO_2 is increasing has risen from about one ppm per year in 1960s to two ppm for 2000s and it is still continuing to increase every year. At present, atmospheric carbon increases by about 4.5 Gt per year on average.

TOP 25 EMITTING COUNTRIES BY TOTAL FOSSIL-FUEL CO_2 EMISSIONS

(Source: IEA 2011)

(1) China, (2) United States of America, (3) India, (4) Russian Federation, (5) Japan, (6) Germany, (7) Islamic Republic of Iran, (8) Canada, (9) Korea (10) United Kingdom, (11) Saudi Arabia, (12) Mexico, (13) Australia, (14) Italy, (15) Indonesia, (16) South Africa, (17) France, (18) Brazil, (19) Poland, (20) Spain, (21) Ukraine, (22) Turkey, (23) Chinese Taipei, (24) Thailand and (25) Kazakhstan

Table 39.1: CO_2 Emissions from Fossil Fuel Combustion in 2009 of Top Ten Countries and World (Source: IEA 2011)

Country	Total CO_2 Emissions (million metric tons of CO_2)		Per Capita Emission (ton CO_2/person)
	Sectoral Approach	Reference Approach	
China	6,877.2	7,085.0	5.14
USA	5,195.0	5,290.0	16.90
India	1,585.8	1,630.0	1.37
Russian Federation	1,532.6	1,528.6	10.80
Japan	1,092.9	1,092.9	8.58
Germany	750.2	755.1	9.16
Islamic Republic of Iran	533.2	549.1	7.31
Canada	520.7	501.3	15.43
Korea	515.5	518.1	10.57
United Kingdom	465.8	476.7	7.54
World	28,999.4	29,549.3	4.29

Table 39.2: Total Carbon Emissions of Top Ten Countries and World (Source: CDIAC 2011)

Country	Total Carbon Emissions (million metric tons of carbon)		
	2008	2009	2010
China	1,917.6	2,035.4	2,247.5
USA	1,546.9	1,438.3	1,497.9
India	475.2	515.8	564.5
Russian Federation	466.0	435.3	460.6
Japan	329.5	290.8	310.5
Germany	214.5	200.1	208.0
Islamic Republic of Iran	146.8	152.4	156.7
Canada	148.4	138.3	141.4
South Korea	138.9	140.7	153.6
United Kingdom	142.6	129.6	134.5
World	8,748.0	8,626.4	9,138.8

How will the world's climate change in the coming century due to increase in global CO_2 emissions? The answer to this question depends on how human societies develop in terms of demographic and economic development, technological change, energy supply and demand and land use change. OhUntil recently, industrialized countries have emitted the large majority of anthropogenic greenhouse gases. However shares of developing countries are rising very rapidly and are projected to continue to do so. The continued growth of global emissions of carbon dioxide and their likely adverse effects on global warming are focusing debate on the contribution of various countries to total emissions and the comparability of efforts across countries in mitigating these emissions.

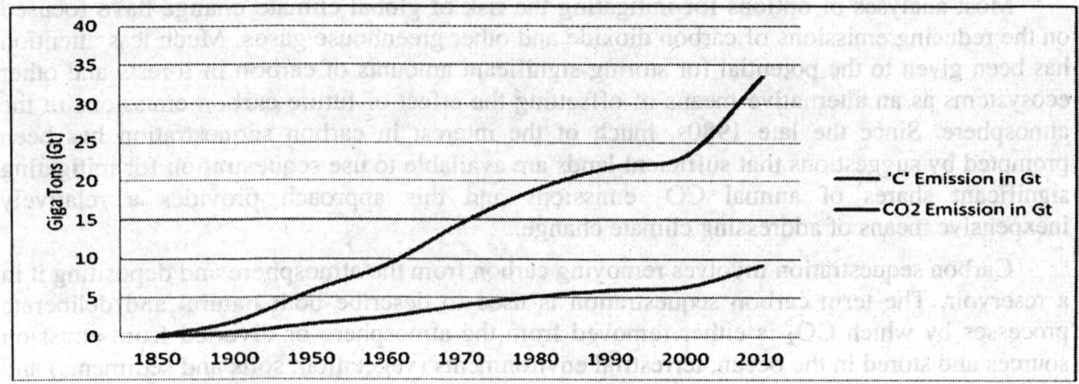

Fig. 39.2: Trend in Global Carbon and CO_2 Emissions from Fossil Fuel Combustion (Source: CDIAC)

In 2010, total global CO_2 emissions reached 33.0 billion tons, a 30 per cent increase from 2000 and 45 per cent more than 1990, the base year of the Kyoto Protocol. In 1990, global emissions were 22.7 billion tons, an increase of 45 per cent on the 1970 level of 15.5 billion tons. The twenty years from 1970 to 1990 and the twenty years from 1990-2010 thus saw the same 45 per cent growth in global emissions. The continuing large variation in national emission growth trends in 2010 resulted in 54 per cent of global emissions by developing countries and 43 per cent from the mature industrialised countries. The top six emitting countries, including the EU-27 block, are responsible for 70 per cent of total global emissions and the top 25 emitting countries more than 80 per cent of total emissions. Increased energy efficiency, nuclear energy and the growing contribution of renewable energy are not compensating for the globally increasing demand for power and transport, which is strongest in developing countries.

NATIONAL SCENARIO OF CO_2 EMISSIONS

India emits more than five per cent of global CO_2 emissions, and emissions continue to grow. India's 2008 total fossil fuel CO_2 emissions rose 8.1 per cent over the 2007 level to 475 million metric tons of carbon. CO_2 emissions have almost tripled between 1990 and 2009. It has been projected that CO_2 emissions in India will increase by almost 2.5 times between 2008 and 2035. Large shares of these emissions are produced by the electricity and heat sector, which represented 54 per cent of CO_2 in 2009, up from 40 per cent in 1990. CO_2 emissions in the transport sector accounted for only nine per cent of total emissions in 2009, but transport is one of the fastest growing sectors.

With the world's second largest population and over 1.1 billion people, India's per capita emission rate for 2009 of 1.37 metric tons of CO_2 is well below the global average (4.29). India's per capita emissions in 2035 will, however, still be well below those in the OECD member countries today. In terms of CO_2 emission per unit of GDP, India has continuously improved the efficiency of its economy and reduced the CO_2 emissions per unit of GDP by 16 per cent between 1990 and 2009. India aims to further reduce emissions intensity of GDP by 20-25 per cent by 2020 compared with the 2005 level.

CARBON SEQUESTRATION

Most analyses of options for mitigating the risk of global climate change have focused on the reducing emissions of carbon dioxide and other greenhouse gases. Much less attention has been given to the potential for storing significant amounts of carbon in forests and other ecosystems as an alternative means of offsetting the effect of future carbon emissions in the atmosphere. Since the late 1980s, much of the interest in carbon sequestration has been prompted by suggestions that sufficient lands are available to use sequestration for mitigating significant shares of annual CO_2 emissions and this approach provides a relatively inexpensive means of addressing climate change.

Carbon sequestration involves removing carbon from the atmosphere and depositing it in a reservoir. The term carbon sequestration is used to describe both natural and deliberate processes by which CO_2 is either removed from the atmosphere or diverted from emission sources and stored in the ocean, terrestrial environments (vegetation, soils and sediments) and geologic formations. Before human-caused CO_2 emissions began, the natural processes that make up the global carbon cycle maintained a near balance between the uptake of CO_2 and its release back to the atmosphere. However, existing CO_2 uptake mechanisms or carbon sinks are insufficient to offset the accelerating pace of emissions related to human activities. Carbon can be sequestered in three major ways: oceanic, geological and terrestrial/ biological.

OCEANIC CARBON SEQUESTRATION

The world's oceans are the primary long term sink for human-caused CO_2 emissions, currently accounting for a global net uptake of about two gigaton of carbon annually. This uptake is not a result of deliberate sequestration, but occurs naturally through chemical reactions between seawater and CO_2 in the atmosphere. While absorbing atmospheric CO_2, these reactions cause the oceans to become more acidic. Many marine organisms and ecosystems depend on the formation of carbonate skeletons and sediments that are vulnerable to dissolution in acidic waters. Laboratory and field measurements indicate that CO_2-induced acidification may eventually cause the rate of dissolution of carbonate to exceed its rate of formation in these ecosystems. The impacts of ocean acidification and deliberate ocean fertilization on coastal and marine food webs and other resources are poorly understood. Scientists are studying the effects of carbon sequestration on these important environments.

GEOLOGIC CARBON SEQUESTRATION

Geologic sequestration begins with capturing CO_2 from the exhaust of fossil fuel power plants and other major sources. The captured CO_2 is piped one to four kilometers below the land surface and injected into porous rock formations. Compared to the rates of terrestrial or biological carbon uptake, geologic sequestration is currently used to store only small amounts of carbon per year. Much larger rates of sequestration are envisioned to take advantage of the potential permanence and capacity of geologic storage.

The permanence of geologic sequestration depends on the effectiveness of several CO_2 trapping mechanisms. After CO_2 is injected underground, it will rise buoyantly until it is trapped beneath impermeable barrier or seal. In principle, this physical trapping mechanism, which is identical to the natural geologic trapping of oil and gas, can retain CO_2 for thousands to millions of years. Some of the injected CO_2 will eventually dissolve in ground water and some may be trapped in the form of carbonate minerals formed by chemical reactions with the surrounding rock. All of these processes are susceptible to change over time following CO_2 injection. Scientists are studying the permanence of these trapping mechanisms, economic costs and environmental risks and the chance of geologically sequestered CO_2 to leak back to the atmosphere. Geological sequestration, which has a fairly large technological potential, has not yet been demonstrated on a scale anticipated to mitigate CO_2 emissions.

TERRESTRIAL OR BIOLOGICAL CARBON SEQUESTRATION

Terrestrial carbon sequestration, also termed biological sequestration, is typically accomplished through forest and soil conservation practices that enhance the storage of carbon (such as restoring and establishing new forests, wetlands and grasslands) or reduce CO_2 emissions (such as reducing agricultural tillage and suppressing forest fires). These practices are implemented to meet a variety of land management objectives. Existing terrestrial carbon storage is susceptible to disturbances such as fire, diseases and changes in climate and land use. Terrestrial carbon sequestration is a form of indirect sequestration whereby ecosystems are maintained, enhanced, or manipulated to increase their ability to store carbon.

Biological carbon sequestration transfers carbon (from CO_2) in the atmosphere to biomass through photosynthesis and ultimately stores it in plants (foliage, wood and roots) and soils. Biological sequestration encompasses various ways of using agricultural and forest land to enhance the natural storage of atmospheric CO_2. Examples include planting or preserving trees, altering crop production practices, planting vegetation in areas prone to soil erosion and changing the way grazing lands are managed. When forests, croplands and grazing lands sequester carbon, they are referred to as "carbon sinks".

Biological carbon sequestration increases the net fixation of atmospheric CO_2 by terrestrial vegetation with emphasis on enhancing physiology and rate of photosynthesis of vascular plants. It retains carbon in plant materials and enhancing the transformation of carbon to soil organic matter. It reduces the emission of CO_2 from soils caused by heterotrophic oxidation of soil organic carbon and also increases the capacity of deserts and degraded lands to sequester carbon. There are two main reservoirs available for biological carbon sequestration: soil and forest.

SOIL CARBON SEQUESTRATION: Soil carbon sequestration is the storage of carbon in soils. Although oceans store most of the Earth's carbon, soils contain approximately 75 per cent of the carbon pool on land, which is three times more than the amount stored in living plants and animals. It is the process of transferring carbon dioxide from the atmosphere into the soil through crop residues, leaf fall and other organic solids and in a form that is not immediately reemitted. This transfer or sequestration of carbon helps to offset emissions from fossil fuel combustion and other carbon-emitting activities while enhancing soil quality and long term productivity. Soil carbon sequestration can be accomplished by management systems that add high amounts of biomass to the soil, cause minimal soil disturbance, conserve soil and water, improve soil structure and enhance soil fauna activity. Continuous no-till crop production is a prime example.

Carbon is added to the soil as litter when plants die and decompose. The primary way that carbon is stored in the soil is as soil organic matter (SOM). The amount of carbon found in SOM ranges from 40 to 60 per cent by mass. SOM is a complex mixture of carbon compounds, consisting of decomposing plant and animal tissue, soil biota (protozoa, nematodes, earthworms, fungi and bacteria), and carbon associated with soil minerals. Carbon can remain stored in soils for millennia or be quickly released back into the atmosphere. Climatic conditions, natural vegetation, soil texture and drainage all affect the amount and length of time carbon is stored. Globally, the top soil stores approximately 1,500 Gt as organic carbon and an additional 900-1,700 Gt as inorganic carbon and exchanges 60 Gt of carbon per year with the atmosphere.

FOREST CARBON SEQUESTRATION: In 1992, over 180 countries joined in signing the United Nations Framework Convention on Climate Change, agreeing in principle to stabilize GHG levels in the atmosphere. Since that time, attention has been given to ways to decrease or at least decelerate the flow of carbon from fossil fuels to the atmosphere. Additionally, it may be possible to increase the rate at which ecosystems remove CO_2 from the atmosphere and store the carbon in plant material, decomposing detritus and organic soil. In essence, forests and other highly productive ecosystems can become biological scrubbers by removing (sequestering) CO_2 from the atmosphere. Forest growth naturally sequesters carbon and the carbon remains in the wood after it is processed into a product. Forest carbon sequestration is a practice to provide sink zone for atmospheric carbon dioxide in term of biological accumulation by increase in forest cover.

CARBON SEQUESTRATION IN FOREST

CARBON AND FORESTS

The process of photosynthesis combines atmospheric carbon dioxide with water, subsequently releasing oxygen into the atmosphere and incorporating the carbon atoms into the cell of plants. Trees, unlike annual plants that die and decompose yearly, are long-lived plants that develop a large biomass, thereby capturing large amounts of carbon over a growth cycle of many decades. Thus, a forest ecosystem can capture and retain large volumes of carbon over long periods.

Forests operate both as vehicles for capturing additional carbon and as carbon reservoirs. A young forest, when growing rapidly, can sequester relatively large volumes of additional carbon roughly proportional to the forest's growth in biomass. An old-growth forest acts as a reservoir, holding large volumes of carbon even if it is not experiencing net growth. Thus, a young forest holds less carbon, but it is sequestering additional carbon over time. An old forest may not be capturing any new carbon but can continue to hold larger volumes of carbon as biomass over long periods of time. Managed forests offer the opportunity for influencing forest growth rates and providing for full stocking, both of which allow for more carbon sequestration. Forest systems operate on a scale of many decades and centuries, rather than annually or over a few years as would be the case with most crops and non-tree vegetation.

FORESTS AS CARBON SINK

The world's forests store more than 650 Gt of carbon, 44 per cent in the biomass, 11 per cent in dead wood and litter, and 45 per cent in the soil. Globally carbon stocks are decreasing as a result of the loss of forest area, however, the carbon stock per hectare has remained almost constant for the period 1990-2010. FAO (2010a) estimates that the world's forests store 289 Gt of carbon in their biomass alone. While sustainable management, planting and rehabilitation of forests can conserve or increase forest carbon stocks, deforestation, degradation and poor forest management reduce them. Carbon stocks in the world's forest biomass decreased by an estimated 0.5 Gt annually during the period 2005-2010. It estimates an average carbon stock of 161.8 tons per hectare in the vegetation of the world's forests for 2010. Each cubic metre of growing stock equals different amounts of biomass and carbon (in biomass) in different regions. Globally, each cubic metre of growing stock equals, on an average, 1 ton of above-ground biomass, 1.3 tons of total biomass and 0.7 ton of carbon in biomass.

Table 39.3: Trends in Total Carbon Stocks in World's Forests, 1990-2010 (Source: FRA 2010a)

Category	Carbon Stock (Gt)				Carbon Stock (tons/ha)			
	1990	2000	2005	2010	1990	2000	2005	2010
Carbon in biomass	299.224	293.843	291.299	288.821	71.8	71.9	71.7	71.6
Carbon in deadwood	34.068	33.172	32.968	32.904	8.2	8.1	8.1	8.2
Carbon in litter	38.855	38.748	38.825	38.984	9.3	9.5	9.6	9.7
Carbon in soil	300.425	295.073	293.232	291.662	72.1	72.2	72.2	72.3
Total carbon stock	672.571	660.836	656.323	652.371	161.4	161.8	161.6	161.8

India's forest and tree cover accounts for about 23.4 per cent of the total geographical area of the country. Over the past decades, national policies of India aimed at conservation and sustainable management of forests have transformed India's forests into a net sink of CO_2. India's forests serve as a major sink of CO_2, the estimates show that the annual CO_2 removals by India's forest and tree cover is enough to neutralize 11.25 per cent of India's total GHG emissions (CO_2e) at 1994 levels and to neutralize 9.31 per cent of total annual emissions of 2000 (Jagdish Kishwan et al 2009).

Table 39.4: Carbon Stock in Indian Forests (Source: FRA 2010b)

Category	Carbon Stock (million metric tons)			
	1990	2000	2005	2010
Carbon in above ground biomass	1,691	1,808	1,989	2,129
Carbon in below ground biomass	532	569	626	671
Sub-total: Living biomass	2,223	2,377	2,615	2,800
Carbon in dead wood	25	25	26	26
Carbon in litter	75	77	80	80
Sub-total: Dead wood and litter	100	102	106	106
Carbon in soil	3,751	3,836	3,972	4,017
Total Carbon Stock	6,074	6,315	6,693	6,923

From 1995 to 2005, the carbon stocks stored in Indian forests and trees have increased from 6,245 to 6,662 million metric tons registering an annual increment of 38 million metric tons of carbon or 138 million metric tons of CO_2e (MoEF 2009a). FAO (2010b) estimates that the India's forests store 6,923 million tons of carbon, 40 per cent in the living biomass, 1.5 per cent in dead wood and litter, and 58 per cent in the soil. The carbon stock for the period 2006–30 is projected to increase from 8.79 to 9.75 GtC with forest cover becoming more or less stable, and new forest carbon accretions coming from the current initiatives of afforestation and reforestation programme.

FOREST CARBON SEQUESTRATION STRATEGIES

Afforestation and reforestation of marginal cropland and pastures which include planting trees on land previously used for other purposes could result in substantial gains in carbon storage in biomass and soils comparing to reforestation (planting trees on land recently devoted to forestry). Carbon storage can also be improved by changing silvicultural practices. Since these practices are usually developed and applied for purposes other than carbon sequestration, it may be difficult to quantify the magnitude of increased total carbon storage when practices change. For instance, increasing timber growth will not necessarily increase biomass growth and soil carbon storage. Conversion of forest land to non-forest use usually means permanent loss of all or a substantial part of live biomass and reduction of organic matter in soils and in the forest floor.

Some carbon may be sequestered in wood products if the harvested biomass is utilized. Carbon dioxide and other GHGs are emitted when the remaining biomass and organic matter is burned or decomposed. Protecting and conserving forests should maintain or increase carbon pools in the short term. Carbon in wood and paper products remains sequestered and is emitted to varying degrees depending on how products are made, used and disposed. Sequestration in products and uses can be increased by altered processing methods, shifts in products used, end-use durability and landfill management. Sequestration in forests and products can be maximized by coordinated understanding of forest ecosystems and product utilization.

Table 39.5: Carbon Sequestration Practices (Richards et al 2006)

Land Type	**Strategy or Objective**		
	Conservation of Carbon Stocks	**Expansion of Carbon Stocks**	**Offsite Sequestration or Emissions Reduction**
Forest	Modified harvesting practices, Preventing deforestation, Change to sustainable forest management, Fire suppression and management	Reforestation, Modified management e.g., fertilization, improve stocking, biotechnology, species mix, extended rotations	Wood fuel substitution, Expanded wood products, Extended wood product life, Displace concrete/steel, Recycling wood and paper products
Crop	Soil erosion and fertility management, Water management, Maintain perennial crops, Residue management	Afforestation, Agroforestry, Improved cropping systems, Improved nutrient and water management, Conservation tillage, Crop residue	Biofuels substitutions, Fertilizer substitution or reduction, Other bioproducts substitution

		management, Restoration of eroded soils, Conversion to grass or other permanent vegetation	
Grazing	**Improved grazing systems**	**Afforestation, Change in species mix including woody species, Restoration of riparian zones, Fertilization, Irrigation**	**Livestock dietary changes**

The Intergovernmental Panel on Climate Change (IPCC) estimated that it may be possible, over the course of the next 50–100 years, to remove between 40 and 80 Gt of the carbon by sequestering it through agroforestry. Thus Agroforestry would be one of the interesting areas of research in land-use related to carbon sequestration, for various reasons. First, the surface involved is considerable and the rate of carbon gain is relatively high, 0.2–3.1 ton per hectare per year. Second, it can mitigate the important CO_2 emission resulting from deforestation. Third, it could provide a sustainable system from technical, ecological and economic points of view.

ASSESSMENT OF CARBON SEQUESTRATION IN FOREST ECOSYSTEM

Monitoring and verifying carbon storage can be expensive depending on the level of scientific validity needed. A system of cost-effective methods is necessary for monitoring and verification on a commercial basis for three types of land use: forest plantations, managed natural forests and agroforestry. It assesses changes in four main carbon pools: above ground biomass, below ground biomass, soils and forest litter. It aims to assess the net change in each pool for land use project and non-project (or pre-project) areas over a specified time period. Carbon monitoring efforts require specialized equipment, methods and trained personnel and most carbon assessment activities are likely to be performed infrequently – once every two to five years.

Carbon assessment system involves the following components (MaCDicken 1997)

- Baseline determination of pre-project carbon pools in biomass, soils and forest litter
- Establishment of fixed area permanent sample plots or carbon inventories for periodic measurement of changes in carbon pool in land use project areas.
- Assessment of carbon stock in above ground biomass, below ground biomass, soils and forest litter.
- Plotless vegetation survey methods *(quarter point or quadrat* sampling) to measure carbon stored in non-project areas or areas with sparse vegetation
- Calculation of the net difference in carbon accumulated in project and non-project land uses
- Use of satellite images as gauges of land-use changes, and as base maps for a microcomputer based geographic information system

- Software for calculating minimum sample size, assigning sample unit locations (either in a systematic grid or randomly), determining the minimum spacing for plots and optimizing site-specific monitoring plans
- Computer modelling of changes in carbon storage for periods between field measurements
- A database of biomass partitioning (roots, wood and foliage) for selected species.

CARBON INVENTORY: For carbon inventory, stratified random sampling generally yields more precise estimates. Each stratum can be defined by vegetation type, soil type, or topography. Useful tools for defining strata include satellite images, aerial photographs and maps of vegetation, soils or topography. These should be combined with ground measurements for accurate result. A GIS can automatically determine stratum size and the size of exclusions or buffer zones. Permanent plot locations can be selected either randomly or systematically

ESTIMATION OF PLANT BIOMASS: For this, diameter of all woody vegetation above the minimum diameter at breast height (e.g. >2 cm), is measured. Subsamples of smaller diameter (< 2 cm) woody vegetation and herbaceous plants are taken using small quadrats or circular plots. Individual dbh values for each plot are converted to biomass using single-entry biomass table or general biomass equations. Where single-entry tables do not provide adequate estimates of biomass, double-entry tables based on dbh and height are used. When weight and biomass tables do not exist for certain tree species, it is needed to develop and construct weight and biomass tables using a minimum of 30 well-selected trees or with a mean tree technique. Regression analysis should be performed for biomass of each component.

PARTITIONING AND WEIGHING

Each individual tree or shrub should be harvested and measured for diameter, length, height and divided into major components. Individual components (stem, branches, leaves, roots etc.) should be divided into several size classes for convenient handling and sub-sampling. Sub-samples should be taken to determine moisture content and specific gravity. The best way is to partition and weigh only one tree at a time and measure green weight in the field. To calculate moisture content, sub-samples should be taken, dried at 80^0 C and re-weighed.

NON-DESTRUCTIVE METHOD

Although harvesting is the preferred way to develop weight tables, it is not always possible due to conservation or regeneration considerations. For biomass tables from non-destructive samples, calculate stemwood volumes and convert them to biomass using specific gravity for wood and expansion factors for canopy biomass.

ESTIMATION OF CARBON STOCK IN TREES, SOIL, LITTER AND HERBACEOUS VEGETATION

Total tree biomass is converted to total carbon using conversion factors or standard formula. Carbon analysis in samples of soil, herbaceous vegetation and forest litter is usually carried out in laboratories using Walkley-Black procedure for determination of soil organic carbon. Other methods include dry combustion and wet combustion.

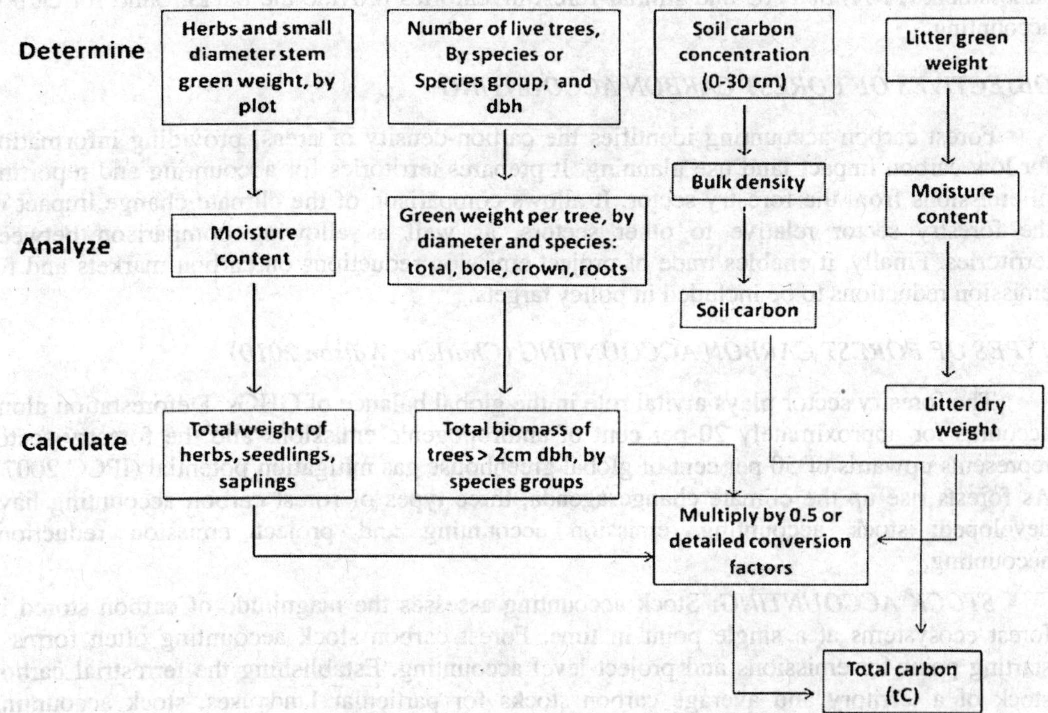

Fig. 39.3: Flow Diagram for Estimation of Carbon Sequestration in Forest Ecosystem (MaCDicken 1997)

FOREST CARBON ACCOUNTING

WHAT IS FOREST CARBON ACCOUNTING?

Carbon accounting is the practice of making scientifically robust and verifiable measurements of GHG emissions. Forests play an important role in the global carbon balance. As both carbon sources and sinks, they have the potential to form an important component in efforts to combat global climate change. Accounting for the carbon within forest ecosystems and changes in carbon stocks resulting from human activities is a necessary step towards the better representation of forests in climate change policy at regional, national and global scales. Although characteristics of forests have been recorded for numerous historical purposes, accounting for carbon is a more recent addition to forest inventories. This follows the growing need to quantify the stocks, sources and sinks of carbon in the context of anthropogenic impacts on the global climate.

Historically, forest inventories recorded stand structure, age, growth rate, biomass accumulation and the wood densities of tree species. These have served both commercial purposes, such as determining merchantable timber volumes and use in the paper and pulp industry, as well as national or regional plannning purposes, such as creating forest and land use inventories for land-use permits, land-use plans and agricultural expansion. In combination with substantial body of forest science researach literature, the Forest Resource Assessment (FRA) of FAO and similar forest inventories provide the background for carbon accounting.

OBJECTIVES OF FOREST CARBON ACCOUNTING

Forest carbon accounting identifies the carbon-density of areas, providing information for low carbon impact land use planning. It prepares territories for accounting and reporting of emissions from the forestry sector. It allows comparison of the climate change impact of the forestry sector relative to other sectors, as well as allowing comparison between territories. Finally, it enables trade of project emission reductions on carbon markets and for emission reductions to be included in policy targets.

TYPES OF FOREST CARBON ACCOUNTING (Charlene Watson 2010)

The forestry sector plays a vital role in the global balance of GHGs. Deforestation alone accounts for approximately 20 per cent of anthropogenic emissions and the forestry sector represents upwards of 50 per cent of global greenhouse gas mitigation potential (IPCC 2007). As forests rise up the climate change agenda, three types of forest carbon accounting have developed: stock accounting, emission accounting and project emission reductions accounting.

STOCK ACCOUNTING: Stock accounting assesses the magnitude of carbon stored in forest ecosystems at a single point in time. Forest carbon stock accounting often forms a starting point for emissions and project level accounting. Establishing the terrestrial carbon stock of a territory and average carbon stocks for particular land uses, stock accounting allows carbon-dense areas to be prioritised in regional land use planning. An early form of forest carbon accounting, emissions and emission reductions accounting have evolved from the principles for stock accouting.

EMISSIONS ACCOUNTING: Emissions accounting assesses the net greenhouse gas emissions to the atmosphere resulting from forests. Emissions accounting is necessary to asssess the scale of emissions from the forestry sector relative to other sectors. It also aids realistic goal-setting for GHG emissions targets. Under the UNFCCC and the Kyoto Protocol, countries are mandated to undertake some land use, land use change and forestry (LULUCF) carbon accounting. With the significant portion of developing country emissions arising form the LULUCF sector, the forestry sector is likely to play a prominent role in climate change stragegies in these countries.

PROJECT EMISSION REDUCTIONS ACCOUNTING: Emission reductions accounting assesses the decrease in emissions from project or policy activities, often so that they can be traded. Carbon accounting for forestry project emission reductions is required for both projects undertaken under the flexible mechanisms of the Kyoto Protocol and the voluntary carbon markets. Both necessitate good carbon accounting to ensure that emissions reductions are real, permanent and verifiable. For projects to generate tradable emission reductions, accouting methods between countries, regions and projects must be standardised in both developed and developing countries.

PRINCIPLES OF FOREST CARBON ACCOUNTING (Charlene Watson 2010)

BEST METHODS OF ACCOUNTING: The adopted practice in forest carbon accounting must be appropriate. In particular, transparency in methods and accuracy and precision in accouting promotes better understanding, legitimacy and trust which are critical for both public and political acceptance of resultant estimates. A basic knowledge of the principles underlying forest carbon accounting is also beneficial. Understanding biomass dynamics and flows between carbon pools in forest ecosystems enables more effective accounting.

ESTIMATION OF BIOMASS STOCK: Forest biomass is organic matter resulting from primary production through photosynthesis minus consumption through respiration and harvest. Assessment of biomass provides information on the structure and functional attributes of a forest and is used to estimate the quantity of timber, fuel and fodder components. With approximately 50 per cent of dry forest biomass comprised of carbon, biomass assessment also illustrates the amount of carbon that may be lost or sequestered under different forest management regimes. Carbon is lost to the atmosphere as CO_2. To convert carbon in biomass to CO_2, the tons of carbon are multiplied by the ratio of the molecular weight of carbon dioxide to the atomic weight of carbon (44/12). Estimating the biomass density of forest components is the first step in forest carbon accounting.

APPROACHES TO EMISSION ACCOUNTING: Although many natural processes lead to emissions and removals of GHGs (fires, insect attacks and local climate variability), anthropogenic activities such as slash and burn, fire management and harvesting have accelerated the release of GHGs from forests. These forest management practices affect the balance of emissions into the atmosphere through biomass fluctuation, soil and litter disturbance. The purpose of emissions accounting is to quantify the exchange of GHGs between the atmosphere, terrestrial vegetation and soils through photosynthesis, respiration, decomposition and combustion. There are two main approaches to emissions accounting: the inventory approach and the activity-based approach, which are supported under IPCC guidance based on the assumption that the flows of GHGs to or from the atmosphere are equal to changes in carbon stocks in the biomass and soils.

Inventory approach measures the difference in carbon stocks averaged between two points in time. It is also called periodic accounting or stock-difference approach. Measurement of stock change in this way can cover large areas and a variety of species and site conditions. The inventory-based system also captures non-linear changes in carbon stocks, for example biomass accumulation through growth. However, relying on the addition of carbon pools and assessments conducted in this way often leaves out smaller biomass components such as leaf biomass, ground vegetation and litter

$$\Delta C = \sum (C_{t2} - C_{t1})/(t_2 - t_1)$$

Where,

ΔC = carbon stock change in tons C per year

C_{t1} = carbon stock at time t_1 in tons C

C_{t2} = carbon stock at time t_2 in tons C (Source: IPCC 2003)

The activity-based approach estimates the net balance of additions and removals from a carbon pool. It is also called the gain-loss or flux approach. This method is useful where individual carbon pools are difficult to measure and is less susceptible to short-term variation in carbon stocks.

$$\Delta C = \sum [A \times (C_I - C_L)]$$

Where,

A = area of land in ha

C_I = rate of gain of carbon in tons C per ha per year

C_L = rate of loss of carbon in tons C per ha per year (Source: IPCC 2003)

ACCOUNTING FOR EMISSION REDUCTIONS: It is most commonly required at the project level but are also relevant when policy targets must be met. Where forestry carbon projects generate emission reductions, these can be traded as offsets either under Kyoto Protocol or on the voluntary carbon markets. It requires and understanding of a number of supplementary principles: the complexities of baseline establishment, demonstration of addionality, issues of leakage and the permanence of emissions reductions. These principles are blamed for the limited demand and limited inclusion of the forestry sector in carbon trading mechanisms to date.

PRACTICE OF FOREST CARBON ACCOUNTING

The practice of forest carbon accounting requires clear identification of the accounting boundary in both space and time. Stratifying the forest into areas with similar carbon characteristics further improves the accuracy of carbon accounting. Data for accounting can be gathered from a variety of sources including existing secondary data, remotely sensed data and primary data through field surveys. The amount of data from each source depends on the quality of the source as well as the trade-offs that must be made between accounting accuracy and costs of resources and time. All forest carbon accounting estimates contain uncertainty. Practitioners should identify, minimise where possible, and quantify this uncertainty through statistical analysis, published information and expert judgement.

GUIDANCE AND TOOLS FOR FOREST CARBON ACCOUNTING

Forest carbon accounting guidance from the Intergovernmental Panel on Climate Change (IPCC) has become the primary source of information for methods, accounting equations and parameters. A number of tools for forest carbon accounting have emerged which vary in terms of geological coverage, forestry activities and the carbon pools accounted for, as well as the level of data input required.

Regional Tools and Models from Developed Countries:

USA (COLE, The Carbon On-Line Estimator, CTCC, FORCARB, LMS), United Kingdom (CARBINE, C-Flow, C-Sort), Australia (CAMfor), Europe (EFI-SCEN)

Tools and Models for Wider Geographical Areas:

Canada (CBM-CFS3), CO2FIX, Graz/GORCAM, CENTURY, BIOME-BGC, AR, TARAM, ENCO-FOR

Tool for Commercial and Service Companies:
Bilan Carbone

CARBON FOOTPRINT AND CARBON TAXATION

Carbon footprint is the total amount of greenhouse gases that are emitted into the atmosphere each year by a person, family, building, organization or company. A person's carbon footprint includes greenhouse gas emissions from fuel that an individual burns directly, such as by heating a home or riding in a car. It also includes greenhouse gases that come from producing the goods or services that the individual uses, including emissions from power plants that make electricity, factories that make products and landfills where trash gets sent. A carbon footprint is a greenhouse gas emissions inventory which captures all business activities resulting in emissions. It can assist organisations to identify and target resource inefficiencies and be used to report the impact of products and services on clients and stakeholders. It deals with greenhouse quantification and carbon management strategies within various industry sector.

Traditonally, industry has used Return on Investment (ROI) to inform business decisions. Establishing a price on carbon changes the variables within the ROI equation and introduces accountablitity to non-financial metrics. Therefore, ROI strategeis should consider the return in terms of carbon cost savings and this should introduce a price incentive for lower emissions goods and services. Electricity consumption, natural gas and LPG consumption, transport fuels, waste to landfill, flights, refrigerant consumption, etc. are counted in the carbon footprint data informing the cost forecasts.

A carbon tax is a specific tax on the consumption of goods which cause carbon dioxide emissions. It mandates a specific price on carbon and revenue generated can be spent on environmental inititatives. A carbon tax sets a price on carbon emissions, but it cannot gurantee the level of emission reduction. As countries have existing tax regimes, a carbon tax may be more straightforward to administer within an individual country. However, it would be challenging to implement internationally.

CARBON TRADING AND CARBON MARKET

CARBON FINANCE

Carbon financing can be defined as financial resources provided to projects generating (or expected to generate) green house gas emission reductions in the form of the purchase of such emission reductions. In simple term, carbon finance is a purchase contracts whereby one party pays another party in exchange for a given quantity of Green House Gas (GHG) emission reductions.

WHAT IS CARBON TRADING?

A series of workshops and conferences related to global warming, climatic changes and GHGs emission finally concluded with adding new chapter to human society in the name of carbon trading. Carbon trading is emissions trading specifically for carbon dioxide calculated in tons of carbon dioxide equivalent (tCO_2e) and currently makes up the bulk of emissions trading. Emission trading is a market-based approach used to control pollution by providing economic incentives for achieving reductions in the emissions of pollutants. The concept of carbon trading came into existence as a result of increasing awareness about the need for pollution control. It was formalized in the Kyoto Protocol, an agreement made to set target for reduction of emission of green house gases (GHG) from 1990 levels by about 5 per cent in the period of 2008-2012.

Carbon trading emerges out as an asset to minimize the cost of reducing emission up to desirable level. At today's perspective, carbon trading is a business of buying and selling of the carbon credits achieved either by carbon sequestration or reducing the emission of GHGs lower than emission allowance. This trading can be done in between the nations, states, even industries. The carbon market can be divided into two: the regulatory (compliance) market and the voluntary market.

COMPLIANCE MARKETS: It generates and trade green house gas emission reductions known as certified emission reductions (CERs), emission reduction units (ERUs), assigned amount units (AAUs) and removal units (RMUs) that are regulated and directly initiated under the Kyoto Protocol.

VOLUNTARY MARKETS: It generates and trade greenhouse gas emission reductions that are not regulated or directly initiated by the Kyoto Protocol and known as verified emission reductions (VERs).

CARBON CREDIT

A carbon credit is a generic term for any tradable certificate or permit representing the right to emit one ton of carbon dioxide or the mass of another greenhouse gas with a carbon dioxide equivalent to one ton (tCO_2e). Carbon credits are a tradable permit scheme and they provide a way to reduce greenhouse gas emissions by giving them a monetary value (based on the market carbon price e.g., about $5 to $10).

International treaties such as the Kyoto Protocol set quotas on the amount of greenhouse gases that countries can produce and countries in turn, set quotas on the emissions of businesses. Businesses that are over their quotas must buy carbon credits for their excess emissions, while businesses that are below their quotas can sell their remaining credits. Credits can be exchanged between businesses or bought and sold in international markets at the prevailing market price.

UNITS OF EMISSION TRADE

Emissions reduction projects or carbon projects can be created by a national government or by an operator within the country. In reality, most of the transactions are not performed by national governments directly, but by operators who have been set quotas by their country. The government can set a limit for GHGs emission within the country. It can issue permits also called allowances or quotas to emit green house gases.

ASSIGNED AMOUNT UNITS (AAUs): Each Annex I Party that ratifies the Kyoto Protocol has a GHG emissions limitation commitment for 2008-2012, which is its "assigned amount." If a country's emissions are lower than that amount, it may sell the unused units.

CERTIFIED EMISSION REDUCTIONS (CERs): Carbon credits are measured in units of certified emission reductions (CERs). For trading purposes, one allowance or certified emission reduction (CER) is considered equivalent to one metric ton of CO_2 emissions reduction. An emission permit is a certificate that gives the holders the right to emit a certain amount of greenhouse gases to the atmosphere. Green house gas reduction of any clean development mechanism (CDM) project is measured according to internationally agreed methods and quantified in standard units called certified emission reductions (CERs). These are expressed in tons of carbon dioxide (CO_2) equivalents.

CERs can be used by an Annex I Party to help meet its emissions limitation commitment under the Kyoto Protocol. The credits issued for sink enhancements achieved by afforestation or reforestation projects under the CDM are either temporary CERs (tCERs), or long-term CERs (lCERs) that are subject to provisions to protect against possible reversals of the sink enhancements.

EMISSION REDUCTION UNITS (ERUs): These are credits issued for emission reductions or removals achieved by a project under the Joint Implementation (JI) mechanism as defined in Article 6 of the Kyoto Protocol. ERUs can be used by an Annex I Party to help meet its emissions limitation commitment under the Kyoto Protocol. Each ERU equals 1 ton of carbon dioxide equivalent (CO_2e) reduced.

REMOVAL UNITS (RMUs): These are credits issued for net sink enhancements achieved by eligible activities under Articles 3.3 and 3.4 of the Kyoto Protocol. RMUs can be used by an Annex I Party to help meet its commitment under the Kyoto Protocol. Each RMU equals 1 metric ton of CO_2e.

VERIFIED EMISSION REDUCTION (VERs): Green house gas reduction outside Kyoto Protocol is measured according to internationally agreed methods and quantified in standard units called verified emission reductions (VERs). These are also expressed in tons of CO_2e.

KYOTO MECHANISMS

The concept of carbon emission activities in one country being used by other countries as credits is well recognized. Institutions are being developed to allow these multi-country activities and transactions to be made. It also leads to the evolution and rapid growth of a whole new class of investment and securities related to environmental impact mitigation or environmental improvement including proposals for emission trading. It comprises,

- Joint Implementation (JI): A way to earn credits as ERUs by investing in emission reduction projects in developed countries that have taken on a Kyoto target.
- The Clean Developmental Mechanism (CDM): A way to earn credits as CERs by investing in emission reduction projects in developing countries.
- International Emissions Trading (IET): It will permit developed countries that have taken on a Kyoto target to buy and sell emission credits among themselves.

Under Joint Implementation, one Annex B country is allowed to purchase carbon credits by undertaking carbon reducing activities in another Annex B country. Clean Development Mechanisms is designed to allow some types of multi-country purchases and trades between Annex B and non-Annex B countries.

JOINT IMPLEMENTATION (JI): Article 6 of the protocol allows a system under which advanced countries jointly implement an emissions reduction project. Developed countries are allowed to obtain emission credits for projects undertaken in another developed country. JI project are a means by which firm in developed countries can invest in other developed countries, including those with economies in transition in central and Eastern Europe, in ways that reduce or avoid GHG emissions. Credit will be earned in the form of emission reduction unit. Joint implementation implies that an industrialized country pay for measures to reduce emissions in another industrialized country. This will give the buyer the right to emit more domestically, while the seller will be required to emit correspondingly less. JI can be a tool to transfer environmentally sound technology.

THE CLEAN DEVELOPMENT MECHANISM (CDM): Article 12 of the protocol allows firms in the developed countries to invest in projects in developing countries that reduce or avoid net GHGs emissions. This promises to be an important new avenue through which government and private corporations will transfer clean technologies and promote sustainable development. Credit will be earned in the form of certified emission reduction. The CDM allows industrialized countries to acquire emissions credit (the right to emit greenhouse gases) by paying for emissions reduction measures in developing countries that do not have emission targets. These measures must also contribute to sustainable development in the recipient country. Detail rules and regulations to ensure that the emissions reduction measures meet all the requirements are worked out.

The CDM has the potential to assist the start of a transition towards sustainable energy in developing countries and, in the long-term, benefit the climate. CDM projects must demonstrate that they add "real, measurable and long term benefits related to the mitigation of climate change" and are meant to contribute to sustainable development in developing countries. Whether or not these goals are reached depends very much on the rules for the CDM.

INTERNATIONAL EMISSION TRADING (IET): Article 17 of the protocol permits the trading of emission credits among countries. Countries with surplus credits can sell these credits to countries with capped emissions. It provides for countries with emission reduction commitments to buy and sell part of their assigned amount of CO_2 emissions among themselves. Reporting rules, a verification system and guidelines are being developed for IET. The idea is to foster a free market in emission credits and reward countries that are most efficient at meeting their target.

EMISSIONS TRADING SYSTEM

There are three main approaches to emissions trading system: cap and trade, baseline and credit and offset

CAP AND TRADE SYSTEM: A regulatory body would first need to set an overall limit on emissions - the 'emissions cap'- the total amount of a pollutant that the participants in the programme are allowed to emit in a given period (e.g. emission of a number of tons of the pollutant per year). Allowances equal to all of the emissions permitted under the cap are then distributed. The way in which allowances are distributed is a key for emissions trading system design and there are two types of distribution: free or by auction. Once allowances are distributed, they may be traded freely.

Defined groups of emitters would be allocated either for free or via auctioning a proportion of the total allowed amount, which they could then trade among one another in the form of permits. Thus emitters that took measures to reduce their pollution below the level represented by permits they held could sell their unused permits to other emitters that were emitting at the level higher than that represented by their permits. Participants that emitted beyond their limits without the necessary permits would face stiff penalties. This system would guarantee that the overall environmental objective was met because total emissions would be limited under the cap.

BASELINE AND CREDIT SYSTEM: Participants in a baseline and credit programme have to earn credits before they can begin trading. First, an emission baseline is defined for each participant by the regulator and the baseline often varies with the level of output. Each participant then makes reductions and monitors or calculates its actual emissions using specified procedures.

In this credit trading system, companies would be allowed to certify emissions reductions below a certain voluntary or imposed level as "tradable credit" that could then be sold to emitters that needed them to meet their own target level of emissions. Baseline and credit trading could be used in conjugation with a cap and trade system to enable sources not regulated by the cap to gain credit for reduction in greenhouse gases caused by their new initiatives. However, it is anticipated that credit creation would be limited to certain sectors and activities.

Emissions trading involving SO_2 and NO_x emissions are called downstream programmes. When emissions are related to the characteristics of a product, they can be regulated prior to their release to the atmosphere. Programmes involving this type of control are upstream programmes. By reducing lead content or choosing low carbon content fuels, it is possible to regulate emissions prior to the point of release. A hybrid programme is one which combines elements of both downstream and upstream ones.

CARBON OFFSETS: Carbon emission credits have been suggested in the energy sector as a vehicle for improving the efficiency of carbon mitigating activities. A similar instrument, carbon offsets, is also being considered under voluntary market. Offsets are greenhouse gas reductions achieved by non-regulated market participants. Greenhouse gas mitigation achieved by non-regulated parties can be purchased as offsets by a regulated party (e.g. large electric power plants) to meet the required cap. Offset opportunities relevant to farms include methane capture from farming operations, end-use efficiency for natural gas, propane or heating oil and afforestation.

The concept is that carbon offsets could be added to the stock of emission credits, thus expanding the number of credits in the system and thereby lowering costs and improving efficiency. Indeed, Kyoto protocol offers a variant of this possibility under the JI and perhaps under the CDM. In recent years, some electrical utilities have expanded their capacity and offset the increased carbon emissions through forestry activities. Carbon offset credits are being considered for protecting forests that otherwise would be destroyed, creating new forests and reducing carbon emissions from the some current practices, e.g. low-impact logging. Such activities would need some method of carbon certification by some certifying firms. Through voluntary offsets market, companies, individuals and events buy emission reductions to reduce their carbon footprint

EMISSION TRADING SCHEME

The UK Emissions Trading Scheme (UK ETS) was the world's first economy-wide greenhouse gas emissions trading scheme. The UK Emissions Trading Scheme was a voluntary emissions trading system created as a pilot prior to the mandatory European Union Emission Trading Scheme which it now runs in parallel with. It ran from 2002 and it closed to new entrants in 2009.

The European Union Emission Trading Scheme (EU ETS) also known as the European Union Emissions Trading System, is the largest multi-national, greenhouse gas emissions trading scheme in the world launched in 2005. It is one of the EU's central policy instruments to meet their cap set in the Kyoto Protocol. Since February 2007, seven U.S. states and four Canadian provinces have joined together to create the Western Climate Initiative (WCI), a regional greenhouse gas emissions trading system. In USA, Regional Greenhouse Gas Initiative (RGGI) program was launched in 2009 with the aim to reduce the carbon "budget" of each state's electricity generation sector to 10 per cent below their 2009 allowances by 2018. California's cap and trade system was launched in 2011 to minimize climate change pollution, becoming the first state in the nation with a cap and trade program. The New Zealand Emissions Trading Scheme (NZ ETS) is a national all-sectors all-greenhouse gases uncapped emissions trading scheme. Many countries are in the process of setting up their own emission trading schemes.

EMISSION AND CARBON MARKETS

Carbon emissions trading have been steadily increasing in recent years. According to the World Bank's Carbon Finance Unit, global carbon market stalled at $142 billion for the year 2010. As CDM and AAU market segments declined, the dominance of the European Union Allowances (EUAs) market became more pronounced in 2010 which accounted for 84 per cent of global carbon market value. While the international regulatory environment remains uncertain, national and local initiatives have noticeably picked up and may offer the potential to collectively overcome the international regulatory gap. Low-carbon initiatives have gained increasing traction in developing economies such as Brazil, China, India and Mexico.

Table 39.6: Carbon Market Evolution Values in $ Billion (Source: World Bank 2011)

Year	EU ETS Allowances	Other Allowances	Primary CDM	Secondary CDM	Other Offsets	Total
2005	7.9	0.1	2.6	0.2	0.3	11.0
2006	24.4	0.3	5.8	0.4	0.3	31.2
2007	49.1	0.3	7.4	5.5	0.8	63.0
2008	100.5	1.0	6.5	26.3	0.8	135.1
2009	118.5	4.3	2.7	17.5	0.7	143.7
2010	119.8	1.1	1.5	18.3	1.2	141.9

In carbon trading, allowances or CERs can be sold privately or in the international market at the prevailing market price. These trade and settle internationally and hence allow allowances to be transferred between countries. Each international transfer is validated by the

UNFCCC. Each transfer of ownership within the European Union is additionally validated by the European Commission.

Climate exchanges have been established to provide a spot market in allowances, as well as futures and options market to help discover a market price and maintain liquidity. Carbon prices are normally quoted in Euros per ton of carbon dioxide or its equivalent (CO_2e). Other greenhouse gasses can also be traded, but are quoted as standard multiples of carbon dioxide with respect to their global warming potential. These features reduce the quota's financial impact on business, while ensuring that the quotas are met at a national and international level.

Currently there are six exchanges trading in carbon allowances: the Chicago Climate Exchange, European Climate Exchange, NASDAQ OMX Commodities Europe, PowerNext, Commodity Exchange Bratislava and the European Energy Exchange. NASDAQ OMX Commodities Europe listed a contract to trade offsets generated by a CDM carbon projects called certified emission reductions (CERs). Many companies now engage in emissions abatement, offsetting and sequestration programs to generate credits that can be sold on one of the exchanges. Carbon credits at Commodity Exchange Bratislava are traded at special platform - Carbon place.

CLIMATE CHANGE MITIGATION EFFORTS IN INDIA

NATIONAL CDM AUTHORITY

Planning Commission constituted a Working Group under the Chairmanship of Dr. R. Mandal, Advisor (E&F), to prepare a National Action Plan for Operationalising CDM in India. The National Plan was ready in December 2003 with its recommendations to establish a Designated National Authority (DNA) for providing the legal teeth to the operationalisation of CDM in India. The ministry of Environment and Forests (MoEF) of Government of India is responsible for the climate change and CDM issues in India. It established the Designated National Authority (DNA) in December 2003 as the National CDM Authority (NCDMA) for the purpose of protecting and improving the quality of environment in terms of the Kyoto Protocol.

The NCDMA is chaired by the Secretary of MoEF. The other members are the Secretary of Ministry of External Affairs Secretary, Secretary of Ministry of Finance, Secretary of Department of Industrial Policy and Promotion, Secretary of Ministry of Non-conventional Energy Sources, Secretary of Ministry of Power, Secretary of Planning Commission, Secretary of DST and MoEF Joint Secretary of Climate Change. The Member-Secretary of the NCDMA is the Climate Change Director of MoEF.

NCDMA duties include approval, registration and certification of CDM projects. The National Clean Development Mechanism (CDM) Authority receives projects for evaluation and approval as per the guidelines and general criteria laid down in the relevant rules and modalities pertaining to CDM in addition to the guidelines issued by the Clean Development Mechanism Executive Board and Conference of Parties serving as Meeting of Parties to the United Nations Framework Convention on Climate Change. The key role of NCDMA is to create National CDM Fund and also to monitor the CDM projects for sustainability. It promotes & markets CDM in the country.

CDM CELL AND CDM PROMOTION AGENCY

MoEF also established CDM Cell and CDM Promotion Agency as per the recommendation of National Action Plan for Operationalising CDM in India. The administrative and regulatory roles pertaining to CDM would be played by the CDM Cell, while the other facilitating roles such as promotion and marketing are envisaged to be played by the CDM Promotion Agency, which could include private sector/NGO. The member secretary of the DNA is responsible for day-to-day activities of the CDM Cell and corordinating CDM Promotion Agency.

NATIONAL ACTION PLAN FOR CLIMATE CHANGE

Government of India released National Action Plan on Climate Change (NAPCC) on 30 June 2008. The NAPCC addresses the urgent and critical concerns of the country through a directional shift in the development pathway, including through the enhancement of the current and planned programmes presented in the Technical Document. The National Action Plan on Climate Change identifies measures that promote our development objectives while also yielding co-benefits for addressing climate change effectively. It outlines a number of steps to simultaneously advance India's development and climate change-related objectives of adaptation and mitigation. The National Action Plan on Climate Change (NAPCC) encompasses a range of measures. It focuses on eight missions, which are as follows:

NATIONAL SOLAR MISSION: The NAPCC aims to promote the development and use of solar energy for power generation and other uses, with the ultimate objective of making solar competitive with fossil-based energy options. It also includes the establishment of a solar research center, increased international collaboration on technology development, strengthening of domestic manufacturing capacity and increased government funding and international support.

NATIONAL MISSION FOR ENHANCED ENERGY EFFICIENCY: The NAPCC recommends mandating specific energy consumption decreases in large energy-consuming industries, with a system for companies to trade energy-saving certificates, financing for public-private partnerships to reduce energy consumption through demand-side management programs in the municipal, buildings and agricultural sectors, and energy incentives, including reduced taxes on energy-efficient appliances.

NATIONAL MISSION ON SUSTAINABLE HABITAT: The NAPCC also aims at promoting energy efficiency as a core component of urban planning by extending the existing Energy Conservation Building Code, strengthening the enforcement of automotive fuel economy standards and using pricing measures to encourage the purchase of efficient vehicles and incentives for the use of public transportation. The NAPCC also emphasizes on waste management and recycling.

NATIONAL WATER MISSION: The NAPCC sets a goal of a 20 per cent improvement in water use efficiency through pricing and other measures to deal with water scarcity as a result of climate change.

NATIONAL MISSION FOR SUSTAINING THE HIMALAYAN ECOSYSTEM: This particular mission sets the goal to prevent melting of the Himalayan glaciers and to protect biodiversity in the Himalayan region.

GREEN INDIA MISSION: The NAPCC also aims at afforestation of 6 million hectares of degraded forest lands and expanding forest cover from 23 to 33 per cent of India's territory. The overarching objective of the National Mission for a Green India is to increase forest and tree cover in 5 million hectares and improve quality of forest cover in another 5 million hectares. Thus, the mission will help in improving ecosystem services in 10 million ha of land, and increase flow of forest based livelihood services and income of about 3 million forest dependent households.

NATIONAL MISSION FOR SUSTAINABLE AGRICULTURE: The NAPCC aims to support climate adaptation in agriculture through the development of climate-resilient crops, expansion of weather insurance mechanisms, and agricultural practices.

NATIONAL MISSION ON STRATEGIC KNOWLEDGE FOR CLIMATE CHANGE: To gain a better understanding of climate science, impacts, and challenges, the plan envisions a new Climate Science Research Fund, improved climate modeling, and increased international collaboration. It also encourages private sector initiatives to develop adaptation and mitigation technologies through venture capital funds.

PART 9:
EVENTS AND ORGANIZATIONS

40 INTERNATIONAL CONFERENCES, MEETINGS AND CONVENTIONS

THE IUCN WORLD CONSERVATION CONGRESS

The IUCN World Conservation Congress is the world's largest and most important conservation event. In the past, Congresses were held every two years, then every three, and now they are held every four years. The Congress aims to improve how we manage our natural environment for human, social and economic development and to debate major sustainable development issues, propose solutions and facilitate the sharing of information and experiences. The first IUCN World Conservation Congress was held in 1948 at Fontainebleau, France. 23rd World Conservation Congress will be held from 6 - 15 September 2012 in Jeju, Republic of Korea.

UNITED NATIONS CONFERENCE ON THE HUMAN ENVIRONMENT

The first United Nations Conference on the Human Environment (UNCHE) was held in Stockholm, Sweden from 5-16 June 1972. Representatives from 113 countries were present, as well as representatives from many international non-governmental organizations, intergovernmental organizations, and many other specialized agencies. This was the first United Nations conference on the environment as well as the first major international gathering focused on human activities in relationship to the environment, and it laid the foundation for environmental action at an international level. The conference acknowledged that the goal of reducing human impact on the environment would require extensive international cooperation, as many of the problems affecting the environment are global in nature. Following this conference, the United Nations Environmental Programme (UNEP) was launched (1972) in order to encourage United Nations agencies to integrate environmental measures into their programs.

The UNCHE emphasized that defending and improving the environment must become a goal to be pursued by all countries. The Stockholm Declaration and Action Plan defined principles for the preservation and enhancement of the natural environment, and highlighted the need to support people in this process. The conference also approved an environmental fund and an action program, which involved many specific recommendations for addressing such problems as global climatic change, acid rain, habitat destruction, marine pollution, population growth, the dumping of toxic wastes and the preservation of biodiversity. Most important result of the conference was the precedent it set for international cooperation in addressing environmental degradation. The nations attending agreed that they shared responsibility for the quality of the environment, particularly the oceans and the atmosphere.

Some of the specific issues addressed were the role which industrialized countries should have in the process of protecting the environment, stating that industrial countries should help to close the gap between them and underdeveloped countries while keeping their own priorities and the protection and improvement of the environment in mind. The conference developed a long set of recommendations to act as goals to pursue its mission. Recommendations included that governments communicate about environmental issues that have international implications (such as air pollution), that governments give attention to the training of those who plan, develop, and manage settlement areas, and that agencies work together to address many issues, such as access to clean water and population growth. However, it was the pending environmental problems that dominated the meeting and led to wider public environmental awareness.

WORLD CLIMATE CONFERENCE

The world climate conferences are a series of international meetings, organized by the World Meteorological Organization (WMO), about global climate issues. The first two conferences largely focused on climate change (global warming) in addition to climate research and forecasting. The third World Climate Conference's theme was "Climate prediction and information for decision-making". Recognizing the growing evidence for the widespread effects of climate variability and change, it aims to spur global action now for countries to cope with both current and future changes.

12-23 February 1979	: First World Climate Conference at Geneva, Switzerland (WCC-1)
29 October - 7 November 1990	: Second World Climate Conference at Geneva, Switzerland (WCC-2)
31 August - 4 September 2009	: Third World Climate Conference at Geneva, Switzerland (WCC-3)

UN CONFERENCE ON NEW AND RENEWABLE SOURCES OF ENERGY

Deeply concerned at the disastrous ecological consequences which might result in the near future from the excessive use of wood as the sole source of energy for the vast majority of people in a large number of developing countries, United Nations decided to convene an international conference. UN Conference on New and Renewable Sources of Energy, convened in Nairobi, Kenya, from 10-21 August 1981, marked the international community's first major attempt to develop a strategy for the use of alternative energies. The conference examined alternative forms of energy and adopted the Nairobi Programme of Action for the Development and Utilization of New and Renewable Sources of Energy as a blueprint for national and international action. The Nairobi Programme identified five broad areas for concentrated action: energy assessment and planning; research, development, and demonstration; transfer, adaptation, and application of mature technologies; information flows; and education and training. Endorsing the Nairobi Programme later that year, the General Assembly set up an interim committee to launch immediate implementation and in 1992, established the Committee on the Development and Utilization of New and Renewable Sources of Energy, open to the participation of all states as full members.

UNEP'S STOCKHOLM +10 CONFERENCE

The UN Environment Programme organized a special Stockholm +10 conference in Nairobi in 1982. The attendees agreed to a declaration expressing serious concern about the present state of the environment and established an independent commission to craft a global agenda for change.

THE UNITED NATIONS CONFERENCE ON ENVIRONMENT AND DEVELOPMENT (1992)

The United Nations Conference on Environment and Development (UNCED), also known as the Earth Summit, took place in Rio de Janeiro, Brazil, from 2-14 June 1992. It was held twenty years after the United Nations Conference on the Human Environment (UNCHE) took place in Stockholm, Sweden. Government officials from 178 countries and, non-governmental organizations participated in this event to discuss solutions for global problems such as poverty, war, and the growing gap between industrialized and developing countries. The central focus was the question of how to relieve the global environmental system through the introduction to the paradigm of sustainable development. This concept emphasizes that economic and social progress depends critically on the preservation of the natural resource base with effective measures to prevent environmental degradation.

The results of the UNCED included the Rio Declaration enunciating 27 principles of environment and development, Agenda 21, and a Statement of principles for the Sustainable Management of Forests, which were all adopted by consensus (without vote) by the conference. The institutional innovation resulting from the conference included an agreement on the operating rules for the Global Environmental Facility (GEF), United Nations Convention on Biological Diversity, and the establishment of the United Nations Commission on Sustainable Development (CSD) on the basis of an Agenda 21 recommendation. The United Nations Framework Convention on Climate Change (UNFCCC) and United Nations Convention on Biological Diversity were products of independent, but concurrent, negotiating processes that were opened for signatures at UNCED.

THE RIO DECLARATION ON ENVIRONMENT AND DEVELOPMENT: It is a set of 27 legally non-binding principles designed to commit governments to ensure environmental protection and responsible development and intended to be an Environmental Bill of Rights, defining the rights of people to development, and their responsibilities to safeguard the common environment. It established the Precautionary principle and the principle of common but differentiated responsibilities. The Declaration recognizes that the only way to have long-term social and economic progress is to link it with environmental protection and to establish equitable global partnerships between governments and key actors of civil society and the business sector.

The Declaration includes many progressive approaches such as the polluter pays principle (the polluter bears the costs of pollution) and the precautionary principle (carry out environmental assessments to identify adverse impacts and eliminate any potential harms from a project before it is started). It advocates that today's development shall not undermine the resource base of future generations and that developed countries bear a special responsibility due to the pressure their societies place on the global environment and the technologies and financial resources they command.

AGENDA 21: It is the international plan of action to sustainable development which outlines key policies for achieving sustainable development that meets the needs of the poor and recognizes the limits of development to meet global needs. Agenda 21 has become the blueprint for sustainability and forms the basis for sustainable development strategies. It attempts to define a balance between production, consumption, population, development, and the Earth's life-supporting capacity. It addresses poverty, excessive consumption, health and education, cities and agriculture; food and natural resource management and several more subjects. Its 40 chapters are broken up into four sections:

- Social and economic dimensions: developing countries; poverty; consumption patterns; population; health; human settlements; integrating environment and development.
- Conservation and management of resources: atmosphere; land; forests; deserts; mountains; agriculture; biodiversity; biotechnology; oceans; fresh water; toxic chemicals; hazardous, radioactive and solid waste and sewage.
- Strengthening the role of major groups: women; children and youth; indigenous peoples; non-governmental organizations; local authorities; workers; business and industry; farmers; scientists and technologists.
- Means of implementation: finance; technology transfer; science; education; capacity-building; international institutions; legal measures; information.

STATEMENT OF PRINCIPLES FOR THE SUSTAINABLE MANAGEMENT OF FORESTS: The Statement of Forest Principles was the first global agreement concerning sustainability of forest management. Although it was not a legally binding contract, all signatories are expected to practice reforestation and forest conservation; they were also to develop programs to find economic and social substitutions for forestry.

AGENDA 21, CHAPTER 11 COMBATING DEFORESTATION: Agenda 21 is a comprehensive plan of action to be taken up globally, nationally and locally by organizations of the United Nations system, governments, and major groups from every area in which humans have an impact on the environment. The four main principles of chapter 11 on combating deforestation are:

- Sustaining the multiple roles and functions of all types of forests, forestlands and woodlands.
- Enhancing the protection, sustainable management and conservation of all forests, and the greening of degraded areas, through rehabilitation, afforestation and reforestation.

- Promoting efficient utilization and assessment to recover the full valuation of the goods and services provided by forests, forestlands and woodlands.
- Establishing and /or strengthening capacities for the planning, assessment and systematic observation of forests and related programmes, projects and activities, including commercial trade and processes.

THE WORLD SUMMIT ON SUSTAINABLE DEVELOPMENT

The World Summit on Sustainable Development (WSSD) was organized by the United Nations (UN) and took place from 26 August to 4 September 2002, at the Sandton Convention Centre in Johannesburg, South Africa. ioThis marked the 30th anniversary of the United Nations Conference on the Human Environment (UNCHE) which took place in Stockholm, Sweden in 1972, and the10th anniversary of the United Nations Conference on Environment and Development (UNCED) which took place in Rio de Janeiro, Brazil in 1992. The 2002 summit is also known as "Rio+10", "Johannesburg Summit 2002" and "Earth Summit 2002"

The summit focused the world's attention toward meeting difficult challenges, including improving people's lives and conserving natural resources, with ever-increasing demands for food, water, shelter, sanitation, energy, health services and economic security. The summit was called in large part to assess world developments in sustainability since the 1992 Earth Summit, and to evaluate the effectiveness of Agenda 21 and other agreements reached at the 1992 meeting. The five topic areas which were to be the key points of discussion at the summit: (1) Water and sanitation, (2) Energy, (3) Human health, (4) Agricultural productivity, and (5) Biodiversity and ecosystem management. The major innovation at Johannesburg was the so-called 'Type 2' agreements. These were informal agreements involving non-state parties, sometimes amongst themselves and sometimes with individual governments.

The summit addressed and emphasized the role of sustainable forest management in overall sustainable development and, more specifically, in poverty reduction, improvements in food security and access to safe drinking water and affordable energy. One of the key commitments of the summit was to accelerate implementation of the IPF/IFF proposals for action by countries and by the Collaborative Partnership on Forests, and intensify efforts on reporting to the United Nations Forum on Forests.

THE JOHANNESBURG DECLARATION: It was a principal outcome of the Summit. The declaration is a collection of general political statements, reaffirming a commitment to agreements made at the Rio de Janeiro summit and at the Stockholm Summit on the Human Environment 30 years prior. International cooperation, decreasing world poverty, special attention for developing nations, empowering women, and maintaining biodiversity, among other things, are outlined as key points to building a sustainable future. The document is meant to serve as a contract for the participants of the summit, binding them to the outlined agreements.

A Plan of Implementation laid down more specific goals for the nations and organizations that participated in the summit. Some of these goals include:

- The establishment of a solidarity fund to wipe out poverty. This fund would be sustained by voluntary contributions; however, developed nations are urged to dedicate 0.7% of their national income to this cause.
- Cutting in half by 2015 the proportion of the world's population living on less than a dollar a day. This is a reaffirmation of a UN Millennium Summit goal.
- Cutting in half by 2015 the number of people who lack clean drinking water and basic sanitation
- Substantially increase the global share of renewable energy
- Cut significantly by 2010 the rate at which rare plants and animals are becoming extinct
- Restore (where possible) depleted fish stocks by 2015, and
- Halving the number of people suffering from hunger.

UNITED NATIONS MILLENNIUM+5 SUMMIT

In September 2000, 150 heads of state met at the United Nations and ratified the UN Millennium Declaration. In September 2005, the UN hosted a Millennium+5 Summit to evaluate the progress towards the goals spelled out in the document. Leaders also resolved to meet a number of Millennium Development Goals (MDGs), which include halving the proportion of people living in poverty and hunger by 2015, ensuring primary schooling for all children, reversing the spread of HIV/AIDS, malaria and other major diseases, sustainable development and environmental protection.

THE UNITED NATIONS CLIMATE CHANGE CONFERENCE

The Conference of the Parties (COP) is the "supreme body" of the United Nations Framework Convention on Climate Change (UNFCCC), that is, its highest decision-making authority. It is an association of all the countries that are Parties to the Convention. The COP is responsible for keeping international efforts to address climate change on track. It reviews the implementation of the Convention and examines the commitments of Parties in light of the Convention's objective, new scientific findings and experience gained in implementing climate change policies. A key task for the COP is to review the national communications and emission inventories submitted by Parties. Based on this information, the COP assesses the effects of the measures taken by Parties and the progress made in achieving the ultimate objective of the Convention.

Since the UNFCCC entered into force, the parties have been meeting annually in Conferences of the Parties (COP) to assess progress in dealing with climate change, and beginning in the mid-1990s, to negotiate the Kyoto Protocol to establish legally binding obligations for developed countries to reduce their greenhouse gas emissions. From 2005 the Conferences have met in conjunction with Meetings of Parties of the Kyoto Protocol (MOP), and parties to the Convention that are not parties to the Protocol can participate in Protocol-related meetings as observers.

LIST OF UNITED NATIONS CLIMATE CHANGE CONFERENCES

1995: COP 1, Berlin, Germany (First UNFCCC Conference of Parties)

1996: COP 2, Geneva, Switzerland

1997: COP 3, Kyoto, Japan (The Kyoto Protocol on Climate Change)

1998: COP 4, Buenos Aires, Argentina

1999: COP 5, Bonn, Germany

2000: COP 6, The Hague, Netherlands

2001: COP 6, Bonn, Germany

2001: COP 7, Marrakech, Morocco

2002: COP 8, New Delhi, India

2003: COP 9, Milan, Italy

2004: COP 10, Buenos Aires, Argentina

2005: COP 11/MOP 1, Montreal, Canada

2006: COP 12/MOP 2, Nairobi, Kenya

2007: COP 13/MOP 3, Bali, Indonesia

2008: COP 14/MOP 4, Poznan, Poland

2009: COP 15/MOP 5, Copenhagen, Denmark

2010: COP 16/MOP 6, Cancun, Mexico

2011: COP 17/MOP 7, Durban, South Africa

2012: COP 18/MOP 8, Qatar

UNITED NATIONS SUMMIT ON CLIMATE CHANGE

Nearly 100 world leaders participated in an historic Summit on Climate Change in New York on 22 September 2009 to mobilize political will and strengthen momentum for a fair, effective, and ambitious climate deal in Copenhagen, Denmark in December 2009. It is an unprecedented gathering of over heads of government and state, brought together by the UN secretary-general Ban Ki-moon to "get leaders moving" in the quest for a global deal to control greenhouse gas emissions and prevent dangerous global warming.

The message from the Summit was to assure that the Copenhagen deal must be comprehensive and ensure:

- Enhanced action to assist the most vulnerable and the poorest to adapt to the impacts of climate change;
- Ambitious emission reduction targets for industrialized countries;
- Nationally-appropriate mitigation actions by developing countries with the necessary support;
- Significantly scaled-up financial and technological resources; and

- An equitable governance structure.

Many leaders underlined that a climate change response should be placed in the broader context of sustainable development. They stressed the need to shift their economies on to low-emissions paths and build climate- resilient societies. There was clear recognition that action on climate change can be consistent with developing country priorities for poverty eradication and sustainable development.

COMMONWEALTH FORESTRY CONGRESS

The Commonwealth is a voluntary association of 54 countries that support each other and work together towards shared goals in democracy and development. Commonwealth Forestry Congress serves as a forum for governments, universities, civil society and the private sector to exchange views and experiences and to formulate recommendations on forestry problems. The Commonwealth Forestry Conferences, held every four years, bring together leading industry experts, policy makers, practitioners and researchers. The Conference has a long history of stimulating innovative solutions to the major topical issues faced around the Commonwealth. The very first Commonwealth Forestry Congress held in London in 1920. 18th Commonwealth Forestry Conference held in Edinburgh from 28 June - 2 July 2010.

THE WORLD FORESTRY CONGRESS

The first World Forestry Congress was held in Rome in 1926 and has generally taken place every six years since then. Gatherings serve as a forum for governments, universities, civil society and the private sector to exchange views and experiences and to formulate recommendations for implementation at national, regional and global levels. The Congress also provides an opportunity for the sector to produce an overview of the state of forests and forestry in order to discern trends, adapt policies and raise awareness of issues among decision-makers, the public and other parties concerned.

Its functions are advisory, not executive, and participants attend the Congress in their personal capacity. The implementation of recommendations is a matter solely for those to whom they are addressed in light of their own particular circumstances - governments, international organizations, scientific bodies and forest owners, among others. The outcomes are brought to the attention of the FAO Conference which may consider endorsing, through a resolution, any declaration emanating from the Congress. These prestigious events produce high quality papers, are recognized for their insightful deliberations and are known for reaching broad consensus on authoritative, although not binding, policy and technical advice to governments and international organizations.

FAO Council selects the host country in accordance with established criteria. The Department of Agriculture, Forestry and Fisheries (DAFF) of South Africa has been awarded to host the Fourteenth World Forestry Congress at the International Convention Centre, Durban in September 2015.

THE COMMITTEE ON FORESTRY SESSIONS

The Committee on Forestry (COFO) is the highest FAO Forestry statutory body. The biennial sessions of COFO (held at FAO headquarters in Rome, Italy) bring together heads of forest services and other senior government officials to identify emerging policy and technical issues, to seek solutions and to advise FAO and others on appropriate action. Other international organizations and, increasingly, non-governmental groups participate in COFO. Participation in COFO is open to all FAO member countries. The first meeting of the FAO Committee on Forestry (COFO) held in 1972 and Twentieth Session of COFO held from 4-8 October 2010. The twenty first session of COFO will be held in 24-28 September 2012 in Rome, Italy.

UNITED NATIONS FORUM ON FORESTS SESSIONS

United Nations Forum on Forests (UNFF) was established by Economic and Social Council of the United Nations (ECOSOC) in October 2000 as part of a new International Arrangement on Forests. It has the main objective to promote the management, conservation and sustainable development of all types of forests and to strengthen long-term political commitment

UNFF Organizational Meeting	: 12 February 2001 in New York
UNFF1 Session	: 11 - 22 June 2001 in New York
UNFF2 Session	: 4 - 15 March 2002 in New York
UNFF3 Session	: 26 May - 6 June 2003 in Geneva
UNFF4 Session	: 3 - 14 May 2004 in Geneva
UNFF5 Session	: 16 - 27 May 2005 in New York
UNFF6 Session	: 13 - 24 February 2006 in New York
UNFF7 Session	: 16 - 27 April 2007 in New York
UNFF8 Session	: 20 April - 1 May 2009 in New York
UNFF9 Special Session	: 30 October 2009 in New York
UNFF9 Session	: 24 January - 4 February 2011 in New York
UNFF10 Session	: 8 - 19 April 2013 in Istanbùl, Turkey

NON-LEGALLY BINDING INSTRUMENT ON ALL TYPES OF FORESTS

Following intense negotiations, the Seventh Session of the UNFF adopted the landmark Non-Legally Binding Instrument on All Types of Forests (NLBI) on 28 April 2007. The instrument is considered a milestone, as it is the first time Member States have agreed to an international instrument for sustainable forest management. The instrument is expected to have a major impact on international cooperation and national action to reduce deforestation, prevent forest degradation, promote sustainable livelihoods and reduce poverty for all forest-dependent peoples. The NLBI was adopted by the UN General Assembly on 17 December 2007. The purpose of this instrument is:

- To strengthen political commitment and action at all levels to implement effectively sustainable management of all types of forests and to achieve the shared global objectives on forests
- To enhance the contribution of forests to the achievement of the internationally agreed development goals, including the Millennium Development Goals, in particular with respect to poverty eradication and environmental sustainability
- To provide a framework for national action and international cooperation

THE ASIA-PACIFIC FORESTRY COMMISSION SESSIONS

The Asia-Pacific Forestry Commission (APFC) is one of six FAO Regional Forestry Commissions that cover the world's major geographic regions. The APFC is a forum for advising and taking action on key forestry issues. It focuses on issues pertinent to Asia and the Pacific, a region characterized by its diversity and rapid changes. The first session of the Asia-Pacific Forestry Commission held in Bangkok, Thailand from 9-17 October 1950. 24th session of the Asia-Pacific Forestry Commission held in Beijing, China from 7-11 November 2011.

WORLD CONGRESS OF AGROFORESTRY

Agroforestry is now receiving increasing attention as a sustainable land-management option the world over because of its ecological, economic, and social attributes. World Congress of Agroforestry is being organized to evaluate progress, assess available options, and design strategies in agroforestry for the future. The congress objective is to provide a global forum for agroforestry professionals to share knowledge, experiences, and ideas.

It promotes to plan future strategies in agroforestry research, education and training, and development. First World Congress of Agroforestry held from 27 June to 02 July 2004 in Orlando, Florida, USA and the second congress held from 23-28 August 2009 in Nairobi, Kenya

THE UN-REDD PROGRAMME

The UN-REDD Programme is the United Nations Collaborative initiative on Reducing Emissions from Deforestation and forest Degradation (REDD) in developing countries. The Programme was launched in September 2008 to assist developing countries prepare and implement national REDD+ strategies, and builds on the convening power and expertise of the Food and Agriculture Organization of the United Nations (FAO), the United Nations Development Programme (UNDP) and the United Nations Environment Programme (UNEP). The Programme currently supports 42 partner countries spanning Africa, Asia-Pacific and Latin America, of which 16 are receiving support to National Programme activities.

The UN-REDD Programme brings together technical teams from around the world to help develop analyses and guidelines on issues such as measurement, reporting and verification (MRV) of carbon emissions and flows, ensuring that forests continue to provide multiple benefits for livelihoods and the environment, and supporting the engagement of

Indigenous Peoples and Civil Society at all stages of the design and implementation of REDD+ strategies.

Reducing Emissions from Deforestation and Forest Degradation (REDD) is an effort to create a financial value for the carbon stored in forests, offering incentives for developing countries to reduce emissions from forested lands and invest in low-carbon paths to sustainable development. "REDD+" goes beyond deforestation and forest degradation, and includes the role of conservation, sustainable management of forests and enhancement of forest carbon stocks.

INDIAN FORESTS CONGRESS

The UN General Assembly has proclaimed the year 2011 as the International Year of Forests. Consequent to the UN General Assembly's adoption of the 'Millennium Development Goals', the global community has come together in a collective search for sustainable solutions to address the world's development challenges. In order to have the synergetic effect of all the efforts presently undertaken by different stakeholders in our country, the Ministry of Environment and Forests organized the First Indian Forest Congress in New Delhi from 22-25 November 2011. Indian Council of Forestry Research and Education (ICFRE) had been entrusted with the task of organizing the first congress. The central theme of this four days Congress is 'Forests in a Changing World'. This Congress worked for a collective search for sustainable solutions to address development and environmental challenges and discussed advances on various aspects related to forestry. The main themes are forests in society, forestry in an expanding economy, expanding frontiers of forestry sciences, forest biodiversity and landscapes and forests and climate change.

UNITED NATIONS CONFERENCE ON SUSTAINABLE DEVELOPMENT (RIO+20)

Rio+20 - the short name for the United Nations Conference on Sustainable Development took place in Rio de Janeiro, Brazil on 20-22 June 2012 to mark the 20th anniversary of the 1992 United Nations Conference on Environment and Development (UNCED), in Rio de Janeiro, and the 10th anniversary of the 2002 World Summit on Sustainable Development (WSSD) in Johannesburg. It was an historic opportunity to define pathways to a safer, more equitable, cleaner, greener and more prosperous world for all. Twenty years after the 1992 Earth Summit in Rio, where countries adopted Agenda 21 - a blueprint to rethink economic growth, advance social equity and ensure environmental protection - the UN again brought together governments, international institutions and major groups to agree on a range of smart measures that can reduce poverty while promoting decent jobs, clean energy and a more sustainable and fair use of resources. Rio+20 was a chance to act to end poverty, address environmental destruction and build a bridge to the future.

At the Rio+20 Conference, world leaders, along with thousands of participants from governments, the private sector, NGOs and other groups, came together to shape how we can reduce poverty, advance social equity and ensure environmental protection on an ever more crowded planet to get to the future we want. The objective of the Conference was to secure renewed political commitment for sustainable development, assess the progress to date and the remaining gaps in the implementation of the outcomes of the major summits on sustainable development, and address new and emerging challenges The Conference focused on two themes: (a) a green economy in the context of sustainable development poverty eradication; and (b) the institutional framework for sustainable development. The Rio+20 have highlighted seven areas which need priority attention/critical issues and these include decent jobs, energy, sustainable cities, food security and sustainable agriculture, water, oceans and disaster readiness.

The UNCSD together with its partners prepared a series of Rio+20 Issues Briefs. The purpose of the Rio+20 Issues Briefs was to provide a channel for policymakers and other interested stakeholders to discuss and review issues relevant to the objective and themes of the conference. The major Rio+20 Issues Briefs were:

Issues Brief 1 - Trade and Green Economy, UNCSD Secretariat and UNCTAD

Issues Brief 2 - Options for Strengthening IFSD: Peer Review, UN-DESA

Issues Brief 3 - IFSD: Issues related to an intergovernmental body on SD, UN-DESA

Issues Brief 4 - Oceans, UN-DESA

Issues Brief 5 - Sustainable Cities, UN-DESA

Issues Brief 6 - Current Ideas on Sustainable Development Goals and Indicators, UN-DESA

Issues Brief 7 - Green jobs and social inclusion, UN-DESA

Issues Brief 8 - Reducing Disaster Risk and Building Resilience, UN-DESA

Issues Brief 9 - Food Security and Sustainable Agriculture, UN-DESA

Issues Brief 10 - Regional, national and local level governance for sustainable development, UN-DESA

Issues Brief 11 - Water, UN-DESA

Issues Brief 12 - Science and Technology for Sustainable Development, UN-DESA

Issues Brief 13 - Sustainable, Low Carbon Transport in Emerging and Developing Economies, UN-DESA

Issues Brief 14 - Population dynamics and sustainable development, UN-DESA

Issues Brief 15 - Migration and sustainable development, UN-DESA

INTERNATIONAL CONVENTIONS

INTERNATIONAL CONVENTION FOR THE REGULATION OF WHALING

The International Convention for the Regulation of Whaling is an international environmental agreement was established in order to provide for the proper conservation of whale stocks and thus make possible the orderly development of the whaling industry. It governs the commercial, scientific, and aboriginal subsistence whaling practices of fifty-nine member nations. It was signed by 15 nations in Washington, D.C. on 2 December 1946 and took effect on 10 November 1948. Its protocol (which represented the first substantial revision of the convention and extended the definition of a "whale-catcher" to include helicopters as well as ships) was signed in Washington on 19 November 1956. The convention is a successor to the International Agreement for the Regulation of Whaling, signed in London on 8 June 1937, and the protocols for that agreement signed in London on 24 June 1938, and 26 November 1945.

The objectives of the agreement are the protection of all whale species from overhunting, the establishment of a system of international regulation for the whale fisheries to ensure proper conservation and development of whale stocks, and safeguarding for future generations the great natural resources represented by whale stocks. The primary instrument for the realization of these aims is the International Whaling Commission which was established pursuant to this convention. The commission has made many revisions to the schedule that makes up the bulk of the convention. The Commission process has also reserved for governments the right to carry out scientific research which involves killing of whales.

CONVENTION ON FISHING AND CONSERVATION OF LIVING RESOURCES OF THE HIGH SEAS

The Convention on Fishing and Conservation of Living Resources of the High Seas is an agreement that was designed to solve through international cooperation the problems involved in the conservation of living resources of the high seas, considering that because of the development of modern technology some of these resources are in danger of being overexploited. It was opened for signature on 29 April 1958 and entered into force from 20 March 1966. At present, there are 38 parties to the convention.

CONVENTION ON INTERNATIONAL TRADE IN ENDANGERED SPECIES OF WILD FAUNA AND FLORA

As a result of growing global environmental awareness, and growing concern about the threats that international trade posed for many species, the idea for a multilateral convention concerning trade in endangered species was formulated in the 1960s under the umbrella of IUCN - The World Conservation Union. This culminated in the drafting of the Convention on International Trade in Endangered Species of Wild Fauna and Flora (CITES), which was signed and adopted by 85 countries during a 1973 conference in Washington, DC. The treaty came into force in 1975 and 175 nations had become Parties. It is an international agreement between governments concerning the international trade in specimens of wild animals and plants and its impacts. Over 30,000 species are protected contributing to sustainable development and achievement of MDGs under this landmark agreement, and more are being added on a regular basis.

CITES is aimed at protecting certain endangered species of wild fauna and flora from overexploitation through international trade, via a system of import/export permits. The convention focuses exclusively on trade and is premised on the view that the control of international markets will contribute to the preservation of endangered species. CITES operates by listing endangered species in one of its three appendices, which constitute different levels of control in international trade. Numerous forest animal species are included in the appendices, but at present only 16 tree species have been listed, mainly species used for timber. Listing of timber producing species of major commercial importance has become increasingly controversial in recent years, because some parties think this may have the potential to lead to the introduction of unfair trade restrictions. While listing in CITES appendices will help draw attention to the need to take action for better management and conservation of given species, it is clear that listing, as such, does not conserve species.

CITES has 25 Articles which describe the legal and scientific definitions, fundamental principles, broad legal guidelines to regulate trade, enforcement measures and administrative aspects. The CITES Secretariat is obliged under Article 11 to convene a Conference of Parties (COPs) at least biennially. COP 10 of CITES recognized in 1997 "that commercial trade may be beneficial to the conservation of species and ecosystems when carried out at levels that are not detrimental to the survival of the species in question"; however, the COP also noted "that some timber species may be under threat because of detrimental levels of use and international trade". In 1994, CITES formed the Timber Working Group to study implementation problems resulting from the inclusion of timber species in CITES and to review the definition of terms and units used to describe parts and derivatives of timber trees in trade. CITES has also instituted a Mahogany Working Group and a Plants Committee.

RAMSAR CONVENTION ON WETLANDS

It is an intergovernmental treaty, which provides the framework for national action and international cooperation for the conservation and wise use of wetlands and their resources. The Convention was adopted in 1971 and entered into force in 1975. The objective of the Ramsar Convention is the conservation and wise use of wetlands through national action and international cooperation. More than a thousand sites, totalling over 80 million hectares, are designated as wetlands of international importance. Some of these sites contain forest ecosystems, such as mangroves, although the exact number cannot be established, as forests as such are not identified under the convention.

Originally, the Ramsar Convention focused exclusively on preserving wetlands as waterfowl habitat, but it has developed extensively over time to embrace the broader implications of wetland destruction. Its COP has taken several decisions to acknowledge the importance of wetlands not only for their waterfowl, but also for their biological diversity and other ecological and environmental functions. More emphasis has been placed on the concept of sustainable development, including for example participation of local communities and indigenous peoples. Since 1999, the criteria for designating a wetland to be of international importance have included the following:

- It supports vulnerable, endangered or critically endangered species or threatened ecological communities
- It supports populations of plant and/or animal species important for maintaining the biological diversity of a particular biogeographic region

CONVENTION ON THE PREVENTION OF MARINE POLLUTION BY DUMPING OF WASTES AND OTHER MATTER

The Convention on the Prevention of Marine Pollution by Dumping of Wastes and Other Matter, commonly called as London Convention (LC), is an agreement to control pollution of the sea by dumping and to encourage regional agreements supplementary to the Convention. The main objective of the London Convention is to prevent indiscriminate disposal at sea of wastes that could be liable for creating hazards to human health; harming living resources and marine life; damaging amenities; or interfering with other legitimate uses of the sea. The 1972 Convention extends its scope over "all marine waters other than the internal waters" of the States and prohibits the dumping of certain hazardous materials. It further requires a prior special permit for the dumping of a number of other identified materials and a prior general permit for other wastes or matter.

The Convention was called for by the United Nations Conference on the Human Environment (June 1972, Stockholm), the treaty was drafted at the Intergovernmental Conference on the Convention on the Dumping of Wastes at Sea (13 November 1972, London) and it was opened for signature on 29 December 1972. It entered into force on 30 August 1975 when 15 nations ratified. Now there are 78 Contracting Parties to the Convention. International Administration of the Convention functions through Consultative Meetings held at International Maritime Organization (IMO) headquarters in London.

CONVENTION FOR THE PROTECTION OF THE WORLD CULTURAL AND NATURAL HERITAGE (WORLD HERITAGE CONVENTION)

It is an international treaty by which countries recognize that the sites located on their national territory constitute a world heritage for whose protection it is the duty of the international community as a whole to co-operate. The Convention was adopted in 1972 at Paris and entered into force in 1981. The World Heritage Convention (WHC) aims to establish a system of collective protection of cultural and natural heritage of outstanding universal value. States recognize their duty to ensure the protection, conservation and transmission to future generations of their cultural and natural heritage. Forests can be considered as natural heritage, defined as "natural sites or precisely delineated areas of outstanding universal value from the point of view of science, conservation or natural beauty" (Article 2).

WHC establishes the World Heritage List containing world heritage sites. Intergovernmental Committee for the Protection of the World Cultural and Natural Heritage, 1999 sets out that the natural heritage sites listed should be outstanding examples representing significant on-going ecological and biological processes in the evolution and development of ecosystems and communities of plants and animals or contain the most important and significant natural habitats for *in-situ* conservation of biological diversity. At present, almost 936 sites are listed as world heritage sites. Of those, 41 sites fall in the category of tropical forests, representing 30.6 million hectares. Other types of forest have also been designated in the World Heritage List.

THE MAN AND THE BIOSPHERE PROGRAMME (MAB)

The Man and the Biosphere (MAB) Programme is an Intergovernmental Scientific Programme aiming to set a scientific basis for the improvement of the relationships between people and their environment globally. Launched in the early 1970s by United Nations Educational, Scientific and Cultural Organization (UNESCO), it proposes an interdisciplinary research agenda and capacity building that target the ecological, social and economic dimensions of biodiversity loss and the reduction of this loss.

The MAB Programme develops the basis within the natural and social sciences for the rational and sustainable use and conservation of the resources of the biosphere and for the improvement of the overall relationship between people and their environment. It predicts the consequences of today's actions on tomorrow's world and thereby increases people's ability to efficiently manage natural resources for the well-being of both human populations and the environment. Sub-programmes and activities focus on specific ecosystems: mountains; drylands; tropical forests; urban systems; wetlands; and marine, island and coastal ecosystems.

THE CONVENTION ON THE CONSERVATION OF MIGRATORY SPECIES OF WILD ANIMALS

The Convention on the Conservation of Migratory Species of Wild Animals (also known as CMS or the Bonn Convention) aims to conserve terrestrial, marine and avian migratory species throughout their range. It is an intergovernmental treaty, concluded under the aegis of the United Nations Environment Programme, concerned with the conservation of wildlife and habitats on a global scale. Since the Convention's entry into force, its membership has grown steadily to include over 116 Parties from Africa, Central and South America, Asia, Europe and Oceania. The Convention was signed and adopted in 1979 in Bonn (hence the name) and entered into force in 1983.

As the only global convention specializing in the conservation of migratory species, their habitats and migration routes, CMS complements and co-operates with a number of other international organizations, NGOs and partners in the media as well as in the corporate sector. Migratory species threatened with extinction are listed on **Appendix I** of the Convention. CMS Parties strive towards strictly protecting these animals, conserving or restoring the places where they live, mitigating obstacles to migration and controlling other factors that might endanger them. Besides establishing obligations for each State joining the Convention, CMS promotes concerted action among the Range States of many of these species. Migratory species that need or would significantly benefit from international co-operation are listed in **Appendix II** of the Convention. For this reason, the Convention encourages the Range States to conclude global or regional Agreements.

CONVENTION ON LONG-RANGE TRANSBOUNDARY AIR POLLUTION

The convention opened for signature on 13 November 1979 and entered into force on 16 March 1983. The Convention, which now has 51 Parties, identifies the Executive Secretary of the United Nations Economic Commission for Europe (UNECE) as its secretariat. Since 1979 the Convention on Long-range Trans boundary Air Pollution has addressed some of the major environmental problems of the UNECE region through scientific collaboration and policy negotiation. The Convention has been extended by eight protocols that identify specific measures to be taken by Parties to cut their emissions of air pollutants.

The aim of the Convention is that Parties shall endeavour to limit and, as far as possible, gradually reduce and prevent air pollution including long-range transboundary air pollution. Parties develop policies and strategies to combat the discharge of air pollutants through exchanges of information, consultation, research and monitoring.

The Parties meet annually at sessions of the Executive Body to review ongoing work and plan future activities including a work plan for the coming year. The three main subsidiary bodies - the Working Group on Effects, the Steering Body to EMEP and the Working Group on Strategies and Review - as well as the Convention's Implementation Committee, report to the Executive Body each year.

Currently, the Convention's priority activities include review and possible revision of its most recent protocols, implementation of the Convention and its protocols across the entire UNECE region (with special focus on Eastern Europe, the Caucasus and Central Asia and South-East Europe) and sharing its knowledge and information with other regions of the world.

CONVENTION FOR THE CONSERVATION OF ANTARCTIC MARINE LIVING RESOURCES

The Convention on the Conservation of Antarctic Marine Living Resources, also Commission on the Conservation of Antarctic Marine Living Resources (CCAMLR), is part of the Antarctic Treaty System. The Convention was opened for signature on 1 August 1980 and entered into force on 7 April 1982 by the Commission for the Conservation of Antarctic Marine Living Resources, headquartered in Tasmania, Australia. The goal is to preserve marine life and environmental integrity in and near Antarctica.

It was established in large part to concerns that an increase in krill catches in the Southern Ocean could have a serious impact on populations of other marine life which are dependent upon krill for food. 31 nations were signatory parties on the convention. In 1985, CCAMLR set up the Ecosystem Monitoring Programme (CEMP) to further monitor the effects of fishing and harvesting of species in the area.

THE VIENNA CONVENTION FOR THE PROTECTION OF THE OZONE LAYER

In 1977 the United Nations Environment Programme (UNEP) concluded a World Plan of Action on the Ozone Layer, which called for intensive international research and monitoring of the ozone layer, and in 1981, UNEP's Governing Council authorized UNEP to draft a global framework convention on stratospheric ozone protection. The Vienna Convention, concluded in 1985, is a framework agreement in which States agree to cooperate in relevant research and scientific assessments of the ozone problem, to exchange information, and to adopt "appropriate measures" to prevent activities that harm the ozone layer. The obligations are general and contain no specific limits on chemicals that deplete the ozone layer.

The Vienna Convention for the Protection of the Ozone Layer outlines states responsibilities for protecting human health and the environment against the adverse effects of ozone depletion. It was adopted in 1985 and entered into force in 1988. It established the framework under which the Montreal Protocol was negotiated. The ultimate objective of the Convention is to protect human health and the environment against adverse effects resulting from human activities which modify or likely to modify the ozone layer and urges the Parties to take appropriate measures in accordance with the provisions in the Convention and its Protocols which are in force for that Party.

MONTREAL PROTOCOL ON SUBSTANCES THAT DEPLETE THE OZONE LAYER

The Montreal Protocol on Substances that Deplete the Ozone Layer (MP) was adopted in 16 September 1987 as an international treaty to eliminate the production and consumption of ozone-depleting chemicals and entered into force in 1989. It is a landmark international agreement designed to protect the stratospheric ozone layer. The treaty was originally signed in 1987 and substantially amended four times in 1990 (London Amendment), 1992 (Copenhagen Amendment), 1997 (Montreal Amendment) and (Beijing Amendment) 1999. The Montreal Protocol stipulates that the production and consumption of compounds that deplete ozone in the stratosphere-chlorofluorocarbons (CFCs), halons, carbon tetrachloride, and methyl chloroform--are to be phased out by 2000 (2005 for methyl chloroform). Scientific theory and evidence suggest that, once emitted to the atmosphere, these compounds could significantly deplete the stratospheric ozone layer that shields the planet from damaging UV-B radiation.

In 1995, the United Nations named 16 September as the International Day for the Protection of the Ozone Layer. Today, all UN recognized nations have ratified the treaty and continue to phase out the production of chemicals that deplete the ozone layer while searching for ozone-friendly alternatives.

BASEL CONVENTION

The Basel Convention on the Control of Transboundary Movements of Hazardous Wastes and Their Disposal, usually known simply as the Basel Convention, is an international treaty that was designed to reduce the movements of hazardous waste between nations, and specifically to prevent transfer of hazardous waste from developed to less developed countries (LDCs). It does not, however, address the movement of radioactive waste. The Convention is also intended to minimize the amount and toxicity of wastes generated, to ensure their environmentally sound management as closely as possible to the source of generation, and to assist LDCs in environmentally sound management of the hazardous and other wastes they generate. The Convention was opened for signature on 22 March 1989, and entered into force on 5 May 1992. At present, there are 175 parties to the Convention.

INTERNATIONAL CONVENTION ON CIVIL LIABILITY FOR OIL POLLUTION DAMAGE

The International Convention on Civil Liability for Oil Pollution Damage, often referred to as the CLC Convention, is an international maritime treaty that was adopted in 1992 to ensure that adequate compensation would be available where oil pollution damage was caused by maritime casualties involving oil tankers (i.e. ships that carry oil as cargo). The convention introduces strict liability for shipowners for an instance of oil pollution. If a ship carries more than 2000 tons of oil in cargo, CLC requires shipowners to maintain insurance or other financial security sufficient to cover the maximum liability for one oil spill. At present, 127 countries, representing 96.6% of the world fleet, are contracting parties to the CLC Protocol 1992.

THE CONVENTION ON BIOLOGICAL DIVERSITY

The Convention on Biological Diversity (CBD) was adopted in 1992 at Rio de Janeiro and entered into force on 29 December 1993. It has 3 main objectives: the conservation of biological diversity, the sustainable use of the components of biological diversity, and the fair and equitable sharing of the benefits arising out of the utilization of genetic resources. A large part of the world's terrestrial biological diversity is found in forests and forest ecosystems are estimated to contain 70 per cent of the world's plant and animal species. Since its adoption, CBD has considerably expanded its horizon to include forests within its purview. At present, there are 193 Parties with 168 Signatures.

In 1995, COP-2 adopted a statement on biological diversity and forests and stressed that forests have a crucial role in maintaining global biological diversity. Subsequently, a Work Programme for Forest Biological Diversity was adopted by COP-3 in 1996. The Work Programme focuses on the research, cooperation and development of technologies necessary for the conservation and sustainable use of forest biological diversity of all types of forests. In 1998, COP-4 decided to consider forests one of the three priority themes for COP-6 in 2002. It also established an Ad Hoc Technical Expert Group on Forest Biological Diversity. This group has a mandate to provide advice on scientific research and development, to review available information on status, trends and threats to forest biological diversity and to suggest actions for the conservation and sustainable use of forest biological diversity. It has had two meetings and has the potential to become an interesting forum for developing an inclusive forest biodiversity policy framework.

CBD has influenced the global dialogue on forests also through its leading role in support of recognition of the traditional forest-related knowledge of indigenous people and forest-dependent people in the IPF/IFF process. CBD certainly includes articles on indigenous and traditional forest-related knowledge, but at the same time it also presents the risk for indigenous people to be seen as a "resource" for biological diversity, i.e. as people who can provide knowledge on components of biological diversity, rather than as people who hold legal, social, cultural and economic rights in relation to it. Another influential link with forests is provided by the Global Environment Facility (GEF), which provides funding for projects under CBD that support activities related to biological diversity, including forest biological diversity. Furthermore, CBD established in 2000 the Global Taxonomy Initiative, which is also of relevance to forests.

THE UNITED NATIONS FRAMEWORK CONVENTION ON CLIMATE CHANGE

The United Nations Framework Convention on Climate Change (UNFCCC or FCCC) is an international environmental treaty produced at the United Nations Conference on Environment and Development (UNCED). The objective of the treaty is to stabilize greenhouse gas concentrations in the atmosphere at a level that would prevent dangerous anthropogenic interference with the climate system. The treaty itself set no mandatory limits on greenhouse gas emissions for individual countries and contains no enforcement mechanisms. In that sense, the treaty is considered legally non-binding. Instead, the treaty provides for updates (called "protocols") that would set mandatory emission limits. The principal update is the Kyoto Protocol, which has become much better known than the UNFCCC itself. The UNFCCC was opened for signature on 9 May 1992, after an

Intergovernmental Negotiating Committee produced the text of the Framework Convention as a report following its meeting in New York from 30 April to 9 May 1992. It entered into force on 21 March 1994 and now UNFCCC has 194 parties.

Since the UNFCCC entered into force, the parties have been meeting in Conferences of the Parties (COP) to assess progress in dealing with climate change. The COP is seen as the "supreme body" of the Convention. The Convention on Climate Change sets an overall framework for intergovernmental efforts to tackle the challenge posed by climate change. It recognizes that the climate system is a shared resource whose stability can be affected by industrial and other emissions of carbon dioxide and other greenhouse gases. Under the Convention, governments:

- gather and share information on greenhouse gas emissions, national policies and best practices
- launch national strategies for addressing greenhouse gas emissions and adapting to expected impacts, including the provision of financial and technological support to developing countries
- cooperate in preparing for adaptation to the impacts of climate change

Parties to UNFCCC are classified as:

- Annex I countries: industrialized countries and economies in transition
- Annex II countries: developed countries which pay for costs of developing countries. They comprise the OECD members, excluding those that were economies in transition in 1992
- Non Annex I countries: Developing countries.

THE UNITED NATIONS CONVENTION TO COMBAT DESERTIFICATION

Desertification, along with climate change and the loss of biodiversity were identified as the greatest challenges to sustainable development during the 1992 Rio Earth Summit. Adopted in 1994, the United Nations Convention to Combat Desertification (UNCCD) is the sole legally binding international agreement linking environment and development to sustainable land management and entered into force in 1997. The Convention addresses specifically the arid, semi-arid and dry sub-humid areas, known as the drylands, where some of the most vulnerable ecosystems and peoples can be found. In the 10-Year Strategy of the UNCCD (2008-2018) that was adopted in 2007, Parties to the Convention further specified their goals: "to forge a global partnership to reverse and prevent desertification/land degradation and to mitigate the effects of drought in affected areas in order to support poverty reduction and environmental sustainability".

The Convention's 194 parties work together to improve the living conditions for people in drylands, to maintain and restore land and soil productivity, and to mitigate the effects of drought. The UNCCD is particularly committed to a bottom-up approach, encouraging the participation of local people in combating desertification and land degradation. The UNCCD secretariat facilitates cooperation between developed and developing countries, particularly around knowledge and technology transfer for sustainable land management.

As the dynamics of land, climate and biodiversity are intimately connected, the UNCCD collaborates closely with the other two Rio Conventions; the Convention on Biological Diversity (CBD) and the United Nations Framework Convention on Climate Change (UNFCCC), to meet these complex challenges with an integrated approach and the best possible use of natural resources. The permanent secretariat of the UNCCD is located in Bonn, Germany.

1996 PROTOCOL

On 17 November 1996, a special meeting of the Contracting Parties adopted the "1996 Protocol to the Convention on the Prevention of Marine Pollution by Dumping of Wastes and Other Matter-1972" which is to replace the 1972 Convention, subject to ratification. In line with UNCED's Agenda 21, the 1996 Protocol reflects the global trend towards precaution and prevention with the Parties agreeing to move from controlled dispersal at sea of a variety of land-generated wastes towards integrated land-based solutions for most, and controlled sea disposal of few, remaining categories of wastes or other matter.

Among the most important innovations brought by the 1996 Protocol is the codification of the precautionary approach and the polluter pays principle. Reflecting these principles, the Protocol embodies a major structural revision of the Convention, the so called "reverse list" approach. Now, instead of prohibiting the dumping of certain (listed) hazardous materials, the Parties are obligated to prohibit the dumping of any waste or other matter that is not listed in Annex 1 (the reverse list) of the 1996 Protocol. Dumping of wastes or other matter on this reverse list requires a permit. Parties to the Protocol are further obligated to adopt measures to ensure that the issuance of permits and permit conditions for the dumping of reverse list substances comply with Annex 2 (the Waste Assessment Annex) of the Protocol.

STOCKHOLM CONVENTION ON PERSISTENT ORGANIC POLLUTANTS

The Stockholm Convention on Persistent Organic Pollutants (POPs) is a global treaty to protect human health and the environment from chemicals that remain intact in the environment for long periods, become widely distributed geographically, accumulate in the fatty tissue of humans and wildlife, and have adverse effects to human health or to the environment. Exposure to Persistent Organic Pollutants (POPs) can lead serious health effects including certain cancers, birth defects, dysfunctional immune and reproductive systems, greater susceptibility to disease and even diminished intelligence. Given their long range transport, no one government acting alone can protect is citizens or its environment from POPs. In response to this global problem, the Stockholm Convention, which was adopted in 2001 and entered into force in 2004, requires Parties to take measures to eliminate or reduce the release of POPs into the environment. The Convention is administered by the United Nations Environment Programme and is based in Geneva, Switzerland. At present, there are 177 parties to the convention.

THE KYOTO PROTOCOL

The Kyoto Protocol is an international agreement linked to the United Nations Framework Convention on Climate Change. The major feature of the Kyoto Protocol is that it sets binding targets for 37 industrialized countries and the European community for reducing greenhouse gas (GHG) emissions. This amount to an average of five per cent against 1990 levels over the five-year period 2008-2012. The major distinction between the Protocol and the Convention is that while the Convention encouraged industrialized countries to stabilize GHG emissions, the Protocol commits them to do so. The only remaining signatory not to have ratified the protocol is the United States. Other United Nations member states which did not ratify the protocol are Afghanistan, Andorra and South Sudan. In December 2011, Canada renounced the Protocol.

Recognizing that developed countries are principally responsible for the current high levels of GHG emissions in the atmosphere as a result of more than 150 years of industrial activity, the Protocol places a heavier burden on developed nations under the principle of "common but differentiated responsibilities." The Kyoto Protocol was adopted in Kyoto, Japan, on 11 December 1997 and entered into force on 16 February 2005. The detailed rules for the implementation of the Protocol were adopted at COP 7 in Marrakesh in 2001, and are called the "Marrakesh Accords". The Kyoto Protocol is generally seen as an important first step towards a truly global emission reduction regime that will stabilize GHG emissions, and provides the essential architecture for any future international agreement on climate change. By the end of the first commitment period of the Kyoto Protocol in 2012, a new international framework needs to have been negotiated and ratified that can deliver the stringent emission reductions.

MARRAKESH ACCORDS

The Marrakesh Accords are the aggregate decisions of the Conference of the Parties (COP) from Decision 2/CP.7 through to Decision 24/CP.7 inclusive of the COP in its seventh session, held at Marrakesh, Morocco from 29 October to 10 November 2001. These decisions were adopted by the Conference of the Parties serving as the meeting of the Parties (COP/MOP) at its first meeting in Montreal in November 2005.

ASIA-PACIFIC PARTNERSHIP ON CLEAN DEVELOPMENT AND CLIMATE

The Asia-Pacific Partnership on Clean Development and Climate, also known as APP, was an international, voluntary, public-private partnership among Australia, Canada, India, Japan, the People's Republic of China, South Korea, and the United States announced on 28 July 2005 at an Association of South East Asian Nations (ASEAN) Regional Forum meeting and launched on 12 January 2006 at the Partnership's inaugural Ministerial meeting in Sydney. As of 5 April 2011, the Partnership formally concluded although a number of individual projects continue. The conclusion of the APP and cancellation of many of its projects attracted almost no media comment.

Foreign, Environment and Energy Ministers from partner countries agreed to co-operate on the development and transfer of technology which enables reduction of greenhouse gas emissions that is consistent with and complementary to the UN Framework Convention on Climate Change and other relevant international instruments, and is intended to complement but not replace the Kyoto Protocol., Ministers agreed to a Charter, Communique and Work Plan that "outline a ground-breaking new model of private-public task forces to address climate change, energy security and air pollution."

Member countries account for over 50 per cent of the world's greenhouse gas emissions, energy consumption, GDP and population. Unlike the Kyoto Protocol, which imposes mandatory limits on greenhouse gas emissions, the Partnership engages member countries to accelerate the development and deployment of clean energy technologies, with no mandatory enforcement mechanism. This has led to criticism that the Partnership is worthless, by other governments, climate scientists and environmental groups. Proponents, on the other hand, argue that unrestricted economic growth and emission reductions can only be brought about through active engagement by all major polluters, which includes India and China, within the Kyoto Protocol framework neither India nor China are yet required to reduce emissions. Canada became the 7th member of the APP at the Second Ministerial Meeting in New Delhi on 15 October 15 2007.

41 INTERNATIONAL AND NATIONAL FOREST ORGANIZATIONS

Present literatures pertaining to the research and education have shown that traditional forestry organizations emphasize sustainable forest management as an overall approach, balancing the social and environmental benefits of forestry with economic values for society. Simultaneously environmental benefits provided by forests such as carbon sequestration and climate change mitigation, biodiversity and landscape protection are frequently addressed by forest and other policy areas such as agriculture, environment and energy. Comprehensive overview of international and national forest organizations and groups are presented here with a view to provide updated information to the students, teachers, scientists, foresters and policy makers in the field of forestry and natural resources. The present state of international and national forest organizations are categorized and presented under five groups: intergovernmental organizations, non-governmental organizations, international research institutes, regional institutes and Indian forestry research and education organizations.

INTERGOVERNMENTAL ORGANIZATIONS

THE FOOD AND AGRICULTURE ORGANIZATION OF THE UNITED NATIONS

The Food and Agriculture Organization of the United Nations (FAO) is a specialized UN agency leading international efforts to defeat hunger and poverty. Serving both developed and developing countries, FAO acts as a neutral forum where all nations meet as equals to negotiate agreements and debate policy. FAO helps developing countries and countries in transition to modernize and improve agriculture, forestry and fisheries practices, ensure good nutrition, achieve food security for all, better the lives of rural populations and contribute to the growth of the world economy.

In 1943, forty four governments gathered in Hot Springs, Virginia, United States for United Nations Conference on Food and Agriculture committed themselves to found a permanent organization for food and agriculture. First session of FAO Conference in Quebec City, Canada on 16 October to 1 November 1945 established FAO as a specialized United Nations agency. FAO is composed of seven departments including Forestry Department. It has its headquarters in Rome, Italy and present in over 130 countries.

FAO FORESTRY DEPARTMENT

The FAO Forestry Department helps nations manage their forests in a sustainable way. The Organization's approach balances social, economic and environmental objectives so that present generations can reap the benefits of the earth's forest resources while conserving them to meet the needs of future generations. The FAO Forestry department supports a number of statutory bodies specifically designed to provide neutral forum for discussion in the field of forestry.

Unasylva is FAO's international journal of forests, forestry and forest industries published quarterly. The State of the World's Forests (SOFO) reports are published biennially since 1995. It reports on the status of forests, recent major policy and institutional developments and key issues concerning forest sector.

COMMITTEE ON FORESTRY

The Committee on Forestry (COFO) is the principal body among the FAO statutory bodies. COFO was established in 1971 as a standing committee of the FAO and its first session held in 1972. The biennial sessions of COFO brings together heads of forest services and other senior government officials to identify emerging policy and technical issues, to seek solutions and to advise FAO and others on appropriate action.

The agenda of COFO has evolved somewhat over time, with an increasing emphasis on technical and forest policy issues of international importance which includes review of FAO programmes in forestry, the work/recommendations of Regional Forestry Commissions, the State of the World's Forests (SOFO), and the World Forestry Congress, the role of forests in achieving the Millennium Development Goals, wildland fire and forests and bio-energy.

REGIONAL FORESTRY COMMISSIONS

Six Regional Forestry Commissions were established by the FAO Conference between 1947 and 1959 viz. African Forestry Wildlife Commission - AFWC (1959), Asia-Pacific Forestry Commission - APFC (1949), European Forestry Commission - EFC (1947), Latin American and Caribbean Forestry Commission - LACFC (1948), Near East Forestry and Range Commission - NEFRC (1953), North American Forest Commission - NAFC (1958). The Commissions consider both policy and technical issues and play a key role in the international arrangement on forests, serving as a link between global dialogue at the Committee on Forestry (COFO) and the United Nations Forum on Forests (UNFF) and national implementation.

UNITED NATIONS COMMISSION ON SUSTAINABLE DEVELOPMENT

The United Nations Commission on Sustainable Development (CSD) was established by the UN General Assembly in December 1992 to ensure effective follow-up of United Nations Conference on Environment and Development (UNCED), also known as the Earth Summit. The Commission is responsible for reviewing progress in the implementation of Agenda 21 and the Rio Declaration on Environment and Development; as well as providing policy guidance to follow up the Johannesburg Plan of Implementation (JPOI) at the local, national, regional and international levels. The JPOI reaffirmed that the CSD is the high-level forum for sustainable development within the United Nations system.

As a functional commission of the UN Economic and Social Council (ECOSOC), CSD has 53 member States (about one third of the members are elected on a yearly basis). Each session of the CSD elects a Bureau, comprised of a Chair and four vice-Chairs. The CSD meets annually in New York, in two-year cycles, with each cycle focusing on clusters of specific thematic and cross-sectoral issues, outlined in its new multi-year programme of work.

IPF AND IFF PROCESS (1995-2000)

The Intergovernmental Panel on Forests (IPF) and the Intergovernmental Forum on Forests (IFF) represent five years of international forest policy dialogue.

INTERGOVERNMENTAL PANEL ON FORESTS: The Intergovernmental Panel on Forests (IPF), established in 1995 by the Commission on Sustainable Development (CSD) for two years (1995-97) to provide a forum for forest policy deliberations. Deliberations by the IPF were on the following issues:

- Implementing the forest-related decisions of the United Nations Conference on Environment and Development (UNCED) at the national and international levels
- International cooperation in financial assistance and technology transfer
- Scientific research, forest assessment and the development of criteria and indicators for sustainable forest management
- Trade and environment in relation to forest products and services
- International organizations and multilateral institutions and instruments, including appropriate legal mechanisms

INTERGOVERNMENTAL FORUM ON FORESTS: Subsequently, in 1997, ECOSOC established the Intergovernmental Forum on Forests (IFF), for three years (1997-2000). Deliberations by the IFF's programme were on the following issues:

- Facilitating the implementation of the proposals for action (see below) of the Intergovernmental Panel on Forests and reviewing, monitoring and reporting on progress in the management, conservation and sustainable development of all types of forest
- Considering matters left pending and other issues arising from the programme elements of the IPF process such as financial resources, transfer of environmental sound technologies and other issues including deliberations on international arrangements and mechanisms on forests
- International arrangements and mechanisms to promote the management, conservation and sustainable development of all types of forests

INTERAGENCY TASK FORCE ON FORESTS

In July 1995, to support the IPF/IFF and to assist governments implement the proposals for action, an informal, high-level Interagency Task Force on Forests (ITFF) was created, made up of eight international organizations. ITFF was created to coordinate the inputs of international organizations to the forest policy process.

UNITED NATIONS FORUM ON FORESTS

In October 2000, the Economic and Social Council of the United Nations (ECOSOC) established the United Nations Forum on Forests (UNFF), a subsidiary body with the main objective to promote the management, conservation and sustainable development of all types of forests and to strengthen long-term political commitment to this end based on the Rio Declaration, the Forest Principles, Chapter 11 of Agenda 21 and the outcome of the IPF/IFF Processes and other key milestones of international forest policy. The Forum has universal membership and is composed of all Member States of the United Nations and specialized agencies.

THE COLLABORATIVE PARTNERSHIP ON FORESTS

The Collaborative Partnership on Forests (CPF) was established in April 2001, following the recommendation of ECOSOC. This innovative partnership of 14 major forest-related international organizations, institutions and convention secretariats, works to support the work of the UNFF and its member countries and to foster increased cooperation and coordination on forests. Member organizations of the Collaborative Partnership on Forests:

- Center for International Forestry Research (CIFOR)
- Food and Agriculture Organization of the United Nations (FAO)
- International Tropical Timber Organization (ITTO)
- International Union of Forest Research Organizations (IUFRO)
- Convention on Biological Diversity (CBD)
- Global Environmental Facility (GEF)
- United Nations Convention to Combat Desertification (UNCCD)
- United Nations Forum on Forests (UNFFS)
- United Nations Framework Convention on Climate Change (UNFCCC)
- United Nations Development Programme (UNDP)
- United Nations Environment Programme (UNEP)
- World Agroforestry Centre (ICRAF)
- World Bank
- World Conservation Union (IUCN)

THE UNITED NATIONS ENVIRONMENT PROGRAMME

The United Nations Environment Programme (UNEP) is an international organization established in 1972 to coordinate United Nations environmental activities, assisting developing countries in implementing environmentally sound policies and practices. It was founded as a result of the United Nations Conference on the Human Environment in June 1972 and has its headquarters in the Gigiri neighborhood of Nairobi, Kenya. UNEP also has six regional offices and various country offices. The UNEP's mission is to provide leadership and encourage partnership in caring for the environment by inspiring, informing and enabling nations and peoples to improve their quality of life without compromising that of the future generations.

THE INTERGOVERNMENTAL PANEL ON CLIMATE CHANGE

The Intergovernmental Panel on Climate Change (IPCC) is the leading international body established in 1988 for the assessment of climate change and hosted at WMO headquarters in Geneva. It was established by the United Nations Environment Programme (UNEP) and the World Meteorological Organization (WMO) to provide the world with a clear scientific view on the current state of knowledge in climate change and its potential environmental and socio-economic impacts. The UN General Assembly endorsed the action by WMO and UNEP in jointly establishing the IPCC. The IPCC is an intergovernmental body and it is open to all member countries of the United Nations (UN) and WMO. Currently 195 countries are members of the IPCC.

The IPCC is a scientific body. It reviews and assesses the most recent scientific, technical and socio-economic information produced worldwide relevant to the understanding of climate change. It does not conduct any research nor does it monitor climate related data or parameters. Thousands of scientists from all over the world contribute to the work of the IPCC on a voluntary basis. Review is an essential part of the IPCC process, to ensure an objective and complete assessment of current information.

THE UNITED NATIONS DEVELOPMENT PROGRAMME

The United Nations Development Programme (UNDP) is the United Nation's global development network, an organization advocating for change and connecting countries to knowledge, experience and resources to help people build a better life. It is based on the merging of the United Nations Expanded Programme of Technical Assistance, created in 1949, and the United Nations Special Fund, established in 1958. UNDP was established in 1965 by the General Assembly of the United Nations and has its headquarters in New York.

UNDP's mandate is to fight poverty. As the development organization of the United Nations, UNDP works to achieve this through supporting development projects in a sustainable way. The overarching goal of reducing poverty has been more concretely defined through the Millennium Development Goals. UNDP's network links and coordinates global and national efforts to reach these goals and helps developing countries attract and use aid effectively.

ORGANISATION FOR ECONOMIC CO-OPERATION AND DEVELOPMENT

The Organisation for European Economic Cooperation (OEEC) was established in 1947 to run the US-financed Marshall Plan for reconstruction of a continent ravaged by war. Encouraged by its success and the prospect of carrying its work forward on a global stage, Canada and the US joined OEEC members in signing the new OECD Convention on 14 December 1960. The Organisation for Economic Co-operation and Development (OECD) was officially born on 30 September 1961, when the Convention entered into force. Other countries joined in, starting with Japan in 1964. Today, 34 OECD member countries worldwide regularly turn to one another to identify problems, discuss and analyze them, and promote policies to solve them. The mission of the Organisation for Economic Co-operation and Development (OECD) is to promote policies that will improve the economic and social well-being of people around the world. It has its headquarters in Paris, France.

UNITED NATIONS INTERNATIONAL STRATEGY FOR DISASTER REDUCTION

The International Strategy for Disaster Reduction (UNISDR) is the secretariat of the International Strategy for Disaster Reduction (ISDR) established in 1999 and has its headquarters in Geneva, Switzerland. It is the successor to the secretariat of the International Decade for Natural Disaster Reduction established in 1990 with the purpose of ensuring the implementation of the International Strategy for Disaster Reduction. ISDR aims at building disaster resilient communities by promoting increased awareness of the importance of disaster reduction as an integral component of sustainable development with the goal of reducing human, social, economic and environmental losses due to natural hazards and related technological and environmental disasters.

GLOBAL FIRE MONITORING CENTER

The Global Fire Monitoring Center (GFMC) was inaugurated at the FAO meeting on Public Policies Affecting Forest Fires (Rome, October 1998) and is financed by the Foreign Office of Germany. Since 1998, GFMC is hosted by the Fire Ecology Research Group, a subdivision of the Biochemistry Department of the Max Planck Institute for Chemistry (Freiburg University, Germany). The GFMC is an activity of the ISDR, chairing an interagency "Working Group on Wildland Fire" under the auspices of the ISDR Inter-agency Task Force for Disaster Reduction. The GFMC Provides a global portal for wildland fire documentation, information and monitoring.

THE FORESTRY WORKFORCE NETWORK

The Forestry Workforce Network (FORWORKNET) is an international network of more than 300 individuals and institutions in some 70 countries interested in forestry workforce issues. The primary functions of FORWORKNET are to enable its members to communicate directly with each other and to open new opportunities for international exchange and cooperation. Specifically, the network endeavours to keep members abreast of new developments, new publications, data and research results, forthcoming events, new technologies affecting forest workers.

THE WORLD TRADE ORGANIZATION

The World Trade Organization (WTO) is an organization that intends to supervise and liberalize international trade. The organization officially commenced on January 1, 1995 under the Marrakech Agreement, replacing the General Agreement on Tariffs and Trade (GATT), which commenced in 1948. The World Trade Organization (WTO) is the only global international organization dealing with the rules of trade between nations. At its heart are the WTO agreements, negotiated and signed by the bulk of the world's trading nations and ratified in their parliaments. The goal is to help producers of goods and services, exporters and importers conduct their business.

At Marrakech in April 1994, Ministers adopted a Decision on Trade and Environment calling for the establishment of a Committee on Trade and Environment (CTE) at the WTO. Three items of the CTE's work plan are relevant to the discussion on forestry:

- Environmental benefits of removing trade restrictions and distortions
- Eco-labelling and certifications of sustainable forest management and forest products
- Certain elements of the Agreement on Trade-Related Intellectual Property Rights

G8 ACTION PROGRAMME ON FORESTS

Since 1975, the head of state of the major industrial democracies also called Group of Eight (France, United States, United Kingdom, Russian Federation, Germany, Japan, Italy, Canada) have been meeting annually to deal with the major economic and political issues facing their domestic societies and the international community as whole. The G8 Action Programme on Forests was launched in 1998 at the Birmingham Summit to address the five issues of particular importance in achieving sustainable forest management: monitoring and assessment, national forest programmes, protected areas, private sector and illegal logging.

WORLD BANK GROUP – FORESTS AND FORESTRY

The World Bank Group's (WBG) mission is to fight poverty and improve the living standards of people in the developing world and has its headquarters in Washington D.C. It is a development bank established in 1944 and now with a closely associated group of five development institutions. It provides loans, policy advice, technical assistance and knowledge sharing services to low and middle income countries to reduce poverty.

THE WORLD BANK FORESTS TEAM: It takes a multi-sectoral approach to forest management that takes into account the impact on forests and forest people of activities, policies and practices outside the forest sector. The Forests Team is comprised of professionals housed in both Agriculture and Rural Development Department and Environment Department, as well as expertise from regional staff and external partners. Furthermore, the Forests Team is part of the Bank's Natural Resources Management Thematic Group which encompasses Land Resources Management, Natural Resource Management Institutions and Irrigation and Drainage.

THE FOREST POLICY AND STRATEGY OF THE WORLD BANK: It was approved in October 2002 and has been implemented to help attain the goal of poverty reduction while safeguarding the environmental values intrinsic to sustainability. The strategy is built on three pillars: harnessing the potential of forests to reduce poverty; integrating forests into sustainable economic development; protecting global forest values

PROGRAMME ON FORESTS: Programme on Forests (PROFOR) is an independent programme hosted at the World Bank within the Environmentally and Socially Sustainable Development Network. It is a multi-donor partnership formed to pursue a shared goal of enhancing forest's contribution to poverty reduction, sustainable development and protection of environmental services. Initially established in 1997 to implement the outcomes of the Intergovernmental Panel on Forests (IPF), PROFOR was relocated from the United Nations Development Programme to the World Bank in 2002.

THE GLOBAL ENVIRONMENT FACILITY

The Global Environment Facility (GEF) unites 182 countries in partnership with international institutions, civil society organizations (CSOs) and the private sector to address global environmental issues while supporting national sustainable development initiatives. Today the GEF is the largest public funder of projects to improve the global environment. An independently operating financial organization, the GEF provides grants for projects related to biodiversity, climate change, international waters, land degradation, the ozone layer and persistent organic pollutants. The Global Environment Facility Secretariat and the Evaluation Office are located in Washington, DC, USA.

The Global Environment Facility was established in October 1991 as a $1 billion pilot program in the World Bank to assist in the protection of the global environment and to promote environmental sustainable development. The United Nations Development Programme, the United Nations Environment Programme and the World Bank were the three initial partners implementing GEF projects. In 1994, GEF was restructured and moved out of the World Bank system to become a permanent, separate institution. Since 1994, however, the World Bank has served as the Trustee of the GEF Trust Fund and provided administrative services. The GEF was entrusted to become the financial mechanism for the UN Convention on Biological Diversity, UN Framework Convention on Climate Change, Montreal Protocol of the Vienna Convention on Ozone Layer Depleting Substances, Stockholm Convention on Persistent Organic Pollutants and the United Nations Convention to Combat Desertification.

EUROPEAN FOREST INSTITUTE

European Forest Institute (EFI) is an international organization established by European States and so far 23 European States have ratified the Convention on EFI. EFI has over 120 Associate and Affiliate Member organizations in more than 30 countries, five Regional Offices and one Project Centre. Its headquarters is in Joensuu, Finland and Policy Advice office is located in Barcelona, Spain. EFI's mission is to promote, conduct and cooperate in research of forestry and forest products at the pan-European level and to make the results of the research known to all interested parties, notably in the areas of policy formulation and implementation, in order to promote conservation and sustainable management of forests in Europe. The five Regional Offices of EFI are:

- **Mediterranean Regional Office (EFIMED)**, located in Barcelona, Spain established in 2007
- **Central European Regional Office and the Observatory for European Forests (EFICENT-OEF)** established in 2009 and it has distributed locations in Germany, France and Switzerland
- **Atlantic European Regional Office (EFIATLANTIC)**, located in Bordeaux, France established in 2009
- **Central-East European Regional Office (EFICEEC)**, located in Vienna, Austria established in 2010
- **North European Regional Office (EFINORD)**, located in Copenhagen, Denmark established in 2010

NON-GOVERNMENTAL ORGANIZATIONS

INTERNATIONAL UNION FOR CONSERVATION OF NATURE

International Union for Conservation of Nature (IUCN) helps the world find pragmatic solutions to our most pressing environment and development challenges. It was founded in October 1948 as the International Union for the Protection of Nature (IUPN) and considered world's first global environmental organization. The organization publishes the IUCN Red List, compiling information from a network of conservation organizations to rate which species are most endangered. IUCN is the world's oldest and largest global environmental organization comprising more than 1,200 member organizations including 200+ government and 900+ non-government organizations. It has its headquarters in Gland, Switzerland.

Conserving biodiversity is central to the mission of IUCN. It demonstrates how biodiversity is fundamental to addressing some of the world's greatest challenges: tackling climate change, achieving sustainable energy, improving human well-being and building a green economy. IUCN work is framed by a Global Programme, developed with and approved by IUCN member organizations every four years. Current IUCN global programme runs from 2009 to 2012. It has five priority areas of work-biodiversity, climate change, sustainable energy, human well-being and green economy. It has six commissions viz. Commission on Education and Communication (CEC), Commission on Environmental, Economic and Social Policy (CEESP), Commission on Environmental Law (CEL), Commission on Ecosystem Management (CEM), Species Survival Commission (SSC) and World Commission on Protected Areas (WCPA)

WORLD WIDE FUND FOR NATURE

World Wide Fund for Nature (WWF) was conceived on 29 April 1961 under the name of World Wildlife Fund and its head office was opened on 11 September 1961 in Morges, Switzerland. Its establishment marked with the signing of the founding document called Morges Manifesto that lays out the formulation ideas of its establishment. WWF was established as an international fundraising organization to work in collaboration with existing conservation groups and bring substantial financial support to the conservation movement on a worldwide scale.

In 1986, the organization changed its name to World Wide Fund for Nature, to better reflect the scope of its activities. For 50 years, WWF has been protecting the future of nature. The world's leading conservation organization, WWF works in 100 countries and is supported by 1.2 million members in the United States and close to 5 million globally. WWF's mission is to conserve nature and reduce the most pressing threats to the diversity of life on Earth. The Goals and Strategies of WWF include, by 2020, WWF will conserve 15 of the world's most ecologically important regions by working in partnership with others to:

- Protect and restore species and their habitats
- Strengthen local communities' ability to conserve the natural resources they depend upon
- Transform markets and policies to reduce the impact of the production and consumption of commodities

- Ensure that the value of nature is reflected in decisions made by individuals, communities, governments and businesses
- Mobilize hundreds of millions of people to support conservation

THE INTERNATIONAL INSTITUTE FOR SUSTAINABLE DEVELOPMENT

The International Institute for Sustainable Development (IISD) is a Canadian-based, public policy research institute that has a long history of conducting cutting-edge research into sustainable development. It was established in 1990 and has its head office in Winnipeg, Manitoba, Canada. The institute champions global sustainable development through innovation, research and relationships that span the entire world. It contributes to the international negotiations on the sustainable management of forests with targeted communications aimed at improving the transparency and clarity of the discussions in a variety of forum viz. UNFF, CSD, WCFSD.

WORLD COMMISSION ON FORESTS AND SUSTAINABLE DEVELOPMENT

Following the Earth Summit in 1992, it was agreed that solutions to forest degradation are likely to be more political than technical. Accordingly, the Inter Action Council, a group of some 30 former heads of government and states, decided to establish an independent commission called World Commission on Forests and Sustainable Development (WCFSD). IISD also hosts the Secretariat of World Commission on Forests and Sustainable Development. The objectives are:

- to increase awareness of the dual function of world forests in preserving natural environment and contributing to economic development
- to broaden the consensus on the data, science and policy aspects of forest conservation and management
- to build confidence between North and South on forest matters with emphasis on international cooperation
- to achieve policy reforms aimed at reconciling economic and environmental objectives for sustainable management of global forests

WORLD RESOURCES INSTITUTE - GLOBAL FOREST WATCH

World Resources Institute (WRI) was launched June 3, 1982 as a center for policy research and analysis addressed to global resource and environmental issues. It has its headquarters in Washington D.C. WRI works with governments, companies and civil society to build solutions to urgent environmental challenges. WRI's transformative ideas protect the earth and promote development because sustainability is essential for meeting human needs and fulfilling human aspirations in the future. Global Forest Watch is an initiative of the World Resources Institute. Global Forest Watch seeks to catalyze changes in forest use to meet human needs and better protect forest ecosystems.

FOREST STEWARDSHIP COUNCIL

Forest Stewardship Council (FSC) is an independent, non-governmental, not-for-profit organization established to promote the responsible management of the world's forests.

Established in 1993 as a response to concerns over global deforestation, FSC is a pioneer forum where the global consensus on responsible forest management convenes and through democratic process effects solutions to the pressures facing the world's forests and forest-dependent communities. FSC is locally represented in more than 50 countries around the world.

INTERNATIONAL RESEARCH INSTITUTES

INTERNATIONAL TROPICAL TIMBER ORGANIZATION

The International Tropical Timber Organization (ITTO) is an intergovernmental organization promoting the conservation and sustainable management, use and trade of tropical forest resources. Its members represent about 80% of the world's tropical forests and 90% of the global tropical timber trade. ITTO was established under the auspices of the United Nations in 1986 amidst increasing worldwide concern for the fate of tropical forests and has its headquarters in Yokohama, Japan. While almost everyone was alarmed at the rate of deforestation occurring in many tropical countries, there was also considerable agreement that the tropical timber trade was one of the keys to economic development in those same countries.

ITTO's origins can be traced back to 1976 when the long series of negotiations that led to the first International Tropical Timber Agreement (ITTA) began at the fourth session of the United Nations Conference on Trade and Development (UNCTAD) as part of that organization's Programme for Commodities. The eventual outcome of these negotiations was the ITTA, 1983, which governed the Organization's work until 31 December 1996, when it was superseded by the ITTA, 1994. Negotiations for a successor to this agreement were concluded in 2006, again under the auspices of UNCTAD. The ITTA, 2006 entered into force on December 7, 2011.

INTERNATIONAL INSTITUTE FOR ENVIRONMENT AND DEVELOPMENT

International Institute for Environment and Development (IIED) is one of the world's most influential international development and environment policy research organizations. It was founded in 1971 by merging the concept and cause of sustainable development and has the headquarters in London, UK. IIED is one of the world's most influential policy research organizations working at the interface between development and environment. IIED contributes to many international policy processes and frameworks, including the Intergovernmental Panel on Climate Change, the Millennium Ecosystem Assessment and the UN conventions on climate change and biological diversity.

THE INTERNATIONAL UNION OF FOREST RESEARCH ORGANIZATIONS

The International Union of Forest Research Organizations (IUFRO) is a non-profit, non-governmental international network of forest scientists, which promotes global cooperation in forest-related research and enhances the understanding of the ecological, economic and social aspects of forests and trees. It has its headquarters in Austria. IUFRO was founded as the International Union of Forest Experiment Stations in 1892. IUFRO focuses on science-based management, conservation and sustainable development of the world's forest resources for the benefit of present and future generations.

IUFRO has identified three strategic goals and related objectives for the period 2011-2014. IUFRO addresses Research and Institutional Goals at the same time. The Research Goals strongly focus on the following six thematic areas: Forests for People, Forests and Climate Change, Forest Bioenergy, Forest Biodiversity Conservation, Forest and Water Interactions and Resources for the Future. IUFRO's three strategic goals are

- To strengthen research and expand IUFRO's capacity for interdisciplinary cooperation
- To strengthen coordination within the scientific community and increase visibility of science-based research findings
- To further strengthen IUFRO's work at the science-policy interface

INTERNATIONAL INSTITUTE FOR APPLIED SYSTEMS ANALYSIS

Founded in 1972, the International Institute for Applied Systems Analysis (IIASA) is a non-governmental international research organization that conducts policy-oriented research into problems that are too large or too complex to be solved by a single country or academic discipline. It is independent and completely unconstrained by political or national self-interest and located in Laxenburg near Vienna in Austria. IIASA concentrates its research efforts within three core research themes: Energy and Climate Change; Food and Water; Poverty and Equity. IIASA's overall theme in the forest area is Global Change and Forests. It addresses the question of how to manage forest sector in order to harmonize geo and bio-physic functions with socioeconomic development

CENTRE FOR INTERNATIONAL FORESTRY RESEARCH

Centre for International Forestry Research (CIFOR) is a nonprofit, global facility dedicated to advancing human wellbeing, environmental conservation and equity. It conducts research that enables more informed and equitable decision making about the use and management of forests in less-developed countries. It was established in May 1993 and has its headquarters in Bogor, Indonesia. CIFOR is one of 15 Centers within the Consultative Group on International Agricultural Research (CGIAR).

CIFOR's research and expert analysis help policy makers and practitioners shape effective policy, improve the management of tropical forests and address the needs and perspectives of people who depend on forests for their livelihoods. In implementing its research, CIFOR is collaborating with other Centers within the CGIAR - the World Agroforestry Centre, the International Center for Tropical Agriculture (CIAT) and Bioversity International - to conduct the CGIAR Consortium Research Programme *Forests, Trees and Agroforestry* (also known as CRP6). CRP6 aims to achieve a clear objective: enhancing the management and use of forests, agroforestry and tree genetic resources across the landscape, from forests to farms.

WORLD AGROFORESTRY CENTRE

The World Agroforestry Centre (formerly the International Centre for Research in Agroforestry-ICRAF) is part of the alliance of the Consultative Group on International Agricultural Research (CGIAR) centres dedicated to generating and applying the best available knowledge to stimulate agricultural growth, raise farmer's incomes and protect the environment. It was established in 1971 and has five regional offices with its headquarters in Nairobi, Kenya. It conducts research in agroforestry, in partnership with national agricultural research systems with a view to developing more sustainable and productive land use. The focus of its research is countries/regions in developing world, in particular the tropics. The World Agroforestry Centre is guided by the broad development challenges: poverty alleviation that entails enhanced food security and health, improved productivity with lower environmental and social costs and resilience in the face of climate change and other external shocks.

SMITHSONIAN TROPICAL RESEARCH INSTITUTE OR THE CENTER FOR TROPICAL FOREST SCIENCE

The Center for Tropical Forest Science (CTFS) is a global network of forest research plots committed to the study of tropical and temperate forest function and diversity. The multi-institutional network comprises more than forty forest research plots across the Americas, Africa, Asia and Europe, with a strong focus on tropical regions. CTFS monitors the growth and survival of about 4.5 million trees of approximately 8,500 species. CTFS plots involve hundreds of scientists from more than 75 institutions worldwide and CTFS has its headquarters in Panama. Ecologists at the Smithsonian Tropical Research Institute established the first plot on Barro Colorado Island (BCI), Panama in 1980.

THE UNEP WORLD CONSERVATION MONITORING CENTRE

The UNEP World Conservation Monitoring Centre (UNEP-WCMC), Cambridge is a collaboration between the United Nations Environment Programme and WCMC, a UK-based charity. The international community has acknowledged the role of UNEP-WCMC and has to carry out many critical global functions. In particular, UNEP's Governing Council has recognized the Centre as UNEP's specialist biodiversity and assessment arm providing a range of biodiversity-related services to UNEP, Governments, Multilateral Environmental Agreements and their constituent party states and other bodies in the NGO and private sectors.

INTERNATIONAL FORESTRY RESOURCES AND INSTITUTIONS RESEARCH NETWORK

International Forestry Resources and Institutions (IFRI) Research Network is a unique group of researchers working on forest governance and livelihoods. It focuses on data collection and analysis across multiple international settings at the local-level, using both social and ecological data collected at several points in time. It aims to explain how institutions shape forest governance outcomes. Founded in 1992 at Indiana University, the network is currently housed at the School of Natural Resources and Environment at the University of Michigan, USA. It is built around links among fourteen Collaborative Research Centers in 12 countries.

REGIONAL INSTITUTES

THE COMMONWEALTH SCIENTIFIC AND INDUSTRIAL RESEARCH ORGANIZATION

The Commonwealth Scientific and Industrial Research Organization (CSIRO) is Australia's national science agency and one of the largest and most diverse research agencies in the world. It was founded in 1916 originally as the Advisory Council of Science and

Industry. CSIRO's corporate headquarters are located at Limestone Avenue in the Australian Capital Territory in Campbell, Australia. CSIRO's research is performed by the 13 Divisions. CSIRO is involved in more than 740 research activities, working with leading scientific organizations in over 80 countries with partners and customers ranging from foreign governments, small companies to large multi-nationals and international foundations.

THE INTERNATIONAL INSTITUTE OF TROPICAL FORESTRY

The International Institute of Tropical Forestry (IITF) is part of the United States Department of Agriculture (USDA) Forest Service that was founded in 1939. The IITF is located in Rio Piedras, Puerto Rico on the grounds of the University of Puerto Rico, Agricultural Experimental Station. The Institute is dedicated to tropical forestry on an international level. Within the Forest Service's mission of caring for the land and serving people, the IITF's mission is to develop and exchange knowledge critical to sustaining tropical ecosystem benefits for human kind

OTHER REGIONAL FORESTRY INSTITUTES

- The Oregon Forest Resources Institute (OFRI) based in Oregon, US
- The Ontario Forest Research Institute (OFRI) located in Sault Ste. Marie, Ontario, Canada
- The Centre for Forest Research (CFR) with headquarters at Montreal, Canada
- The Finnish Forest Research Institute (METLA) with headquarters at Vantaa, Finland
- The Federal Research Centre for Forestry and Forest Products, Hamburg, Germany
- The Forest Research Institute Malaysia (FRIM) located in Kepong, near Kuala Lumpur, Malaysia
- The Iwokrama International Centre for Rain Forest Conservation and Development established by Guyana and the Commonwealth
- The Norwegian Forest and Landscape Institute with headquarters at As, Norway
- The Research Institute for Nature and Forest (INBO) with headquarters at Brussels Belgium
- New Zealand Forest Research Institute Limited (SCION) with headquarters at Rotorua, New Zealand
- Sukachev Institute of Forest, Krasnoyarsk, Russia
- Korea Forest Research Institute (KFRI), Seoul, Korea

FOREST RESEARCH AND EDUCATION ORGANIZATIONS IN INDIA

THE MINISTRY OF ENVIRONMENT & FORESTS

The Ministry of Environment & Forests (MoEF) is the nodal agency in the administrative structure of the Central Government for the planning, promotion, co-ordination and overseeing the implementation of India's environmental and forestry policies and programmes. The Ministry also serves as the nodal agency in the country for the United Nations Environment Programme (UNEP), South Asia Co-operative Environment Programme (SACEP), International Centre for Integrated Mountain Development (ICIMOD) and for the follow-up of the United Nations Conference on Environment and Development (UNCED). The Ministry is also entrusted with issues relating to multilateral bodies such as the Commission on Sustainable Development (CSD), Global Environment Facility (GEF) and of regional bodies like Economic and Social Council for Asia and Pacific (ESCAP) and South Asian Association for Regional Co-operation (SAARC) on matters pertaining to the environment. The broad objectives of the Ministry are:

- Conservation and survey of flora, fauna, forests and wildlife
- Prevention and control of pollution
- Afforestation and regeneration of degraded areas
- Protection of the environment and
- Ensuring the welfare of animals

SUBORDINATE OFFICES OF MoEF

- Forest Survey of India, Dehradun
- Botanical Survey of India, Kolkatta
- Zoological Survey of India, Kolkatta
- Indira Gandhi National Forest Academy, Dehradun
- Directorate of Forest Education, Dehradun
- National Institute of Animal Welfare, Faridabad
- National Zoological Park, New Delhi
- National Museum of Natural History, New Delhi

AUTONOMOUS ORGANIZATIONS OF MoEF

- Govind Ballabh Pant Institute of Himalayan Environment & Development, Almora
- Indian Council of Forestry Research and Education (ICFRE), Dehradun
- The Indian Institute of Forest Management (IIFM), Bhopal
- Indian Plywood Industries Research and Training Institute (IPIPTI), Bangalore
- Wildlife Institute of India (WII), Dehradun

AUTHORITIES OF MoEF

- Central Zoo Authority, New Delhi
- National Biodiversity Authority, New Delhi
- National Ganga River Basin Authority, New Delhi

- National Tiger Conservation Authority, New Delhi

BOARDS OF MoEF

- Animal Welfare Board of India, Chennai
- Central Pollution Control Board, Delhi
- National Afforestation and Eco-development Board, New Delhi

INDIAN COUNCIL OF FORESTRY RESEARCH AND EDUCATION

Indian Council of Forestry Research and Education (ICFRE), Dehradun is an autonomous body of Ministry of Environment and Forests, Government of India established in the year 1988. ICFRE is an apex body in the national forestry research system and has been undertaking the holistic development of forestry research through need based planning, promoting, conducting and coordinating research, education and extension covering all aspects of forestry. The Council deals with the solution based forestry research in tune with the emerging issues in the sector, including global concerns such as climate change, conservation of biological diversity, combating desertification and sustainable management and development of resources. Topical research by the Council enhances public confidence in the ability of forest managers and researchers to successfully handle challenges related to natural resource management.

ICFRE has eight Regional Research Institutes and four Research Centres located in different bio-geographical regions of the country to cater the forestry research needs of the nation.

REGIONAL RESEARCH INSTITUTES UNDER ICFRE

- Forest Research Institute (FRI), Dehradun was established in 1906
- Institute of Forest Genetics and Tree Breeding (IFGTB), Coimbatore was established in 1988
- Institute of Wood Science and Technology (IWST), Bangalore was established in 1988 but its origin goes back to 1938 when, the then Government of Mysore had set up a Forest Research Laboratory (FRL) at Bangalore.
- Tropical Forest Research Institute (TFRI), Jabalpur came into existence in 1988, although its origin goes back to 1973 when a Regional Centre of Forest Research Institute, Dehradun was established at Jabalpur
- Rain Forest Research Institute (RFRI), Jorhat was established in 1988
- Arid Forest Research Institute (AFRI), Jodhpur was established in 1988
- Himalayan Forest Research Institute (HFRI), Shimla was established as High Level Conifer Regeneration Research Centre during 1977. The institute made its humble beginning from this Centre at the time of re-organization of forestry research in ICFRE during 1998
- Institute of Forest Productivity (IFP), Ranchi was established in 1993

ADVANCED RESEARCH CENTRES UNDER ICFRE

- Centre for Social Forestry and Eco-Rehabilitation (CSFER), Allahabad was established in 1992 as an advanced centre under the umbrella of ICFRE and presently, it is a branch of Forest Research Institute (FRI), Dehradun.
- Centre for Forestry Research and Human Resource Development (CFRHRD), Chhindwara came into existence in 1995. It was declared in 1996 as a satellite Centre of Tropical Forest Research Institute, Jabalpur
- Forest Research Centre (FRC), Hyderabad was initially established in 1997 as Advanced Centre for Biotechnology and Mangrove Forests functioning under the administrative control of Institute of Wood Science and Technology Bangalore. It was formally inaugurated as FRC in 2000
- Advanced Research Centre for Bamboo and Rattans (ARCBR), Aizawl was established in 2004

INDIRA GANDHI NATIONAL FOREST ACADEMY

Established in May 1987, Indira Gandhi National Forest Academy (IGNFA), Dehradun became the successor institute of the erstwhile Indian Forest College. It's the premier institution offering professional training services to the Indian Forest Service and probationers. The main objective is to provide proper professional knowledge of forestry, managerial skills and to help the probationers follow the fundamental ethos of the All India Services. There are Compulsory Training Courses held in the academy for the IFS officers, sponsored by the Ministry of MoEF. The Academy also conducts five Skill Upgradation courses, which is promoted by State Forest Service (SFS) and cadres belonging to different states are trained under this program.

DIRECTORATE OF FOREST EDUCATION

Previously Directorate of Forest Education (DFE), Dehradun was a part of the FRI Colleges, Dehradun and remained responsible for professional and technical level training and education in the country. After the reorganization of FRI & Colleges and creation of an autonomous ICFRE, this Directorate was de-linked in 1991 and functions directly under the Ministry of Environment and Forests. The directorate is responsible for controlling, coordinating and managing the entire regular training courses of State Forest Service and Forest Range Officers in the country. It also organizes short-term special refresher courses for the service officers. The colleges, which are under the administrative control of the Directorate are:

- State Forest Service College, Dehradun (Uttaranchal)
- State Forest Service College, Coimbatore (Tamil Nadu)
- State Forest Service College, Burnihat (Assam)
- Eastern Forest Rangers College, Kurseong (West Bengal)

WILDLIFE INSTITUTE OF INDIA

Wildlife Institute of India (WII), an autonomous Institution of Ministry of Environment and Forests, Government of India, was established at Dehradun in 1982 with a mandate to train government and non-government personnel, carry out research and advise on matters of conservation and management of wildlife resources. WII is an internationally acclaimed Institution, which offers training program, academic courses and advisory in wildlife research and management. The Institute is actively engaged in research across the breadth of the country on biodiversity related issues. It was initially established under the ministry of Agriculture and Corporation in 1982. In April 1986, WII became an autonomous Institute of the Ministry of Environment and Forests, comprising a 35-member WII society headed by the Union Minister of Environment and Forests, Government of India.

INDIAN INSTITUTE OF FOREST MANAGEMENT

The Indian Institute of Forest Management (IIFM), Bhopal is a sectoral management institute which constantly endeavors to evolve knowledge useful for the managers in the area of Forest, Environment and Natural Resources Management and allied sectors. Since its establishment in the year 1982, IIFM is working as an autonomous organization under the Ministry of Environment and Forests. The Objective of the institute is to provide training in management and its related subjects to the officers from the IFS, Forest departments, Forest Development Corporations and Forest Related Industries with a view to inculcate professionalism in forestry management. The activities of IIFM fall under four realms, which are education, training, research and consultancy, related to forest management.

INDIAN PLYWOOD INDUSTRIES RESEARCH AND TRAINING INSTITUTE

Indian Plywood Industries Research and Training Institute (IPIPTI), Bangalore which was started in 1962 as a Co-operative Research Association of the Plywood Industry, has evolved into a dynamic and creative organization dedicated in developing new environment friendly technologies for wood based industry. It became an autonomous organization of the MoEF when the Government of India in the year 1990 transferred it from the Ministry of Industry. The institute is recognized internationally as a centre of expertise in the mechanical wood industries and is well equipped with the modern facilities of research and training in saw milling and plywood. The Institute frequently conducts research to enlarge the service life of wood and products, economize wood utilization by reducing wastage in conversion as well as in application.

CENTRE FOR ENVIRONMENT EDUCATION

Centre for Environment Education (CEE) was created in recognition of the importance of environmental education in India's overall environment and development strategy. The result of a unique partnership between government and a non-governmental institution, CEE was established as a Centre of Excellence in 1984 by the Ministry of Environment and Forests (MoEF) in Ahmedabad, Gujarat. CEE now works in the field of environment education and action for a wide range of sectors, target groups and geographical areas.

FOREST SURVEY OF INDIA

Forest Survey of India (FSI) is a premier national organization under the union Ministry of Environment and Forests, responsible for assessment and monitoring of the forest resources of the country regularly. In addition, it is also engaged in providing the services of training, research and extension. FSI was established in Dehradun on 01 June 1981. It started as an organization called Pre-Investment Survey of Forest Resources (PISFR) in 1965 as FAO/UNDP/GOI Project. The main objective of PISFR was to ascertain the availability of raw material for establishment of wood based industries in selected areas of the country. National Commission on Agriculture (NCA) report in 1976 recommended for the creation of a National Forest Survey Organization for a regular, periodic and comprehensive forest resources survey of the country leading to creation of FSI. After a critical review of activities undertaken by FSI, Government of India redefined the mandate of FSI in 1986 in order to make it more relevant to the rapidly changing needs and aspirations of the country.

NATIONAL RESERCH CENTRE FOR AGROFORESTRY

The National Research Centre for Agroforestry (NRCAF), Jhansi as unit of ICAR was established in 1988. It undertakes basic and applied research for developing and delivering technologies based on sustainable agroforestry practices on farms, marginal and wastelands for different agroclimatic zones in India.

CENTRAL ARID ZONE RESEARCH INSTITUTE

To arrest the land degradation process and for scientific and sustainable management of the resources, Desert Afforestation Station was established in 1952 at Jodhpur. This was later expanded into Desert Afforestation and Soil Conservation Station in 1957, and finally upgraded to Central Arid Zone Research Institute (CAZRI) in 1959 under Indian Council of Agricultural Research (ICAR), New Delhi. The CAZRI operates through Six Divisions, located at the headquarters in Jodhpur. There are four Regional Research Stations located in different agro-climatic zones to work on location-specific problems. CAZRI focuses on development of sustainable farming systems in the arid ecosystem.

INDIAN INSTITUTE OF REMOTE SENSING

Indian Institute of Remote Sensing (IIRS), Dehradun under Indian Space Research Organization, Department of Space, Govt. of India is a premier training and educational institute set up for developing trained professionals in the field of remote sensing, geoinformatics and GPS technology for natural resources, environmental and disaster Management. Formerly known as the Indian Photo-Interpretation Institute (IPI), the IIRS was set up in 1966 under the patronage of the Survey of India to provide training on remote sensing technology.

NATIONAL REMOTE SENSING CENTRE

National Remote Sensing Center (NRSC), a constituent of the Indian Space Research Organization (ISRO), located at Hyderabad with five regional remote sensing centres located at Bengaluru, Nagpur, Kolkata & Jodhpur. NRSC at Hyderabad has been converted into a full-fledged centres of ISRO since September 2008. Earlier, NRSC was an autonomous body called National Remote Sensing Agency (NRSA) under Department of Space (DOS). National Remote Sensing Centre's mandate is to provide Earth observation data from space and aerial platforms to users, develop technologies for the management of natural resources, support disaster monitoring and management and capacity building.

STATE FOREST RESEARCH INSTITUTE

State Forest Research Institute (SFRI), Jabalpur the first state level forest research institute in the country, came into existence on 27 June 1963, as per the recommendations of the 10th Silvicultural Conference held at Dehradun in 1961. The mandate of the institute is to provide scientific technical support to the M.P. forest department and various forestry related institutions, as well as other stakeholders and forestry sector as a whole, in the state.

KERALA FOREST RESEARCH INSTITUTE

Established in 1975, the Kerala Forest Research Institute (KFRI) is an institution of the Government of Kerala under its Science and Technology Department. The Institute is dedicated to research in tropical forestry, environment and biodiversity conservation. In 2003, when the State owned R & D institutions were re-organized to constitute the Kerala State Council for Science and Technology (KSCSTE), KFRI also became part of the latter. The Main Campus of KFRI is situated at Peechi, Thrissur District.

OTHER STATE FOREST RESEARCH INSTITUTES

State Forest Research Institute, Kolapakkam, Chennai

State Forest Research Institute, Kanpur, Uttar Pradesh

State Forest Research and Training Institute (SFRTI), Raipur, Chhattisgarh

State Forest Research Institute, Itanagar, Arunachal Pradesh

State Forest Research Institute, Janipura, Jammu (J&K)

FORESTRY EDUCATION INSTITUTIONS IN INDIA

The first systematic program of forestry education was started in 1763 at Master Schools in Germany at Harting and Cotta. Master Schools spread widely from these two places and German foresters were invited to many countries to take charge of forest management and to educate foresters. The first forestry school was established in Austria by Prince Schwarzenberg and Prince Liechtenstein, the two largest owners of forest land. This was followed by the State owned forestry schools on the same pattern of military academics in various countries viz. 1790 at Munich (Bovaria), 1803 at St. Petersburg (Russia), 1813 at Mariabrunn (Austria), 1825 at Nancy (France), 1828 at Stockholm (Sweden), 1848 at Turin (Italy), 1878 in Russia and later in Poland and the Scandinavian countries.

Thirteenth session of the FAO Advisory Committee on Forestry Education held in Mexico City during June, 1985 discussed the status of forestry education in developing countries and attention was drawn to the imbalance between university education and technical and vocational training and the need to train foresters at all levels. It also discussed the necessity to adjust forestry training to the real needs of the world.

The scientific forestry started in India is as early as 1864 to train the forestry professionals for the forest management. Teaching of forestry, research and extension in India had been under the purview of Federal and State Forest Services till 1985. In view of the shortage of trained manpower in the Forestry sector, Indian Council of Agriculture Research (ICAR) considered the expansion of Forestry education and research in selected State Agricultural Universities during the VII Plan period (1985-90) with the objective of getting professionally trained students in forestry on sustainable basis. Forestry education at the University level in the country was started from the State Agricultural Universities (SAUs) in the year 1985 with the inception of B.Sc. Forestry as a four year degree program. ICAR selected 13 State Agricultural Universities of India for starting B.Sc. Forestry degree programme during the academic year 1985-86.

The starting of forestry degree programme in the Universities was more or less a consequence of the general awareness to the environmental and forestry related problems. The media played an active role in highlighting the problems of deforestation, wasteland formation, environmental degradation as the consequence of development projects. Now there are many State Agricultural Universities (SAUs), Traditional Universities, Deemed Universities and private educational institutions in the country, recognized by the UGC, ICAR and ICFRE, have been offering B.Sc. Forestry programme. M.Sc. and Ph.D. in forestry/agroforestry with specialization of particular topic related to forestry have also been started in many institutions.

LIST OF FORESTRY ORGANIZATIONS PROVIDING SPECIALIZED EDUCATION AND TRAINING

- Indian Institute of Forestry Management (IIFM), Bhopal on forest management
- Wildlife Institute of India (WII), Dehradun on wildlife research and management
- Indian Plywood Industries Research and Training Institute (IPIRTI), Bangalore on plywood manufacturing and mechanical wood processing
- Institute of Paper Technology (IPT), Saharanpur on pulp and paper technology

LIST OF UNIVERSITIES, COLLEGES AND DEPARTMENTS OFFERING FORESTRY, AGROFORESTRY AND WILDLIFE EDUCATION

- ASPEE College of Horticulture and Forestry, Navsari of Navsari Agricultural University, Gujarat
- Forest Research Institute-Deemed University, Dehradun, Uttarakhand
- Forest College Research Institute (FC&RI), Mettupalayam of Tamil Nadu Agricultural University
- College of Forestry, Nauni, Solan of Dr.Y.S. Parmar University of Horticulture and Forestry, H.P.

- College of Forestry and Environment, Allahabad of Sam Higginbottom Institute of Agriculture Technology and Sciences (Formerly Allahabad Agricultural Institute) - Deemed University, U.P.
- College of Forestry and Hill Agriculture, Ranichauri of G.B. Pant University of Agriculture and Technology, Uttarakhand
- College of Forestry, Akola of Dr. Panjabrao Deshmukh Krishi Vidyapeeth, Maharashtra
- College of Forestry, Dapoli of Dr. Balasaheb Sawant Konkan Krishi Vidyapeeth, Maharashtra
- College of Forestry, Ponnampet of University of Agricultural Sciences-Bangalore, Karnataka
- College of Forestry, Sirsi of Unviersity of Agricultural Sciences-Dharwad, Karnataka
- College of Forestry, Thrissur of Kerala Agricultural University
- College of Horticulture and Forestry, Jhalawar of Maharana Pratap University of Agriculture and Technology, Rajasthan
- College of Horticulture and Forestry (CH&F) Pasighat, Arunachal Pradesh of Central Agricultural University
- College of Forestry & Hill Agriculture, Tehri Garhwal of Uttarakhand University of Horticulture and Forestry
- Faculty of Forestry, Kanke, Ranchi of Birsa Agricultural University, Ranchi
- Faculty of Forestry, Shalimar, Srinagar of Sher-e-Kashmir University of Agricultural Sciences and Technology, Kashmir
- Division of Agroforestry of Sher-e-Kashmir University of Agricultural Sciences and Technology, Jammu
- Department of Forestry, Nirjuli of North Eastern Regional Institute of Science and Technology (NERIST) Deemed University, Arunachal Pradesh
- Department of Forestry and Natural Resources, Ludhiana of Punjab Agricultural University
- Department of Forestry of Kumaun University, Uttarakhand
- Department of Forestry, Bhubaneswar of Orissa University of Agriculture and Technology
- Department of Forestry, Hisar of CCS Haryana Agricultural University
- Department of Forestry, Kanpur of C.S. Azad University of Agriculture and Technology, U.P.
- Department of Forestry, Raipur of Indira Gandhi Agricultural University, Chhattisgarh
- Department of Forestry, Srinagar (Garhwal) of H.N.B. Garhwal Universiy, Uttarakhand

- Department of Forestry, Wildlife and Environment Science, Bilaspur of Guru Ghasidas Vishwavidyalaya University, Chhattisgarh
- Department of Forestry, Cooch Behar of Uttar Banga Krishi Vishwavidyalaya, West Bengal
- Department of Wildlife Science of Aligarh Muslim University, U.P.
- Department of Agroforestry, Jhansi of Bundelkhand University, U.P.
- School of Studies (Forestry & Wildlife), Bastar of Bastar University, Chhattisgarh
- Doon (PG) College of Agriculture Science and Technology (DCAST), Dehradun, Uttarakhand
- Dolphin (PG) Institute of Biomedical & Natural Sciences, Dehradun, Uttarakhand.

BIBLIOGRAPHY

GENERAL REFERENCES

Publications, Technical Reports and Website of Ministry of Environment and Forests, GOI, New Delhi

Publications, Technical Reports and Website of Indian Council of Forestry Research and Education, Dehradun

Publications, Technical Reports and Website of Ministry of Water Resources, GOI, New Delhi

Publications, Technical Reports and Website of Ministry of Tribal Affairs, GOI, New Delhi

Publications, Technical Reports and Website of DoLR (Department of Land Resources), Ministry of Rural Development, GOI, India

Publications, Technical Reports and Website of Planning Commission, GOI, New Delhi

Publications, Technical Reports and Website of Forest Survey of India, Dehradun

Publications, Technical Reports and Website of National Remote Sensing Centre, Hyderabad

Publications, Technical Reports and Website of Wildlife Institute of India, Dehradun

Publications, Technical Reports and Website of Indian Institute of Forest Management, Bhopal

Publications, Technical Reports and Website of FAO (Food and Agricultural Organization of the United Nations)

Publications, Technical Reports and Website of Global Forest Resources Assessment (FRA)

Publications, Technical Reports and Website of IPCC (Intergovernmental Panel on Climate Change)

Bor NL 2010 A Manual of Indian Forest Botany, Vedam Books International

Buckman HO and Brady CN 1960 Nature and Properties of Soils, A College text book of Edaphology, 6th Edition, The Macmillan Company, USA.

Chaturvedi AN and Khanna LS 1982 Forest Mensuration, Dehradun: International Book Distributors

Evans J 1992 Plantation Forestry in the Tropics, Oxford Press

Gulhati NK 1981 Glossary of Forestry Terms, International Book Distributors

Hosette BB and Venkateshwarlu M 2001 Trends in Wildlife Biology Conservation and Management, Volume 1 and 2, New Delhi, India: Daya Publishing House

Hosetti BB 2003 Wildlife Management in India, Jaipur, India: Pointer Publishers

Jha LK and Sen-Sarma PK 1994 Forest Entomolgy, Chapter 20: Control of Forest insect pests, New Delhi: Ashish Publishing House

Khanna LS 1982 Forest Protection, Dehradun: Khanna Bandhu Publication, India

Khanna LS 1984 Principles and Practices of Silviculture, Dehradun: Khanna Bhandu

Mehta T 1981 A Hand Book of Forest Utilization, Dehradun: International Book Distributors

Negi SS 1986 Handbook of Social Forestry, Dehradun: International Book Distributors

Negi SS 1993 Manual for Wildlife Management in India, Dehradun: International Book Distributors

Negi SS 1994 Indian Forestry through the Ages, New Delhi: Indus Publishing Co.

Negi SS 2000 Joint Forest Management: People's Participation in forestry, Dehradun: International Book Distributors

Negi SS 2001 Forest Policy and Laws, Dehradun: International Book Distributors

Negi SS 2005 Hand Book of Forestry, Dehradun: International Book Distributors

Ram Prakash 2006 Forest Management, Dehradun: International book distributors

Saharia VB 1982 Wildlife in India, Dehradun, India: Natraj Publisher

Smith DM 1962 The Practice of Silviculture, New York: John Wiley

Tejwani RG 1994 Agroforestry in India, New Delhi: Oxford & IBH Publishing Co. Pvt. Ltd.

Tewari DN 1995 Agroforestry for increased productivity, sustainability and poverty alleviation, Dehradun, India: International Book Distributors

Thakur ML 2000 Forest Entomology (Ecology and Management), Chapter XIX: Forest Insect Control, Dehradun: Sai Publishers

REFERENCES

Abod SA 1995 Forest Plantations Current Issues, Paper presented at the International Short Course on Sustainable Tropical Forest Management, UPM

Agragamee 1997 Mandibishi Mahila Mandal, hill brooms and TDCC - a case of a government against its people, In (ed) SR Hiremath Forest lands and forest produce: as if people mattered, Dharwad, Karnataka, India, National Committee for Protection of Natural Resources (NCPR)

AgriLORE 2012 Innovations in Technology Mediated Learning: An Institutional Capacity building in Using Re-Usable Learning Objects in Agro-horticulture, NAIP Project, website: http://agropedialabs. iitk.ac.in/agrilore/?q=node/2666

Aldo Leopold 1933 Game Management, New York: Scribner Publication (printed in USA)

Anonymous 1961 100 years forestry, Delhi: Manager of Publications, 2 and 381 pages

Anonymous 1996 Studies on seed collection, processing and standardization of storage practices in Silvicultural crops of economic importance, Final Report, ICAR Adhoc Scheme, Tamil Nadu Agricultural University, Coimbatore, Tamil Nadu

Anonymous 2001 Proceedings of the JFM Nodal Officers Meeting, 5th December 2001, New Delhi. http://www.rupfor.org/jfmcell/Proceedings-nodalofficersmeet.doc

Avery TE and Burkhart HE 1983 Forest Measuremen, McGraw-Hill Book Company

Baharuddin HJ and Simula M 1994 Certification for all Timber and Timber Products, ITTO, Yokohama, Japan

Bahuguna VK 1999 Forest Fire Prevention and Control Strategies in India, *International Forest Fire News* **20** 5-9

Bahuguna VK and Satyendra Singh 2002 Fire Situation in India, *International Forest Fire News* **26** 23-27

Bahuguna VK, Unnikrishnan KP and Dhandiyal UC 1989 Studies on the performance of Philippines and Malaysian provenance of *Albizia falcataria* (L) Feberg at nursery stage under north Indian moist tropical conditions, *Indian Forester* **115** 209-215

Bakshi BK 1973 Forest Pathology - Principles and Practice in Forestry

Bakshi BK and Boyce JS 1959 *Polyporus shoreae* root rot in sal, *Indian Forester* **83** 656-658

Bakshi BK and Sujan Singh 1970 Heart rots in trees, *International Review of Forest Research*, Academics Press, New York **3** 197-251

Barnett TP 2001 Detection of anthropogenic climate change in the world's oceans, *Science* **292** 270-274

BCFT (British Commonwealth Forest Terminology) 1953 Part I, London: Empire Forestry Association

Beeson CFC 1941 The Ecology and control of Forest Insects of India and the neighboring Countries, Dehradun: Vasantha press, (Reprint in 1961) 761 page

Bene JC, Beall HW and Cole A 1977 Trees, Food and People, IDRC, Ottawa

Benson Jonathan 2012 Woodworker's Guide to Veneering and Inlay, Fox Chapel Publishing, website: *http://www.jonathanbenson.com/*

Bentzer BG 1993 Strategies for clonal forestry with Norway spruce, In Clonal Forestry II: Conservation and Application (eds) MR Ahuja and WJ Libby, New York: Springer-Verlag, 120-138 pages

Bhardwaj SP and Chakraborty AK 1994 Studies on time of seed collection, sowing and presowing seed treatments of *Terminalia bellirica* Roxb and *Terminalia chebula* Retz., *Indian Forester* **120** 430-439

Bier JE 1959a The relation of bark moisture to the development of canker diseases caused by native, facultative parasites, II Fusarium canker on black cottonwood *Canadian Forest Botany* **37** 1140-1142

Bier JE 1959b The relation of bark moisture to the development of canker diseases caused by native , facultative parasites, II Fusarium canker on Poplar. *Canadian Forest Botany* **37** 781-788

Blasing TJ 2012 Recent Greenhouse Gas Concentrations, Published in CDIAC website: http://cdiac.ornl.gov/pns/current_ghg.html

Boisvenue C and Running SW 2006 Impacts of climate change on natural forest productivity-evidence since the middle of the 20th century, *Global Change Biology* **12** 862-882

Bolund P and Hunhammar S 1999 Ecosystem services in urban areas, *Ecological Economics* **29** (2) 293-301

Boyce JS 1961 Forest pathology, 3rd Editon, New York: Mc Graw-Hill Book Co. Inc., 572 page

Brack CL 2002 Pollution mitigation and carbon sequestration by an urban forest, *Environmental Pollution* **116** 195-200

Brown AA and Davis KP 1973 Forest Fire: Control and Use, New York: McGraw Hill

Bruce D and Schumacher F 1930 Forest Mensuration, New York: McGraw-Hill Book Co.

Bruijnzeel LA and Critchley WRS 1994 Environmental impacts of logging, IHP Humid tropic programme series 7(UNESCO)

Bryant RL 1996 Romancing Colonial Forestry: the Discourse of forestry as progress in British Burma, *The Geographical Journal* **162** (2) 169-178

BSI (Botanical Survey of India) 2010 Plant Discoveries 2009: New Genera, Species and New Records-GOI, MoEF, BSI

Budowski G 1976 Environmental Conservation **3** 27-31

Burley J and Wood PJ 1976 A manual on species and provenance research with particular reference to the tropics, London: Commonwealth Forestry Institute, 226 page

Burns RM and Honkala BH 1990 Silvics of North America, Agric. Hand book, USDA Forest Service, Washington DC (2 volumes), 654 page

CBD (Convention on Biological Diversity) 1992 CBD adopted in 1992 at Rio de Janeiro and entered into force on 29 December 1993

CCC (Civilian Conservation Corps) 2009 Forest Mensuration (Chapter 8), In: Civilian Conservation Corps Forestry, Publication of CCC Educational Program, Office of Education, United States Department of the Interior

CCSP 2008 The Effects of Climate Change on Agriculture, Land Resources, Water Resources, and Biodiversity in the United States, A Report by the U.S. Climate Change Science Program and the Subcommittee on Global Change Research, Washington DC, USA: U.S. Environmental Protection Agency

CDIAC (Carbon Dioxide Information Analysis Center) 2011 Preliminary CO_2 Emissions 2010, United States: Carbon Dioxide Information Analysis Center, Oak Ridge National Laboratory

Ceballos-Lascurain H 1987 The future of ecotourism, *Mexico Journal* (**January**) 13-14

Chakraborty S 2004 Biodiversity, Jaipur: The Diamond Printing Press, Pointer Publishers, India 136 page

Champion HG and Seth SK 1968 A Revised Survey of the Forest Types of India, Government of India Press (Reprinted in 2005 by Natraj Publishers, Dehradun)

Charlene Watson 2010 Forest Carbon Accounting: Overview and Principles, Report of UNDP, UNEP & UNEP Risoe Centre: CDM Capacity Development in Eastern and Southern Africa

Chatrubuj Shau 1998 Primitive Tribes of India (An Ethnographic Profile), New Delhi: Published by Prabhat Kuma Sharma for Sarup & Sons

Chatterjee PN and Singh P 1968 *Celosterna scarabrator* Fab. (Lamiidae: Coleoptera), a new pest of *Eucalyptus* and its control, *Indian Forester* **94** (11) 826-830

Chaturvedi AN and Khanna LS 1982 Forest Mensuration, Dehradun: International Book Distributors

Chaturvedi AN and Khanna LS 2000 Forest Mensuration and Biometrics (3rd ed.), Dehradun: Khanna Bandhu Publisher

Chaudhary P and Tewari VP 2011 Urban forestry in India: development and research Scenario, *Interdisciplinary Environmental Review* **12** (1) 80-93

Chaudhary P, Bagra K and Singh B 2011 Urban Greenery Status of Some Indian Cities: A Short Communication, *International Journal of Environmental Science and Development* **2** (2) 1-4

Chepil WS and Woodruff NP 1965 The Physics of Wind Erosion and Its Control, *Advances in Agronomy* **15** 211-301

Chetan Kumar 2002 Community involvement in Forest Fire Prevention and Control: Lessons from Joint Forest Management (JFM), *International Forest Fire News* **26** 28-31

Chittaranjan Kumar Paty (ed) 2007 Forest Government and Tribe, Paper presented at a national conference held at Chaibasa during 2-3 mrach 2005, New Delhi: Concept publishing company

Chopra K and Kadekodi GK 1997 Natural Resource Accounting in the Yamuna Basin: Accounting for Forest Resources-Monograph, Delhi: Institute of Economic Growth, University of Enclave

Chris Brack 2007a Lecture Notes on Forest Mensuration Australian National University, website: http://www. sres. anu.edu.au/associated/mensuration/

Chris Brack 2007b RWG2 (1992) Code of Forest Mensuration Practice, website: http://sres-associated.anu.edu. au /mensuration/rwg2/code/index.htm

Christopher Biermann 1993 Handbook of Pulping and Papermaking, San Diego: Academic Press

Conway JA 1975 Proceedings of 1st International conference on "Stored Product Entomology", Savannah, Georgia, 554-566

Cramer W, Bondeau A, Woodward FI, Prentice IC, Betts RA, Brovkin V, Cox PM, Fisher V, Foley J and Friend AD 2001 *Global Change Biology* **7** 357-373

Curtis JT and Cottam G 1956 Plant Ecology Work Book: Laboratory field reference manual, Minnesota: Burgers Publication Co., 193 page

Damodaran A 1992 Economics of Forest Conservation and Sustainable Development, Proceedings of the Seminar on Socio-Economic Research in Forestry, Thrissur, 308 page

Damodaran A 1997 Countering Forest Encroachment with JFM: A Simulation Exercise, Working Paper No. WPS/9712, Bangalore: Indian Institute of Plantation Management

Damodaran A 2002 Tree Plantations and Biodiversity in the Indian Context, Working Paper No. IIPM/WPS/WTO Cell/05, Bangalore: Indian Institute of Plantation Management

Delaware ENVIROTHON 2012 website: http://www.delawareenvirothon.org/

Dharmalingam C and Masilamani P 1997 Radiography technique for testing the quality of teak (*Tectona grandis* Linn.f) seeds, *Bangladesh Journal of Forest Science* **26** 51-55

Dhyani SK 2009 Challenges and scope of Agroforestry research in India, In (eds) OP Chaturvedi, A Venkatesh et al., Agroforestry-Natural Resource sustainability, livelihood and climate moderation, 13-25 pages

Dickinson NM 1996 Metal resistance of some trees, In (ed) I Glimmersveen, *Heavy metals and trees*, Edinburgh, UK, Institute of Chartered Foresters (ICF), 85-92, 177-178 pages

Dimond AE 1955 Pathogenesis in wilt diseases, *A Review of Plant Physiology* **6** 329-350

Ditlevsen B, Shrestha NB and Robbins AMJ 1988 Tree improvement: An outline and plan of action for Nepal, HMGN/EEC/ODA National Tree Seed Project

Dixon JA, Louise FS, Richard A, Carpenter and Paul B Sherman 1986 Economic Analysis of Environmental Impacts, London: Earth Scan Publications Limited

DoLR (Department of Land Resources) Government of India website: http://dolr.nic.in/

DoLR and NRSC 2010 Wastelands Atlas of India, Department of Land Resources, Ministry of Rural Development, GOI, New Delhi and National Remote Sensing Centre, Hyderabad

DoLR and NRSC 2011 Wastelands Atlas of India: Change Analysis based on Temporal Satellite Data of 2005-06 and 2008-09, Department of Land Resources, Ministry of Rural Development, GOI, New Delhi and National Remote Sensing Centre, Hyderabad

Duarirasu P, Parthiban KT, Umesh Kanna S, Sekhar C, Vennila S and Rajendran P 2011 A Compendium on Forestry Statistics, Shri Garuda Publication, ISBN: 978-93-80769-12-7

Dumroese RK, Landis TD, Barnett JP and Burch F 2005 Forest Service Nurseries: 100 years of ecosystem restoration, *Journal of Forestry* **103** 241-247

Duncan EA, van Deventer F, Kietzka JE, Lindley RC and Denison NP 2000 The applied subtropical Eucalyptus clonal programme in Mondi forests, Zululand coastal region, In *Proceedings of the International Union of Forest Research Organizations (IUFRO) Working Party, Forest Genetics for the Next Millennium,* Durban, South Africa, 95-97 pages

Edward Mohr and Van Baren FA 1954 Tropical Soils: A Critical study of Soil Genesis as related to climate, Rock and vegetation, New York: Interscience pub. Inc.

Emmanuel CSJK 2007 Forest genetics and tree improvement: what we have achieved? *Envis Forestry Bulletin* **7** (1) 40-58

Emmanuel CSJK, Kapoor ML and Sharma VK 1992 Three decades of forest genetics and tree improvement, *Indian Forester* **115** 489-499

Eslyn WE 1959 Radiographical determination of decay in living trees by means of the thulium x-ray unit, *Forest Science* **5** 37-47

Evans J (ed) 2009 Planted Forests: Uses, Impacts and Sustainability, Wallingford: CABI

Evans J 1992 Plantation Forestry in the Tropics, Oxford Press

Evans J 1997 The sustainability of wood production in plantation forestry, Voluntary paper submitted to the Eleventh World Forestry Congress, Antalya, Turkey

Evans J 2000 Sustainability of Productivity in successive rotations, Proceedings of the International Conference on Timber Plantation Development

Evans J and Turnbull 2009 Plantation Forestry in the Tropics, The Role, Silviculture and Use of Planted Forests for Industrial, Social, Environmental and Agroforesry Purposes, Oxford: Oxford University Press

FAO (Food and Agricultural Organization) 1978 Eight World Forestry Congress, Forest for People, Jakarata Declaration, website: http://www.fao.org/docrep/meeting/011/ag398e/ag398e12.htm

FAO (Food and Agricultural Organization) 1992 Mixed and pure plantations in the tropics and subtropics, FAO Forestry Paper 103, Rome: FAO

FAO (Food and Agricultural Organization) 2006 Global Forest Resources Assessment 2005, Rome: FAO

FAO 1993a The Challenge of Sustainable Forest Management - what future for the world's forests?: Technical Paper, FAO, Rome

FAO 1993b Sustainable management of tropical moist forest for wood: Technical Paper, FAO Rome

FAO 1993c Policy, legal and institutional aspects of sustainable forest management: Technical Paper, FAO Rome

FAO 1999 Beyond Sustainable Forest Management: Opportunities and Challenges for improving Forest Management in the next Millennium: Technical Paper, FAO Rome

FAO 2008 Community based adaptation in action, a case study from Bangladesh, Project Summary Report

FAO 2012 Non-wood forest products, website: http://www.fao.org/forestry/nwfp/en/

Faulkner R 1975 Seed Orchard, *Forestry Commission Bulletin* **54** 149 page

FDA 2007 Guidelines for forest management planning, Forest Development Authority, Liberia (www.fda.gov.lr)

Foley JA, Levis S, Prentice IC, Pollard D and Thompson SL 1998 *Global Change Biology* **4** 561-579

Forests and Rangelands 2012 website: http://www.forestsandrangelands.gov/

FRA (Global Forest Resources Assessment) 2010a Global Forest Resources Assessment 2010, Rome: Food and Agricultural Organization of the United Nations

FRA (Global Forest Resources Assessment) 2010b Global Forest Resources Assessment 2010: Country Report for India, Rome: Food and Agricultural Organization of the United Nations

FRA 2000 Global Forest Resources Assessment 2000, Rome: Food and Agricultural Organization of the United Nations (FAO)

Francis Cairncross 2006 Time to get Stern on climate change, The First Post (Dated 30 October 2006): http://www.thefirstpost.co.uk/index.php?menuID=2&subID=1055

FSI (Forest Survey of India) 1987 India State of Forest Report-1987, Ministry of Environment and Forests, GOI

FSI (Forest Survey of India) 2009 State of Forest Report-2009, Dehradun : Forest Survey of India, Ministry of Environment and Forests

FSI (Forest Survey of India) 2011 India State of Forest Report-2011, Dehradun: Forest Survey of India, Ministry of Environment and Forests

Gartland KMA and. Davey MR 1995 *Agrobacterium* Protocols, In Methods in Molecular Biology, Humana Press, 44 page

Gartland KMA, Crow1 RM, Fenning TM and Gartland JS 2003 Genetically modified trees: production, properties, and potential, *Journal of Arboriculture* **29** (5) 259-265

Gary W Luck, Gretchen C Daily and Paul R Ehrlich 2003 Population Diversity and Ecosystem Services, *Trends in Ecology and Evolution* **18** (7)

Gautam NC and Narayan LRA 1988 Wastelands in India, Mathura: Pink Publishing House

GFU 2008 List of underutilized plants, website: http://www.underutilized-species.org/

Gillet NP, Weaver AJ, Zwiers FW and Flannigan MD 2004 Geophysics Research Letter **31**

Goldammer JG 2000 Overview of Fire and Smoke Management Issues and Options in Tropical Vegetation, *The Global Fire Monitoring Center/ Fire Globe website*

Government of India 1991 Status Paper on Forestry Situation, New Delhi: Ministry of Environment and Forests

Government of India 2002 Joint Forest Management: A Decade of Partnership, New Delhi: RUPFOR, Ministry of Environment and Forests

Gowda K, Sridhara MV and Rajan S 2007 Planning and management of parks and green areas: the case of Bangalore metropolitan area, *Management of Environmental Quality* **19** (3) 270–282

Grahn P and Stigsdotter UA 2003 Landscape planning and stress, *Urban Forestry and Urban Greening* **2** 001-018

Graudal L and Kajaer ED 2000 Priorities and strategies for tree Improvement, Denmark: DANIDA-Forest Seed Center, 17 page

Griffin DM 1958 Influence of pH on the incidence of damping-off, *Trans. Br. Mycol. Soc.* **41** 483-490

Grimm NB, Faeth SH, Golubiewski NE, Redman CL, Wu J, Bai X and Briggs JM 2008 Global change and the ecology of cities, *Science* **319** (5864) 756-760

Guariguata MR, Rheingans R and Montagnini F 1995 Early woody invasion under tree plantations in Costa Rica: Implications for forest restoration, *Restoration Ecology* **3** (4) 252-260

Gubbi S 2003 Fire, fire burning bright!, Bangalore: Deccan Herald, India. Accessed from http://wildlifefirst.info/images/wordfiles/fire.doc

Gulati SC and Sharma S 2000 Population pressure and deforestation in India, Population Research Centre, Institute of Economic Growth, University Enclave, Delhi-7

Gulhati NK 1981 Glossary of Forestry Terms, International Book Distributors

Gupta BN 1994 Non-Wood Forest Products in Asia, RAPA Publication, Bangkok, 1994/28, FAO

Gupta HK 2006 Joint Forest Management: Policy, Participation and Practices in India, Dehradun: IBD, 255-287 pages

Gurudev singh B, Arjunan MC, Manimuthu L, Mahadevan NP and Masilamani P 2001 Teak seed research problems and prospects, In Genetics and Silviculture of Teak (eds) AK Mandal and SA Ansari, Dehradun: International Book Distributors, 41-53 pages

Hadker N, Sharma S, David S and Muraleedharan TR 1997 Willingness-to-pay for Borivali National Protected Area: Evidence from a Contingent Valuation, *Ecological Economics* **21** (2) 105-122

Hall CM and Weiler B 1992 Introduction: What's special about special interest tourism? In Special Interest Tourism (eds.) B Weiler and CM Hall, London: Belhaven Press, 1-14 pages

Hannah Jaenicke 1999 Good Tree Nursery Practices: Practical Guidelines for Research Nurseries, Nairobi, Kenya: ICRAF

Hans Jenny 1941 Factors of Soil Formation, New York and Londaon: Mc Graw-Hill Book company, Inc.

Haripriya GS 2000 Carbon Budget of Indian Forest Ecosystem, Mumbai: Indira Gandhi Institute of Development Research

Hartley C 1939 The clonal variety for tree planting: Asset or liabilities? *Phytopathology* **29** 9

Hedegart T 1975 Seed collection of teak, In FAO / Danida training course on forest seed collection and handling II Rome: FAO, 274-279 pages

Helgerson 2012 Forest Ecosystem: Structure and Function, Module 2, Cooperative Extension,Washington State University, Skamania County

Honey M 2008 Ecotourism and Sustainable Development: Who owns Paradise? Washington DC: Island Press, 29-31 pages

Hopkins WG and Huner NPA 2004 Introduction to Plant Physiology, 3rd edn. J. Wiley, Hoboken, NJ

Hughes JF and Willan RN 1976 Policy planning and objectives, In A manual on species and provenance research with particular reference to the tropics, (eds) J Burley and PJ Wood), London: Commonwealth Forestry Institute, 4-14 pages

Husch B, Beers TW and Miller CI 2003 Forest Mensuration (4th ed.), New Jersey, Canada: John Wiley and Sons Inc.

ICFRE (Indian Council of Forestry Research and Education) 1990 Indian Forest Utilization Vol I and Vol II, Dehradun: ICFRE, India

ICRISAT 1976 Farming Systems Annual Reports 1975-1984, Hyderabad: ICRISAT, India

IEA (International Energy Agency) 2011 CO_2 Emissions from Fuel Combustions Highlights, Paris: International Energy Agency

IIED (International Institute for Environment and Development) 2003 Valuing Forests: A Review of Methods And Applications in Developing Countries, London: Environmental Economics Programme, WC1H 0DD, UK

Indianetzone 2012 The largest free encyclopedia on India, website: http://www.indianetzone.com/

IPCC (Intergovernmental Panel on Climate Change) 2003 Good Practice Guidance for Land Use, Land-Use Change and Forestry, Hayama, Japan: Intergovernmental Panel on Climate Change, IPCC/IGES

IPCC (Intergovernmental Panel on Climate Change) 2007 Climate Change 2007: The Physical Science Basis, Cambridge, UK: Cambridge University Press

ITTO 1992 Guidelines for the Sustainable Management of Natural Tropical Forests, Policy Development Series No. 1, International Tropical Timber Organization, Yokohama

ITTO 2005 Revised ITTO Criteria and Indicators for the Sustainable Management of Tropical Forests including Reporting Format, ITTO Policy Series No 15. ITTO, Yokohama, Japan

IUCN 1988 Guidelines for Re-introductions, Prepared by the IUCN/SSC Re-introduction Specialist Group, IUCN, Gland, Switzerland and Cambridge, UK 10 page

IUCN 2004 Protected Area Categories, By Kevin Bishop, Nigel Dudley, Adrian Phillips and Sue Stolton, *Parks* **14** (3)

IUFRO 2012 Non-wood forest resource management, website: http://www.iufro.org/

Jackson JK 1987 Manual of Afforestation in Nepal, Nepal-UK Forestry Research Project, Kathmandu, Nepal: Forestry Survey and Research Office

Jadeyegowda M 2000 Effect of Farmyard Manure on Maize and fingermillet based intercropping system in BRHills, M.Sc.Thesis submitted to University of Agricultural Sciences, Bangalore

Jagdish Kishwan, Rajiv Pandey, and Dadhwal VK 2009 India' Forest and Tree Cover: Contribution as a Carbon Sink, *ICFRE Technical Paper* 130 ICFRE BL-23, Dehradun: Indian Council of Forestry Research and Education

Jambulingam R 1990 Recent development in research on *Casuarina* in Tamil Nadu: Variation in populations, Advances in *Casuarina* research and utilization, Proceedings of the Second International *Casuarina* workshop, Cairo, Egypt, 15-20

Jerry Leech 2007 Handbook: An Australian Standard for Valuing Commercial Forests-Version 1, Association of Consulting Foresters of Australia in conjunction with Forest and Wood Products Research and Development Corporation

Jhala YV, Qureshi Q, Gopal R and Amin R 2009 Field Guide: Monitoring Tigers, co-predators, prey and their habitats, Third Edition, Technical Publication of National Tiger Conservation Authority, New Delhi and Wildlife Institute of India, Dehradun

Jhala YV, Qureshi Q, Gopal R and Sinha PR 2011 Status of the Tigers, co-predators and prey in India-2010, National Tiger Conservation Authority, New Delhi and Wildlife Institute of India, Dehradun, TR2011/003 pp-302

Jim CY and Chen WY 2006 Recreation-amenity use and contingent valuation of urban green spaces in Guangzhou, China, *Landscape and Urban Planning* **75** 81-96

JPA 2011 Top Ten Paper Producing Countries in 2010, Japan Paper Association, website: http://www.jpa.gr.jp/en/

Julia Szulecka 2011 Historical Paradigm Shifts in Tropical Forest Plantations, Conference of International Research on Food Security, Natural Resources Management and Rural Development, University of Bonn

June Seve 2001 The economic environment for plantations in developing countries: making plantation forestry a viable land use option, Proceedings of the International Conference on Timber Plantation Development, Rome: Food and Agricultural Organization

Kadekodi GK 2001 Valuation of Natural Resources: What Have We Learnt from Indian Experience? *Indian Journal of Agricultural Economics* Conference Number **56(3)** 285 – 312

Kandya AK 1987 Proceedings of International Symposium on Forest Seed problems in Africa (ed) SK Kamra and RD Ayling, Harare, Zimbabwe

Khanna LS 1982 Forest Protection, Dehradun : Khanna Bandhu Publication, India, 81-125 pages

Khongsak P 2005 Seed Production Areas: Guidelines for Establishment and Management Practices (MS4), Canbera: CSIRO, Australia

Khosla PK and Kohli RK 1987 Social Forestry for Rural Development, Solan: ISTS, 1-177 pages

King KFS and Chandler MT 1978 The Wastelands, Nairobi: ICRAF, 85 page

Kleinschmit J, Khurana DK, Gerhold HD and Libby WJ 1993 Past, present, and anticipated applications of clonal forestry, In Clonal Forestry II: Conservation and Application (eds) MR Ahuja and WJ Libby, New York: Springer-Verlag, 9- 41 pages

Konijnendijk CC 2003 A decade of urban forestry in Europe, *Forest Policy and Economics* **5** (2) 173–186

Kramer PJ and Kozlowski T 1960 Physiology of Trees, McGraw Hill publishers

Krantz BA 1981 Water conservation, management, and utilization in semi-arid lands, In: Advances in Food Producing Systems for Arid and Semi-arid Lands, Part A, University of California, 339-378 pages

Krasowski MJ 2012 Producing Planting Stock in Forest Nurseries, Forest and Forest Plants vol III, *Encyclopedia of Life Support System* (EOLSS), www.eolss.net/Sample-Chapters/C10/E5-03-05-07.pdf

Kuchelmeister G 1998 Urban forestry in the Asia-Pacific Region - status and prospects, Asia-Pacific Forestry ,Sector Outlook Study Working Paper Series No. 44. Rome, FAO, (website: http://www.fao.org/forestry/FON/FONS/outlook/Asia /APFSOS /44/ Apfsos 44.htm)

Kuchelmeister G 1999 Urbanization in developing countries - time for action for national forest programs and international development cooperation for the urban millennium, Forest Policy Research Forum: The Role of National Forest Programs to Ensure Sustainable Forest Management, 14-17 June 1999, Joensuu, Finland

Kulkarni N and Joshi KC 1998 Insect pests of forest tree seeds : their economic impact and control measures, *Journal of Tropical Forest Science* **10** 438-455

Laarman JG and Durst PB 1987 Nature travel in the tropics, *Journal of Forestry* **85** (5) 43-46

Lal P 2000 Integrated development of farm forestry plantations and wood based industries, *Indian Forester* **130** (1) 71-78

Lal P 2010 Clonal Forestry in India, *Indian Forester*, **136** (1) 17-37

Lamb D and Lawrence P 1993 Mixed species plantations using high value rainforest trees in Australia, In: (eds) H Lieth and M Lohmann, Restoration of Tropical Forest Ecosystems. Kluwer Academic Publishers, Holand 101-108 pages

Leakey R 1996 Definition of agroforestry revisited, *Agroforestry Today* **8**(1) 5-7

Leakey RRB 1987 Clonal forestry in the tropics - a review of developments, strategies and opportunities, *Commonwealth Forestry Review* **66** 61-75

Lerner IM 1958 The Genetic Basis of Selection, New York: John Wiley & Sons Inc. 298 page

Libby WJ 1976 Introductory remarks, In Proceedings of IUFRO Joint Meeting on Advanced Generation Breeding, 1-9 page

Libby WJ 1987 Testing for Clonal Forestry, *Annals of Forestry* **13** (1/2) 61-76

Lillesand TM and Kiefer RW 1994 Remote Sensing and Image Interpretation, Third edition, John Wiley and Sons Inc.

Lundgren BO 1987 Institutional aspects of agroforestry research and development, In (eds) HA Steppler and PKR Nair, Agroforestry: A decade of development, Nairobi: ICRAF, 43-51 pages

Lundgren BO and Raintree JB 1982 Sustained Agroforestry, In (ed) B Nestel, Agricultural Research for Development: Potentials and Challenges in Asia, Hague, Netherlands: ISNAR, 37-49 pages

MA (Millennium Ecosystem Assessment) 2005 Ecosystems and Human Well-being: Biodiversity Synthesis, Washington DC: World Resources Institute, web page: http://www.maweb.org/en/index.aspx

MaCDicken K 1997 A Guide to Monitoring Carbon Storage in Forestry and Agroforestry Projects, Arlington, United States: Winrock International

Madhu Verma 2000 Economic Valuation of Forests of Himachal Pradesh, Bhopal: Indian Institute of Forest Management (IIFM)

Malhi Y and Wright J 2004 Spatial patterns and recent trends in the climate of tropical forest regions, Philosophical Transactions of the Royal Society of London, Series B-Biological Sciences

Malhotra KC, Deb D, Dutta M, Vasula TS, Yadav G and Adhikari M 1991 Role of non-timber forest produce in village economy: A household survey in Jamboni Range, Midnapore, West Bengal, Calcutta: Indian Institute of Biosocial Research and Development (IBRAD)

Manmohan Yadav, Kotwal PC and Menaria BL 2007 Forest Certification: A Tool for Sustainable Forest Management, IIFM, Bhopal

Mannering JV and Fenster CR 1983 What is conservation tillage? *Journal of Soil and Water Conservation* 38(3) 141-143 (Special Edition on Conservation Tillage)

Mapier I and Robbins AMJ 1989 Forest Seed and Nursery practice in Nepal, Nepal-U.K. forestry research project, Kathmandu, Nepal: Forestry research division, Department of Forestry and plant research

Martin-Trillo M and Martinez-Zapater JM 2002 Growing up fast: Manipulating the generation time of trees, *Curr. Op. Biotechnolgy* **13** 151–155

Masilamani P 1992 Production, Processing and Storage technology for seeds of *Cassia siamea* Lamk, *Hardwickia binata* Roxb and *Prosopis juliflora* swartz DC, M.Sc.(Ag.) Thesis, Tamil Nadu Agricultural University, Coimbatore, Tamil Nadu.

Masilamani P, Annadurai K and Marimuthu R 1998 Influence of season of fruit collection on germination and seedling vigour in Neem (*Azadirachta indica)*, Paper presented in IUFRO Seed Symposium on "Recalcitrant seeds" Kuala Lumpur, Malaysia, October 12-15.

Masilamani P, Bhaskaran M, Chinnusamy C and Annadurai K 2000 Influence of age of mother tree on germination and initial vigour of neem (*Azadirachta indica), Journal of Tropical Forest Science* **12** 188-190

Masilamani P, Dharmalingam C and Annadurai K 1999a Seed source variation in seed and seedling attributes of Teak (*Tectona grandis* Linn.f), Paper presented in National Symposium on "Forestry towards 21st Century" at Tamil Nadu Agricultural University, Coimbatore, Abst.85 page

Masilamani P, Gurudev Singh B and Manimuthu L 1999b Collection methods and treatments to enhance Teak germination, Paper presented in National Symposium on "Forestry towards 21st Century" at Tamil Nadu Agricultural University, Coimbatore, Abst. 131 page

Masilamani P, Parthiban KT, Rajeswari E and Tilak M 1995 Effect of seed microflora on germination and vigour of Honey mesquite (*Prosopis juliflora* Swartz DC), Paper presented in "National Seminar on seed pathology" held at University of Mysore, March 23-25. Abst. 25 page

Mehta T 1981 A Hand Book of Forest Utilization, Dehradun: International Book Distributors

Menninger EA 1962 Flowering trees of the world for tropics and warm climates, New York: Hearthside Press, 40 page

Miller RW 1997 Urban Forestry: Planning and Managing Urban Green Spaces, Second Ed., New Jersey: Prentice Hall

Ministry of Tribal Affairs 2010 Statistical Profile of Scheduled Tribes in India, New Delhi: Statistics Division, Ministry of Tribal Affairs, GOI

Ministry of Tribal Affairs 2012 Ministry of Tribal Affairs, GOI, website: http://tribal.nic.in/index.asp

Mittal RK and Sharma MR 1981 Seed mycoflora of *Cassia fistula* L., *Indian Journal of Forestry* **4** 70

MoEF (Ministry of Environment and Forests) 1988 National Forest Policy (1988), Resolution No. 3-1/86-FP. New Delhi, India, Department of Environment, Forests and Wildlife

MoEF (Ministry of Environment and Forests) 2009a India's Forest and Tree Cover: Contribution as a Carbon Sink, New Delhi: Ministry of Environment and Forests

MoEF (Ministry of Environment and Forests) 2011 Annual Report of Ministry of Environment and Forests 2010-11, New Delhi: MoEF, GOI

MoEF (Ministry of Environment and Forests) Government of India website: http://moef.nic.in/index.php

MoEF (Ministry of Environment and Forests) 1999 Annual Report of MoEF-1999, New Delhi: MoEF, GOI

MoEF (Ministry of Environment and Forests) 2009 India's Fourth Report to the Convention on Biological Diversity, New Delhi: Ministry of Environment and Forest, GOI

Mongenstern EK, Holst MJ, Teich AH and Yeatman CW 1975 Plus tree selection: Review and Outlook, Ottawa: Department of the Environment, Canadian Forestry Service, Publication no. **1347**, 72 page

Montagnini F, Gonzalez E, Rheingans R and Porras C 1995 Mixed and pure forest plantations in the humid neotropics: Comparison of early growth, pest damage and establishment costs, *Commonwealth Forestry Review* **74** (4) 306-314

Montane J (ed) 1994 La Foret Landaise, Editions Privat, Toulouse, France

Morab SG 1977 The Soligas of Biligiriraganan Hills, Calcutta, Anthropological Survey of India, 35-47 pages

Morales DJ, Micha FR and Weber RC 1983 Two methods of evaluating trees on residential sites, *Journal of Arboriculture* **9** (1) 21-24

MPSG 2012 IUCN Medicinal plant special group, website: http://mpsg.org/

MPSMFPCF 2012 MFP Resources and their marketing, M.P. State Minor Forest Produce (T & D) Co-op. Fed. Ltd, website: http://www.mfpfederation.com/

MSU (Michigan State University) 2012 Forest Ecology, Rapid River Schools, Michigan State University Extension

Murali KS et al 2000 Status of Participatory Forest Management in India: An Analysis, In (eds) KS Murali, K Indu, Murthy and Ravindranath NH 2002 Joint forest management in India and its ecological impacts, *Environmental Management and Health* **13** (5) 512-528

Murthy MSR, Badarinath KVS, Biswadip Gharai, Rajshekhar G and Roy PS 2006 *International Forest Fire News* **34** 72-77

Myers N 1988 Threatened Biotas: "Hot spots" in Tropical Forests, *The Environmentalist* **8** 1-20

Myers N 1990 The Biodiversity Challenge: Expanded Hot Spots Analysis, *The Environmentalist* **10** 243-256

Myers N, Mittermeier RA, Mittermeier CG, da Fonseca GAB and Kent J 2000 Biodiversity Hotspots for Conservation Priorities, *Nature* **403** 853-858

Naik ST, Suryanarayana V and Neelannavar TN 1998 Survey and management of forest nursery diseases in Uttar Kannada, Annual Meeting and Symposium on Integrated disease management and crop loss assessment, Indian phytopathological, Society (South Zone) Dec. 10-12, Dept. of plant Pathology, UAS GKVK, Bangalore-65

Nair KK 1992 Prevention of Forest Fires, *National Seminar on Forest Fires Report*-1992, Session II

Nair KSS, Sudheendrakumar VV, Varma RV and Chacko KC 1985 Studies on the seasonal incidents of teak defoliators and the effect of defoliation on volume increment of teak, Peechi, Kerela: KFRI Report No.30, 85page

Nair PKR 1979 Intensive multiple cropping with coconuts in India: Principles, programmes and prospects, Verlag Paul Parey, Berlinand Hamburg, Germany

Nair PKR 1985 Classification of agroforestry systems, *Agroforestry Systems* **3** 97-128

Nair PKR 1993 An Introduction to Agrofoestry, Kluwer Academic Publishers, Dordrecht, Netherlands

Nanda KK 1970 Investigations on the use of Auxins in Vegetative Reproduction of Forest Plants, Final Report of PL 480-Research Project A7FS-11 (FG In 255), 1-215 pages

National Tiger Conservation Authority (Project Tiger), Ministry of Environment and Forests, GOI website: http://www.projecttiger.nic.in/

Negi SS 1986 Handbook of Social Forestry, Dehradun: International Book Distributors

Negi SS 1996 An Introduction to Forest Pathology, Dehradun: IBD

Negi SS 2000 Joint Forest Management: People's Participation in forestry, Dehradun: IBD, 1-132 pages

Negi SS 2001 Forest Policy and Laws, Dehradun: IBD, 1-62 pages

Negi SS 2005 Hand Book of Forestry, Dehradun: International Book Distributors

Nemani RR, Keeling CD and Hashimoto H 2003 Climate-driven increases in global terrestrial net primary production from 1982 to 1999, *Science* **300** 1560-1563

Nienstaedt H 1953 Tannin as a factor in the resistance of chestnut (*Castanca* spp.) to the chestnut blight fungus, *Endothia parasitica* (Murr.) A. & A., *Phytopathology* **43** 32-38

NMPB (National Medicinal Plants Board) 2012 National Medicinal Plants Board, Ministry of Health and Family Welfare, GOI, website: http://nmpb.nic.in/index.php

Norman W Hudson 1987 Soil and Water Conservation in Semi Arid Areas, FAO Land and Water Development Division, Rome: Food and Agriculture Organization of the United Nations

Noss RF 2001 Beyond Kyoto: Forest Management in a Time of Rapid Climate Change, *Conservation Biology* **15**(3) 578–590

NRCAF 2007 Perspective plan vision-2025, NRC for Agroforestry, Jhansi, India

NTFP 2012 NTFP Information Resources, website: http://www.ntfp.org

Nyland RD 1996 Silviculture: Concepts and Applications, New York: McGraw Hill

Ohba K 1993 Clonal Forestry with sugi *(Cryptomeria japonica),* In Clonal Forestry II: Conservation and Application (eds) MR Ahuja and WJ Libby, New York: Springer-Verlag, 66-90 pages

Otegbeye GO 2001 Forestry mating designs and progeny testing: principles, methods and application, In Forest Genetics and tree Breeding

Owens JN and Lund HG 2006 Forest and Forest Plants, In: Encyclopedia of Life Support Systems (EOLSS), Oxford, UK: EOLSS Publishers, 324 page

Pal M 1991 Research Paper, In Forest seeds, (eds) P Khullar, RC Thapliyal, BS Beniwal, RK Vakshasys and A Sharma, Dehradun: ICFRE, 343-370 pages

Paramathma M, Parthiban KT and Neelakantan KS 2004 Strategies for Improvement and Utilization of Tree Borne Oilseeds, Mettupalayam: Forest College and Research Institute (FC & RI) Publications, TNAU

Park YD, Ronis DH, Boe AA and Cheng ZM 1995 Plant regeneration from leaf tissue of four North Dakota genotypes of potato (*Solanum tuberosum* L.), *American Potato Journal* **72** 329-338

Parmathama M 2002 Seed orchards, In Textbook of Forest Tree Breeding, (eds) C Surendern, RN Sehgal and M Parmathama), New Delhi: ICAR, 120-146 pages

Parthiban KT, Umesh Kanna S and Durairasu P 2011 Dendro biomass based power generation, Technical Bulletin, Mettupalayam: Forest College and Research Institute (FC & RI) Publications, TNAU

Parvez Ahmed 2005 Study of Alternatives for meeting the demand of raw material by wood based industries, *Indian Forester* 609-633

Pathak PS, Miranda SM and El-Swaify SA 1985 Improved rainfed farming for semi-arid tropics-implications for soil and water conservation, In (eds) E1-Swaify, Moldenhauer and Lo, Soil Erosion and Conservation, Soil Conservation Society of America, 338-354 pages

Pathak PS, Pateria HM and Solanki KR 2000 Agroforestry system in India-A diagnosis and Design Approach, NRCAF, ICAR, New Delhi, 184 page

Patric Purser 1999 Timber Measurement Manual: Standard Procedure for the Measurement of Round Timber for Sale purposes in Ireland, Purser Tarleton Russell Ltd., Funded by Forest Service in the Department of Marine and Natural Resources, Republic of Ireland

Patrick BD, Thomas R Waggener, Thomas Enters and Tan Lay Cheng 2001 Forests out of bounds: Impacts and effectiveness of logging bans in natural forests in Asia-pacific, Rome: Food and Agriculture Organization

Planning Commission 2001 Report of the Task Force on Greening India for livelihood security and sustainable development, New Delhi: Planning Commission, GOI

Planning Commission 2005 Report of the Task Group on Development of Scheduled Castes and Scheduled Tribes, New Delhi: Planning Commission, GOI

Planning Commission Government of India website: www.planningcommission.nic.in

Prabhakar VK 1998 Social and Community Forestry, New Delhi: Satish Garg, 90-106 pages

Prasad R 1999 Joint forest management in India and the impact of state control over non-wood forest products, In Non-wood Forest products and income generation, *Unsaylva* - No. 198

Prasad R 1999 Management of tropical forestry seeds for wasteland afforestation, In Seed and Nursery Technology of Forest Trees, (eds) DGW Edwards and SC Naithani, New Delhi: New Age International (P) Limited, Publications, 1-14 pages

Prasad R and Kandya AK 1992 Handling of forestry seeds in India, New Delhi: Associated publishing company, 45-144 pages

Prasad R and Saxena NC 1996 Role of the state in NTFP marketing exchanges, Bangalore: Actionaid

Prasad R, Shukla PK and Bhatnagar P 1996 Leaves from the forest: a case study of tendu leaves in Madhya Pradesh, Jabalpur, Madhya Pradesh: Centre for Environment and Sustainable Development

Putz FE and Pinard MA 1993 Reduced-Impact Logging as a Carbon-Offset Method, *Conservation Biology* **7** 755–757

Quijada R 1980 Vegetative Propagation Methods, Report on the FAO/DANIDA Training Course on Forest Tree Improvement, Merida, Venezuela, 140-145 pages

Race D and Desmond H 2001 Forestry Out-grower Scheme: A Global View, *Forest Plantation Thematic Papers* **FP/11** Rome: FAO

Raghubanshi AS, Rai LC, Gaur JP and Singh JS 2005 Invasive Alien Species and Biodiversity in India, *Current Science* **88** (4) 25

Rajeeva 1989 An Introduction to the Tribal development in India, Dehradun: International Book Distributors

Raju G et al 1993 Development of People's Institution for Management of Forests, Ahmedabad: VIKSAT

Ram Prakash 2006 Forest management, Dehradun: International book distributors

Randhawa HS, Sharma HL, Jagmeet K and Rattan GS 1986 The acceleration of germination of *Cassia fistula*: Fungi associated with rotting seeds and abnormal seedling, *Indian Forester* **112** 524-527.

Raunkiaer C 1934 The Life Forms of Plants and Statistical Plant Geography, Oxford: Clarendon Press, 632 page

Raven PH 1987 The scope of the plant conservation problems worldwide, In *Botanic Gardens and the World Conservations Strategy* (eds) B Bramwell, O Hamann, VV Heywood and H Synge, London: Academic Press

Ravindranath NH et al (eds) 2000 Joint Forest Management and Community Forestry in India: An Ecological and Institutional Assessment, New Delhi: Oxford & IBH Publishing Co. Pvt. Ltd., 25-58 pages

Ravindranath NH, Joshi NV, Sukumar R and Saxena A 2006 Impact of climate change on forests in India, *Current Science* **90** (3) 354-361

Revi A 2008 Climate change risk: An adaptation and mitigation agenda for Indian cities, *Environment and Urbanization* **20** (1) 207-229

Richards KN, Sampson N and Brown S 2006 Agricultural and Forestlands: US Carbon Policy Strategies, Report for the Pew Center of Global Climate Change

Rishbeth J 1963 Stump protection against *Fomes annosus* & Inoculation with *Peniophora gigantean, Ann. Appl. Biol.* **52** 63-77

RISI 2011 List of top hundred paper industry, RISI, website: http://www.risiinfo.com/

Robinson RA 1973 Horizontal resistance, *Pl. Pathology* **52** 483-501

Roonwal ML 1990 Field-ecological observations on the *Ailanthus excelsa* defoliator *Eligma narcissus indica* (Lepidoptera: Noctuidae) in Peninsular India, *Indian Journal Forestry* **13** (2) 81-84

Rose R, Campbell SJ, Landis TD (eds) 1990 Proceedings, target seedling symposium, combined meeting of the western forest nursery associations. Fort Collins (CO): USDA Forest Service, Rocky Mountain Forest Range Experiment Station, General Technical Report RM-200

Roshetko JM, Tolentino EL, Carandang WM, Bertomeu M, Tabbada A, Manurung GES and Yao CE 2010 Tree Nursey Source Book: Options in Support of Sustainable Management, ICRAF and Winrock International , Bogor, Indonesia

Rosvall O, Lindgen D and Mullin TJ 1998 Sustainability, robustness and efficiency of a multi-generation breeding strategy based on within-family selection, *Silvae Genetica* **47** 307-321

Russel TG, Alan EH, Theresa BJ and Jonalea RT 1999 The Effects of Thinning and similar Stand Treatments on Fire Behaviour in Western Forests, Gen. Tech. Rep. PNW-GTR-463, USDA Forest Service

Russell JH and Loo-Dinkins JA 1993 Distribution of testing effort in cloned genetic tests, *Silvae Genetica* **42** 98-104

SAC (Space Application Centre) 2007 Desertification and Land Degradation Atlas of India, Ahmedabad: Space Application Centre, Indian Space Research Organization

Sadguru Foundation 2012 N.M. Sadguru Water Development Foundation, website http://www.nmsadguru.org/

Sanchez 1995 Defining Agroforestry System, *Agroforestry*

Sara J. Scherr and Satya Yadav 1996 Land Degradation in the Developing World: Implications for Food, Agriculture and the Environment to 2020, IFPRI Food Agriculture and the Environment Discussion Paper 14, Washington DC: International Food Policy Research Institute

Sargent MS and Carter KS (ed) 1999 Managing Michigan Wildlife: A Landowners Guide, Michigan United Conservation Clubs, East Lansing, MI

Sargos J 1997 Histoire de la Foret Landaise, Deuxieme edition, L'Horizon chimerique, Bordeaux, France

Sarin M 1993 From Conflict to Collaboration: Local Institutions in Joint Forest Management, JFM Working Paper No. 14, Society for Promotion of Wastelands Development and Ford Foundation, New Delhi

Sarkar RM (ed) 2006 Land and Forest Rights of the Tribals Today, New Delhi: Serial Publication

Saxena NC 1997 Saga of Participatory Forest Management, Bogor, Indonesia: CIFOR

Schell LM and Ulijaszek SJ 1999 Urbanism, Health and Human Biology in Industrialized Countries, Cambridge: Cambridge University Press

Schmidt L 2000 *Guide to handling of tropical and subtropical forest seed*, Denmark: Danida forest seed centre publications, 143-182 pages

Schreiner EJ 1960 Objectives of pest resistance improvement in forest trees and their possible attainment, Seattle, USA: Proceedings of V World Forestry Congress 2 721-727

Schulze ED, Beck E and Muller-Hohenstein K 2002 Plant Ecology, Berlin: Springer-Verlag

Sen-Sarma PK and Thakur ML 1985 Pest management in Indian forestry, *Indian Forester* **111** (11) 956-964

Sharma PD 2009 Ecology and Environment, Meerut, India: Rastogi Publications

Sharma S, Bhattacharya S and Garg A 2006 Greenhouse gas emissions from India: A perspective, *Current Science* **90** (3) 326-333

Shashua-Bar L, Potchter O, Bitan A, Boltansky D and Yaakov Y 2010 Microclimate modelling of street tree species effects within the varied urban morphology in the Mediterranean city of Tel Aviv, Israel, *International Journal of Climatology* **30** (1) 44-57

Shaw DV and Hood JV 1985 Maximizing gain per effort by using clonal replicates, *Theoretical and Applied Genetics* **71** 392-399

Sheikh AB and Tahir Siddiqui M 1999 Current Issues in Tropical Plantation Forestry – A Malaysian Perspective, *Indian Forester* **December** 1228-1238

Singh NB, Kumar D, Rawat GS and Shrivastva SK 1999 Improvement of *Populus deltoides* in India II-future strategies, *Indian Forester* **125** (3) 341-354

Singh P and Bhandari RS 1988 Insect pests of bamboo and their control, *Indian Forester* **114** (10) 670-683

Singh P, Rawat DS, Misra RM, Fasih M, Prasad G and Tyagi BDS 1983 Epidemic defoliation of poplar and its control in Tarai Central Forest Division, Uttar Pradesh, *Indian Forester* **109** (9) 675-693

Singh UM 1997 Fighting Forest Fires, *Yojana* **41**(8) 63-72

Singh VS, Pandey DN and Chaudhary P 2010 Urban Forests and open green spaces: lessons for Jaipur, Rajasthan, India, RSPCB Occasional Paper No. 1/2010. http\:www.rpcb.nic.in doi 15.04.2012

Sivaramakrishnan K, Kundu A and Singh B 2005 Handbook of Urbanization in India: An analysis of trends and processes, New Delhi: Oxford University Press

Sivasamy M 1991 Net reproductive effort, seed production spectrum and quality polymorphism as influenced by ecological factors in *Azadirachta indica* A. Juss., *Dessis indica* L. and *Ailanthus excelsa* Roxb. Seeds. Ph.D. Thesis, Tamil Nadu Agricultural University, Coimbatore, Tamil Nadu

Smith DM 1962 The Practice of Silviculture, New York: John Wiley

Smith LW and Wood LW 1964 website: http://www.fpl.fs.fed.us/documnts/misc/miscpub_6409.pdf

Smith ST 1997 Environmental Tax Design, In (eds) Timothy O' Riordan, Eco Taxation, UK: Earthscan Publication, 21-36 pages

Sociology Guide 2012 Sociology Guide – A Student's Guide to Sociology, website: www.sociologyguide.com

Sojostrom E 1993 Wood Chemistry: Fundamentals and Applications, Academic Press

Somasundaram HN and Kibe RV 1990 The Soliga - Tribe and its Stride, 3-5 pages

SPWD 1992 Joint Forest Management: Concepts and Opportunities, Proceedings of the National Workshop at Surajkund, August 1992

Srivastava J, Lambert J and Vietmeyer N 1996 Medicinal Plants: An Expanding Role in Development, World Bank Technical Paper 320, Washington DC: World Bank

Stebbing EP 1914 Indian Forest Insects, (reprinted 1989), Bisen Singh and Mohindra, Dehra Dun

Stein BA and Flack SR 1996 America's Least Wanted: Alien species invasions of US ecosystems, Arlington, Virginia: The Nature Conservancy

Stern N 2006 Stern Review on The Economics of Climate Change (pre-publication edition), Executive Summary, London: HM Treasury

Stewart BA, Dusek DA and Musick JT 1981 A management system for the conjunctive use of rainfall and limited irrigation of graded furrows, *Soil Science Society of America Journal* **45** 413-419

Sudarshan H 1998 Traditonal Medicine and health care system of Soligas, In (eds) KN Ganeshaiash and R Uma Shaankaer, Biligiri Ranga Swamy Wild Life Sanctuary, Natural History, Biodiversity and Conservation, Vivekananda Girijana Kalyana Kendra, Chamarajanagar, 17-19 pages

Sudha P and Ravindranath NH 2000 A study of Bangalore urban forest, *Landscape and Urban Planning* **47** (1-2) 47–63

Suresh, KK, Swaminathan C and Rai RSV 1994 Nursery efficiency by use of presprouted seeds in *Hardwickia binata* Roxb., *Indian Journal of Foresetry,* **17** 356-358

Swaminathan C, Srimathi P and Sivagnanam K 1994 Effect of tree age on germination and seed and pod quality in *Ciba pentandra* (L.), *Indian Journal of Forestry* **17** 46-48

Swaminathan MS 1982 Rainfall and dryland agriculture, Proceedings of Symposium on Rainwater and Dryland Agriculture, New Delhi: IARI

Taubenbock H, Wegmann M, Roth A, Mehl H and Dech S 2009 Urbanization in India: Spatiotemporal analysis using remote sensing data, *Environment and urban systems* **33** (3) 179-188

Tejwani RG 1994 Agroforestry in India, New Delhi: Oxford & IBH Publishing Co. Pvt. Ltd.

Tewari DN 1994 A Monograph on *Dalbergia sissoo* Roxb., Dehradun: International Book Distributors, India, 316 page

Tewari DN 1995 Agroforestry for increased productivity, sustainability and poverty alleviation, Dehradun, India: International Book Distributors

Tewari KM 1983 Social Forestry in India, Natraj Publishers, 1-243 pages

Thakur ML Sushikumar and Negi A 1989 Chemical control of termites in *Eucalyptus* hybrids, *Indian Forester* **115** (10) 733-743

Thakur S, Sharma K and Badiyala SD 2000 Propagation and Conservation of *Terminalia chebula*, In Man and Forest (eds) RK Kohli, HP Singh, SP Vij, KK Dhir, DR Batish and DK Khurana, DNES, IUFRO, ISTS, Chandigarh, India 427-432 pages

Thapliyal RC and Rawat MS 1982 Tropical pines, Seminar cum Workshop, Koraput: Orrisa, 10-13

Thomas Enters, Patrick B. Drust, Grahame B. Applegate, Peter CS, Kho, Gray Man 2002 Applying reduced impact logging to advance sustainable forest management, International conference proceedings, Rome: Food and Agriculture Organization

Thurston E 1909 Castes and tribes of Southern India, **No.6,** 379-386 pages

TIES (The International Ecotourism Society) 1990 Definition of Ecotourism.

Tiwari SC 1992 Modern Ecology, Printmendia, Srinagar Garhwal, Uttarakhand, India

Turnbull JW 2002 Tree Domestication and the History of Plantations, In (ed) VR Squires, The Role of Food, Agricultre, Forestry and Fisheries in Human Nutrition Vol-II

Tyrvainen L 1999 *Monetary valuation of urban forest amenities in Finland,* Doctoral dissertation, Research Paper No. 739, Vantaa, Finland, Finnish Forest Research Institute

Ulrich RS 1984 View through a window may influence recovery from surgery, *Science* **224** 420-421

Umesh Kanna S, Parthiban KT, Sekar I, Vidhayavathi A, Devanand PS, Anbu PV and Durairasu P 2011 Value Addition of agroforestry plantation residues through briquetting technology for energy purpose, BENWOOD Conference: Short Rotation Forestry and Agroforestry: An Exchange of experience between CDM Countries and Europe, June 20-22, 2011, 93-98 pages

UN 2008 The General Assembly of the United Nations, Resolution 62/98

UNCED 1992 Earth Summit - Rio Declaration & Forest Principles

UNEP 1996 Sharing of Experiences on Incentive Measures for Conservation and Sustainable Use, Note by the Executive Secretary, Conference of the Parties to the Convention on Biological Diversity, Third Meeting, Beunos Aires

UNFF 2006 Global Objectives on Forests, United Nations Forum on Forests

UNGA (United Nations General Assembly) 2008 Resolution adopted by the General Assembly [on the Report of the Second Committee (A/62/419-Part I)] 62/98. Non-Legally Binding Instrument on All Types of Forests-A/RES/62/98

Uniyal PL, Chamola BP and Semwal DP 2010 The Plant Wealth of Uttarakhand-First Edition, New Delhi: Jagdamba Publishing Co. 436 page

USEPA (United States Environmental Protection Agency) 2012 Climate impacts on forests, web page: http://www.epa.gov/climatechange/impacts-adaptation/forests.html

USGCRP (2009) Global Climate Change Impacts in the United States, (eds) TR Karl, JM Melillo and TC Peterson, United States Global Change Research Program, New York, NY, USA: Cambridge University Press

Vaartaja O 1952 Forest humus quality and light conditions as factors influencing damping off, *Phytopathology* **42** 501-506

Valentine PS 1992 Review: Nature-based tourism, In Special Interest Tourism (eds.) B Weiler and CM Hall, London: Belhaven Press, 105-127 pages

Van Buijtenen JP 1969 Progress and problem in forest tree selection, Proceedings of 10th Southern Conference on Forest Tree Improvement, Houston, Texas, 17-16 pages

Van Buijtenen JP, Donovon GA, Long EM, Lowe WJ, McKinley CR, Robinson JF and Woessner RA 1981 Introduction to practical forest tree improvement, Circular no. 207, Texas Forest Service, 22 page

Van der Plank JE 1963 Plant Diseases: Epidemics and Control, Acadcmic Press, New York and London, 349 page

Varma RV 1986 Seasonal incidence and possible control of important insect pests in plantations of *Ailanthus triphysa*, Peechi, Kerela: KFRI Research Report No. **39**, 32page

Velayutham M and Bhattacharya T 2000 Soil resource management, In: (eds) JSP Yadav and GB Singh Natural, Resource Management for Agricultural Production in India, Indian Society of Soil Science, New Delhi

Venkatesh A, Vanangamudi K and Umarani R 2000 Effect of seedling grades on seedling vigour of *Acacia nilotica* ssp. *indica*. *My Forest* **36** (2) 91-97

Venkatesh A, Vanangamudi K and Umarani R 2001 Influence of salt stress on seed germination and vigour of *Acacia nilotica* (L.) Willd. ssp. *indica* Benth, *My Forest* **37** (4) 687-695

Venkatesh A, Vanangamudi K and Umarani R 2003 Effect of seedling grade on growth and survival of pungam (*Pongamia pinnata*), *Journal of Tropical Forest Science*, **15** (1) 231-233

Vidhyavathi A and Swaminathan LP 2007 Valuation of Use Value of Forest Resources: Application of Contingent Valuation, *International Journal of Ecology and Development* **7** 74-91

Vinod Kumar 1995 Nursery and plantation practices in Forestry, Jodhpur: Scientific Publishers, 515 page

W.E.D'arcy 1898 Preparation of working plans in India, Office of the Superintendent of Government Printing, India

Wastelands Atlas of India 2011 Wasteland Change Analysis based on Temporal Satellite Data of 2005-06 and 2008-09, DoLR, Ministry of Rural Development, GOI, New Delhi and NRSC, Hyderabad

WCED 1987 Our common future, World Commission on Environment and Development, New York: Oxford University Press.

Webb R 1998 Urban and peri-urban forestry in South-East Asia: a comparative study of Hong Kong, Singapore and Kuala Lumpur, Rome: FAO

Wellendorf H 1991 Tree Improvement strategies, Lecture note no. D-10, Denmark: Danida Forest Seed Centre, 10 page

Westerling AL, Hidalgo HG, Cayan DR, Swetnam TW 2006 *Science* **313** 940-943

White TL, Adams WT and Neale DB 2007 Forest Genetics, Oxford: CABI Publishing, UK

Wight P 1993a Ecotourism: Ethics or eco-sell? *Journal of Travel Research* **XXXI** (3) 3-9

Wight P 1993b Sustainable ecotourism: Balancing economic, environmental and social goals, within an ethical framework, *Journal of Tourism Studies* **4** (2) 54-66

Wikipedia 2012 Forest Tribes of India website: www.wikipedia.org

Williams JGK, Kubelik AR, Livak KJ, Rafalski JA and Tingey SV 1990 DNA polymorphism amplified by arbitrary primers are useful as genetic markers, *Nucleic Acid Research* **18** 6531-6535

Williams JR 1975 Sediment yield prediction with universal equation using runoff energy factor, In: Present and prospective Technology for predicting sediment yields and sources, ARS-S-40.USDA, Agricultural Research Service, Washington, DC: USDA, 118-124

Wilson EO 1989 Threats to Biodiversity, *Scientific American* **(Sept)** 106-116

Wischmeier WJ and Smith DD 1978 Predicting Rainfall Erosion loss: A Guide to Conservation Planning, Agri. Hand book No. 537, Washington, DC: USDA

Wison MA and Laarman JG 1988 Nature tourism and enterpraise development in Ecuador, *World Leisure and Recreation* **29/30** (1) 22-27

Working Plan Code 2011 Ministry of Environment and Forest, New Delhi: MoEF, GOI

Working plan of Doda Forest Division (2004-2014) Doda Forest Division Working plan prepared by working plan circle, Department of Forest, Wildlife and Remote Sensing, Jammu /Srinagar

World Bank 2011 State and Trends of the Carbon Market 2011, Washington DC: Carbon Finance Unit of World Bank

Wormald TJ 1992 Mixed and pure forest plantations in the tropics and subtropics, FAO, No:103, 166 page

Wright JW 1976 Introduction to Forest Genetics, New York: Academic Press, 463 page

Yadav BK 2012a Forest Mensuration, Forestry Nepal: Gateway to Forestry Information in Nepal, website: http://www. forestrynepal.org /notes/mensuration

Yadav BK 2012b Nursery, Forestry Nepal: Gateway to Forestry Information in Nepal, website: http://www. forestrynepal.org /notes/silviculture/nursery/1

Young A 1989 Agroforestry for Soil Conservation, Wallingford, UK: CAB International

Zabala NQ 1993 Mass Vegetative Propagation of *Dipterocarp* Species, Rome: FAO, website: http://www.fao. org/docrep/006/ad222e/ad222e00.htm

Zobel BJ and Talbert BJ 1984 Applied Forest Tree Improvement, New York: John Wiley & Sons

Zsuffa L, Sennerby-Forsse L, Weisgerber H and Hall RB 1993 Strategies for clonal forestry with poplars, aspens, and willows, In Clonal Forestry II: Conservation and Application (eds) MR Ahuja and WJ Libby, New York: Springer-Verlag, 91-119 pages

Zycha H and Dimitri L 1962 Erfahrungcn mit cinem Gcriit zur Fauleermittlung an stehenden Stammen, *Forslwiss. Z'-nlBl.* **81** 222-230